Soil and Water Conservation
for Productivity and Environmental Protection

Fourth Edition

Frederick R. Troeh
Professor Emeritus of Agronomy
Iowa State University

J. Arthur Hobbs
Professor Emeritus of Agronomy
Kansas State University

Roy L. Donahue
Late Professor Emeritus of Soil Science
Michigan State University

Prentice Hall
Upper Saddle River, New Jersey 07458

Library of Congress Cataloging in Publication Data
Troeh, Frederick R.
Soil and water conservation for productivity and environmental protection / Frederick R. Troeh, J. Arthur Hobbs, Roy L. Donahue.—4th ed.
p. cm.
Includes bibliographical references.
ISBN 0-13-096807-2 (alk. paper)
1. Soil conservation. 2. Water conservation. I. Hobbs, J. Arthur (James Arthur) II. Donahue, Roy Luther III. Title.

S623.T76 2004
631.4'5—dc21 2002192987

Editor in Chief: Stephen Helba
Executive Editor: Debbie Yarnell
Editorial Assistant: Jonathan Tenthoff
Managing Editor: Mary Carnis
Production Editor: Emily Bush, Carlisle Publishers Services
Production Liaison: Janice Stangel
Director of Manufacturing and Production: Bruce Johnson
Manufacturing Buyer: Cathleen Petersen
Creative Director: Cheryl Asherman
Cover Design Coordinator: Miguel Ortiz
Marketing Manager: Jimmy Stephens
Cover Photo: Numerous terraces help to control erosion on this cropland near Miniato in northern Italy. Courtesy of F. R. Troeh

---

Pearson Education LTD.
Pearson Education Australia PTY, Limited
Pearson Education Singapore, Pte. Ltd.
Pearson Education North Asia Ltd.
Pearson Education Canada, Ltd.
Pearson Educación de Mexico, S.A. de C.V.
Pearson Education—Japan
Pearson Education Malaysia, Pte. Ltd.

10 9 8 7 6 5 4 3 2 1
ISBN 0-13-096807-2

Contents

Preface

Soil and water have always been vital for sustaining life, and these resources are becoming more limiting and crucial as population increases. The importance of conserving soil productivity and protecting the quality of both soil and water is becoming clear to more people than ever before. Declining productivity and increasing pollution could spell disaster for all residents of the Earth. The soil and water resources of the planet are finite and are already under intensive use and misuse. Environmental degradation is becoming painfully evident, and increasing numbers of people are demanding that steps be taken to not only reduce the amount of current degradation but also to amend some of the previous damage.

Soil and water conservation deals with the wise use of these important resources. Wise use requires knowledge, understanding, and value judgments. The hazards posed by erosion, sedimentation, and pollution, and the techniques needed to conserve soil and maintain environmental quality are all treated in this book. Situations and examples are drawn from many places to constitute a cross-section of the soils, climates, and cultures of the world. The scope includes agricultural, engineering, mining, and other uses of land. Soil and water are recognized as essentials for everyone's life.

This fourth edition continues the use of foot-pound-second units as the principal units of measurement. Metric units are usually included in parentheses and are presented as the principal or only units where they are the units generally used in the United States. The fourth edition has been updated throughout with many citations to the literature published since the third edition was printed. Significant new material has been added, and certain sections have been expanded. The trend toward computerizing the soil-loss equations is emphasized in Chapter 6. The rapidly growing use of no-till cropping is recognized with an expanded treatment in Chapter 9.

The former chapters "Vegetating Mining and Construction Sites" and "Vegetating Other Areas of High Erosion Hazard" have been combined in one chapter titled "Vegetating Drastically Disturbed Areas." This and several smaller changes helped to consolidate similar topics and make the material flow more smoothly. The increased emphasis on water conservation initiated in the third edition is continued in this edition.

Much of this book can be read and understood by anyone with a good general education. Some parts, however, necessarily assume an acquaintance with basic soil properties such as texture, structure, water-holding capacity, and cation exchange capacity. These topics are covered in any introductory soil science textbook and one of these should be consulted if the reader lacks this background. The system of soil taxonomy used in the United States is followed in this book. An explanation of that system can also be found in modern introductory soils textbooks.

The broad collective background of the authors in soil science and soil conservation in the United States and abroad has been reflected in each edition and carries forward into this new edition. Much credit goes to Dr. Roy Donahue for having originated this project, enlisted his co-authors, and contributed enthusiastically to all three previous editions in spite of his advancing age. However, his death in 1999 made it necessary to change the handling of revisions for this edition. Suggestions were obtained from several users of the present text, and Dr. Troeh accepted the responsibility of incorporating these suggestions along with new material from the literature into the text. Dr. Hobbs contributed by reviewing all of the material and making valuable suggestions.

The authors' experience has been supplemented by extensive use of excellent libraries to locate appropriate literature including journals, books, and publications from government agencies. Many colleagues have also contributed valuable suggestions and some have thoughtfully reviewed the manuscript of one or more chapters dealing with subject matter in which they were especially well qualified. The helpful assistance of the following persons is gratefully acknowledged:

Paul L. Brown, ARS-USDA, Northern Plains Soil and Water Research Center, Bozeman, Montana (deceased)

Lee Burras, Associate Professor of Agronomy, Iowa State University

J. Brian Carter, Oklahoma State University

Julian P. Donahue, Assistant Curator, Entomology, Natural History Museum, Los Angeles County, California

George R. Foster, ARS-USDA, National Sedimentation Laboratory, Oxford, Mississippi (retired)

Paula M. Gale, Associate Professor, Plant and Soil Science, University of Tennessee at Martin

Harold R. Godown, NRCS-USDA (retired)

Robert Gustafson, Botanist, Natural History Museum, Los Angeles County, California

Lawrence J. Hagen, ARS-USDA, Northern Plains Area Wind Erosion Research Unit, Kansas State University, Manhattan, Kansas

Walter E. Jeske, NRCS-USDA, Washington, D.C.

John M. Laflen, Laboratory Director and Research Leader, USDA–ARS National Soil Erosion Research Laboratory, Purdue University, Lafayette, Indiana

Rattan Lal, Agronomy Department, Ohio State University

John Malcolm, USAID, Washington, D.C.

Gerald A. Miller, Associate Dean of Agriculture and Professor of Agronomy, Iowa State University

John A. Miranowsky, Professor of Economics, Iowa State University

Basil Moussouros, former Minister for Agriculture, Government of Greece

Kenneth R. Olson, Professor of Crop Sciences, University of Illinois

Gerald W. Olson, former Professor of Soil Science, Cornell University (deceased)

G. Stuart Pettygrove, Department of Land, Air, and Water Resources, University of California at Davis

Durga D. Poudel, Assistant Professor of Soil Science, University of Louisiana at Lafayette

Kenneth G. Renard, ARS-USDA, Southwest Watershed Research Center, Tucson, Arizona (retired)

E. L. Skidmore, ARS-USDA, Northern Plains Area Wind Erosion Research Unit, Kansas State University, Manhattan, Kansas

Barbara M. Stewart, NRCS-USDA, Des Moines, Iowa

Gene Taylor, formerly U. S. Congress from 7th District of Missouri (deceased)

Glen A. Weesies, NRCS-USDA, National Soil Erosion Research Laboratory, Purdue University

D. Keith Whigham, Professor of Agronomy, Iowa State University

C. M. Woodruff, Professor Emeritus, Department of Agronomy, University of Missouri (retired)

Frederick R. Troeh
Ames, Iowa

J. Arthur Hobbs
Winnipeg, Canada

CHAPTER

1

CONSERVING SOIL AND WATER

Soil and water are vital resources for the production of food, fiber, and other necessities of life. Food and fiber are renewable resources—a fresh crop can be grown to replace what is consumed. The soil that produces these renewable resources is essentially nonrenewable. Water can recycle, but its supply is limited, and it is frequently the limiting factor for crop production.

Strong reactions occur when there are shortages of food products or other consumer items. Prices of coffee and sugar, for example, have increased dramatically when a significant part of the world's crop was damaged, often by unfavorable weather, temporarily decreasing the supply. Such situations arise suddenly and require adjustments in the lives of many people. The more gradual changes resulting from persistent processes such as soil erosion may escape attention despite their fundamental importance. The long-term loss of productivity caused by soil erosion should be of greater concern than temporary shortages.

The purpose of soil conservation is not merely to preserve the soil but to maintain its productive capacity while using it. Soil covered with concrete is preserved, but its ability to produce crops is lost in the process. Intensive cropping uses the soil but often causes erosion on sloping land. Land needs to be managed for long-term usefulness as well as for current needs; that is, its use should be sustainable. Scarred landscapes, as shown in Figure 1–1, tell a sad story of waste and ruin where long-term principles have been sacrificed for short-term gain.

Soil erosion is often more detrimental than might be supposed from the amount of soil lost. The sorting action of either water or wind removes a high proportion of the clay and humus from the soil and leaves the coarse sand, gravel, and stones behind. Most of the soil fertility is associated with those tiny particles of clay and humus. These components are also important in microbial activity, soil structure, permeability, and water storage. Thus, an eroded soil is degraded chemically, physically, and biologically.

Degradation of soil and water resources is a worldwide problem that takes many forms (Napier et al., 2000). It is especially severe in developing countries where people are struggling to eke out an existence and are more concerned with survival than with conservation. Each situation is different and calls for its own distinct solution.

Figure 1–1 The amount of soil eroded by gullies eating their way into a landscape is spectacular but is often exceeded by sheet erosion around the gullies. (Courtesy USDA Natural Resources Conservation Service.)

1–1 NEEDS INCREASING WITH TIME

The demand for plant and animal products increases with time as population increases and standards of living are raised. People now consume more food than do all other land animals combined (Deevey, 1960). Their needs place an increasing load on soil productivity—a load that can severely strain the ecosystem (Wöhlke et al., 1988). Plants can be grown without soil by hydroponics and by sand or gravel culture, but the expense is high and the scale is small. Even seafood is used on a much smaller scale than are soil products.

Until recent decades, production increases came mostly by using more land. New frontiers were opened, forests were cut, prairies were plowed, and deserts were irrigated. It was suggested that one hectare (2.5 acres) of cropland per person was needed to maintain a satisfactory standard of living. A continually expanding land base maintained approximately that much area for a long time. Of course, the best land was chosen first, so the average suitability of the land declined even while the area per person was maintained.

The one-hectare-per-person rule is no longer supported. Most countries now have more people than hectares of cropland; the world average is declining and will soon be down to 0.1 ha per person (Lal, 1999). Production depends on soil, crop, climate, and management as well as land area. One hectare per person may not be adequate in some places, but it is enough to support ten or twenty people in other places.

In recent decades, the land base has been relatively constant. Most of the good cropland is already in use. Irrigation has been increasing and may continue to increase, but much of the newly irrigated land comes from previously rainfed cropland. The small areas of new cropland being added each year are offset by new roads and buildings on former cropland. Ryabchikov (1976) estimates that people already are using 56% of Earth's land surface, 15% of it intensively. Much of the rest is covered by glaciers, bare rock, steep slopes, desert, or other conditions that make it unsuitable for crop production.

Increased production is now obtained mostly by using present cropland more intensively. New crop varieties and increased fertilization are important factors producing higher yields. More intensive cropping systems increase row crops and grain crops at the expense of forage crops. Multiple cropping has increased, and the rest period in the slash-and-burn system has been reduced or eliminated in many tropical areas. The effect of these changes on soil erosion has been mixed. Fertilization and multiple cropping increase plant cover on land and reduce erosion. The replacement of forage crops with row crops and grain crops and the shortening of rest periods in slash-and-burn tend to increase erosion.

1–2 EROSION PROBLEMS

Erosion occurs in many forms as a result of several causes. Anything that moves, including water, wind, glaciers, animals, and vehicles, can be erosive. Gravity pulls soil downslope—either very slowly as in soil creep or very rapidly as in landslides.

1–2.1 Intermittent Erosion

Erosion can be uniform and subtle. Sheet erosion, for example, removes layer after layer a little at a time until a lot of soil has escaped almost undetected. Most erosion, though, is intermittent and spotty. Surface irregularities concentrate the erosive effect of either wind or water in certain spots. Cavities may be blown out by wind or gullies cut by water. The pattern is usually spotty, as illustrated in Figure 1–2.

Generally, more than half of the annual soil loss in an area occurs in only a few storms during which rain and wind are intense and plant cover is at a minimum. Weeks, months, or even years may pass without much soil being lost. The loss from a single ferocious storm sometimes exceeds that of an entire century.

The spotty and intermittent nature of erosion complicates the interpretation of erosion measurements. A field with an average soil loss of 4 tons/ac (9 mt/ha) annually is within the accepted tolerable rate for most deep soils if the loss is evenly distributed. But if most of the loss comes from part of the field eroding at 40 tons/ac, that part of the field is being ruined by erosion. Furthermore, crops on adjoining areas may be suffering damage from sedimentation, as shown in Figure 1–3. An average over time is equally deceptive. The benefits of having only small soil losses for nine years are wiped out if severe loss during the tenth year completely destroys a crop and carries away the topsoil that produced it.

1–2.2 Accelerated Erosion

The normal rate of erosion under natural vegetation is in approximate equilibrium with the rate of soil formation. A particular set of conditions maintains sufficient soil depth to insulate the underlying parent material from weathering just enough so that soil is formed as

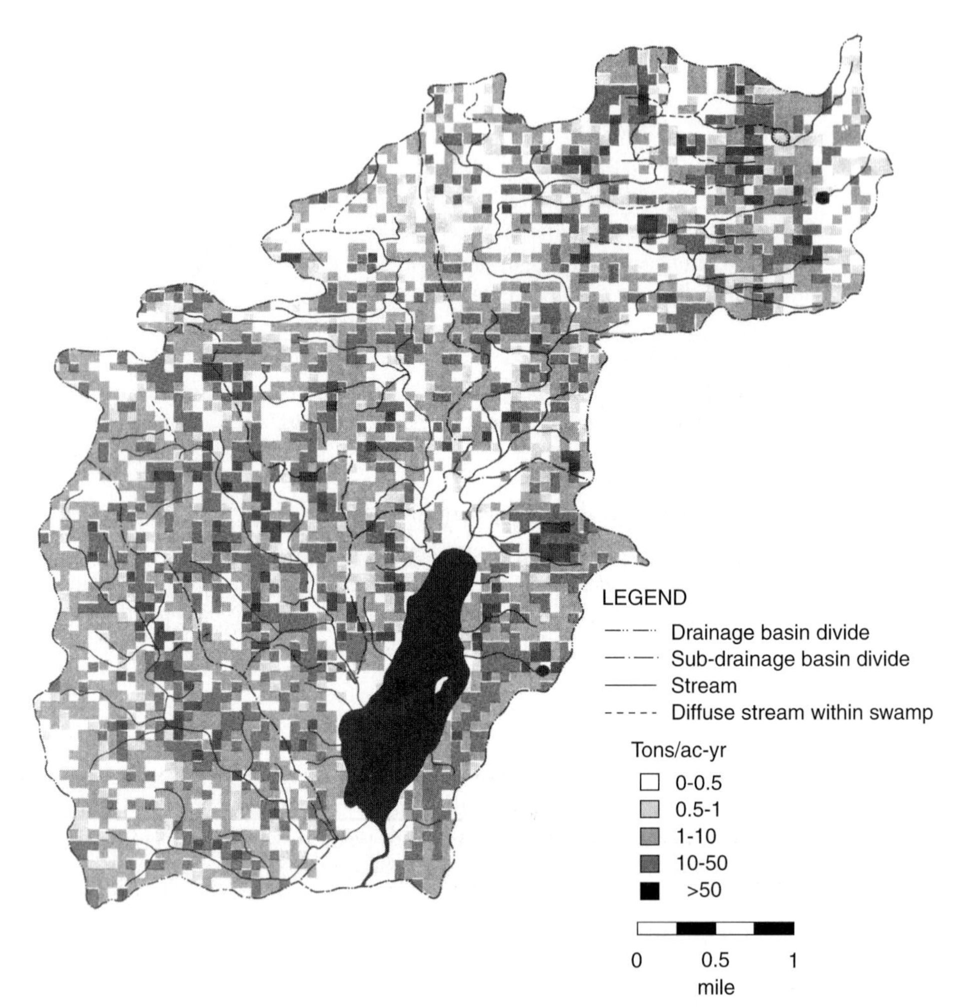

Figure 1–2 Estimated annual soil-loss rates by 10-acre (4-ha) cells in the Lake Canadarago drainage basin, New York. (From Kling and Olson, 1975.)

fast as it is lost. Deviations from equilibrium cause the soil to get either thicker or thinner until a new equilibrium is established. Precise data on rates of geologic (natural) erosion and soil formation are difficult to obtain but are thought to average about 0.5 ton/ac (1 mt/ha) annually (see Section 3–8).

Tilling cropland, grazing pasture or rangeland, or cutting trees nearly always increases the rate of soil erosion. Loss of soil cover reduces protection and may accelerate soil loss by a factor of 10, 20, 50, or 100 times. Formation of new soil cannot keep pace with greatly accelerated erosion rates, so the soil becomes progressively thinner, sometimes until little or no soil remains. The quality of the remaining soil generally deteriorates, not only because the soil has less depth but also because its physical, chemical, and biological properties become less favorable for plant growth (Lal et al., 1999). Islam and Weil (2000) suggest that microbial biomass, specific respiration rate, and aggregate stability are good indicators of soil quality.

Figure 1–3 Sediment from the higher areas covered and killed the crop in the foreground in this Iowa field. (Courtesy USDA Natural Resources Conservation Service.)

Accelerated erosion reduces the amount of plant growth a soil is able to support. The productive potential is reduced even if the actual production is maintained or increased by the use of fertilizer and other management techniques. A shallower soil, with its reduced capacity for storing water and plant nutrients and its generally poorer structure and aeration, cannot match the productive potential of the uneroded soil.

1–2.3 An Old Problem in a New Setting

Cultivated fields, overgrazed pastures, and cutover forestlands have suffered from erosion since the dawn of civilization in all parts of the world. The eroded soil becomes sediment that covers bottomlands and sometimes becomes so thick that it buries both fields and cities. The result becomes an archaeologist's treasure when a famous city such as Babylon is uncovered centuries after its inhabitants lost a frustrating battle with sediment eroded from nearby hills.

Gullies, sand dunes, and other obvious signs of erosion have caused concern since the beginning of agriculture. Impressive terrace systems were built thousands of years ago to stop erosion. Even so, entire soil profiles have been lost by sheet erosion, gullies have dissected hillsides, and sand dunes have drifted across anything in their path, such as the sidewalk shown in Figure 1–4. Many millions of acres of formerly productive land have been abandoned because of erosion damage.

In recent years, a new concern has been added to the age-old problems of erosion and deposition. Dust clouds and muddy water signify air and water pollution. Soil particles carry plant nutrients and other chemicals that contaminate water. Erosion has become an environmental problem that must be remedied for the sake of clean air and water. This new

Figure 1–4 Sand from a nearby beach drifted into the city and blocked this sidewalk in Montevideo, Uruguay. (Courtesy F. R. Troeh.)

concern has added an urgency to erosion control that should have been recognized earlier. Similarly, the increasing amounts of waste materials such as sewage sludge and mine tailings that are being spread on soil as a means of disposal generate concerns of soil pollution, especially by accumulation of heavy metals in the soil (Dinel et al., 2000). Soil and water pollution concerns are addressed in Chapters 16 and 17.

An increasing part of the impetus for soil conservation, especially that which is legally mandated, stems from environmental concerns. The early stages of pollution control concentrated on point sources such as sewage systems and smokestacks. Current efforts are beginning to include nonpoint sources such as soil erosion. Soil conservation practices must be used along with other pollution controls to protect the environment.

1–2.4 A Concern for All People

Eroded soil and the chemicals it carries are matters of concern because a degraded environment harms everyone's health and enjoyment. Polluted water, for example, is unsafe for drinking, swimming, and many other uses. It can kill fish; moreover, the surviving fish may impair the health and reproductive capacity of birds that eat them. Both the fish and the birds may be made unfit for human food.

Erosion adds to the cost of producing food and other soil products, thereby increasing the cost of living. With worldwide trade and emergency relief programs, the effects of reduced production in any major area spread through the world markets. Eswaran et al., (2001) point out that soil erosion, soil compaction, and plant nutrient depletion are worldwide problems that add billions of dollars per year to the cost of food production. In extreme conditions, ruined land must be taken out of production, and the increased load placed on the remaining land drives up production costs. Installing expensive erosion control practices also adds to production costs, but these practices help assure that production will continue.

Soil conservation legislation should be of concern to all voters, even those not directly affected by it. Government may provide too little, too much, or even the wrong kind of control in an effort to bring about effective soil conservation. Tax funds are used to pay the public's share of conservation costs. The public needs to understand and support the principles of soil and water conservation and environmental protection.

1–3 OBSTACLES TO CONSERVATION

Conservation is difficult to oppose, yet easy to overlook or ignore. Too many people give lip service to conservation but leave the application to someone else. Reasons for inaction include economic and aesthetic obstacles, insecurity and uncertainty, ignorance, and apathy.

1–3.1 Economic Obstacles

Major decisions are usually based largely on economic considerations. How much will it cost? What returns can be expected? Will the cost be repaid in a short time, in a long time, or not at all? Much reluctance to apply conservation practices is based on economics. The people who must spend money to conserve their soil are not the only ones who suffer if the soil is eroded or benefit if it is conserved. Often the persons most affected live someplace downstream or downwind or will live at a later time. People are commonly reluctant to spend their money for unknown beneficiaries; some are unwilling to spend money to conserve soil for their own future benefit.

Conservation practices vary greatly in costs, returns, and effectiveness. The easiest practices to promote are those like a good fertilizer program that will both conserve soil and return a profit within a short time. Longer-term practices such as liming and soil drainage may be recognized as desirable for some time before any action is taken. The time lag is still longer for terracing and other practices whose high investment costs require many years to repay. Least popular of all are practices such as changing to a less intensive land use with lower probable returns.

The economic value of many conservation practices is further complicated by benefits that accrue to persons other than those who install the practices. Reduced erosion generally means there will be less air and water pollution and probably less flood damage in downstream areas. Consideration of "externalities" shows that many conservation practices are economically desirable for society as a whole even though their costs exceed the on-farm benefits (Stonehouse and Protz, 1993). The farmer should not be the only one involved in the decision nor the only one involved in paying for such practices. This kind of situation may be resolved by governmental involvement in the form of laws and cost sharing for conservation practices.

1–3.2 Aesthetic and Cultural Obstacles

A great deal of pride can be involved in certain agricultural traditions. Straight rows, for example, are considered a mark of skill. Years ago, young farm workers were instructed "Don't look back!" because a tug on the reins would turn the horses and make a crooked row. Straight rows are appealing, but they cause erosion on hilly land by providing channels for runoff water to erode. Contour tillage is often the solution, but it must overcome tradition.

Figure 1–5 Much hand labor is used in areas where people are barely able to subsist by tilling the land. (Courtesy F. Botts, Food and Agriculture Organization of the United Nations.)

Many farmers take pride in plowing so that crop residues are completely covered. Unfortunately, this practice exposes the soil to the impact of rain, runoff, and wind. Conservation tillage reduces erosion by leaving residues on the surface. This concept is now widely recognized, but it still must overcome tradition before many of its critics will accept it.

1–3.3 Insecurity, Uncertainty, and Small Holdings

Many people in developing countries can barely eke out a living from their land by hard work such as the hand tillage shown in Figure 1–5. They know that traditional management has kept them and their predecessors alive, and that they have nothing to spare for gambling on a new method. It is difficult for them to change their techniques even for immediate benefits such as higher yields and less soil loss. It is still more difficult for them to adopt a practice that requires an investment, especially if the benefits are delayed or distributed over several years. The establishment of conservation practices under such conditions requires a reliable guarantee that these people will not starve to death if the new practice fails (Napier and Sommers, 1993).

Short-term tenancy prevents the adoption of many desirable practices. A one-year contract, or even a five-year contract, does not give the renter enough time to benefit from the sizable investment of money and labor required to install long-term conservation practices. Theoretically, the landowners should be willing to invest in sound long-term practices, but many owners are too far removed from the land to realize what practices are needed. Short-term tenancy makes it easy for both tenants and owners to overlook problems, even when those problems reach critical stages.

Small holdings are a common problem in developing countries. They may need conservation structures that cross several property lines or even need to be applied to an entire watershed to be effective (Pandey, 2001). Often, much of the benefit would go to people living downstream in the form of flood control and pollution prevention. Individuals cannot be expected to apply such practices.

1–3.4 Ignorance and Apathy

Most erosion occurs so gradually and subtly that its effects are easily overlooked until long after preventive action should have been taken. Even rills (small erosion channels) in a field are often ignored because tillage operations can smooth the surface again. Unproductive subsoil exposed on the shoulder of a hill is overlooked if the rest of the field remains productive. Even people who work with land often are unaware of how many tons of soil are being lost each year, of how costly these losses are, and of how short the useful life expectancy may be for a rapidly eroding soil.

Many people are apathetic about future needs and have short-term viewpoints regarding the use of soil and other resources. Land that was ruined in the past is unavailable now, and land that is ruined now is lost to future generations. Reduced productivity of eroded but usable land is even more important because it is more widespread.

Erosion-control practices needed to prevent environmental pollution often are not installed or are long postponed because of indifference. Some landowners claim the right to use their land as they please even if it is being ruined and even if the sediment is damaging other people's property. Public opinion and environmental considerations have provided the impetus for laws restricting the rate of soil erosion allowable under certain conditions.

1–4 CONSERVATION VIEWPOINT

The need for soil conservation has been clear enough to catch the attention of both modern and ancient people. For example, the people of ancient Rome, India, Peru, and several other places valued soil enough to build terraces that still stand today, such as those shown in Figure 1–6. Terrace walls were built of stones left on eroded hillsides; then laborers carried soil in baskets on their backs or heads from the foot of the hill up to the terraces to make level benches. The Chinese still carry out similar laborious projects, but most modern conservation structures are built with the aid of machines.

Concern for the land is the most important characteristic of a soil conservationist. Those who have such concern will find a way to conserve their soil and water; those who lack concern often neglect to use even the most obvious and inexpensive means of conservation. Conservation efforts, therefore, include education and persuasion aimed at convincing more people to care for their land.

Several organized groups now exist to promote soil and water conservation. The Natural Resources Conservation Service of the U.S. Department of Agriculture helps people install conservation practices; several other agencies assist their efforts. Employees of the Natural Resources Conservation Service work in cooperation with local Soil Conservation Districts that have their own national association. Interested individuals can become members of the Soil and Water Conservation Society, and there are many other groups at national, state, and local levels that advocate conservation of natural resources.

Figure 1–6 These terraces in Bolivia were built hundreds, or perhaps thousands, of years ago and are still protecting the soil from erosion. (Courtesy F. R. Troeh.)

Conservationists take a long-term view regarding the use of resources. Some land has been used for several thousand years and is still productive. All land needs to be used in ways that will maintain its usefulness. The objective of soil conservation has been stated as "the use of each acre of agricultural land within its capabilities and the treatment of each acre of agricultural land in accordance with its needs for protection and improvement."

1–5 CONSERVATION TECHNIQUES

The practices used for conserving soil and water are many and varied. Some practices are expensive and some only require new habits; some are permanent and some are temporary; some are limited to very specific conditions whereas others are widely useful, although none have universal application. The amount of soil and water saved varies from one practice to another and from one set of circumstances to another.

1–5.1 Land Use and Management

One of the first items a soil conservationist considers is the use of land within its capabilities. Some land is suited for intensive cropping, especially where the soil is deep, level, fertile, well drained, and has favorable texture and structure. Other land is so steep, shallow, stony, or otherwise limited that it is suitable only for wildlife or other nondisruptive uses. Most land is suitable for some uses but unsuitable for others.

Land use can be broadly classified into cropland, pastureland, woodland, wildlife and recreational land, and miscellaneous use. Each broad class can be subdivided several

Figure 1–7 A dense growth of bluegrass in this lawn provides excellent protection against erosion. (Courtesy F. R. Troeh.)

times. For example, cropland may be used for cultivated row crops, small grain crops, or hay crops. The soil exposure to erosive forces declines from cropland to pasture and woodland and then to wildlife land. These latter uses are therefore considered to be progressively less intense. Nonagricultural classes may parallel agricultural ones. For example, lawn grasses, as shown in Figure 1–7, might be roughly equivalent to a similar growth of pasture grasses.

Management can alter the erosive effects of land use. Row crops, for example, can be grown in wide or narrow rows that may or may not follow contour lines. The time of exposure to the elements between the harvesting of one crop and the protective growth of the next varies considerably. The soil may or may not be protected by crop residues or by special cover crops during periods when the main crop is not on the land. These variables have considerable effect on the amount of erosion that is likely to occur.

Variations also occur with other types of land use. Pasture, for example, may have grasses and legumes that were selected to provide good ground cover and forage, or it may have whatever happens to grow. The number of livestock may be limited to what the pasture can readily support, or overgrazing may kill much of the vegetation. Extreme over- and under-use may occur in the same pasture if the animals spend too much time in one area. Also, both soil and vegetation may be damaged by trampling if livestock are allowed to graze when the soil is too wet.

1–5.2 Vegetative and Mechanical Practices

Conservation techniques are often divided into vegetative and mechanical practices. There is no good reason for always favoring one type over the other; both include a wide variety of methods for protecting soil against erosive forces. Often, the best approach is to use a combination of vegetative and mechanical practices.

Vegetative practices include techniques that provide denser vegetative cover for a larger percentage of the time. Changing to less intensive land use usually reduces erosion considerably. The problem is that less intensive land use is usually less profitable. A crop rotation provides a compromise by using a series of different crops, some providing more income and some giving more soil protection. Crops grown for the purpose of protecting soil between other crops are known as cover crops.

Choices of land use, crop rotations, and cover crops need to be accompanied by good management practices that help each crop grow well. Good seed planted at the right time in a proper seedbed helps get the crop off to a good start. Adequate fertilizer and lime where needed promote vigorous growth. Narrow row spacing allows a row crop to provide better soil cover sooner. These management techniques generally improve both yield and erosion control.

Special vegetation is needed in critical places. Grassed waterways can prevent the formation of gullies. Windbreaks can direct air currents away from erodible land. Various forms of strip cropping reduce water erosion, wind erosion, and pollution. Appropriate plantings in odd corners, steep slopes, or other problem areas provide food and cover for wildlife as well as erosion control. Disturbed areas such as roadbanks and mine spoils need special plantings.

Vegetation can limit erosion to geologic rates (the rate of erosion under native vegetation defines the geologic rate for a particular setting). Grasses, trees, and other plants are nature's tools for controlling erosion. Although geologic rates are usually quite slow, they occasionally are as sudden and rapid as a landslide. Sometimes the rate of erosion should be reduced below the geologic rate by providing more than the natural amount of protection. More often, some increase above the geologic rate is permissible.

Mechanical methods broaden the choice of vegetation and allow higher-income crops to be grown even though the crops provide less soil protection. Contour tillage, for example, often reduces erosion to half of that resulting from straight-line tillage. Tillage systems that leave more crop residues on the soil surface reduce erosion markedly. The ultimate in reduced tillage, a no-till system, is an excellent means of conserving soil. Its use is expanding rapidly, partly because modern herbicides are helping to make it practical to reduce or eliminate tillage. Additional erosion control can be achieved by building terrace systems, such as those shown in Figure 1–8, to hold soil in the field. Soil movement may occur between terraces, but the soil caught in terrace channels will not pollute a stream. Of course, the channels must be cleaned periodically as a part of terrace maintenance.

Various structures made of concrete, wood, metal, or other sturdy material limit erosion by controlling water flow. Critical points occur where water must drop to a lower elevation. The water may be conducted through a pipeline, down a flume or chute, or over a drop structure. Pilings, riprap, or other bank protection may be used to keep a stream from meandering to a new location.

Mechanical methods of erosion control tend to be either very inexpensive or very expensive. Conservation tillage saves fuel, time, and money by reducing the number of trips and the total amount of work done on the soil. Contour tillage may require more planning and layout, and it generally adds some inconvenience in the form of short rows, but the fuel requirement for working across the slope is usually slightly less than that for up and down the slope.

Figure 1–8 Terraces such as these hold the soil on the field rather than letting it erode away. (Courtesy USDA Natural Resources Conservation Service.)

Tillage changes require no new investments unless new equipment is needed. However, these inexpensive practices are short-lived and must be repeated for each new crop. Most long-lasting mechanical methods of erosion control involve expensive structures such as terraces, dams, and drop structures. The earthmoving and concrete work required are costly. Expensive structures are usually justified by many years of usefulness and increased flexibility of land use.

1–5.3 Conserving Soil and Water Together

Soil and water conservation are so interrelated that they must be accomplished together. There are very few techniques that conserve one but not the other.

Both soil and water can be conserved by protecting the soil from raindrops that would puddle on the surface and produce a crust. Plant material intercepting raindrops helps maintain permeability so that water can infiltrate instead of running off. The soil acts as a reservoir that conserves water. Reducing both splash and runoff conserves soil.

Contouring, contour strip cropping, rough surfaces created by tillage, and terracing all increase infiltration by holding water on the land. Any runoff that occurs is slower and carries less soil. Streams fed by seepage and slow runoff have more uniform flow and lower flood peaks than would occur from unprotected watersheds.

Reducing erosion reduces the rate at which streams, ponds, and lakes fill with sediment. Reservoir capacities are thus maintained for recreation, flood control, power generation, and irrigation. Keeping sediment out of the water also lowers the supply of plant nutrients in the water and thereby reduces unwanted growth of algae and other vegetation. Soil particles absorb pollutants that are best kept out of water by keeping the soil on the land. The control of nonpoint sources of water pollution therefore centers on conserving soil.

Wind erosion control is more closely related to water conservation than it might seem. Water conservation is very important for plant growth in dry climates. Anything that slows runoff and helps get more water into the soil or keeps it there by reducing evaporation provides more water for plant growth. Improved plant growth in turn helps reduce wind erosion as well as water erosion.

1–6 CHOOSING CONSERVATION PRACTICES

Soil and water conservation is too complex to be solved by only one approach. Each situation needs to be analyzed to determine what problems and potentials exist and what alternatives are available.

1–6.1 Soil Properties That Influence Conservation

Many soil properties influence soil and water conservation, but a few deserve special emphasis because they strongly influence runoff and erosion control. Soil topography, depth, permeability, texture, structure, and fertility are worth consideration in relation to conservation.

Topography includes the gradient, length, shape, and aspect (direction) of slopes. These features control the concentration or dispersion of erosive forces such as runoff water and wind. Topography also influences the practicality of erosion control practices such as contouring, strip cropping, and terracing. These practices may be very helpful on long, smooth slopes but impractical on rolling topography with short, variable slopes.

Soil depth, the nature and thickness of soil horizons, and the type of underlying rock material all affect the rate of soil formation. The tolerable rate of erosion is much lower for shallow soils over hard bedrock than for deep soils underlain by loess or other unconsolidated material. Subsoils with high clay contents or other unfavorable properties need a covering of topsoil to support plant growth. Deep soil is favorable for water storage and plant growth. Where the soil is shallow, it may be impossible to smooth the land for drainage and irrigation or to move soil to build terraces and ponds.

Soil permeability helps determine how much water will run off and cause erosion. Soil permeability is most commonly limited by a soil surface puddled by raindrops or traffic, plowsoles or other compact layers, heavy subsoils with small water passages, frozen soil, and bedrock or cemented layers. Restrictive layers near the soil surface require little water to saturate the overlying soil and cause runoff to begin. Soil permeability also influences the functioning of subsurface drainage systems and septic tank drain fields.

Soil texture and structure influence soil permeability and erodibility. Clay can bind soil either into a solid mass or into structural units with pore space between them. Individual clay particles are difficult to detach from soil but can be moved long distances after they are detached. Sand particles are easily detached from sandy soils, but fast-moving water is

Figure 1–9 Soil survey reports such as these contain soil maps, descriptions, and interpretations for various uses such as soil and water conservation. (Courtesy F. R. Troeh.)

required to transport them. Silty soils are often the most erodible by water, because the silt particles are too large to stick together well and are small enough to be transported readily. Silt particles are small enough, however, to resist detachment by wind unless they are knocked loose by something else, such as moving sand particles.

Soil fertility is important to soil conservation because plant cover helps protect the soil. Vigorous growth produced on a fertile soil provides more complete cover and better protection than sparser growth. Fertilizer and lime are therefore important for soil conservation.

1–6.2 Maps for Conservation Planning

Most of the soil properties discussed in the preceding section can be mapped. Topographic shapes and elevations are shown by contour lines. Soil depth, texture, structure, and many other properties are considered in naming the soil series shown on soil maps. Slope gradient and past erosion are also generally indicated on soil maps.

Conservation plans are based on soil maps. The soil map units are classified on the basis of the intensity of land use for which they are suited and the treatment they need. Soil maps are often colored to make important features stand out for planning. The soil maps are published along with descriptions of the soils and interpretations for various uses in soil survey reports such as those shown in Figure 1–9.

1–6.3 Considering Alternatives

Any piece of land could be used and managed in a variety of ways. Some ways would cause disastrous damage or monetary loss, but several satisfactory ways usually remain after unsuitable uses are eliminated. For example, a field might be used for pasture or hay production without any special practices, for a crop rotation without excessive erosion if contour strip cropping and conservation tillage were used, or for intensive row crops if terraces were built and conservation tillage used. Increased intensity of use normally requires additional conservation practices to protect the land. Economic factors and personal preference are usually considered when a choice must be made from alternatives such as these.

Many choices depend on the type of agriculture being practiced. Growing hay or pasture on part or on all of one's land implies that the forage will be fed to livestock. Building terraces to permit more row crops fits a cash-crop operation. Cover crops can be used to protect the soil between the trees in an orchard. Irrigation makes it possible to grow a wide variety of crops in arid climates, and soil drainage permits previously wet areas to be cropped.

1–7 CARING FOR THE LAND

Conservationists see the possession and use of land as a stewardship. The land that one person has now was previously someone else's and will soon pass to others. Its condition should be as good when passed on as when it was received. The owner has a responsibility to society for the way the land is used and the care it receives. The authority of governmental units to tax land, to place restrictions on its use, and to require that access and some other rights be granted to others indicates that ownership is not absolute.

Soil and water conservation attitudes and practices are needed everywhere. Even the best land is subject to damage if it is abused. Good land, fair land, and poor land are all useful if they receive proper care. People need constant reminders not to choose short-term exploitation over long-term productivity.

The use and care of agricultural land are stressed throughout this book, but the conservation needs of nonagricultural land must not be overlooked. Erosion on a construction site is often more rapid than in any nearby field. Excess traffic, especially by off-road use of motorcycles, four-wheel drive, all-terrain, or other vehicles, can start a gully. Modified versions of agricultural practices may control erosion in these and many other circumstances. Vegetation and mechanical structures can be adapted to a wide variety of situations.

Large areas of land are devoted to raising grass for livestock or trees for wood products, recreation, and wildlife use. These uses can provide excellent protection for the land, but they can also be abused. Overgrazing a pasture degrades the vegetation and exposes the soil to erosive forces (Herrick et al., 1999). Similarly, road building, logging operations, or fire can open a forested area to landslides and gullying (Elliot et al., 1999).

Population growth makes good land stewardship more crucial. "How many people can the Earth support?" is a pertinent question (Brown and Kane, 1995). Increasing pressure from higher population densities makes the conservation task both more important and more difficult. The need for population control has become obvious enough to cause many programs to be developed for that purpose, such as the family planning center shown in Figure 1–10. These programs and soil conservation practices are both needed, literally, for the salvation of the world.

Figure 1–10 Family planning centers such as this one on Mauritius, an island in the Indian Ocean, are helping reduce birth rates and control population. This island has 850,000 people living in an area of 720 mi^2 (1865 km^2). (Courtesy P. Morin, Food and Agriculture Organization of the United Nations.)

SUMMARY

Erosion of the soil resource often goes unnoticed. The loss in productive capacity is usually worse than the tonnage indicates because erosion sorts the soil and removes the most fertile parts. Soil conservation seeks ways to use the soil without losing it.

Production increases formerly came mostly by cultivating more land. Now, increased production must be obtained by increasing yields and intensity of land use because new land is hard to find.

Erosion is so intermittent and spotty that averages fail to reveal much of the serious damage done to soil and crops. Cultivated fields, overgrazed pastures, and cutover forest lands have suffered from erosion since the dawn of civilization, and sediment has been polluting streams and burying fields and cities. Erosion control efforts such as terrace systems have been in use for thousands of years but have been inadequate to prevent the loss of millions of acres of land. The contribution of erosion to air and water pollution has become a major concern in recent years. Erosion, pollution, and soil conservation are costly to everyone.

The installation of conservation practices is costly, and people are reluctant to abandon traditional methods. Subsistence agriculture, short-term tenancy, ignorance of erosion problems, and apathy are additional obstacles, but people who care about the land find ways to conserve their soil and water.

Many different techniques are available for conserving soil and water. The first requirement is to select an appropriate use within the land capability. Good management and conservation practices come next. Protective practices may be vegetative, mechanical, or a combination of the two. The effectiveness of vegetative practices depends on the density of the vegetation and the percentage of time it covers the land. Permanent vegetation, such as grassed waterways, windbreaks, or other plantings, can provide protection for vulnerable sites. Mechanical methods such as conservation tillage and water-control structures permit the growth of higher-income crops.

Soil and water conservation must be accomplished together. Contouring, terracing, and protecting the soil surface against crusting all increase infiltration and conserve both soil and water. Keeping sediment out of water reduces pollution and lengthens the life of reservoirs. Water conservation in dry climates increases plant growth and reduces wind erosion.

Soil properties such as topography, depth, permeability, texture, structure, and fertility influence the erodibility of soil and the best choice of conservation practices. Topographic maps and soil maps identify many of these properties and are useful for conservation planning.

Good stewardship of land requires passing it on to others in good condition for continued productivity. Both agricultural and nonagricultural lands need soil and water conservation. Population growth makes land stewardship increasingly important.

QUESTIONS

1. In what ways can average rates of erosion be misinterpreted?
2. Why should a factory worker living in an apartment be concerned about erosion?
3. Why do people fail to adopt new methods of erosion control?
4. Why would one build expensive terraces to control runoff and erosion that could be controlled by inexpensive vegetative methods?
5. What influence has increased environmental concern had on soil conservation?
6. Why are different techniques needed to conserve soils of sandy, silty, and clayey textures?
7. What information useful for conservation planning can be shown on maps?
8. Why do some people do a much better job of soil and water conservation than others?

REFERENCES

Ashby, J. A., J. A. Beltrán, M. del Pilar Guerrero, and H. F. Ramos, 1996. Improving the acceptability to farmers of soil conservation practices. *J. Soil Water Cons.* 51:309–312.

Brown, L. R., and H. Kane, 1995. Reassessing population policy. *J. Soil Water Cons.* 50:150–152.

Deevey, E. S., Jr., 1960. The human population. *Sci. Am.* 203(3):195–204.

Dinel, H., T. Paré, M. Schnitzer, and N. Pelzer, 2000. Direct land application of cement kiln dust- and lime-sanitized biosolids: Extractability of trace metals and organic matter quality. *Geoderma* 96:307–320.

Elliot, W. J., D. Page-Dumroese, and P. R. Robichaud, 1999. The effects of forest management on erosion and soil productivity. Ch. 12 in R. Lal (ed.), *Soil Quality and Soil Erosion.* CRC Press, Boca Raton, FL, 329 p.

ESWARAN, H., R. LAL, and P. F. REICH, 2001. Land degradation: An overview. In E. M. Bridges, I. D. Hannam, L. R. Oldeman, F. W. T. Penning de Vries, S. J. Scherr, and S. Sombatpanit (eds.), *Response to Land Degradation.* Science Pub., Inc., Enfield, NH, 510 p.

GALL, G. A. E., and G. H. ORIANS, 1992. Agriculture and biological conservation. *Agric. Ecosystems Environ.* 42:1–8.

GREENLAND, D. J., 1994. Soil science and sustainable land management. Ch. 1 in J. K. Syers and D. L. Rimmer (eds.), *Soil Science and Sustainable Land Management in the Tropics.* Cab International, Wallingford, U.K., 290 p.

HERRICK, J. E., M. A. WELTZ, J. D. REEDER, G. E. SCHUMAN, and J. R. SIMANTON, 1999. Rangeland soil erosion and soil quality: Role of soil resistance, resilience, and disturbance regime. Ch. 13 in R. Lal (ed.), *Soil Quality and Soil Erosion.* CRC Press, Boca Raton, FL, 329 p.

ISLAM, K. R., and R. R. WEIL, 2000. Soil quality indicator properties in mid-Atlantic soils as influenced by conservation management. *J. Soil Water Cons.* 55:69–78.

KLING, G. F., and G. W. OLSON, 1975. *Role of Computers in Land Use Planning.* Information Bull. 88, Cornell Univ., Ithaca, NY, 12 p.

LAL, R., 1999. Soil quality and food security: The global perspective. Ch. 1 in R. Lal (ed.), *Soil Quality and Soil Erosion.* CRC Press, Boca Raton, FL, 329 p.

LAL, R., D. MOKMA, and B. LOWERY, 1999. Relation between soil quality and erosion. Ch. 14 in R. Lal (ed.), *Soil Quality and Soil Erosion.* CRC Press, Boca Raton, FL, 329 p.

LEE, L. K., 1996. Sustainability and land-use dynamics. *J. Soil Water Cons.* 51:295.

NAPIER, T. L., and D. G. SOMMERS, 1993. Soil conservation in the tropics: A prerequisite for societal development. In E. Baum, P. Wollf, and M. A. Zöbisch (eds.), *Acceptance of Soil and Water Conservation: Strategies and Technologies.* DITSL, Witzenhausen, Germany, 458 p.

NAPIER, T. L., S. M. NAPIER, and J. TVRDON, 2000. Soil and water conservation policies and programs: Successes and failures: A synthesis. Ch. 38 in T. L. Napier, S. M. Napier, and J. Tvrdon (eds.), *Soil and Water Conservation Policies and Programs: Successes and Failures.* CRC Press, Boca Raton, FL, 640 p.

NOWAK, P. J., 1988. The costs of excessive soil erosion. *J. Soil Water Cons.* 43:307–310.

PANDEY, S., 2001. Adoption of soil conservation practices in developing countries: Policy and institutional factors. In E. M. Bridges, I. D. Hannam, L. R. Oldeman, F. W. T. Penning de Vries, S. J. Scherr, and S. Sombatpanit (eds.), *Response to Land Degradation.* Science Pub., Inc., Enfield, NH, 510 p.

RYABCHIKOV, A. M., 1976. Problems of the environment in a global aspect. *Geoforum* 7:107–113.

STONEHOUSE, D. P., and R. PROTZ, 1993. Socio-economic perspectives on making conservation practices acceptable. In E. Baum, P. Wollf, and M. A. Zöbisch (eds.), *Acceptance of Soil and Water Conservation: Strategies and Technologies.* DITSL, Witzenhausen, Germany, 458 p.

TWEETEN L., 1995. The structure of agriculture: Implications for soil and water conservation. *J. Soil Water Cons.* 50:347–351.

WÖHLKE, W., G. HENGYUE, and A. NANSHAN, 1988. Agriculture, soil erosion, and fluvial processes in the basin of the Jialing Jiang (Sichuan Province, China). *Geojournal* 17:103–115.

CHAPTER

2

SOIL EROSION AND CIVILIZATION

The earliest cultivated fields were small, with tillage likely restricted to more level, productive soils, and with short periods of cultivation followed by long periods of fallow (shifting cultivation). Erosion losses were not serious until populations multiplied, permanent settlements were established, and the cultivated acreage was expanded to meet the increased need for food. Sharpened sticks and stones were no longer adequate tools for clearing land and raising crops (Warkentin, 1999), so hoes and wooden plows were invented (see Chapter 9). More intensive and widespread cropping increased erosion, decreased soil productivity, destroyed considerable land, and added sediment to streams, lakes, and reservoirs.

2–1 ORIGIN OF AGRICULTURE

Human beings were hunters and gatherers until relatively recent times. It is impossible to determine where crops were first cultivated, but relics from old village sites indicate the location and age of early tillage.

Archaeologists in 1946 uncovered an ancient village at Jarmo in northern Iraq (Braidwood and Howe, 1960). One relic found in the "dig" was a stone hand sickle. Other stone implements found at the village site could have been used for tilling the soil and for weeding growing crops. This village was occupied about 11,000 B.C.E. and is considered to be the earliest known site of cultivated agriculture. Other prehistoric villages were found in the area dating from 11,000 to 9500 B.C.E.

These villages were located on upland sites with friable, fertile, easily tilled, silt loam soils. The progenitors of modern domestic wheats and barleys are believed to have been among the grasses native to the area. Wild beans, lentils, and vetches were among the indigenous legumes. Rainfall probably amounted to 18 to 20 in. (450 to 500 mm) annually.

Villages that were occupied during the period 9500 to 8800 B.C.E. were also excavated on lowland sites in the southern part of Iraq, near the Tigris and Euphrates rivers. The climate of this area is considerably drier than that farther north. Crop production by dryland farming methods was nearly impossible, so irrigation had to be invented and used. The record is clear that water from the rivers was used to produce crops. An abundant food supply enabled large cities, such as Babylon, to develop.

2–2 EROSION IN THE CRADLE OF CIVILIZATION

Mesopotamia, an ancient country located between the Tigris and Euphrates rivers in what is now part of Iraq, is sometimes called the "cradle of civilization." It is supposed to be the locale of the biblical Garden of Eden and the Tower of Babel. Lowdermilk (1953) recounted the early history of this area. The following is a condensed version of his account.

Long ago, this area was covered by a great flood that wiped out the previous civilization and left a thick deposit of brown alluvium over the whole area. After the flood subsided, the area was resettled and known as Sumer. Kish was its capital city. Kish must have been a magnificent city in its day, but eventually its buildings were abandoned and buried by a thick layer of sand and silt eroded from the nearby hills. This sediment preserved the ruins until they were excavated by archaeologists in the early twentieth century.

Babylon, one of the most famous cities of ancient times, succeeded Kish as the capital of Mesopotamia. King Nebuchadnezzar was proud of having built the city of Babylon. He also boasted of building a canal and irrigation system and of cutting down huge cedar trees from Mount Lebanon to erect magnificent palaces and temples. He did not know that cutting down the forest and allowing sheep and goats to overgraze the hillsides would cause massive erosion. Nor did he realize that erosion would cause the downfall of Babylon by filling its irrigation canals with silt faster than slaves could clean it out. Babylon, too, was abandoned and its buildings were buried under about 4 m (13 ft) of erosional debris (Juo and Wilding, 2001). It, too, was later excavated by archaeologists.

The Sumerians of Mesopotamia numbered about 25 million at the peak of their power and prestige. By the 1930s, Iraq, a major part of ancient Mesopotamia, had a population of about four million. What happened to cause the area to lose most of its population? Lowdermilk suggests that the primary reason was that it was weakened by the failure of its agriculture as the canals silted full and the soils became saline.

Irrigated fields of the southern, lowland region were not damaged by soil removal, rather they were ruined by sediment from eroded land above them. Demand for food forced cultivation higher up the steep slopes in the watershed to the north. Sheep and goats overgrazed the hill pastures, and trees were felled indiscriminately for lumber and fuel. These practices denuded the watershed, causing severe erosion and erratic river flow. Large sediment loads, carried by the rivers in the sloping areas where flow was rapid, settled out in the more level areas, and salts from the irrigation water accumulated in the soils. Much sediment was deposited in the irrigation canals and ditches. In time, human labor was insufficient to cope with removal, so sections of irrigated land were abandoned. Drifting, windblown soil from the sparsely covered, abandoned lands filled the remaining irrigation structures. Eventually, the whole irrigated area had to be abandoned. Large urban populations could no longer be supported, and the area became a virtual desert. The area is very badly gullied now, and much of the original soil is gone. This was the first of many failed attempts to develop a permanent, productive, cropland agriculture.

2–3 EROSION IN MEDITERRANEAN LANDS

Communities on the trade routes between Mesopotamia and Egypt developed systems of arable agriculture shortly after tillage was first used. It is probable that the knowledge of cultivation and irrigation was carried from Mesopotamia to these countries by traders and other travelers. Unfortunately, much of the detailed history and knowledge was lost when the library at Alexandria burned in the third century C.E. (Warkentin, 1999). It was the greatest library of its time, with over a half million manuscripts.

2–3.1 Soil Productivity in Egypt

Egypt had a dry climate. Only the narrow Nile River floodplain could be cultivated. Natural flooding provided a type of irrigation with no canals. Little soil eroded because the floodplain was level.

The Nile River rises in the mountains and tablelands of Ethiopia far to the south of the irrigated Egyptian floodplain. Extensive forest cutting and cultivation of steep slopes in the upper reaches of the watershed caused considerable erosion. The coarser sediments were deposited mostly in the Sudan where the river left the high country and entered the more level plain. Sediment carried into Egypt was fine textured and fertile. The annual deposit helped maintain soil productivity. The system developed in the lower Nile valley was the first successful attempt to develop a permanently productive, cultivated agriculture with irrigation. But the High Aswan Dam, completed on the Nile River 600 mi (1000 km) south of Cairo in 1970, has upset the precarious balance that kept the floodplain productive (Thoroux, 1997). The new dam regulates the flow of the river, reduces major flooding, and traps most of the revitalizing silt in Lake Nasser, behind the dam. It was the annual flood with its regular fertile silt deposit that was largely responsible for the maintenance of the delta's productivity. What will happen now? Egypt and the world are anxiously watching to see what the final answer will be.

Bedouins have a long history of grazing livestock on the uplands above the valley of the Nile. The sparse vegetation in this desert area makes it subject to wind erosion, especially if overgrazed, and requires flocks to range over a large area. Some Bedouins have small areas of cropland in favorable sites. Briggs et al. (1998) discuss the reasons why some Bedouins choose to crop sites in the channel of a small stream (which is dry most of the time but stores water in the soil when there is runoff), in the floodplain adjacent to a major stream, or on the edge of Lake Nasser. The problem is that any of these sites is subject to flooding during wet periods and to drought at other times.

2–3.2 Erosion in Israel, Lebanon, Jordan, and Syria

The ancient lands of Canaan, Phoenicia, and Syria have long had sedentary populations and established, cultivated agricultures. Cultivation was restricted initially to gently sloping lowland areas, with flocks and herds utilizing the steeper-sloping lands as range. The gradual encroachment of cultivation on the steeper lands and the reduction of protective native cover by overgrazing and timber harvesting increased runoff, erosion, and sedimentation.

The Phoenicians were among the first people to experience severe erosion on steep cultivated slopes. They found that bench terraces made by constructing stone walls on the contour and leveling the soil above them reduced water erosion and made irrigation on steeply sloping land possible. The ancient terraces in this region are still being cultivated successfully. Large soil losses have occurred from nonterraced sloping land through the years.

Extremely severe erosion has occurred on more than a million acres of rolling limestone soils between Antskye (Antioch), now in Turkey, and Allepo, in northern Syria. From 3 to 6 ft (1 to 2 m) of soil has been removed. Half of the upland soil area east of the Jordan River and around the Sea of Galilee has been eroded down to bedrock. Much of the eroded material was deposited in the valleys, where it is still being cultivated. Even these floodplain soils are subject to erosion, however, and gullies are cutting ever deeper into them. Archaeologists found the former city of Jerash buried on a floodplain under as much as 13 ft (4 m) of erosional debris.

Some arid regions, such as the Sinai Peninsula, have been so severely overgrazed that the land is cut by extremely large gullies, despite low rainfall. Winds severely eroded the soil between the gullies, but as stones of various sizes were exposed, a closely fitted desert pavement formed that now prevents further wind damage. (See Figure 5–1.) The productive capacity of the soil has greatly deteriorated. As a result, population in many sections has been drastically reduced by starvation and emigration. Some areas that formerly produced food for export to the Roman and Greek empires now cannot produce enough to feed the small indigenous population.

Some authorities blame the decline in productivity, particularly in the drier sections, on a change of climate, but agricultural scientists are convinced that the climate has not changed sufficiently to account for this decline. They believe that soil loss due to water and wind erosion is the root of the problem (Le Houerou, 1976).

In spite of the damage done to the soils in this region, there is still hope for cultivation agriculture. Burgeoning population has forced the Israelis to produce as much food as possible locally. They are cultivating much more land than was ever tilled in their country previously, and with good results in spite of the dry climate. They are employing excellent farming practices, including the use of terraces and other soil and water conservation measures.

2–3.3 Erosion in Northern Africa and in Southern Europe

Knowledge of cultivation spread westward from Egypt and the Middle East to northern Africa and to Greece, Italy, and other parts of Europe.

Erosion in Northern Africa. Tunisia and Algeria, on the Mediterranean seacoast, with annual precipitation of about 40 in. (1000 mm), produced an abundance of crops in the early Roman era. There was substantial production even inland, where precipitation was much lower.

Carthagineans, residents of present-day Tunisia, were excellent farmers; cultivation techniques were advanced for the times, and yields were good. They used very careful water conservation methods and had extensive irrigation works. Water-spreading techniques were employed in many of the drier regions. Relics of grain-storage structures and olive-oil presses attest to the farmers' expertise and to the region's productivity. Grain produced in excess of local needs was exported.

Serious deterioration took place over the centuries despite the high initial productivity of the soils. Winter rainfall caused erosion on bare soils, and wind erosion often destroyed soils left without cover during drier parts of the year. After the fall of Rome,

Carthage was attacked by desert dwellers from the south (herders mostly), and the region's agriculture declined. Vegetative deterioration caused by neglect and overgrazing further reduced soil productivity. Much formerly productive land lost all of its topsoil as it was eroded down to a stony desert pavement. Desert encroached onto productive cultivated fields; the potential for food production was drastically reduced.

Soils in many areas that once supported large populations now provide food for only a few hundred people. Lowdermilk (1953) mentioned the ruined city of El Jem on the plains of Tunisia whose amphitheater could accommodate 65,000 people. He wrote that in the late 1930s there were fewer than 5000 inhabitants in the whole district surrounding the city's ruins. El Jem's cultivation agriculture was destroyed after only a short period of successful production. Local wars and invasion played a part in the massive decrease in production, but soil erosion was the main cause of the deterioration and ultimate demise of a productive system.

Erosion in Greece and Italy. The Greeks originally were a pastoral people. The upland areas of their country were covered with forests; the productive lowland soils were used for grazing. Farming gradually replaced flocks and herds. Productive valley soils were cultivated first but, as population increased, food needs demanded production from land higher up the hillsides.

Old Greek agricultural literature describes a few special soil-management practices, such as multiple cultivations, fallowing for one to several years, and deep plowing, which help to maintain productivity or at least reduce the rate of its decline. Soil erosion and soil deterioration, however, were rarely mentioned. Apparently they used no special erosion-control practices. More than 3 ft (1 m) of soil was washed from the surface of extensive areas, and in places the soil was eroded to bedrock. Severe gullying occurred on the steeper slopes, and lower-lying fields were buried under unproductive erosion debris.

Increased food needs and declining production from their soils caused the Greeks to exploit the grain-producing potential of their colonies in Italy, in northern Africa, and on Crete and smaller islands in the Mediterranean. The cheap grain imported from the colonies caused Greek farmers to shift to the growth of more profitable olives, grapes, and vegetables. Some areas that had not been too seriously impoverished continued to produce food grain to meet the country's emergency needs resulting from acts of war, piracy, and bad storms.

Italy was a colony of Greece for several centuries before Rome established its own empire. Initially, little land was cultivated, but when tillage was introduced both dryland and irrigation farming methods were used. Italy eventually became a granary for Greece.

When Rome first became independent of Greece, it produced all its food needs locally. Agriculture was a prominent and respected vocation. Many Roman authors assembled knowledge of successful farming methods used at home and abroad. Widespread use was made of fallow; the legume crop, alfalfa (lucerne), was highly recommended as a valuable forage and as a soil-fertility-improving crop. Other legumes were recommended and used also.

As Roman population increased, demands for food also increased. The level bottomland areas were insufficient to produce what was needed, so cultivation moved farther up the hillsides. Erosion from the uplands accelerated and became even more serious as forests were felled to supply timber for ships and for fuel. Denuded soils and huge gullies resulted. Dedicated and industrious farmers terraced and contoured much of their land to reduce erosion. Figure 2–1 shows some terraces currently in use in northern Italy.

Sediment still washed into streams and was deposited in irrigation works, making regular removal necessary. When the empire deteriorated as a result of wars and invasions,

Figure 2–1 Numerous terraces help to control erosion on this cropland near Miniato in northern Italy. (Courtesy F. R. Troeh.)

labor for sediment removal became scarce, and irrigated fields were progressively abandoned. Sedimentation in the major river channels caused frequent and destructive floods. As swamps developed close to the streams, malaria and other diseases increased in severity, forcing people to move to higher ground and to intensify farming on the steeper slopes. This caused still larger soil losses.

With growing populations, deteriorating soils, and smaller yields, Rome had to depend on imported grain, particularly from Carthage, Libya, and Egypt. Cheap imported grains and declining soil productivity forced farmers to abandon some fields and to switch to the production of olives, vegetables, and grapes on others, as shown in Figure 2–2.

Italian soils, despite severe damage, were more durable than those of Greece, and recuperated after the fields were abandoned. Reasonable yields can be produced with modern technology on most soils, but erosion is a continuing threat.

2–4 EROSION IN EUROPE

Western Europe, north of the major mountain ranges (Alps, Jura, and Pyrennes), has a forest climax vegetation. Rainfall is abundant in most areas. Local agriculture was generally improved when Roman methods were introduced.

2–4.1 Erosion in the United Kingdom

The Celts, pre-Roman inhabitants of Britain, had a well-developed, arable agriculture. They cultivated fields across the slope. Gully erosion was not a serious problem because rainfall was gentle. Washed sediment gradually built up on the downhill side of each field as a result

Figure 2–2 This vineyard near Florence, Italy, has widely spaced rows that allow room for soil-conserving grain and hay crops. Hillside ditches provide drainage below each row of grapes. (Courtesy F. R. Troeh.)

of sheet erosion and of tillage-induced downhill movement (Bennett, 1939). The Celtic agricultural methods were altered but not replaced by Roman techniques.

The Saxons, who later came to Britain from mainland Europe, introduced a system of long, narrow, cultivated fields, mainly on the more level lowlands. Most of the old Celtic fields were then abandoned, many permanently.

Erosion has been recognized in Great Britain since the middle 1800s, but it has not been widespread, and annual losses generally are not large. The extent and magnitude of erosion has been studied on a national scale since the late 1970s. Most measured losses are in the 1–2 t/ac-yr (2–4 mt/ha-yr) range, although some fields have lost up to 20 t/ac-yr (45 mt/ha-yr) under unusual conditions. Most erosion is blamed on shifts from ley farming to straight cereal production, with little or no animal manure or green manure returned to the soil. Overgrazing by sheep has also been blamed for some instances of erosion.

Soil erosion was more severe in Scotland than in England because more steeply sloping areas were cultivated, soils were sandier, and amounts of rainfall were greater there (Arden-Clarke and Evans, 1993). The detrimental effects of erosion were recognized early in Scotland, and conservation practices were developed. Contour ridges were recommended and used, and a predecessor of the graded terrace was developed.

2–4.2 Erosion in France, Germany, and Switzerland

Erosion became extremely severe in many hilly areas of central Europe as steep slopes were cleared of forest cover and cultivated. More than a thousand years ago, the farmers in what is now France returned some cultivated land to forest and developed bench terraces. Some terraces were constructed on slopes as steep as 100% (45°). The soils on the benched areas

Figure 2–3 Steep areas in the Swiss Alps are either left in forest or grazed judiciously. (Courtesy F. R. Troeh.)

were turned deeply every 15 to 30 years as the soils became "tired." These terraced areas are still being used. Lowdermilk (1953) suggested that the Phoenicians were responsible for these developments. Appropriate land use and careful husbandry reduced erosion losses over most of the region. The steep lands at high elevations are now used only for forest and pasture, as shown in Figure 2–3. Moderate slopes at lower elevations produce grain crops, as in Figure 2–4. Mainly truck crops are grown on level bottomlands, as shown in Figure 2–5.

Wind erosion is a serious menace in Europe also. Sandy soils subject to wind damage are found along seacoasts and inland where water-laid or glaciated coarse deposits occur. Systems of revegetation have been developed to hold the sands and prevent drifting in most areas where wind erosion poses a threat. These sandy soils are generally used for pasture or forest and are rarely cultivated.

2–4.3 Soil Reclamation in the Low Countries

The Low Countries include most of The Netherlands and a part of Belgium. They occur on a generally flat plain with little land more than 150 ft (50 m) above sea level. Instead of losing land by soil deterioration, arable area has been increased by reclaiming land from the sea. This is a very expensive method of acquiring land, but it is justified by the extreme population pressure of the area. The new land is obtained by building dikes such as the one shown in Figure 2–6; pumping the salt water out, originally with windmills such as the one in Figure 2–7, now with large electric pumps; and using river water for reclamation and irrigation (Note 2–1). This new land is a mixture of sandy soils, clayey soils, and peat. Water erosion was never serious on these lands, and wind erosion has been well controlled by protective vegetation on sandy sites. The soils in this area are generally more productive now than they were when first reclaimed.

Figure 2–4 Upland areas in Switzerland are cropped where the soils are favorable and the slopes are not too steep. Pasture and woodland are interspersed with the cropland. (Courtesy F. R. Troeh.)

Figure 2–5 Level alluvial soils in Swiss valleys are used for vegetables, such as the lettuce in this field, and other high-value crops. (Courtesy F. R. Troeh.)

Figure 2–6 Dike protecting lowlands in the southern part of The Netherlands. (Courtesy F. R. Troeh.)

Figure 2–7 Windmill near Kapelle, The Netherlands. Windmills were the original power source for pumping water to reclaim the polders. (Courtesy F. R. Troeh.)

NOTE 2–1
RECLAMATION OF THE POLDERS

The people of The Netherlands and Belgium have been reclaiming land from the sea for hundreds of years to help meet the needs of dense populations. Most of the reclaimed land was near sea level and much of it was reclaimed in relatively small tracts. The polders form a much larger and better-known reclamation project than any of the others.

The general plans for large-scale reclamation were drafted about 1890 by Cornelis Lely, but initiation was delayed until the 1930s. The Zuiderzee, an arm of the North Sea that reached deep into the Netherlands, was cut off and converted into a body of freshwater called the Ijsselmeer. Five polders covering about 545,000 ac (220,000 ha) have been reclaimed within the Ijsselmeer by the following procedures:

1. The Afsluitijk, a barrier dam 20 mi (30 km) long and nearly 325 ft (100 m) wide, was completed in 1932 to form the Ijsselmeer. River inflow gradually converted it to a freshwater lake. Excess river water is emptied into the ocean through sluice gates.
2. An area within the Ijsselmeer was surrounded by an inner dike, and the water was pumped out to form a polder.
3. Rushes were planted in the freshly drained polder to control weeds, use up water, and help aerate the soil.
4. Trenches, ditches, and canals were dug for drainage and irrigation. The rushes were burned and a crop of rape was planted the first year and wheat the second year. Both rape and wheat tolerate the initial saline conditions and are good soil conditioners.
5. After three to five years, the land was dry enough to replace the trenches with drain tile. Farmsteads were built, and the units leased to selected farmers.

The inland border around the basin that contains the polders is a hilly area. Much of the cropland there is covered with highly erodible loess deposits. Erosion has been a problem ever since they began to clear the forests decades ago, but it became much more serious after about 1975. The population has increased greatly. This has caused much additional urbanization with runoff from extensive paved surfaces and an accompanying intensification of agriculture (De Roo, 2000).

2–4.4 Erosion in Eastern Europe

Some sheet erosion occurs in Hungary, the Czech republic, Slovakia, and Poland, but gully erosion is not severe. This is because most of the rains are gentle.

Polish information (Ryszkowsk, 1993) indicates that cultivation began in that country between 5500 and 3000 B.C.E., when the area warmed up after the glaciers receded. There was very little erosion, however, until large numbers of trees were felled during the iron age (200–400 C.E.). About 20% of the country's land was being cultivated at the end of the tenth century. Cultivated acreage continued to increase as population increased until about 1500 C.E., when half the land was being farmed. Erosion was still rather insignificant at that time, but it became more serious at the end of the eighteenth century when row crops, especially potatoes, became more common. At present, 39% of the cultivated land is being damaged by water erosion and about 11% is subject to wind erosion, but erosion rates are low compared to those in most other areas of the world.

A type of agriculture was developed throughout this region that was uniquely adapted to forest soils. It improved rather than destroyed soil productivity (Jacks and Whyte, 1939). Adapted crop rotations are used, animal manure and crop residues are returned to the soils, and judicious rates of lime and fertilizers are now being applied. Cultivated crops in most of the area are produced only on gentle slopes. Grain and other agricultural products are imported when local production is not sufficient to meet national needs. The main conservation needs of these soils are the maintenance of soil fertility and the improvement of drainage.

2–5 EROSION IN RUSSIA AND ASSOCIATED NATIONS

Russia is the world's largest country, spanning eastern Europe and northern Asia. It encompasses a wide range of climate, topography, and soils. Much of its land, especially in Siberia (the Asiatic part) is too cold for use as cropland. Much of this area is forested, so erosion has not been a serious problem. It is thinly populated and has few roads, schools, and cultural institutions (Denisova, 1995).

Russia has a long, proud history in soil science, but the emphasis has been on soil classification, epitomized in the heritage of V. V. Dokuchaev, the 19th century "father of soil science." Soil erosion problems received little emphasis under the centralized planning system of the former Soviet Union. Erosion is receiving more attention recently along with other conservation problems such as soil pollution, salinization, and desertification. A new map has been prepared at a scale of 1 to 2.5 million (Kastanov et al., 1999). It shows that the most serious erosion problems in the northern part (north of 60° N latitude, dipping to 55° N latitude near the Ural Mountains), in the area of reindeer pastures and localized farming, are caused by runoff from snowmelt. Snowmelt becomes less significant farther south and rainfall erosion becomes the more important factor as the amount of cropland increases. The importance of water erosion diminishes and wind erosion increases in the drier southern part.

Severe erosion in Russia has a long history. Excessive tree cutting on forested slopes and overgrazing and cultivation of steep and semiarid lands have caused serious water and wind erosion in both European and Asian regions. Cultivation of "new lands" in southern Siberia after World War II increased the extent of submarginal, cold, and dry farming areas. Severe wind erosion occurred. Some of these areas proved so erodible that cultivation was abandoned.

Most of the Russian people live east of the Ural Mountains in the European part. As far as agriculture is concerned, the climate ranges from too cold in the north to too dry in the south. The best soils are south of Russia in Ukraine (part of the former Soviet Union but now a separate nation), but drought is a hazard in that area.

The Soviet Union was known for its large collective farms and centralized planning. Each collective farm had a quota it was supposed to produce for each of several different crops, even if some of them were not well suited to the area. Fields on the collective farms were large, and localized areas of severe erosion were largely ignored. Crop yields were often low, partly because of unfavorable weather and partly because the top-heavy system inhibited individual initiative. Much higher yields were obtained from small household plots where individuals were permitted to manage them personally. These small plots represented only 1.5% of the Soviet Union's agricultural land but contributed about 25% of the value of agricultural output, largely through a heavy emphasis on livestock production (Lerman, 2002).

The breakup of the Soviet Union in the 1990s led to land reform and much confusion. The amount of land in household plots has doubled to 3% of the agricultural land, and private farms have been formed on another 7%. The remaining 90% is still in large farms, but ownership has been transferred from the state to collectives that include all the people involved with each farm. Even so, the 10% of the land that is in private hands receives the best care and produces about 50% of the agricultural output (Lerman, 2002).

The southern part of the former Soviet Union has become several countries that struggle with their own problems, including wind erosion, salinization, desertification, and a shortage of irrigation water. For example, Uzbekistan has much land that was devoted to raising cotton under the Soviet system, but cotton requires large quantities of irrigation water and contributes to salinization (Craumer, 1995). They need to install drainage systems, shift to other crops that will use the water more efficiently, and probably abandon some of the poorest land.

2–6 EROSION IN ASIA

China, countries of south-central Asia (India, Pakistan, Bangladesh, and Sri Lanka), and of southeastern Asia have suffered catastrophic soil erosion. Cultivation began in China and India soon after the Sumerians developed it in Mesopotamia. This long history of land utilization is responsible for very severe erosion damage, although extreme population pressure has also caused severe exploitation and considerably increased soil losses.

2–6.1 Erosion in China

China is the third largest country in the world (slightly smaller than Canada and a little larger than the United States). It has the largest population of any nation, and its numbers are still increasing. It has abundant natural resources, but much of its land is too dry and too steep to be farmed safely. In fact, only about 13.5% is now thought to be suitable for cultivation. Annual precipitation ranges from near zero in the desert areas in the north and northwest to more than 60 in. (1600 mm) in the subtropical areas in the southeast. Rainfall tends to be erratic and is often intense.

Long-term historical records in China provide a basis for calculating the rate of sedimentation in the deltas of its major rivers (Saito et al., 2001). Until about 1000 C.E., the Huang He (Yellow River) delta is estimated to have received 100 million metric tons (110 million short tons) of sediment per year, mostly silt from the loess plateau, but it now receives approximately a billion metric tons each year, or ten times as much. Farther south, the Changjiang (Yangtze River) delta received about 240 million metric tons of sediment per year prior to 0 C.E. but now receives approximately 600 million metric tons each year. Together, these two rivers contribute about 10% of the sediment delivered annually to the world's oceans. Add to that the sediment deposited on floodplains and irrigated fields along the way and that carried by other streams. The annual losses from the loess plateau in the mid-1980s amounted to about 2.2 billion metric tons (2.4 billion short tons) (Dazhong, 1993).

These large increases in erosion rates reflect the activities of China's growing population. The loess plateau receives about 200 mm (8 in.) of precipitation in its northwestern part and about 600 mm (24 in.) in its southern part. It was originally covered with much grassland and considerable forest, especially in the southern part. Most of the land has been cleared, the level land first, followed by steeper and steeper slopes as the population increased. Large areas have been dissected by gullies as deep as 185 m (600 ft). Some areas between the gullies have lost entire soil profiles by sheet and rill erosion.

As erosion continued and accelerated, soil productivity decreased due to losses of topsoil, organic matter, and plant nutrients. Crop yields declined also, making it necessary to cultivate still more fragile land. And so the vicious circle continued.

The Chinese people, because of their long acquaintance with soil erosion, have developed considerable expertise in methods of controlling its ravages. Early records of soil conservation appeared in 956 B.C.E. The major control mechanism that was developed was the contour terrace. Terraces in farm fields were described in writings about the Tang dynasty (760 C.E.). Modern conservation research was initiated in the 1930s. Significant efforts toward soil conservation have been made since then, including installation of large numbers of terraces (often built by hand), reforestation, and replanting grasslands (Shi and Shao, 2000). These practices have reduced the erosion rate, but much land needs more protection. The rivers still run yellow with silt.

Serious erosion has occurred in other areas of China as well. Despite more rugged terrain and greater rainfall in southern China, water erosion has been less than on the loess plateau. These southern soils were less erodible, they were better protected by native forest vegetation, and crop cover was more adequate on cultivated fields. Bench terraces were used extensively on the steeper cultivated slopes; contouring was employed on less steeply sloping land.

Over the last century, however, significant changes have taken place. The Chinese Civil War from 1910 to 1940 and the village-centered, small-scale iron-smelting program during the 1950s led to large-scale deforestation. Much of the land that formerly was adequately protected by forest vegetation is now almost bare. The result has been that these deeply weathered soils, on steep slopes and under a high rainfall regime, have been severely damaged by water erosion. In addition, the sediment from the hills has buried many of the valley rice fields and has choked streams and reservoirs (Biot and Xi, 1993).

Wind erosion has also been severe in many sections of northern and western China where semiarid and arid climates prevail (Zhibao et al., 2000). Soil drifting has been most severe in desert areas and on cultivated fields during droughts in semiarid areas. An estimated 500 million ha (1.2 billion ac) have been degraded by wind erosion.

The lost soil that threatens the ability of the Chinese in the hilly lands to produce food becomes sediment that threatens millions of people on the plains in eastern China. The rivers level out in the east and the resulting sedimentation elevates the streambeds, reduces channel capacities, and causes periodic flooding. This is particularly true of the Hwang He (Yellow River). About 4000 years ago, the Chinese started constructing levees along the river. A system of double dikes about 400 mi (600 km) long has been built. The inner dikes are close to the river channel and the outer ones are about 7 mi (10 km) on each side of the inner levee system. River-borne sediment continues to raise the river channel within the levees about 4 in. (10 cm) a year (Dazhong, 1993). In many places now the river flows more than 40 to 50 ft (15 m) above the floodplain and delta. Floodwater used to break through the levees occasionally, laying waste to cultivated fields and homes, drowning more than three hundred thousand people in the twentieth century alone (Zich, 1997), and causing hundreds of thousands more to die of starvation. However, conservation efforts in the watershed have now reduced sedimentation sufficiently so that there has been no serious flooding in the delta for three decades (Dazhong, 1993).

2–6.2 Erosion in South-Central Asia

South-central Asia is a very old and geologically stable area with very diverse soils. The topography ranges from very steep slopes in the mountains and hills to level alluvial plains. Most soils are Oxisols and Ultisols in the higher rainfall sections, but a large belt of Vertisols has formed on basalt rock on the Deccan Plateau. Aridisols and Regosols are found in drier areas. It is a tropical and subtropical region with rainfall ranging from less than 4 in. (100 mm) in the northwestern desert to more than 400 in. (10,000 mm) annually in the western Ghats and along the border between Bangladesh and Assam. The region has monsoonal wet summers and dry winters.

Native vegetation controlled erosion over the subcontinent except in the desert and in some semiarid areas. When these soils were first cultivated, soil erosion and soil deterioration were negligible because shifting cultivation was employed with the recuperative periods ten or more times as long as the cultivation periods. As the population has increased, especially over the last 150 years, fallow periods have been shortened. Now, continuous cropping is the norm, giving the soil little chance to recuperate. Large, contiguous areas of cultivated land increase the potential for erosion also. Not all crops and cropping systems cause rampant erosion. Well-managed plantation crops, such as tea and coffee, provide good ground cover and thus help to reduce erosion.

On hilly terrain, the wholesale destruction of forest cover, beginning in the early 1800s and accelerating after 1852 when the Forest Land Settlement Act was promulgated, also greatly increased the potential for accelerated erosion. And with the deforested hills used for grazing by cattle and especially by goats, soon little vegetative cover was left to protect the soil from rain and torrents of runoff water (Khoshoo and Tejwani, 1993).

The most severely eroded area in south-central Asia is in the hills south of the Himalayas in what is now Pakistan. Livestock herders, woodcutters, and farmers combined to denude the slopes. As a result, torrents of water rush downhill during the rainy season, causing severe sheet erosion and forming deep gullies. Deterioration is not restricted to land from which soil is removed. Floods inundate fields and villages, and transported sediments are deposited on good soil, causing severe crop and soil damage.

Erosion is not the only factor causing soil deterioration in this region. Fields on more level plains and alluvial soils have not eroded seriously, but many have been ruined by excessive salt buildup.

Major obstacles that must be faced in developing plans to control erosion in this region are the tremendous demand for food, excessive livestock numbers, the general abject poverty of the people, and the great difficulty of communicating effective conservation information to millions of small farmers.

2–7 EROSION IN THE AMERICAS

There are two major areas in the Americas that have long histories of continuous cultivation: Peru and the Mayan area of Central America. In other areas of the Americas, population pressure did not force the people to use continuous cultivation until after the European settlers arrived. With increasing immigration, the number of fields and their sizes increased. Farm implements and animal power made it possible to cultivate larger fields and to incorporate crop residues thoroughly into the soil. Erosion increased and soil deterioration followed.

2–7.1 Erosion in Ancient America

In Peru, the Incas were expert soil and water conservationists. They developed bench terraces to control water and soil loss when the population became too large for adequate food production from the level bottomland soils. Relatively large areas were covered with these terraces, and many of them are still being cultivated. As shown in Figure 2–8, the face of each bench was a stone wall generally 3 to 10 ft (1 to 3 m) high, with some as high as 50 ft (15 m). Rocks, gravel, and other nonsoil materials were placed behind the wall to within a few feet (about a meter) of the level surface of the bench. This last meter was filled with

Figure 2–8 Rock-faced bench terraces constructed by the Incas at Machu Picchu, Peru (the "Lost City of the Incas"). The dark channel in the foreground carried irrigation water to the terrace. (Courtesy F. R. Troeh.)

Figure 2–9 Peruvian fields on a steep mountainside south of Cuzco. Hand tillage is used to raise grain crops on these steep slopes. (Courtesy F. R. Troeh.)

good soil, which was carried from the level bottomland, or more frequently from a considerable distance. They also installed much broader terraces in the stream valleys to prevent flooding and reduce soil erosion.

The Incas probably had the world's most effective erosion-control structures. Modern-day farmers are unwilling to pay the price in money, energy, and labor to emulate them. Peruvian farmers recently have cut the forests, overgrazed the range, and greatly reduced the length of the recuperative fallow between periods of cultivation. They are farming on exceedingly steep slopes, as shown in Figure 2–9. They use manure and even excavated tree roots for cooking fires because fuel is scarce. These practices have all combined to increase erosion greatly in a relatively short time.

The Maya of Central America specialized in corn production. Initially only the level lowlands were cultivated; eventually cultivation spread to forested slopes also. Many unprotected sloping sites were damaged so severely that the soils lost much of their productive capacity, and the streams—which were necessary to the continued existence of the population—were filled with sediment. Recently, archaeologists (Matheny and Gurr, 1979) found a complex system of Mayan bench terraces, dams and other water-diverting devices, and underground water-storage cisterns and walk-in wells in southeastern Mexico. These practices and structures apparently reduced soil and water losses.

Beach and Dunning (1995) also have discovered evidence of sophisticated Mayan conservation practices in sections of Guatemala. Modern-day Maya and recent immigrants have no knowledge of these important conservation measures; they have reverted to shifting cultivation, taking two or three crops, then abandoning the field for long periods of recuperation under forest vegetation. This system cannot support a large population such as that of the Mayan civilization at its peak.

Beach and Dunning propose that an intensive study of these old conservation methods should be undertaken so that the Mayan conservation ethic could be reintroduced to the people of the region and to others in similar circumstances around the world.

2–7.2 Erosion in the Amazon Rain Forest

Much of the uncultivated arable land of the world is located in tropical areas. One large block of this land is found in the Amazon Valley. Many believe that this land can be cultivated without difficulty, but this is not true. Soil deterioration from fertility loss and erosion is a constant threat.

Until relatively recently, the populations of most tropical regions have remained in balance with food supplied by a system of shifting cultivation. This is a permanently productive system, and it was common in the Amazon Valley.

Over the last few decades, the local population in the Amazon region has increased and, in addition, entrepreneurs have acquired large blocks of forested land for cultivation or for mining or other commercial activities. Too often, these efforts are undertaken with little concern for their long-term effects on soils and other environmental factors. Extensive denuded areas, from whatever cause, generally erode rapidly and spectacularly. In addition, valuable hardwood logs are burned or otherwise destroyed, climax vegetation is lost, perhaps forever, and carbon dioxide is released to the atmosphere where it contributes to global warming.

To date, the Brazilian government appears not to consider the situation serious. Steps need to be taken, however, to control these losses before it is too late.

2–7.3 Erosion in the United States

The influx of new settlers from all over Europe and the population buildup in the United States, particularly after 1800, provided labor and a market for expanded crop production. Crops were grown as monocultures, not as mixtures. These crops, particularly those grown in widely spaced rows, left the fields open to the beating action of the rain and to the force of runoff water for long periods each year. Cotton and tobacco production in the southern states permitted extensive erosion. Check-rowed corn grown in northern states (with plants placed on a grid so they could be cultivated both lengthwise and crosswise) was nearly as destructive. Rainfall intensity was much greater in the new land than it was in Europe, and far greater damage was done to the soil. With little knowledge of how to reduce the losses and with an abundance of new land available, a philosophy of unconcern was adopted. A few individuals developed new techniques to reduce erosion losses and to maintain productivity. Several of these innovations are still in use. Most landowners tried to cope with land deterioration, without much success, or they moved west to start again.

The western frontier vanished at the end of the nineteenth century; since then the nation has had to depend on existing farms. The productive capacity of many eastern farms has been greatly depleted; severe soil deterioration has occurred on many "new" farms in central and western parts also. Water erosion was the major hazard in the humid sections; wind erosion was a serious menace in drier regions.

The Dust Bowl of the 1930s brought wind erosion to the attention of the nation. It eroded millions of acres in an area of the southern Great Plains extending about 400 miles

(650 km) northward from the Texas panhandle and reaching about 300 miles (500 km) across, with its center near Liberal, Kansas (Hurt, 1981). This is a semiarid area that can grow good crops during favorable years. Much of the land was planted to wheat as part of the agricultural base needed to win World War I. Farm prices declined after the war was over and ultimately triggered the Great Depression. Farmers responded to the price declines by cropping more land. A severe drought cycle parched the land beginning in 1932 and lasted for about five years. Dust clouds obscured the skies and were carried eastward to the Atlantic coast and beyond. Lawmakers in Washington, D.C., could watch them roll by when they passed the law to establish the Soil Conservation Service in 1935. Crops were wiped out, land values plummeted, and farmers went broke. An estimated 165,000 people moved out of the area between 1930 and 1936 (Drache, 1996). Their plight became the theme for the novel *Grapes of Wrath* that was published in 1939. It won a Pulitzer prize for its author, John Steinbeck (1902–1968) and was made into a powerful movie. Another moving, personal account of this tragic period was a book called *An Empire of Dust* (Svobida, 1940) that was reissued in 1986 as *Farming the Dust Bowl* and became the basis for a Public Broadcasting System documentary, *Surviving the Dust Bowl.*

Drought occurs periodically in the Great Plains on about a 20-year cycle (Thompson, 1990), and the dry years in the 1930s struck when a large expanse of land had been tilled for wheat and other crops. Drought in the 1950s caused less severe wind erosion in the same general area, and others since then have been still less serious as tillage has been reduced and more of the land has been protected by perennial vegetation.

Loss of fertility was serious also. Limited applications of nutrients were made in humid areas prior to 1930; essentially no fertilizer or manure was applied in the drier regions. Summer fallow helped to stabilize crop production in semiarid areas, but it increased soil exposure to erosive rains and winds, and contributed to soil salinity (saline seeps).

Erosion surveys in the middle 1930s showed the following: 12% of U.S. cropland essentially ruined, 12% severely damaged, 24% with one-half to all of its topsoil lost, another 24% with measurable erosion loss, and only 28% unaffected by erosion. The survey showed that 3 billion tons (2.7 billion metric tons) of soil was washed off the fields, pastures, and forests of the United States annually. This is equivalent to the loss of the furrow slice 7 in. (17 cm) thick from 3 million ac (1.2 million ha) of land each year. The severe dust storm of May 12, 1934, that started in southeastern Colorado, southwestern Kansas, and the panhandles of Texas and Oklahoma, carried an estimated 200 million tons (185 million metric tons) of soil about 2000 mi (3000 km) to New York City and Washington, D.C., and out to sea.

Progressive farmers and research and extension personnel gradually came to realize the importance of reducing soil losses. They pressured Congress in the late 1920s and early 1930s to establish government agencies to combat erosion. These agencies are discussed in Chapter 19. National awareness prompted the passage of the Soil Conservation Act in 1935. This act established the Soil Conservation Service as a research and promotional body. Soil conservation activities in the United States have done much to reduce soil degradation, but erosion is still serious in many areas.

2–8 EROSION IN AUSTRALIA

Australia was thinly populated prior to European immigration. The aborigines were hunters and gatherers. They did not grow crops and had no grazing livestock (Hallsworth, 1987).

Cultivation and introduction of Western-style livestock began approximately 200 years ago. These introductions caused erosion to increase. Rapid increase in the wild rabbit population after their introduction in the mid-1800s further depleted vegetative cover and permitted increased erosion (Edwards, 1993). Water erosion became especially severe in southeastern Australia, where precipitation falls mostly in the summer and where intense thunderstorms are common. Forest cutting in Queensland also increased water erosion. Overgrazing dryland vegetation in southern and western Australia increased wind erosion.

After about 150 years of exploitive agriculture, concern about soil deterioration surfaced. Development and promotion of conservation measures were started in the early 1930s. Individual farmers began to develop and adopt soil improvement practices and state and Commonwealth governments initiated research, promotion, and financial programs to control soil losses.

Soil erosion and other forms of soil degradation continue to plague the nation, but soil productivity losses are being reduced and practices for improvement continue to be promoted.

2–9 EROSION IN AFRICA

Africa has a very wide range of climates and soils. In more temperate regions, animal agriculture was dominant with only a few, scattered fields. In tropical regions, animal agriculture was restricted in the lowlands by parasites, mainly trypanosomes carried by tsetse flies, so crop agriculture developed. Arable agriculture in the tropics always involved shifting cultivation.

Until recently, populations in most African countries have been in balance with the food supplied by animals and by shifting cultivation. This system broke down when the population increased dramatically. A much larger cultivated acreage is now needed and the fallow period must be greatly curtailed.

2–9.1 Erosion in Southern Africa

Erosion was not a serious problem in southern Africa until gold discovery accelerated European immigration in the late nineteenth century. Prior to that, population density was low, and livestock numbers were moderate. Native forage plants and forests thrived, and cropping was not intense. Even in the deserts and on the desert fringes, wind erosion was serious only intermittently. Immigration and a rapidly increasing native population expanded the need for cultivated land. Erosion quickly became severe in many areas and was exacerbated by steadily increasing numbers of livestock owned by native populations. Rainfall in the region is generally erratic. This increases erosion potential.

A recent study in the Kalahari Sandveld area of north-central Botswana (Sele et al., 1996) has shown that a long-term, slow aridification process has accompanied the successive wet/dry climatic cycles in the region for more than 10,000 years. This trend has changed little over the last 70 years. However, marked desertification (as indicated by less adequate water supply, less total biomass production, less palatable forage and browse, more bare soil, and more wind erosion) has been observed in the area recently. This change has accompanied dramatic increases in human and livestock numbers but no appreciable increase in cultivated land in the area, which remains a negligible acreage. Sele et al. (1996) blame the desertification on the synergy of low precipitation, overpopulation, and overexploitation, not on slow, long-term aridification.

Most countries in southern Africa have dual farming sectors—a commercial sector dominated by Europeans, and a communal sector operated entirely by native peoples. Erosion was apparent and became serious first on the commercial farms. With increasing native populations and enlarged flocks and herds confined to relatively small, poor-quality areas, however, erosion soon became very serious on the communal lands also. In many communal areas, overgrazing by livestock was responsible for more erosion than was cultivation. Serious consequences of erosion were first recognized in the 1920s and efforts were made by individual commercial farmers to initiate control programs. By the 1930s, government assistance was being demanded and provided. Quality of land, density of population, availability of capital, control over resources, and production of food and export revenue clearly favored the commercial sector in the battle against erosion and in getting the assistance and direction provided by government. Native sectors were left largely unassisted. With independence from European political dominance, some redresses are being made, but erosion losses from communal lands continue at very high levels. The developing countries in this region will have great difficulty funding suitable control programs without external assistance (Whitlow, 1988).

2–9.2 Erosion in Eastern Africa

Many east African countries have extensive areas with high elevations and temperate climates. Originally these, and more tropical areas, were occupied and farmed by native peoples without serious erosion taking place. European immigrants took over much of the arable lands in the temperate regions in the late nineteenth and early twentieth centuries, leaving drier and more tropical areas for the native population.

Erosion incidence increased after this partition, first on the "white" farms, but soon on communal lands also. Production of corn and cotton on more sloping land, overgrazing, and unrestricted felling of forest trees caused erosion to increase in Kenya. Increased cotton production was primarily responsible for erosion in Uganda. The desert enlarged in both countries in the early 1930s as a result of expanded cultivation on marginal land in the semiarid desert fringes and from overgrazing as livestock numbers increased.

Ethiopian highlands form the upper watershed of the Nile River. The original forest cover protected this area from erosion. Excessive forest cutting and overgrazing have denuded the land over the last several centuries. Whereas 75% of Ethiopia was forest covered, now only about 4% is protected in this way. Loss of forest cover caused rapid runoff and soil loss, and the highlands, particularly the Amhara Plateau, are now very severely eroded.

2–9.3 Erosion in the Sahara and Sahel Regions

The Sahara Desert is the largest desert in the world (Mann, 1993), extending about 3000 miles (5000 km) across northern Africa, in a band about 1000 miles (1600 km) wide south of the Mediterranean coastal areas discussed in Section 2–3.3. It is in the band where global air circulation causes descending air currents and aridity. It is best known for its large sand dunes, a few widely spaced oases, and a very sparse nomadic population. It also has large areas of desert pavement and some mountainous areas. There are no rivers or even major streams between the Nile River in Egypt and the western coastal rivers in Morocco because the annual precipitation of most of the area averages an inch (2.5 cm) or less. Most of the erosion is geologic wind erosion, and there is little that people can do to change it.

The Sahel is a transitional band that crosses Africa south of the Sahara Desert. It typically receives between 10 and 20 in. (250 to 500 mm) of precipitation per year. Like other semiarid lands, its precipitation is erratic and inadequate for many purposes. The drier northern part is used mostly for raising livestock, and the amount of cropland increases to the south. Food demand exceeds the capacity of the land to produce, so marginal lands are cultivated and even small areas of poor pasture are grazed. The drought problem is aggravated by poor communication and transportation systems. The problem caught worldwide attention in the 1970s, as thousands of people starved to death.

Severe wind and water erosion have occurred, and the northern part of the Sahel has become drier and drier. This process has been called desertification, as the land becomes part of the Sahara Desert. Various causes have been proposed, such as a long-term change in climate. The reduction in vegetative cover resulting from overgrazing and excessive cropping is a more likely reason (Sircar, 1993). Less vegetation not only leads to increased runoff and erosion but also to reduced transpiration. Reduced transpiration may lower the relative humidity and the probability of precipitation, but this effect is probably not measurable because the weather is highly variable (droughts of one or two years and longer-term dry periods lasting a decade or more are both common). It should be noted that these conditions also occur elsewhere and that desertification has occurred in other places, including parts of northern India, eastern Europe, and western United States.

2–9.4 Erosion in Western Africa

Prior to the beginning of the twentieth century, soils in western Africa did not suffer greatly from accelerated erosion. The population was relatively small, and needed food was produced from small fields in a system of shifting cultivation with short crop cycles and long fallow cycles. During the 20th century, however, population in the region increased rapidly, cultivated acreage expanded, the length of the cropping cycle increased, and the duration of the fallow cycle was curtailed.

Accelerated erosion began with these changes and became more severe in the northern savanna areas first, but erosion followed quickly in the forested areas in the south as well (Lal, 1993). With further population increases, and especially with the development of large cities, a system of continuous cultivation has developed around urban centers and elsewhere. Lack of roads from farms to markets and from villages to urban centers demands that vast areas in the vicinity of the cities must be cultivated every year. Local farmers cannot let land lie idle

Figure 2–10 Ironstone (hardened plinthite) in a tropical soil in Ghana exposed by the erosion of about 6 in. (15 cm) of overlying soil. (Courtesy Henry Obeng.)

long enough to replenish its productive capacity. As a result, erosion has increased, huge gullies have developed (Osuji, 1984), and soil productivity and crop production have decreased.

It is not enough to add fertilizers to continuously cultivated land in this region. Fertilizers are scarce and very high priced. Many nitrogen fertilizers cause soil acidification. This can have very serious consequences on the fragile soils in areas where liming materials are scarce.

Another cause of productivity loss in moister regions of west Africa and other tropical areas is the irreversible dehydration of hydrous iron oxides. When forest cover is removed for extended periods, the temperature and moisture regimes of the soils change and plinthite (soft iron concretions) in the soil hardens into ironstone pebbles or an ironstone layer that reduces or prevents water and root penetration. The resulting poor plant cover allows the friable soil above the ironstone to be washed away (see Figure 2–10).

2–10 EXPANDING INTEREST IN CONSERVATION

For centuries, only farm workers and farmland owners seemed aware of soil erosion. Only they appeared to have any interest in doing anything to reduce it or to ameliorate the damage it did to their soils and to their individual farms. Nonfarm people considered erosion a farmer's problem, if they considered it at all. Eventually some nonfarm, even urban, people grasped the idea that erosion is harmful to the land and eventually will hurt everyone. Some of these were lay people; others were employed in research or administrative positions in agriculture. These, together with some interested farmers, began to promote the idea of erosion control, and the soil conservation movement was born. At first the movement was concerned only with soil damage and its prevention, but gradually it became apparent that erosion had a much wider impact. Sections of the general public have recently become aware of their own environments and they are interested in all things that affect them. Their current interest in erosion includes the damage their environment can suffer when soils erode—soil loss, loss of soil productivity, sedimentation, water and air pollution, and more. Some organized environmental groups are becoming very active, some even militant. They believe it is time to force lagging farmers and landowners to reduce erosion and thereby reduce environmental degradation.

Soil erosion is a serious problem wherever there are relatively large populations of people and animals and where rainfall and wind are conducive to soil detachment and transportation. If appropriate techniques were selected and used worldwide, excessive soil losses could be reduced to tolerable levels, and soil productivity could be maintained indefinitely for the benefit of all the world's people. This is not being done, even though the tools and methods to do the job are at hand. More than knowledge is necessary. "Human desires, motives, and emotions, including greed and the drive for self preservation, are as crucial to the cause of soil erosion as rainfall (and wind) erosivity and soil erodibility" (Miller et al., 1985, p. 25).

SUMMARY

Accelerated erosion is older than recorded history. It has been a common cause of soil deterioration all over the world. Soil erosion has been severe from the earliest civilizations in the Middle East to the most recently cultivated areas in the Americas, Australia, and Africa. Soil deterioration has often been so great that the land was abandoned because it was no longer productive.

People have frequently failed to develop effective systems to maintain soil productivity over long periods. Soil deterioration resulting from erosion has caused the decline of many civilizations. A few civilizations have developed and used conservation measures that reduced soil deterioration and maintained soil productivity at satisfactory levels for many centuries. In many "new" lands, colossal productivity losses occurred in very short periods. This emphasized the extremely urgent need to develop and use effective conservation measures.

QUESTIONS

1. Point out where soil deterioration and sedimentation have played a major role in the decline of early civilizations.
2. What positive and negative effects did the Romans have on soil conservation methods in Europe?
3. How and why have people reclaimed land from the sea?
4. Why does China have gullies 600 ft deep and rivers 50 ft above their floodplains?
5. What kind of erosion prevails in northern Russia?
6. Explain why most European settlers in the United States failed to take adequate precautions to reduce soil erosion and deterioration as they moved into and across the country.
7. What caused the Dust Bowl? Why hasn't it happened again?

8. What causes desertification?
9. Describe the forces that have impinged on the stable farming systems found in many tropical countries and caused the systems to break down.

REFERENCES

ARDEN-CLARKE, C., and R. EVANS, 1993. Soil erosion and conservation in the United Kingdom. In D. Pimentel (ed.), *World Soil Erosion and Conservation.* Cambridge University Press, Cambridge, U.K., p. 193–215.

BEACH, T., and H. P. DUNNING, 1995. Ancient Maya terracing and modern conservation in the Peten rain forest of Guatemala. *J. Soil Water Cons.* 50:138–145.

BENNETT, H. H., 1939. *Soil Conservation.* McGraw-Hill, New York, 993 p.

BIOT, Y., and L. X. XI, 1993. Assessing the severity of the problem and the urgency for action. In E. Baum, P. Wolff, and M. A. Zoebisch (eds.), *Acceptance of Soil and Water Conservation: Strategies and Technologies.* DITSL, Witzenhausen, Germany, 458 p.

BRAIDWOOD, R. J., and B. HOWE, 1960. *Prehistoric Investigations in Iraqi Kurdestan.* Univ. of Chicago Press, Chicago, 184 p.

BRIGGS, J., I. D. PULFORD, M. BADRI, and A. S. SHAHEEN, 1998. Indigenous and scientific knowledge: The choice and management of cultivation sites by bedouin in Upper Egypt. *Soil Use and Management* 14:240–245.

CRAUMER, P., 1995. *Rural and Agricultural Development in Uzbekistan.* Royal Institute of International Affairs, 48 p.

DAZHONG, W., 1993. Soil erosion and conservation in China. In D. Pimentel (ed.), *World Soil Erosion and Conservation.* Cambridge University Press, Cambridge, U.K., p. 63–85.

DENISOVA, L. N., 1995. *Rural Russia Economic, Social and Moral Crisis.* Nova Science Pub., Commack, NY, 259 p.

DE ROO, A. P. J., 2000. Applying the LISEM model for investigating flood prevention and soil conservation scenarios in South-Limburg, the Netherlands. Ch. 2, p. 33–41 in J. Schmidt (ed.), *Soil Erosion: Application of Physically Based Models.* Springer, Berlin, 318 p.

DRACHE, H. M., 1996. *History of U.S. Agriculture and Its Relevance to Today.* Interstate Pub., Inc., Danville, IL, 494 p.

EDWARDS, K., 1993. Soil erosion and conservation in Australia. In D. Pimentel (ed.), *World Soil Erosion and Conservation.* Cambridge University Press, Cambridge, U.K., p. 147–169.

HALLSWORTH, E. G., 1987. Soil conservation down under. *J. Soil Water Cons.* 42:394–400.

HURT, R. D., 1981. *The Dust Bowl: An Agricultural and Social History.* Nelson-Hall, Chicago, 214 p.

JACKS, G. V., and R. O. WHYTE, 1939. *Vanishing Lands.* Doubleday, New York, 332 p.

JUO, A. S. R., and L. P. WILDING, 2001. Land and civilization: An historical perspective. In E. M. Bridges, I. D. Hannam, L. R. Oldeman, F. W. T. Penning de Vries, S. J. Scherr, and S. Sombatpanit (eds.), *Response to Land Degradation.* Science Pub., Inc., Enfield, NH, 510 p.

KASTANOV, A. N., L. L. SHISHOV, M. S. KUZNETSOV, and I. S. KOCHETOV, 1999. Problems of soil erosion and soil conservation in Russia. *Eurasian Soil Sci.* 32:83–90.

KHOSHOO, T. N., and K. G. TEJWANI, 1993. Soil erosion and conservation in India (status and policies). In D. Pimentel (ed.), *World Soil Erosion and Conservation.* Cambridge University Press, Cambridge, U.K., p. 109–143.

LAL, R., 1993. Soil erosion and conservation in West Africa. In D. Pimentel (ed.), *World Soil Erosion and Conservation.* Cambridge University Press, Cambridge, U.K., p. 7–25.

LE HOUEROU, H. N., 1976. Can desertization be halted? In *Conservation in Arid and Semiarid Zones.* FAO Conservation Guide 3. FAO, Rome, p. 1–15.

LERMAN, Z., 2002. The impact of land reform on the rural population. Ch. 2, p. 42–67 in D. J. O'Brien and S. K. Wegren (eds.), *Rural Reform in Post-Soviet Russia.* Johns Hopkins Univ. Press, Baltimore, 430 p.

LOWDERMILK, W. C., 1953. *Conquest of the Land through Seven Thousand Years.* USDA, SCS Agric. Inf. Bull. 99.

MANN, H. S., 1993. Deserts in India and Africa and related desertification problems. Ch. 2, p. 5–19 in A. K. Sen and A. Kar (eds.), *Desertification and its Control in the Thar, Sahara & Sahel Regions.* Scientific Pub., Jodhpur, India. 478 p.

MATHENY, R. T., and D. L. GURR, 1979. Ancient hydraulic techniques in the Chiapas Highlands. *Am. Sci.* 64(4):441–449.

MILLER, F. P., W. D. RASMUSSON, and L. D. MEYER, 1985. Historical perspectives of soil erosion in the United States. In R. F. Follett and B. A. Stewart (eds.), *Soil Erosion and Crop Productivity.* Am. Soc. Agron., Madison, WI, p. 23–48.

OSUJI, G. E., 1984. The gullies of Imo. *J. Soil Water Cons.* 39:246–247.

RYSZKOWSK, L., 1993. Soil erosion and conservation in Poland. In D. Pimentel (ed.), *World Soil Erosion and Conservation.* Cambridge University Press, Cambridge, U.K., p. 217–232.

SAITO, Y., Z. YANG, and K. HORI, 2001. The Huanghe (Yellow River) and Changjiang (Yangtze River) deltas: A review on their characteristics, evolution and sediment discharge during the Holocene. *Geomorphology* 41:219–231.

SELE, F., S. RINGROSE, and W. MATHESON, 1996. Desertification in north central Botswana: Causes, processes, and impacts. *J. Soil Water Cons.* 51:241–248.

SHI, H., and M. SHAO, 2000. Soil and water loss from the Loess Plateau in China. *J. Arid Environ.* 45:9–20.

SIRCAR, P. K., 1993. Droughts in India and Africa. Ch. 3, p. 21–38 in A. K. Sen and A. Kar (eds.), *Desertification and its Control in the Thar, Sahara & Sahel Regions.* Scientific Pub., Jodhpur, India, 478 p.

SVOBIDA, L., 1940. *An Empire of Dust.* Caxton Printers, Caldwell, ID, 203 p.

THOMPSON, L. M., 1990. Impact of global warming and cooling on midwestern agriculture. *J. Iowa Acad. Sci.* 97:88–90.

THOROUX, P., 1997. The imperial Nile delta. *National Geographic.* National Geographic Society, Washington, D.C., 191(1):2–35.

WARKENTIN, B. P., 1999. The return of the "other" soil scientists. *Can. J. Soil Sci.* 79:1–4.

WHITLOW, R., 1988. Soil conservation history in Zimbabwe. *J. Soil Water Cons.* 43:299–303.

ZHIBAO, D., W. XUNMING, and L. LIANYOU, 2000. Wind erosion in arid and semiarid China: An overview. *J. Soil Water Cons.* 55:430–444.

ZICH, A., 1997. China's three gorges. *National Geographic.* National Geographic Society, Washington, D.C., 192(3):2–33.

CHAPTER

3

Geologic Erosion and Sedimentation

Geologic erosion processes are natural. They include slow but persistent processes of water and wind erosion punctuated by catastrophic events such as glaciation, landslides, and floods, but they exclude accelerated erosion resulting from tillage, traffic, overgrazing by livestock, cutting of trees, or other human influence. Understanding geologic erosion improves one's comprehension of erosion processes and underscores the importance of controlling accelerated erosion.

3–1 THE GREAT LEVELER

Erosion is known as the "great leveler" because it wears down high places and fills up low places. Valleys are created and the hills between them are eroded. The theoretical end result of long-continued erosion is a rather featureless plain.

Geologic erosion is very old and very persistent. It has worn mighty mountain systems down to stubs. Entire sea basins have been filled with sediment, then lifted and eroded into new forms. Earth's landscapes are shaped and reshaped by the action of geologic forces, especially erosion and sedimentation.

The nature of sedimentary deposits depends on the source of the material, the transporting agent, and the environment of deposition. Geologists learn much about the climate, vegetation, and landscapes of the past by studying sediment and rock layers.

3–2 ROCK TYPES

The Earth is gradually releasing heat from its interior. This heat is presumed to originate from the decay of radioactive elements. This type of heat source gradually diminishes with time, so it must have been stronger in the past than it is now. Projecting this idea backward over a long period of time indicates that the entire Earth was once molten. The original crust was formed of igneous rocks that solidified as the Earth cooled. Weathering forces caused the surface of the early rocks to crumble into small particles. The leveling action of erosive forces has been active ever since. Sedimentary and metamorphic rocks now cover about 85% of the land surface, but the underlying "basement" rocks are igneous (Note 3–1).

NOTE 3–1
ROCK TYPES

Rocks are broadly classified as igneous, sedimentary, or metamorphic depending on how they were formed. Igneous rocks form by solidification when magma (molten rock material) cools. The rate of cooling determines the grain size. Magma forms a fine-grained rock when it erupts to the surface as a lava flow that cools quickly. Lava rocks are extrusive in contrast to the slower-cooling, coarser-grained intrusive rocks. A simple classification of four important igneous rock types is as follows:

	Extrusive (fine-grained)	Intrusive (coarse-grained)
Rocks containing quartz crystals	Rhyolite	Granite
Rocks lacking quartz crystals	Basalt	Gabbro

Basalt and granite are the most common igneous rocks. Several other types are defined as intermediates or extremes of the above types or as having special mineral contents.

Sedimentary rocks form where transported material is deposited. Soft sedimentary rocks are classified by transporting agent and grain size. Some important examples are:

Alluvium: material deposited by water
Colluvium: material moved downslope by gravity
Glacial till: material deposited by glaciers
Loess: silty material deposited by wind

Hard sedimentary rocks include a cementing agent or process in their formation. Some important examples are:

Sandstone: cemented sandy alluvium
Shale: clayey material compressed enough to bond together
Limestone: primarily precipitated calcium carbonate

Heat and pressure change other rocks into metamorphic rocks by making them recrystallize and form new, usually larger, crystals. The material does not melt—that would result in an igneous rock. Metamorphism may be either low grade (relatively small changes produced) or high grade. Some low-grade metamorphic rocks are easily related to an earlier rock type, but some are more general. The metamorphosed rock is often harder than the original.

Marble:	metamorphosed limestone
Slate:	metamorphosed shale
Quartzite:	metamorphosed sandstone with a high quartz content
Schist:	low-grade metamorphic rock high in small flakes of mica
Gneiss:	high-grade metamorphic rock with elongated grains and a general banded appearance

Many sedimentary deposits have accumulated to thicknesses of several miles (usually the area gradually sinks under the added weight, so the surface may remain near the same level). The pressure and heat of deep burial change the deposits into metamorphic rocks that later may be re-exposed at the surface by uplift and erosion.

3–3 PROCESSES THAT ELEVATE LAND

Erosion has worked long and persistently. It would have eliminated all dry land long ago, but fortunately there are equally potent and persistent counteracting forces. The net result is a balance that maintains a nearly constant amount of land. The processes that elevate land masses can be classified into three groups: deposition, lava flows, and uplift.

3–3.1 Deposition

Sediments that gradually fill a pond, lake, or other basin often produce fertile level land. Gray colors are common in sediments deposited in water. Some of these sediments include organic layers where the residues of lush vegetation fall into shallow water and are preserved. Bogs are formed where thick layers of organic materials fill a low area. Low, wet areas can become part of a smooth valley floor by the combined action of erosion on the high points and deposition in low areas.

Much of the material removed by young, actively eroding streams is deposited downstream where the stream floods across a valley. Small streams entering a large valley often deposit sediment in alluvial fans that slope down to merge with the valley floor, as does the one shown in Figure 3–1. Sediments deposited in valleys are commonly several feet thick, but some accumulate to thicknesses of a few miles. A large deposit can cause the area to sink and continue receiving sediment.

Figure 3–1 An alluvial fan is formed where a small stream flowing on a steep gradient enters a large valley and is slowed by the flatter slope. Orchards are grown on alluvial fans in this Idaho scene because the fan soils are well drained, but the bottomland soils are wet. (Courtesy F. R. Troeh.)

3–3.2 Lava Flows

Molten rock flowing out on the land, as shown in Figure 3–2, is an obvious way to increase elevation. Lava flows several feet (a few meters) thick are common. Most lava areas have multiple flows resulting in cumulative thicknesses ranging up to thousands of feet (several hundred meters). Some lava rock is covered by or layered with ash that spewed out of a volcano and was spread across large areas by wind. Loess layers and alluvial deposits may also occur on or between lava flows. The largest area of lava rock on land is the Columbia River Basalt, covering large areas of Washington, Oregon, and Idaho in the northwestern United States.

The ocean basins are lined with basalt (the most common type of lava rock). Undersea lava flows have built up mountains that reach miles above the seafloor and form islands where they protrude above sea level. The Hawaiian Islands are an outstanding example of thick lava accumulations surrounded by deep ocean.

3–3.3 Uplift

Slow, persistent uplift is the principal process that elevates land masses and offsets erosion. Both upward and downward movements occur, resulting in the warping and twisting evidenced by contorted rock layers, but the upward movement is dominant in the uplands.

Figure 3–2 The edge of a basalt lava flow in Oregon. This flow is about 50-ft (15-m) thick. (Courtesy F. R. Troeh.)

Uplift is an expression of *isostasy* (the tendency to equalize pressure). Even solid rock is able to flex and move across dimensions of many miles. Pressure inequalities produced by erosion are balanced by internal shifts in the Earth's crust. Eroded areas are lifted up again, and depositional areas gradually sink. Thus, uplift and erosion continually reactivate each other and produce stresses in the Earth's crust. Sometimes the stresses are released suddenly in an earthquake.

Density differences in rock types cause unequal elevations to produce equal pressures. Basalt dominates in ocean basins and has an average density of about 3.0 g/cm^3. Granitic rock in the cores of large mountain ranges has an average density of about 2.7 g/cm^3. The 10% difference in density permits the lighter-weight rocks in the mountains to "float" above the denser rocks. Like an iceberg, the mountain exposes only about 10% of its total mass. The other 90% is submerged and provides buoyancy, as shown in Figure 3–3. The steep slopes, exposure to severe weather, and a relatively high incidence of earthquakes cause mountain lands to be high-risk areas for both geologic and human-related hazards (Hewitt, 1997). Earthquakes, avalanches, landslides, volcanic eruptions, glaciation, and floods are examples of processes that can produce disasters in the mountains. Road building, tunneling, mining, and logging increase the hazards.

Continental masses are believed to be shifting slowly around the globe in a process known as *plate tectonics* that is widening the Atlantic Ocean at the expense of the Pacific. The subcontinent of India illustrates a dramatic type of uplift deduced within this framework. India appears to have split from Africa and drifted across the ocean to its present position. The "collision" of India with Asia pushed the Himalayan Mountains up as high as 29,023 ft (8846 m), making them the highest mountains in the world (Frank et al., 1987).

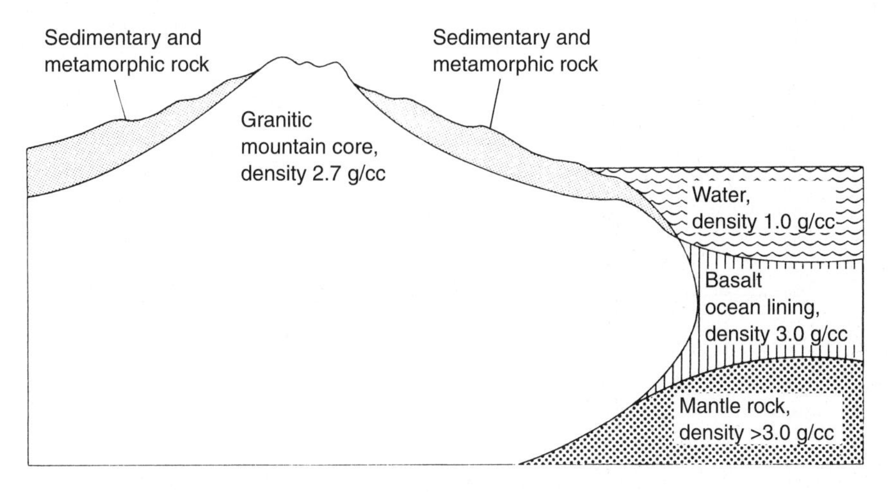

Figure 3–3 Land areas, especially mountains, protrude above sea level because they are formed of rock with a lower density than that lining the oceans.

3–4 LANDSCAPE DEVELOPMENT

Erosion, deposition, lava flows, and uplift play major roles in shaping landscapes. Usually the building processes such as uplift and lava flows are dominant during the early stages, but erosional processes take over and are dominant over longer periods of time. The results are seen as hills and valleys, mountains, plateaus, and plains. All landscapes, whether simple or picturesque, have been affected by these age-old forces. While the geologic history of some landscapes is fairly easily interpreted, that of others is very complex.

Some landforms are controlled by single agents of erosion or deposition in ways that produce close approximations of geometric shapes. River floodplains, for example, come close to being plane surfaces. The flow of a small stream into the side of a larger valley produces a nearly uniform slope profile in every direction it moves. The result is a fan-shaped area with contour lines approximating segments of concentric circles. The effect continues even though the slope gradient decreases with distance.

Erosional surfaces resembling alluvial fans are called *pediments.* A classic example of a pediment is located in the area of Gila Butte, Arizona. As shown in Figure 3–4, the pediment surface comes close to fitting a mathematical equation. The floodplain and other adjoining landforms do not fit the equation (Troeh, 1965).

3–4.1 Geomorphology

"The study of the Earth's surface forms, and of the processes that shape them, constitutes the field of geomorphology" (Butzer, 1976, p. 7). Geomorphology and soil science are so interrelated that neither can be understood without some consideration of the other. As Butzer says, "Soils modify the gradational process and, in turn, the external

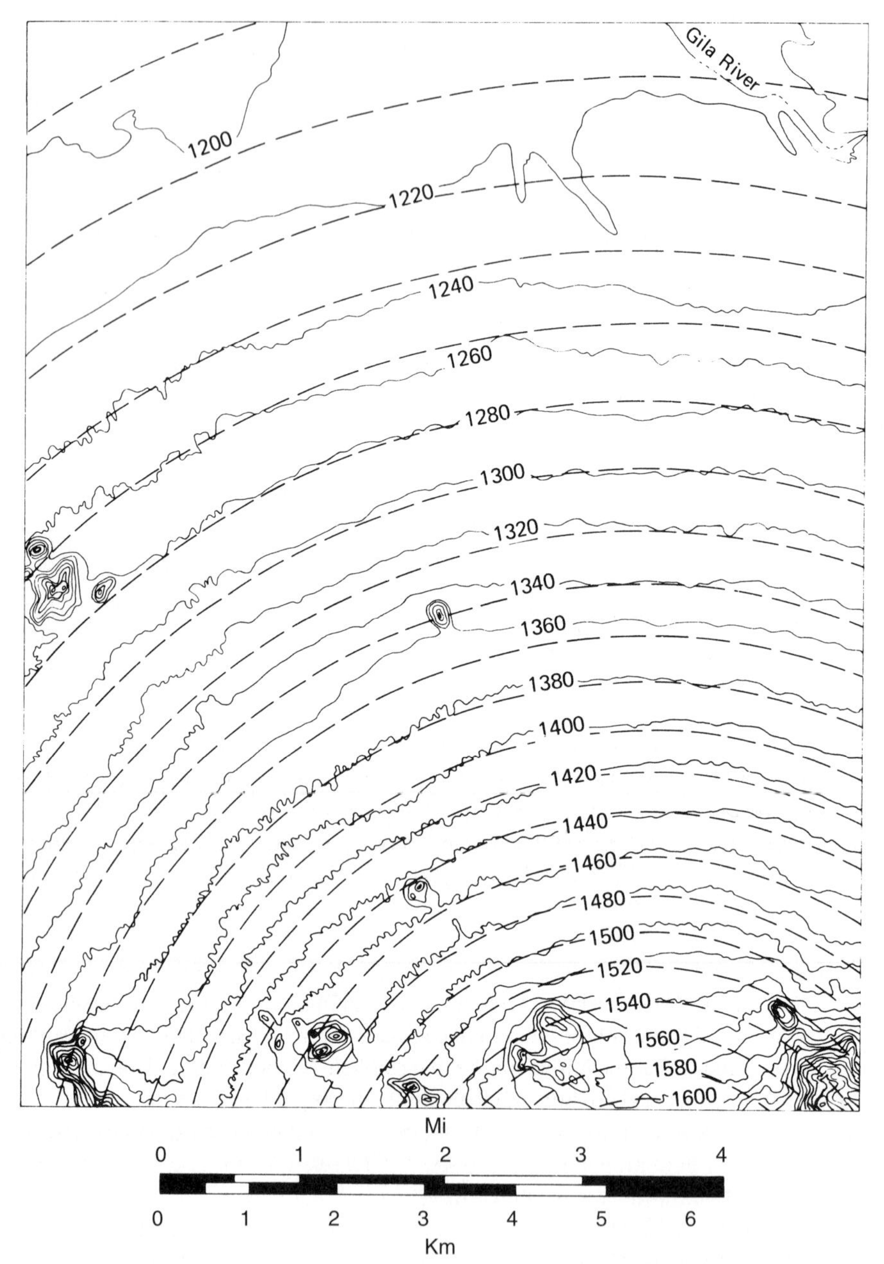

Figure 3–4 The contour lines of the area around Gila Butte, Arizona, are closely approximated by a series of concentric circles drawn to fit the equation $Z = 1707\text{ ft} - 0.0192R + (0.000{,}000{,}176/\text{ft})R^2$. Inserting a radial distance (R) in feet gives the corresponding elevation (Z) in feet above mean sea level. (Courtesy F. R. Troeh.)

agents of erosion and deposition affect the soil mantle and its development." Some principles of geomorphology are included in this section to help clarify the subject of landscape development.

3–4.2 Uniformitarianism

Credit for the early development of modern geomorphic principles is given to James Hutton (Bloom, 1998). Hutton presented his ideas in a paper in 1785 and in his book, *Theory of the Earth,* in 1795. The underlying theory that Hutton developed and used to explain how the Earth received its form is called the principle of uniformitarianism. Previous theories had supposed that landscapes were shaped primarily by sudden nonrecurrent events. Hutton argued that past processes are still at work and that one can study present processes to explain past events. Uniformitarianism is summed up in the statement, "The present is the key to the past." Due allowance must be made for variations in rates and importance of various processes, but strange and unknown processes need not be invoked.

3–4.3 Structure, Process, Climate, and Stage

The principles of geomorphology were further clarified by W. M. Davis in the 1890s. Davis theorized that erosion phenomena resulted in a "geographical cycle" of landform development. According to Davis, the nature of any landform is a function of structure, process, and stage of development in the geographical cycle (Bloom, 1998).

Structure. Structure includes all aspects of the physical nature and arrangement of the rocks in which the landscape formed. The hardness of each rock material; the presence or absence of layers; and any tilting, folding, or faulting are part of structure. Grain size, degree of cementation, and permeability are also included because they influence the rate and pattern of erosion. The shape of erosional landscapes depends largely on structural factors that determine which rocks are most erodible.

Process. Process refers to the combined action of the agents building and eroding a landscape. Water erosion, wind erosion, and mass movement remove material from some areas and deposit it elsewhere. Glaciers add their own dramatic touch. Soil movement by plants and animals should not be overlooked (Gabet, 2000). Human activity, however, causes accelerated erosion and is excluded from geologic erosion.

Rock weathering and soil formation are vital predecessors of the erosion process because they help produce fragments small enough to be moved. Erosion rates are limited in many places by the rate of production of loose material.

Erosive agents sometimes leave identifiable impressions on the landscape. A hillside scar and debris in the valley mark a landslide long after the event. Scratches on the bedrock of central Canada remain today as evidence of glacial scour that occurred more than ten thousand years ago. Moraines and other glacial features in northern United States show how far the glaciers reached. Old bison trails show conspicuously on aerial photos of certain areas in the Great Plains (Clayton, 1975).

Climate. Climate exerts a strong influence on geomorphic processes. Arid regions typically have angular topography (Note 3–2), and humid regions usually have rounded features because of the combination of weathering, erosion, and deposition working on their landscapes. Cold climates with heavy snowfall produce glaciers and glacial landscapes. A trained observer can learn much about the climate simply by observing landscapes (see Figure 3–5).

NOTE 3–2
ARID LANDSCAPES

It is often assumed that arid landscapes are shaped by wind. Wind is an active agent in arid regions, but other factors such as the sparse vegetation and the distribution of water in both time and space should not be overlooked.

Arid regions have bare soil exposed to the direct impact of rainfall because the bunchgrasses, shrubs, cacti, and other native plants cover only a fraction of the area. Also, the relatively low organic matter contents produce weaker soil aggregates, slower permeability, and less resistance to erosion than would be likely under more humid conditions.

Rainfall in arid regions can be intense even though it is infrequent. Runoff can accumulate quickly and cause a local "flash flood" because there is not enough vegetation to slow the water flow. A literal wall of water comes rushing down a streambed that was dry only a moment before. The stream may empty onto a broad flat area where the water spreads out and infiltrates into the alluvium deposited by countless flash floods of ages past. These flats are a marked contrast to the adjacent steep angular hills. Some flat areas completely surround one or more hills.

Local runoff gives the flats more water than adjacent hills, part of it coming from nearby mountains that receive more precipitation than the lower areas. Rock material within a few feet of the flats is often moist enough to weather and erode faster than the drier rock above. Wind and water erode these softened lower rock layers and produce angular topography with unique landforms such as natural bridges and balanced rocks. Wind erosion abrades most at low levels because the wind carries more and heavier particles at low levels than at higher levels.

Stage. Landscapes pass through stages of *youth, maturity,* and *old age.* When water erosion first cuts a set of steep, narrow valleys in an upland, the landscape is said to be in youth even though the relatively undisturbed areas between the valleys represent an older stage of a preceding geomorphic process. Maturity is reached when the new cycle has molded the entire area. Maturity is a long stage that is often divided into early and late substages. A mature landscape typically has many valleys and is very hilly. Old age is reached when the hills are nearly worn away and erosion has slowed.

(A)

(B)

(C)

Figure 3–5 The shape of a landscape tells much about the climate of the area: (A) an angular arid landscape in Arizona (Courtesy F. R. Troeh.); (B) a semiarid landscape on the island of Kauai, Hawaii (Courtesy USDA Natural Resources Conservation Service.); (C) a rounded humid-region landscape in New York. (Courtesy F. R. Troeh.)

Figure 3–6 Glacial ice still covers major areas between 60° and 90° North and South latitudes and once covered much more extensive areas in Europe, Asia, and the Americas. (Courtesy F. R. Troeh.)

Multiple Cycles. Landscape development is often interrupted by environmental changes before old age is reached. Any of several factors such as climatic change, uplift, or the cutting of a barrier by erosion may initiate a new cycle of landscape development. A landscape may show the effects of two or more cycles of development long after such changes occur.

Climatic changes have caused glaciers across vast areas of North America, Europe, and Asia to come and go several times during the last million years. At their peak, glaciers covered 30% of the Earth's land surface as shown in Figure 3–6, although they presently cover only about 10% of the land (Bloom, 1998). Each glacial and interglacial stage had marked effects on landscapes both within and beyond the glacier-covered areas. Glacial deposits have been cut by valleys, filled by a later glacier, and then cut by new valleys in different directions. The glacial processes combined with associated water and wind erosion and deposition have produced very complex landscapes. Kemp and Derbyshire (1998) indicate that the deep loess deposits in China include more than 30 paleosols. They suggest that each paleosol represents a relatively warm, moist period with intensified summer monsoons affecting the loess area, whereas the loess between the paleosols represents cooler, drier periods with dominant winds from the north or northwest adding new layers of silt.

Climatic changes can produce dramatic differences in landscape development even without glaciation. A small change in precipitation or temperature might influence the rate and pattern of erosion by altering the vegetation. Major climatic changes may cause angular topography to shift toward roundness, or vice versa.

Renewed uplift rejuvenates old landscapes. Greater elevation differences result in faster erosion and deeper valleys with steeper slopes. Uplift in one area may be accompanied by downwarp (sinking) in another area that will then receive deposits. The intervening area is tilted to a new angle that shifts the position of streams and their erosional and depositional areas. The record of such changes is often preserved in buried surfaces, sloping and dissected terraces, and other landscape variations.

Buried Landscapes. Land surfaces may appear quite stable when viewed over a time period of a few years or decades, but they are subject to great changes during geologic time. Geologists have determined that mountain chains are pushed up, worn down, and pushed up again several times during the course of their existence (Opie, 1993). Lower areas are covered by sediment and/or by lava flows.

Rock layers are studied to learn the geologic history of an area. Fossils are used to identify the age of the various rock layers present at any particular site. The presence of rock layers of a particular geologic era indicates deposition at that time, whereas the absence of such rock layers indicates a period of uplift and erosion. Tilting and warping of layers reveal periods of violent shifts within the crust of the Earth. After a period of erosion, a new period of deposition smooths the topography and produces an *unconformity* where the new material rests on the eroded surface.

Some rock layers are massive and impermeable, whereas others such as sand and gravel deposits are porous and permeable. A buried permeable layer sandwiched between two impermeable layers commonly fills with water and may serve as an aquifer. Some aquifers are localized, but others extend across large areas. One of the largest is the Ogallala aquifer underlying portions of several states in the Great Plains from South Dakota to Texas (discussed in more detail in Chapter 15). This aquifer represents a buried piedmont (discussed in Section 3–5.1) extending eastward from the Rocky Mountains and buried by subsequent glaciation and other events. It was filled with water during glacial times but has since lost most of that water, first through natural drainage and recently through pumping for irrigation (Opie, 1993).

3–4.4 Stream Systems

Streams cut valleys in all but the most arid landscapes. Most streams are integrated into river systems that carry runoff water to the oceans. Exceptions occur where the water all infiltrates and flows underground and where the water accumulates in a low area such as the Dead Sea or Great Salt Lake and escapes by evaporation. Such exceptions are called *closed drainage systems* in contrast to *open drainage systems* that flow into the oceans.

Some streams originated from water flowing down the initial slope of the area when it was first exposed to erosion. Such streams are called *consequent streams* because they are a consequence of the original slope.

Figure 3–7 A dendritic stream pattern is one that "branches like a tree" and has streams flowing in all directions. (Courtesy F. R. Troeh.)

Tributaries flowing into consequent streams are called *subsequent streams* because they developed later. Subsequent streams flow into the sides of consequent valleys, and headward erosion cuts their valleys across the original slope of the area. Tributaries to subsequent streams are called *secondary consequent* if they flow in the same direction as the consequent stream and *obsequent* if they flow in the opposite direction.

The preceding terminology works well on *rectangular* stream patterns where all streams tend to meet at right angles. Tilted rock layers of varying erodibility can produce such a pattern. Most of the streams that flow in valleys cut into the softest rock strata are subsequent streams and are parallel to each other. The connecting links are consequent streams that flow down the tilt-produced slope. Parallel fault lines are a less common cause of rectangular patterns. A *trellis* pattern is a rectangular pattern with a large preponderance of streams in one direction.

Dendritic stream patterns are the most common type. Directional terminology is not very useful in describing dendritic systems because the branching streams may flow in any direction. Dendritic stream patterns on a map resemble the branching of a tree, as illustrated in Figure 3–7. They develop where there are thick rock masses with uniform degrees of erodibility.

Other patterns also occur. For example, a mass of rock may be pushed up in the form of a dome and develop a *radial* drainage pattern from the water running off on all sides. *Complex* patterns include elements of several patterns.

Streams may also be classified into orders. *First-order streams* are the first identifiable channels formed where runoff water begins to concentrate into streams. A *second-order stream* is formed where two first-order streams meet, a *third-order stream* is formed by the union of two second-order streams, and so on. A fifth- or sixth-order stream is probably a major river.

3–4.5 Development of Valleys

Stream erosion produces valleys. Valleys are the growing parts of a landscape, whereas hills are remnants that have not yet been worn away. Observers may see the hills, but they need to study the valleys to understand how the landscape formed.

A valley begins where water collects from a large enough area to form a stream. The stream cuts a channel in the most erodible soil and rock in a low part of the landscape. The early development is often so rapid that a gully forms, eroding its way toward the source of the water.

A stream is a dynamic entity that does three things as it forms its valley:

1. It erodes downward toward a *base level* determined by the elevation of the stream or area into which it empties.
2. It erodes headward toward its main source of water, thus lengthening the valley.
3. It picks up tributaries as smaller streams enter the sides of the main valley and produce their own smaller valleys.

Streams initially develop where water accumulates in the concave parts of landscapes, but their erosional effects permit them to shift locations with time and move to the areas where the rock is easiest to erode. Such shifting can even produce an inversion of the topography (Pain and Ollier, 1995). For example, a lava flow in a valley may be so resistant that new valleys form on each side of it, leaving the old valley as a ridge between them.

3–4.6 Mass Wasting

Much material that a stream removes from its valley reaches the stream by *mass wasting*. This broad term includes all processes by which gravity moves soil downslope. The movement may be rapid and dramatic, as in landslides and mudflows, or it may be measured in fractions of millimeters per year (Troeh, 1975). Slow movement across a gently sloping surface is called *solifluction* or *soil creep*. Water promotes movement by seepage pressure and added weight but is not the actual transporting agent (Bloom, 1998). The slope need not be very steep because saturated soils on gentle slopes can become almost fluid. Saturated soil is especially common during the thawing period where wet springs follow winters that freeze the soil to a significant depth.

Two competing theories, often called *downwearing* and *backwearing*, describe how mass wasting progresses with time.

3–4.7 Downwearing

W. M. Davis developed the downwearing theory in the 1890s as part of his geographical cycle idea discussed in Section 3–4.3. Davis suggested that the hills in a landscape are gradually worn down by erosion and the steep valley slopes of youth gradually become more gentle as the landscape ages. The end result, if the landscape remained stable long enough, would be such gentle slopes that erosion would cease. The landscape would then be a nearly featureless, poorly drained surface called a *peneplain*. Some argue that present-day landscapes

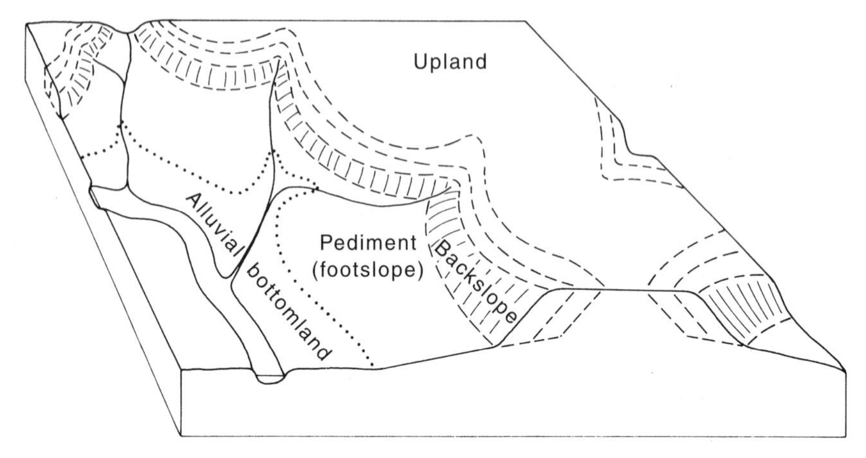

Figure 3–8 A backwearing landscape gradually erodes an upland as the backslope recedes in a manner described by Penck as "parallel retreat of slopes." A sloping pediment surface forms between the backslope and the bottomland. (Courtesy F. R. Troeh.)

are not old enough to be called peneplains, but surfaces such as the Pampas of Argentina certainly are "almost a plain." The Pampas occupy some 300,000 square miles (777,000 km^2) of land that is so flat it is often described as featureless. Lateral migration (Band, 1987) of the Paraná River appears to have been a major factor in the formation of the northern Pampas. The Paraguayan landscape east of the river is hilly and contrasts sharply with the Pampas to the west of the river.

3–4.8 Backwearing

Davis delivered a set of lectures in Germany challenging the leading European geomorphologist, Albrecht Penck. Penck's son, Walther, answered Davis in his book, *Die Morphologische Analyse,* published in 1927. Walther's untimely death prevented him from seeing his book in print and from developing his concepts further.

Penck's lengthy development of the concept of backwearing (also called *parallel retreat of slopes*) made geomorphologists realize that the subject of slope retreat deserved more study. Penck argued that a valley develops two kinds of slopes—a relatively steep backslope (the valley wall) and a flatter footslope between the backslope and the bottomland along the stream, as shown in Figure 3–8.

Penck reasoned that after an equilibrium is reached, the slope gradients should remain constant as the hill gradually wastes away. Flat hilltops would not wear down as Davis had suggested but would be worn away from the sides. The end result is similar to the peneplain concept but is called a *pediment* to distinguish between the two theories. The term "pediment" can be used to describe an area smaller than a peneplain and is useful at an earlier stage.

The pediment surface begins as an enlarging footslope. It is normally mantled with soil material that is slowly moving across the pediment from the backslope area to the stream. The soil on the pediment is considered to be moving *en masse* as a form of mass wasting in addition to any water transport occurring at the soil surface.

Many efforts have been made to combine the ideas of Davis and Penck. Some have suggested that the rounded topography of humid regions may best fit Davis' downwearing concept, whereas backwearing better explains the angular topography of arid regions.

3–4.9 Soils on Eroding Landscapes

The concepts of downwearing and backwearing have important implications for soil development. Downwearing implies constant erosion of the entire landscape with the highest areas eroding fastest so that the landscape becomes smoother. The soils on hilltops should therefore be younger and less developed than those in lower areas.

The backwearing theory predicts stronger development in the hilltop soils than elsewhere, because the hilltops should remain undisturbed until removed by eroding slopes. The shallowest and least-developed soils, apart from floodplains, should occur on the backwearing slopes. Material accumulating on footslopes should produce deep, moderately developed soils.

Soil scientists usually favor the backwearing theory because the soil development pattern outlined in the preceding paragraph occurs on many landscapes. A composite theory dominated by backwearing but including some downwearing loss from the hilltops might have the most support. Most hilltops lose some soil by water erosion (especially from the shoulder areas), and wind erosion can occur anywhere. Furthermore, chemical erosion leaches material from hilltop soils in humid regions. Clément (1993) indicates that solution processes remove more material in forested landscapes in the Canadian Appalachians than is lost by physical erosion on slopes up to 8 to 10° (14 to 18%). It should be realized, of course, that this is a humid region and that the forest vegetation gives it excellent protection against physical erosion.

3–5 SEDIMENTARY LANDFORMS

Material eroded from one place is deposited elsewhere, so the wearing away of one area causes another to be built up. Deposition sometimes converts water areas into land by filling lakes or ponds. Even a shallow sea may become land through deposition and uplift. The many gray-colored limestone, sandstone, and shale formations composed of materials that accumulated on seafloors are evidence that vast amounts of erosion, deposition, and uplift have taken place. The term "sedimentation" as used in this book includes the depositional process and related factors as they influence the nature of the deposited sediment.

3–5.1 Water Deposits

Water sorts materials by particle size as it transports them. The coarsest materials are either left in place or occasionally moved for short distances. Progressively finer particles are deposited as the water flow reaches lower, flatter areas and moves at slower velocities (see Chapter 4 for more details).

Alluvium. Deposits left by flowing water are called *alluvium.* Alluvial deposits occur in floodplains, terraces, fans, piedmonts, and deltas. Alluvial land is desirable for many purposes because it tends to have smooth topography, deep, fertile, permeable soils, and more readily available water than most other land. Such land is in demand for farming, road building and other construction activities, and many other uses.

Floodplains are narrow or nonexistent in young valleys but become broader and flatter as the landscape ages. Heavy runoff flooding across these areas and depositing a fresh layer of alluvium is a natural process (Baker, 1994). The deposition produces a nearly flat surface except where stream channels cut the floodplain. Each flood may rearrange material deposited earlier and cut new channels while filling old ones. The stream channel itself meanders gradually across its floodplain, occasionally making sudden major shifts. Although natural, these shifts are distressing and costly to landowners.

Flood hazard should be taken into account whenever construction on floodplains is considered. Protective works such as levees and channel dredging may prevent small floods but are often inadequate to protect against major floods. Actually, there is no known way to control a major river permanently, largely because flood-prevention efforts cause rivers to raise the level of their beds by depositing sediment that should have been deposited on floodplains. The result is an ever-increasing flood hazard such as that experienced by the Chinese along the Yellow River and by the Americans along the Mississippi River. Staying off the floodplain during the flood season is a nuisance, but it is a successful long-term strategy, as the Egyptians demonstrated for thousands of years in the Nile Valley before the Aswan Dam was built (Chapter 2, Section 2–3.1). Now they are trapping sediment in the reservoir instead of receiving its benefits on the floodplain.

Alluvium in river valleys is usually only a few feet thick, although some very thick exceptions exist. Thick alluvium occurs mostly in downwarp areas. The weight of the sediment helps cause the area to sink, and the lowered position increases the likelihood that the area will trap more sediment. The magnitude such deposits can reach is illustrated in Note 3–3.

NOTE 3–3
SEDIMENTARY DEPOSITS

Some sedimentary deposits become very large. The Idaho Formation will serve as an example. This body of sand and silt occupies an area 100 mi (160 km) long and 30 mi (50 km) wide in southern Idaho and eastern Oregon. Sediments washed from the Rocky Mountains were deposited in a gradually sinking area that has been called the Snake River Downwarp. A well drilled to a depth of 5000 ft (1500 m) was still in the formation; in fact, it penetrated only 30% of the formation's probable thickness as estimated from the slope of its layers. Samples taken from the well indicated that the sediments were all deposited on dry land even though most of the formation is now below sea level.

Sedimentary deposits are common in most parts of the world, though most are smaller than the Idaho Formation. Some deposits remain loose, but many are cemented into sandstone and shale. The nature of the rock indicates the depositional environment. Red colors characterize land deposits where oxidizing conditions prevailed; gray colors and fossils of water creatures indicate that the sedimentation occurred under water.

Terraces are formed when the base level of a stream changes and its channel is cut too deep for normal flooding to occur. The stream may have cut through a barrier to a lower base level, or uplift may have increased the slope of the stream channel. Either cause leaves the former floodplain higher than the stream and its new floodplain. The old floodplain becomes a terrace, but its origin is recognizable by its flat surface bordered by a slope down to the new floodplain on one side and a slope up to another terrace or the upland above on the other side. Terraces are subject to erosion as the new floodplain enlarges and to dissection where tributary streams cross them.

Alluvial fans, also called *alluvial cones,* are named for their shapes. They line the sides of many valleys where tributary streams erode the adjoining uplands. Suddenly free to spread out, a stream slows down, loses energy, and deposits much of its load of sediment. The deposit forms the shape of a fan radiating downward from the point of entry into the main valley. The depositing stream often disappears by soaking into highly permeable fan deposits. The upper part of the fan is normally the steepest and coarsest textured. The gradient decreases and the texture becomes finer with distance, until the fan blends into the terrace or floodplain below.

A *piedmont* is a plain formed at the base of a mountain range. It forms in a manner similar to that of alluvial fans but on a grander scale. Many streams blend their sediment and form a smooth surface sloping away from the mountains. A piedmont surface is shaped like a pediment but has a great thickness of accumulated sediment, in contrast to the thin covering of soil characteristic of a pediment. Similar depositional surfaces in arid regions where the coalescing alluvial fans merge into the level area of an undrained basin are called *bajadas.*

Deltas form where streams empty into bodies of standing water. Their shape, as seen on a map, normally resembles the Greek capital letter delta, Δ. Deltas have nearly level surfaces that are partly above and partly below water level. A small delta may be only a few feet across, but some large ones continue for hundreds of miles. Small deltas that form in lakes are quite sandy because the finer sediments are carried farther into the lake. Rivers flowing into the ocean produce finer-textured deltas because salt water flocculates clay particles and makes them settle out. Such sediment represents the richest part of the soils from the source area and produces a fertile soil. The internal drainage is poor because of the delta's low elevation. Much rice is grown on deltas in warm climates.

Peat bogs are another way that nature slowly fills lakes and ponds. Organic residues from vegetation growing in, on, and around a water body accumulate in the water, because decomposition is very slow where oxygen is deficient. What decomposition does occur releases nutrients that add to those carried by inflowing water; these nutrients promote more plant growth. Over time, layers of peat several feet or tens of feet (a few meters) thick may accumulate. Burial of such layers beneath new layers of rock leads to pressure and heat that convert old deposits of peat into coal, oil, and natural gas.

Bottom Deposits. Sand and gravel are seldom carried far into a body of water, but fine sediments and chemical precipitates may cover the entire area and gradually form deposits on most lake bottoms and seafloors. Sediment is sometimes dominant, but chemical precipitates dominate at other times and places.

Lacustrine deposits, also called lake-laid clays, are composed mostly of fine sediment carried past deltas into the main body of water. The texture of the deposit may vary from silt or sand during seasons of high runoff to fine clay during cold or dry seasons, thus producing a layered effect such as that shown in Figure 3–9. These layers are called *varves*

Figure 3–9 Varved clay deposit in Michigan. Each varve is about 1 mm thick and is composed of a thin, sandy layer deposited during the summer thaw and a coating of silt and clay deposited while the glacial lake was frozen. The scale is in centimeters. (Courtesy Roy L. Donahue.)

and are sometimes counted to estimate the age of the deposit. The depositional process usually levels the surface, so the varves lie horizontally.

Glaciers crossing hilly topography produced many small lakes whose locations are marked by deltas and lacustrine deposits. Deltaic deposits high on hillsides show the former water level and have been used as sources of gravel for road construction. The lacustrine deposits are at lower elevations and produce fine-textured soils that are poorly drained. They are erodible and, when wet, subject to small slips and even landslides.

Limestone is the most common sedimentary rock. It normally contains some silt and clay sediments but is composed mostly of consolidated calcium carbonate that is formed as a result of chemical precipitation triggered by the growth of algae and other plant and animal life. Conditions that result in the formation of limestone are common on the floors of shallow seas, as the water in seas and some lakes is nearly saturated with calcium carbonate. Large formations of limestone have been uplifted and converted into dry land. Limestone rock is subject to solution erosion in humid regions but is quite resistant to weathering in arid regions. Subsurface solution erosion forms caverns in limestone bedrock. Chalk, a saltwater deposit, and marl, a freshwater deposit, are less consolidated calcium carbonate deposits (and less resistant to weathering) than limestone.

Evaporite deposits form where salts are concentrated by evaporation. An ideal situation for evaporites occurs where a warm, dry climate evaporates water from a shallow bay that receives an input of salty seawater. Another evaporite situation occurs where rainfall in mountain areas seeps down into dry, closed basins such as Death Valley, the Great Salt Lake, or the Dead Sea. Large deposits of sodium chloride have been formed as evaporites. Other salts also precipitate, sometimes in relatively pure form. Potassium chloride from evaporite deposits in western Canada, southwestern United States, and elsewhere is the principal source of potassium fertilizer. Borax deposits in Death Valley provide much of the world's commercial boron. Sodium nitrate from the Atacama Valley in Chile was an early source of nitrogen fertilizer. All of these salts are too soluble for their deposits to form in humid environments.

3–5.2 Wind Deposits

Wind sorts materials by particle size even more effectively than water does. Wind deposits are relatively free of layering and of textural variations in any particular vicinity. Wind deposits are widespread, but much of their area is not very thick. A few deposits attain thicknesses of 600 ft (200 m) or more, but even these thin to less than 100 ft (30 m) within a short distance and continue to thin with increasing distance from the source until they become too mixed with other materials to be identified. Thick wind deposits bury the underlying landforms and have a steep topography of their own resulting from a combination of depositional and erosional effects. Most of the area is mantled with a silty deposit averaging only a few feet thick and conforming approximately to the shape of the buried surface.

Aeolian Sands. Wind normally does not move large particles, so stone fragments are left in place. Desert areas often accumulate a surface layer of gravel, one pebble thick, as the wind carries away the finer soil particles. This gravel layer, called *desert pavement* or *lag gravel,* prevents further wind erosion. The sand component of the soil is deposited nearby, often in the form of dunes.

The difference between a stable surface and one that can be blown into sand dunes is largely a matter of vegetative cover. Drought and blowing sand make it difficult to maintain and more difficult to establish vegetation on sand dunes. Climate is also a factor—additional precipitation increases the likelihood of stabilizing vegetation being established. Thinner, smoother deposits of aeolian sands are easier to stabilize than sand dunes, especially if they have a water table within reach of plant roots so they can support a thick stand of vegetation.

Loess. Loess is wind-deposited material dominated by silt-size particles. Loess deposits are the most extensive form of wind deposits because silt particles are more detachable than clay and easier to transport than sand. Most wind deposits change from aeolian sands to loess within one or a few miles of the source.

Guy Smith (1942), in his classic study on Illinois loess, found three distinct trends:

1. Loess deposits are thickest near the source and become thinner with increasing distance.
2. The average particle diameter decreases from coarse silt to fine silt, and the clay percentage increases with increasing distance from the source.
3. The calcium carbonate percentage decreases with increasing distance from the source.

The decreases in deposit thickness and in particle size are direct results of the additional energy required to transport particles longer distances. The decreased calcium carbonate content is attributed to the thin portions being deposited more slowly and leached more thoroughly than the thicker portions during the period of deposition.

Soils formed in loess resist wind erosion because they lack the sand particles that would move first and knock silt particles loose. But loess deposits and the soils formed in them are susceptible to both water erosion and mass movement. Falling raindrops and flowing water can readily detach silt particles, which are easily transported in runoff water. Gully erosion is relatively common in loess materials, and the gullies often cut straight down to the bottom of the deposit. The nearly vertical sides and flat bottoms of such gullies are described as U-shaped. U-shaped gullies help to identify loess deposits because most other gullies are V-shaped.

When an exposed loess surface becomes saturated with water, it is subject to mass movement ranging from small slumps to large landslides. Landslides were a common problem in loess roadbanks until road builders learned to make vertical banks in loess. Loess has a natural tendency to cleave along nearly vertical planes that are more stable than sloping surfaces, as shown in Figure 3–10. Friction between the flat silt particles and cementation by calcium carbonate have been suggested as reasons for vertical stability. Also, while vertical surfaces stay relatively dry in a storm, sloping surfaces may become saturated with water—an unstable condition for loess.

Figure 3–10 A nearly vertical loess bank in western Iowa. (Courtesy F. R. Troeh.)

Windblown Clay. Wind deposits contain less clay than silt and sand. Clay particles stick together and are hard to erode, but bouncing sand particles knock loose some clay, much of which is mixed with silt in the downwind portions of loess deposits. Clay also may be picked up and carried with drifting snow.

Wind formed extensive clay deposits in southern Australia during the Pleistocene. More recent clay deposits occur near the Gulf coast of Texas and Mexico, and in Senegal and Algeria (Bowler, 1973). In an arid environment, material containing 20 to 77% clay can become strongly aggregated and so loose between aggregates that the wind can cause it to drift like sand. Bowler (1973) indicates that clay dunes form only on the downwind side of seasonally exposed mud flats around shallow bodies of saline water.

Volcanic Ash. Volcanoes occasionally spew out large quantities of ash that blanket the landscape. The ash is much like a thin loess deposit but has more angular particles and more clay. When saturated with water, the thicker ash, cinders, and other deposits on the flank of a volcano are subject to a form of debris flow called a *lahar.* Some ash layers can be traced to a specific event and thus become a geologic time marker. The explosion that formed Crater Lake in Oregon by blowing the top off Mount Mazama is a prime example. The resulting layer of volcanic ash can be identified across much of Oregon. The ash from the 1980 eruption of Mount St. Helens in the state of Washington is a more recent example of the same phenomena (Karowe and Jefferson, 1987). Volcanic ash soils have high clay contents and are very sticky when wet and hard when dry. Some of them have wide cracks during dry seasons along with the self-swallowing action characteristic of Vertisols.

3–6 MASS MOVEMENT DEPOSITS

Landslides, soil slips, solifluction, and other forms of mass movement produce accumulations of soil and rock materials in footslope positions below valley walls, cliffs, and other steep slopes. This unsorted material is called *colluvium.* Its characteristics vary according to the nature of the source material. Usually a certain amount of churning has occurred that makes a colluvial deposit relatively uniform with depth, though it may be interlayered with sorted alluvial deposits. Being a depositional site makes the soils that form in such materials relatively deep and commonly quite moist and fertile. Some of them are not tillable, however, if the material in the source slopes provides enough rock fragments to make the deposit excessively stony.

Catastrophic mass movements such as landslides are inherently intermittent events separated by long intervals of stability. Less obviously, slower movements such as solifluction may also be intermittent in their effects as they respond to changes in climate and vegetation. For example, Eriksson et al. (2000) were able to distinguish two major colluvial deposits on slopes in the Irangi Hills in central Tanzania. They related the first set of deposits to a climatic change from dry to wet conditions, which took place during the Late Pleistocene. They used a recently developed technique called optically stimulated luminescence to date these deposits between 15,000 and 11,000 years before present. A long period of stability allowed mature soils to develop in this colluvium until the second colluvial deposit occurred. Its dates, beginning about 900 years ago and continuing to the present, indicate that it is likely associated with human influence such as the development of settlements and the introduction and intensification of agriculture in the area.

3–7 GLACIAL LANDSCAPES

Glaciers produce dramatic effects on landscapes. Moving ice picks up soil and stones, then deposits them far from their source. Stones held in the ice at the bottom of a glacier scratch and gouge the bedrock. Areas such as central Canada were scoured by the ice. Glacial movement that followed the length of valleys in New York deepened the Finger Lakes so their bottoms are below sea level. The fjords of Norway, Greenland, Canada, and Alaska had a similar glacial origin in coastal areas and are now flooded with seawater. The moving ice also steepened valley walls near the lakes and fjords. Other landscapes where the ice crossed pre-existing valleys often had the hilltops scraped off and the valleys filled, thus smoothing the topography.

Much material picked up by glaciers crossing Canada was dumped in hilly areas where the ice melted in the United States. Ice margins were marked by *terminal moraines,* forming irregular ridges that protrude a few feet or tens of feet (several meters) above the adjacent landscape. A thinner *ground moraine* (also called a till plain) extends back over the glaciated area and is commonly marked by lines of low arcing hills. These hills represent end moraines formed by brief advances of the ice during its waning phase. Many former valleys were filled with glacial deposits, so present landscapes may be quite different from the preglacial ones.

Glacial influence extended beyond the area actually covered by ice. Valleys leading away from glaciers were enlarged by torrents of meltwater laden with debris that was deposited as sediment on broad, flat floodplains. Some of these floodplains remain today as oversized floodplains for present streams; others are terraces or outwash plains on upland areas completely separated from major streams. The broad floodplains of major rivers such as the Missouri and Mississippi became source areas for wind erosion that produced some of the world's major loess deposits.

3–8 RATE OF GEOLOGIC EROSION

Measured or estimated geologic erosion rates can serve as an important reference for evaluating rates of accelerated erosion. Ruhe (1969) gives examples of erosion rates in selected Iowa landscapes. Carbon-14 dating was used to determine how long it took to build sediment deposits in closed basins. The amount of sediment divided by the number of years and the area of land draining into the basin provided an average erosion rate ranging from 0.07 to 8.76 in. (2 to 222 mm) per 1000 years before agriculture entered the area. One area changed from a presettlement rate of 6 in. (150 mm) per 1000 years to 60 in. (1500 mm) per 1000 years after settlement. Each inch of soil depth represents 150 tons/ac (336 mt/ha) of soil if the bulk density is 1.3 g/cm^3. Thus Ruhe's sample data are equivalent to 0.01 to 1.3 tons/ac (0.02 to 2.9 mt/ha) of geologic erosion per year and 9 tons/ac (20 mt/ha) of accelerated erosion per year.

Most landscapes do not provide closed basins suitable for using the above method to compute long-term average erosion rates. A relatively new method described by Bierman et al. (2001), however, has more widespread application. Cosmic rays constantly produce certain nuclides (^{3}He, ^{10}Be, ^{21}Ne, ^{26}Al, and ^{36}Cl) in soils and rocks to a depth of about 2 m (6 or 7 ft). Each of these nuclides gradually decays at its own rate. For example, the shortest half-life of the group is that of ^{26}Al, about 700,000 years. Measuring the concentrations of one or more of these nuclides at various depths provides data that can be used to calculate the length of time that the material has been close enough to the surface to be exposed to cosmic radiation and thereby to calculate average long-term erosion rates. A comparison of the concentrations of two or more of these nuclides in buried sediments can be used to estimate how long the materials have been buried. Bierman et al. (2001) provide data for a site in the mountains near the Oregon coast with erosion rates from 36 to 139 m per million years. An area in central Texas averaged 29 m/million years, and a site in Australia yielded results of 14 and 18 m/million years. For comparison to the data in the preceding paragraph, 1 m/million years = 1 mm/1000 years = 0.04 in./1000 years. On the basis of a bulk density of 1.3 g/cm^3, these data convert to average annual geologic erosion rates ranging from 0.08 to 0.81 tons/ac (0.18 to 1.8 mt/ha).

A rate of 0.5 ton/ac-yr (1 mt/ha-yr) may be considered typical for geologic erosion from gently sloping soils. Actual rates vary widely from essentially zero for thousands of years in the most stable areas to thousands of tons of soil and rock moved by a landslide in a moment's time. Water and wind have caused remarkable geologic erosion that is more puzzling and only slightly less sudden than a landslide. For example, the origin of the "channeled scablands" in the state of Washington (Figure 3–11) was unknown for a long time. Finally, it was explained on the basis of a sudden release of water from glacial Lake Missoula in northern Idaho and eastern Montana (Baker, 1978). Torrents of water crisscrossed hundreds of square miles of land, eroding channels to bedrock but leaving scattered islands of soil. This event was brief, but its effects, including gravel bars 100 ft (30 m) high and soil islands surrounded by channels of bare rock, are still conspicuous.

Except for occasional catastrophic events such as those mentioned, geologic erosion rates are determined by such factors as climate, vegetation, slope, and soil material. The effect of precipitation is of interest because the lowest rates of geologic erosion are from moderate precipitation. The rate is higher in very wet climates because high precipitation has great erosive force. Arid climates also have relatively high rates because there is too little vegetation to protect the soil against wind erosion and the occasional heavy rains that occur in such areas.

SUMMARY

Erosion, "the great leveler," persistently removes material from uplands and fills low areas with sediment. The process is usually interrupted before it can reach the end result—a featureless peneplain. All land would have been worn down to sea level if not for the opposing forces—deposition, lava flows, and uplift. The most important of these forces is uplift caused by isostatic adjustments.

Geomorphology is the study of landforms and the processes that shape them. Geomorphologists apply the principle of *uniformitarianism* to landscapes and consider the effects of *structure, process,* and *stage.* Much landscape development occurs as streams cut downward, lengthen headward, and pick up tributaries. Streams cut valleys that gradually enlarge as the hills erode. Two theories have been developed to explain erosion by mass wasting. The *downwearing* theory supposes that higher areas wear down and slopes become less steep until a *peneplain* is formed. According to *backwearing,* a parallel retreat of slopes erodes the hillsides but not the hilltops. The surface formed as a slope retreats is

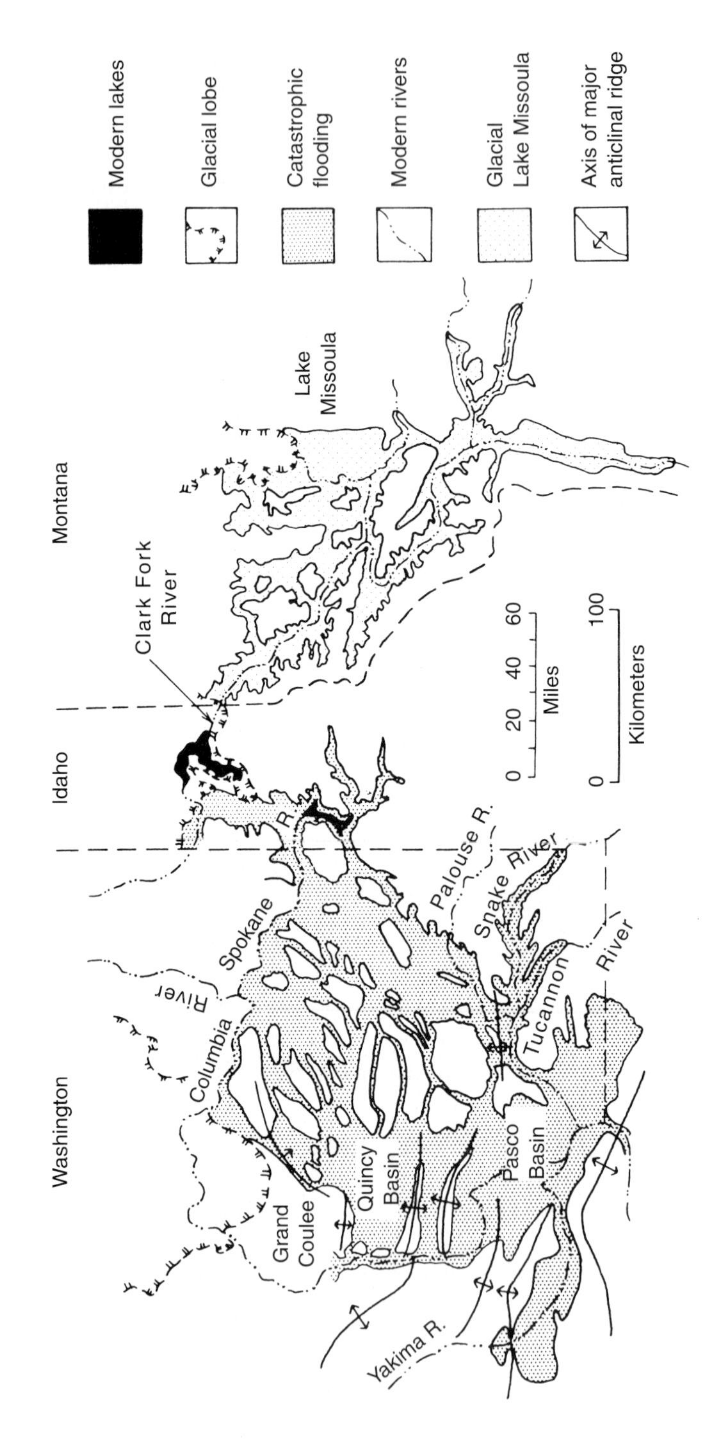

Figure 3–11 The channeled scablands were formed when an ice blockage broke in northern Idaho and released water from Lake Missoula in Montana as a torrent that raged across eastern Washington. (Courtesy Victor R. Baker, *Science* 202:1255, Dec. 22, 1978. Copyright 1978 by the American Association for the Advancement of Science.)

called a *pediment.* Most soil scientists favor the backwearing theory because older soils are usually found on hilltops and younger soils on hillsides.

Water deposits *alluvium* in fans and on floodplains; remnants of former floodplains become terraces when a new floodplain forms at a lower level. Water sorts sediment by particle size—coarse material settles near the source and finer material is carried into quieter waters. *Deltas* form where a stream enters a lake or sea. *Lacustrine deposits* high in clay accumulate on lake bottoms. Calcium carbonate precipitates as limestone, chalk, or marl on the bottom of many bodies of water, especially on shallow seafloors. More soluble salts precipitate from briny lakes in arid regions.

Wind also sorts the material it erodes. Stones are left in place and sand is deposited in nearby dunes, but silty *loess* deposits blanket larger areas. *Volcanic ash* also blankets large areas but in a much shorter time.

Unsorted material deposited by mass movement at the base of a slope is called *colluvium.* It may be interlayered with alluvium.

Glaciers scour soil and rock in their source areas and deposit the material where they melt. Glacial margins are marked by hilly *moraines* and the area between is covered by *ground moraine.* Much of the area covered by glaciers is smoothed as the hills are eroded and the low places filled.

A rate of 0.5 ton/ac (1 mt/ha) per year is considered typical for geologic erosion from gently sloping soils.

QUESTIONS

1. Discuss the relative importance of erosion and deposition in forming steep and level landscapes.
2. Why is the principle of uniformitarianism important to geomorphologists and soil scientists?
3. What combinations of structure, process, and stage produce hilly landscapes?
4. Why are some hills angular, whereas others are rounded?
5. What difference does it make whether the dominant process of landscape development is downwearing or backwearing?
6. Why is an alluvial fan steepest near the top and flatter in its lower parts?
7. What evidence is there that wind is able to sort particles by size?
8. Why did glacial action deepen some valleys and fill others?

REFERENCES

BAKER, V. R., 1978. The Spokane flood controversy and the Martian outflow channels. *Science* 202:1249–1256.

BAKER, V. R., 1994. Geomorphological understanding of floods. *Geomorphology* 10:139–156.

BAND, L. E., 1987. Lateral migration of stream channels. In *Geomorphological Models: Theoretical and Empirical Aspects.* Catena Supp. 10, Catena Verlag, Cremlingen, W. Germany, p. 99–110.

BIERMAN, P., E. CLAPP, K. NICHOLS, A. GILLESPIE, and M. W. CAFFEE, 2001. Using cosmogenic nuclide measurements in sediments to understand background rates of erosion and sediment transport. Ch. 5, p. 89–115 in R. S. Harmon and W. W. Doe III, (eds.), *Landscape Erosion and Evolution Modeling.* Kluwer Academic Pub., NY, 540 p.

BLOOM, A. L., 1998. *Geomorphology: A Systematic Analysis of Late Cenozoic Landforms.* 3rd ed. Prentice-Hall, Englewood Cliffs, NJ, 482 p.

BOULTON, G. S., 1987. Progress in glacial geology during the last fifty years. *J. Glaciol.* Special Issue:25–32.

BOWLER, J. M., 1973. Clay dunes: Their occurrence, formation, and environmental significance. *Earth Sci. Rev.* 9:315–338.

BUTZER, K. W., 1976. *Geomorphology from the Earth.* Harper & Row, New York, 463 p.

CLAYTON, L., 1975. Bison trails and their geologic significance. *Geology* 3:498–500.

CLÉMENT, P., 1993. Erosion balance and slope evolution under continental humid temperate climate: An Appalachian example (Quebec, Canada). *Catena* 20:303–315.

DAVIS, W. M., 1922. Peneplains and the geographical cycle. *Geol. Soc. Am. Bull.* 23:587–598.

DEARING, J. A., and I. D. L. FOSTER, 1993. Lake sediments and geomorphological processes: Some thoughts. Ch. 2 in J. McManus and R. W. Duck (eds.), *Geomorphology and Sedimentology of Lakes and Reservoirs.* John Wiley & Sons, New York, 278 p.

ERIKSSON, M. G., J. M. OLLEY, and R. W. PAYTON, 2000. Soil erosion history in central Tanzania based on OSL dating of colluvial and alluvial hillslope deposits. *Geomorphology* 36:107–128.

FRANK, W., A. BAUD, K. HONEGGER, and V. TROMMSDORFF, 1987. Comparative studies on profiles across the northwest Himalayas. In J. P. Schaer and J. Rodgers (eds.), *The Anatomy of Mountain Ranges.* Princeton Univ. Press, Princeton, NJ, 298 p.

GABET, E. J., 2000. Gopher bioturbation: Field evidence for non-linear hillslope diffusion. *Earth Surf. Process. Landforms* 25:1419–1428.

HARDWICK, P., and J. GUNN, 1995. Landform-groundwater interactions in the Gwenlais Karst, South Wales. Ch. 5 in A. G. Brown (ed.), *Geomorphology and Groundwater.* John Wiley & Sons, New York, 213 p.

HEWITT, K., 1997. Risk and disasters in mountain lands. Ch. 16, p. 371–408 in B. Besserli and J. D. Ives (eds.), *Mountains of the World: A Global Priority.* Parthenon Pub. Group, NY, 495 p.

KAROWE, A. L., and T. H. JEFFERSON, 1987. Burial of trees by eruptions of Mount St. Helens, Washington: Implications for the interpretation of fossil forests. *Geol. Mag.* 124:191–204.

KEMP, R. A., and E. DERBYSHIRE, 1998. The loess soils of China as records of climatic change. *Eur. J. Soil Sci.* 49:525–539.

OPIE, J., 1993. *Ogallala: Water for a Dry Land.* Univ. of Nebraska Press, Lincoln, 412 p.

PAIN, C. F., and C. D. OLLIER, 1995. Inversion of relief—A component of landscape evolution. *Geomorphology* 12:151–165.

PENCK, W., 1927. *Die Morphologische Analyse.* J. Engelhorns Nachf., Stuttgart, 283 p.

RUHE, R. V., 1969. *Quaternary Landscapes in Iowa.* Iowa State Univ. Press, Ames, IA, 255 p.

SMITH, G. D., 1942. *Illinois Loess: Variations in Its Properties and Distribution.* Ill. Agric. Exp. Sta. Bull. 490, p. 139–184.

TROEH, F. R., 1965. Landform equations fitted to contour maps. *Am. J. Sci.* 263:616–627.

TROEH, F. R., 1975. Measuring soil creep. *Soil Sci. Soc. Am. Proc.* 39:707–709.

CHAPTER

4

WATER EROSION AND SEDIMENTATION

Movement of soil by water occurs in three stages; individual grains are detached from the soil mass, detached grains are transported over the land surface, and soil grains are deposited on new sites. Soil removal and deposition occur to some degree in nearly all locations, but detachment and transportation are of major concern on uplands, and deposition is most important on lowland sites and in streams and lakes. Only about a quarter of the eroded soil material actually leaves the nearby area (den Biggelaar et al., 2001). Even the sediment that rivers carry to the oceans is deposited mostly in deltas near the coast.

Abundant vegetative cover limits erosion to a slow rate typical of geologic erosion. Cultivation, overgrazing, logging, mining, construction activity, and fire, which reduce or destroy vegetation, cause accelerated erosion. Even arid regions may suffer severe water erosion from infrequent, intense rainstorms on areas where their normal sparse cover has been damaged by overgrazing or other misuse.

4–1 TYPES OF WATER EROSION

Most water erosion is classified as sheet erosion, rill erosion, gully erosion, or streambank erosion. The classification is based on the nature and extent of soil removal. A related process, solution erosion, is usually handled separately or ignored.

4–1.1 Sheet Erosion

Sheet erosion is the removal of a thin layer of soil over an entire soil surface. Raindrop splash and surface flow cause sheet erosion, with splash providing most of the detaching energy and flow providing most of the transporting capacity. Sheet erosion is insidious because it is difficult to see. The first sign is when subsoil color begins to show, as cultivation mixes surface soil and subsoil. It is most apparent on upper portions of convex slopes (see Figure 4–1).

Figure 4–1 The hilltops in this Iowa field are light-colored because the dark-colored topsoil has been lost by sheet erosion. (Courtesy F. R. Troeh.)

4–1.2 Rill and Ephemeral Gully Erosion

Runoff water tends to concentrate in streamlets as it passes downhill. This water is more turbulent and has greater scouring action than sheet flow, and it cuts small channels by removing soil from the edges and beds of the streamlets. These small channels frequently occur between crop rows and along tillage marks, but some channels follow the slope across plant rows and break through tillage ridges as they pass downhill. The crop-row channels are called *rills* (see Figure 4–2). The channels that follow the slope are called *ephemeral gullies,* because they tend to form repeatedly in the same places in fields, and, if not managed carefully, they can grow into full-fledged gullies. Normal cultivation smooths the surface where these small channels form. Accordingly, long-term soil loss effects are similar to those resulting from sheet erosion.

4–1.3 Gully Erosion

Erosion channels too large to be erased by ordinary tillage are called *gullies* (Figure 4–3). Deep, relatively straight-sided channels develop where the soil material is uniformly friable throughout the profile. The channels in deep loess soils are U-shaped with almost vertical walls. Broad V-shaped channels often develop where friable surface soils overlie cohesive, tight, nonerodible subsoils.

Gullies are described as *active* when their walls are free of vegetation and *inactive* when they are stabilized by vegetation. Gullies are also classified as small, medium, and large according to depth, with medium-sized gullies measuring 3 to 15 ft (1 to 5 m) deep.

Gully erosion sometimes expands by a process that has been called internal erosion. Water enters the soil in cracks or other large passageways and flows for a considerable distance beneath the surface. The underground flow erodes soil along the way and often

Figure 4–2 Rills in a cultivated Vertisol in India. (Courtesy Roy L. Donahue.)

Figure 4–3 A gully eroding uphill from a cultivated field into a grassed area near Bruxelles, Manitoba, Canada. The equipment in the background is smoothing the gully to make it into a grassed waterway. (Courtesy Manitoba Department of Agriculture.)

produces a deposit where it exits on a slope or in a gully. Enlargement of the underground channel eventually causes the surface to cave in and converts it into a visible gully.

4–1.4 Streambank Erosion

Removal of soil material from the sides of running streams, *streambank erosion,* is usually greatest along the outside of bends, but inside meanders may be scoured intensively during severe floods. Streambank erosion affects relatively small areas, but it often removes the entire soil profile of very productive soils.

Streams that are "unloaded" pick up sediment from their beds and banks. Thus, streambank and bed erosion are increased when the sediment load brought into the streams is reduced as a result of conservation measures on uplands or when upstream sediments are caught in reservoirs or other traps.

4–1.5 Solution Erosion

Solution erosion is more likely to be a component of geologic erosion than of accelerated erosion. Water that percolates through a soil into the water table carries soluble materials with it. The amount is generally small, but the process is persistent, and it gradually removes weathering products. Even relatively stable soil minerals are subject to this process over very long periods of time. Solution erosion can be the principal means of soil loss on level areas with good cover in a humid region.

4–2 EROSION DAMAGE

Water erosion damages soil in many ways. Soil is lost, plant nutrients are removed, texture changes, structure deteriorates, productive capacity is reduced, and fields are dissected. Sediments pollute streams and lakes and pile up on bottomlands, in stream channels, and in lakes and reservoirs. Estimates of the annual damage caused by water erosion in the United States are $500 million to $1.2 billion in on-site damages (Colacicco et al., 1989) and $3.4 billion to $13 billion in off-site damages (Clark et al., 1985).

4–2.1 Soil Loss

The most apparent damage caused by water erosion is the removal of soil from eroding surfaces. Surface soil is generally more friable, more permeable to water, air, and roots, and higher in organic matter and fertility than subsoil, so loss of surface soil is critical. While geologic erosion from land covered with perennial vegetation, either grass or trees, amounts to only a fraction of a ton per acre annually, erosion from bare, cultivated fields may exceed 200 tons/ac (450 mt/ha).

Several means have been devised to measure soil loss rates. Much of the data used as a basis for the soil loss prediction equations discussed in Chapter 6 was obtained by weighing the sediment in runoff intercepted at the base of a slope and relating it to the rainfall and other conditions that produced the runoff. Often the process was facilitated by using a rainfall simulator to create the desired amount of runoff rather than waiting for whatever rainfall nature might provide. The data were expanded to an annual basis, usually by means of a soil loss prediction equation.

Longer-term soil loss rates have sometimes been estimated where a measurable depth of soil has been lost in a known period of time. For example, a farmer may remember that a large stone or other marker that now protrudes 6 in. (15 cm) above the soil was level with the surface when the field was first farmed 50 years ago. Since 6 in. of an average soil are equal to about 900 tons of soil per acre (2000 mt/ha), this example would indicate an average annual soil loss rate of 18 t/ac (40 mt/ha). A similar approach has sometimes been used to estimate rates of sedimentation from the depth of cover over a buried marker.

Another approach that has somewhat wider application has been developed in recent years based on the concentration of a contaminant in the soil. Cesium 137 (^{137}Cs) originating mostly from nuclear testing during the 1950s and 1960s is the marker that has been most used for this purpose. The depth distribution of ^{137}Cs in the soil profile(s) is determined and compared to the amounts measured in uneroded soil nearby or calculated from other fallout measurements in the same general area. An appropriate computer model is needed to satisfactorily interpret such data into erosion rates (Yang et al., 1998). A related approach is based on lead 210 (^{210}Pb) (Walling and He, 1999). Soils naturally contain a small amount of ^{210}Pb produced by the decay of radon (^{226}Ra) that comes from uranium (^{238}U) in the soil, but they also receive deposits of ^{210}Pb by deposition of decay products from atmospheric ^{222}Ra. The concentration of ^{210}Pb at various depths can therefore indicate rates of erosion or sedimentation.

Fly ash from the Illinois Central railroad provided a basis for Hussain et al. (1998) to measure erosion patterns between Chicago and Cairo, Illinois. They compared cropped sites to uncultivated sites and found that the cultivated areas averaged 47% less fly ash in the upper 22.5 cm (9 in.) of the soil profile, thus indicating that an average of 10.6 cm (0.47 22.5 cm) of soil had been lost in the 142 years from 1855 to 1997. This would be about 10 mt/ha (4.5 t/ac) annually.

4–2.2 Plant Nutrient Losses

Bennett and Chapline (1928), in one of the earliest reports on erosion, stated that more than 43 million tons (40 million mt) of N, P, and K were lost in the soil that was washed each year from the fields, grasslands, and forests of the United States. Only a small proportion of this quantity was immediately available to plants, but this amount of available and potential nutrients was nearly 65 times the amount of fertilizer N, P, and K applied in the United States in 1934 (Bennett, 1939, p. 10). U.S. fertilizer use increased rapidly after World War II. It peaked at nearly 24 million tons (nearly 22 million mt) in 1981. Since that time, consumption has ranged mostly from 19 to 22 million tons (17 to 20 million mt) of N, P, and K per year (Taylor, 1994). These applications have not counteracted erosion losses because they have not been large enough to replace nutrients sold off the farm in crops plus those lost by erosion. Fertilizer applications have increased crop yields; they have even raised soil test values. But the losses of plant nutrients in solution and in eroded sediments still are very important.

4–2.3 Textural Change

Water erosion makes sandy soils even sandier by moving the finest particles considerable distances and leaving the coarser particles close by. Medium- and fine-textured soils are usually well aggregated. Their textures are not altered seriously by erosion because the water sorts aggregates, not individual soil particles. Generally, small and large aggregates have similar textural compositions.

Erosion may remove the entire surface horizon. The new surface soil, which is really the exposed subsoil, generally is finer textured, contains less organic matter, and has poorer structure than the original surface soil. This makes seedbed preparation more difficult, increases the amount of runoff, and poses problems at other stages of crop production.

4–2.4 Structural Damage

Water erosion affects soil structure in three ways. Loss of surface soil generally exposes a less granular and less permeable subsoil; raindrops disintegrate aggregates on the surface and produce a compact surface crust; and percolating rainwater carries suspended soil grains through the soil surface, plugging pores, and reducing infiltration and permeability rates. These changes cause increased runoff, which increases erosion and reduces the amount of water stored. Vandervaere et al. (1998) found that crusted soils in Niger produced runoff sooner and had five times as much runoff as similar soils without a crust.

4–2.5 Productivity Loss

Losses of soil material and nutrients, and deteriorating structure reduce productive potential. The magnitude of this reduction depends on properties of surface soil and of subsoil. Reductions are largest when subsoil is shallow and infertile, or fine textured, compact, and intractable. Often the reduction in productive potential is masked by increased inputs and better management. Early studies showed that grain and cotton yield losses on artificially truncated profiles averaged 77% on several soil conservation stations when no fertilizer or manure was used. Where adequate fertilizer was applied, corn yields were reduced by about 80% in years with poor rainfall distribution, but comparable yields were obtained on truncated and normal soils when rains were adequate. In another study, crops were grown on farmers' fields on soils that had experienced differing erosion severity. Yields were superior where little or no erosion had occurred and were reduced progressively as erosion was more severe (Table 4–1).

TABLE 4–1 EFFECT OF EROSION SEVERITY ON CROP YIELD

			Crop yield (bu/ac)		
Crop	State	No. of farms	Slight erosion	Moderate erosion	Severe erosion
Corn	Wisconsin	8	80	67	60
Grain	Wisconsin	11	65	50	43
Grain	Minnesota	5	36	30	23

Source: Hays et al., 1949.

More recent field trials on soils with differential natural erosion severity, but receiving uniform tillage practices and fertilizer applications, showed that erosion was associated with smaller, but still significant, yield reductions. Mokma and Sietz (1992) studied the effect of erosion severity on corn yields in Michigan during a five-year period. The plots were moldboard plowed and were seeded to corn each year. Fertility differences in the plots were masked by high fertilizer-application rates. The slightly eroded plots produced the most grain, though not significantly more grain than the moderately eroded plots. The severely eroded plots, on average, produced 21% less grain than the slightly eroded plots. The authors attributed the lower production to less available moisture in the severely eroded soil.

Weesies et al. (1994) studied the effect of erosion on corn and soybean grain yields on three Indiana Corn Belt soils over a ten-year period. Severe erosion reduced yields significantly—corn yield showed a 9–18% reduction and soybean yield a 17–24% reduction—compared to yields on areas of similar soil with only slight erosion. Yields on moderately eroded phases were lower than on slightly eroded phases of the same soil, but differences were not statistically significant.

Larson et al. (1985) developed a mathematical model to calculate indirectly the effect of erosion on soil productivity index by including soil properties that influence productivity and that are known to be affected by erosion. Seventy-five important north-central U.S. soils were examined. The study predicted that the loss of 10 in. (25 cm) of soil will reduce the index for one-third of the soils by more than 10%, and one-eighth by more than 20%. A 20-in. (50-cm) loss will reduce the index for one-half of the soils by more than 10%, one-third by more than 20%, and one-sixth by more than 30%. The index of four soils (5% of the total) will be reduced more than 40%. Current rates of erosion for these soils range from 0 to 50 tons/ac-yr (0 to 105 mt/ha-yr), or from 0 to 1/3 in. (0 to 8.5 mm) per year on a depth basis.

4–2.6 Field Dissection

An eroding field can be farmed as a unit as long as erosion channels are small enough for ordinary machinery to cross them. Reduced cultivatable-land area, lower yields, and higher production costs combine to reduce net farm income when a field is cut by larger gullies.

4–2.7 Engineering Structure Damage

Erosion damages buildings, roads, bridges, and other engineering structures. Foundations are undermined by washing, landslides, and soil creep. Road ditches and culverts can cause gullies that affect road safety (Figure 4–4). Bridge approaches, footings, pilings, and supports are structurally weakened or destroyed.

4–2.8 Water Pollution

Soil sediment is the greatest single pollutant of surface water, on a volume basis. The muddying of streams and lakes reduces their value for home and industrial use, for recreation,

Figure 4–4 Erosion damage to the ditch along a roadway in Kansas. (Courtesy USDA Natural Resources Conservation Service.)

and for fish and wildlife. Erosion also contaminates streams when fertilizers and pesticides are dissolved in runoff water or adsorbed on eroded soil.

4–2.9 Sedimentation

Rich bottomland soils owe their productivity to sediments eroded from the surface of upland soils, but they can be harmed if subsequent erosion deposits subsoil or other less productive material on them. Lower-lying areas within a field can also be damaged in this way (see Figure 4–5). Such deposition may be apparent from the topography, from the color and structure of the sediment, or from chemical measurements. Wallbrink et al. (1999) propose a method of relating sediment to its original soil depth position based on its content of three radionuclides: ^{7}Be, ^{210}Pb, and ^{137}C. These three radionuclides normally have different distributions with depth in a soil profile, so their concentrations relative to each other can be used as a depth indicator.

Sediment is also deposited in stream channels, lakes, and reservoirs. This changes the aquatic environment and affects plant and animal life. For example, sediment in fish spawning areas may ruin these sites for fish propagation. Sedimentation raises streambeds, reducing the depth and capacity of the channels. This causes navigational problems and more severe flooding. Sedimentation of lakes and reservoirs reduces their capacity, value, and life expectancy. In extreme cases, sedimentation changes an aquatic habitat into a terrestrial one.

Figure 4–5 Sediment deposited at the foot of a slope in a cultivated field near Alexander, Manitoba, Canada. Several rills and a small gully were subsequently eroded through the sediment. (Courtesy J. A. Hobbs.)

4–3 AGENTS ACTIVE IN WATER EROSION

Two major agents are active in water erosion: falling raindrops and running water. Both derive the energy needed to detach and transport soil grains from the force of gravity. Water also adds to the soil weight and reduces soil strength when gravity causes soil to roll or slide downhill *en masse.* While spectacular and significant where it occurs, rapid mass movement of soil is much less important overall than the more widespread effects caused by water transport and by the very slow mass movement described in Chapter 3 (Section 3–4.8).

4–3.1 Falling Raindrops

In the early years of erosion-control activity, it was assumed that overland flow caused water erosion. Now, a substantial part of soil detachment, and even transportation, is credited directly to raindrops.

Energy of Falling Raindrops. The kinetic energy of a falling body can be calculated from the equation

$$E = 1/2mv^2$$

where E = kinetic energy, ergs (ft-lb)

m = mass of falling body, g (lb)

v = velocity of fall, cm/s (ft/s)

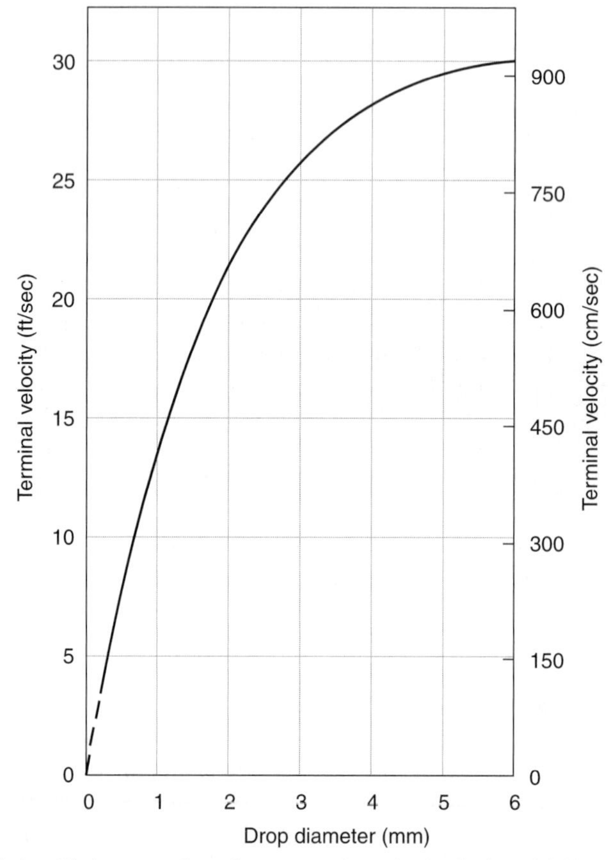

Figure 4–6 Relationship between drop diameter and terminal velocity of falling raindrops. (Modified from Wischmeier and Smith, 1958.)

Air friction limits drop fall to a maximum known as *terminal velocity* that is related to drop mass. Laws (1941) and Gunn and Kinzer (1949) studied the relationship between drop diameter, from 0.2 to 6.0 mm, and terminal velocity. Figure 4–6 presents a graphic form of their data as developed by Wischmeier and Smith (1958). These data show that larger drops have higher terminal velocities. Thus, larger drop size affects kinetic energy in two ways—by larger mass and faster fall. Laws (1941) found that smaller drops reached terminal velocity in a shorter distance, but that all drops are at terminal velocity after falling 30 ft (9 m) or less.

Rainstorm Intensity and Energy. Laws and Parsons (1943) found that average drop size increased as storm intensity increased up to 6 in./h (150 mm/h), but a range of sizes from 0.25 to 7 mm was present in each storm. Data from several sources show that drop size decreases slightly at high intensity. Hudson (1971, p. 52–53) concluded that its maximum is at about 80 to 100 mm/hr (3 to 4 in./hr); McGregor and Mutchler (1977, p. 138) showed a 3-mm maximum mean drop size between 2 and 3 in./hr (50 to 75 mm/hr).

Wischmeier and Smith (1958) used Laws and Parsons' data to develop the relationship

$$e = 916 + 331\ log_{10}I$$

where e = total energy, ft-ton/ac-in. for each inch of rainfall

I = rainfall intensity, in./h (up to 3 in./h)

Table 6–1, developed from this equation, is presented in Chapter 6.

Wind and Raindrop Energy. Strong winds add a horizontal-velocity component to raindrop fall. Wind partially reduces air resistance by moving air horizontally away from one side of the drop. This reduced resistance and the actual force of the wind combine to accelerate the drop. Smith and Wischmeier (1962) suggested that the velocity of wind-driven rain can be estimated by multiplying the drop's terminal velocity in still air by the secant of the angle between vertical and the direction of fall in the wind. They calculated that 3-mm median-drop-size rain falling in a wind at a 30° angle of inclination has a velocity 17% greater and kinetic energy 36% greater than the same rain falling vertically. Lyles (1977) calculated that the kinetic energy of a 2-mm drop in a 20-mi/h (32-km/h) wind was 2.75 times that of a similar drop falling in still air. Wind tends to break large drops into smaller ones, but the increased energy makes windblown drops much more erosive than drops falling in still air.

Vegetation and Raindrop Energy. Wollny, in 1890, first reported that vegetative cover affected rainfall and that growing crops intercepted up to 45% of the raindrops (Baver, 1939). Shaw (1959) showed that a growing corn crop intercepts and holds 0.13 to 0.19 in. (3 to 5 mm) of moisture from a storm that exceeds 0.3 in. (7.5 mm) of rainfall. This water never reaches the ground. Drops that are temporarily intercepted by plants have much of their energy dissipated at the plant surface. Some break into smaller drops that also have reduced energy. It should be noted, however, that when raindrops are intercepted by tree leaves, the water tends to accumulate and fall in large drops. These can gain high energy if they fall from sufficient height. Remember that all drops reach maximum velocity within 30 feet.

Work of Raindrops. Raindrops release energy when they strike a surface. This energy does three kinds of work: it breaks peds and clods into smaller aggregates and individual particles, it moves soil grains to new locations as water splashes back into the air, and it compacts and puddles the surface layer of soil. The first is the basis for the detaching capacity of the raindrops; the second is the source of the transporting potential; and the third reduces the soil's infiltration rate, causing more water to run off and erode the soil.

Raindrops and the Erosion Process. Ellison (1944, 1947a, 1947b) showed that falling raindrops are the main detaching agent in sheet erosion. He also showed that raindrops break many aggregates into smaller aggregates and into individual particles and detach them from the soil mass. Roose (1977) demonstrated that raindrops are more responsible for sheet erosion than is sheet flow.

A wide range of grain sizes is loosened by raindrop splash. Certainly, all material 2 mm in diameter and smaller can be detached directly. Ellison (1944) found that 10-mm pebbles, when partly submerged in surface flow, were moved by raindrops. He also found splashed particles 2 ft (60 cm) or more above the land surface. They moved horizontally as much as 5 ft (1.5 m). Heavy rains can splash more than 100 tons/ac (225 mt/ha) on bare, highly detachable soil (Ellison, 1947a).

Raindrops falling on level, bare soil, in the absence of wind, splatter equally in all directions; soil carried out of any area is matched by movement into the area. Although there is a lot of soil movement, there is no net soil loss and no measurable erosion. When raindrops fall on sloping land, more of the rebounding splash goes downhill than uphill. The distances individual splashes travel is greater downhill also. Thus, slope can cause soil movement downhill even if no runoff takes place. Ellison (1944) found 75% of the splash was downhill on a 10% slope. Wind causes a similar directional splash even on level surfaces. Wind-driven raindrops hit the soil at an angle, and water and soil are carried downwind by splash.

Duley and Kelly (1939) noted the development of a thin compacted layer at the surface of the soil, where unimpeded raindrops beat on bare soil. When compact layers developed, the soil's infiltration rate declined rapidly during the storm, and runoff increased. They proved that it was the thin compacted layer that produced the reduction in infiltration rate. Crop residues on the soil surface eliminated or greatly reduced the compacted layer.

4–3.2 Running Water

Runoff is recognized as an important cause of erosion. The most common cause of runoff is rain falling faster than soil can absorb it. Other causes are snowmelt and irrigation.

Runoff is classified as prechannel, or sheet, flow and channelized flow. The depth of sheet flow seldom exceeds 0.1 to 0.2 in. (3 to 5 mm) unless surface vegetative mulch causes it to be slightly deeper. Channelized flow depth varies from about 0.2 in. (5 mm) in small rills to 10 ft (3 m) or more in large streams. The erosiveness of runoff water depends on its depth, velocity, turbulence, and abrasive material content.

Energy of Running Water. Kinetic energy and erosive force of running water are related to the quantity and velocity of flow, just as raindrop energy is controlled by mass and velocity of fall. Water passing across the land in very thin films moves at relatively low velocity in a smooth, non-mixing manner called *laminar flow*. It is incapable of initiating soil movement. Thin sheets of water are nonerosive unless they are churned by falling raindrops. Water velocity generally increases with the depth of the water layer but seldom exceeds 5 ft/s (1.5 m/s) even in a gully. Runoff water moving at this speed, if it had laminar flow, would have only 1/28 of the energy of an equal mass of 3-mm raindrops falling at a terminal velocity of 26 ft/s (8 m/s).

Turbulence and Energy. Fast-flowing fluids move in an irregular manner with random oscillations in direction and sudden changes in both horizontal and vertical velocity. This irregular motion is called *turbulence*. Turbulence develops as depth of flow and velocity increase until the whole flow becomes turbulent. Kinetic energy of a stream and its erosive capacity are both increased dramatically by full-flow turbulence.

Energy of Transported Material. Clear water has limited erosive energy. Soil grains saltating (jumping along the bed) in the stream are about 2.65 times as dense as water and carry about 2.65 times as much energy as an equal volume of water traveling at the same velocity. Similarly, soil aggregates that are about 1.3 times as dense as water carry about 1.3 times as much energy as the same volume of water traveling at the same velocity. Energy differential is greater when grains move faster than water that is in contact with the streambed (the common case). Thus, a stream carrying abrasive material has greater power than clear water to break up aggregates and clods on the bed and put them into motion. Such a stream can even abrade bedrock.

Being denser than water, transported material tends to fall and would soon settle out if the water were not turbulent. Turbulence lifts falling soil particles where its eddies happen to be moving upward and causes them to bounce off the bottom where eddies are directed downward. Either way, energy is transferred from the water to the particles, and any given water flow has a limited sediment transport capacity (Zheng et al., 2000). A fully loaded stream of water may still pick up new particles, but it will deposit a similar amount of sediment at the same time.

A stream of water has more abrasive power when it is lightly loaded with sediment than when it is heavily loaded. Merten et al. (2001) found evidence of two causes for this phenomenon. One cause is the use of energy to keep the particles in suspension, thus reducing the turbulent energy of the water. The other cause is the blanket of sediment moving slowly along the bottom of the stream. This blanket protects the underlying surface from fast-moving particles in the main stream.

Stream Depth and Energy. Depth of water moving across a field is controlled by rainfall intensity, soil infiltration rate, and type of flow. Most soils can absorb all the water from a 0.1 in./h (2 or 3 mm/h) rain, so no runoff will occur. Runoff commences when rainfall intensity exceeds soil infiltration rate. Greater rainfall intensity increases thickness of runoff films, runoff velocity, and erosive energy.

Water films at the top of a slope are thin even in intense rainstorms because there is little water movement from above. Water films are thicker farther down the slope because of water accumulating from above. However, with increasing thickness, the water moves at a faster rate and this limits flow thickness.

Water is more likely to flow in channels farther down the slope. Runoff from relatively broad areas is funneled into rills and gullies, causing appreciably thicker layers, higher velocities, turbulence, and more energy to erode soil than sheet flow has.

Slope, Runoff, and Erosion. Four features of slope affect velocity and amount of runoff and, hence, erosiveness: slope gradient, slope length, slope shape, and slope aspect.

Slope gradient (steepness) is measured in units of vertical fall either per single horizontal unit (decimal fraction) or per 100 horizontal units (percent). Increasing slope gradient increases the speed of water moving downhill and the erosive force of the flowing water, as explained in Note 4–1.

NOTE 4–1
SLOPE STEEPNESS AND FORCE OF RUNOFF

Force of a body moving on a frictionless inclined plane is defined by the equation

$$F = mg \sin \theta$$

where F = force, dynes

m = mass, grams

g = force of gravity, cm/s^2

θ = gradient angle, degrees

Increasing the gradient angle increases sin and therefore increases the value of F. If land slope is measured in units of vertical fall per unit of distance along the land surface, rather than along the true horizontal, the slope so determined is numerically equal to the sine of the gradient angle, or

$$F = mg \times \text{gradient}$$

where "gradient" is a decimal fraction. Thus, a unit volume of runoff water exerts twice the force on a 10% (0.1) frictionless slope as on a similar 5% (0.05) slope. When friction is considered in calculating the force, the equation becomes

$$F = mg \sin \theta - fmg \cos \theta$$

where f is the coefficient of friction (dimensionless). Sin θ increases but cos θ decreases as the gradient, θ, increases. Consequently, the force of the water, taking friction into account, increases proportionately faster than sin θ increases, or faster than the gradient, expressed as a decimal fraction, increases.

Runoff from medium- and fine-textured soils usually increases with increasing slope; runoff from sandy soils does not always increase on steeper slopes. Erosion, however, always increases with slope steepness.

Slope length is the distance from the crest of a knoll or hill to the point where deposition of transported material starts, or where runoff enters a natural or prepared waterway. Slope length in a terraced field is the distance from the ridge top of one terrace to the bottom of the channel of the terrace immediately below. Slope length has a variable effect on runoff and erosion. Runoff losses per unit area are usually (though not always) greatest on short slopes. There is generally more erosion on longer slopes in spite of less runoff per unit area. Longer slopes increase the amount of erosion when rainfall intensity is high, or the permeability of the soil is low, or both.

Slope shape across the slope (along the contour lines) can be straight, concave, or convex, and the downhill direction also can be straight, convex (increasingly steeper gradient downhill), or concave (gradients progressively less steep downhill). Many slopes are

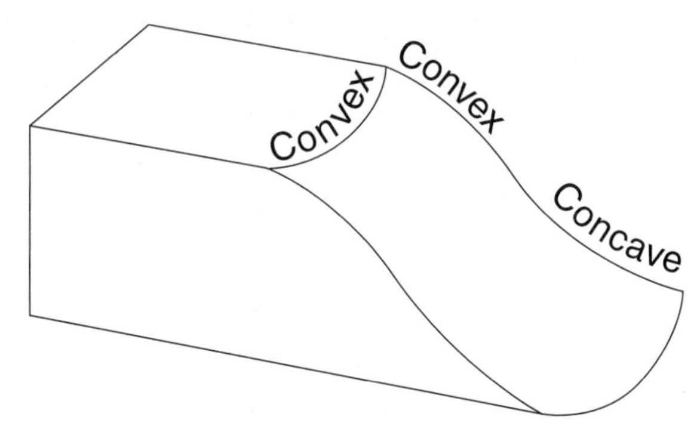

Figure 4–7 Diagram of a hypothetical hillside showing convex, concave, and complex slopes. (Courtesy F. R. Troeh.)

convex at the top and concave at the bottom (complex), as illustrated in Figure 4–7. Convex slopes (in either the contour or the slope direction) cause water to flow away from an area, making the site drier than it otherwise would be, whereas concave slopes cause water to accumulate and make the soils wetter and deeper (Troeh, 1964). Slope-shape effect on runoff and erosion is complex. Runoff velocity is slow and soil movement is minimal near the top of convex slopes because runoff volume is small and slope is gentle. Water movement is faster lower on the slope because the gradient is steeper and more runoff water accumulates. Soils tend to be shallow on convex slopes; soil replacement from the flatter slope above is slower than is the loss to steeper slopes below.

Nearly all natural slopes in humid regions are complex, having convex slopes at the top and concave slopes at the bottom. Runoff generally slows as it moves down the concave part of the slope because the slope gradient is decreasing. This causes the water depth to increase, gives more time for water to infiltrate, and causes sedimentation to occur. Thus, the convex upper part of the slope is a relatively dry, erosive environment, and the concave lower part is a relatively moist environment with a deep soil.

Slope aspect is the direction the slope faces. A limited study on the effect of slope aspect on runoff and erosion was conducted by Wollny and his associates in Germany about a century ago. They found smaller runoff but greater erosion losses on south-facing slopes (Baver, 1939). The major effects of slope aspect apparently result from its influence on the angle at which the sun's rays strike the land. Slope aspect has minimal effect at the equator, but the influence increases toward the poles. Slopes that face the afternoon sun (south- or west-facing slopes in the northern hemisphere and north- or west-facing slopes in the southern hemisphere) are warmer and have higher evaporation during the growing season. Water storage is reduced on these sites. As a result, there is less plant growth and less soil organic matter, especially in dry climates. Reduced vegetation usually results in increased erosion. North- and east-facing slopes in the northern hemisphere are usually noticeably cooler, more moist, better vegetated, and less eroded.

Surface Condition. Small depressions in the soil surface and plant material, living or dead, reduce runoff volume and velocity. This reduces the energy of runoff water and the amount of erosion. A smooth, bare surface offers the least possible frictional resistance; a pitted soil covered with dense vegetation presents the ultimate in resistance to water movement. Manning's formula (see Note 4–2) predicts quantitatively the effect of surface roughness on flow velocity.

NOTE 4–2
MANNING'S FORMULA

The most common method for estimating flow velocity in an open channel employs Manning's formula:

$$V = 1.5\frac{R^{2/3}S^{1/2}}{\eta}$$

where V = average velocity of flow, ft/sec

R = hydraulic radius, ft

S = land slope, ft/ft

η = coefficient of surface roughness

Hydraulic radius (R) is related to depth of flow, but it is not depth, not even average depth. It is defined as

$$R = \frac{A}{P}$$

where A = cross-sectional area of flow, ft^2

P = wetted perimeter, ft

The method for calculating the hydraulic radius is shown below.

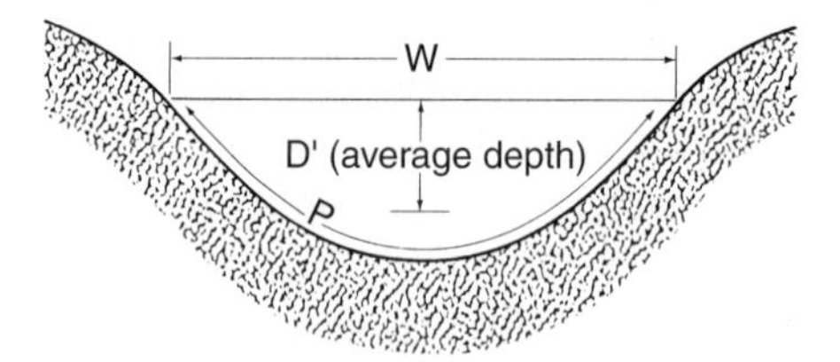

The value of R is equal to average depth in sheet flow but is less than average depth in channel flow.

The surface roughness coefficient (η) ranges from 0.017 for a smooth, straight, bare, earthen ditch to 0.300 for a dense, uniform stand of grass such as a 10 in. (25 cm) tall stand of bermudagrass. Selected values of η are shown in Table 4–2.

TABLE 4–2 SELECTED VALUES FOR COEFFICIENT OF ROUGHNESS (η) FOR USE IN MANNING'S FORMULA

	Values of η		
Surface condition	Minimum	Design	Maximum
Ditches			
Earth, straight and uniform	.017	.022	.025
Winding, sluggish ditches and gullies	.022	.025	.030
Ditches with rough, stony beds, and vegetated (weedy) banks	.025	.035	.040
Earth bottom, rubble on sides	.028	.032	.035
Natural stream channels			
Clean, straight banks, no rifts, pools	.025		.033
Clean, straight banks, some weeds and stones	.030		.040
Winding, some pools, but clean	.033		.045
Winding, some pools, some weeds and stones	.035		.050
Sluggish river channels, either weedy or with deep pools	.050		.080
Very weedy channels	.075		.150
Vegetated waterways (prepared)			
Dense, uniform bermudagrass, 10-in. tall[a]	.040	.040	.300
Dense, uniform bermudagrass, 2.5-in. tall	.034		.110

[a]Other sod grasses slightly less rough.

Source: NRCS Engineers' Handbook.

Runoff and the Erosion Process. Both detaching and transporting capacity of runoff are small when the flow is in thin films and the velocity is low, but they increase as the runoff becomes channelized because flow depth and velocity increase. Prechannel flow by itself has little or no capacity to detach and transport soil particles. Raindrops that fall into prechannel flow detach large numbers of soil grains. Falling drops also cause sheet flow to be turbulent, thus increasing its carrying capacity. Soil grains detached by raindrops can be transported long distances when turbulent prechannel flow carries them to a channel.

Raindrops cannot easily detach grains deep below the surface of running water, but velocities of such streams give them detaching and transporting powers. Erosive energy of soil particles carried in runoff must be added to the energy of clear water. Erosiveness of the particles generally exceeds that of the clear stream. A stream's carrying capacity is related to the third power of its velocity and to its turbulence. Even when a stream carries its maximum burden, the composition of the sediment load constantly changes. Gravity causes some soil particles to drop out of suspension as other particles are picked up. The larger and heavier grains drop out quickly; the smallest remain in suspension for very long periods.

Sand can be deposited on land from fast-moving waters; silt will settle out in quantity only as the stream's velocity is markedly reduced, such as when the stream gradient levels off; and clay settles out in quantity only when the water is still or when the clay flocculates as the stream enters a body of salt water.

4–3.3 Gravity

Gravity moves soil directly. Soil movements caused by gravity are known as *landslides, mudflows, slips, slumps, soil creep,* and *surface creep.* An *avalanche* (strictly speaking) is a mass of moving snow, but it may strip all vegetation and soil down to bedrock and deposit them in a heap at the bottom of the slope.

In most cases of gravity-induced movement, the topography must be steeply sloping. Soil creep, which occurs on slopes of only a few percent, especially in footslope positions, is a notable exception.

Land in native vegetation, even on steep slopes, is ordinarily in equilibrium with its environment. Soil movement is usually extremely slow. If soil cover is destroyed or greatly reduced by fire, overgrazing, logging, cultivation, surface mining, or construction, the soil loses both the water extracting potential and the structural binding power of plant roots. If an excess of moisture then develops, due to heavy rains or rapid snowmelt (especially while the subsoil is still frozen), soil weight increases, normal friction between the semiviscous soil mass and the underlying material is reduced, and the mass slowly or rapidly slides downhill. Sometimes the soil slumps down and out, leaving a depression at the top and forming a hump above the former surface at the bottom. This sort of movement occurs on the north sides of hills on the steep cultivated fields in the Palouse region of eastern Washington and northern Idaho. It is also common on steep roadbanks.

Along the Pacific slopes of the Cascades and the Coastal Range, in the Intermountain Region, and in other hilly and mountainous areas of the United States and other countries, a watery soil mass called a mudflow may pass down a valley like a very slowly moving river. Mudflows also occur in permafrost regions when upper soil layers thaw and become saturated by rain or snowmelt. There the mudflows move either downslope or vertically into solution caverns where ice has melted.

4–4 SOIL PROPERTIES AND SOIL ERODIBILITY

Differences in soil erodibility are obvious and were noted in the early years of the conservation movement. It has been difficult, though, to determine what soil properties actually make soils erodible.

Erosion detaches individual soil grains from the soil mass and carries them away in raindrop splash or running water. Soil erodibility, therefore, must be related to the soil's detachability and transportability. Any property that makes soil detachment or soil transportation difficult reduces soil erodibility. Soil texture and structure certainly affect the size of soil grains exposed to erosive elements. Runoff must occur for rapid erosion to take place, so soil properties that affect infiltration rate and permeability must also affect the rate of erosion. Infiltration rate and permeability are also influenced in a major way by texture and structure.

Early studies by Middleton and his associates (1930, 1932, 1934) related the results of many chemical and physical analyses on soils from the early soil erosion stations and from other sites to assessments of soil erodibility made by scientists in the field. Properties and combinations of properties related to aggregate size, structural stability, and soil permeability appeared to be most closely correlated with soil erodibility. Wischmeier et al.

(1971) showed that soil texture, organic matter content, soil structure, and soil permeability were the soil properties most closely correlated with soil erodibility. They developed a nomograph based on these properties to predict erodibility (see Figure 6–1 in Chapter 6).

4–4.1 Soil Texture

Sand particles are easy to detach because they lack cohesiveness, but they are difficult to transport because they are relatively large and heavy. Clay particles tend to stick together and are difficult to detach, but are easily carried great distances once separated from the soil mass. Silty soils are frequently well aggregated, but the aggregates break down readily when wetted, and the individual particles are easily transported.

Infiltration rate and permeability to water are related, in part, to texture. Water moves rapidly through macropores but slowly through micropores. Large pores between sand particles permit rapid water movement. Fine to very fine pores common in medium- and fine-textured soils such as the loams, clay loams, and clays restrict water movement. While total porosity of the fine-textured soils is nearly always greater than that of coarse-textured soils, the individual pores in the fine soils are usually much smaller, and both infiltration and permeability are slower. Therefore, a moderate rainstorm generally produces more runoff and erosion from the finer-textured soils than from sandy ones. Silty soils such as those formed in loess are very susceptible to water erosion.

4–4.2 Soil Structure

Large, stable aggregates make a soil difficult to detach and transport and make it more permeable to water. While soils high in clay usually have low permeabilities and low infiltration rates, a well-aggregated clay soil permits faster water movement than a poorly-aggregated clay.

Factors that influence the size and stability of aggregates include texture, cations on the exchange complex, type of clay mineral, organic-matter content, cementing materials other than clay and organic matter, and cropping history.

Texture. Sand has a weakening and loosening effect on structure. Clay is a cementing and aggregating agent. The higher the clay content, up to about 40%, the larger and more stable the aggregates. Clay contents above 40% promote development of very small aggregates that erode easily, especially where surface soils freeze and thaw frequently during winter. Some soil aggregates, particularly those high in silt and very fine sand, are relatively unstable. Raindrops destroy these aggregates, and the fine grains flow into and plug surface pores to produce a dense compact layer at the soil surface. Slow infiltration into this compacted layer causes increased runoff and erosion.

Type of Cation on the Exchange Complex. If the cation-exchange complex is occupied mainly by H^+ or di- or trivalent cations, the colloid will be flocculated, and individual soil particles will aggregate. Large, stable aggregates resist both detachment and transportation. On the other hand, soil colloids with large amounts of Na^+ and K^+ or with very large amounts of Mg^{2+} on the exchange sites will deflocculate. Deflocculated colloids prevent aggregate formation and cause low permeability and high erodibility.

Type of Clay Mineral. Aggregation of soils is influenced by the type of clay mineral. Tropical and subtropical soils, which are high in hydrous oxides of iron and aluminum and in the 1:1-type lattice clay, kaolinite, tend to be better aggregated than soils high in the 2:1-type lattice clays, smectite and illite.

Organic-Matter Content. Soil structure improves and the individual aggregates become more stable as organic-matter content increases. This is accompanied by increased permeability and decreased runoff and erosion. These effects are mostly related to decomposable organic matter such as dying roots and fresh plant residues incorporated into the soil (Gale et al., 2000). The direct cause is probably the high degree of microbial activity producing sticky exudates and hyphae that bind the soil particles into aggregates. Too much old, stable organic matter may cause soils to be very erodible because of the small size and low density of granules formed. This loose condition makes exposed organic soils subject to severe wind erosion even though their topographic position usually protects them from water erosion.

Water Repellency. Recent studies have shown that most soil aggregates have some degree of water repellency produced largely by coatings of extracellular polysaccharides and other exudates from roots and microbes (Hallett et al., 2001). These coatings increase aggregate stability, but they reduce the water infiltration rate.

Cementing Agents. Secondary lime is a cementing agent and helps to hold particles in aggregates. Some iron compounds bond clays and other soil grains together in stable forms in many strongly leached, temperate-region soils and in numerous tropical soils. These soils may be quite resistant to erosion.

Cropping History. Soils plowed from native vegetation, cultivated pasture, or meadow resist erosion. They tend to have excellent structure and relatively large, stable aggregates. Roots permeating the aggregates and large amounts of incorporated crop residue add to stability. Actively decomposing plant materials help to develop resistant structure; humus, which resists further decomposition, is less effective. Aggregates become less stable and more subject to breakdown, and erodibility increases unless decomposable organic material is added to the soil regularly and in abundance. Intense cultivation decreases organic matter content and aggregate stability by promoting decomposition and erosion.

4–4.3 Freezing and Thawing

Freezing and thawing influences soil erodibility through its effects on soil structure and permeability. The shifting of soil particles resulting from ice formation helps produce soil aggregates, and organic bonding agents probably diffuse to the contact points. Aggregate stability is maximized after two or three freeze-thaw cycles (Lehrsch, 1998). This type of action loosens compacted layers such as crusts and plowpans, making them easier for plant roots to penetrate and increasing their permeability to air and water. Soil compaction is therefore a much more serious problem in warm climates than it is where the soils freeze and thaw one or more times each year.

The thawing of frozen soil coupled with rainfall and snowmelt commonly produces a layer of saturated soil above an impermeable frozen layer. Saturated soil is very weak and erodible, so soil erosion can easily occur under these conditions. More than half of the annual soil loss from many temperate-region soils occurs while the soil is thawing (Froese et al., 1999). Also, both vehicle and animal traffic need to be kept away from such areas while the soil is saturated with water. The hazardous period is easily identified by watching ponded areas in the fields. The water disappears quickly when the frost is gone.

4–5 VEGETATION AND WATER EROSION

Vegetation limits the erosive action of raindrops on soil. Plant material intercepts raindrops and slows down runoff. An historic early erosion study at Columbia, Missouri, showed clearly that crops differ significantly in their effect on water erosion (Table 4–3). These results show that runoff and erosion are most severe on bare soil (fallow). Continuous corn, planted in intertilled rows, was the next most erosive cropping system, followed by continuous wheat, and then by the rotation of corn, wheat, and clover. Continuous bluegrass gave the most protection of any system tested.

The density of vegetative cover is as important as the kind of cover. Zöbisch (1993) found in Kenya that the critical ground cover for protection against erosion was 40%. Erosion was largely controlled by cover densities of 40%. Erosion occurred at an increasing rate as cover density was reduced below 40%.

Kramer and Weaver (1936) demonstrated conclusively that plant top growth was significantly more effective than roots in reducing water erosion. Duley and Kelly (1939) showed that dead plant material on the soil surface is as effective as thick perennial vegetation in maintaining infiltration rate and in reducing runoff and erosion. The concept of *stubble mulch tillage,* an early conservation tillage practice, developed from this early research finding.

Even relatively narrow belts of dense vegetation reduce the amount of erosion. Owens et al. (1996) studied the effect of fencing cattle out of pastures immediately adjacent to streams. They found that preventing cattle from grazing right to the streambank

TABLE 4–3 EFFECT OF CROPPING SYSTEM ON RUNOFF AND EROSION AT COLUMBIA, MISSOURI, 1918–1931

	Average runoff		Average annual erosion	Time to erode a 7-in. layer
Cropping system	inches	percent	(tons/ac)	(yrs)
Continuous fallow	12.3	30.5	41.4	24
Continuous bluegrass	4.8	12.0	0.3	3043
Continuous wheat	9.4	23.3	10.1	100
Continuous corn	11.9	29.4	19.7	50
Corn, wheat, clover rotation	5.6	13.8	2.8	368

Source: Modified from Miller and Krusekopf, 1932.

reduced annual sediment concentration by 50% and reduced soil loss from the area by 40%. Similarly, Robinson et al. (1996) showed that grass filter strips along streams in cultivated Iowa fields promoted infiltration, reduced runoff volume, and decreased runoff sediment concentration. Seventy percent of the sediment was removed in the first 10 ft (3 m) of the strip, 85% in 30 ft (9 m).

4–6 TRAFFIC AND WATER EROSION

Bare, smooth, compacted soil surfaces facilitate runoff and erosion. Foot traffic by wild and domestic animals and humans, and vehicular traffic such as that involving bicycles and motorized equipment, kills vegetation and smoothes and packs the soil surface on the trail, path, or road. As a result, water erosion is common where trails, paths, and roads traverse sloping ground, unless major steps are taken to control runoff water.

In the past, erosion has been common on nearly all roads and rights-of-way, but better design and more careful construction now reduces erosion losses on all major and most minor highways in developed countries. In the developing world, steps are also being taken to reduce erosion losses on their major roads. However, erosion is still a menace on many municipal roads and on or beside many roads and trails on private lands, on trails in government-controlled areas such as national and state parks and recreational areas, even on the shoulders of access and cross-over roads in the interstate highway system in the United States. It takes only a few passes up a sloping site in the semiarid southwest of the United States or on the fragile tundra soils in the north to leave marks that will last indefinitely and permit water to run along them and deepen into ruts and gullies as time goes on. In areas more favorable for plant growth, it will usually take more traffic to initiate erosion and gully formation, but be assured, erosion will accompany traffic on all trails or roads unless preventative treatments are provided.

A study in the Petroglyph National Monument in New Mexico (Gellis, 1996) showed that 50 gullies had formed along a 20-mile-long (32-km) basalt escarpment. Thirty of these gullies were connected by surface drainage to primitive dirt roads and foot and bicycle trails at the top and up the sides of the escarpment. In the area closest to a population center, where trails and roads were most abundant, gullies were more frequently encountered. Ten of the gullies must have developed recently because they could not be found on four-year-old aerial photos. Rainfall during that four-year period was not unusually intense. All storms recorded in the area of the new gullies had a recurrence interval of two years or less for the rainfall intensity measured. These results show the sensitivity of these fragile soils to even relatively mild traffic development.

4–7 WATER EROSION AND POLLUTION

In the past, major concern about erosion was centered on damage to the soil and its productivity. Increasing emphasis now is being placed on the pollution that erosion sediments cause on land and particularly in water. Concern about agricultural chemicals in runoff and eroding sediments is growing. Pollutants that enter streams in diffuse patterns from agricultural activities over wide areas constitute *nonpoint source* pollution.

While soil conservationists rate various soils as having "tolerable limits" of soil loss ranging from 1 to 5 tons/ac-yr (2 to 11 mt/ha-yr), many people view even "tolerable level" erosion as excessive. They are deeply concerned about fertilizers, pesticides, and animal wastes that enter streams and groundwater. Water pollution and its control are discussed in Chapter 17.

4–8 WATER EROSION AND SEDIMENTATION

Sedimentation is a part of the erosion process. Repeated detachment, transportation, and deposition move soil from the highest uplands to ocean beds. Not all sediment comes from cultivated lands or upland sites. Bottomlands, even stream banks and beds, lose material to erosion. Using every practical means to control erosion on cultivated land, range, and forest will reduce but not prevent sedimentation. For example, the Missouri River in the central United States was given the name "Big Muddy" by the western pioneers at a time when there was little or no cultivation along its banks.

Sedimentation is both beneficial and harmful. Alluvial soils are among the world's most productive soils; they develop in sediments eroded from rich surface soils on the uplands. Accelerated erosion removes surface soil initially, but soon carries less productive subsoil. Subsoil deposited on fertile bottomland soils reduces their productivity. Serious floods in the midwestern United States in 1950 and again in 1993 deposited 2 ft (60 cm) or more of coarse, unproductive sand on thousands of acres of highly productive alluvial soil. Erosion from denuded mountain slopes often deposits several feet of soil, stones, and other coarse material on the land at the foot of the slopes.

Sedimentation damages all types of vegetation. Even large trees may be killed. Highways, railroads, commercial buildings, and residences may be covered by layers of flood-borne sediment.

Sedimentation is a continuing process. Sediment from upland areas and streambeds is deposited on the bed when flow velocity decreases. This raises the level of the river bed and reduces channel capacity so that subsequent floodwater overtops the banks and causes increased damage.

Levees have been built to control river flow. Their effectiveness is seldom permanent because sedimentation raises channel beds in the levee systems so much that many river beds are actually above much of the surrounding land. Severe storms eventually cause overflow that results in extreme losses on the alluvial plains the levees were built to protect.

Siltation of stream channels forms shallow areas and sandbars that must be cleared from navigable streams. Dredging operations are expensive and may be needed at frequent intervals.

Most alluvial soils have water tables at shallow depths that feed seepage water into the nearby stream or river. As sediments raise river beds, water tables also rise. Higher water tables reduce crop growth by reducing depth of well-aerated soil. Some of these areas become swamps with no commercial value, though they may have value as wetlands for wildlife.

Sedimentation of reservoirs is costly also. Silt and coarse clay carried by streams are deposited when velocity slows as they enter lakes and reservoirs. The finest clay passes through the lakes and out the spillways. Excessive sedimentation and a short useful life are

Figure 4–8 This reservoir in Kansas originally had a surface area of 20 ac (8 ha) but has been filled with sediment that has changed most of the area into mud flats. The remaining water is only about 3 ft (1 m) deep. (Courtesy USDA Natural Resources Conservation Service.)

likely where stream gradient is steep, catchment area soils are erodible, watershed area is small (less than 100 mi^2 or 250 km^2), and the ratio of watershed area to volume of storage is less than 80 mi^2 : 1 ac-in. (200 km^2 : 1 ha-cm). When a reservoir is filled with sediment, its value for recreation, flood control, and/or irrigation is gone forever. Figure 4–8 shows a reservoir destroyed by sediment.

4–9 PRINCIPLES OF WATER-EROSION CONTROL

Water erosion occurs when conditions are favorable for the detachment and transportation of soil material. Climate, soil erodibility, slope gradient and length, and surface and vegetative conditions influence how much erosion will take place. Methods for predicting the effects that variations of each of these factors have on erosion loss are presented in Chapter 6.

Many different practices have been developed to reduce water erosion. Not all practices are applicable in all regions. However, the principles of water-erosion control are the same wherever serious water erosion occurs. These principles are:

1. Reduce raindrop impact on the soil.
2. Reduce runoff volume and velocity.
3. Increase the soil's resistance to erosion.

Management practices that put one or more of these principles into effect will help to control water erosion. These practices are discussed in later chapters.

SUMMARY

Water erosion occurs wherever rainfall strikes bare soil or runoff water flows over erodible and insufficiently protected soil. The principal forms of water erosion are *sheet erosion, rill erosion, gully erosion, solution erosion,* and *streambank erosion*. Erosion damages the upland areas from which soil and plant nutrients are removed and also the bottomlands on which sediments are deposited. It washes out crops on uplands, buries them on depositional sites, pollutes water, fills in river channels and reservoirs, and contributes to flooding.

The erosiveness of rainfall is influenced by the total amount of rain, the size of the drops, and their velocity of fall. The erosiveness of runoff is influenced by both the volume and the velocity of flow and by the abrasiveness of particles carried by the water. *Flow volume* is directly related to rainfall intensity and duration and inversely related to soil infiltration rate and permeability. *Flow velocity* is related to flow thickness, slope gradient, and surface condition.

Water also acts as a lubricant, aiding gravity in mass soil movement. *Landslides* and *mudflows* occur in some areas with steeper topography. Mudflows are most common in winter rainfall areas where cultivated soils are nearly saturated and lack adequate anchoring roots.

Soil properties, especially texture and structure, influence the ease or difficulty with which soil grains are detached and transported, and also influence infiltration and percolation rates. Vegetation, both living and dead, intercepts raindrops and reduces the energy they release at the soil surface. It also slows the passage of runoff water over the soil surface, thus reducing its energy.

Erosion removes valuable soil from the uplands and reduces the productive potential of the soils. Eroded materials are deposited on lower-lying lands. *Sedimentation* is both beneficial and harmful, and part of it is natural. Sedimentation is most damaging when it raises the level of a stream channel or fills a reservoir with soil material. Eroded soil is an important *nonpoint-source water pollutant* because of its large bulk and the chemicals it carries.

Water erosion and sedimentation can be controlled by reducing the energy of the erosive agents, usually rainfall and runoff water, or by increasing the soil's resistance to erosion.

QUESTIONS

1. In what ways is water erosion important to the nonfarming segment of the world's population?
2. Compare the appearance and effects of rill erosion to sheet erosion and to ephemeral gully erosion.
3. How do raindrops cause soil loss?
4. Compare the kinetic energy of a 1-mm raindrop to that of a 5-mm raindrop.
5. Describe briefly how runoff depth, land slope, and surface condition affect the erosiveness of running water.
6. How does soil material carried by a stream influence the ability of the stream to erode its channel?
7. How does soil structure influence the amount of soil lost by raindrop splash and runoff water?
8. Describe circumstances where freezing and thawing make a soil less subject to erosion and other circumstances when they make soil more subject to erosion.
9. How does vegetation reduce soil loss that is caused by water erosion?

REFERENCES

BAVER, L. D., 1939. Ewald Wollny—A pioneer in soil and water conservation research. *Soil Sci. Soc. Amer. Proc.* (1938) 3:330–333.

BENNETT, H. H., 1939. *Soil Conservation.* McGraw-Hill, New York, 993 p.

BENNETT, H. H., and W. R. CHAPLINE, 1928. *Soil Erosion a National Menace.* USDA Circ. 33.

CLARK, E. H. II, J. A. HAVERKAMP, and W. CHAPMAN, 1985. *Eroding Soils: The Off-Farm Impacts.* The Conservation Foundation, Washington, D.C., 252 p.

COLACICCO, D., T. OSBORN, and K. ALT, 1989. Economic damages from soil erosion. *J. Soil Water Cons.* 44:35–39.

DEN BIGGELAAR, C., R. LAL, K. WIEBE, and V. BRENEMAN, 2001. Impact of soil erosion on crop yields in North America. In D. L. Sparks (ed.), *Advances in Agronomy* 72:1–52, Academic Press, San Diego.

DULEY, F. L., and L. L. KELLY, 1939. *Effect of Soil Type, Slope, and Surface Condition on Intake of Water.* Nebraska Agr. Exp. Sta. Res. Bull. 112, Lincoln, NE.

ELLISON, W. D., 1944. Studies of raindrop erosion. *Agric. Eng.* 25:131–136, 181–182.

ELLISON, W. D., 1947a. Soil erosion studies: II. Soil detachment hazard by raindrop splash. *Agric. Eng.* 28:197–201.

ELLISON, W. D., 1947b. Soil erosion studies: V. Soil transportation in the splash process. *Agric. Eng.* 28:349–351.

FROESE, J. C., R. M. CRUSE, and M. GHAFFARZADEH, 1999. Erosion mechanics of soils with an impermeable subsurface layer. *Soil Sci. Soc. Am. J.* 63:1836–1841.

GALE, W. J., C. A. CAMBARDELLA, and T. B. BAILEY, 2000. Root-derived carbon and the formation and stabilization of aggregates. *Soil Sci. Soc. Am. J.* 64:201–207.

GELLIS, A. C., 1996. Gullying at the Petroglyph National Monument, New Mexico. *J. Soil Water Cons.* 51:155–159.

GUNN, R., and G. D. KINZER, 1949. The terminal velocity of fall for water droplets. *J. Meteorol.* 6:243–248.

HALLETT, P. D., T. BAUMGARTL, and I. M. YOUNG, 2001. Subcritical water repellency of aggregates from a range of soil management practices. *Soil Sci. Soc. Am. J.* 65:184–190.

HAYS, O. E., A. G. MCCALL, and F. G. BELL, 1949. *Investigations in Erosion Control and the Reclamation of Eroded Land at the Upper Mississippi Valley Conservation Experiment Station near LaCrosse, Wisconsin, 1933–1943.* USDA Tech. Bull. 973.

HUDSON, N., 1971. *Soil Conservation,* Cornell Univ. Press, Ithaca, NY, 320 p.

HUSSAIN, I., K. R. OLSON, and R. L. JONES, 1998. Erosion patterns on cultivated and uncultivated hillslopes determined by soil fly ash contents. *Soil Sci.* 163:726–738.

KRAMER, J., and J. E. WEAVER, 1936. *Relative Efficiency of Roots and Tops of Plants in Protecting the Soil from Erosion.* Cons. Dept. Bull. 2, Univ. Nebraska, Lincoln.

LARSON, W. E., F. J. PIERCE, and R. H. DOWDY, 1985. Loss in long-term productivity from soil erosion in the United States. In S. A. El Swaify, W. C. Moldenhauer, and A. Lo (eds.), *Soil Erosion and Conservation.* Soil Cons. Soc. Am., Ankeny, IA, p. 262–271.

LAWS, J. O., 1941. Measurements of fall velocity of water drops and raindrops. *Trans. Am. Geophys. Union* 22:709–721.

LAWS, J. O., and D. A. PARSONS, 1943. The relation of raindrop-size to intensity. *Trans. Am. Geophys. Union* 24:452–459.

LEHRSCH, G. A., 1998. Freeze-thaw cycles increase near-surface aggregate stability. *Soil Sci.* 163:63–70.

LYLES, L., 1977. Soil detachment and aggregate disintegration by wind driven rain. In *Soil Erosion: Prediction and Control.* Soil Cons. Soc. Am., Ankeny, IA, p. 152–159.

MCGREGOR, K. C., and C. K. MUTCHLER, 1977. Status of the R factor in northern Mississippi. In *Soil Erosion: Prediction and Control.* Soil Cons. Soc. Am., Ankeny, IA, p. 135–142.

MERTEN, G. H., M. A. NEARING, and A. L. O BORGES, 2001. Effect of sediment load on soil detachment and deposition in rills. *Soil Sci. Soc. Am. J.* 65:861–868.

MIDDLETON, H. E., 1930. *Properties of Soils Which Influence Soil Erosion.* USDA Tech. Bull. 178.

MIDDLETON, H. E., C. S. SLATER, and H. G. BYERS, 1932. *The Physical and Chemical Characteristics of the Soils from the Erosion Experiment Stations.* USDA Tech. Bull. 316.

———*The Physical and Chemical Characteristics of the Soils from the Erosion Experiment Stations.* Second Report. USDA Tech. Bull. 430.

MILLER, M. F., and H. H. KRUSEKOPF, 1932. *The Influence of Systems of Cropping and Methods of Culture on Surface Runoff and Soil Erosion.* Mo. Agr. Exp. Sta. Res. Bull. 177, Columbia, Missouri.

MOKMA, D. L., and M. A. SIETZ, 1992. Effects of soil erosion on corn yields on Marlette soils in south-central Michigan. *J. Soil Water Cons.* 47:325–327.

OWENS, L. B., W. M. EDWARDS, and R. W. VAN KEUREN, 1996. Sediment losses from a pastured watershed before and after stream fencing. *J. Soil Water Cons.* 51:90–94.

ROBINSON, C. A., M. GHAFFARZADEH, and R. M. CRUSE, 1996. Vegetative filter strip effects on sediment concentration in cropland runoff. *J. Soil Water Cons.* 51:227–230.

ROOSE, E. J., 1977. Use of the universal soil loss equation to predict erosion in West Africa. In *Soil Erosion: Prediction and Control.* Soil Cons. Soc. Am., Ankeny, IA, p. 60–74.

SHAW, R. H., 1959. Water use from plastic-covered and uncovered corn plots. *Agron. J.* 51:172–173.

SMITH, D. D., and W. H. WISCHMEIER, 1962. Rainfall erosion. In *Advances in Agronomy,* Vol. 14, Academic Press, New York, p. 109–148.

TAYLOR, H. H., 1994. *Fertilizer Use and Price Statistics, 1960–1993.* E.R.S. Stat. Bul. 893.

TROEH, F. R., 1964. Landform parameters correlated to soil drainage. *Soil Sci. Soc. Am. Proc.* 28:808–812.

VANDERVAERE, J.-P., M. VAUCLIN, R. HAVERKAMP, C. PEUGEOT, J.-L. THONY, and M. GILFEDDER, 1998. Prediction of crust-induced surface runoff with disc infiltrometer data. *Soil Sci.* 163:9–21.

WALLBRINK, P. J., A. S. MURRAY, and J. M. OLLEY, 1999. Relating suspended sediment to its original soil depth using fallout radionuclides. *Soil Sci. Soc. Am. J.* 63:369–378.

WALLING, D. E., and Q. HE, 1999. Using fallout lead-210 measurements to estimate soil erosion on cultivated land. *Soil Sci. Soc. Am. J.* 63:1404–1412.

WEESIES, G. A., S. J. LIVINGSTON, W. D. HOSTETER, and D. L. SCHERTZ, 1994. Effect of soil erosion on crop yield in Indiana: Results of a 10-year study. *J. Soil Water Cons.* 49:597–600.

WISCHMEIER, W. H., C. B. JOHNSON, and R. V. CROSS, 1971. A soil erodibility nomograph for farmland and construction sites. *J. Soil Water Cons.* 26:189–193.

WISCHMEIER, W. H., and D. D. SMITH, 1958. Rainfall energy and its relations to soil loss. *Trans. Am. Geophys. Union* 39:285–291.

YANG, H., Q. CHANG, M. DU, K. MINAMI, and T. HATTA, 1998. Quantitative model of soil erosion rates using ^{137}Cs for uncultivated soil. *Soil Sci.* 163:248–257.

ZHENG, F., C. HUANG, and L. D. NORTON, 2000. Vertical hydraulic gradient and run-on water and sediment effects on erosion processes and sediment regimes. *Soil Sci. Soc. Am. J.* 64:4–11.

ZÖBISCH, M. A., 1993. Erosion susceptibility and soil loss on grazing lands in some semiarid and subhumid locations in eastern Kenya. *J. Soil Water Cons.* 48:445–448.

CHAPTER

5

Wind Erosion and Deposition

Wind erosion is the process of detachment, transportation, and deposition of soil material by wind. Like water erosion, it is a cause of serious soil deterioration that has both geologic and human-caused components. Loess deposits provide evidence that much of the geologic component came from outwash plains associated with continental glaciation, from large desert areas, and from beaches. The human-caused component can also be very important. USDA data indicate that more soil is lost by wind erosion than by water erosion in the Great Plains region of the United States. Furthermore, this soil may carry agricultural chemicals with it and they can be deposited on sensitive crops in downwind areas (Clay et al., 2001).

> *The basic causes of wind erosion are few and simple. Wherever (1) the soil is loose, finely divided, and dry, (2) the soil surface is smooth and bare, and (3) the wind is strong, erosion may be expected.*
>
> *By the same token, whenever (a) the soil is compacted, kept moist, or made up of stable aggregates or clods large enough to resist the force of the wind, (b) the soil surface is roughened or covered by vegetation or vegetative residue, or (c) the wind near the ground is somewhat reduced, erosion may be curtailed or eliminated. (Chepil, 1957)*

Wind erosion is usually considered to be a problem of dryland regions, but even in humid areas, wind can cause severe damage to sandy soils, particularly along seacoasts, to muck soils, and to medium- and fine-textured soils that are stripped of their vegetative cover. Wind erosion can occur even on irrigated land when the soil becomes dry between irrigations or during seasons when the land is not being irrigated. It is a worldwide problem that is particularly serious in the United States and Canada, in the drier parts of Argentina, Bolivia, and Peru, in parts of the former USSR, in Middle East countries, in China, India, and Pakistan, in Africa both north and south of the equator, and in Australia.

In North America, wind erosion has been most damaging in the Great Plains states and in the Canadian prairie provinces, but it is also important around the Great Lakes, in eastern Washington and Oregon, in southeastern coastal areas, and along the Atlantic seaboard. It was most widespread and most damaging in the 1930s. Since then, research has provided a better understanding of this phenomenon, and better techniques for its control have been developed. The U.S. Natural Resources Conservation Service (formerly the Soil Conservation Service), the Canadian Prairie Farm Rehabilitation Agency, local conservation districts, and agricultural universities have promoted conservation measures. Governments have subsidized some conservation practices to promote their adoption.

Increased use of conservation measures has reduced the effects of severe droughts and strong winds so that soil losses, although often serious, have not again reached the levels encountered in the 1930s. In spite of this, the U.S. Department of Agriculture estimated that wind erosion in 1992 caused excessive soil losses on more than 116 million ac (47 million ha) in the United States. More than 45% of this area was cropland, more than 52% was rangeland, and less than 1% was in pasture (NRCS, 1994).

5–1 TYPES OF SOIL MOVEMENT

Wind carries soil in three ways:

1. *Suspension.* Soil particles and aggregates less than 0.05 mm in diameter (silt size and smaller) are kept suspended by the turbulence of air currents. Suspended dust does not drop out of the air in quantity unless rain washes it out or the velocity of the wind is drastically reduced.
2. *Saltation.* Intermediate-sized grains, approximately 0.05 to 0.5 mm in diameter (very fine, fine, and medium sand sizes), move in a series of short leaps. The jumping grains gain a great deal of energy and may knock other grains into the air or bounce back themselves. These saltating grains are the key to wind erosion. They drastically increase the number of both smaller and larger grains that move in suspension and in surface creep.
3. *Surface creep.* Soil grains larger than 0.5 mm in diameter cannot be lifted into the wind stream, but those smaller than about 1 mm may be bumped along the soil surface by saltating grains.

Aggregates, clods, and particles larger than 1 mm in diameter remain in place on the eroding surface and form a protective covering, often called *desert pavement* or *lag gravel,* as shown in Figure 5–1.

5–2 EROSION DAMAGE

Wind-erosion damage includes loss of soil depth, textural change, nutrient and productivity losses, abrasion, air pollution, and sedimentation. Precise measurements are scarce and difficult to make, but qualitative observations are abundant and were first documented more than 2000 years ago. For example, dust storms were noted in China as early as 205 B.C.E. (Zhibao et al., 2000). Wind erosion equations (Chapter 6, Sections 6-5 through 6-8) have been devised to estimate the rate of soil loss when actual measurements are not available.

Figure 5–1 Desert pavement in New Mexico. The finer soil particles have been blown away, leaving a heavy, continuous gravel cover. (Scale is in inches and centimeters.) (Courtesy F. R. Troeh.)

5–2.1 Loss of Soil

Annual losses higher than 300 tons/ac (700 mt/ha) have been estimated for highly erodible, bare, sandy soils. An entire furrow slice (about 1000 tons/ac) could be blown away in three or four years at this rate if soil was removed uniformly from the entire surface. Actual losses are usually less than this because land is seldom left bare and unprotected for a whole year, but greater losses have occurred. Plow layers from many recently tilled farm fields in the Great Plains were blown away in single dust storms in the 1930s. Some fields in western Canada, such as the area in Figure 5–2, lost 1 ft (30 cm) of soil in a single year.

A relatively new method of measuring soil loss since the 1950s is based on measurements of cesium 137 (^{137}Cs). Atmospheric fallout of this radioactive element, associated with nuclear-weapons testing, provided a tracer for studies of soil erosion and sedimentation. Ping et al. (2001) used this approach to estimate wind erosion rates. They calculated 40-year average erosion rates of 84 mt/ha-yr (38 tons/ac-yr) for shrub coppice dunes, 70 mt/ha-yr (31 tons/ac-yr) for semi-fixed dunefields, 30.5 mt/ha-yr (14 tons/ac-yr) for farmland, and 22 mt/ha-yr (10 tons/ac-yr) for grasslands in the Qinghai-Tibet Plateau in China.

5–2.2 Textural Change

Wind winnows soil much as it sifts chaff from threshed grain. Fine soil grains are carried great distances in suspension, saltating grains move to the fence rows or other barriers at the edges of fields, and coarser grains stay where they are or move relatively short distances within the eroding field.

Figure 5–2 More than a foot (30 cm) of fine sandy loam soil was removed by wind erosion from this field in southwestern Manitoba. The concrete structure on the left is a geodetic survey marker. (Courtesy Canada-Manitoba Soil Survey.)

The winnowing action of wind can coarsen textures in soils developed from glacial till, mixed residuum, and other materials having a wide range of particle sizes. Texture changes were reported during the erosion period of the 1930s in North America (Chepil, 1946; Daniel, 1936; Moss, 1935). Lyles and Tatarko (1986) confirmed texture coarsening of western Kansas soils between 1948 and 1984.

The largest changes occur in the more erodible sandy soils. Medium- and fine-textured soils suffer less from texture change. In these soils, the coarser grains left in eroding fields and the finer grains in dunes are generally aggregates rather than particles. These have the same texture as the whole soil had before erosion. Many silt loam "dunes" were deposited around buildings in the Great Plains during the 1910–1914 erosion period. The textures of these old dunes and of associated cultivated fields are the same.

5–2.3 Nutrient Losses

Colloidal clay and organic matter are the seat of most of the soil's fertility. Colloidal material lost in dust storms contains a lot of fertility. Lyles and Tatarko (1986) showed that ten soils in the western half of Kansas had lost an average of one-fifth of their 1948 organic-matter content by 1984. Fertility loss is particularly severe in coarse-textured soils that become coarser as erosion progresses, but it is also important in medium-textured soils that lose surface soil but do not change texture. Fertility loss also occurs by crop removal and by oxidation of organic matter, so it is difficult to determine how much of it should be attributed to wind and water erosion.

5–2.4 Productivity Losses

Soils become less productive as winds erode them. Soils developed from glacial till and other mixed-textured material lose productivity mostly because of lowered nutrient content and reduced water-holding capacity. Soils developed from loess or other relatively uniform material lose productivity because of loss of friable, productive surface soil and the exposure of more clayey, less permeable, less fertile subsoil material.

Loss of soil depth reduces crop production by making the root zone shallower and/or less favorable for root growth. Larney et al. (2000) found that grain yields from their plots on three soil types in the northern Great Plains were reduced by an average of 53% by the removal of 20 cm of topsoil. They also found that heavy applications of manure were more effective than mineral nitrogen and phosphorus fertilizers for restoring short-term soil productivity. Lyles (1977) suggested a method for estimating productivity decline based on the prediction of surface soil loss and a known relationship between depth of surface soil and crop yield, but this relationship undoubtedly depends on the type of soil. Also, it may be partly incremental instead of gradational. Abrupt changes may occur when certain critical values are passed. For example, a small change in soil depth causes a comparable change in water storage capacity that influences plant survival during dry periods.

5–2.5 Abrasion

Soil grains carried by wind have etched automobile windows and sandblasted paint on houses, cars, and machinery. Soil particles sift into bearing surfaces in machinery and accelerate wear. This type of damage is costly, but it is insignificant compared to the damage done to young, growing plants.

Plants seldom suffer permanent damage solely from the flogging action of high winds, but severe damage is done, especially to young plants, when the erosive wind carries abrasive soil material. Damage ranging from delayed growth and reduced yield to actual death has been inflicted on cotton, sorghum, wheat, soybeans, sunflower, alfalfa, a number of native grasses, and several vegetables. A relatively short exposure to soil blast will reduce final yields; plants are killed when the exposure is long enough. Figure 5–3 shows a field of wheat in southwestern Kansas destroyed by severe wind erosion.

5–2.6 Air Pollution

The presence of soil particles in the air has long been noted. Early Greek writers mentioned dust storms that probably originated in the Sahara Desert. Most dust originates in deserts or in dryland areas temporarily bared of vegetative cover by overgrazing, cultivation, or fires.

Atmospheric dust causes discomfort some distance from its source. Near the source, discomfort is much greater, and a dust storm can be fatal to travelers caught in it.

In addition to the physical hazards produced by dust in the air, chemical hazards also may be encountered. Wind-blown dust can carry toxic substances (Holmes et al., 1993). Mine-spoil banks and waste-disposal sites often contain unusual chemicals. Some areas are accidentally, some purposely, contaminated with chemicals. Some of these chemicals are benign, while others, such as polychlorinated biphenyls (PCBs), are extremely toxic. Cultivated fields and some permanently vegetated areas are regularly or intermittently treated with agrochemicals. Many of these contaminants are adsorbed on the surface of the fine soil particles and are carried into the air by erosive forces. These chemicals add to the inconvenience and the hazard of living in, and breathing, dust-laden air.

Dust that results from farming activities seldom causes death directly, but it can and does cause accidents and respiratory ailments that sometimes prove fatal. Dust in the air can

Figure 5–3 The wheat crop in this field in southwestern Kansas was destroyed by a windstorm on February 10, 1976. (Courtesy USDA Natural Resources Conservation Service.)

reduce visibility dangerously on highways and at airfields. Dust may also carry pathogens that cause skin disorders.

5–2.7 Deposition (Sedimentation)

Suspended dust is carried long distances and deposited as a thin film over exposed surfaces. It constitutes a nuisance that can be demoralizing to those who try to keep surfaces dust-free, but it generally does no great physical damage. Some soils may even gain from added nutrients and organic matter.

Reynolds et al. (2001) suggest that aeolian dust contributes to soil formation, furnishes essential nutrients for plants to invade hostile environments, and has considerable influence on arid-land ecosystems. They used particle-size and mineralogical analyses to estimate that soils on the Colorado Plateau are composed of as much as 20 to 30% aeolian dust and that much of it came from areas more than 60 miles (100 km) away. Variations in the orientation of magnetite particles allowed them to infer a time line for the deposits and this, coupled with changes in mineralogy, made it possible to estimate how much of the deposition had occurred since farming began in the upwind areas.

Measurements of airborn dust in the Southern High Plains plateau in Texas and New Mexico indicate a relatively stable level of background dust with sudden peaks representing specific dust events (Stout, 2001). The background dust was unrelated to wind, but the

Figure 5–4 Wind erosion sediments have buried three snow fences erected to protect the road at the right of this picture. The surface of the dunes over the snow fences is 3 to 10 ft (1 to 3 m) above the former ground level. (Courtesy USDA Natural Resources Conservation Service.)

peaks were well correlated with wind velocities higher than 4 m/s (9 mph) when measurements were taken downwind from areas of dry soil with little surface cover.

Efforts have been made to develop prediction equations for the amount of dust in the air. These equations consider not only the rate of soil loss from source areas but also the wind direction, distance from the source area, and elevation profiles (Saxton et al., 2000).

Saltating soil material does not travel far, but it causes considerable physical damage. Many farm fences were buried during the 1930s by soil that settled behind tumbleweeds that were trapped by the fences. Sometimes two fences had to be built successively on top of the original one to keep livestock confined and cropland protected, or to control soil and snow drifts (see Figure 5–4). Drainage and irrigation ditches have also been plugged with blowing soil. Land leveled for irrigation as well as ordinary farm fields have been made hummocky by drifting soil. An uncultivated area can be a source of sediment that can fill corrugations or even furrows, thus making it difficult or impossible to maintain these forms of surface irrigation (Chapter 15, Section 15-5.1).

Crops can be buried by drifting soil, particularly when they are planted in furrows. Young plants are most likely to be damaged, but even mature plants on the windward edges of fields next to eroding areas can be completely covered. Sand dunes can move into windbreaks and tree shelterbelts, eventually killing them if the drifts get too deep. Highways and other engineering works can be covered with blown soil. Huszar and Piper (1986) surveyed off-site costs of wind erosion in New Mexico and reported annual costs of over $465 million. This dwarfs the $10 million on-site costs claimed annually for the state.

Sand deposits are expensive to remove. The Santa Fe Railway found this out when the John Martin Dam was built in eastern Colorado. The mainline track had to be relocated from the river valley to the uplands south of the river, passing through a very sandy region.

Blowouts were common during the 1930s and sand frequently covered the right-of-way. Soil drifting was controlled only after livestock numbers were reduced and the area was revegetated with the help of large government input.

5–3 EROSIVENESS OF SURFACE WIND

Moving air possesses energy. The higher the velocity of the wind, the higher the energy level and the more erosive the wind.

5–3.1 Velocity of Wind Near the Ground

Standard wind velocity is measured at a fixed height, usually 30 ft (9 m), above the ground. Height is important because velocity of even a steady wind increases dramatically above the ground surface. The height of measurement (z) therefore must be specified along with wind velocity (u).

Wind velocity over a bare surface is zero at a height (Z_o) slightly above the average height of a bare soil surface but below the tops of soil irregularities (see Figure 5–5A). Velocity approaches zero considerably above the soil surface in a vegetated area, as shown in Figure 5–5B. Here D represents the zero plane displacement caused by the vegetative cover and Z_o is the roughness parameter. The velocity gradient behaves as if velocity were zero at ($D + Z_o$), (top dashed line in Figure 5–5C), but air still moves slowly and erratically through the crop below this point, as indicated by the lower dotted line. The velocity gradients within 15 ft (5 m) of soil surfaces plot as straight lines on a semilog graph as shown in Figure 5–5D, if ($z - D$) is used for height plotted on the log scale.

Measurement of wind velocity profiles shows that the height Z_o, where velocity becomes zero, is the same on a specific bare surface for all wind speeds. Similarly, the height ($D + Z_o$) is constant for a specific vegetated surface as long as the crop is not bent over by stronger winds.

Friction Velocity and Erosive Power of Wind. Friction velocity (u^*) controls the erosive power of wind (Note 5–1). It is related to the velocity profile and to the drag exerted by wind on the soil surface. Over a bare soil, it increases in direct proportion to the wind velocity, measured at a specific height, until the drag on the soil surface (surface shear stress) begins to cause erosion. As soil starts to erode, some of the wind's energy is used to transport soil grains so velocity decreases.

NOTE 5–1
FRICTION VELOCITY OF WIND

Friction velocity, u^*, is not an actual velocity, but it has the same units as velocity, l/t. It is defined by the following equation:

$$u^* = \frac{\sqrt{\tau_o}}{\rho}$$

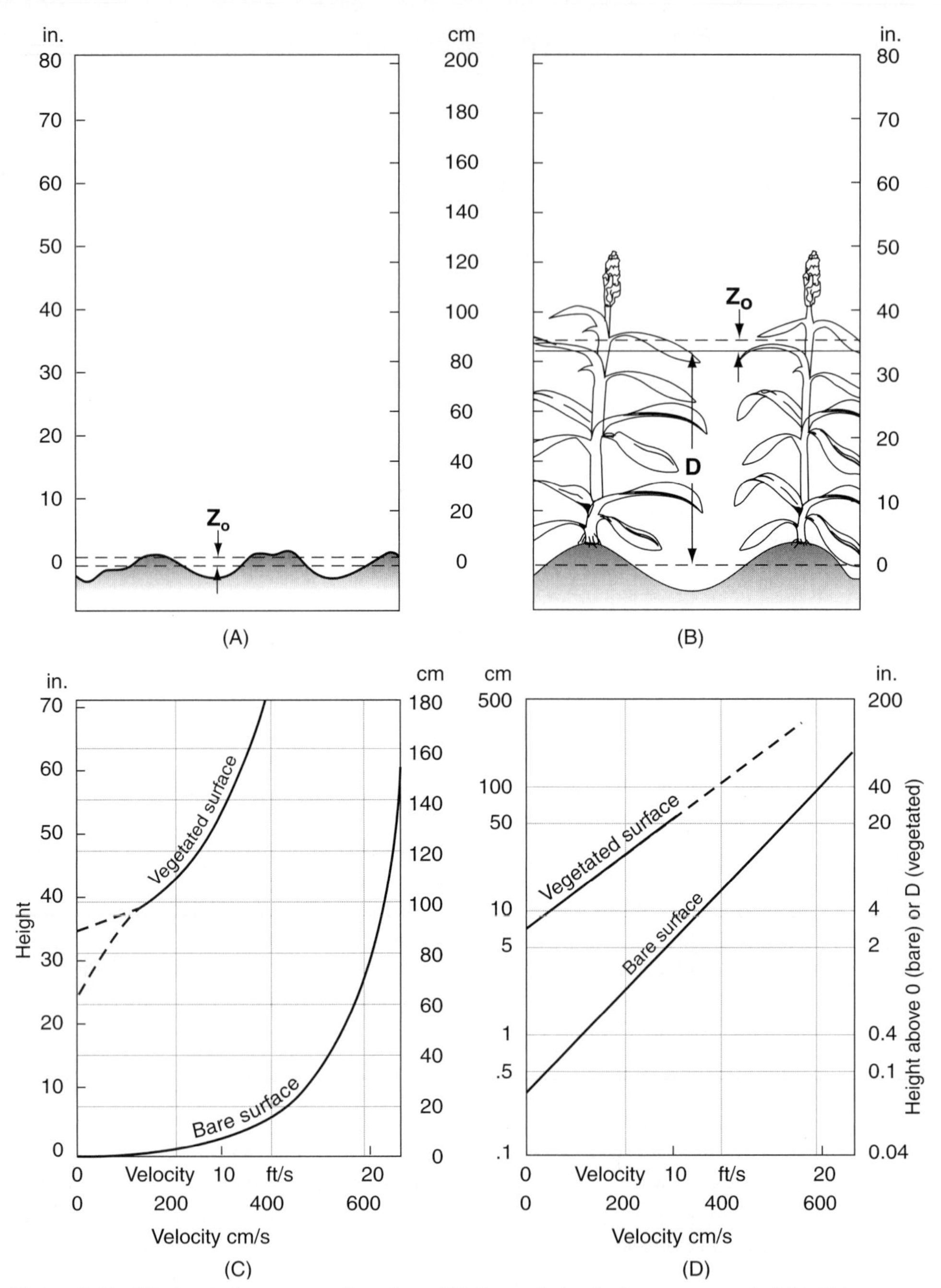

Figure 5–5 Wind velocity near a soil surface. (A) Zero wind velocity occurs at a height (Z_o) above the average height of the soil surface but below the high points. (B) A crop or other vegetative cover raises the level where wind velocity extrapolates to zero by a distance (D), making (D + Z_o) equal to about 70% of the height of the vegetation. (C) Velocity profiles above a bare surface, as in (A), and above and within a vegetated surface, as in (B), under the influence of the same "free" velocity wind (same velocity at 1600$^+$ ft [500$^+$ m]). (D) Velocity gradients plot as straight lines when a logarithmic scale is used for the height.

where τ_o = surface shear stress, dynes/cm^2

ρ = air density, g/cm^3

Wind velocity is related to friction velocity as indicated in the following equation:

$$u_z = \frac{u^*}{k}\ln\left(\frac{z - D}{Z_o}\right) + \phi_o$$

where uz = mean wind velocity, cm/s, at height z, cm

k = von Karman's constant < 0.4

D = zero plane displacement, cm (Figure 5–5B)

Zo = effective roughness height (roughness parameter), cm (Figure 5–5A)

fo = integral adiabatic influence function (usually zero for highly turbulent flow)

Friction velocity, u^*, is proportional to the mean wind velocity, u_z, over any noneroding surface because *k, D,* and Z_o are constant for each surface (unless vegetative cover height is reduced by bending with increasing wind), and *z* can be made constant by always measuring wind velocity at the same height. Friction velocity will be different for bare and vegetated surfaces because vegetation changes values of Z_o and *D.* Movement of soil grains changes u^* by absorbing wind energy.

For a given friction velocity, shear force is greater over rough surfaces, which therefore should be more erodible than smooth ones. Actually, rough surfaces usually reduce erosion because elements of surface roughness absorb much of the drag and leave only a small residual force to strike erodible soil grains.

5–3.2 Wind Turbulence

Wind strong enough to cause erosion is always turbulent, with eddies moving in all directions at a variety of velocities. Turbulence increases with increases in friction velocity, with increasing surface roughness, and with pronounced changes in surface temperature. It is also more pronounced close to the soil surface than higher in the wind stream (Chepil and Milne, 1941).

Air turbulence was once considered to be the major factor initiating movement of grains in saltation, but other factors are now known to play a more significant role (Section 5–4.1). Turbulence is important in keeping soil grains suspended in air.

5–3.3 Wind Gustiness

Wind velocity fluctuates widely and frequently. Wind tunnel studies show that soil composed of a mixture of erodible and nonerodible components will stabilize if wind velocity is constant. Nonerodible components eventually blanket the surface and protect it from further loss. Variable wind velocity prevents the soil surface from stabilizing completely. Higher-velocity gusts cause a stabilized surface to start eroding again, with erosion continuing until gust velocity drops below that required to cause erosion.

5–3.4 Prevailing Wind Direction

Winds from any direction can cause erosion. Changes in wind direction affect the erosion process in two major ways. First, soil surfaces become stabilized against the erosive force of the wind by developing surface patterns with protected areas behind nonerodible grains and clods that resist the wind. If the direction of the wind changes sufficiently, this pattern loses its ability to protect the surface and erosion may start again. A shift in direction of as little as 30° will allow this to happen. Second, if erosive winds are predominantly from one direction or from both of two opposite directions, it is possible to reduce soil losses by placing barriers, such as furrows, crop strips, and windbreaks, perpendicular to the prevailing wind direction. If, on the other hand, erosive winds show no seasonal or annual prevailing direction, wind barriers are of little practical value. For example, the prevailing direction of wind at Dodge City and Wichita, Kansas, is from the north for six months and from the south for six months. Each month, more than twice as much erosive wind force occurs parallel to the prevailing direction as occurs perpendicular to it. East-west directional barriers are therefore quite effective. In contrast, Douglas, Arizona, has prevailing winds from the east for three months, the northwest or north-northwest for two months, the southwest and south-southwest for six months, and the south for one month, and in no month is much more than half the erosive wind force from the prevailing direction. Barriers are of little value against such variable winds.

5–4 INITIATION OF SOIL MOVEMENT BY WIND

Slow-velocity winds do not have enough energy to pick up soil particles. The *threshold velocity* required to cause wind erosion varies, depending on soil and surface properties, from about 8 to 30 mi/hr (3.6 to 13.5 m/s), measured 6 in. (15 cm) above the soil surface. Soil movement begins with saltating sand grains unless there is an external agent such as vehicle or animal traffic.

5–4.1 Saltation

Saltation was recognized as a factor in wind erosion by King (1894) long before the mechanism that causes it was understood. Free (1911) suggested that turbulence was responsible for starting soil grains moving in saltation. However, calculations showed that a velocity in excess of 10.4 mi/hr (4.65 m/s) would be necessary for turbulence to lift a soil cube 0.5 mm on a side (Chepil and Milne, 1939). Measurements showed, however, that sand grains larger than 0.8 mm in diameter were saltating in a wind with a velocity of only 5 mi/hr (2.25 m/s) at 1/4 in. (5 mm) height. Therefore, some factor(s) other than turbulence must be involved in initiating saltation.

Bagnold (1937) suggested that sand particles rise into the air in saltation because stationary particles lying on the soil surface are knocked into the air by the impact of de-

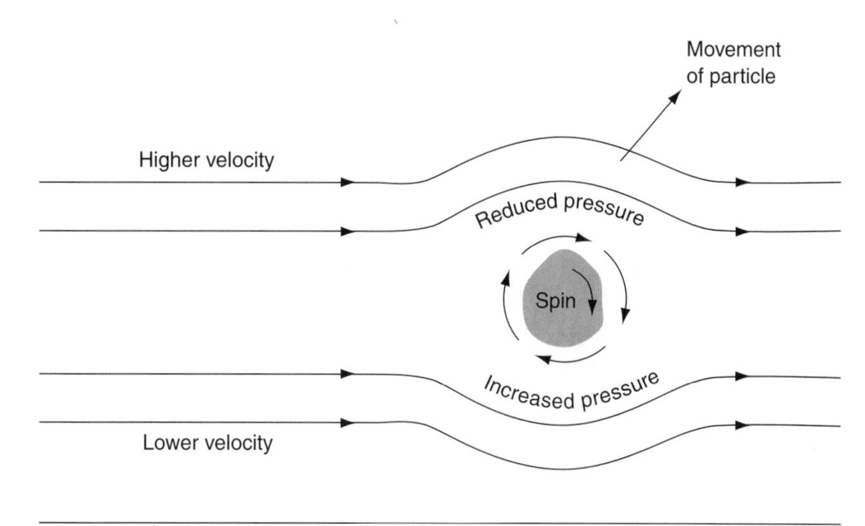

Figure 5–6 A spinning sand grain in a moving airstream is lifted by increased air pressure below and reduced pressure above.

scending particles, or that saltating particles returning to the soil surface carom back into flight. Neither of these suggestions explains how particles first get into movement. Chepil (1945a) studied the possibility that soil grains might be bounced into the airstream by irregularities on their surfaces or on the ground. He decided, however, that the nearly vertical takeoffs he observed could not be explained by this mechanism. He then examined the possibility that particle spin and the steep velocity gradient close to the soil surface might initiate saltation. Differences in the velocity of fluid flow over the top and bottom surfaces of an object set up pressure differences on these surfaces. A zone of lower pressure develops where flow is more rapid; higher pressure develops where flow is slower. The lift on airplane wings is a result of this principle.

Sand particles rolling along the bed of a wind tunnel increase the flow velocity of air on top of the grain and decrease it on the bottom. The rapid increase in wind velocity with height close to the surface contributes to a large velocity differential between the top and bottom of the spinning grains, as shown in Figure 5–6. Chepil felt that the pressure differential was sufficient to force the particles steeply upward into the wind stream.

Chepil watched sand grains on the smooth, wooden floor of his wind tunnel and observed that they started to roll along the bed. After a relatively short distance, without contacting other grains or striking any visible irregularity, the particles suddenly jumped almost vertically into the air. The angle of ascent generally ranged between 75 and 90°. By this time they had a very significant spin, later measured at 200 to 1000 revolutions per second (Chepil, 1945a). Ascending grains were carried downwind by the increasingly rapid velocity of the wind currents through which the grains rose. They continued downwind with increasing velocity even as they fell back toward the tunnel surface. Acceleration brought about by the wind seemed to match the acceleration of gravity, because the path back to the bed, as shown in Figure 5–7, was nearly a straight line with an impact angle generally between 6 and 12°.

Chepil's explanation of saltation based on spinning particles and pressure differences is now generally accepted with the understanding that it is probably augmented by surface irregularities, turbulence, and perhaps venturi effects.

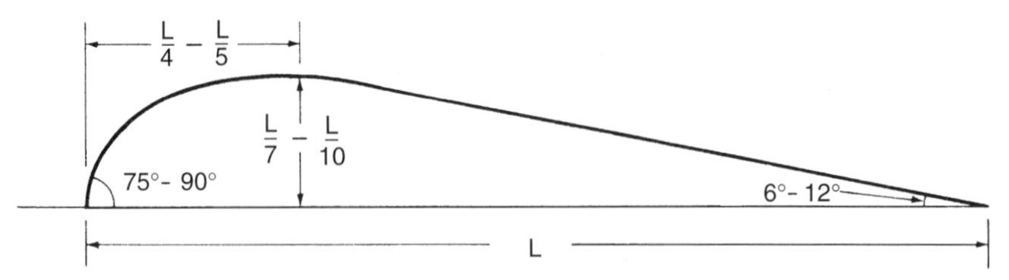

Figure 5–7 Path of a sand grain in saltation.

5–4.2 Avalanching

The increasing rate of erosion as the wind blows farther across a field is called *avalanching*. Chepil (1946) suggested the following three major reasons for avalanching:

1. Accumulation of eroded material from previous storms. Erosion often results from successive high winds coming from the same prevailing direction. Soil farther downwind in such fields is likely to be more erodible simply because it has been eroded previously and deposited there. The farther leeward the wind moves, the more abrasive material it is likely to carry.

2. Increasing quantity of saltating particles. Saltating particles constantly drop back to the soil surface and, in the process, detach new soil grains and abrade clods, peds, and aggregates. Abrasion, and consequently erosion, accelerates as these new materials are added to the wind stream.

3. Smoother surfaces. Wind moving across a field smooths the surface by cutting humps and filling hollows. Wind flows faster across smoother surfaces, and faster winds are more erosive.

5–5 WIND AND THE EROSION PROCESS

Wind has power to detach and transport soil grains. After transportation, particles and aggregates come to rest short or long distances from their origin. Coarse sand particles roll only a short distance and saltating sand particles move in short jumps, but silt particles may travel many miles and tiny clay particles may be carried around the world.

5–5.1 Detaching Capacity of Wind

The detaching capacity of wind is related to its friction velocity, or shear stress, and to the size of the erodible grains.

$$D = f(u'^{*})^2$$

where D = detaching capacity, g/cm^2-s

u'^{*} = friction velocity over an eroding surface, cm/s

A sharp velocity gradient near the soil causes grains that protrude higher into the wind stream to be struck by stronger wind force. Larger particles stick up higher, but their larger mass requires more force to detach them. Any particle from 0.05 to 0.5 mm in diameter can

Figure 5–8 Abrasion of soil cylinders 3 in. (7.5 cm) in diameter and 2.5 in. (6.2 cm) tall by dune sand carried in a wind blowing from left to right with a friction velocity of 2 ft/sec (61 cm/s). Soils from left to right are fine sandy loam, loam, light silt loam, heavy silt loam, and silty clay. (Courtesy USDA-ARS Northern Plains Wind Erosion Research Unit.)

be detached if the wind is strong enough, but those from 0.1 to 0.15 mm are the easiest to detach of any grains that move in saltation. Silt- and clay-sized grains (< 0.05 mm) and very coarse sand and gravel sizes (> 1.0 mm) cannot be separated from the soil mass even by strong winds if they are free of saltating particles. Fine particles can only move in saltation if they are bound together into aggregates of the right size.

Winds containing abrasive material can detach both smaller and larger grains. These abrasive materials not only detach erodible grains from the soil, they also abrade nonerodible clods, detaching small erodible grains from them. If the wind continues long enough, whole clods may be disintegrated (see Figure 5–8).

5–5.2 Transporting Capacity of Wind

Transport capacity of wind is related to wind velocity, but not to soil-grain size. Greater numbers of smaller-sized grains can be picked up, but the total weight of material that a specific wind can carry remains relatively constant.

Early work by Chepil (1945a) showed that the proportion of material in suspension, saltation, and creep depended on the aggregate- and particle-size composition of the soil. Minor quantities of suspended material were found over very coarse-textured soils and over strongly aggregated, fine-textured soils. Quantities of creeping particles were relatively large. Suspension was greater and creep noticeably smaller over dusty, silty, and fine sandy soils. In every soil Chepil studied, the amount of material in saltation was always greater than that in suspension and creep combined. Amounts in suspension ranged from 3 to 38%, in saltation from 55 to 72%, and in creep from 7 to 25% of the moving soil.

Bagnold (1941) and Chepil (1945b) noted that the rate of dune sand and soil movement by wind (weight of material moving past a unit width, normal to the direction of movement, per unit time) was related to the third power of the friction velocity:

$$q = f\frac{\rho}{g}u'^{*3}$$

where q = rate of soil movement, g/(cm width)-s

ρ = air density, g/cm^3

g = gravitational constant, 980 cm/s^2

u'^{*} = friction velocity over an eroding surface, cm/s^2

The relationship between amount of soil removed from a unit area and erosive wind force probably has greater significance than that between rate of loss and wind force. Chepil and Woodruff (1963) suggest that

$$X = f(u^*)^5$$

where X = the transportation capacity, g/cm^2

This relationship is influenced by a number of factors, but the total soil removal appears to be proportional to the fifth power of the friction velocity.

5–5.3 Soil Deposition

The distance that soil is transported from its original site depends on wind velocity and on size and weight of particles and aggregates. Dust, which is kicked up by saltating particles and carried in suspension by the wind, is moved far away from the original location. This cloud of dust, however spectacular it may be, contains only a small part of the full soil load carried by the wind. Nevertheless, it is a very important loss because it contains the finer, more fertile elements from the soil—the clay and the humus.

Silt and very fine sand particles are the main components of loess deposits. As discussed in Chapter 3 (Section 3–5.2), loess mantles areas downwind from source areas. The deposit is thickest and contains more very fine sand near the source and becomes gradually thinner and finer textured as the transport distance increases. Major loess deposits can be traced for tens or hundreds of miles (or kilometers) before they become too thin to detect.

Saltating grains usually remain in the vicinity of the eroding field. They come to rest in the lea of cultivation ridges, are piled up in the field as small dunes behind clumps of vegetation, or are trapped by vegetation in downwind fencerows. Coarser saltating grains are deposited on the windward side of the dunes, finer ones to the leeward ("leesands"). Dune formation is a complex process with the shape of the dunes depending on wind speed, constancy of direction, topographic features, and vegetation. Certain types of shrubs (especially varieties of mesquite) can survive even if they are mostly covered by sand. Such shrubs form the core for semi-stabilized dunes called *nabkas* or *coppiced* dunes. Langford (2000) describes nabkas as large as 130 ft (40 m) across and 14 ft (4.3 m) deep. Branches protruding from the upwind side and broken branches littering the adjacent area indicate that the nabka is migrating slowly downwind. Nabkas typically occur in groups with the areas between them covered by desert pavement or by low-growing shrubs and grasses. Langford calls other areas *sand sheets* where the vegetation and sand deposits are more uniform.

Individual grains moving in surface creep travel only a very short distance from their original location. Lag materials are not moved by wind, but they may lose some bulk by abrasion.

5–6 FACTORS AFFECTING WIND EROSION

High-velocity winds do not always cause soil drifting, and erosive winds do not cause the same amount of erosion in all situations. Factors that influence the amount of erosion that wind will cause are the soil's resistance to erosion, surface roughness, rainfall, land slope (hummocks), length of exposed area, and vegetative cover.

5–6.1 Soil Resistance

The major factor that makes soil resist wind erosion is size (mass) of individual soil grains. If the mass is sufficient, particles, aggregates, or clods are nonerodible. Large grains also protect and stabilize erodible grains in their wind shadow. Chepil (1950) called this "the governing principle of surface roughness." This is roughness caused by soil cloddiness, not by mechanical ridging.

Dry Aggregate Size Distribution. Grains larger than 1 mm in diameter resist wind erosion; those between 1 and 0.5 mm (coarse sand size) are erodible only in very high-velocity winds; those less than 0.5 mm in effective diameter are highly erodible. The more nonerodible grains the surface contains, the less erodible the soil is.

The proportion of nonerodible grains present on or near the soil surface affects the ease and speed with which soil starts to move, how long erosion will continue before a nonerodible blanket forms to halt it, and the total amount of soil that will be lost. A relatively smooth soil containing nonerodible clods may become rough as it erodes and thus reduce erosion losses. The likelihood of a field forming a protective surface condition can be assessed by determining the size distribution of stable clods in the top inch (2.5 cm) of soil. The relationship between proportion of nonerodible clods and soil erodibility is shown in Figure 5–9.

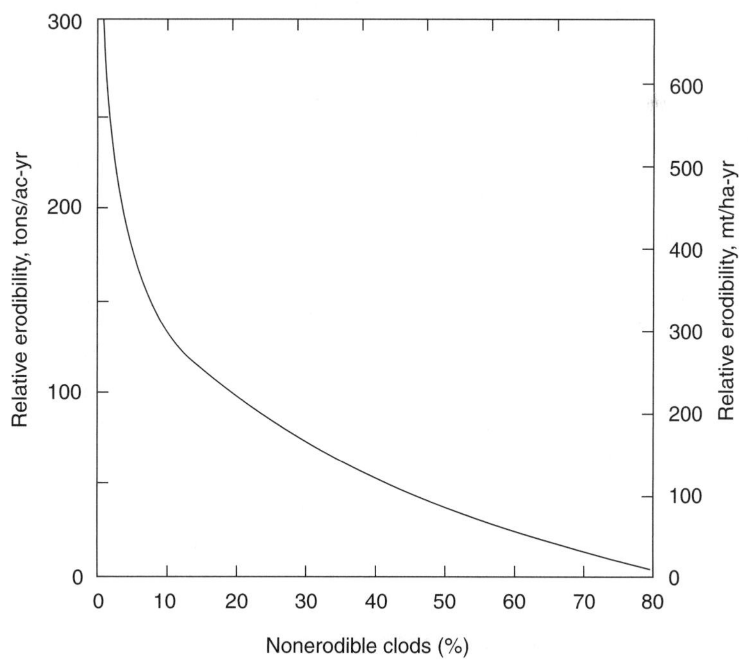

Figure 5–9 Relationship between percentage of nonerodible clods in a soil and its relative erodibility (*I*). (Modified from Soil Science Society of America Proceedings, Volume 29, p. 602–608, 1965 [Woodruff and Siddoway], by permission of the Soil Science Society of America.)

A smooth, bare, infinitely wide field with only 2% nonerodible clods could lose 250 tons/ac (560 mt/ha) in a year under the high winds common in the central Great Plains. A similar field with 40% nonerodible clods would lose 56 tons/ac-yr (125 mt/ha-yr) under the same conditions.

Mechanical Stability of Structural Units. The presence of nonerodible structural units does not, by itself, prevent soil erosion. Aggregates and clods that are easily abraded do not resist erosion for long. Accordingly, aggregate stability is an important attribute also.

Factors Affecting Aggregate Size and Stability. Soil properties such as texture, organic matter, exchangeable cations, and free calcium carbonate influence aggregate size and stability.

Coarse-textured soils do not contain enough clay to bind the sandy particles into structural units. Clayey soils develop aggregates and clods, but weathering, especially freezing and drying while frozen, breaks them down. Chepil (1953) claimed that a clay content of about 27% is best for clod development. Clay content less than 15% almost precluded a stable cloddy condition. The presence of large amounts of fine and medium sand influences soil erodibility directly because these sizes can saltate. Very coarse sand and gravel help reduce soil erodibility because they are too large to be moved by most winds.

Soil organic matter is often associated with high levels of aggregation and with structural stability, but soils with very high organic-matter levels are often more erodible than those with moderate contents. Wind erosion is a major hazard on organic soils that are drained and cropped.

Chepil (1955) found that additions of cereal straw and legume hay increased nonerodible dry clods and reduced erodibility while the materials were undergoing active decomposition (about six months). As decomposition slowed down, initial cementing materials lost their aggregating ability, and microbial fibers disintegrated. Nonerodible-clod size decreased and soil erodibility increased, especially on high-residue treatments. Often a high proportion of highly erodible small- to medium-sized aggregates developed.

Applications of lime to acid soils in humid areas often improve soil structure. This is not true for soils in arid and semiarid regions. These are rarely calcium deficient. Their surface soils have an abundance of calcium, and their subsoils invariably contain free calcium carbonate. In fact, shallow calcium carbonate horizons in drier areas may be mixed with surface soil by cultivation. Chepil (1954) showed that the presence of as little as 1% free calcium carbonate in a soil generally caused clod disintegration and increased erodibility. Despite this relationship, when calcareous B or C horizons are exposed by erosion, surface crusts often develop that effectively protect soils from further erosion.

5–6.2 Surface Ridges and Roughness

Surface ridges produced by tillage reduce erosion. Effectiveness of ridges depends on height, lateral frequency, shape, and orientation relative to the direction of the wind. Ridges reduce wind velocity near the ground and trap eroding soil grains in the furrows between ridge crests. Bielders et al. (2000) used hoes to make ridges 8-in. (20-cm) high and 5 ft (1.5 m) apart in Niger. Soil ridges reduced wind erosion by 57%, and the reduction was 87% when millet stover residues were included to stabilize the soil. However, the ridges had to be oriented across the wind direction; little or no reduction in erosion occurred when the wind blew parallel to the ridges. They noted that ridge orientation may be a problem where both wind and water erosion are likely to occur.

Wind tunnel measurements by Batt and Peabody (1999) showed that ridges across the wind direction were the most effective form of surface roughness for reducing wind erosion, but that any erosion-resistant surface roughness was helpful. Ridges, clods, stubble, and gravel all reduced wind erosion caused by both low-velocity and high-velocity winds. They evaluated a normalized roughness factor by dividing the height difference between high and low points by the average spacing between high points (H/S) and found that the amount of soil loss decreased linearly as the H/S factor increased from 0 to 0.12. The soil loss rate remained nearly constant at about 10% of the flat surface value for H/S factors greater than 0.12.

5–6.3 Rainfall

Rain moistens surface soil, and moist soil is not eroded by wind. Studies show soil erodibility decreases—slowly at first, then more rapidly—as a soil is moistened from the air-dry condition to the wilting point. Chepil (1956) worked with four Great Plains soils and found they became nonerodible at moisture contents ranging from 0.82 to 1.16 times the water content at 15-atm tension. Bisal and Hsieh (1966) studied three Canadian soils and found moisture contents from 0.32 to 1.46 times that at 15-atm tension prevented soil drifting. Unfortunately, the direct effect of moisture on soil erodibility is transitory; it takes only a very thin layer of dry surface soil to permit erosion to start, even if moisture is abundant immediately below. Wind soon reduces the moisture content of surface layers. Some sandy soils can begin to drift 15 or 20 minutes after an intense shower.

Rainfall also reduces erosion indirectly by increasing plant growth. Crop response to rainfall is extremely important because plant cover controls wind erosion best (see Section 5–6.6).

Rain can also increase wind erosion. This happens when raindrops break exposed wind-resistant grains, detach erodible grains from the soil mass, and smooth soil surfaces so they are more susceptible to erosion.

5–6.4 Knoll Slopes

Over long slopes, short slopes not exceeding 1.5%, and level land, the velocity gradient and friction velocity are reasonably constant for a given wind. Over hummocky topography, where slopes are relatively short, layers of higher wind velocity move closer to the soil surface as they pass over knoll crests (see Figure 5–10). Z_o is a relatively constant height above the surface, so the shorter vertical distance to the higher-velocity flow makes the friction velocity greater over the knolls. This makes the wind's erosive force much greater on the crests than on level land or on long slopes. Chepil et al. (1964) calculated probable increases in erosion on the crests and upper slopes of relatively short 3, 6, and 10% slopes.

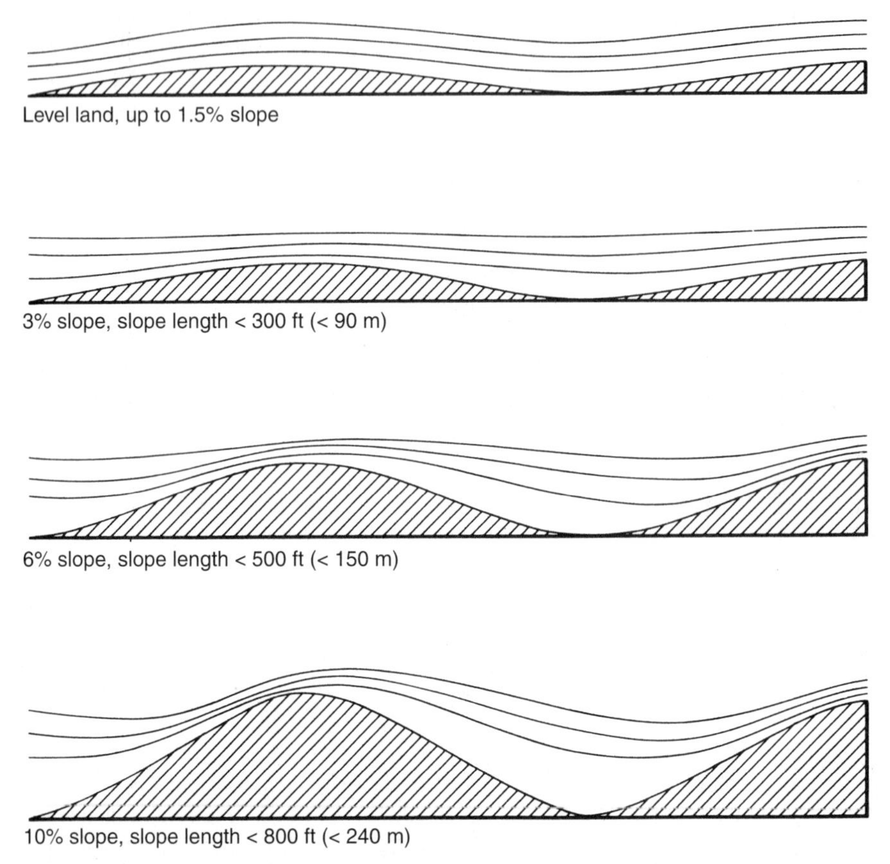

Figure 5–10 Lines of equal wind velocity over different land slopes. If the top line in each diagram represents 14 mi/hr (625 cm/s), this velocity is reached 1.0, 0.6, 0.32, and 0.18 ft (30.5, 18.3, 9.8, and 5.5 cm) above the knoll crest on the 1.5%, 3%, 6%, and 10% slopes, respectively. (Modified from the *Journal of Soil and Water Conservation*, Volume 19, p. 179–181, 1964 [Chepil et al.]).

They assumed that the zone of 14-mi/hr (625-cm/s) wind was found approximately 1.0, 0.6, 0.32, and 0.18 ft (30.5, 18.3, 9.8, and 5.5 cm) above the crests of knolls with side slopes of 1.5, 3, 6, and 10%, respectively. Their calculated values are presented in Table 5–1.

5–6.5 Length of Exposed Area

Soil drifting increases substantially with increasing length of the eroding strip. Wind starts to pick up soil grains close to the windward side of an erodible field. It continues to add to its load as it passes over the field until it can carry no more. The wind may continue to pick up other soil grains as it travels farther, but it also drops some of its load because its carrying capacity is finite.

The transport capacity of wind at a specific friction velocity is similar for all soils, but the distance the wind must travel across a field to pick up its full load depends on soil

TABLE 5–1 RELATIVE AMOUNTS OF EROSION FROM LEVEL (1.5% SLOPE) AND FROM SHORT-SLOPING (HUMMOCKY) LAND

	Relative amounts of erosion	
Slope (%)	Crests	Upper slopes
1.5 (level)	100	100
3.0	150	130
6.0	320	230
10.0	660	370

Source: From the *Journal of Soil and Water Conservation,* Volume 19, p. 179–181 (Chepil et al., 1964).

erodibility. The more erodible the soil, the shorter the distance required to reach its load capacity. Distance needed to acquire maximum load varies from less than 180 ft (55 m) for a structureless fine sand to more than 5000 ft (1500 m) for a cloddy, medium-textured soil (Chepil and Woodruff, 1963). Many fields are not wide enough for the wind to pick up its maximum load.

5–6.6 Vegetative Cover

The most effective way to reduce wind erosion is to cover the soil with a protective mantle of growing plants or with a thick mulch of crop residue. Barriers of plant material increase the thickness of the blanket of still air ($D + Z_o$) next to the soil.

The protection that plant cover provides is influenced by plant species (amount of vegetative cover and time of year when cover is provided), plant geometry and population, and row orientation. Crop residues left on the surface, especially if tall and dense, offer almost as much protection as a comparable amount of growing plants. Standing residues provide more protection than the same amount of flattened residues. One anomaly is that a sparse cover over less than 10% of the surface can increase the turbulence of windgusts and actually increase wind erosion to more than that from bare soil (Sterk, 2000).

A complete cover of growing plants offers maximum protection, but individual plants and rows of plants across the direction of the wind also reduce ground-level wind velocity and erosion. This is apparent where isolated weeds in fallow fields trap saltating grains and where soil piles up around wind barriers such as field shelterbelts.

Barriers are effective because air is a fluid. As air moves toward a porous barrier, part of it is pushed over or around the barrier. Air that is not deflected builds up pressure that forces it through the barrier at a fast rate (funneling effect). This air immediately slows down as it spreads out to occupy all the space behind the barrier. Wind speed returns to normal only when the deflected air returns to its initial position in the windstream.

5–7 WINDBREAKS AND SHELTERBELTS

Windbreaks and shelterbelts are groups of trees and shrubs planted across the prevailing winds for the purpose of moderating the winds, reducing wind erosion, trapping dust and snow, reducing evaporation, increasing relative humidity, ameliorating the environment for livestock and wildlife, reducing fuel costs in heating and cooling a home, and enhancing the environment for people. *Windbreaks* are small groups of trees and shrubs planted to protect livestock and people. *Shelterbelts* are extensive groups of trees and shrubs planted primarily to protect fields from erosive winds. Windbreaks and shelterbelts can be seen in Figure 5–11.

5–7.1 Windbreak and Shelterbelt Effectiveness

Windbreaks reduce winter fuel bills, protect cattle and other livestock, and reduce wind erosion. They can also provide food and habitat for wildlife. During the early days of the promotion of windbreaks, the Lake States Forest Experiment Station conducted an experiment at Holdrege, Nebraska, on the home fuel saved by the proper establishment of a windbreak. Two identical homes were compared, one with a windbreak and one without. The home with a windbreak used about 23% less fuel.

Windbreaks and shelterbelts are desirable and almost essential for economical production of range livestock in northern states. The University of Montana researched this relationship and reported that during a mild winter, tree-protected cattle gained 35 lb (16 kg) more or, during a severe winter, lost 10 lb (4.5 kg) less than cattle that were not protected by a windbreak or shelterbelt (Stoeckeler and Williams, 1949).

A windbreak or shelterbelt that is well designed and properly oriented will create a protected area with wind speed less than half of that in the open, as shown in Figure 5–12. Windbreak effectiveness is significant as far leeward as 15 or 20 times the height of the windbreak and windward for about twice its height. Dense windbreaks give the most reduction in wind velocity near the windbreak, but more open ones are effective for a greater distance downwind. The most effective windbreaks have porosities of about 50%. Any reduction in wind speed is important in decreasing wind erosion because the amount of soil eroded is proportional to the fifth power of the wind speed (Section 5–5.2). Thus, the amount of wind erosion is approximately halved when the wind velocity is reduced to 87% of what it would have been without the windbreak.

5–7.2 Trees and Shrubs for Windbreaks

Suitable trees and shrubs for windbreak and shelterbelt plantings that are adapted to conditions in the northern Great Plains are listed in Table 5–2, and those adapted for the southern Great Plains are in Table 5–3. Selection of the most appropriate species depends partly on the number of rows to be included in the windbreak. Traditional windbreaks have had seven rows, with the tallest trees flanked by shorter trees and shrubs. One row of trees should be evergreens for greater winter protection.

Seven-row windbreaks effectively divert the wind upward but they occupy a lot of land area. This may be justified if the trees chosen can be marketed for wood products when they mature. Otherwise, a smaller windbreak with fewer rows may be preferred. The most effective wind diversion with three-, five-, and seven-row windbreaks occurs with the tallest trees in the second, fourth, or fifth row downwind, respectively. The most efficient use of land for windbreaks is achieved with single-row windbreaks. These are usually composed of evergreens (often spruce trees) so they will provide both summer and winter protection, and they must have branches near the ground so the wind cannot blow under them.

Figure 5–11 Windbreaks around homes and farmsteads ameliorate the weather for people and livestock, and shelterbelts along field borders protect crops against wind erosion in North Dakota (above) and on the High Plains of Texas (below). (Courtesy USDA Natural Resources Conservation Service.)

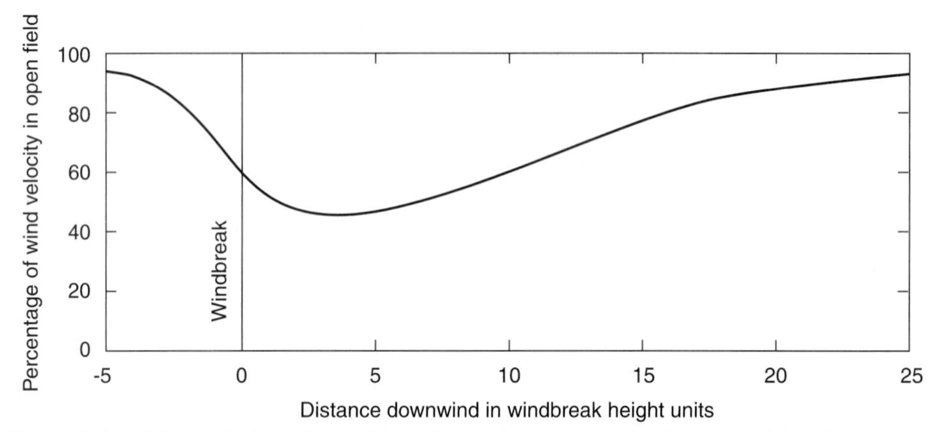

Figure 5–12 Wind velocities about 16 in. (40 cm) above ground level in the vicinity of a windbreak, as percentages of what they would be without the windbreak. The curve represents average results of data from several sources. Many data vary ±10% from the curve depending on width, height, and porosity of the windbreak, wind velocity, and topographic features. (Courtesy USDA Natural Resources Conservation Service.)

TABLE 5–2 TREE AND SHRUB SPECIES ADAPTED FOR PLANTING WINDBREAKS AND SHELTERBELTS IN THE NORTHERN GREAT PLAINS

Trees	Shrubs
Austrian pine	American plum
Black Hills spruce	Amur maple
Bur oak	Arnold hawthorn
Colorado blue spruce	Buffaloberry
Cottonwood	Caragana (Siberian peashrub)
Eastern red cedar	Common chokecherry
Green ash	Cotoneaster
Hackberry	Hanson hedgerose
Robusta poplar	Late lilac
Rocky Mountain juniper	Nanking cherry
Scotch pine	Saskatoon serviceberry
Siberian crabapple	Redosier dogwood
White spruce	Silverberry
White willow	Skunkbush sumac
	Tartarian honeysuckle
	Western sandcherry

Source: Soil Conservation Service, 1984a and 1984b.

Tree spacing is also important and should be proportioned to the ultimate size of the trees. A mature windbreak should still be open enough to permit some air movement through the windbreak, but there must be no major gaps. Missing trees should be replaced promptly because wind speeds through a gap are likely to be about 120% of what they would be without the windbreak. Also, windbreaks should end in sheltered or vegetated areas because wind velocities reach about 120% of normal around the ends. Windbreaks should be fenced to keep livestock from eating the vegetation that is needed to form a good barrier near the ground.

TABLE 5–3 TREE AND SHRUB SPECIES ADAPTED FOR PLANTING WINDBREAKS AND SHELTERBELTS IN THE SOUTHERN GREAT PLAINS

Trees	Shrubs
Crack willow	American plum
Desert willow	Buckthorn
Eastern red cedar	Caragana (Siberian peashrub)
Hackberry	Chickasaw plum
Honeylocust (thornless)	Cotoneaster
Kentucky coffeetree	Lilac
Mulberry	Redbud
One-seed juniper	
Osage-orange	
Pecan	
Russian olive	
Shortleaf pine	
Sycamore	

Source: Selected from *Trees,* 1949 Yearbook of Agriculture, p. 848–849, 1949, Stoeckeler and Williams.

5–7.3 Prairie States Forestry Project

The U.S. Congress authorized the establishment of the Prairie States Forestry Project in 1934 following the drought and dust-storm years of the early 1930s. During the years from 1935 to 1942, 218 million trees and shrubs were planted on 31,000 farms and ranches to establish 20,000 mi (32,000 km) of windbreaks and shelterbelts in the ten Great Plains states from Texas to the Canadian border.

Farmers and ranchers in the Great Plains were pleased to have the new plantings of trees and shrubs because it seemed a sign of government concern during the crisis years of drought and economic depression.

5–7.4 Recent Trends

During recent decades, farms and ranches have become larger, tractors and equipment larger and more powerful, and large center-pivot irrigation systems are being used in many areas. Many farmers, most economists, and some technical agriculturists claim that management systems such as strip cropping, stubble mulching, minimum tillage, and clod tillage eliminate the need for windbreaks and shelterbelts. Many existing tree belts are being destroyed, mostly by newcomers to the Great Plains. They refuse to believe that wind erosion and dust storms are hard facts of life in the area.

Many conservationists, however, believe that the new management practices are necessary and, furthermore, that plantings of trees and shrubs as windbreaks should be increased tenfold. So, while some windbreaks and shelterbelts in the Great Plains are being destroyed, new ones are being planted. The Forest Service and the Natural Resources Conservation Service offer assistance in establishing trees and shrubs. Recent soil survey reports list the tree and shrub species recommended for planting as windbreaks and shelterbelts on each soil map unit. Cost-sharing for planting trees in the Great Plains is available through the Farm Service Agency. Twenty-four plant materials centers, operated

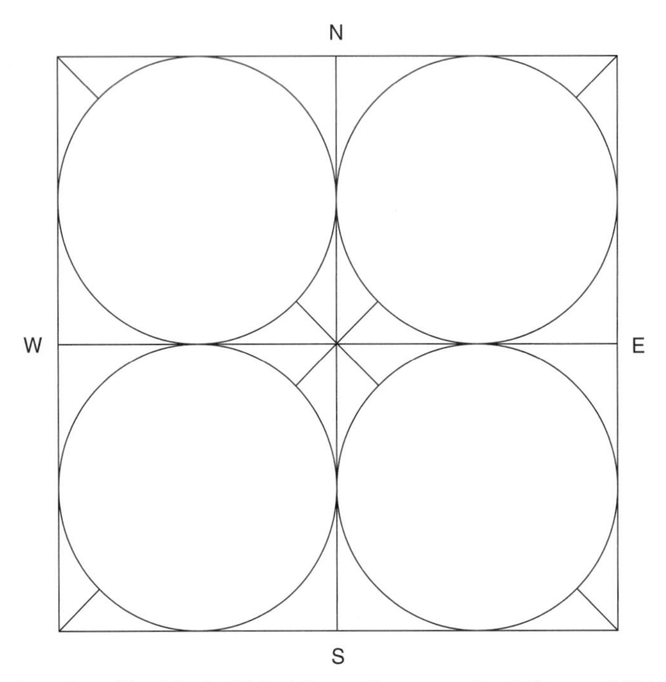

Figure 5–13 A section of land in the United States (1 square mile, 640 ac, or 259 ha) has room for four standard center-pivot irrigation systems. Each circle irrigates about 133 ac (54 ha) of land. Shelterbelts can be planted around the outside and across the middle of the section and, if necessary, along the diagonals. Such shelterbelts reduce wind velocities near the tree lines and improve the distribution of the irrigation water. (Courtesy USDA Natural Resources Conservation Service.)

throughout the United States by the USDA Natural Resources Conservation Service, are constantly searching for, testing, and releasing new plants for use as windbreaks and shelterbelts in arid and semiarid regions and for water erosion control in humid regions.

5–7.5 Irrigation and Shelterbelts

Thousands of windbreaks and shelterbelts have been removed because they interfered with the establishment of large center-pivot irrigation systems. These systems are carried by wheels traveling in concentric circles as described in Chapter 15. The tracks for the wheels must be smooth and nothing should obstruct the moving irrigation pipe; hence, no shelterbelts. Shelterbelts are needed, however, to calm the winds so irrigation water can be applied evenly.

An example of a compromise solution to the irrigation-shelterbelt dilemma is presented in Figure 5–13. Four of the usual 133-ac (54-ha) irrigation circles can be arranged in one section (a square mile containing 640 ac or 259 ha) of land and still have a shelterbelt located every half mile (0.8 km). The nine odd-shaped areas outside the circles can also be planted to trees, or they may be seeded to grass or nonirrigated crops. Such an arrangement will adequately protect the land within about 500 to 1000 ft (150 to 300 m) downwind and 100 ft (30 m) upwind from the shelterbelts if the trees are 50-ft (15-m) tall. However, this may still leave as much as 2000 ft (600 m) of land unprotected in a typical center-pivot system on a quarter section of land (2640 × 2640 ft).

5–8 PRINCIPLES OF WIND-EROSION CONTROL

Wind erosion occurs whenever conditions are favorable for detachment and transportation of soil material by wind. Soil erodibility, surface roughness (ridging), climatic conditions (wind velocity and humidity), length of exposed surface, and vegetative cover influence how much erosion will take place. Little can be done to change the climate in an area, but it is usually possible to alter one or more of the other factors.

Many successful practices for reducing wind erosion have been developed by farmers based on their observations, and by research scientists as a result of their studies. These practices are not universally successful; some work well in one region, others are better in another. Regardless of the relative success of individual techniques, the aims or principles of soil-drifting control are the same in all areas where wind erosion occurs. The aims are to reduce wind velocity near ground level below the threshold velocity, to remove abrasive material from the windstream, and to reduce soil erodibility. Any practice that accomplishes one or more of these goals will reduce the severity of wind erosion.

SUMMARY

Wind erosion has both geologic- and human-induced components, and both are erratic over time and place. Wind erosion is a common phenomenon on most soils in dryland regions and on coarse-textured and high organic-matter soils in humid areas. It damages the land that loses soil by reducing soil depth and removing fine particles. Slow deposition may contribute to soil fertility, but rapid deposition damages the land where sediment is deposited. Flying soil grains bury or abrade crops and damage buildings and equipment. Dust in the air is a health and safety hazard.

Soil moves in the wind in three general forms: the finest grains move in *suspension,* intermediate sizes in *saltation,* and the coarsest erodible grains in *surface creep.* Finer and coarser sizes are difficult for even strong winds to dislodge, but intermediate-sized grains are relatively easy to move. Wind erosion begins with saltation, and the saltating particles start the smaller grains moving in suspension and cause larger grains to move in surface creep.

Wind velocity, turbulence, gustiness, and direction all affect the severity of wind erosion. Forces that start particles saltating include pressure differences caused by spinning soil grains and steep wind-velocity gradient near the soil augmented by air turbulence and irregularities on soil grains and land surfaces. These lift or bounce susceptible particles a short distance vertically, where they are accelerated by the wind. When they drop back to the soil surface, they rebound higher and faster, or they knock other particles into motion.

Friction velocity represents the velocity gradient over the soil surface and is responsible for wind's erosive power. Erosive force and detaching capacity of a wind are proportional to the square of the friction velocity, the carrying capacity is proportional to the cube of the friction velocity, and the amount of material eroded by wind from a unit area is proportional to the fifth power of the friction velocity.

Factors other than wind also affect wind erosion. Soil cloddiness and stone content, amount and nature of field ridging, amount of rainfall, topography (smooth or hummocky), length of erodible area in the direction of the wind, and amount, nature, and distribution of vegetative cover influence the amount of soil movement by wind.

Windbreaks are used to reduce the wind velocity around farmsteads, livestock, and people. Shelterbelts are used similarly to protect fields from wind erosion for a distance upwind about twice the height of the trees and downwind 15 to 20 times the height of the trees. Appropriate tree species may produce useful wood products and wildlife habitat as well as erosion protection.

There are three principal ways of reducing wind erosion. They are to reduce the velocity of the wind close to the soil, remove abrasive material (saltating grains) from the windstream, and increase the soil's resistance to wind erosion.

QUESTIONS

1. Name and describe the three types of movement of soil particles in the wind.
2. Briefly describe the types of damage to soils, crops, and structures that can be caused by wind erosion.
3. List the characteristics of wind that influence the amount of soil that is eroded and describe how each characteristic affects the erosion process.
4. What physical forces appear to be responsible for starting soil movement by wind?
5. Describe how each of the following soil properties affects the erodibility of soils: texture, structure, organic-matter content, lime content, and moisture content.
6. Describe each of the mechanisms by which vegetation reduces the amount of wind erosion.
7. Why are some shelterbelts being removed while others are being planted in the Great Plains?
8. What are the three principles of wind erosion control and why is each effective?

REFERENCES

BAGNOLD, R. A., 1937. The transport of sand by wind. *Geog. J.* 89:409–438.

BAGNOLD, R. A., 1941 (1973 reprint). *The Physics of Blown Sand and Desert Dunes.* Chapman & Hall, London, 265 p.

BATT, R. G., and S. A. PEABODY II, 1999. Entrainment of fine sand particles from rough surfaces at high wind speeds. *Trans. ASAE* 42:79–88.

BIELDERS, C. L., K. MICHELS, and J.-L. RAJOT, 2000. On-farm evaluation of ridging and residue management practices to reduce wind erosion in Niger. *Soil Sci. Soc. Am. J.* 64:1776–1785.

BISAL, F., and J. HSIEH, 1966. Influence of moisture on erodibility of soil by wind. *Soil Sci.* 102:143–146.

CHEPIL, W. S., 1945a. Dynamics of wind erosion: I. Nature of movement of soil by wind. *Soil Sci.* 60:305–320.

CHEPIL, W. S., 1945b. Dynamics of wind erosion: III. The transport capacity of the wind. *Soil Sci.* 60:475–480.

CHEPIL, W. S., 1946. Dynamics of wind erosion: V. Cumulative intensity of soil drifting across eroding fields. *Soil Sci.* 61:257–263.

CHEPIL, W. S., 1950. Properties of soil which influence wind erosion: I. The governing principle of surface roughness. *Soil Sci.* 69:149–162.

CHEPIL, W. S., 1953. Factors that influence clod structure and erodibility of soil by wind: I. Soil texture. *Soil Sci.* 75:473–483.

CHEPIL, W. S., 1954. Factors that influence clod structure and erodibility of soil by wind: III. Calcium carbonate and decomposed organic matter. *Soil Sci.* 77:473–480.

CHEPIL, W. S., 1955. Factors that influence clod structure and erodibility of soil by wind: V. Organic matter at varying stages of decomposition. *Soil Sci.* 80:413–421.

CHEPIL, W. S., 1956. Influence of moisture on erodibility of soil by wind. *Soil Sci. Soc. Am. Proc.* 20:288–292.

CHEPIL, W. S., 1957. Dust bowl: Causes and effects. *J. Soil Water Cons.* 12:108–111.

CHEPIL, W. S., and R. A. MILNE, 1939. Comparative study of soil drifting in the field and in a wind tunnel. *Sci. Agric.* 19:249–257.

CHEPIL, W. S., and R. A. MILNE, 1941. Wind erosion of soil in relation to roughness of surface. *Soil Sci.* 52:417–431.

CHEPIL, W. S., F. H. SIDDOWAY, and D. V. ARMBRUST, 1964. Wind erodibility of knolly terrain. *J. Soil Water Cons.* 19:179–181.

CHEPIL, W. S., and N. P. WOODRUFF, 1963. The physics of wind erosion and its control. In *Advances in Agronomy,* Vol. 15. Academic Press, New York, p. 211–302.

CLAY, S. A., T. M. DESUTTER, and D. E. CLAY, 2001. Herbicide concentration and dissipation from surface wind-erodible soil. *Weed Sci.* 49:431–436.

DANIEL, H. A., 1936. The physical changes in soils of the southern High Plains due to cropping and wind erosion and relation between sand plus silt over clay ratios in these soils. *J. Am. Soc. Agron.* 28:570–580.

FREE, E. E., 1911. *The Movement of Soil Material by the Wind.* USDA Bur. Soils Bull. 68, Washington, D.C.

HOLMES, G., B. R. SINGH, and L. THEODORE, 1993. *Handbook of Environmental Management and Technology.* John Wiley & Sons, Inc., New York, 651 p.

HUSZAR, P. C., and S. L. PIPER, 1986. Estimating the off-site costs of wind erosion in New Mexico. *J. Soil Water Cons.* 41:414–416.

KING, F. H., 1894. *Destructive Effects of Winds on Sandy Soils.* Univ. Wis. Bull. 42, Madison, WI.

LANGFORD, R. P., 2000. Nabkha (coppice dune) fields of south-central New Mexico, U.S.A. *J. Arid Environ.* 46:25–41.

LARNEY, F. J., B. M. OLSON, H. H. JANZEN, and C. W. LINDWALL, 2000. Early impact of topsoil removal and soil amendments on crop production. *Agron. J.* 92:948–956.

LYLES, L., 1977. Wind erosion: Processes and effects on soil productivity. *Trans. Am. Soc. Agr. Eng.* 20:880–884.

LYLES, L. and J. TATARKO, 1986. Wind erosion effects on soil texture and organic matter. *J. Soil Water Cons.* 41:191–193.

MOSS, H. C., 1935. Some field and laboratory studies of soil drifting in Saskatchewan. *Sci. Agric.* 15:665–679.

NATURAL RESOURCES CONSERVATION SERVICE, 1994. *1992 National Resources Inventory.* USDA, Washington, D.C. (*http:/www.nrcs.usda.gov.text.nritable.12*)

PING, Y., D. ZHIBAO, D. GUANGRONG, Z. XINBAO, and Z. YIYUN, 2001. Preliminary results of using ^{137}Cs to study wind erosion in the Qinghai-Tibet Plateau. *J. Arid Environ.* 47:443–452.

REYNOLDS, R., J. BELNAP, M. REHEIS, P. LAMOTHE, and F. LUISZER, 2001. Aeolian dust in Colorado Plateau soils: Nutrient inputs and recent change in source. *Proc. Natl. Acad. Sci. USA* 98:7123–7127.

SAXTON, K., D. CHANDLER, L. STETLER, B. LAMB, C. CLAIBORN, and B.-H. LEE, 2000. Wind erosion and fugitive dust fluxes on agricultural lands in the Pacific Northwest. *Trans. ASAE* 43:623–630.

SOIL CONSERVATION SERVICE, 1984a. *Farmstead and Feedlot Windbreaks.* Tech. Guide Notice ND-39, North Dakota. Standards and Specifications, Section IV, 380–1.

SOIL CONSERVATION SERVICE, 1984b. *Field Windbreaks.* Tech. Guide Notice ND-39, North Dakota. Standards and Specifications, Section IV, 392–1.

STERK, G., 2000. Flattened residue effects on wind speed and sediment transport. *Soil Sci. Soc. Am. J.* 64:852–858.

STOEKELER, J. H., 1962. *Shelterbelt Influence on Great Plains Environment and Crops.* USDA Prod. Res. Rep. 62, Washington, D.C.

STOECKELER, J. H., and R. A. WILLIAMS, 1949. Windbreaks and shelterbelts. In *Trees.* 1949 Yearbook of Agriculture, Washington, D.C., p. 191–199.

STOUT, J. E., 2001. Dust and environment in the southern High Plains of North America. *J. Arid Environ.* 47:425–441.

WOODRUFF, N. P., and F. H. SIDDOWAY, 1965. A wind erosion equation. *Soil Sci. Soc. Am. Proc.* 29:602–608.

ZHIBAO, D., W. XUNMING, and L. LIANYOU, 2000. Wind erosion in arid and semiarid China: An overview. *J. Soil Water Cons.* 55:430–444.

CHAPTER

6

PREDICTING SOIL LOSS

The need to evaluate erosion losses and to determine the effectiveness of control measures became apparent as soon as field workers started promoting conservation. In the 1930s, research officers of the U.S. Soil Conservation Service (SCS) began to assess levels of erosion and to study quantitatively the effects of soil physical characteristics and of control measures on the erosion process. Eventually, soil loss prediction equations were devised for both water and wind erosion. Field technicians and others use these prediction equations to estimate soil losses on specific sites. If the estimated losses exceed tolerable limits, the equations are used to ascertain what farming-system changes will reduce the erosion losses to acceptable levels. Often several suitable options are found, from which the farmer can select the one that suits the circumstances best. The equations are refined as new information becomes available, as farming practices change, and as new applications requiring erosion prediction develop.

6–1 TOLERABLE SOIL LOSS

The concept of soil loss tolerance was introduced by SCS research officers (Smith, 1941), based initially on estimates of rates of soil loss that could be sustained without reducing the level of soil organic matter. Different bases were suggested by other workers, such as loss rates that would permit sufficient soil depth to remain for optimum crop growth, or rates that could be permitted and still not lose fertility or have gully formation. Wischmeier and Smith (1978) defined soil loss tolerance as "the maximum level of soil erosion that will permit a high level of crop productivity to be sustained economically and indefinitely."

In 1961 and 1962, a group of American scientists assigned soil loss tolerance (T) values to a large group of U.S. soils. These T values ranged from 2 to 6 tons/ac-yr (4 to 13 mt/ha-yr). Later it was agreed that 5 tons/ac-yr (11 mt/ha-yr) should be the maximum rate and that some soils are so fragile that a rate of only 1 ton/ac-yr (2 mt/ha-yr) should be added. This range of T values, from 1 to 5 tons/ac-yr, is still used in the United States.

Four major factors affect the rate of erosion that can occur without permanent loss of soil productivity: depth of soil, type of parent material, relative productivity of surface soil and subsoil, and amount of previous erosion. The thicker the soil (A + B horizons) and the greater the thickness of material permeable to plant roots, the faster erosion can occur without irreparable loss of productive capacity. Unconsolidated, fertile parent material such as glacial till or loess is more quickly converted into soil than is bedrock. Where surface soil is notably more fertile and productive than subsoil and parent material, loss of even small quantities of surface soil will seriously reduce productivity. A soil that has already suffered serious erosion cannot stand further losses as well as soils not previously damaged.

Johnson (1987), Lal (1985), and others have criticized accepted tolerable loss levels. This is based on their belief that soil is not rebuilt, even under the best conditions, as fast as these rates require. Two U.S. studies indicate that A-horizon soil can develop fairly rapidly from B-horizon material or from unconsolidated parent material. An Iowa study showed that the top 4 in. (10 cm) of 24 in. (61 cm) of a subsoil fill was transformed into surface soil, indistinguishable from normal topsoil in the area, in 100 to 125 years (Hallberg et al., 1978). In an Indiana study, Kohnke and Bertrand (1959) found that 1 in. (2.5 cm) of soil indeed developed from glacial till in 58 years at a 1-in. depth, but it took about 700 years for an inch of soil to develop if buried 24 in. (61 cm), and at 40 in. (1 m) it took more than 1000 years.

To maintain soil depth, new soil must develop at the bottom of the profile, not at the top. Thus, even at the 1 ton/ac (2.2 mt/ha) annual rate (cumulative rate of 1 in./143 yr), the lost soil will not be fully replaced by the soil development rate found in Indiana until the soil depth is reduced to approximately 3 in. This makes one wonder if assigned T values really are tolerable.

There are also some who feel that, on occasion, other factors may restrict tolerable losses more than soil productivity damage alone. Pierce et al. (1984) described two kinds of T values. One (T_1) measures the effect of erosion on soil physical properties that in turn influence inherent productivity, and another (T_2) involves social goals, such as reducing air and water pollution. Lower tolerable limits also may be necessary to prevent damage to property and crops by sediment and wind-driven soil (Hayes, 1965). For example, some crops are extremely sensitive to abrasion damage. Production of the more sensitive plants would require that soil losses be kept lower than the usual tolerable limits.

No controversy is complete without hearing from the opposing side. There are those who argue that the T values are too low for practical purposes because they often indicate that a soil should not be used in a way that has already proven to be highly profitable. Some say that a lessening of the thickness of a deep deposit of loess, alluvium, or glacial till does little damage to productivity. Others claim that certain soils produce little of value when they are restricted to uses that limit the rate of erosion to their T values, so the crops that can be produced now are more valuable than the soil.

It should also be recognized that wind and water are not the only agents that cause soil loss. Logically, the total annual soil loss should be compared to the T value, but in practice most soil loss calculations are based on the one dominant erosion process or, occasionally, on a sum of the estimates for water and wind erosion. The subsidence of an organic soil following drainage and its gradual loss by decomposition added to losses by wind erosion generally make it impossible to farm an organic soil without gradually losing it. The surface may drop 1 to 2 in. (2 to 5 cm) per year—a loss rate far in excess of any T value. The only way known to stop such loss of organic soil is to keep it saturated with water, but that prevents it from being used as cropland.

There are disagreements, too, on the importance of soil movement within a field. Some argue that a set of terraces controls erosion because any soil movement that occurs is caught in the terrace channels and little or no soil leaves the field. Others suggest that the soil caught in the terrace channels indicates a need for better management of the land above each terrace. Tillage erosion is another type of soil movement that is usually ignored, but repeated plowing can remove soil from convex topography (such as a hilltop) at a rate considerably faster than the T value (Lindstrom et al., 2000). This soil accumulates in concave areas or in backfurrows at the edge of the field. Other tillage implements have a similar effect, but to a lesser degree. The actual amount of movement depends on the implement, the depth of tillage, and the slope gradient and shape (Gerontidis et al., 2001).

6–2 THE UNIVERSAL SOIL LOSS EQUATION (USLE)

The first equations for predicting soil losses from water erosion were developed in the midwestern United States. Zingg (1940), Smith (1941), Browning et al. (1947), and Musgrave (1947), all SCS research officers, developed equations that initially involved only slope steepness and length. Then the effects of crops and special management practices were added, later inherent soil erodibility was included, and finally a rainfall erosiveness factor was introduced.

The comprehensive soil loss prediction equation now known as the Universal Soil Loss Equation (USLE) was developed by scientists of the Agricultural Research Service (ARS) and the Soil Conservation Service (SCS) in the U.S. Department of Agriculture (USDA) and scientists of Purdue University under the leadership of Walter H. Wischmeier. The new prediction equation took definite form late in the 1950s with initial field use starting almost immediately. It was developed to predict long-term average annual soil losses from sheet and rill erosion on uniform cultivated fields in the eastern half of the United States.

The equation was published in handbook form by Wischmeier and Smith (1965). The handbook contained a description of the equation accompanied by the tables and graphs necessary for its use in the field. The equation rapidly gained acceptance and was very widely used. Conservationists in SCS field offices and others interested in soil conservation used the USLE to predict soil loss by water erosion on farmers' fields in their areas. If the analysis showed that excessive erosion was occurring, the equation was used to assess the likely losses from alternate systems. It became an excellent conservation and extension tool. The equation is

$$A = R \times K \times LS \times C \times P$$

where A = estimated average annual soil loss, tons/ac-yr

R = rainfall and runoff factor, 100s of ft-tons/ac-yr

K = soil-erodibility factor, soil loss per unit of rainfall erosivity from bare fallow on a 9% slope 72.6-ft (22.1-m) long, tons (of soil)/100 ft-tons (of rainfall)

LS = slope length and steepness factor, dimensionless

C = cover-management factor, dimensionless

P = supporting-practice factor, dimensionless

TABLE 6–1 KINETIC ENERGY, *e*, IN FT-TONS/AC, PER INCH OF PRECIPITATION AS INFLUENCED BY RAINFALL INTENSITY, *I*, IN IN./HR, BASED ON THE EQUATION $e = 916 \text{ T } 331 \log_{10} I$

	Rainfall intensity, *I* (in./hr)									
I (in./hr)	0.00	0.01	0.02	0.03	0.04	0.05	0.06	0.07	0.08	0.09
0	000	254	354	412	453	485	512	534	553	570
0.1	585	599	611	623	633	643	653	661	669	677
0.2	685	692	698	705	711	717	722	728	733	738
0.3	743	748	752	757	761	765	769	773	777	781
0.4	784	788	791	795	798	801	804	807	810	814
0.5	816	819	822	825	827	830	833	835	838	840
0.6	843	845	847	850	852	854	856	858	861	863
0.7	865	867	869	871	873	875	877	878	880	882
0.8	884	886	887	889	891	893	894	896	898	899
0.9	901	902	904	906	907	909	910	912	913	915
	0.0	0.1	0.2	0.3	0.4	0.5	0.6	0.7	0.8	0.9
1	916	930	942	954	964	974	984	992	1,000	1,008
2	1,016	1,023	1,029	1,036	1,042	1,048	1,053	1,059	1,064	1,069
3	1,074[a]									

[a]All intensities greater than 3.0 in./hr also have $e = 1{,}074$.

Source: Wischmeier and Smith, 1978.

6–2.1 Components of the USLE

The USLE is an empirical equation based on measurements rather than theory. Each of its five factors includes the influence of two or more subfactors. Combining the subfactors into a factor may be complex, but combining the five factors with each other is simply a matter of multiplication.

Rainfall and Runoff Factor (*R*). Wischmeier (1959) found that the total energy (E) and the maximum 30-minute intensity (I_{30}) of rainstorms were the characteristics most closely correlated with the amounts of erosion produced. The product of the energy value and the intensity (in inches but used as a dimensionless number) was designated as EI_{30} or R. Using information from recording rain gauge charts and energy/intensity data (see Table 6–1), but excluding data from storms that produced 0.5 in. (13 mm) or less of rain, Wischmeier and his associates analyzed 4000 location-years of rainfall records to arrive at a picture of the varying erosiveness of the rainfall over the eastern half of the United States. These data were presented in an iso-erodent map. EI_{30} values ranged from less than 50 in the dry regions in the western Great Plains to more than 500 along the northern Gulf Coast and in parts of Florida.

Soil Erodibility Factor (*K*). K represents inherent soil erodibility. It is the rate of soil loss on a standard plot (9% slope, 72.6-ft [22.1-m] long, kept fallow by periodic tillage up and down hill). A few K values were obtained from studies on the soil conservation experiment stations—from fallow plots and from row-crop plots, corrected for vegetative cover. Many were obtained from small plots using a rainfall simulator.

Slope Length and Steepness Factor (*LS*). *LS* is the ratio of soil loss expected per unit area from a particular field's slope conditions compared to what would occur from a 9% slope 72.6-ft (22.1-m) long. This ratio is calculated from equations developed from those devised first by Zingg (1940).

Cover-Management Factor (*C*). Another new concept in the USLE was that the changing protective effect of crops and crop management practices could be programmed over the crop year (Wischmeier, 1960). To do this, the crop year was divided into five different *cropstage periods*: Rough Fallow, Seedling, Establishment, Growing and Maturing Crop, and Residue or Stubble. Each cropstage period for a particular crop and crop management practice was assigned a soil loss ratio—a ratio of soil loss with the crop and/or the practice on the land to the soil loss when the land was under fallow conditions with periodic cultivation up- and downhill.

These soil loss ratios were developed from analyses of thousands of plot-years of runoff and soil loss data. The values include the effects of all factors of crop production that affect crop growth and erosion reduction.

Supporting-Practice Factor (*P*). Special practices are frequently needed in addition to the protection provided by the crop and normal crop- and soil-management practices. Most common special practices for cropland are contour cultivation, contour strip cropping, and terracing. The *P* factor indicates the fractional amount of erosion that occurs when these special practices are used compared with what would occur without them.

Contour Cultivation. *P* values for contour cultivation were based entirely on slope gradient, assuming slopes were not too long. Additional measures may be needed to reduce soil losses on longer slopes. Factor values range from 0.5 to 0.9.

Contour Strip Cropping. By definition, contour strip cropping includes contour strips of sod in cultivated fields. Where half the total length of the slope is in sod strips, the strip cropping *P* factor is 0.50. This factor multiplied by the contouring *P* factor brings the total *P* for contour strip cropping down to values ranging from 0.25 to 0.45. Where one-fourth of the field is in sod strips, the *P* factor is 0.75, bringing the total contour strip-cropping factor value to between 0.38 and 0.68. No benefit is credited to strip cropping that contains only cultivated annual crops.

Terraces. The only conservation benefit Wischmeier and Smith credited to terraces was changing the effective length of the slope. Therefore, the terrace effect was credited under the Slope Length and Steepness Factor, not under the Supporting Practices Factor.

6–2.2 Acceptance of the Universal Soil Loss Equation

The USLE was developed primarily to help field workers assess current erosion levels and, where necessary, plan changes in management to reduce excessive rates of erosion on cultivated farm fields in the eastern half of the United States. Several calculations were needed for each field proposal developed. The time that was required to make these calculations might have been expected to slow early acceptance, but the new tool was taken up with enthusiasm by SCS administrators and by many of the staff in the SCS state offices. It was adopted as an official tool to be used in the development of conservation farm plans by staff in SCS field offices. Other government staff members and private individuals, interested in conservation and in agricultural extension, also adopted the USLE.

Its use spread, not only throughout the eastern half of the United States, but also to other states and to several other countries. With widespread use, especially use outside of the eastern half of the United States, the equation was found to need adjustments to meet new needs. For example, McCool et al. (1974) used the USLE in the Pacific Northwest region of the United States. Much less intensive rains are the rule there. Also, the USLE did not consider rain falling on frozen ground and runoff from snow melt in developing *R* values. In addition, exceedingly steep slopes complicated determination of satisfactory *LS* values.

Excluding storms of less than 0.5 in. (12.7 mm) from the *R* value calculations greatly reduced the number of storms to be calculated and made only a small difference in the results for the humid portion of the United States. The smaller storms can be important elsewhere, however. Evans and Nortcliff (1978) pointed out that thunderstorms producing as little as 0.3 in. (7.4 mm) of total rainfall cause soil erosion in the United Kingdom. Evans and Kalkanis (1976) used the equation in California to predict erosion not only on cropped lands but on pastures and haylands. They found it necessary to modify the *R* factor so that it would fit the hydrological conditions in three different regions of the state. They also had to include snow melt in *R* for about one-third of the state and had to adjust the *LS* factor to suit local conditions.

Different problems were encountered in the semiarid sections of the southwestern states. Sediment in the streams, as well as erosion from the soil surfaces, needed to be predicted. Renard et al. (1974) suggested that "a channel erosion term (*Ec*) was needed to explain different sediment yields from watersheds with different channel characteristics." A partial pavement of stones provided protection to the soil surface in many areas and needed to be accounted for in the *C* factor.

Soil and climatic conditions in Hawaii vary widely; many crop species are very different from those grown on the mainland. Local SCS officials were confused about how to develop a method for assessing erosion and ranking the benefits of conservation measures quantitatively (Brooks, 1974). SCS representatives and University of Hawaii personnel met with Wischmeier to find a solution. As a result of these discussions, the Hawaiians assembled the climatic, soil, and other information that was required, and in 1975 issued two SCS handbooks that described a method for predicting soil erosion losses in Hawaii.

Over the years some discrepancies showed up. Some of the data proved not to be quite as universal as was originally supposed. The equation was used outside of its target area and for purposes for which it had not been designed. Adjustments were made to it from time to time to overcome many of these problems, but by the mid-1970s it was obvious that the equation needed to be overhauled. Its structure needed to be modified so that it could serve the needs of its users better. A modified version was developed and published in a new handbook (Wischmeier and Smith, 1978).

6–2.3 Modified USLE

The basic prediction equation of the modified USLE was not changed. It was still an empirically based method that continued to predict soil losses caused by sheet and rill erosion. It did not predict losses by accelerated flow in small field gullies. But some basic data were modified. Changes were made in the way some of the factors were prescribed and calculated.

A section in the handbook's appendix titled "*Conversion to Metric System*" gives tables on the relationship between rainfall intensity and kinetic energy, both in foot-ton units and in metric ton-meters per centimeter of rain. Explicit instructions are provided for developing *EI* values from rain-gauge charts with measurements in mm. A conversion factor for transforming *K* values to metric is provided. Minor modifications for *LS* values may be needed only occasionally; these are discussed. As *C* and *P* values are dimensionless, no modifications are needed to convert these values to metric equivalents.

***R* Values.** The major change in the Rainfall-Erosion Index (*R*) portion of the equation was the expansion of the index to cover the entire United States, excluding Alaska. This was done by analyzing many years of rainfall data from the states in the western half of the country and from Hawaii. Records of additional stations in the eastern part were also examined. New iso-erodent maps were drawn up for continental states and for Hawaii, and somewhat patchy information on the monthly distribution of *EI* over the year was developed for eight additional states. *EI* values on the new continental map range from less than 20 to more than 550. For the Hawaiian Islands, the range is from less than 125 to more than 450.

A method was devised to take into account the effect of thaw and snow melt on *R* values in more northern regions, especially in the Pacific Northwest.

***K* Values.** The new handbook contains information on a soil erodibility nomograph (Figure 6–1) developed by Wischmeier et al. (1971) for determining *K* values from five soil parameters, % silt + very fine sand, % other sand, % organic matter, structure, and permeability. This nomograph offers a means of determining *K* values when no actual measurements are available and for mixed materials such as the altered soils found on construction sites, mine-spoil banks, and other areas. A few representative *K* values from the handbook are reproduced in Table 6–2.

***LS* Values.** The equations for calculating *LS* factor values were changed slightly in the modified version of the equation. A new table for use in selecting *LS* values for known slope conditions is included in the handbook, as is a graph. Some features of *LS* have proven difficult to standardize. Liu et al. (2000) analyzed data for slopes in the loess plateau of China with gradients up to 60% and concluded that the proper exponents for developing the *LS* equations are more sensitive to rainfall intensity than they are to slope length, and that the original USLE values represent their data better than the revised values.

A major addition is the discussion of a method to determine *LS* values for multisegment slopes (Foster and Wischmeier, 1974). A table is included that can be used in calculating an *LS* value for slopes that have from two to five different slope segments. This table can also be used to obtain combined $K \times LS \times C$ values where from two to five different soils, crop densities, and slope conditions are found in a single field.

***C* Values.** In the new handbook, the number of *Cropstage Periods* was increased from five to six. The breaks between most Cropstage Periods coincide with the cover the crop gives, rather than with a fixed time period after a particular field activity or between two field activities.

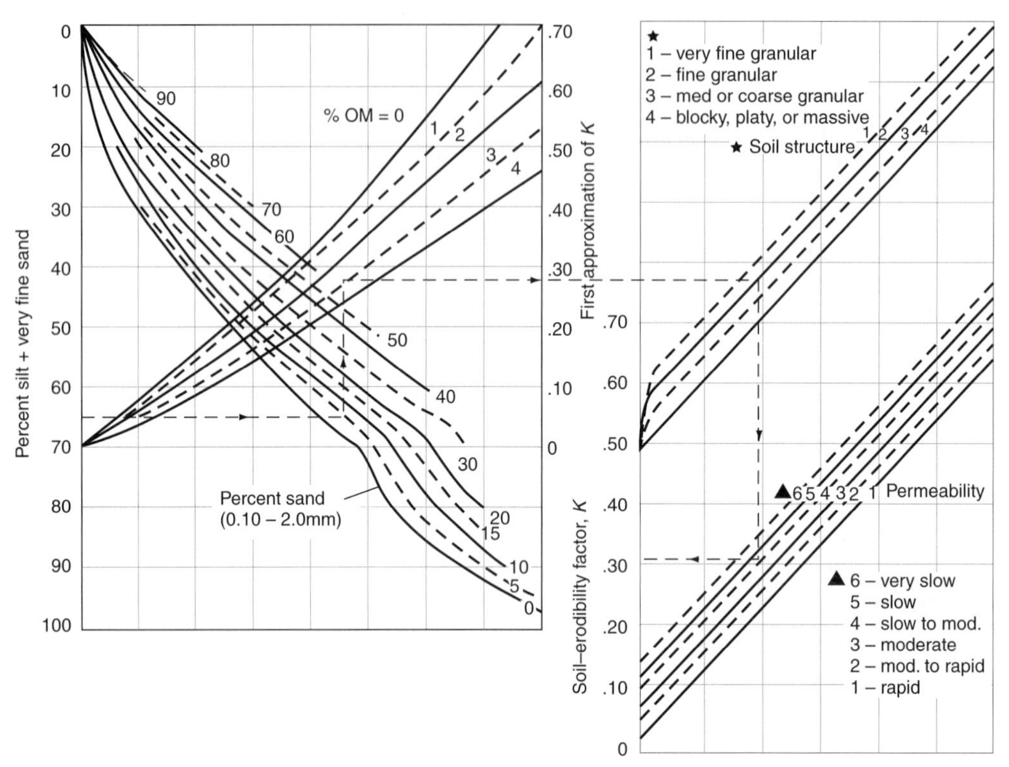

Figure 6–1 A nomograph to determine the soil-erodibility factor, *K*, from percent silt plus very fine sand (0.002 to 0.1 mm), percent sand (0.1 to 2.0 mm), percent organic matter, soil structure, and soil permeability. The dashed line shows how the nomograph is used to obtain a *K* value of 0.31 for a soil having 65% silt + very fine sand, 5% sand, 2.8% organic matter, fine granular structure, and slow-to-moderate permeability. This same sequence of properties must always be used to obtain *K* values from the nomograph. (From the *Journal of Soil and Water Conservation,* Volume 26, p. 189–193, 1971 [Wischmeier et al.]).

Soil Loss Ratios for many additional crops and cropping practices were provided in the modified version. Much more information is given on reduced tillage options, including no-till. Soil loss ratios for a few selected crops and cropping systems are presented in Table 6–3.

The modified USLE also provided *C* values for construction areas, idle land, pasture and rangeland, for undisturbed forest land, for woodland that is grazed, burned, or selectively harvested, and for forest land that has had post-harvest site preparation for re-establishment.

***P* Values.** *P* values for contour cultivation and contour strip cropping were not changed in the modified USLE. Stripping without a sod crop was still not credited with any beneficial effect on water erosion.

P values for terracing were changed. In addition to being credited with reducing slope length, terraces were also credited with causing sediment deposition in the terrace system because of their influence on the speed of water flow along the terrace channel. While this may not greatly reduce impoverishment of the soils on the slope, some of this soil can be saved, and certainly it is not carried off the field to pollute streams and be deposited on lands lower

TABLE 6–2 COMPUTED SOIL ERODIBILITY (*K*) AND TOLERABLE SOIL LOSS (*T*) VALUES FOR SOILS ON EROSION RESEARCH STATIONS

Soil type	Location	K^a (tons/100ft-ton)	T^b (tons/ac)
Albia gravelly loam	Beemerville, NJ	0.03	–
Austin clay	Temple, TX	0.29	2
Bath flaggy silt loam with surface stones >2 in. removed	Arnot, NY	0.05[c]	3
Boswell fine sandy loam	Tyler, TX	0.25	5
Cecil clay loam	Watkinsville, GA	0.26	4
Cecil sandy clay loam	Watkinsville, GA	0.36	3
Cecil sandy loam	Clemson, SC	0.28[c]	3
Cecil sandy loam	Watkinsville, GA	0.23	3
Dunkirk silt loam	Geneva, NY	0.69[c]	3
Fayette silt loam	LaCrosse, WI	0.38[c]	5
Freehold loamy sand	Marlboro, NJ	0.08	4
Hagerstown silty clay loam	University Park, PA	0.31[c]	4
Honeoye silt loam	Marcellus, NY	0.28[c]	3
Ida silt loam	Castana, IA	0.33	5
Keene silt loam	Zanesville, OH	0.48	4
Lodi loam	Blacksburg, VA	0.39	–
Mansic clay loam	Hays, KS	0.32	4
Marshall silt loam	Clarinda, IA	0.33	5
Mexico silt loam	McCredie, MO	0.28	3
Ontario loam	Geneva, NY	0.27[c]	–
Shelby loam	Bethany, MO	0.41	5
Tifton loamy sand	Tifton, GA	0.10	4
Zaneis fine sandy loam	Guthrie, OK	0.22	4

[a]Metric units in mt/MJ are 0.1317 times as large as these.

[b]*T* values were obtained from Soil Survey soil descriptions. Metric units in mt/ha are 2.24 times as large as these.

[c]Evaluated from continuous fallow. All others computed from row crop data.

Source: Adapted from Wischmeier and Smith, 1978.

down the slopes. Entrapment of sediment in terrace channels is predicted to range from 80 to 95%, depending on the type of terrace and slope.

6–2.4 Acceptance of the Modified USLE

The modified version of USLE was accepted readily. It was integrated into the operations of SCS field offices in all states. With its new features, it was tested in many countries outside of North America. For example, Ministry of Agriculture officials in Kenya asked the SCS for assistance in testing the use of the modified USLE in their country. Ulsaker and Onstad (1984) measured natural rainfall and erosion during a 13-month period, compared the correlations between 15 individual erosivity factors, and measured erosion rates. The single erosivity factor that correlated best was runoff volume, but EI_{30} was nearly as good. They were able to establish that Wischmeier and Smith's *P* factor values could be transferred to Kenya. A few cover and management practices tested indicate that at least some of the *C* factor values established in the United States have validity in Kenya also (Onstad et al., 1984).

TABLE 6–3 RATIO OF SOIL LOSS FROM CROPLAND TO CORRESPONDING LOSS FROM CONTINUOUS FALLOW

Line no.	Cover, crop sequence, and management	Spring residue lb	Cover after planting %	soil loss ratio for cropstage period and canopy cover[a] F %	SB %	1 %	2 %	3:80 %	3:90 %	3:96 %	4L %
Corn after corn, grain sorghum, small grain, or cotton in meadowless rotation											
	Moldboard plow, conv. till										
1	Res. left, spring TP	4500	—	31	55	48	38	—	—	20	23
3	Res. left, spring TP	2600	—	43	64	56	43	32	25	21	37
5	Res. left, fall TP	HP[b]	—	44	65	53	38	—	—	20	—
9	Res. removed, spring TP	HP	—	66	74	65	47	—	—	22	56
	Reduced tillage systems										
27	No till, plant in crop residue	4500	80	—	5	5	5	—	—	5	15
43	Conservation tillage	4500	30	—	21	18	15	—	—	13	21
Corn in sod-based systems											
	Moldboard plow, conv. till										
A[c]	1st year corn, spring TP	HP	—	8	22	19	17	—	—	10	14
B[c]	2nd year corn, spring TP	HP	—	22	44	38	32	—	—	18	20
Soybeans after corn											
123	Spring TP, RdL, conv. tillage	HP	—	33	60	52	38	—	20	17	38
Small grains after grain crops											
129	Spring seeded, disked residues	4500	70	—	12	12	11	7	4	2	7
142	Fall seeded, RdL, conv. tillage	HP	—	31	55	48	31	12	7	5	4
Small grains after summer fallow											
146	Conv. tillage, after sm. grain	200	10	—	70	55	43	18	13	11	12
149	Cons. tillage, after sm. grain	1000	50	—	26	21	15	8	7	6	8
155	Cons. tillage, after row crop	1000	30	—	40	31	24	13	10	8	4
Established meadows											
—	Grass and legume meadow[d]	HP	—	—	—	—	—	—	—	—	0.4

[a]F= Rough Fallow, S= Seedbed, 1= Establishment, 2= Development, 3= Maturation, % canopy cover, 4= Residue.

[b]HP = high productivity—3- to 5-ton yields, 4500 lb. of spring residue.

[c]Calculated from information in Tables 5 and 5D of Wischmeier and Smith, 1978.

[d]From Table 5B.

Source: Modified from Wischmeier and Smith, 1978.

Hussein (1986) could not determine accurate EI_{30} values in the usual way from Iraqi records. Instead, he used an equation developed by Arnoldus (1977). He obtained reasonable values in this way and prepared an iso-erodent map. He also obtained monthly contributions to annual EI_{30} values by using sums of squares of monthly rainfall. With these data as a first approximation, he predicted sheet and rill erosion for fields in northern Iraq using his adaptation of the modified USLE.

Yu et al. (2001) proposed that rainfall records that lack some of the detail needed for calculating EI_{30} can be used by supplementing them either with detailed data collected over a period of 20 to 30 months or with data from another site with a similar climate.

A similar problem occurred in studies of soil erosion in Northern Rhodesia (now Zimbabwe). Rainfall records were not plentiful, and EI_{30} values from available records correlated poorly with observed and measured erosion. Elwell (1978) solved this problem by developing an independent method of estimating soil loss. His equation was

$$Z = K \times C \times X$$

Z is the estimated average annual soil loss (mt/ha-yr). *K* is the soil-erodibility factor (mt/ha-yr). It is a combination of a climate subfactor (*E*) defined as the total energy of the average yearly rainfall and a soil subfactor (*F*). *C* is a crop factor obtained from the percent of rain that strikes vegetation before falling to the soil surface. *X* is a slope factor that is similar to Wischmeier and Smith's, but standard slope conditions are different: length is 30 m, gradient 4.5%. This estimator (Soil Loss Estimator for Southern Africa or SLEMSA) gave satisfactory predictions both on arable and on grazing land.

After some years of use, the modified USLE began to show some significant weaknesses. In spite of the fact that it was still the paramount instrument for predicting soil loss by water erosion both in the United States and abroad and was being used very widely in planning conservation activities, the USDA-ARS decided, in 1985, to develop an improved prediction model. The new model was to include new data that would be handled in new and different ways and would incorporate corrections for the weaknesses that were becoming apparent in the modified USLE. The new program was to be called the Revised Universal Soil Loss Equation (RUSLE).

6–3 REVISED UNIVERSAL SOIL LOSS EQUATION (RUSLE)

RUSLE was developed under the leadership of ARS with scientists from other government agencies cooperating (Renard et al., 1991; Renard et al., 1994; Renard et al., 1997). It was developed as a paper-based program, but because of the increasing availability of personal computers at the time the first prototype was field tested, the program was also adapted for computer use.

6–3.1 Changes Made in Revising USLE

The basic USLE has been retained, but nearly every factor is evaluated by new techniques, often using new databases. The *R* factor was expanded to cover mainland United States (Figure 6–2) and Hawaii. *R* factors in Hawaii vary from less than 100 to 2000, with the higher values occurring at high elevations, because these receive the most rainfall. Both *K* and *C* are varied seasonally by climatic data (monthly precipitation and temperature, frost-free period, and 15-day distribution of *EI*). The slope (*LS*) factor has been revised. The cover-management (*C*) factor has been expanded into a continuous function with four subfactors: prior land use, surface cover, crop canopy, and surface roughness. The conservation-practice (*P*) factor has been expanded to include data for rangeland. The major differences between the modified USLE and RUSLE are set out in Table 6–4.

6–3.2 Release and Acceptance of RUSLE

RUSLE was first released by USDA-ARS in 1992. The Soil Conservation Service (now Natural Resources Conservation Service [NRCS]) decided that same year to adopt the new RUSLE and incorporate it into its Field Office Computing System (FOCS). By 1997, RUSLE was being used in about half the conservation districts in the United States. In additional states, data are being gathered that will be used to develop local databases for the *C* factor.

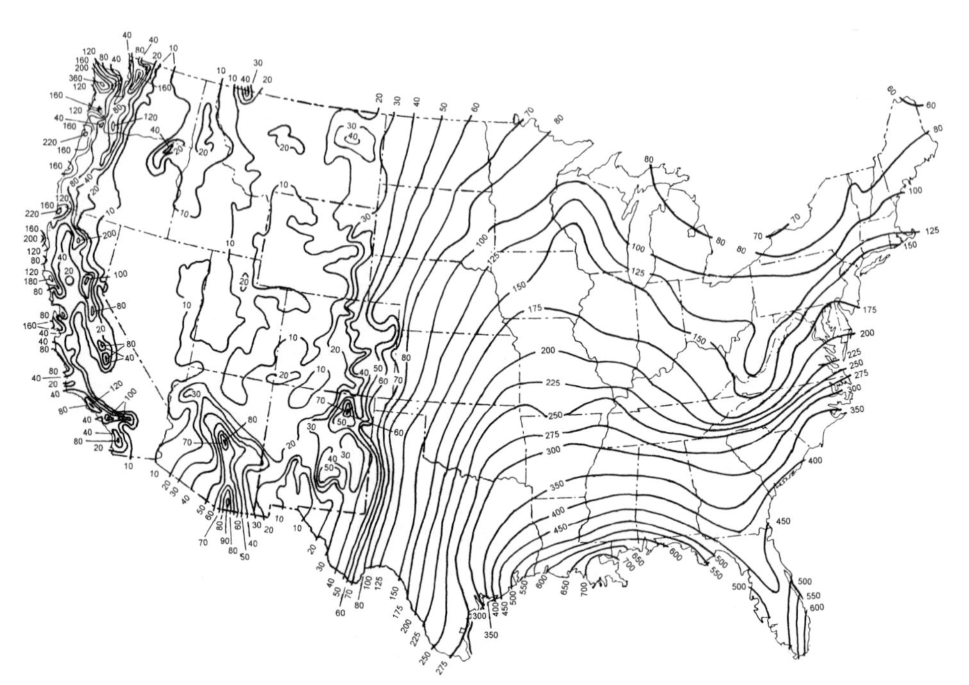

Figure 6–2 Isoerodent map (*R* values) in units of 100 ft-ton/ac-yr for the conterminous United States. (Modified from Renard et al., 1997.)

6–3.3 Training NRCS Personnel to Use RUSLE

Rather than undertake the responsibility for training NRCS employees in the use of the new equation and software, ARS negotiated a Cooperative Research and Development Agreement (CRADA) with the Soil and Water Conservation Society (SWCS) in 1992. (SWCS is a not-for-profit, private organization headquartered in Ankeny, Iowa.) The then current version of RUSLE was turned over to SWCS. By signing the CRADA, SWCS agreed to assume many responsibilities (Kautza et al., 1995). Among other things, SWCS

1. developed training materials that were used to transfer a knowledge of the RUSLE technology to the trainers of the ultimate users of the software
2. worked with NRCS and other user groups as they incorporated RUSLE into their activities
3. served as liaison between ARS and the user groups
4. conducted training workshops for prospective trainers of the technology
5. accepted responsibility for merchandising RUSLE software and manuals.

6–3.4 Updating RUSLE

As SWCS began its training-development program, it became evident that further improvement of the software was necessary. SCS specialists and ARS scientists analyzed

TABLE 6–4 SUMMARY OF THE DIFFERENCES BETWEEN THE MODIFIED USLE AND RUSLE

Factor	Modified USLE	RUSLE
R	Based on long-term average rainfall conditions for geographic areas in the U.S.	Generally the same as USLE in the eastern U.S. Values for western states (Montana to New Mexico and west) are based on data from more weather stations and thus are more precise for any given location. RUSLE computes a correction to *R* to reflect the effect of raindrops striking water ponded on flat surfaces.
K	Based on soil texture, organic-matter content, permeability, and other factors inherent to soil type.	Same as USLE but adjusted to account for seasonal changes such as freezing and thawing, soil moisture, and soil consolidation.
LS	Based on length and steepness of slope, regardless of land use.	Refines USLE by assigning new equations based on the ratio of rill to interrill erosion, and accommodates complex slopes.
C	Based on cropping sequence, surface roughness, and canopy cover, which are weighted by the percentage of erosive rainfall during the six crop stages. Lumps these factors into a table of soil loss ratios, by crop and tillage scheme.	Uses the subfactors prior land use, canopy cover, surface cover, surface roughness, and soil moisture. Refines USLE by dividing each year in the rotation into 15-day intervals, calculating the soil loss ratio for each period.
		Recalculates a new soil loss ratio every time a tillage operation changes one of the subfactors. RUSLE provides improved estimates of soil loss changes as they occur throughout the year, especially relating to surface and near-surface residue and the effects of climate on residue decomposition.
P	Based on installation practices that slow runoff and thus reduce soil movement. *P* factor values change according to slope ranges with some distinction for various ridge heights.	*P* factor values are based on hydrologic soil groups, slope, row grade, ridge height, and the 10-year, single-storm erosion-index value. RUSLE computes the effect of strip-cropping based on the transport capacity of flow in dense strips relative to the amount of sediment reaching the strip.
		The *P* factor for conservation planning considers the amount and location of deposition.

Source: Renard et al., 1994.

more data and modified the software. Improvements to the model resulted in SWCS RUSLE Versions 1.03 (January 1994) and 1.04 (June 1994).

Work on a new Version 2.0 began in 1993. There were three major reasons for developing Version 2.0 (Yoder and Lown, 1995):

1. The earlier versions were not making full use of the USLE erosion-predicting power. For example, *K, C,* and *P* factors were averaged separately over the crop season and the resulting averages were multiplied to get a year-long *A* value, even though they all change over the growing season. A more accurate way to handle the changes is to obtain the interactions of these elements (mathematical products of *K* and *C* and *P*) as they change individually over the season and then average these products to obtain a seasonal value.

2. The computer software produced after the paper-structured program was developed was awkward to use, and internal movement during execution caused technical and computer problems.

3. Many of the countries that want to use the new RUSLE technology work in metric units. It is far simpler to include this capability in a new computer-based technology than to try to introduce changes in the versions already completed.

Some very important changes are made in RUSLE Version 2.0. This version uses the same RUSLE technology and many of the same databases. Rather than use an independent-factor approach as was done in former versions, Version 2.0 asks users to describe fully the situation in the field for which erosion losses are to be predicted. With the complete description keyed in, the program calculates the erosion rate. In addition, the individual time-varying factors mentioned previously (*K, C,* and *P*) are multiplied together, automatically, throughout the year and the individual products averaged to give seasonal $K \times C \times P$ values.

Version 2.0 also has the capability to calculate soil loss values for slopes with different soils, management schemes, even crops. Version 2.0 is a Windows™-based program. It takes advantage of the extra resources that operating system provides and the more powerful computers that are required to run Windows™.

6–3.5 Predicting Soil Losses Caused by Water Erosion—A Specific Example

USLE and the early versions of RUSLE were solved using paper tables and graphs. Many NRCS district offices still use this paper model of RUSLE for predicting soil losses on cooperators' farms. This example will generate a prediction the paper way, even though computer programs with data banks instead of tables and graphs are available now.

The factors described in the earlier sections of this chapter and the data contained in the figures and tables can be used to predict soil loss. This example will calculate the average annual soil loss from a field planted to continuous corn on a Marshall silt loam soil in southwestern Iowa. The field has a slope gradient of 8% and a slope length of 250 ft (75 m). Corn yields average 125 bu/ac (78.6 quintals/ha), and about 4500 lb/ac (5045 kg/ha) of crop residue remains on the field in the spring. The land is turn plowed in early April, and the seedbed is prepared subsequently by disking and harrowing. No special erosion-control practices are employed.

Many tables of values for the time-dependent variables *K, C,* and *P* have been computer generated for a variety of conditions. Selected tables are reproduced in this section. Rainfall values in mountainous areas such as western United States and Hawaii vary greatly with elevation and cannot be shown accurately on a small map, such as Figure 6–2. Local *R* values, 24-hour storm *EI* values, and appropriate tables for *K* and *C* values are best obtained from state or district NRCS offices. Another alternative is to purchase the manual and software for computer solutions.

An *R* value for southwestern Iowa (170 hundred ft-ton/ac-yr) is obtained from Figure 6–2. The most recent *K* value for Marshall silt loam, obtained from the NRCS state office in Des Moines, Iowa (0.32 ton soil loss/100 ft-ton of *R*), is adjusted to 0.26 as prescribed in Table 6–5.

An *LS* value of 1.62 for row-cropped fields on 8% gradient and 250-ft slope length is obtained from Table 6–6B.

Data for calculating a *C* value are found in Table 6–7. This is just one set of tables from a total of 12 sets that were developed to evaluate *C* values for Climatic Zone 99 (Shenandoah, Iowa). This particular set of seven tables is for Mulch Till preparation,

TABLE 6–5 AVERAGE ANNUAL *K*-FACTOR ADJUSTMENTS CLIMATIC ZONE: 99 (SHENANDOAH, IA)

Current K	RUSLE adjusted K	Current K	RUSLE adjusted K
0.02	0.02	0.28	0.24
0.05	0.05	0.32	0.26
0.10	0.08	0.37	0.30
0.15	0.12	0.43	0.35
0.17	0.15	0.49	0.40
0.20	0.17	0.55	0.46
0.24	0.20	0.64	0.52

Source: USDA NRCS, Technical Guide, Section 1, Erosion Prediction, 1996.

TABLE 6–6A VALUES FOR TOPOGRAPHIC FACTOR (*LS*) FOR LOW-RATIO RILL TO INTERRILL EROSION, SUCH AS FOR RANGELAND

Slope	Slope length (ft)											
(%)	25	50	75	100	150	200	250	300	400	600	800	1000
0.5	0.08	0.08	0.08	0.09	0.09	0.09	0.09	0.09	0.09	0.09	0.09	0.09
1	0.13	0.13	0.14	0.14	0.15	0.15	0.15	0.15	0.16	0.16	0.17	0.17
2	0.21	0.23	0.25	0.26	0.27	0.28	0.29	0.30	0.31	0.33	0.34	0.35
3	0.29	0.33	0.36	0.38	0.40	0.43	0.44	0.46	0.48	0.52	0.55	0.57
4	0.36	0.43	0.46	0.50	0.54	0.58	0.61	0.63	0.67	0.74	0.78	0.82
5	0.44	0.52	0.57	0.62	0.68	0.73	0.78	0.81	0.87	0.97	1.04	1.10
6	0.50	0.61	0.68	0.74	0.83	0.90	0.95	1.00	1.08	1.21	1.31	1.40
8	0.64	0.79	0.90	0.99	1.12	1.23	1.32	1.40	1.53	1.74	1.91	2.05
10	0.81	1.03	1.19	1.31	1.51	1.67	1.80	1.92	2.13	2.45	2.71	2.93
12	1.01	1.31	1.52	1.69	1.97	2.20	2.39	2.56	2.85	3.32	3.70	4.02
14	1.20	1.58	1.85	2.08	2.44	2.73	2.99	3.21	3.60	4.23	4.74	5.18
16	1.38	1.85	2.18	2.46	2.91	3.28	3.60	3.88	4.37	5.17	5.82	6.39
20	1.74	2.37	2.84	3.22	3.85	4.38	4.83	5.24	5.95	7.13	8.10	8.94

Source: Modified from USDA NRCS, Technical Guide, Section 1, Erosion Prediction, 12/96.

Medium Production (125 bu/ac) fields. From the first section of the table, *Corn after,* the value 0.20 for *Spring Primary Tillage—Plow* occurs in the first column of the first row. The value for P is 1.0 because no special practices are employed. Therefore

$$A = 170 \times 0.26 \times 1.62 \times 0.20 \times 1.0 = 14.3 \text{ tons/ac-yr}$$

This predicted soil loss is almost three times the tolerable rate for Marshall silt loam ($T = 5$ tons/ac-yr, as shown in Table 6–2).

Several changes can be made to reduce the rate of soil loss. Reducing tillage will leave more crop residue on the soil surface for protection; contour cultivation or strip cropping can be employed; terraces can be installed; a crop rotation with a lower C value can be initiated; or some combination of these conservation practices can be devised that will reduce the rate of soil loss to the tolerable level.

Using a mulch tillage scheme to ensure 30% ground cover would reduce the C factor value to 0.11, as shown in Table 6–7 for *Corn after Corn,* with *Spring Primary Tillage,* and

TABLE 6–6B VALUES FOR TOPOGRAPHIC FACTOR (*LS*) FOR MODERATE-RATIO RILL TO INTERRILL EROSION, SUCH AS FOR ROW-CROPPED AGRICULTURAL AND OTHER MODERATELY CONSOLIDATED SOIL CONDITIONS WITH MODERATE COVER

Slope	Slope length (ft)											
(%)	25	50	75	100	150	200	250	300	400	600	800	1000
0.5	0.08	0.08	0.08	0.09	0.09	0.09	0.09	0.09	0.10	0.10	0.10	0.10
1	0.12	0.13	0.14	0.14	0.15	0.16	0.17	0.17	0.18	0.19	0.20	0.20
2	0.19	0.22	0.25	0.27	0.29	0.31	0.33	0.35	0.37	0.41	0.44	0.47
3	0.25	0.32	0.36	0.39	0.44	0.48	0.52	0.55	0.60	0.68	0.75	0.80
4	0.31	0.40	0.47	0.52	0.60	0.67	0.72	0.77	0.86	0.99	1.10	1.19
5	0.37	0.49	0.58	0.65	0.76	0.85	0.93	1.01	1.13	1.33	1.49	1.63
6	0.43	0.58	0.69	0.78	0.93	1.05	1.16	1.25	1.42	1.69	1.91	2.11
8	0.53	0.74	0.91	1.04	1.26	1.45	1.62	1.77	2.03	2.47	2.83	3.15
10	0.67	0.97	1.19	1.38	1.71	1.98	2.22	2.44	2.84	3.50	4.06	4.56
12	0.84	1.23	1.53	1.79	2.23	2.61	2.95	3.26	3.81	4.75	5.56	6.28
14	1.00	1.48	1.86	2.19	2.76	3.25	3.69	4.09	4.82	6.07	7.15	8.11
16	1.15	1.73	2.20	2.60	3.30	3.90	4.45	4.95	5.86	7.43	8.79	10.02
20	1.45	2.22	2.85	3.40	4.36	5.21	5.97	6.68	7.97	10.23	12.20	13.99

Source: Modified from USDA NRCS, Technical Guide, Section 1, Erosion Prediction, 12/96.

TABLE 6–6C VALUES FOR TOPOGRAPHIC FACTOR (*LS*) FOR HIGH-RATIO RILL TO INTERRILL EROSION, SUCH AS FRESHLY PREPARED CONSTRUCTION AND SIMILAR HIGHLY DISTURBED SITES WITH LITTLE OR NO COVER

Slope	Slope length (ft)											
(%)	25	50	75	100	150	200	250	300	400	600	800	1000
0.5	0.07	0.08	0.08	0.09	0.09	0.10	0.10	0.10	0.11	0.12	0.12	0.13
1	0.10	0.13	0.14	0.15	0.17	0.18	0.19	0.20	0.22	0.24	0.26	0.27
2	0.16	0.21	0.25	0.28	0.33	0.37	0.40	0.43	0.48	0.56	0.63	0.69
3	0.21	0.30	0.36	0.41	0.50	0.57	0.64	0.69	0.80	0.96	1.10	1.23
4	0.26	0.38	0.47	0.55	0.68	0.79	0.89	0.98	1.14	1.42	1.65	1.86
5	0.31	0.46	0.58	0.68	0.86	1.02	1.16	1.28	1.51	1.91	2.25	2.55
6	0.36	0.54	0.69	0.82	1.05	1.25	1.43	1.60	1.90	2.43	2.89	3.30
8	0.45	0.70	0.91	1.10	1.43	1.72	1.99	2.24	2.70	3.52	4.24	4.91
10	0.57	0.91	1.20	1.46	1.92	2.34	2.72	3.09	3.75	4.95	6.03	7.02
12	0.71	1.15	1.54	1.88	2.51	3.07	3.60	4.09	5.01	6.67	8.17	9.57
14	0.85	1.40	1.87	2.31	3.09	3.81	4.48	5.11	6.30	8.45	10.40	12.23
16	0.98	1.64	2.21	2.73	3.68	4.56	5.37	6.15	7.60	10.26	12.69	14.96
20	1.24	2.10	2.86	3.57	4.85	6.04	7.16	8.23	10.24	13.94	17.35	20.57

Source: Modified from USDA NRCS, Technical Guide, Section 1, Erosion Prediction, 12/96.

30% cover. This would bring the predicted erosion level down to 6.9 tons/ac-yr (15 mt/ha-yr), still above the tolerable level.

$$A = 170 \times 0.26 \times 1.62 \times 0.11 \times 1.0 = 6.9 \text{ tons/ac-yr}$$

For reduced tillage to reduce erosion losses to the tolerable level on this farm, at least 50% ground cover would have to be maintained on the surface **every year.** This would require instituting a no-till system. With a minimum 50% cover on the soil surface every year, the

TABLE 6–6D VALUES FOR TOPOGRAPHIC FACTOR (*LS*) FOR THAWING SOILS WHERE MOST EROSION IS CAUSED BY SURFACE FLOW

Slope	Slope length (ft)											
(%)	25	50	75	100	150	200	300	400	500	600	800	1000
0.5	0.05	0.07	0.09	0.10	0.12	0.14	0.17	0.20	0.22	0.24	0.28	0.31
1	0.08	0.11	0.14	0.16	0.20	0.23	0.28	0.32	0.36	0.40	0.46	0.51
2	0.14	0.20	0.25	0.29	0.35	0.41	0.50	0.58	0.64	0.71	0.82	0.91
3	0.21	0.29	0.36	0.42	0.51	0.59	0.72	0.83	0.92	1.02	1.17	1.31
4	0.27	0.38	0.47	0.54	0.66	0.77	0.94	1.08	1.20	1.33	1.53	1.71
5	0.33	0.47	0.58	0.67	0.82	0.94	1.16	1.34	1.50	1.64	1.89	2.11
6	0.40	0.56	0.69	0.79	0.97	1.12	1.38	1.59	1.79	1.95	2.25	2.51
8	0.52	0.74	0.91	1.05	1.28	1.48	1.81	2.09	2.33	2.56	2.96	3.31
10	0.62	0.88	1.08	1.25	1.53	1.77	2.16	2.50	2.79	3.06	3.54	3.95
12	0.70	0.98	1.21	1.39	1.71	1.97	2.41	2.78	3.10	3.41	3.94	4.40
14	0.76	1.08	1.32	1.53	1.87	2.16	2.64	3.05	3.41	3.74	4.31	4.82
16	0.82	1.17	1.43	1.65	2.02	2.33	2.86	3.30	3.69	4.04	4.67	5.22
18	0.88	1.25	1.53	1.82	2.16	2.50	3.06	3.53	3.94	4.32	5.00	5.58
20	0.94	1.33	1.63	1.88	2.30	2.66	3.25	3.76	4.19	4.60	5.31	5.94

Source: Modified from Renard et al., 1997.

C value becomes 0.036 (from a no-till table not presented here). The predicted soil loss would then be 2.6 tons/ac-yr (5.8 mt/ha-yr).

To determine the effect of contour cultivation on the predicted level of erosion, a subfactor value for contour cultivation (*Pc*) must be obtained. This is based on the soil's *Cover-Management Condition,* its *Soil Hydrologic Group,* the field's *Ridge Height* category, and an estimate of the *Grade* of the contour rows. The cover-management condition is ascertained by selecting the management condition (Codes 1 to 7) from Table 6–8 that best describes the field condition during the one-quarter of the year when rainfall and runoff are most erosive and the soil is most susceptible to erosion. Because *P* factor effects are approximate, no provision is made to vary the cover-management condition code during the year. The *Soil Hydrologic Group* designation can be obtained from the Soil and Water Section of an appropriate County Soil Survey Report. *Ridge Height* characteristics are described in Table 6–9. The selection (*Very Low, Low,* etc.) is chosen that best describes the conditions during the one-quarter of the year when rainfall and runoff are the most erosive and the soil is most susceptible to erosion.

Conditions set up for the example field suggest that the *Cover-Management Condition* is “5” and the *Ridge Height* classification is “Low.” The *Soil Hydrologic Group* is “B.” A Row Grade of 1% will be assumed for this example, since a field measurement is not available. The *Pc* subfactor value of 0.61 for low ridge heights, 8% slope, and 1% row grade is obtained from Table 6–10. (This table, like Table 6–7, is one set of a whole family of table sets for determining contour-cultivation subfactor values.) With contour cultivation, then

$$A = 170 \times 0.26 \times 1.62 \times 0.20 \times 0.61 = 8.7 \text{ tons/ac-yr}$$

Combining conservation tillage and contouring, the predicted soil loss is

$$A = 170 \times 0.26 \times 1.62 \times 0.11 \times 0.61 = 4.8 \text{ tons/ac-yr}$$

This level of erosion is below the tolerance limit.

TABLE 6–7 RATIO OF SOIL LOSS FROM CROPLAND TO CORRESPONDENT LOSS FROM CONTINUOUS FALLOW, MULCH-TILL, MEDIUM PRODUCTION (CORN—125 BU/AC, SOYBEANS—35 BU/AC); THE COLUMN HEADINGS INDICATE THE PERCENT RESIDUE COVER

	Spring primary tillage							Fall primary tillage						
Corn after	Plow	10	20	30	40	50	60	Plow	10	20	30	40	50	60
Corn	.20	17	.13	.11	.09	.08		.21	.17	.13	.11	.095	.085	
Corn silage w/cover	.26	.20	.16	.14	.125									
Corn silage	.36	.34*						.37	.35*					
Small grain-baled	.33	.26	.22	.19	.17			.34	.27	.23	.20	.18		
Soybeans	.30	.25	.22	.20	.19			.33	.25	.22	.20	.18		
Soybeans w/narrow row	.29	.25	.22	.20	.19			.31	.25	.22	.19	.18		
Meadow	.10	.09	.08	.07				.15	.13	.11	.10			

	Spring primary tillage							Fall primary tillage						
Wide-row soybeans after	Plow	10	20	30	40	50	60	Plow	10	20	30	40	50	60
Corn	.24	.19	.15	.12	.10	.085	.075	.26	.19	.15	.12	.10	.09	
Corn silage w/cover	.26	.22	.18	.15	.13	.12	.11							
Corn silage	.41	.39*						.42	.40*					
Small grain-baled	.37	.30	.25	.18	.16			.43	.32	.27	.23	.21		
Soybeans	.34	.29	.26	.24				.37	.30	.27	.25			
Meadow	.10	.085	.075	.07				.15	.13	.11	.11			

	Spring primary tillage							Fall primary tillage						
Narrow-row soybeans after	Plow	10	20	30	40	50	60	Plow	10	20	30	40	50	60
Corn	.18	.14	.11	.09	.075	.06		.18	.13	.10	.082	.07	.06	
Corn silage w/cover	.22	.18	.14	.12	.11	.10								
Corn silage	.31	.31*						.32	.31*					
Small grain-baled	.28	.23	.20	.18	.17			.34	.25	.21	.19	.18		
Soybeans w/narrow rows	.24	.20	.18	.16	.15			.27	.20	.17	.16			
Meadow	.075	.065	.055	.05										

	Spring primary tillage							Fall primary tillage						
Spring small grain (baled) after	Plow	10	20	30	40	50	60	Plow	10	20	30	40	50	60
Corn	.048		.029	.022	.018	.014	.01	.055	.04	.03	.023	.02	.018	.015
Corn silage	.14	0.11*							.12*					
Soybeans	.08	.065	.054	.048	.045			.095	.07	.058	.051	.046		
Soybeans w/narrow rows	.075	.064	.053	.048	.045	.041		.09	.066	.055	.045	.04		

	Spring primary tillage							Fall primary tillage						
Corn silage after	Plow	10	20	30	40	50	60	Plow	10	20	30	40	50	60
Corn	.23	.18	.14	.12	.10			.24	.18	.14	.12	.10		
Corn silage	.40	.38*						.41	.39*					
Small grain-baled	.36	.30	.25	.22				.37	.32	.26	.23			
Soybeans	.34	.29	.26	.24				.35	.30	.27	.25			
Soybeans w/narrow rows	.34	.29	.26	.24				.36	.29	.25	.23			
Meadow	.11	.09	.08	.075				.15	.13	.12	.11			

	Spring primary tillage							Fall primary tillage						
Oats with meadow after	Plow	10	20	30	40	50	60	Plow	10	20	30	40	50	60
Corn	.022		.012	.009	.007	.005	.004	.028	.018	.013	.01	.008	.006	.005
Corn silage	.089	.08*						.10	.09*					
Soybeans	.04	.03	.026	.024	.022			.055	.038	.031	.028	.025		
Soybeans w/narrow rows	.04	.032	.026	.024	.022			.052	.038	.031	.028	.026		

	Spring primary tillage							Fall primary tillage						
Corn silage with cover after	Plow	10	20	30	40	50	60	Plow	10	20	30	40	50	60
Corn	.22	.17	.13	.10	.09	.08								
Soybeans w/cover	.26	.21	.16	.14	.12	.11								
Soybeans	.31	.24	.21	.20	.19									
Soybeans w/narrow rows	.31	.24	.21	.195	.18									

*Residue levels will normally be below 10% cover.

Source: USDA NRCS Technical Guide, Section 1, Erosion Prediction, 1996.

This gives two known alternatives that would bring about satisfactory reductions in erosion: (a) no-till, and (b) conservation tillage with contour cultivation. In real life, these systems would be explained to the farmer as a basis for deciding the best program for the field.

Other systems could be devised also. A terrace system could be used to reduce erosion from this field. If an open terrace system with 100-ft terrace intervals and 0.2% channel grade were developed, the LS factor value would change to 1.04 (Table 6–6B), and the terrace subfactor value (Pt) would be 0.6 (Table 6–11). For a combination of contour cultivation and terraces (contour cultivation should be used when a field is terraced), the overall P factor value would be

$$P = Pc \times Ps \times Pt = 0.61 \times 1.0 \times 0.6 = 0.37$$

With the original plow, disk, and harrow tillage system, plus contour cultivation and terraces, the predicted soil loss is

$$A = 170 \times 0.26 \times 1.04 \times 0.20 \times 0.37 = 3.4 \text{ tons/ac-yr}$$

TABLE 6–8 DESCRIPTION OF COVER-MANAGEMENT CONDITIONS USED IN RUSLE FOR ESTIMATING *P* FACTOR VALUES

Cover-management condition	Description
Code 1. Established meadow	In this condition, the grass is dense and runoff is very slow, about the slowest under any vegetative condition. When mowed and baled, this condition is condition 2.
Code 2. 1st year meadow, hay	In this condition, the hay is a mixture of grass and legume just before cutting. The meadow is a good stand of grass that is nearing the end of its first year. When mowed and baled, this condition becomes a condition 4 for a short time.
Code 3. Heavy cover and/or very rough	Ground cover for this condition is about 75 to 95%. Roughness would be like that left by a high-clearance moldboard plow on a heavy-textured soil. Roughness depressions would appear to be 7-in. deep or deeper. Vegetative hydraulic-roughness would be similar to that produced by a good legume crop, such as lespedeza, which has not been mowed.
Code 4. Moderate cover and/or rough	The ground cover for this condition is about 40 to 65%. Roughness would be similar to that left by a moldboard plow in a medium-textured soil. Depressions would appear to be about 4- to 6-in. Vegetative hydraulic-roughness would be much like that produced by winter small grain at full maturity.
Code 5. Light cover and/or moderate roughness	Ground surface cover is between 10 and 35%. The surface roughness is like that left by the first pass of a tandem disk over a medium-textured soil that has been moldboard plowed, or that left after a chisel plow cultivated a medium-textured soil at optimum moisture conditions (for tillage). Roughness depressions appear to be about 2- to 3-in. deep. In terms of hydraulic-roughness produced by vegetation, this condition is much like that produced by spring small grain at about three-fourths maturity.
Code 6. No cover and/or minimal roughness	This condition is very much like the condition found in row-cropped fields after the field has been planted and exposed to moderately intense rainfall. Ground cover is 5% or less, and the roughness is that characteristic of a good seedbed for corn or soybeans. The surface is rougher than that of a finely pulverized seedbed prepared for vegetables or grass.
Code 7. Clean-tilled, smooth fallow	This condition is essentially bare, with a cover of 5% or less. The soil has not had a crop grown on it in 6 months or more. Much of the residual effect of previous cropping has disappeared. The surface is smooth, much like the soil surface that develops on a very finely-pulverized seedbed exposed to several intense rainstorms. This condition may be found in fallowed and vegetable fields, or in newly-sown lawns and hay fields.

Source: Adapted from USDA NRCS Technical Guide, Section 1, Erosion Prediction, 9/96.

This, then, is another way of controlling erosion losses, albeit an expensive way.

It should be remembered that erosion predictions for individual fields and farms, made with USLE, RUSLE, and most other equations, are anticipated long-term average losses under normal conditions and average management. There are many farms that are managed in unconventional ways, so there will be cases where predicted losses will be larger or smaller than those actually taking place. Jackson (1988) studied a farm in Ohio that had been operated by the same Amish family in the same way for 150 years. Predicted average soil loss from the Wooster silt loam soil on the farm was 15 tons/ac-yr (36 mt/ha-yr). This is a loss equivalent to about 15 in. (38 cm) of soil depth during that 150-year period. In the mid-1980s, however, the soils of the farm still matched the Wooster silt loam soil description as to horizon depth and other properties. Actual erosion losses must have been near the rate of soil renewal, not at the rate predicted.

TABLE 6–9 DESCRIPTION OF RIDGE HEIGHTS USED IN RUSLE FOR ESTIMATING *P* FACTOR VALUES

Ridge height		
Category	Inches	Descriptive field situation
Very low	0.5–2	Plants not closely spaced, but with a perceptible ridge height.
		No-till planted crop rows.
		Fields that have been rolled, pressed, or dragged after planting.
		Conventionally drilled crops when erosive rains occur during or soon after planting.
		Clean-seeded hay that leaves a very low ridge.
Low	2–3	No-till drilled crops.
		Mulch-tilled row crops.
		Conventionally planted row crops without row cultivation.
		Conventionally drilled small grain crops when erosive rains are uniformly distributed throughout the year.
		Winter small grains when runoff from snow melt occurs during winter and early spring.
		Transplanted crops, widely spaced.
Moderate	3–4	Conventionally (clean) tilled row crops with row cultivation.
		High-yielding, winter small grain crops when erosive rains are concentrated in the late spring after plants have developed a stiff, upright stem.
		Transplanted crops that are closely spaced and/or in narrow rows.
High	4–6	Ridge-tilled crops with high ridges during periods of erosive rains.
Very high	>6	Ridge-tilled crops with very high ridges during periods of erosive rains.
		Hipping, bedding, or ridging with very high ridges during periods of erosive rains.

Source: Adapted from USDA NRCS Technical Guide, Section 1, Erosion Prediction, 9/96.

6–4 COMPUTER-BASED SOIL LOSS PREDICTION MODELS

As computer technology developed and as personal computer availability increased, a number of other computer software programs were developed to predict soil erosion losses for specific purposes. New programs that analyzed the erosion process on either a daily or a small-area basis were also developed. These programs are generally known by their acronyms rather than by the long names those acronyms represent.

6–4.1 soil loss Prediction for Special Purposes

Erosion predictions can be used for a variety of purposes. For example, eroded soil carries plant nutrients with it and thereby contributes to water pollution if it enters a body of water. The emphasis in a program to predict water pollution is on the delivery of nutrients to water that may become polluted rather than on how much soil is lost from an area in a field. Other programs are designed for administrative or other purposes that have their own unique requirements.

ANSWERS (Aerial Non-Point Environmental Response Simulation). Beasley et al. (1980) developed the ANSWERS program to predict movement of nutrients and sediment from agricultural land to lakes and streams. Beasley and Huggins (1982) prepared a computer manual for this program.

TABLE 6–10 ***P* SUBFACTOR FOR CONTOURING (*Pc*)**[a]

10-Year Rainstorm *EI* Value—80
Cover Management Condition—5
Soil Hydrologic Group—B

Very low ridge heights

	Row grade													
Slope	0	1	2	3	4	5	6	8	10	12	14	16	18	20
2	0.56	0.87	1.00											
4	0.50	0.75	0.85	0.93	1.00									
6	0.51	0.71	0.79	0.86	0.91	0.96	1.00							
8	0.55	0.71	0.78	0.83	0.87	0.91	0.94	1.00						
10	0.62	0.74	0.79	0.83	0.86	0.89	0.91	0.96	1.00					
12	0.71	0.79	0.83	0.86	0.88	0.90	0.92	0.95	0.98	1.00				
14	0.81	0.86	0.88	0.90	0.91	0.92	0.93	0.95	0.97	0.99	1.00			
16	0.91	0.93	0.94	0.95	0.96	0.96	0.97	0.97	0.98	0.99	0.99	1.00		
18	1.00	1.00	1.00	1.00	1.00	1.00	1.00	1.00	1.00	1.00	1.00	1.00	1.00	
20	1.00	1.00	1.00	1.00	1.00	1.00	1.00	1.00	1.00	1.00	1.00	1.00	1.00	1.00

Low ridge heights

	Row grade													
Slope	0	1	2	3	4	5	6	8	10	12	14	16	18	20
2	0.50	0.85	1.00											
4	0.39	0.70	0.82	0.92	1.00									
6	0.38	0.63	0.74	0.82	0.89	0.95	1.00							
8	0.40	0.61	0.70	0.77	0.82	0.87	0.92	1.00						
10	0.45	0.62	0.70	0.75	0.80	0.84	0.88	0.94	1.00					
12	0.50	0.64	0.70	0.75	0.79	0.82	0.85	0.91	0.96	1.00				
14	0.57	0.69	0.73	0.77	0.80	0.83	0.85	0.90	0.93	0.97	1.00			
16	0.64	0.73	0.77	0.80	0.82	0.84	0.86	0.90	0.93	0.95	0.98	1.00		
18	0.73	0.79	0.82	0.84	0.86	0.87	0.89	0.91	0.93	0.95	0.97	0.98	1.00	
20	0.81	0.85	0.87	0.88	0.90	0.91	0.91	0.93	0.95	0.96	0.97	0.98	0.99	1.00

Moderate ridge heights

	Row grade													
Slope	0	1	2	3	4	5	6	8	10	12	14	16	18	20
2	0.45	0.84	1.00											
4	0.29	0.65	0.79	0.90	1.00									
6	0.26	0.56	0.69	0.78	0.86	0.94	1.00							
8	0.27	0.53	0.64	0.72	0.79	0.85	0.90	1.00						
10	0.29	0.52	0.61	0.68	0.74	0.79	0.84	0.93	1.00					
12	0.33	0.52	0.60	0.67	0.72	0.76	0.81	0.88	0.94	1.00				
14	0.38	0.55	0.62	0.67	0.71	0.75	0.79	0.85	0.91	0.95	1.00			
16	0.43	0.57	0.63	0.68	0.72	0.75	0.78	0.83	0.88	0.92	0.96	1.00		
18	0.49	0.61	0.66	0.70	0.73	0.76	0.79	0.83	0.87	0.91	0.94	0.97	1.00	
20	0.55	0.65	0.69	0.73	0.75	0.78	0.80	0.84	0.87	0.90	0.93	0.95	0.98	1.00

High ridge height

	Row grade													
Slope	0	1	2	3	4	5	6	8	10	12	14	16	18	20
2	0.42	0.83	1.00											
4	0.21	0.61	0.77	0.89	1.00									
6	0.16	0.50	0.65	0.75	0.85	0.93	1.00							
8	0.16	0.46	0.58	0.68	0.75	0.82	0.89	1.00						
10	0.17	0.43	0.54	0.63	0.70	0.76	0.81	0.91	1.00					
12	0.20	0.43	0.53	0.60	0.66	0.72	0.77	0.85	0.93	1.00				
14	0.23	0.44	0.52	0.59	0.64	0.69	0.74	0.81	0.88	0.94	1.00			
16	0.27	0.45	0.53	0.59	0.64	0.68	0.72	0.79	0.85	0.90	0.95	1.00		
18	0.31	0.47	0.54	0.59	0.64	0.68	0.71	0.77	0.83	0.88	0.92	0.96	1.00	
20	0.36	0.50	0.56	0.61	0.65	0.68	0.71	0.77	0.82	0.86	0.90	0.93	0.97	1.00

[a]This is one set of tables for determining *P* subfactor (contouring) values. Other sets for different combinations of 10-Year Rainstorm *EI* values, Cover Management Conditions, and Soil Hydrologic Groups are also available.

Source: Adapted from USDA NRCS Technical Guides, Section 1, Erosion Prediction, 12/96.

TABLE 6–11 TERRACE SUBFACTOR (*Pt*) VALUES FOR CONSERVATION PLANNING

	Pt subfactor value			
		Open outlets, with terrace grade of		
Terrace interval ft	Closed outlets	0.1–0.3%	0.4–0.7%	0.8%+
<111	0.5	0.6	0.7	1.0
111–140	0.6	0.7	0.8	1.0
141–180	0.7	0.8	0.8	1.0
181–225	0.8	0.8	0.9	1.0
226–300	0.9	0.9	0.9	1.0
>300	1.0	1.0	1.0	1.0

Source: Adapted from USDA NRCS Technical Guide, Section 1, Erosion Prediction, 12/96.

CREAMS (Chemicals, Runoff, and Erosion from Agricultural Management). CREAMS was developed by ARS scientists for use on small watersheds. It predicts simple runoff and also has a component to assess nutrient losses on sediment and in runoff (Knisel, 1980). It requires topographic input, information on soil erodibility, potential runoff, cover management, and climatic data. It was designed for use on a mainframe computer.

EPIC (Erosion Productivity Impact Calculator). The EPIC program was developed by ARS scientists to assess the long-time effect of soil erosion on soil productivity (Williams et al., 1983). Its primary, initial function was to refine data for USDA reports to fulfill the 1977 Soil and Water Resources Conservation Act mandate for a periodic review of the condition of soil, water, and related resources on agricultural land. It was also used to make predictions on productivity losses by erosion and on management needed to control these losses. The model includes components of climate, runoff, erosion, soil moisture, soil chemistry, crop growth, crop and soil management, and economics. It computes erosion from a single point on the landscape, but it does not consider sediment deposited in the field. Initially, it required a mainframe computer, but was reprogrammed for personal computers.

AGNPS (Agricultural Non-Point Source). The AGNPS pollution model was developed by ARS, SCS, the University of Minnesota, and the Minnesota Pollution Control Agency to estimate runoff water quality from agricultural watersheds ranging in size from 5 to 50,000 acres (Young et al., 1989a). It predicts runoff volume, peak rate of flow, eroded and delivered sediment, nitrogen and phosphate in runoff, and oxygen required to decompose organic and inorganic chemicals that are added from feedlots, springs, wastewater treatment plants, and so forth. It can be run on an IBM-compatible personal computer. A user's guide is available (Young et al., 1989b).

CAMPS (Computer-Assisted Management and Planning System). SCS developed the CAMPS program. It was used in SCS field offices to develop conservation plans and engineering designs for conservation measures, compute soil loss, make soil interpretations, and track conservation progress. It was also used to manage workloads, develop public mailings, and prepare reports. It was designed for use on a personal computer. The Natural Resources Conservation Service (successor to SCS) replaced this program in their district offices with the Field Operations Computer System (FOCS).

6–4.2 Process-Based soil loss Prediction Equations

The early soil loss equations were based on empirical relationships derived from the analysis of data with no requirement for a scientific explanation of the processes involved. They averaged values over a year's time or within a few subdivisions of a year so that the necessary calculations could be made on paper. Programmable computers made it possible to speed up the calculations and were applied to the RUSLE equation, but it was still an empirical approach.

A more theoretical analysis of the erosion process became feasible as computers became more powerful. Some of the special-purpose programs in the preceding section are at least partly process-oriented. The programs discussed in this section incorporate this approach throughout their structure by developing models to simulate water flow and soil movement.

WEPP (Water Erosion Prediction Project). The WEPP program was initiated "to develop a new generation of water erosion prediction technology" for use by a wide range of user agencies involved "in soil and water conservation and environmental planning and assessment" (Foster and Lane, 1987). The Agricultural Research Service, Natural Resources Conservation Service, Forest Service, all of the USDA, and the Bureau of Land Management in the U.S. Department of the Interior were involved in the project.

WEPP is the first American model developed for erosion prediction on a broad scale that is not based on USLE technology. It is a fundamental, process-based, daily-simulation model developed to replace the Universal Soil Loss Equation for erosion prediction (Laflen et al., 1991; Lane and Nearing, 1989). It is a model that makes daily computations of the status of the soil and biomass in a field. If it rains, runoff is computed. If runoff occurs, detachment, transportation, and deposition are computed down the slope. It includes a climate generator, a hydrology component, a plant-growth model, and climate, soil, tillage, and plant databases for most conditions in the United States. A preliminary version of the technology was released in 1989. It has been extensively tested on cropland in the United States and has performed well. It has also been tested in Europe, Asia, and Africa. It was released to the general public in 1993.

Because WEPP calculates not only the amount of soil eroded but also the transporting capacity of the runoff, it is able to predict the amount and location of sediment that will be deposited when the water flow slows as slopes begin to flatten (Favis-Mortlock and Guerra, 2000). It has also been modified experimentally to predict changes in erosion patterns that may occur as a result of global warming.

WEPP comes in two versions—a Hillslope Version and a Watershed Version. Both are specifically designed to estimate sediment flow in ephemeral gullies and in terrace channels and grassed waterways; neither can be used for areas with permanent gullies or stream channels. It runs on a personal computer and is available on the Internet by using a World Wide Web browser connected to the following location:

http://topsoil.nserl.purdue.edu/nserlweb/weppmain/wepp.html

The Soil and Water Conservation Society (SWCS) in Ankeny, Iowa, stocks WEPP software, including the CD-ROM disk and a 500-page user's guide.

EUROSEM (European Soil Erosion Model). The EUROSEM program was developed by a team of scientists from ten European countries under the leadership of R. P. C. Morgan and R. J. Rickson (Morgan, 1994). It is a modular-structured, process-based, computerized model that incorporates terms of soil erodibility and surface roughness (surface depression storage) that change with time as a storm progresses. It integrates raindrop-splash and surface-flow effects as soil detachment and movement occur, predicting amounts of removal and sediment deposition. It will predict erosion for individual storms from fields and small watersheds, and, in addition, will evaluate a variety of conservation practices.

A documentation manual that explains the background of the EUROSEM model for potential users is available (Morgan et al., 1992). A user's guide has also been prepared that describes the input data that are needed, how to obtain them, and how to create the necessary input/output files to run the model (Morgan et al., 1991).

Future EUROSEM models are planned that will consider the development of surface armor and crusting during a storm and their effects on erosion. Subroutines will also incorporate the effects of snow melt and stoniness on the erosion process.

GUEST (Griffith University Erosion System Template). The GUEST program focuses on sediment concentration in runoff, beginning with the interaction of raindrops, soil, and any water or vegetation present on the soil surface (Misra and Rose, 1996). It considers the sediment-transporting capacity of flowing water and the shape of the landscape to determine where the water will flow faster and pick up more soil or slow down and deposit sediment.

LISEM (Limburg Soil Erosion Model). LISEM was developed in The Netherlands to consider overland flow for drainage basin planning (Jetten et al., 1996). It places an emphasis on microrelief and surface differences such as the presence of stones or a depression caused by a wheel track. The resulting calculations show that compaction often causes field roads covering 2 to 3% of the area in a drainage basin to account for 25 to 50% of the runoff (De Roo, 2000). Several alternative approaches for considering infiltration differences can be used with this program, depending on the type of input data available and the amount of detail desired.

6–5 THE WIND-EROSION PREDICTION EQUATION (WEQ)

Attempts to use research results to predict wind erosion were made soon after information on the mechanics of wind erosion and the factors that influence the process began to accumulate. Much of the early work on wind-erosion prediction was done by W. S. Chepil, N. P. Woodruff, and A. W. Zingg.

Climate was the first factor shown to be quantitatively related to wind erosion. In the U.S. Great Plains region, the amount of erosion was found to vary directly with the cube of the average March–April wind velocity and inversely with the previous year's rainfall (Zingg et al., 1952). Preliminary relationships between percentage of erodible aggregates, quantity of surface crop residue, soil ridge roughness, and amount of erosion taking place in a wind tunnel were developed as early as 1953. The original wind tunnel equation was modified as new information developed and new factors could be included. Woodruff and Siddoway (1965) published a prediction equation that had already undergone extensive field testing. It is still being refined at the USDA-ARS Northern Plains Area Wind Erosion Research Unit, Manhattan, Kansas, the USDA-ARS Southern Plains Area Wind Erosion Management Unit, Big Spring, Texas, and elsewhere.

6–5.1 Components of the WEQ

The wind-erosion prediction equation (WEQ) is

$$E = f(I', K', C', L', V)$$

where E = predicted soil loss, tons/ac-yr (mt/ha-yr)

I' = soil-erodibility factor, tons/ac-yr (mt/ha-yr)

K' = soil-ridge-roughness factor, dimensionless

C' = climatic factor, dimensionless

L' = width-of-field factor, ft (m)

V = vegetative-cover factor, dimensionless

The factor values in the WEQ often interact with each other. As a result, solutions to WEQ require the use of complex equations and complicated charts and nomographs.

Soil-Erodibility Factor (I'). Soil erodibility (I') is the potential annual soil loss from a wide, unsheltered, isolated field with a bare, smooth, noncrusted surface. Soil-erodibility values in the original equation were based on wind-tunnel data for relative

TABLE 6–12 SOIL-ERODIBILITY (*I*) VALUES IN TONS/AC-YR[a] FOR SOILS WITH VARIOUS PERCENTAGES OF NONERODIBLE CLODS (>0.84 mm diameter) AS DETERMINED BY STANDARD DRY SIEVING

Percentages	Percentages (units)									
(tens)	0	1	2	3	4	5	6	7	8	9
					tons/ac-yr					
0		310	250	220	195	180	170	160	150	140
10	134	131	128	125	121	117	113	109	106	102
20	98	95	92	90	88	86	83	81	79	76
30	74	72	71	69	67	65	63	62	60	58
40	56	54	52	51	50	48	47	45	43	41
50	38	36	33	31	29	27	26	24	23	22
60	21	20	19	18	17	16	16	15	14	13
70	12	11	10	8	7	6	4	3	3	2
80	2	-	-	-	-	-	-	-	-	-

[a]To convert tons/ac-yr to mt/ha-yr, multiply each value by 2.24.

Source: Reproduced from *Soil Science Society of America Proceedings,* Volume 29, p. 602–608, 1965 (Woodruff and Siddoway) by permission of the Soil Science Society of America.

erodibilities and on soil losses measured in the vicinity of Garden City, Kansas, during 1954 to 1956, an excessively dry and windy period (Chepil, 1960). The standard method of assessing the *I* value is based on the content of nonerodible clods (>0.84 mm diameter) in the top inch of soil as determined by means of a rotary sieve. These percentages can be converted into *I* values (soil loss rates) by means of Table 6–12, but this conversion must still be considered an approximation. Scientists at the Wind Erosion Research Management Unit, Big Spring, Texas, and elsewhere are using improved soil-collecting devices to reevaluate field soil losses (Fryrear et al., 1991).

I factor values are modified for knolly topography. See Table 5–1 in Chapter 5 or Figure 6–3 in this chapter for *Is* values. The *Is* value expresses the ratio in percent of soil eroded on a particular slope to a loss on level land. *Is* for a short 3% slope is 150% on the crest of the knoll and 130% on the upper windward slope.

Wind-erosion susceptibility depends partly on degree of surface crusting, but since crusts are transitory, they are seldom evaluated. Therefore, the soil-erodibility factor (I') usually has been considered to be the product of soil erodibility (I) and knoll-steepness factor (Is).

Soil-Ridge-Roughness Factor (K'). Surface roughness results from three elements: cloddiness of surface soil, I'; vegetative cover, V, to be discussed later; and ridges on the soil surface. The last is the element involved in the soil-ridge-roughness factor, K'. Precise evaluation of K' can be made only with a wind tunnel, but Skidmore (1983) developed Kr as an estimate of K' based on measured field roughness. Skidmore's Kr values are shown in Table 6–13. For example, if measured field ridge height is 6 in. (15 cm) and ridge spacing is 30 in. (75 cm) in a downwind direction, Kr is 0.5 (the cell where the 30-in. spacing row and the 6-in. height column intersect).

TABLE 6–13 SOIL RIDGE-ROUGHNESS FACTOR (*Kr*) VALUES

	Ridge height (in.)											
Ridge spacing (in.)	1	2	3	4	5	6	7	8	9	10	11	12
1	0.5	0.8										
2	0.5	0.6	0.8									
4	0.6	0.5	0.7	0.8								
6	0.7	0.5	0.6	0.8								
8	0.8	0.5	0.5	0.6	0.8							
10	0.8	0.6	0.5	0.6	0.8							
12	0.9	0.6	0.5	0.5	0.7	0.8						
14	0.9	0.6	0.5	0.5	0.6	0.8						
16	0.9	0.6	0.5	0.5	0.6	0.7	0.8					
18	0.9	0.7	0.5	0.5	0.5	0.6	0.8					
20	0.9	0.7	0.5	0.5	0.5	0.6	0.8					
22	0.9	0.7	0.6	0.5	0.5	0.6	0.7	0.8				
24	0.9	0.7	0.6	0.5	0.5	0.6	0.7	0.8				
26	0.9	0.8	0.6	0.5	0.5	0.5	0.6	0.8				
28	0.9	0.8	0.6	0.5	0.5	0.5	0.6	0.7	0.8			
30	0.9	0.8	0.6	0.5	0.5	0.5	0.6	0.7	0.8			
32	1.0	0.8	0.6	0.5	0.5	0.5	0.6	0.6	0.8			
34	1.0	0.8	0.6	0.5	0.5	0.5	0.5	0.6	0.7	0.8		
36	1.0	0.8	0.6	0.5	0.5	0.5	0.5	0.6	0.7	0.8		
38	1.0	0.8	0.6	0.6	0.5	0.5	0.5	0.6	0.7	0.8		
40	1.0	0.8	0.7	0.6	0.5	0.5	0.5	0.6	0.7	0.8		
42	1.0	0.9	0.7	0.6	0.5	0.5	0.5	0.6	0.6	0.7	0.8	
44	1.0	0.9	0.7	0.6	0.5	0.5	0.5	0.5	0.6	0.7	0.8	
46	1.0	0.9	0.7	0.6	0.5	0.5	0.5	0.5	0.6	0.7	0.8	
48	1.0	0.9	0.7	0.6	0.5	0.5	0.5	0.5	0.6	0.7	0.8	

Source: From the *Journal of Soil and Water Conservation,* Volume 38, p. 110–112, 1983 (Skidmore).

Climatic Factor (C'). The climatic factor includes both direct effects such as wind velocity and the indirect effects of precipitation and temperature that influence surface soil moisture and plant growth.

The rate of erosion by wind is related to the wind velocity cubed. In developing the WEQ, Chepil et al. (1962) studied the effects of mean annual and mean monthly wind velocities. They suggested that the mean annual velocity should be used where general propensity for wind erosion is the major concern.

The direct effect of *humidity* and *precipitation* on soil cohesiveness and resistance to erosion is evaluated in WEQ. The more frequently soils are moistened by rain, and the longer they remain moist after each storm, the more difficult they are to dislodge. Chepil et al. (1962) assessed soil moisture by the *P-E* Index method (Thornthwaite, 1931). (see Note 6–1)

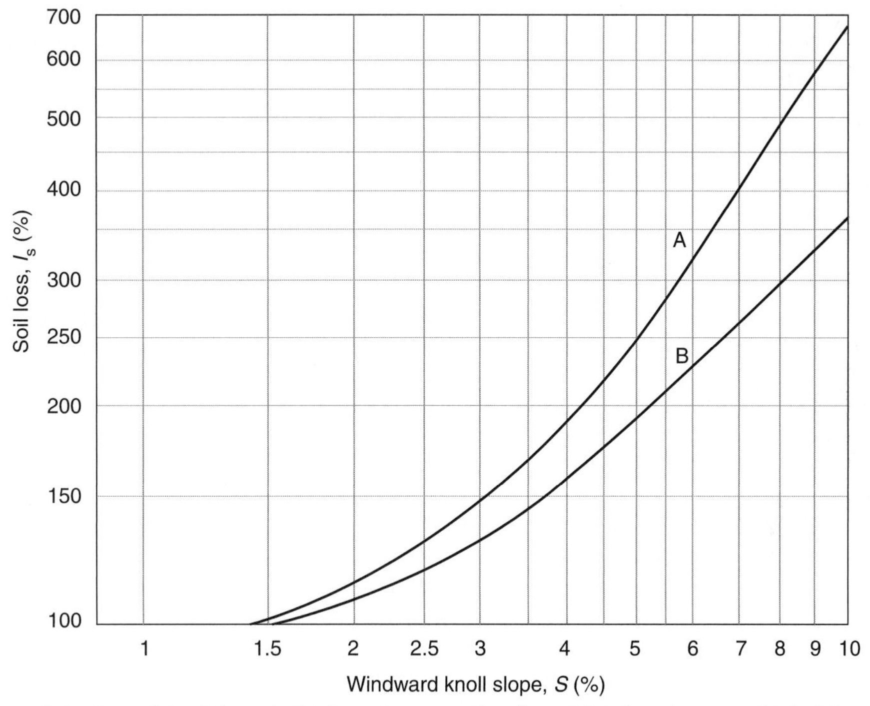

Figure 6–3 Potential soil loss, *Is,* (A) from the crest of knolls and (B) from the upper third of the windward side of slopes,< 500-ft (<150-m) long as percentages of *Is* on level land. (Reproduced from *Soil Science Society of America Proceedings,* Volume 29, p. 602–608, 1965 [Woodruff and Siddoway] by permission of the Soil Science Society of America.)

NOTE 6–1
THORNTHWAITE'S HUMIDITY FACTOR

Thornthwaite (1931) suggested the following complex rainfactor (*P-E* Index) to characterize atmospheric humidity.

where P = monthly precipitation, in.

T = mean monthly temperature, °F

i = months

$$P - E = 115 \sum_{i=1}^{12} \left(\frac{P}{T - 10} \right)_i^{10/9}$$

Thornthwaite used 28.4°F as a minimum mean monthly temperature. Omitting recorded temperature data for colder months made 18.4 the minimum $T - 10$ value. This modification becomes important in higher latitudes where mean temperatures are considerably below freezing several months each year, with consequent reductions in evaporation and with precipitation that falls nonerosively as snow.

P-E is very small in dry regions, and calculated C values would be excessively high there (as high as 1000 in some drier U.S. locations). To reduce these erroneously high values, Chepil used 0.5 in. (13 mm) as a minimum monthly precipitation figure.

The C value for a location is obtained by the use of the equation

$$C = u^3/(P\text{-}E)^2$$

where u = wind velocity and

P-E = Thornthwaite's P-E ratio

The value of the climatic factor (C') for any location is obtained by comparing the C value for that location to the C value for Garden City (2.9) on a percentage scale. Thus:

$$C' = \frac{u^3}{(P - E)^2} \times \frac{100}{2.9} = 34.8 \times \frac{u^3}{(P - E)^2}$$

Chepil et al. (1962) developed a generalized climatic erosiveness map for the western half of the United States. Lyles (1983) calculated the C' factor values for many additional locations in the western states and in Alaska. Figure 6–4 contains a summary of C' values. More detailed C' values can be obtained from the NRCS field offices.

Because values for C' in the WEQ are dimensionless, other methods of expressing moisture or humidity can be substituted for it. For example, the metric equivalent can be used with wind velocity in meters per second, precipitation in mm, temperature in degrees Celsius, and the constant recalculated to match these units.

Skidmore (1986) developed a physically based, wind-erosion climatic factor (CE) that can be used in place of C' to solve the WEQ. This method can be used to predict wind erosion from seasonal or individual events and it avoids the need to exclude low temperatures and low amounts of precipitation. CE is calculated from readily available climatic measurements—wind speed, precipitation, and temperature—combined with calculated solar and net radiation values.

$$CE = \rho \int_R^{\infty} \left[u^2 - \left(u_t^2 + \frac{\gamma'}{\rho \alpha^2} \right) \right]^{3/2} f(u)du$$

where ρ = air density

u = wind velocity at 33 ft (10 m)

u_t = threshold wind speed

γ' = soil cohesive resistance (proportional to relative surface soil water content, i.e. actual moisture content/moisture content at the permanent wilting point)

α = a constant (depending on the von Karman constant, the height at which wind speed is measured, and a surface roughness parameter)

$f(u)$ = wind speed probability function

$R = u_t^2 + \gamma' / \rho \alpha^2$

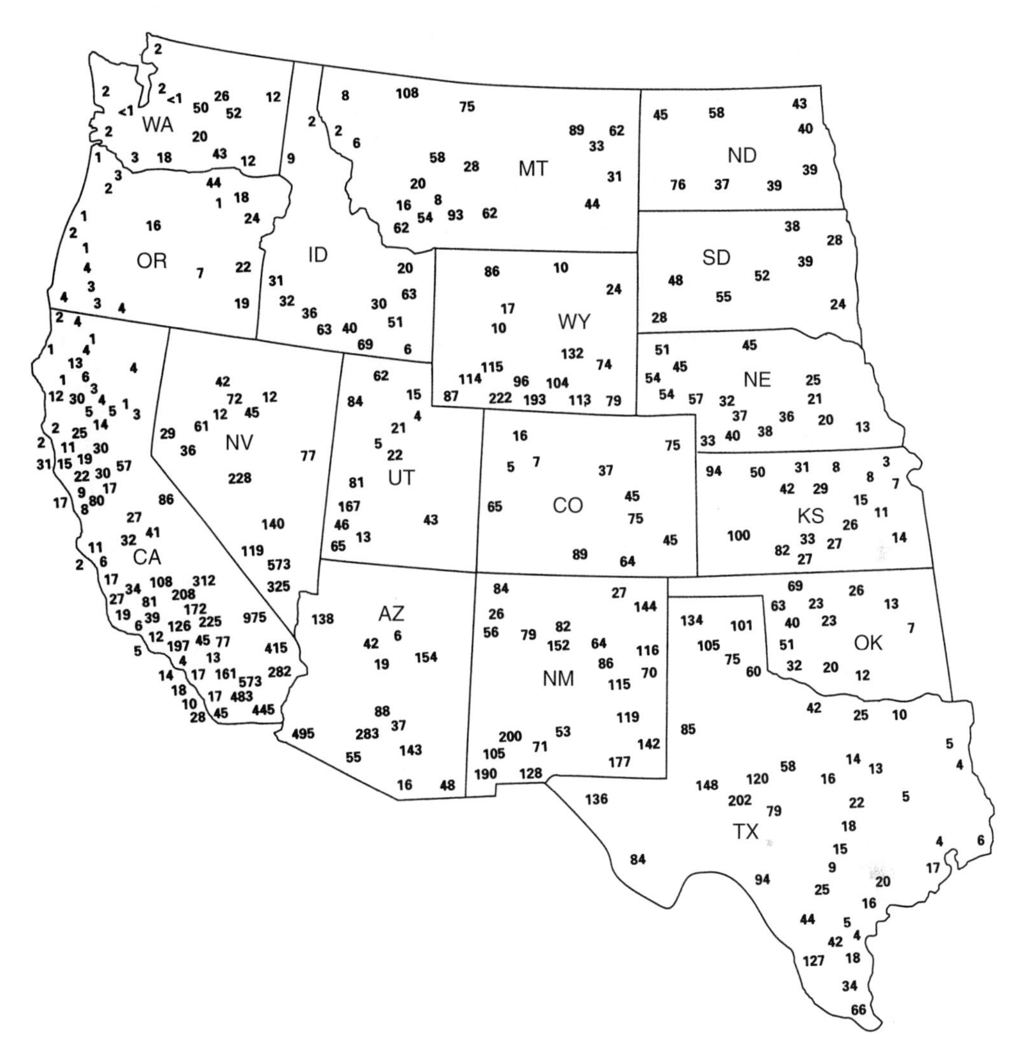

Figure 6–4 Wind-erosion climatic factor C' in percent of the value at Garden City, Kansas. (From the *Journal of Soil and Water Conservation,* Volume 38, p. 106–109, 1983 [Lyles].)

Width-of-Field Factor (L'). The width of a field (Df) is the unsheltered distance in feet in the downwind direction. Df would always be the same if all winds were from a prevailing direction. With variable winds, the unsheltered distance either increases or decreases when the wind changes direction.

The proportion of erosive forces that are parallel to, as opposed to forces perpendicular to, the prevailing wind direction is called the *preponderance factor.* A preponderance of 1.0 means that there is no prevailing wind direction; 2.0 means that there is a prevailing direction and twice as much erosive force is parallel to the prevailing direction as is perpendicular to it. Skidmore (1987) expanded and improved the procedure developed by Skidmore and Woodruff (1968) to use prevailing wind direction and preponderance to calculate wind-erosion direction

TABLE 6–14A WIND-EROSION DIRECTION FACTOR FOR RECTANGULAR FIELDS WITH LENGTH/WIDTH RATIO OF 2

	Angle of deviation (degrees)								
Preponderance	0.0	11.25	22.50	33.75	45.00	56.25	67.50	78.75	90.00
1.0	1.42	1.42	1.42	1.42	1.42	1.42	1.42	1.42	1.42
1.2	1.30	1.31	1.35	1.40	1.42	1.43	1.44	1.46	1.46
1.4	1.20	1.21	1.27	1.36	1.42	1.44	1.48	1.52	1.55
1.6	1.14	1.15	1.22	1.32	1.42	1.46	1.53	1.62	1.66
1.8	1.10	1.11	1.18	1.29	1.42	1.47	1.58	1.72	1.80
2.0	1.07	1.09	1.16	1.28	1.42	1.47	1.62	1.82	1.96
2.2	1.05	1.07	1.14	1.27	1.42	1.48	1.65	1.94	2.00
2.4	1.04	1.06	1.13	1.26	1.42	1.49	1.68	1.97	2.00
2.6	1.03	1.05	1.12	1.26	1.42	1.49	1.70	1.99	2.00
2.8	1.02	1.04	1.12	1.25	1.42	1.50	1.72	2.00	2.00
3.0	1.02	1.04	1.12	1.25	1.42	1.50	1.73	2.00	2.00
3.2	1.01	1.04	1.12	1.25	1.42	1.50	1.74	2.00	2.00
3.4	1.01	1.04	1.12	1.25	1.42	1.50	1.74	2.01	2.00
3.6	1.01	1.04	1.12	1.25	1.42	1.50	1.75	2.01	2.00
3.8	1.01	1.04	1.11	1.25	1.42	1.50	1.76	2.02	2.00
4.0	1.01	1.04	1.11	1.25	1.42	1.51	1.76	2.02	2.00

Source: Modified from *Soil Science Society of America Journal,* Volume 51, p. 198–202, 1987 (Skidmore) by permission of the Soil Science Society of America.

TABLE 6–14B WIND-EROSION DIRECTION FACTOR FOR RECTANGULAR FIELDS WITH LENGTH/WIDTH RATIO OF 4

	Angle of deviation (degrees)								
Preponderance	0.0	11.25	22.50	33.75	45.00	56.25	67.50	78.75	90.00
1.0	1.48	1.48	1.48	1.48	1.48	1.48	1.48	1.48	1.48
1.2	1.30	1.31	1.35	1.40	1.48	1.57	1.66	1.72	1.76
1.4	1.20	1.21	1.27	1.36	1.48	1.65	1.86	2.00	2.00
1.6	1.14	1.15	1.22	1.32	1.48	1.72	1.98	2.24	2.35
1.8	1.10	1.11	1.18	1.29	1.48	1.77	2.08	2.43	2.55
2.0	1.07	1.09	1.16	1.28	1.48	1.82	2.17	2.58	2.78
2.2	1.05	1.07	1.14	1.27	1.48	1.85	2.20	2.74	3.06
2.4	1.04	1.06	1.13	1.26	1.48	1.86	2.38	2.89	3.35
2.6	1.03	1.05	1.12	1.26	1.48	1.87	2.42	3.02	3.58
2.8	1.02	1.04	1.12	1.25	1.48	1.88	2.44	3.15	3.74
3.0	1.02	1.04	1.12	1.25	1.48	1.88	2.45	3.28	3.92
3.2	1.01	1.04	1.12	1.25	1.48	1.89	2.46	3.33	4.00
3.4	1.01	1.04	1.12	1.25	1.48	1.89	2.47	3.35	4.00
3.6	1.01	1.04	1.12	1.25	1.48	1.89	2.48	3.38	4.00
3.8	1.01	1.04	1.12	1.25	1.48	1.89	2.48	3.39	4.00
4.0	1.01	1.04	1.11	1.25	1.48	1.90	2.48	3.41	4.00

Source: Modified from *Soil Science Society of America Journal,* Volume 51, p. 198–202, 1987 (Skidmore) by permission of the Soil Science Society of America.

TABLE 6–14C WIND-EROSION DIRECTION FACTOR FOR RECTANGULAR FIELDS WITH LENGTH/WIDTH RATIO OF 10

	Angle of deviation (degrees)								
Preponderance	0.0	11.25	22.50	33.75	45.00	56.25	67.50	78.75	90.00
1.0	1.48	1.48	1.48	1.48	1.48	1.48	1.48	1.48	1.48
1.2	1.30	1.31	1.35	1.40	1.48	1.57	1.66	1.72	1.76
1.4	1.20	1.21	1.27	1.36	1.48	1.65	1.86	1.99	2.00
1.6	1.14	1.15	1.22	1.32	1.48	1.72	1.98	2.33	2.55
1.8	1.10	1.11	1.18	1.30	1.48	1.77	2.10	2.76	3.08
2.0	1.07	1.09	1.16	1.28	1.48	1.82	2.26	3.18	3.73
2.2	1.05	1.07	1.14	1.27	1.48	1.85	2.34	3.61	4.47
2.4	1.04	1.06	1.13	1.26	1.48	1.86	2.61	4.01	5.22
2.6	1.03	1.05	1.12	1.26	1.48	1.87	2.70	4.38	5.93
2.8	1.02	1.04	1.12	1.25	1.48	1.88	2.77	4.73	6.61
3.0	1.02	1.04	1.12	1.25	1.48	1.88	2.82	5.03	7.28
3.2	1.01	1.04	1.12	1.25	1.48	1.89	2.84	5.20	7.79
3.4	1.01	1.04	1.12	1.25	1.48	1.89	2.86	5.31	8.17
3.6	1.01	1.04	1.12	1.25	1.48	1.89	2.88	5.39	8.54
3.8	1.01	1.04	1.12	1.25	1.48	1.89	2.90	5.46	8.91
4.0	1.01	1.04	1.11	1.25	1.48	1.90	2.91	5.51	9.27

Source: Modified from *Soil Science Society of America Journal,* Volume 51, p. 198–202, 1987 (Skidmore) by permission of the Soil Science Society of America.

factors. These direction factors serve as multipliers for use on actual field lengths to give the median travel distance of the wind across a particular field.

Direction of prevailing winds and preponderance values can be obtained from Skidmore (1987) or from NRCS field offices. Wind-erosion direction factors for a variety of conditions are found in Tables 6–14A-C. The median travel distance (*MTD*) across a field is the dimension of the side closest to the wind direction times the wind erosion direction factor (wdf).

$$MTD = \text{field dimension (ft)} \times \text{wdf}$$

Any distance in a field that is sheltered by a wind barrier (B) must be subtracted from the *MTD.* The distance sheltered by field hedges, tree shelterbelts, and similar barriers is taken to be 10 times the height of the barrier ($10B$). Thus the width of field factor (L') is the difference between (*MTD*) and sheltered distance ($10B$).

$$L' = MTD - 10B$$

Calculation of the effect of L' on soil loss is complex; it depends on the amount of soil being carried in the wind. The nomogram in Figure 6–5 is used to assess the effect of L' on erosion loss by incorporating L', along with soil-erodibility/soil-ridge-roughness erosion estimates ($E_2 = I' \times K'$), and the soil-erodibility/soil-ridge-roughness/climate erosion estimates ($E_3 = I' \times K' \times C'$) in the prediction equation.

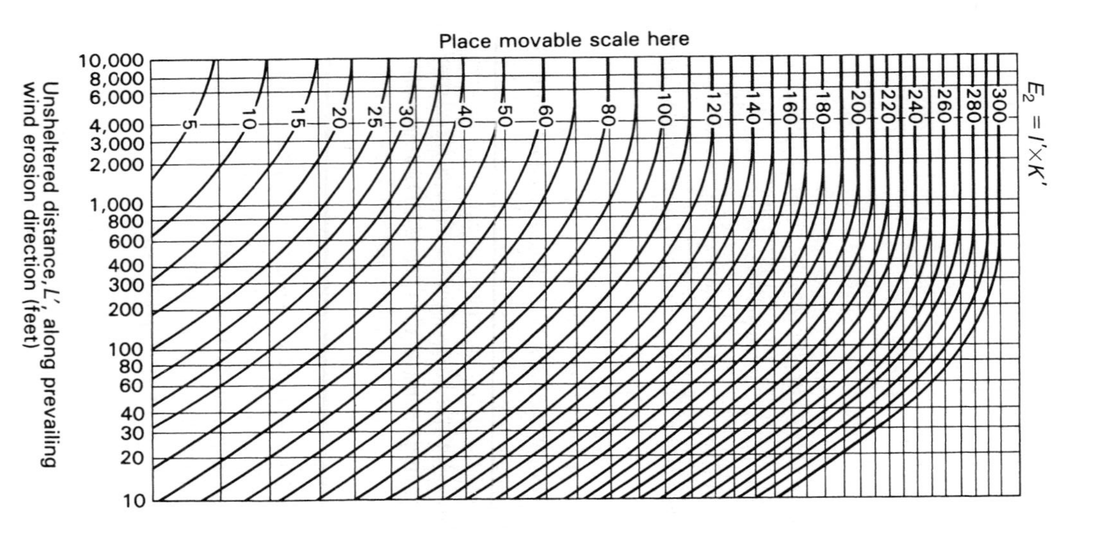

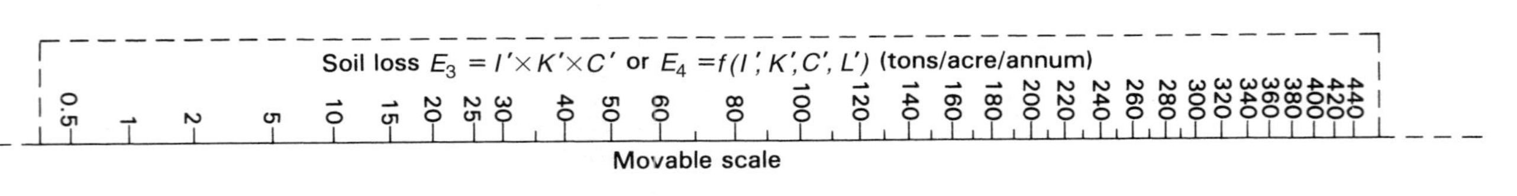

Figure 6–5 Nomogram to determine soil loss, $E_4 = f(I', K', C', L')$, from soil loss $E_2 = I' \times K'$ and $E_3 = I' \times K' \times C'$ and the unsheltered distance L' across the field. In use, a copy of the movable scale is placed along the left side of the graph so that E_3 on the scale is aligned with E_2 on the graph. A line parallel to the curved lines is then followed to its intersection with the unsheltered distance line. The value of E_4 is located by following a horizontal line from the intersection back to the movable scale. For example, an E_3 value of 150 and an E_2 value of 120 combined with an unsheltered distance of 600 ft give an E_4 value of 130. (From *Soil Science Society of America Proceedings,* Volume 29, p. 602–608, 1965 [Woodruff and Siddoway], by permission of the Soil Science Society of America.)

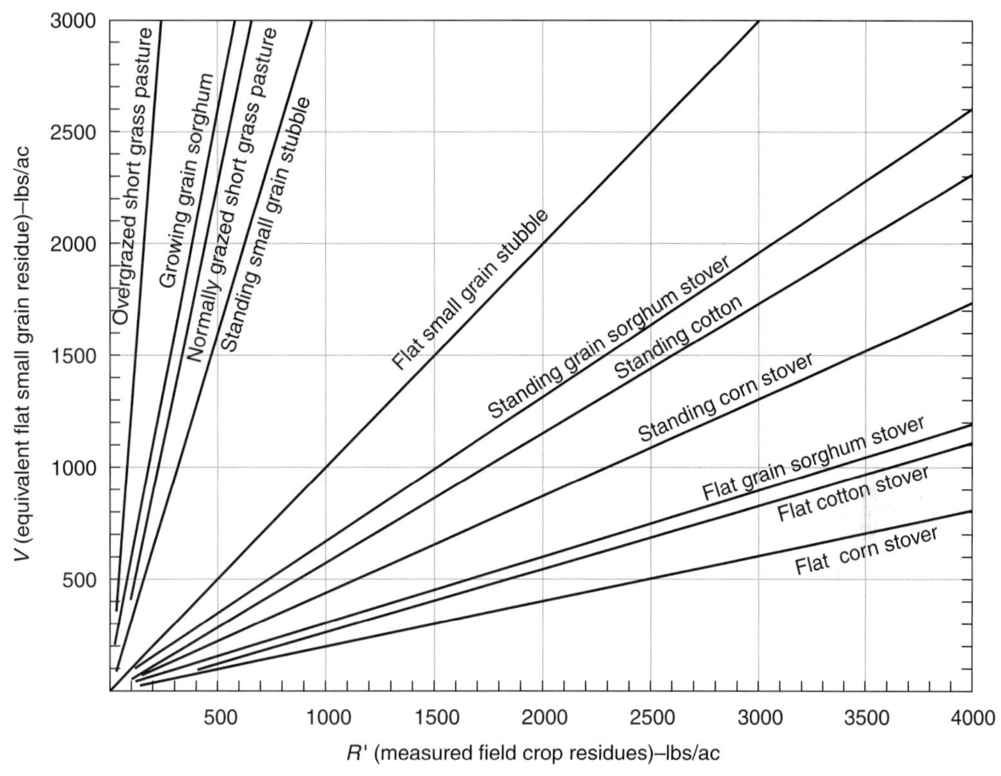

Figure 6–6 Curves to convert weights of living or dead crop material (R') into equivalent amounts of flattened wheat straw (V). (Modified from *Agronomy Journal,* Volume 77, p. 703–707, 1985 [Armbrust and Lyles] by permission of the American Society of Agronomy; *Journal of Range Management,* Volume 33, p. 143–146, 1980 [Lyles and Allison]; and *Transactions of American Society of Agricultural Engineers,* Volume 24, p. 405–408, 1981 [Lyles and Allison].)

Vegetative-Cover Factor (V). The protection offered by vegetation depends on how much dry matter it contains, its texture, and whether it is living or dead, standing or flat. The original work on vegetative protection involved flattened wheat straw. This condition is now the standard for determining V. For use in the equation, dry weights of growing crops or crop residues (R') are converted to equivalent quantities of flat wheat straw (V) by using Figure 6–6.

The effectiveness of V in reducing soil loss also depends on the level of erosion. Figure 6–7 provides the means for relating erosion to V.

The best way to estimate the amount of surface vegetation on a field is to hand pick, dry, and weigh the living plants or surface residue from a unit area. Soil must be removed from plant material before weighing. Results are translated into lb/ac or kg/ha (Whitfield et al., 1962).

Shortcuts to the pick, clean, and weigh method have been suggested. Hartwig and Laflen (1978) describe a method in which the total widths of the various pieces of residue touching one face of a meter stick are recorded. Total length of contact per set

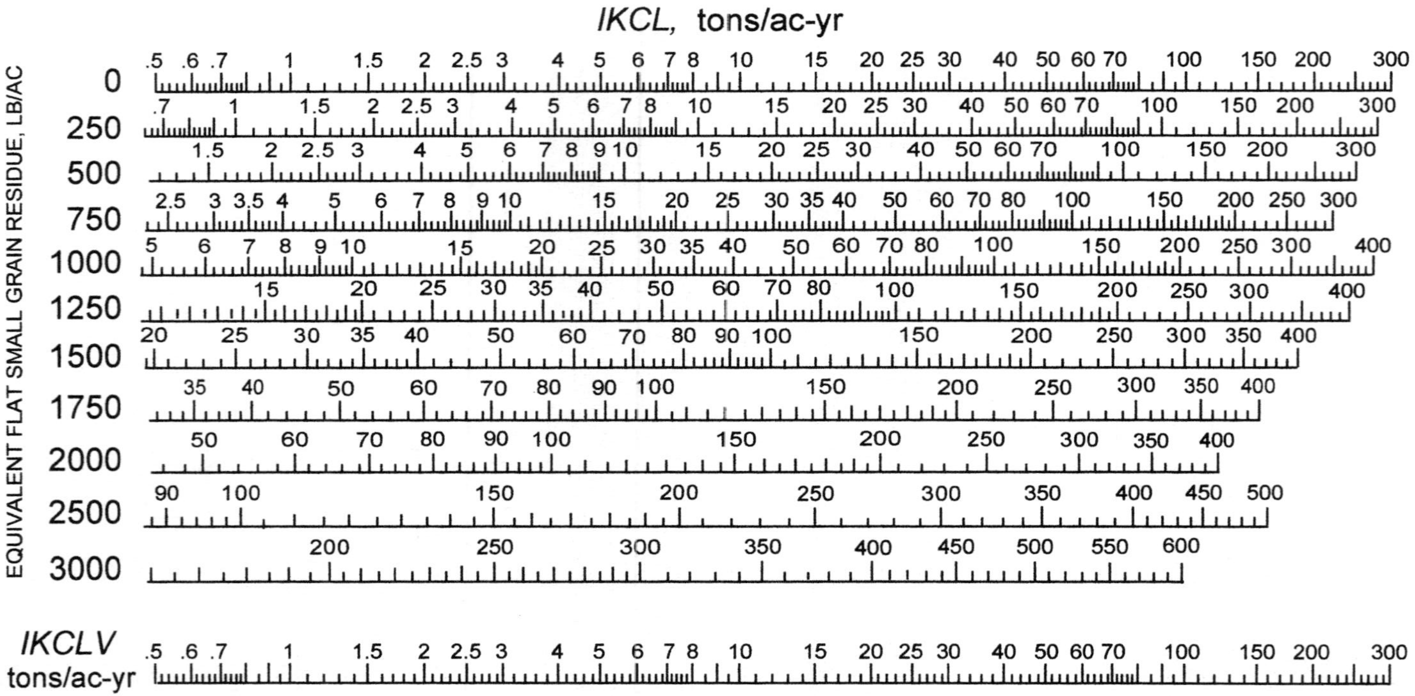

IKCL, tons/ac-yr

EQUIVALENT FLAT SMALL GRAIN RESIDUE, LB/AC

0 — .5 .6 .7 1 1.5 2 2.5 3 4 5 6 7 8 10 15 20 25 30 40 50 60 70 100 150 200 300
250 — .7 1 1.5 2 2.5 3 4 5 6 7 8 10 15 20 25 30 40 50 60 70 100 150 200 300
500 — 1.5 2 2.5 3 4 5 6 7 8 9 10 15 20 25 30 40 50 60 70 100 150 200 300
750 — 2.5 3 3.5 4 5 6 7 8 9 10 15 20 25 30 35 40 50 60 70 80 100 150 200 250 300
1000 — 5 6 7 8 9 10 15 20 25 30 35 40 50 60 70 80 100 150 200 250 300 400
1250 — 15 20 25 30 35 40 50 60 70 80 100 150 200 250 300 400
1500 — 20 25 30 35 40 50 60 70 80 90 100 150 200 250 300 350 400
1750 — 35 40 50 60 70 80 90 100 150 200 250 300 350 400
2000 — 50 60 70 80 90 100 150 200 250 300 350 400
2500 — 90 100 150 200 250 300 350 400 450 500
3000 — 200 250 300 350 400 450 500 550 600

IKCLV tons/ac-yr — .5 .6 .7 1 1.5 2 2.5 3 4 5 6 7 8 10 15 20 25 30 40 50 60 70 100 150 200 300

Figure 6–7 Revised wind erosion calculator scale for finding $E = f(I', K', C', L', V)$ as a function of E_4 and designated values of equivalent flat small grain residue (V) in lb/ac. The E value is determined by entering the graph on the left at the appropriate level of V and following it horizontally across to the value of E_4. A vertical line through the E_4 value will point to the value of E on the bottom scale. (From the *Journal of Soil and Water Conservation,* Volume 38, p. 110–112, 1983 [Skidmore].)

is the percentage of residue cover. Sloneker and Moldenhauer (1977) describe a method in which 50 beads are fastened at equal intervals along a string. The beaded string is stretched across the area to be evaluated. Each bead that touches a "significant" piece of residue is counted as 2% soil cover. Morrison and his associates studied a variety of equipment for estimating ground cover by crops and crop residues (Morrison et al., 1997). They tested five different kinds of measuring lines and four different wheels. Measurement precision and operation time were similar for all devices, except for the simple 50-ft measuring tape that took 50% longer than the other devices. Variability among operators was greater than differences among measuring devices. Count variability increased with all devices as the cover increased. The first preference of the operators was tied between a wheel and a random-bead spacing line.

Sloneker and Moldenhauer (1977) determined that the fraction of surface cover can be translated into residue weight with the following equations:

$$X = -5{,}278 \log (1 - Y) \quad \text{for corn}$$
$$X = -15{,}519 \log (1 - Y) \quad \text{for small grain}$$
$$X = -27{,}813 \log (1 - Y) \quad \text{for soybeans}$$

where X = crop residue, lb/ac

Y = cover as a decimal fraction

6–5.2 Predicting Soil Losses Caused by Wind Erosion—A Specific Example

A 2640-ft by 5280-ft (1/2-mile by 1-mile or 800-m by 1600-m) field with long axis east-west on Ortello fine sandy loam (Wind Erodibility Group 3) near Dodge City, Kansas, will serve as an example. The surface soil (WEG 3) contains 25% nonerodible clods. Several knolls with short 3% slopes are in the field. A crop of grain sorghum was produced in 30-in. (75-cm) rows; 1150 lb/ac (1300 kg/ha) of tall, standing stubble remains in March, when the strongest winds are expected. The average ridge height is 6 in. (15 cm).

The predicted soil loss from the knolls is calculated as follows:

STEP 1: $E_1 = I' = I \times Is$

I for 25% nonerodible clods is 86 tons/ac-yr (190 mt/ha-yr) (Table 6–12). The value for Is on the knolls of a 3% slope is 150% (Figure 6–3).

$$E_1 = 86 \times \frac{150}{100} = 129 \text{ tons/ac-yr}$$

STEP 2: $E_2 = E_1 \times K'$

The soil-ridge-roughness condition, (Kr), for $HR = 6$ and $IR = 30$ is 0.50 (Table 6–13).

$$E_2 = 129 \times 0.50 = 65 \text{ tons/ac-yr}$$

STEP 3: $E_3 = E_2 \times C'$

The C' value for Ford County is about 80 (Figure 6–4).

$$E_3 = 65 \times \frac{80}{100} = 52 \text{ tons/ac-yr}$$

STEP 4: $E_4 = E_3 \times f(L')$

The prevailing wind for March is due north (deviation = 0°). Preponderance is 2.4 (Skidmore and Woodruff, 1968). The wind-erosion direction factor from Table 6–14A is 1.04.

$$MTD = 2640 \times 1.04 = 2746 \text{ ft (837 m)}$$

Because no wind barrier was described, $L' = 2746$ ft. Following the directions in the caption of Figure 6–5 for values of $E_2 = 65$, $E_3 = 52$, and $L' = 2746$ ft shows E_4 to be 50 tons/ac-yr. Note that a shorter value of L' would have increased the difference between the values for E_3 and E_4.

STEP 5: $E = E_4 \times f(V)$

There is no growing crop, so R' is 1150 lb/ac (the sorghum stubble). V from Figure 6–6 is about 750 lb/ac for $R' = 1150$ lb/ac of standing sorghum stubble. The value for E, read from Figure 6–7 using $E_4 = 50$ tons/ac-yr and $V = 750$ lb/ac, is 23 tons/ac-yr.

The tolerable loss for Ortello fine sandy loam is 5 tons/ac-yr. How can the loss be reduced to the tolerable limit? About 750 lb/ac (850 kg/ha) of sorghum grain would be produced along with 1150 lb of stover. This is a low yield, even for a dry area. If grain yield could be raised to 1300 lb/ac or above regularly, stover would increase to about 1950 lb/ac ($V = 1250$ lb/ac), and the predicted soil loss would be reduced to about 6.8 tons/ac. Another possibility would be to split the field in three parts on an east-west axis and use one part for each year in a sorghum-fallow-wheat rotation. This would cut field width to 880 ft (one-third of 2640) and reduce E_4 to 40 tons/ac-yr. Combining the higher yield goal already suggested with the narrower field width would reduce the final E value to about 4.8 tons/ac-yr. Conservation tillage (Chapter 10) would be needed to ensure that sufficient crop residues were produced and maintained on the soil for protection against wind erosion every year.

6–5.3 Field Use of the Wind-Erosion Prediction Equation

The WEQ was developed primarily to help conservation technicians assess wind-erosion conditions and evaluate the changes in management needed to reduce an excessive erosion hazard. Solving the equation is time-consuming. The need to examine several options requires even more time. As a result, SCS specialists and others modified certain factor values and proposed calculation shortcuts.

Establishment of Wind Erodibility Groups (WEGs) was a real timesaver. It allows I values to be assessed without collecting, drying, and dry sieving samples. The nine Erodibility Groups are described in Table 6–15.

TABLE 6–15 WIND-ERODIBILITY GROUP (WEG) CHARACTERISTICS

WEG	Predominant soil textural classes of surface layer	Dry soil aggregates >0.84 mm (%)	Wind erodibility index, *I* (tons/ac-yr)[a]
1	Very fine sands; fine sands, and coarse sands	1	310
		2	250
		3	220
		5	180
		7	160
2	Loamy very fine sands; loamy fine sands, loamy sands, and loamy coarse sands; sapric organic soil materials	10	134
3	Very fine sandy loams, fine sandy loams, sandy loams, and coarse sandy loams	25	86
4	Clays and silty clays; noncalcareous clay loams and noncalcareous silty clay loams with > 35% clay content	25	86
4L	Calcareous loams, calcareous silt loams, calcareous clay loams, and calcareous silty clay loams	25	86
5	Noncalcareous loams and noncalcareous silt loams with < 20% clay content; sandy clay loams and sandy clays; hemic organic soil materials	40	56
6	Noncalcareous loams and noncalcareous silt loams with > 20% clay content; noncalcareous clay loams with < 35% clay content	45	48
7	Silts; noncalcareous silty clay loams with < 35% clay content; fibric organic soil materials	50	38
8	Soils not susceptible to wind	> 80	0

[a]To convert *I* values to mt/ha-yr, multiply each by 2.24.

Source: USDA Natural Resources Conservation Service.

Chepil, Zingg, and other staff in the Wind-Erosion Research Unit collected and published a series of photographs that show the surface-residue and ridge-roughness conditions of a range of field situations (Chepil and Woodruff, 1959). Each photo is labeled for residue weight and ridge roughness. These permit comparisons of field conditions with the photos, and quick evaluations of residue cover and ridge roughness can be made. This replaces residue-sample collection, preparation, and weighing, and ridge height and spacing measurements, thus saving time and labor.

The development of the Fortran IV computer program WEROS (Fisher and Skidmore, 1970) was a big timesaver. A version developed by Skidmore in 1975 was used by SCS central staff to prepare tables of erosion prediction values for use by their conservation field officers. The officers can select values from these tables rather than calculate them. They still need to determine WEG, ridge roughness, field width, and crop residue present, but the tables make most of the usual calculations unnecessary and permit an analysis of the present situation and several alternatives to be made in a matter of minutes.

The wind-erosion equation slide rule (Skidmore, 1983) was another helpful innovation. This instrument also speeded the process of evaluating current conditions and alternatives.

More recently, upgrades of WEQ computer software have been developed. These have increased the availability and adaptation of this technology to conservation officers and others. Cole et al. (1983) added a wind-erosion prediction capability to the Erosion-Productivity Impact Calculator (EPIC) (Williams et al., 1983). It predicted soil losses based on periodic simulation. Skidmore and Williams (1991) upgraded Cole et al.'s part of the EPIC model and adapted it for use on a personal computer. Two new computer prediction models, RWEQ and WEPS, are discussed in Sections 6–7 and 6–8.

6–5.4 Predicting Wind Erosion for Crop Periods

I', K', C', and V factor values are annual figures in the WEQ, but these values actually change with time. Bondy et al. (1980) studied changes in wind energy values over the year. They determined the portion of the wind energy that occurs each month and compiled a table that shows the cumulative monthly percentage of average annual wind energy for 76 weather stations in the 10 Great Plains states. Lyles (1983) developed similar information for the seven western states and Alaska. This was done so that estimates of erosion by periods could be made using the WEQ.

This method had several advantages. Changes with time in I', K', and V could be cataloged, along with changes in C', so times during the crop year that show undue susceptibility to erosion could be pinpointed. If crop periods were employed in the WEQ, crop and residue values for these periods could be used in both the water- and wind-erosion equations. Nearly all the computer programs mentioned in Sections 6–4.2, 6–7, and 6–8 analyze the erosion picture daily or over relatively short time periods.

There are also disadvantages. Calculations for the crop-period method are more complex and take much longer than those for the original method. There are few experimental data to show specifically how the various factors, other than climate, change over time.

6–6 EXPANDED USE OF THE WIND-EROSION PREDICTION EQUATION

The WEQ was developed for use in the semiarid Great Plains region. Soil, climate, and other data from this region were used to develop a practical and effective prediction tool. Scientists and technicians in other regions of the United States immediately became interested in the equation and started to use it.

6–6.1 Use on Cultivated Soil in Humid Regions in the United States

The WEQ has been used in many areas outside the Great Plains. The Wind Erosion Research Unit's portable wind tunnel was taken to some areas to check soils and results; in other areas, the equation was used either without modification or with modifications based only on judgment of field situations.

A wind tunnel study in northwestern Ohio showed soils there were more erodible than Great Plains soils with the same nonerodible clod composition. A new table relating the percentage of nonerodible clods to soil erodibility was set up for that area.

Carreker (1966) studied sandy soils of the coastal plains of southeastern United States and found a relationship between the percentage of coarse and very coarse sand (X) and soil erodibility.

$$I = 174 - 4.64X + 0.03X^2$$

Hayes (1965) compared I values based on dry-sieving results and on the WEG. He found that WEG values were not sufficiently accurate for the very sandy soils in New Jersey. WEG 1 was subdivided subsequently into five subgroups. Hayes used the calculations method for determining Kr. He also pointed out the need to reduce wind-erosion losses to less than 5 tons/ac-yr to avoid damage to some crops.

In all of these studies, either satisfactory estimates of the climatic factor could be obtained from research publications, or sufficient local data were available to evaluate climatic erosiveness. Reasonable estimates of erosion losses were generally obtained.

6–6.2 Use on Rangeland in Western United States

The Bureau of Land Management, other agencies, and individuals are concerned about wind erosion on rangelands and pasturelands. Values for V can be obtained for these assessments by using relationships between range species and flattened wheat straw established by Hagen and Lyles (1988) and Lyles and Allison (1980). Values for other factors are obtained in the normal way.

6–6.3 Use of the WEQ to Predict Air Pollution

Suspended dust is a serious air pollutant. It poses a threat to highway and air traffic and to human health. The WEQ evaluates soil movement of all kinds—surface creep, saltation, and suspension. Suspended dust constitutes a very small fraction of the total amount of soil moved by wind, but it is the important air pollutant.

Efforts have been made to modify WEQ to assess the amount of dust in the air under a variety of natural conditions. Wilson (1975) used I' values based on the WEQ to predict dust emissions in New Mexico. He multiplied the E values he obtained by 0.003 to change eroding soil to suspended dust. The equation was useful in this survey, but great care must be exercised in using WEQ this way; results obtained must be viewed with caution.

6–6.4 Use of the Equation in Other Countries and Continents

Problems arise when the WEQ is used in areas very different from the Kansas area where it was developed. The relationship between percentage of nonerodible clods or WEGs and I is not universally constant, but the initial assumption that must be made in a new area is that the proposed relationship will hold. Values for Kr in new areas can be obtained by measurement, using Table 6–13.

The equation's method of evaluating erosiveness of climate should give reasonable values in other areas with similar climatic conditions. A problem arises in very dry areas because the P-E value used to calculate C' becomes very small and makes C' very large. Using Budyko's "Dryness Ratio" might solve this problem (Skidmore, 1986). Lack of long-term weather records from which C' values can be obtained is likely to be another major problem. Climatic zones probably can be outlined and approximate relationships assigned on the basis of general knowledge. The major limitation resulting from a lack of wind and other weather records is the problem of relating local climatic erosiveness to that of the WEQ standard reference at Garden City, Kansas.

Wind records are also needed to determine L values. Where wind records are not sufficient to develop statistically valid prevailing directions and preponderances, the general direction of the prevailing winds during the major wind-erosion periods should be estimated at least as close as the nearest compass point (22.5°). Maximum travel distance (WL) along the prevailing wind direction can be estimated from field width (FW) and field length (FL) and the declination angle (A) between the prevailing wind direction and side FW, by the method suggested by Cole et al. (1983).

$$WL = FW \sec A \quad \text{where } FW \sec A \leq (FW^2 + FL^2)^{0.5}$$

or

$$WL = FL \csc A \quad \text{where } FW \sec A > (FW^2 + FL^2)^{0.5}$$

It should be recognized that WL is only the same as the median travel distance (MTD, used to calculate L' in Section 6–5.1) of wind along the prevailing direction for declination angles of 0° and 90° to FW. MTD will decrease, slowly at first, then more rapidly as the declination angle separates from 0 or 90° until it is only half WL where WL equals $(FW^2 + FL^2)^{0.5}$, i.e., along the diagonal of the field. The sheltering effect of barriers should be the same worldwide.

Values for R' can be accurately obtained by picking, drying, and weighing the residues or growing crop, or they can be estimated by photo interpretation or by one of the line-transect methods. Surface residue should be translated into flat wheat straw using the information in Figure 6–6. Unevaluated material will have to be translated to standard residue on the basis of its observed effectiveness in relation to small-grain or sorghum residue.

Predictions employing less precise techniques should be useful in assessing potential trouble and recommending needed control methods. Lack of some information required for accurate analyses is no excuse for disregarding the prediction equation altogether.

6–6.5 Upgrading the Wind-Erosion Equation

WEQ held a preeminent position among systems for predicting soil losses from wind action from the late 1960s to the early- and mid-1990s, but it is not a perfect prediction tool. The WEQ is empirical by design and rigid in makeup. New information generally cannot be imbedded in the equation to upgrade its capabilities.

The equation is based primarily on the results of laboratory and wind tunnel studies. Few soil loss measurements were made in the field due to lack of suitable and reliable soil-catching equipment. Early estimates of erosion losses were made by visual observations, measurements of depth of soil removed, or depth of soil deposited in drifts in fence rows, around buildings, and so forth. When the equation was first released, it was the hope, rather than the assured promise, that estimates of field losses would be accurate. Only recently has extensive field research been conducted to measure soil losses accurately under a variety of conditions (Fryrear, 1995).

WEQ assumes that the field situation downwind is homogenous—one I' value (only one soil type or one WEG), one K' value (similar ridging throughout), and one V value (uniform vegetative cover over the whole area). It assumes that field conditions are static—that soil erodibility, ridges, and vegetation, living or dead, do not change over the year in nature or amount from the values selected.

The climatic factor (C'), based as it is on the climate of Garden City, Kansas, so circumscribes the area where the equation can be used successfully that it is difficult to apply it accurately to areas outside of the U.S. Great Plains. Use of Skidmore's climatic factor (CE) in the WEQ would have helped to overcome the provincial nature of the equation (Skidmore, 1986).

In higher latitudes, the equation does not allow for periods when soils are rendered nonerodible by freezing or by a protective snow cover.

There is no factor in the equation that takes into account random surface roughness, caused by whatever means, as opposed to roughness in parallel ridges produced by planting or cultivating equipment.

In spite of these deficiencies, however, the overriding problem of the WEQ is that it just does not account well enough for the complex interactions between the variables that control wind erosion. The WEQ represents a technology that cannot readily be adapted to new conditions in the field or to climates that are much different from that of the Great Plains region of the United States. New knowledge, developed from research, and the increased power and availability of personal computers were seen as avenues for the development of new, flexible, process-based erosion equations.

In 1986, a team of research scientists, mainly from USDA agencies but with representatives from the Environmental Protection Agency and the Bureau of Land Management, was appointed to take the lead in developing a new Wind Erosion Prediction System (WEPS) as a replacement for the WEQ. It was to be a process-based, computerized technology that would reflect the dynamic changes that take place in many of the variables over time. It was anticipated that the new system would be ready for deployment by the early- to mid-1990s.

As an intermediate step between the technologies in WEQ and WEPS, a few agricultural scientists and engineers were requested to incorporate new technology into a Revised Wind Erosion Equation (RWEQ). The new equation was to use state-of-the-art computer, analytical, and programming techniques. Predicted soil losses from RWEQ are verified by the results of many new field measurements of soil movement in wind erosion (Fryrear et al., 2000).

6–7 REVISED WIND EROSION EQUATION (RWEQ)

RWEQ uses data banks for the climatic variables—wind velocity, rainfall, temperature, and solar radiation, and for the surface-soil properties—soil erodibility, soil-oriented roughness, and soil random roughness. Mathematical equations have been designed to estimate the extent and strength of soil crust and to predict soil roughness parallel and perpendicular to the wind. They are also used to predict the rates of soil-roughness decay and to assess rates of plant-residue decomposition. The five factors—wind potential, soil-erodible fraction, soil crust, soil-roughness criteria, and a combined residue factor—together with the field-length factor, are then combined to generate soil loss predictions for periods of 15 days or less (Fryrear et al., 2000).

The developmental RWEQ is being tested against measured field losses at the present time and is showing promise. It should be able to estimate soil erosion for a wide variety of soil, climate, cropping, and tillage conditions and systems. RWEQ is designed for use on personal computers, but it will also be released in a paper model with tables and figures.

An effort is also being made to combine RWEG with a geographic information system (GIS) to predict wind erosion on a regional basis. Zobeck et al. (2000) used this approach in two counties in Texas and concluded that this approach is feasible.

6–8 THE WIND EROSION PREDICTION SYSTEM (WEPS)

The group of scientists that was assigned responsibility for developing a new process-based Wind Erosion Prediction System (WEPS) to replace the WEQ decided on a two-step strategy. First, they planned to develop a wind-erosion research model (WERM) in Fortran 77. It would be a daily-simulated, process-based model, with seven submodels (hydrology, management, soil, crop, decomposition, erosion, and weather) and four databases (climate, soils, management, and crop/decomposition). It would serve as a reference standard for wind-erosion predictions. In the second stage, the development of the WEPS model, the WERM submodels would be reorganized to increase computation speed, databases would be expanded, and a user-friendly input/output section would be added (Hagen, 1991).

The WERM model was developed in 1991 as planned. The WEPS model also was completed and has been released for testing. It is still undergoing improvement as new ideas and new research results are incorporated.

WEPS is a process-based computer model that simulates weather and field conditions. It uses a *Hydrology* submodel to predict water contents, including surface soil drying, along with other factors that affect the soil's susceptibility to wind erosion and the climatic processes that cause wind erosion. Its *Erosion* submodel is activated whenever wind speed exceeds its threshold velocity. It then simulates soil erosion and deposition on a sub-hourly basis (Hagen et al., 1999). It considers loose material entrained by the wind and material abraded from clods and crusts as separate sources and simulates them over time with information from the databases. It is modular in structure, so it can be updated and improved from time to time as new material becomes available and as model use shows the need for change.

Results of simulation are not biased by the climate of any particular region; rather climatic data from any location can be utilized in the prediction process. Snow cover and frozen soils are taken into account, where they occur, in assessing the erodibility of the soil over the year. The erosion condition can be simulated for the duration of a single storm, for several storms, for a season, or for a whole year.

The simulation area in WEPS is a single field or a few adjacent fields. Nonhomogenous sites are simulated by breaking them down into a number of homogenous subsites, based on different soils, tillage, cover, or other conditions affecting the erodibility of the surface soil.

WEPS estimates soil movement, plant abrasion damage, and emissions of PM-10 material (dust), if and when wind speeds exceed the threshold velocity. Because of its modular nature, new features can be incorporated that will make it possible to predict such things as long-term soil productivity changes, physical damage to crops and their cost, the loading of lakes and streams with wind-carried sediments, and the reduction of visibility caused by air-borne dust in the vicinities of airports and on highways. No doubt it will be used to calculate on-site and off-site economic costs of wind erosion as well (Wagner, 1996).

SUMMARY

Erosion cannot be prevented, but it must be limited to tolerable rates. Methods for predicting erosion rates are needed for two main reasons: to ascertain the erosion hazard with present management, and to evaluate alternative methods of crop and soil management. Erosion-prediction equations have been developed for estimating soil losses by both water and wind.

The water-erosion prediction equation (USLE) was designed to predict soil movement by sheet and rill erosion in cultivated fields. It is

$$A = R \times K \times LS \times C \times P$$

where A is the expected average annual rate of erosion in tons/ac-yr based on factors for rainfall erosivity (R), soil erodibility (K), slope length and steepness (LS), cropping management (C), and special erosion-control practices (P). Techniques were developed to calculate values for each factor from information obtained by field inspection and from tables, charts, and equations. The equation became popular immediately and was widely used by Soil Conservation Service field workers to develop effective alternative management practices for erosion control. It was also used by research workers and by individuals in many other agencies and organizations. Many computer programs were developed to speed up and broaden the use of USLE.

Over time, it became apparent that the original equation needed modification. It was first modified, and later revised (Revised Universal Soil Loss Equation—RUSLE) to meet new and expanded needs. Other soil loss equations have been developed, but the USLE/RUSLE has been by far the most used technology. RUSLE can be used for estimating soil losses from pasture, range, and forest land, as well as from cropland.

A new prediction model has been developed (WEPP). It is a process-based (rather than empirical), daily-simulation model. It can be used to estimate erosion losses in ephemeral gullies, and in terrace channels and grassed waterways in addition to losses from sheet and rill erosion. It is expected to be more widely useful than even the USLE/RUSLE models have been.

The wind-erosion prediction equation (WEQ), designed to evaluate the erosion hazard on cultivated lands in the Great Plains region of the United States, is now widely used by conservationists and others in many areas. It is

$$E = f(I', K', C', L', V)$$

where E is the expected average annual rate of erosion in tons/ac-yr as a function of soil erodibility (I'), soil-ridge-roughness (K'), climatic erosiveness (C'), field length (L'), and vegetative cover (V). Techniques are provided for calculating values for each factor using information obtained by field examination along with tables, graphs, and equations.

The WEQ has been used widely by SCS (now NRCS) field workers to determine the current erosion hazard and to evaluate the effectiveness of alternative management and erosion-control programs. It was also used successfully in the Great Plains each fall and winter to predict the likelihood and extent of wind erosion the following spring.

Considerable interest has developed in using the equation on highly erodible soils (sands and mucks) in humid regions.

Modifications have been introduced that have permitted WEQ's use in predicting erosion on perennially vegetated areas and in predicting airborne dust.

A revised form of the WEQ is being developed (RWEQ), as is a new prediction model (WEPS). RWEQ is designed to estimate erosion for a single day or for any longer period. It will predict erosion for a wide variety of soil, climate, and tillage conditions. It has both paper and computer technologies. WEPS is a process-based computer model that uses four databases: climate, soils, management, and crop/decomposition. It will serve as a reference standard for wind-erosion predictions.

QUESTIONS

1. List the four major soil factors that affect the rate of erosion that can be tolerated, and point out why each factor is important.
2. What needs prompted the development of the soil loss prediction equations?
3. What is the water-erosion K factor? Describe three methods that have been used to establish specific K values in the United States.
4. What are the principal advantages of RUSLE, compared to USLE, for predicting soil loss from water erosion in the eastern half of the United States?
5. Briefly describe the specific conditions for which the wind-erosion prediction equation was designed initially.
6. How does ridge roughness affect wind erosion? How was this effect evaluated initially?
7. What major improvements were included in the new prediction models, WEPP and WEPS, that make them potentially more effective than the former models, USLE and WEQ?
8. (a) Describe a farm field and calculate the total erosion losses (water plus wind) predicted for it. (b) If the predicted losses exceed the tolerable amount, propose and justify management changes and control measures for reducing erosion.

REFERENCES

ARMBRUST, D. V., and L. LYLES, 1985. Equivalent wind-erosion protection from selected growing crops. *Agron. J.* 77:703–707.

ARNOLDUS, H. M. J., 1977. Methodology used to determine the maximum potential average annual soil loss due to sheet and rill erosion in Morocco. Annex IV in *Assessing Soil Degradation.* FAO Soils Bull. 34, FAO, Rome.

BEASLEY, D. B., E. J. MONKE, and L. F. HUGGINS, 1980. ANSWERS: A model for watershed planning. *Trans. Am. Soc. Agr. Eng.* 23:938–944.

BEASLEY, D. B., and L. F. HUGGINS, 1982. *ANSWERS Users' Manual.* Publ. EPA-905-82-001. U.S. Environmental Protection Agency, Chicago, 54 p.

BONDY, E. L., L. LYLES, and W. A. HAYES, 1980. Computing soil erosion by periods using wind energy distribution. *J. Soil Water Cons.* 35:173–176.

BROOKS, F. L., 1974. Use of the universal soil loss equation in Hawaii. In *Soil Erosion: Prediction and Control.* Soil Conservation Society of America, Ankeny, IA.

BROWNING, G. M., C. L. PARISH, and J. A. GLASS, 1947. A method for determining the use and limitation of rotation and conservation practices in the control of soil erosion in Iowa. *J. Am. Soc. Agron.* 39:65–73.

CARREKER, J. R., 1966. Wind erosion in the southeast. *J. Soil Water Cons.* 21:86–88.

CHEPIL, W. S., 1960. Conversion of relative field erodibility to annual soil loss by wind. *Soil Sci. Soc. Am. Proc.* 24:143–145.

CHEPIL, W. S., F. H. SIDDOWAY, and D. V. ARMBRUST, 1962. Climatic factor for estimating wind erodibility of farm fields. *J. Soil Water Cons.* 17:162–165.

CHEPIL, W. S., and N. P. WOODRUFF, 1959. *Estimation of Wind Erodibility of Farm Fields.* USDA Prod. Res. Rep. 25.

COLE, G. W., L. LYLES, and L. J. HAGEN, 1983. A simulation model of daily wind erosion soil loss. *Trans. Am. Soc. Agr. Eng.* 26:1758–1765.

DE ROO, A. P. J., 2000. Applying the LISEM model for investigating flood prevention and soil conservation scenarios in South-Limburg, The Netherlands. Ch. 2, p. 33–41 in J. Schmidt (ed.), *Soil Erosion: Application of Physically Based Models.* Springer, Berlin, 318 p.

ELWELL, H., 1978. Modeling soil losses in southern Africa. *J. Agric. Eng. Res.* 23:117–127.

EVANS, R., and S. NORTCLIFF, 1978. Soil erosion in north Norfolk. *J. Agric. Sci.* 90:185–192.

EVANS, W. R., and G. KALKANIS, 1976. Use of the universal soil loss equation in California. In *Soil Erosion: Prediction and Control.* Soil Conservation Society of America, Ankeny, IA.

FAVIS-MORTLOCK, D. T., and A. J. T. GUERRA, 2000. The influence of global greenhouse-gas emissions on future rates of soil erosion: A case study from Brazil using WEPP-CO_2. Ch. 1, p. 3–31 in J. Schmidt (ed.), *Soil Erosion: Application of Physically Based Models.* Springer, Berlin, 318 p.

FISHER, P. S., and E. L. SKIDMORE, 1970. *WEROS: A Fortran IV Program to Solve the Wind Erosion Equation.* USDA, ARS, 41–174, 13 p.

FOSTER, G. R., and L. J. LANE (compilers), 1987. *User Requirements: USDA Water Erosion Prediction Project (WEPP).* NSERL Rep. 1. National Soil Erosion Research Laboratory, USDA-ARS, West Lafayette, IN, 43 p.

FOSTER, G. R., and W. H. WISCHMEIER, 1974. Evaluating irregular slopes for soil loss prediction. *Trans. Am. Soc. Agr. Eng.* 17:305–309.

FRYREAR, D. W., 1995. Soil losses by wind erosion. *Soil Sci. Soc. Amer. J.* 59:668–672.

FRYREAR, D. W., J. D. BILBRO, A. SALEH, H. SCHOMBERG, J. E. STOUT, and T. M. ZOBECK, 2000. RWEQ: Improved wind erosion technology. *J. Soil Water Cons.* 55:183–189.

FRYREAR, D. W., J. E. STOUT, L. J. HAGEN, and E. D. VORIES, 1991. Wind erosion: Field measurements and analysis. *Trans. Am. Soc. Agric. Eng.* 34:155–160.

GERONTIDIS, D. V. S., C. KOSMAS, B. DETSIS, M. MARATHIANOU, T. ZAFIRIOUS, and M. TSARA, 2001. The effect of moldboard plow on tillage erosion along a hillslope. *J. Soil Water Cons.* 56:147–152.

HAGEN, L. J., 1991. A wind erosion prediction system to meet user needs. *J. Soil Water Cons.* 46:106–111.

HAGEN, L. J., and L. LYLES, 1988. Estimating small grain equivalents of shrub-dominated rangelands for wind erosion control. *Trans. Am. Soc. Agric. Eng.* 31:769–775.

HAGEN, L. J., L. E. WAGNER, and E. L. SKIDMORE, 1999. Analytical solutions and sensitivity analyses for sediment transport in WEPS. *Trans. Am. Soc. Agric. Eng.* 42:1715–1721.

HALLBERG, G. R., N. C. WOLLENHAUPT, and G. A. MILLER, 1978. A century of soil development in soil derived from loess in Iowa. *Soil Sci. Soc. Am. J.* 42:339–343.

HARTWIG, R. O., and J. M. LAFLEN, 1978. A meterstick method for measuring crop residue cover. *J. Soil Water Cons.* 33:90–91.

HAYES, W. A., 1965. Wind erosion equation useful in designing northeastern crop protection. *J. Soil Water Cons.* 20:153–155.

HUSSEIN, M. H., 1986. Rainfall erosivity in Iraq. *J. Soil Water Cons.* 41:336–338.

JACKSON, M., 1988. Amish agriculture and no-till: The hazards of applying the USLE to unusual farms. *J. Soil Water Cons.* 43:483–486.

JETTEN, V., J. BOIFFIN, and A. DE ROO, 1996. Defining monitoring strategies for runoff and erosion studies in agricultural catchments: a simulation approach. *Eur. J. Soil Sci.* 47:579–592.

JOHNSON, L. C., 1987. Soil loss tolerance: Fact or myth? *J. Soil Water Cons.* 42:155–160.

KAUTZA, T. J., D. L. SCHERTZ, and G. A. WEESIES, 1995. Lessons learned in RUSLE technology transfer and implementation. *J. Soil Water Cons.* 50:490–493.

KNISEL, W. J. (ed.), 1980. *CREAMS: A field scale model of Chemical, Runoff, and Erosion from Agricultural Management Systems.* Cons. Rep. 26. Science and Education Administration, USDA, Washington, D.C.

KOHNKE, H., and A. R. BERTRAND, 1959. *Soil Conservation.* McGraw-Hill, NY, 298 p.

LAFLEN, J. M., L. J. LANE, and G. R. FOSTER, 1991. WEPP: A new generation of erosion prediction technology. *J. Soil Water Cons.* 46:34–38.

LAL, R., 1985. Soil erosion in its relation to productivity in tropical soils. In S. A. El Swaify, W. C. Moldenhauer, and A. Lo (eds.), *Soil Erosion and Conservation.* Soil Conservation Society of America, Ankeny, IA, p. 237–247.

LANE, L. J., and M. A. NEARING (eds.), 1989. *USDA-Water Erosion Prediction Project: Hillslope Profile Model Documentation.* NSERL Rpt. No. 2. National Erosion Res. Lab., USDA ARS, West Lafayette, IN.

LINDSTROM, M. J., J. A. SCHUMACHER, and T. E. SCHUMACHER, 2000. TEP: A Tillage Erosion Prediction model to calculate soil translocation rates from tillage. *J. Soil Water Cons.* 55:105–108.

LIU, B. Y., M. A. NEARING, P. J. SHI, and Z. W. JIA, 2000. Slope length effects on soil loss for steep slopes. *Soil Sci. Soc. Am. J.* 64:1759–1763.

LYLES, L., 1983. Erosive wind energy distribution and climatic factors for the west. *J. Soil Water Cons.* 38:106–109.

LYLES, L., and B. E. ALLISON, 1980. Range grasses and their small grain equivalents for wind-erosion control. *J. Rge. Mgmt.* 33:143–146.

LYLES, L., and B. E. ALLISON, 1981. Equivalent wind-erosion protection from selected crop residues. *Trans. Am. Soc. Agr. Eng.* 24:405–408.

MCCOOL, D. K., W. H. WISCHMEIER, and L. C. JOHNSON, 1974. Adapting the universal soil loss equation to the Pacific Northwest. Paper No. 74-2523. *Am. Soc. Agr. Eng.,* St. Joseph, MI.

MISRA, R. K., and C. W. ROSE, 1996. Application and sensitivity analysis of process-based erosion model GUEST. *Eur. J. Soil Sci.* 47:593–604.

MORGAN, R. P. C., 1994. The European soil erosion model: An update on its structure and research basis. In R. J. Rickson (ed.), *Conserving Soil Resources: European Perspectives.* CAB International, Wallingford, Oxfordshire, U.K., p. 286–299.

MORGAN, R. P. C., J. N. QUINTON, and R. J. RICKSON, 1991. *EUROSEM: A User Guide.* Silsoe College, Silsoe, Bedford, U.K.

MORGAN, R. P. C., J. N. QUINTON, and R. J. RICKSON, 1992. *EUROSEM: Documentation Manual.* Silsoe College, Silsoe, Bedford, U.K.

MORRISON, J. E., JR., R. W. TICKMAN, D. K. MCCOOL, and K. L. PFEIFFER, 1997. Measurement of wheat residue cover in the Great Plains and Pacific Northwest. *J. Soil Water Cons.* 52:59–65.

MUSGRAVE, G. W., 1947. The quantitative evaluation of factors in water erosion: A first approximation. *J. Soil Water Cons.* 2:133–138.

ONSTAD, C. A., A. M. KILEWE, and L. G. ULSAKER, 1984. An approach to testing the USLE factor values in Kenya. In *Challenges in African Hydrology and Water Resources.* Proceedings of the Harare Conference, July 1984, Int. Ass. Hydrol. Publ. No. 144.

PIERCE, F. J., W. E. LARSON, and R. H. DOWDY, 1984. Soil loss tolerance: Maintenance of long-term soil productivity. *J. Soil Water Cons.* 39:136–138.

RENARD, K. G., G. R. FOSTER, G. A. WEESIES, D. K. MCCOOL, and D. C. YODER (Coords.), 1997. *Predicting Soil Erosion by Water: A Guide to Conservation Planning with the Revised Universal Soil Loss Equation (RUSLE)*. USDA Agric. Handbook 703, 404 p.

RENARD, K. G., G. R. FOSTER, G. A. WEESIES, and J. P. PORTER, 1991. RUSLE: Revised universal soil loss equation. *J. Soil Water Cons.* 46:30–33.

RENARD, K. G., J. M. LAFLEN, G. R. FOSTER, and D. K. MCCOOL, 1994. RUSLE revisited: Status, questions, answers and the future. *J. Soil Water Cons.* 49:213–220.

RENARD, K. G., J. R. SIMANTON, and H. B. OSBORN, 1974. Applicability of the universal soil loss equation to semiarid rangeland conditions in the Southwest. *Proc. Ariz. Sec., Am. Water Resources Assoc., Hydrology Sec., Ariz. Acad. Sci.* 4:18–32.

SKIDMORE, E. L., 1983. Wind erosion calculator: Revision of residue table. *J. Soil Water Cons.* 38:110–112.

SKIDMORE, E. L., 1986. Wind erosion climatic erosivity. *Climatic Change* 9:195–208.

SKIDMORE, E. L., 1987. Wind-erosion direction factors as influenced by field shape and wind preponderance. *Soil Sci. Soc. Am. J.* 51:198–202.

SKIDMORE, E. L., and J. R. WILLIAMS, 1991. Modified EPIC Wind Erosion Model. In J. Hanks and J. T. Ritchie (eds.), *Modeling Plant and Soil Systems.* Agron. Monograph 31. American Society of Agronomy, Madison, WI, p. 457–469.

SKIDMORE, E. L., and N. P. WOODRUFF, 1968. *Wind Erosion Forces in the United States and Their Use in Predicting Soil Loss.* USDA Agric. Handbook 346.

SLONEKER, L. L., and W. C. MOLDENHAUER, 1977. Measuring the amounts of crop residue remaining after tillage. *J. Soil Water Cons.* 32:231–236.

SMITH, D. D., 1941. Interpretation of soil conservation data for field use. *Agric. Eng.* 22:173–175.

THORNTHWAITE, C. W., 1931. Climates of North America according to a new classification. *Geograph. Rev.* 21:633–655.

ULSAKER, L. G., and C. A. ONSTAD, 1984. Relating rainfall erosivity factors to soil loss in Kenya. *Soil Sci. Soc. Amer. J.* 48:891–896.

WAGNER, L. E., 1996. An overview of the wind erosion prediction system. In Conference Proceedings. Int. Conf. on *Air Pollution from Agricultural Operations,* Kansas City, MO, Feb. 7–9, 1996.

WHITFIELD, C. J., J. J. BOND, E. BURNETT, W. S. CHEPIL, B. W. GREB, T. M. MCCALLA, J. S. ROBINS, F. H. SIDDOWAY, R. M. SMITH, and N. P. WOODRUFF, 1962. *A Standardized Procedure for Residue Sampling: A Committee Report.* USDA, ARS, 41–68.

WILLIAMS, J. R., K. G. RENARD, and P. T. DYKE, 1983. EPIC—A new method for assessing erosion's effect on soil productivity. *J. Soil Water Cons.* 38:381–383.

WILSON, L., 1975. Application of the wind erosion equation in air pollution surveys. *J. Soil Water Cons.* 30:215–219.

WISCHMEIER, W. H., 1959. A rainfall-erosion index for a universal soil loss equation. *Soil Sci. Soc. Am. Proc.* 23:246–249.

WISCHMEIER, W. H., 1960. Cropping-management factor evaluations for a universal soil loss equation. *Soil Sci. Soc. Am. Proc.* 24:322–326.

WISCHMEIER, W. H., C. B. JOHNSON, and B. V. CROSS, 1971. A soil erodibility nomograph for farmland and construction sites. *J. Soil Water Cons.* 26:189–193.

WISCHMEIER, W. H., and D. D. SMITH, 1965. *Predicting Rainfall-Erosion Losses from Cropland East of the Rocky Mountains.* USDA Agric. Handbook 282.

WISCHMEIER, W. H., and D. D. SMITH, 1978. *Predicting Rainfall Erosion Losses: A Guide to Conservation Planning.* USDA Agric. Handbook 537.

WOODRUFF, N. P., and F. H. SIDDOWAY, 1965. A wind erosion equation. *Soil Sci. Soc. Am. Proc.* 29:602–608.

YODER, D., and J. LOWN, 1995. The future of RUSLE: Inside the new Revised Universal Soil Loss Equation. *J. Soil Water Cons.* 50:484–489.

YOUNG, R. A., C. A. ONSTAD, D. D. BOSCH, and W. P. ANDERSON, 1989a. AGNPS: A nonpoint source pollution model for evaluating agricultural watersheds. *J. Soil Water Cons.* 44:168–173.

YOUNG, R. A., C. A. ONSTAD, D. D. BOSCH, and W. P. ANDERSON, 1989b. *AGNPS User's Guide, Version 3.50—October 1989.* USDA-ARS, Morris, MN.

YU, B., G. M. HASHIM, and Z. EUSOF, 2001. Estimating the R-factor with limited rainfall data: A case study from peninsular Malaysia. *J. Soil Water Cons.* 56:101–105.

ZINGG, A. W., 1940. Degree and length of land slope as it affects soil loss in runoff. *Agric. Eng.* 21:59–64.

ZINGG, A. W., W. S. CHEPIL, and N. P. WOODRUFF, 1952. *Analysis of Wind Erosion Phenomena in Roosevelt and Currie Counties, New Mexico.* Region VI SCS Albuquerque, NM, M-436.

ZOBECK, T. M., N. C. PARKER, S. HASKELL, and K. GUODING, 2000. Scaling up from field to region for wind erosion prediction using a field-scale wind erosion model and GIS. *Agric. Ecosys.Environ.* 82:247–259.

CHAPTER

7

SOIL SURVEYS AS A BASIS FOR LAND USE PLANNING

Soils are complex and variable. Their chemical and physical properties differ from one location to another on both large and small scales. Any farmer can tell differences in the soils on different parts of the farm, a gardener can do likewise on a smaller scale, and someone who makes a close examination can find differences in soil samples taken only inches or even millimeters apart. Still more differences are noted if one travels to a different area and finds a whole new set of soils. Some of these differences are easily observed, some can be measured by laboratory tests, and some are difficult to evaluate. Soil surveys produce maps and soil descriptions that are a means of compiling information about soils and the landscapes where they occur.

Certain soil differences have a great influence on the suitability of the soil for specific uses; other properties may be more important for other uses. For example, the contents of various plant nutrients are a concern where plants are to be grown, but physical properties like texture and permeability are more important when soil is used to build a road or a dam.

The term *land* generally includes such natural resources as soil, mineral deposits, climate, water supply, location in relation to markets and transportation, vegetative cover, and structures such as buildings and terraces. The soil component is an important part of land but not the only factor to be considered.

Land use planning works on various scales (FAO, 1993) to help make the best use of land with practices that meet current needs and protect land resources so they will still be available in the future. For example, national planning concerned with policies and laws must consider an entire nation. It can benefit from a generalized map that shows the nation on a single sheet. Local planning is concerned with decisions that apply to specific locations and is based on a detailed map of a small area. Intermediate levels also occur at county, state, regional, or district levels.

Many nations have provided for a *land use survey* to make a *land use inventory* to be used as a basis for national land use planning (Davidson, 1992). The work is often done on a sampling basis so that it can be completed within a reasonable period of time.

Observations of soil differences began thousands of years ago and were recorded by Chinese, Greek, Hebrew, Roman, and other scholars (Gardner, 1998). Scientific analyses, however, had to await developments in chemistry, geology, and other sciences that made it possible to analyze soils, and an understanding of the factors that influence soil formation. Most of that knowledge became available during the 19th and 20th centuries.

7–1 SOIL SURVEYS

Most 19th-century soil maps were closely related to geology, and soils were understood mostly as weathered rock. A map of Albany County, New York, completed in 1820 by a lawyer-geologist named Amos Eaton is credited as the first of this kind in the United States (Gardner, 1998). Eaton not only identified soils formed from different kinds of rocks but also distinguished between upland and lowland positions and alluvial deposits versus soils formed in place. Other maps followed, and the work gradually became more sophisticated. Gardner (1998) credits E. W. Hilgard with having considered the influence of vegetation along with geology in his 1860 map of the soils of Mississippi.

In 1894, the U.S. Congress allocated funds within the Weather Bureau to study the relationship of climate to organic life. This study became the Division of Agricultural Soils with Milton Whitney as its chief. The project that is now known as the National Cooperative Soil Survey officially began in 1899 under his direction. Whitney stretched his $16,000 budget that first year by cooperating with state institutions such as experiment stations and geological surveys, thus establishing a precedent that is still in practice (Gardner, 1998).

The first soil surveys were very general. The 1899 maps covered 720,000 ac (291,000 ha) in portions of the Connecticut Valley of Massachusetts and Connecticut, Cecil County in Maryland, Pecos Valley in New Mexico, and Salt Lake Valley in Utah (Whitney, 1900); this increased to 6,557,320 ac (2,653,700 ha) in 1901 (Gardner, 1998). Whitney considered it more urgent to provide maps for large areas rather than to include much detail.

Soil surveys became more detailed with time and were published on progressively smaller scales, moving from an early standard of 1 in. per mile (1:63,360), to 2 in. per mile (1:31,680) and 3.168 in. per mile (1:20,000) to the present norm of 4 in. per mile (1:15,840). The amount of information included has also increased with more detailed soil descriptions and more interpretations for various types of land use. The USDA Natural Resources Conservation Service and the various land grant universities with their agricultural experiment stations are the leaders in the National Cooperative Soil Survey, and a number of other federal, state, and local agencies participate in some places.

Soil surveyors study soils on the landscapes where they occur, delineate soil areas on maps, and describe their characteristics (Figure 7–1). The information needed to classify the soils and evaluate their suitability for various uses is gathered, recorded, and published in soil survey reports. The United States now has more than 1000 soil surveyors working throughout the nation. They have designated more than 100,000 map units and classified them into more than 13,000 soil series. Soil survey reports are available for most of the nation, usually on a county basis. Some counties have two or even three published reports because they needed to be updated with more detailed maps and more descriptive and interpretive information. Soil surveys have proven to be so valuable that they are being made by nations throughout the world.

Figure 7–1 This soil surveyor is showing the farmer how he uses 10% HCl solution to detect the presence of calcium carbonate in soils and rocks. This information helps him classify the soil and make a map that will help natural resource managers make efficient long-term use of the land. (Courtesy USDA Natural Resources Conservation Service.)

7–1.1 Soil Classification

Soil classification developed separately at first but soon became an important tool for organizing the soil information amassed from soil surveys. A Russian named V. V. Dokuchaiev is credited with having introduced the concept that soils are independent natural bodies and published the first scientific system of soil classification in 1879 (Basinski, 1959). K. D. Glinka translated the Russian work into German in 1914, and C. F. Marbut translated it from German into English in 1927. Marbut was the head of the U.S. soil survey program from 1910 to 1935. He used the Russian work as a basis for his own system of soil classification (Marbut, 1935).

Marbut died in 1935, and a revised system of soil classification was published in the 1938 yearbook of agriculture, *Soils and Men* (Baldwin et al., 1938). This was the official system of soil classification in the U.S. until 1965 when the *7th Approximation* of a new system was adopted. This became *Soil Taxonomy* when the Soil Survey Staff published the complete system in 1975. The second edition of this work (Soil Survey Staff, 1999) is the current guide for soil classification in the U.S. and is widely used internationally. It is available from the Superintendent of Documents and is accessible on the World Wide Web at *http://www.statlab.iastate.edu/soils/soiltax/.* A description of the system is included in textbooks designed for a course in introductory soil science.

The basic unit of soil classification is the soil series. Soil series descriptions specify the kind and sequence of soil horizons and the range of characteristics such as the texture, color, and thickness that each horizon may have. Defining characteristics are chosen to relate to how the soil formed and how it should be classified. Soil Taxonomy has six categorical levels wherein soil series are grouped into progressively broader classes as one moves toward the higher categories. The six categories are:

1. Order
2. Suborder
3. Great Group
4. Subgroup
5. Family
6. Series

Soil Taxonomy provides a systematic nomenclature and definitions for the twelve established orders and for their breakdown into suborders, great groups, and subgroups. Family and series names and definitions are determined during the course of each soil survey and integrated into the system through a process known as *soil correlation.*

Many other nations have developed their own systems of soil classification. Some have used the Russian work, the 1938 U.S. Soil Classification System, or Soil Taxonomy as a starting point, and each has been adapted to the needs of the particular nation. Some systems rely entirely, or at least as much as possible, on soil properties that can be described in the field, some include properties that require laboratory analyses, and some depend heavily on theories of soil formation.

The Food and Agriculture Organization (FAO) of the United Nations has developed a system of soil classification that it uses in the various countries where it has soil scientists working. FAO has attempted to make its system useful worldwide in a variety of fields, including agriculture, ecology, geology, and hydrology. FAO has published a World Reference Base for Soil Resources that was endorsed in 1998 at an international meeting of soil scientists (Nachtergaele et al., 2000). This system and the U.S. Soil Taxonomy are the widest used soil classification systems in the world at present.

7–1.2 Using Soil Surveys

The original objective of soil surveys was to predict the probable success of a crop when planted on a soil where it had never been grown before. Another objective was to predict the response of a crop to irrigation. Other uses were soon found. By the 1920s, the Michigan Highway Department was applying soil surveys to plan highways, and North Dakota was applying them as a means of tax assessment (Kellogg, 1933).

Land-grant universities and state and federal agricultural research stations were founded in the last half of the 19th century and grew vigorously. The research stations obtained results from field plots, but there was no assurance that the new ideas could be transferred successfully from experimental plots to farms and ranches. There was a missing link: soil mapping and classification. Soil surveys were needed to determine where each set of experimental results should be applicable to the soils on private land.

Soil surveys made today provide an essential basis for transferring scientific agricultural information from experiment stations to farms and from one farm to another. This includes estimating yields of adapted crops, predicting erosion hazards, evaluating sites for wildlife, locating sand and gravel for construction, and identifying and conserving prime and unique agricultural lands for agricultural use. A modern soil map supplemented by appropriate laboratory tests can be used to make predictions about the suitability or limitations of each soil map unit for:

- Field crop adaptations, estimated yields, and erosion hazard
- Forest tree species suitability
- Pasture grass environment
- Wildlife habitat
- Recreation potential
- Building sites
- A source of sand or gravel for construction
- Suitability for a septic tank drainfield
- Suitability for a farm pond

Consultants from various government agencies and private companies combine information from soil survey reports with research results and information from other sources to make recommendations for the uses in the preceding list. Individual farmers can obtain copies of soil survey reports from the Extension Service or the Natural Resources Conservation Service and use them to make their own management plans.

7–1.3 Kinds of Soil Surveys

There are two general kinds of soil surveys: detailed and reconnaissance. A *detailed soil survey* is one where sufficient field work is done to observe all soil mapping units and to trace all soil boundaries throughout their length. A *reconnaissance soil survey* has only intermittent field observations. Much of the information is interpreted from aerial photos or other maps of the area on a scale of 1:20,000 or 1:15,640.

The current detailed surveys being made in the United States for publication under the National Cooperative Soil Survey are known as *Standard Soil Surveys.* Completed surveys are published in *Soil Survey Reports.* The maps and accompanying reports must meet certain standards to serve the needs of their users.

7–1.4 Soil Survey Reports

Soil maps and their accompanying descriptions, interpretations, and other information are published in soil survey reports, usually on a county basis. As of February 2002, the USDA National Resources Conservation Service listed 3251 survey areas in the 50 states as shown in Table 7–1. Most of the land that has not been surveyed is in Alaska, in mountainous regions, in desert areas, or in swamps.

Soil survey reports in the United States consist of two parts: the text and the soil maps. The text tells how the survey was made, describes the county and the soils, tells how the soils are classified, and discusses their use, management, and conservation.

The first soil maps were line drawings similar to a road map with the soil information superimposed on it. Most of them were a single large sheet that represented an entire

TABLE 7–1 STATUS OF SOIL SURVEYS IN THE UNITED STATES AS OF FEBRUARY 2002

Description	Number	% of U.S. land area
Published Soil Survey	1557	43
Initial Mapping Complete	360	12
Initial Mapping in Progress	133	5
Update Field Work Complete	162	3
Update Field Work in Progress	187	3
Out-of-date Publication	365	8
Maintenance Needed	287	7
Maintenance	38	1
Non Project	162	18

Source: Web site at *ftp://ftp.ftw.nrcs.usda.gov/pub/ams/soils/ssa_small.pdf*

county at a scale of 1 in. = 1 mile (1:63,360). Modern soil survey reports contain two kinds of soil maps: a general soil map and detailed soil maps. The *general soil map* is actually a soil association map made on a large enough scale to be printed on one sheet. This map is useful in county-wide land use planning. The *detailed soil maps* are divided into a number of foldout sheets in the back of the report. They have an aerial photo background, typically at a scale of 4 in. = 1 mile (1:15,840).

7–1.5 Base Maps

Soil surveys are made on base maps that show geographic features such as roads, streams, houses, and other reference points. Early soil surveyors used topographic maps published by the U.S. Geological Survey or other suitable maps if any were available. If no adequate base map was available, they made their own as they mapped the soils, often using a plane table and alidade and sometimes measuring distances by pacing for short distances and by an automobile odometer for longer distances.

The advent of aerial photographs caused a major revolution in base maps for soil surveys as well as for many other uses. From about the 1930s to the 1950s, aerial photographs for most areas of the United States became generally available and were adopted for use as base maps. These aerial photographs are updated periodically and continue to be the best base for drawing soil boundaries. The standard type is a black-and-white photograph with a matte (nonglossy) surface to facilitate marking soil boundaries and symbols in the field with a pencil. A more permanent ink version is prepared later in the office for use by farm planners and others. The published maps in a soil survey report are mosaics made by blending the individual photos into a map of the entire area that is then subdivided into page-size sheets. Currently, soil maps are being digitized so they can be manipulated and printed by computer.

Since the mid-1960s, infrared film and some full color prints have been used in some mapping. These color photos are superior to black-and-white photos where land use and vegetation are to be interpreted but are seldom used as soil survey base maps.

At about the same time as infrared photography became available, scanners (multispectral photographic systems) were developed. This technique permits the mapping of wetlands, dark soils, and well drained soils, but is useful only when the soil is bare of vegetation.

Thermal infrared sensors in aircraft can measure relative soil temperatures. Scanners are most useful at the time of day when the temperatures of well drained soil, poorly drained soil, surface rock, and types of vegetation are expected to be in greatest contrast, usually at about 3 P.M.

Earth Resources Technology Satellites (Landsats 1 and 2) have made four-spectral-band photographs at 18-day intervals for most of the world. Using Landsat maps at scales of 1:500,000 and 1:1,000,000, Klingebiel (1977) reports a technique for making a map that delineates areas that are *good*, *fair*, and *poor* for food and fiber production in Mexico. This type of work has tremendous value for planning national development and identifying needs for more detailed soil surveys by conventional methods.

Although extremely useful as supporting and supplemental information, all air photo data require ground observations for full interpretation. Some reconnaissance-type soil surveys are made with minimal ground observations, but this diminishes the amount of information that can be mapped, and the accuracy is much less than that of surveys made in the field.

7–1.6 Soil Map Units

The individual soil map units are almost all named as phases of soil series, although phases of a class at a higher categorical level are used on some generalized maps. Also, a few miscellaneous units are used for nonsoil areas such as bare rock or gravel bars and for drastically disturbed sites such as mine spoils and made land in and around cities and developments. Some areas are mapped as rough broken land or as intermixed soil series too small in area to show separately.

Phases are used to specify information that is not considered in Soil Taxonomy but has practical significance. Phases specify the soil series plus slope range, erosion status, degree of stoniness, or some other property that affects the use and management of the land even though it does not change the classification of the soil. A soil surveyor draws boundaries on the map to separate the soil map units defined in the local mapping legend. The map units approximate as nearly as possible the taxonomic units known as polypedons. A symbol is placed in each soil area to designate the type of polypedon plus phase designations such as slope and erosion classes.

Polypedons are composed of contiguous pedons of the same soil series. One soil series is differentiated from another by properties inherent in the constituent *pedons* (in Greek, *pedon* = ground). Each pedon represents an individual unit of soil large enough to grow a representative plant. By definition, soil pedons have a minimum surface area of 1 m^2 and may be larger (up to 10 m^2) if necessary to adequately represent the soil. They are characterized by the color, texture, structure, porosity (permeability), consistence of structural units (peds or clods), pH, concretions, clay coatings, organic matter, and root abundance of each horizon and the depth to bedrock or other root-restricting layer. Polypedon field descriptions also record the vegetation, surface drainage, slope class, erosion class, parent materials, stoniness, and land use (Soil Survey Staff, 1993). The field descriptions are supplemented by laboratory analyses of soil samples taken from selected sites.

7–2 SOIL MAP UNIT INTERPRETATIONS

Soil maps and soil descriptions need to be interpreted to make them useful; in fact, they need to be interpreted in several ways to suit the needs of different users. Modern reports include interpretations for agronomic, engineering, and sanitation uses. Interpretations for horticulture, forestry, and rangeland are included where these uses are important.

When the mapping units are used according to the limitations specified, the soils will remain productive, nonpolluting, and useful. Examples of soil map unit interpretations are included in the following sections.

7–2.1 Land Use Capability Groupings

The Natural Resources Conservation Service promotes the use of land within its capabilities and the treatment of land according to its needs for protection and improvement. The land use capability system was designed to help fulfill that purpose. The emphasis is on the soil component of land and the limitations it places on land use.

There are eight *land use capability classes,* as shown in Figure 7–2, differentiated and described on the basis of limitations (hazards) that restrict intensity of use or require special treatment. The first four classes have potential for use as cropland; classes five through seven may be used for pasture, rangeland, or woodland; and class eight is restricted

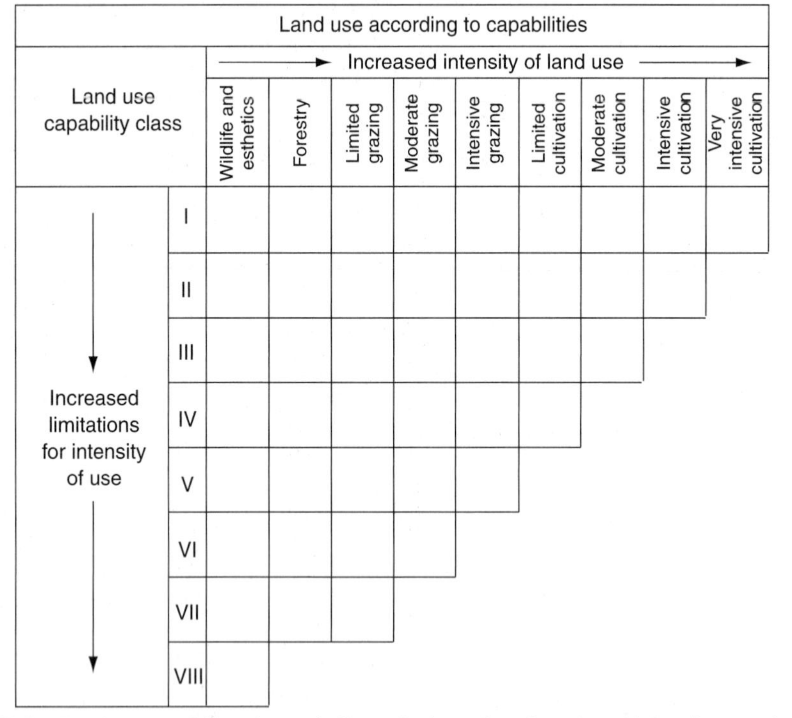

Figure 7–2 Land use capability classes indicate the intensity of use for which soils are suited. Class I is suitable for any use, but other classes are restricted by various limitations and hazards. (Courtesy USDA Natural Resources Conservation Service.)

to noninvasive uses. Maps used for farm planning are commonly colored to make the land use capability classes easy to identify. The standard colors are indicated along with the definitions below:

Class I: Soils that have few or no limitations for cultivation of the crops usually grown in the area—green on maps.

Class II: Soils that have moderate limitations that reduce the choice of adapted crops or that require suitable conservation practices such as erosion control or soil drainage—yellow on maps.

Class III: Soils that have strong limitations that restrict the choice of crops, require intensive conservation practices, or both—red on maps.

Class IV: Soils that have severe limitations that reduce the choice of crops, require very intensive conservation practices, or both—blue on maps.

Class V: Soils that are not likely to deteriorate when used for pasture, rangeland, woodland, wildlife, or esthetics but have limitations such as wetness, stoniness, or climate that make them unsuitable for cultivation—dark green or white on maps.

Class VI: Soils that have strong limitations that make them generally unsuited to cultivation and limit their use primarily to pasture, rangeland, woodland, wildlife, recreation, or esthetics—orange on maps.

Class VII: Soils that have severe limitations that make them unsuited to cultivation and that require careful management when used for pasture, rangeland, woodland, or recreation (an example is shown in Figure 7–3)—brown on maps.

Class VIII: Soils and landforms that have very severe limitations that make them unsuitable for commercial purposes and restrict their use to wildlife, esthetics, recreation, and/or watersheds—purple on maps.

Land use capability subclasses are subdivisions of land use capability classes that designate the *dominant kind of limitation* or *hazard* restricting land use. The subclasses are designated by writing a lower-case letter following the Roman numeral that signifies the land capability class. The four subclasses are:

e: erosion hazards

w: wetness problems

s: soil limitations including excessive shallowness, extremely fine or coarse texture, stoniness, salinity, or sodicity

c: climatic limitations such as excessive coldness or dryness

The relative abundance of these types of limitations in the United States is shown in Figure 7–4. Land Use Capability Class I, by definition, has no hazards limiting its use and therefore has no subclasses. Each of the other seven classes is divided into subclasses. Soil map units are classified into subclasses according to which one of the four types of hazards limits the use of the land to its designated capability class. When two subclasses limit land use equally, only one is used, priority being assigned in the sequence e, w, s, and c.

Land use capability units are divisions of land use capability subclasses (or of Class I) into smaller, more homogeneous groups of soil map units that have similar use potential and management requirements. They have also been called *soil management groups.* Capability units are established because subclasses are too broad and soil map units are

Figure 7–3 Fairmont flaggy silty clay loam in Kentucky on 25% slopes. The soil is a Typic Hapludoll. It is in Land Use Capability Class VII and Subclass e (erosion limitation). Limited grazing is probably the most intensive use this soil can serve without excessive erosion. (Courtesy USDA Natural Resources Conservation Service.)

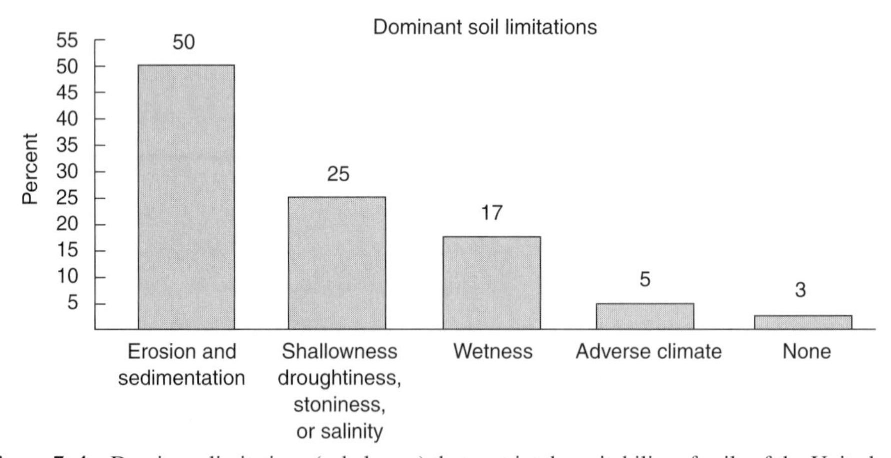

Figure 7–4 Dominant limitations (subclasses) that restrict the suitability of soils of the United States for more intensive use. Erosion and sedimentation limit the most intensive use of half of the soils of the nation. (From Soil Conservation Service, 1971.)

more specific than necessary for many land management considerations. Therefore, soil map units with similar erosion hazards, degrees of wetness, stoniness, textural classes, and crop adaptations are grouped in the same land use capability unit.

Capability units are identified by Arabic numbers added to the symbols for the land use capability class and subclass. An example is IIIe-1. The classes and subclasses are the same everywhere, but the land use capability units are so variable throughout the 50 states that separate definitions and interpretations are established for individual counties.

Other countries have also developed land capability classification systems to suit their needs. For example, the British and Canadian systems each have seven classes that parallel the U.S. system with Class V omitted and with a separate O class for organic soils in the Canadian system (Davidson, 1992). A land suitability classification is used in The Netherlands to evaluate land into classes with ratings of suitability for arable land, grassland, both, or neither (Davidson, 1992). Davidson also discusses land use capability assessments made by the Food and Agriculture Organization of the United Nations for many of the nations where they work. These are adapted to the circumstances of each nation.

7–2.2 Evaluating Rural Land

Soil map interpretations have been used in several states to evaluate rural land. For example, a study in Illinois compared the prices paid in 1509 rural land sales in 33 counties to soil productivity ratings of soil map units. The standard soil survey report and soil map can also be used to evaluate rural land to help equalize taxes.

Taxes on farmland and ranchland should be based on the productive potential of the soils and not on the management skills of the farmers and ranchers. Taxes on rural property that is being farmed should not be based on its potential value for urban development.

An economic analysis of yield potential and costs of production permits calculation of the potential agricultural value of each soil. Assessors equipped with these data and with soil maps can calculate the potential crop income of a farm or ranch. The procedures and the crops being considered vary from state to state, but the essential steps are the following:

1. Calculate the weighting factor to be assigned to each soil map unit.
2. Measure the area of each kind of soil in each land ownership.
3. Sum the products of area × weighting factor.
4. Adjust for buildings, location, and other factors.

Several states and countries now use the above procedure to determine all or part of the assessed valuation for property taxes.

7–2.3 Upgrading Soil Test Recommendations

A soil testing laboratory can make more precise recommendations for soil samples that are collected and labeled on the basis of a soil map unit. Fertilizer and lime recommendations can be adjusted according to field experimental data from the same or a similar soil series. The recommended rates for certain herbicides that depend on organic matter and/or clay contents can be determined from soil maps and the associated descriptions and laboratory data.

The soil testing laboratory at Iowa State University, for example, requests the name of the soil series along with each soil sample so as to adjust the fertilizer recommendation based on known subsoil characteristics. They make adjustments such as recommending higher rates of potassium fertilizer for poorly drained soils and for soils formed in glacial till as compared to those formed in loess. Soil characteristics such as slope, depth of topsoil, and drainage can be supplied instead if the soil series is unknown.

7–2.4 Determining Need for Artificial Drainage

Soil surveys indicate soil series. One criterion for differentiating soil series is internal drainage. The natural internal drainage is classified as either *excessively drained*, *somewhat excessively drained*, *well drained*, *moderately well drained*, *somewhat poorly drained*, *poorly drained*, or *very poorly drained*. The last three soil drainage classes usually require some type of artificial drainage to achieve optimum yields of most upland farm crops (Chapter 14).

In 1994, federal legislation mandated all poorly drained soils to be designated as *hydric* and to be mapped as *wetlands*. Natural wetlands that are not already drained are to be preserved for wildlife and hydrologic benefits (Chapter 14). They are not to be drained for agricultural crops or building sites (Soil Conservation Service, 1994).

7–2.5 Evaluating Woodland and Windbreak Sites

Tree species adaptation and relative growth rates depend on many factors, including soil depth through which roots may grow without physical or chemical hindrance, available water-holding capacity of the soil, soil texture, organic matter, aeration, and depth to the water table. Information about all these factors can be interpreted from the soil map unit. Such information for each county in which forest trees grow naturally, as well as for semiarid areas where windbreaks and shelterbelts are common, can be found in each county soil survey report.

The soil survey reports for humid regions have a section on "Woodland" that itemizes the soil map symbol, woodland group, potential productivity, seedling mortality, plant competition, equipment limitations, and preferred species for planting. The soil survey reports for semiarid areas discuss the suitability of the various soils for windbreak plantings.

7–2.6 Selecting Sites for Wildlife Habitats

Different species of wildlife occupy different habitats. Deer, for example, are favored by patches of woodland intermixed with open areas. Pheasants can feed in fields but need grassy areas such as fencerows for nesting. Beavers need a stream that they can dam to form a pond. Ducks and other waterfowl need open bodies of water, and so forth.

Land owners who want to favor wildlife have many options. Odd corners that might not be profitable to farm can be vegetated with grass, shrubs, or trees that will provide wildlife habitat. Leaving grass strips along fencelines or field borders benefits many birds and small animals. Some farmers plant corn or some other crop in a remote area and leave it for winter food for wildlife. Birdhouses and bird feeders may be placed in fencelines or other suitable sites.

A table included in standard soil survey reports issued since 1965 rates all soil map units on suitability for producing *wildlife habitat elements*. These elements include grain and seed crops, grasses and legumes, wild herbaceous plants, hardwood trees and shrubs, coniferous trees, and wetland food and cover plants. Also, soil map units are rated for suitability to openland, woodland, and wetland wildlife.

7–2.7 Interpretations for Engineering Uses

Engineers place buildings, roads, and other structures on soils that must provide proper support and drainage for the structures. They also use soils as construction materials for roadfills, dams, terraces, and so forth. The most important soil characteristics for engineering uses are particle (grain) sizes, permeability to water, compressibility, shear strength, compaction, drainage, shrink-swell potential, plasticity, soil pH, depth to water table, depth to bedrock, and topography. Engineering test data are presented in tables in soil survey reports in a section called "Engineering Uses of the Soils." The interpretations are based on field testing, laboratory analyses, and estimations of soil properties.

The engineering test data can be used to predict the suitability of each soil map unit as a source of topsoil, sand, gravel, or road fill. They can also be used to evaluate the suitability of each soil map unit for constructing a sanitary landfill, a filter field for a septic tank, a sewage lagoon, a foundation for a house, a farm pond, a highway, or a playground. Other available information important to engineering uses of soils includes flooding hazard, relative wetness, erodibility, and stabilization of construction slopes with adapted vegetation.

7–3 MANAGING LAND

Land is used for many activities and purposes that have a wide variety of soil requirements. Land managers can use soil survey information and interpretations to select sites that are best suited for particular uses and make appropriate management decisions.

7–3.1 Managing Agricultural Land

Farmers and ranchers need to be good land managers to make a profit without causing long-term soil deterioration. Soil survey reports provide useful information for making land-management decisions. They can be used for major decisions such as what land to purchase and what use to make of it. They also provide a basis for designing field layouts, planning conservation practices, and making general management decisions.

Selecting land for a particular enterprise is easier and more precise when soil survey information is used. For example, a rancher choosing land for a cattle ranch in the southern Great Plains would find that the Abilene soils shown in Figure 7–5 are known as Pachic Argiustolls to soil surveyors and deep hardland range soils to local ranchers. The native grasses are mid- and short-prairie grasses. Good grazing management is needed because overgrazing causes blue gramagrass and buffalograss to increase at the expense of the taller grasses. Forage production then declines from its normal 600 lb/ac to 90 lb/ac (700 down to 100 kg/ha) annually (Soil Conservation Service, 1974).

Adapted crop varieties can be selected for farms on the basis of experiment station results when soil survey maps show that the soils are similar. Also, needs for artificial drainage or for erosion-control practices can be identified and serious damage avoided. The

Figure 7–5 A soil profile of the Abilene series (Pachic Argiustoll) in the 25-in. (640-mm) precipitation belt in Texas. This soil supports a good growth of nutritious range grasses. The depth to the whitish calcium carbonate zone at the bottom of the photo is about 3 ft (nearly a meter). (*Note:* the arrow is at 1 meter.) (Courtesy Texas Agricultural Experiment Station.)

general likelihood of needing lime and fertilizer may also be indicated in a soil survey report, but soil samples from the actual field should be tested when specific fertility recommendations are needed.

7–3.2 Precision Agriculture

Farmers have traditionally used entire fields as production units and applied uniform treatments across each field. An approach known as *precision agriculture* or *site-specific agriculture* is currently under development (Stombaugh and Shearer, 2000). It divides a field into small *management units* and uses the global positioning system (GPS) to identify the location of each unit for both testing and treatment purposes. Variables such as seeding rate, fertilizer rates, pesticide application rates, and soil amendment rates that have usually been made on a field basis are adjusted for differences in the management units within the field. Variable rate equipment that can be adjusted as it travels across the field is used to seed the crop, apply the specified rates of nitrogen, phosphorus, and potassium fertilizers, or to apply lime or other soil amendment. This equipment is expensive enough that it is usually owned and operated by a fertilizer dealer or a contractor rather than by an individual farmer.

Precision agriculture using variable rate technology has been shown to have the potential of producing higher yields with less fertilizer in comparison to uniform field rates.

For example, Van Alphen and Stoorvogel (2000) used this type of system for nitrogen fertilization of winter wheat in The Netherlands and reported a saving of 23% in fertilizer input, an increased yield of 3%, and an increased seed density of 4% relative to traditional uniform fertilization. The higher yields and seed density are attributed to parts of the field that were underfertilized when a uniform rate was used. The savings from the parts of the field that had been overfertilized not only reduced the total amount of fertilizer used, but also must have reduced water pollution by nitrate leaching from such areas.

Improved results from precision agriculture can only come from fields where there is sufficient information to adjust the rates appropriately. Opinions differ regarding the best method of obtaining such information, partly because situations differ. One field may include several distinctly different soil series that might best be represented by taking soil samples from each soil map unit in the field with arbitrary subdivisions for any such units that are larger than a specific size. Another field with several areas that reflect past use and management differences should be sampled in a way that separates the effects of those differences.

Grid sampling is another approach that fits well with the management unit concept of precision agriculture. Soil samples are taken at regular intervals, each one representing a specific area. Bongiovanni and Lowenberg-DeBoer (2000) analyzed the results of lime applications in Indiana using 1-ha (2.5-ac) and 0.4-ha (1-ac) areas. They concluded that the 1-ha areas were the most profitable; the added cost of taking more samples was not justified by their data.

Yield monitors mounted on harvesting equipment make it possible to use records from the previous crop to make fertilizer recommendations for the next crop. Long et al. (2000) used a monitor that recorded yield and protein content of spring wheat at one-second intervals in a field in northern Montana. From these data, they calculated how much nitrogen was removed by the crop and how much additional nitrogen would be needed to increase the protein content to the desired level. They compared variable fertilizer rates for the next year's crop based on these data to applying the same total amount of fertilizer at a uniform rate in parallel strips. Yields were the same, but the variable rate produced grain with a higher and more uniform protein content. Ellsworth (2001) used a yield monitor approach combined with a soil map and soil test data to analyze corn yield response to nitrogen in an Iowa field. Compiling these types of information in a digitized data set for the small areas that make up a field produces a *geographical information system* (GIS) (Sylvester-Bradley et al., 1999). Such data are used to adjust fertilizer recommendations for the next crop, and a map of the field is prepared to show how much fertilizer each management unit should receive.

7–3.3 Mechanization in Developing Countries

Developing countries often want to move rapidly from hand-hoe cultivation to animal-powered cultivation to cultivation by tractors. These transitions are not always successful. One reason for the failure of animal-powered farming in humid Africa is trypanosomiasis (African sleeping sickness) carried by several species of tsetse fly (*Glossina*). In humid Africa, many governments want to lead their farmers from the use of the village-made hoes directly into the use of sophisticated tractors.

Rapid mechanization is feasible on some soils. For example, nearly level and fertile Vertisols occur in Ethiopia and Sudan (eastern Africa) and in Ghana and Nigeria (western Africa). Tractors are the most suitable farm power on Vertisols because animals cannot pull plows and other implements through such fine-textured soils.

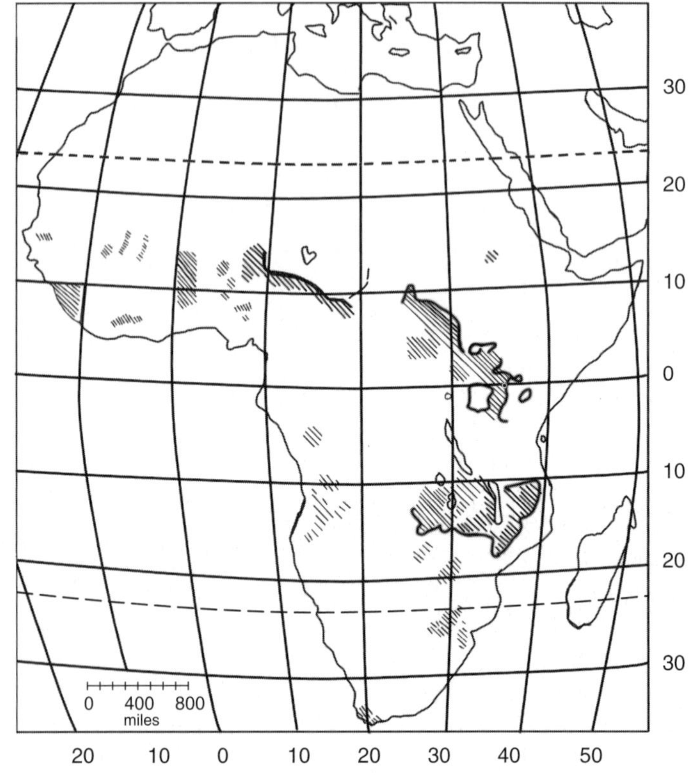

Figure 7–6 The distribution of plinthite (also known as laterite) in Africa. (From Prescott and Pendleton, 1966.)

By contrast, the ferruginous soils of Ghana in western Africa are so steep and irregular in topography, so low in fertility, and so high in plinthite (laterite) that they cannot be cultivated with a tractor (Kline et al., 1969) (Figures 7–6 and 7–7). Care must be used in working with these soils to keep them from drying enough to harden irreversibly into ironstone.

7–3.4 Delineating Nutritional Problem Areas

As early as 1878, President Welch of the Agricultural College at Ames, Iowa, (now Iowa State University of Science and Technology) suggested that there should be a national study of soils in relation to animal and human nutrition. The Annual Report of the USDA Bureau of Animal Industry for 1898 contained a section on the relationship between a bone disease in animals and the forage from pastures on the "noncalcareous" soils of the Gulf Coastal Plains. Later it was discovered that the soils were acid and low in calcium and phosphorus. The bone disease became known as *osteomalacia.* Extreme cases were reported in the 1930s in some northern states when animals ate only native prairie grasses and overwintered on prairie hay. Soils likely to supply only small amounts of phosphorus can be identified on soil maps and checked by chemical tests. Phosphorus percentages in most forages can be increased by a soil application of phosphorus fertilizer or by liming acid soils. The

Figure 7–7 A plinthite soil in Ghana in western Africa. Plinthite, also known as laterite, hardens into ironstone when wetted and dried during many years of high temperature. (*Note:* the depth scale shown is in inches.) (Courtesy Roy Donahue.)

phosphorus fertilizer will also increase forage production and reduce erosion on soils low in available phosphorus.

Forage plants growing on soils that are poorly drained, both on Histosols (peats and mucks) and on wet mineral soils, may contain concentrations of copper and molybdenum that are toxic to cattle and sheep (Kubota, 1975). These soils are readily identifiable on soil maps.

Some nutritional problems, such as those mentioned, relate to specific areas that can be mapped. They occur in places where most of the food is produced locally. People in most developed nations obtain food from a wide area and many soils. A balanced diet is therefore likely to contain adequate nutrients.

TABLE 7–2 FAVORABLE SOIL CHARACTERISTICS FOR USING A SOIL AS A SEPTIC TANK DRAIN FIELD OR AS A SEWAGE LAGOON

Septic tank drain field	Sewage lagoon
Nearly level	Level
Well drained	Poorly drained
As deep as possible to a water table	No possibility of contaminating water supplies
Sandy loam to loam texture	Fine texture (high in clay)
High cation exchange capacity	Structureless
High in organic matter	Low in organic matter
Free of coarse fragments	Free of coarse fragments
pH 6.5 to 8.5	Any pH

7–3.5 Cleansing the Environment

Pollution of the environment is a serious threat to human life. Pollutants discarded in oceans and other bodies of water may not degrade; water may preserve them. Soil is the only biodegrading medium for rational disposal of many kinds of polluting wastes (Chapter 16).

Soils vary in the kinds of pollutants each can absorb and how much of each pollutant they can degrade. Soils that are ideal for use as septic tank drain fields are usually not ideal for sewage lagoons, as may be seen by comparing the columns in Table 7–2. Sanitary landfills requirements are different from both of these columns and even require different soil characteristics for different types of landfills. For example, because a trench-type landfill requires soil manipulation under varying moisture conditions, its soil should have stable structure and should be deep, medium-textured, and high in organic matter.

Table 7–3 rates twelve soil series in Pike County, Illinois, according to the severity of their limitations for use as a disposal site for each of these kinds of polluting wastes. Similar tables can be found in other modern soil survey reports.

7–4 LAND USE PLANNING

Soil survey reports and soil maps have been used in almost every conceivable type of land resource planning, zoning, and general development. One very large project is that for the entire 201-county watershed of the Tennessee Valley, comprising all or parts of seven states. Land use planning is often perceived as having four objectives:

1. Protect current land use
2. Guide future developments
3. Reduce present and future conflicts
4. Avoid pollution (Gold et al., 1989)

An area of stratified, sandy glacial deposits will serve as an example of how soil survey maps are used to plan land use and avoid pollution. Sandy deposits are natural locations for rainwater to recharge aquifers. For this reason, they should be avoided when locating a landfill, a storage lagoon for manure, a salt-storage pile for deicing roads, or a drain

TABLE 7–3 RELATIVE LIMITATION RATINGS, AND THE REASONS FOR THE RATINGS, OF SELECTED SOIL SERIES IN PIKE COUNTY, ILLINOIS, FOR USE IN WASTE DISPOSAL

Soil series	Relative ratings				Reason(s) for ratings
	Septic tank drainfield	Sewage lagoon	Sanitary landfill		
			Trench type	Area type	
Blair	Severe	Severe	Severe	Severe	Wetness
Downs	Moderate	Moderate	Severe	Moderate	Slope, wetness
Drury	Moderate	Moderate	Slight	Slight	Slope, seepage
Elizabeth	Severe	Severe	Severe	Severe	Slope, shallow, stony
Fayette	Moderate	Moderate	Moderate	Slight	Slope, slow permeability
Hamburg	Severe	Severe	Severe	Severe	Steep slope
Rozetta	Moderate	Moderate	Severe	Moderate	Wetness
Seaton	Slight	Moderate	Slight	Slight	Slope
Sylvan	Moderate	Moderate	Severe	Severe	Wetness
Tama	Moderate	Moderate	Severe	Moderate	Wetness
Virden	Severe	Severe	Severe	Severe	Ponding
Worthen	Slight	Moderate	Slight	Slight	Slope, seepage

Note: For soil series with more than one map unit, these ratings are for the unit with the least restrictions (usually the unit with the lowest slope gradient for that soil series).

Source: Struben and Lilly, 1999.

field for a septic tank. Nor should such deep sandy deposits be fertilized heavily with nitrogen fertilizer, animal manures, or sewage sludges.

Stratified drift deposits must be identified and located accurately to avoid their misuse. This can be done with the proper use of soil survey maps or U.S. Geological Survey (USGS) maps. Fortunately, these two kinds of maps are in agreement about 85% of the time. Both kinds of maps must be used because neither kind is available for all areas where deep sand deposits exist. USGS maps are available for 5 to 10% of the United States; as already indicated in Table 7–1, up-to-date soil survey maps are available for about 58%, with another 8% in progress; older soil survey maps are still available for another 16% of the U.S.

7–4.1 Sustainable Land Use

Major uses of nonfederal rural land in the United States are listed in Table 7–4. They include cropland, 28%; pastureland, 9%; rangeland, 27%; forestland, 26%; and other nonfederal rural lands, 10%.

Sustainability is an appropriate test for a land use and management system (FAO, 1995). Is the productivity of the land maintained for both present and future use? Any system that causes land to deteriorate and become less productive with time is not sustainable in that place even if the decline is slow. Conversion to a less intensive use that is sustainable may cause the affected area to produce less profit in the short run, but it will be much more productive in the long run.

Each soil map unit has a T factor that designates the annual rate of erosion that it can tolerate without permanent loss of soil productivity. Map unit characteristics that influence

TABLE 7–4 USE OF NONFEDERAL RURAL LAND IN THE UNITED STATES IN 1982

Land use	Acres (millions)	Hectares (millions)	Percent of total
Cropland	421.4	170.7	28
Pastureland	133.3	54.0	9
Rangeland	405.9	164.4	27
Forestland	393.8	159.5	26
Subtotal	1,354.4	548.6	90
Other nonfederal rural lands	143.2	58.0	10
Grand total	1,497.6	606.6	100

Source: National Research Council, 1986, p. 7.

TABLE 7–5 AVERAGE ANNUAL SHEET, RILL, AND WIND EROSION

Land use	Sheet and rill erosion		Wind erosion	
	tons/ac-yr	mt/ha-yr	tons/ac-yr	mt/ha-yr
Cropland				
All	4.4	9.86	3.0	6.72
Cultivated	4.8	10.75	3.3	7.39
Pastureland	1.4	3.14	0.0	0.00
Rangeland	1.4	3.14	1.5	3.36
Forestland				
Grazed	2.3	5.15	0.1	0.22
Not grazed	0.7	1.57	0.0	0.00

Source: National Research Council, 1986, p. 8.

this T factor are depth of the solum (A + B horizons), kind of parent material, relative productivity, and previous erosion. Soil loss calculations as outlined in Chapter 6 can be used to estimate the soil-loss rate under various use and management options to serve as a basis for choosing land use and management practices that do not degrade the land. Soil-loss rates that exceed the T factor indicate that something needs to be changed to make the practices sustainable.

There is little doubt that a well-managed, fully stocked forest provides maximum protection against soil erosion and sediment loss. Next in rank of soil protection is a thick cover of grass, and least protective is bare soil. A generalized comparison of the average rates of erosion associated with various types of land use in the United States as a whole is shown in Table 7–5. These data show that cropland is eroding much faster than pastureland, rangeland, and forestland in spite of the fact that cropland is generally on the best soils and has the least average erodibility. More detailed data show that much of the cropland erosion is occurring on a relatively small amount of land that is being used beyond its capability.

Nationwide reductions in total erosion and sedimentation could be achieved if all land use capability classes were used within their designated limitations. Human needs dictate that most land must be used as intensively as its limitations permit, and the country needs to reserve its best agricultural land for agricultural purposes. A new national concern for prime and unique agricultural lands recognizes this concept.

Figure 7–8 Most of the soil under this subdivision in Maryland would be classified in Land Use Capability Class I and would be considered prime agricultural land. Building costs were less than elsewhere because the area is nearly level and adequately drained. Who should determine whether it will be used for housing or reserved for agriculture? (Courtesy USDA Natural Resources Conservation Service.)

7–4.2 Prime and Unique Agricultural Lands

There has been increasing concern that the best lands for agricultural use are being irreversibly converted to other uses such as residential and business sites, as illustrated in Figure 7–8. This concern has led to efforts to define and identify prime and unique agricultural lands that should be kept in farmland (Soil Survey Staff, 1993).

Prime farmland is land that has the best combination of physical and chemical characteristics for producing food, feed, forage, fiber, and oilseed crops. The land could be used as cropland, pastureland, rangeland, forestland, or other land, but not as urban built-up land. It has the soil quality, growing season, and moisture supply needed to economically produce sustainable high yields of crops when treated and managed according to modern farming methods. Prime agricultural lands are characterized by these eight parameters:

1. Adequate natural rainfall or adequate good-quality irrigation water to meet normal needs 7 out of 10 years
2. Mean summer temperatures warmer than 59°F (15°C) at a soil depth of 20 in. (50 cm)

3. Lack of excessive moisture, which means (a) no flooding more often than once in two years and (b) water table below rooting zone
4. Soil not excessively acidic or basic (pH between 5.5 and 8.6), not saline, not sodic
5. Permeability at least 0.38 in./hr (1.0 cm/hr) in the upper 20 in. (50 cm) of soil
6. Gravel, cobbles, or stones not excessive enough to interfere with power machinery
7. Soil deep enough to any root-restricting layer to permit adequate moisture storage for crop plants
8. Soil not excessively erodible. The universal soil loss equation K-factor multiplied by slope percentage is 5 or less.

Unique farmland is land other than prime farmland that is used for the production of specific high-value food and fiber crops. It has the special combination of soil quality, location, growing season, and moisture supply needed to economically produce sustainable high quality and/or high yields of a specific crop when treated and managed according to modern farming methods (Johnson, 1975).

Unique agricultural lands include, for example, cranberry bogs, citrus orchards, and rice fields. Three criteria characterize such lands:

1. Adequate soil moisture from whatever source
2. Soil temperatures high enough and a growing season long enough to produce a satisfactory harvest of the selected crop
3. A location that has the favorable attributes needed, such as nearness to market, good air drainage, the proper aspect (direction of slope), favorable relative humidity, and suitable soil temperature

7–4.3 Soil Surveys for Planning and Zoning

Examples of specific uses of soil survey information for planning and zoning will be briefed for a city, two counties, and for "critical areas" in general.

Canfield is a town in northeastern Ohio with a population of less than 5000. Soil surveys were used in establishing subdivision regulations in Canfield as early as 1966. Based on those properties relevant to construction, public health, and erosion and sedimentation hazard, all soil series in Canfield were placed in one of five groups as follows (Soil Conservation Service, 1967):

I: Favorable soils (well drained, 2 to 12% slopes)
II: Steep soils (12 to 50% slopes)
III: Seasonally wet soils (0 to 6% slopes)
IV: Permanently wet soils with high shrink-swell potential (0 to 6% slopes)
V: Restricted soils (flood hazard, nearly level)

Each of the first four soil groups had specific mandatory specifications for foundation construction, whereas Group V was not to be used for buildings without having very costly flood control structures approved in advance.

Figure 7–9 A zoning order based on soil maps would have prevented this house from being located on this area of Linside soil series. The soil is a Fluvaquentic Eutrochrept—the prefix "Fluv-" indicates the flooding hazard. (Courtesy H. C. Porter, Virginia Agricultural Experiment Station, Virginia Polytechnic Institute and State University.)

Black Hawk County in north-central Iowa is a prosperous agricultural and industrial area with a population of 133,000 persons. Before zoning, pressures were tremendous to use prime agricultural lands for housing developments. Prime agricultural land was defined as any land that would produce 6900 lb/ac (7735 kg/ha) or more of corn. Based on this criterion, 68% of the county was rated as "prime." After a soil survey of Black Hawk County was completed in 1973, a public hearing was held on a proposed county order to zone the county. One very controversial section of the order prohibited the use of prime agricultural lands for residential developments. The county board of supervisors passed the order and land developers moved to the surrounding counties. Within a year, however, the adjoining counties started passing similar orders (Vincent, 1977).

Walworth County, Wisconsin, using a 1966 soil survey as fundamental resource information, passed a sanitary code in 1968 and a shoreland and subdivision zoning order in 1971. Orders such as this help prevent disasters such as that shown in Figure 7–9.

Critical areas are those that either possess unique economic, recreational, historic, or cultural values for the nation, or that pose environmental hazards. Examples of national concerns include floodplains, virgin redwood forests, Native American burial grounds, lakeshores, ocean beaches, groundwater recharge areas, unique wilderness areas, waterfowl flyways, prime agricultural land, and historic trails. Soil surveys are as important for delineating critical areas as they are in preparing and enforcing city and county zoning ordinances and orders.

7–4.4 Environmental Impact Statements

The National Environmental Policy Act of (NEPA) 1969 established the Environmental Protection Agency (EPA) in the U.S. and mandated the use of environmental impact statements. NEPA applies to projects that could damage the environment whether they are undertaken by a government agency or require approval or licensing by a government agency (Bregman, 1999). Projects involving major soil disturbances, such as starting a new housing development, building a new highway, opening a surface mine for coal, or siting a power plant, must be evaluated to determine how much environmental damage would be caused and what alternatives are available. Public hearings are held where objections and suggestions can be made. The agency or company must receive a permit that is based in part on an environmental impact statement submitted to the U.S. Council on Environmental Quality before it proceeds with the project.

Sixteen states, the District of Columbia, and Puerto Rico have passed laws requiring that an environmental impact statement also be filed with a designated state agency, and many cities and counties have planning and zoning commissions that must approve all actions that are "unfriendly" to the local environment (Kreske, 1996).

Many other nations also have environmental protection policies. An international organization called OECD (Organisation for Economic Co-operation and Development, 1999) established a Project on Environmental Requirements for Industrial Permitting in 1993. OECD was established in Paris in 1960 with 20 member nations (mostly European nations, the United States, Canada, and Iceland) and had grown to 30 member nations by 2002. The nations in this organization use licensing processes to regulate various industrial and construction activities on the basis of environmental, economic, and territorial criteria.

The essentials of an environmental impact statement include an inventory of air quality, water quality, aquatic life, terrestrial wildlife, people, jobs, transportation, and endangered plant and animal life. Another part of the report consists of predictions of the changes and disturbances that will be caused by the proposed activity. Soil considerations are usually crucial in environmental impact statements because of the wind and water erosion and sedimentation that result from land disturbance (Bregman, 1999).

Teams set up to research and write an environmental impact statement should consist of subject-matter specialists trained in the various environmental areas involved. Any such team needs to have engineers, scientists, and planners who are concerned with soil, water, vegetation, and wildlife in addition to those dealing with construction of the proposed development. A forester should be a member of a study team in a forested or potentially forested area, an urban planner in an urban or potentially urban area, and a dairy specialist where dairies exist or are being considered. Because several specialties are involved, the average environmental impact team consists of 15 to 20 members. The team should be under the direction of an administrator who will ensure that all pertinent factors are considered on their merits without bias.

SUMMARY

Soils are complex and variable. They have differences that influence their use for various purposes. Soils, along with climate, water supplies, roads, buildings, and other structures, are important components of land. Land use planning is applied to areas ranging in size from a small garden to a large nation. Planning needs to be based on maps that show an appropriate amount of detail for the area involved.

Soil surveys in the United States were officially instituted in 1899. Their use has spread to most nations throughout the world. The first surveys were quite general so that a large area could be covered quickly. Since then, they have become gradually more detailed. Some generalized soil surveys, known as reconnaissance surveys, are still made, usually in new areas that have not been mapped before, but most are more detailed standard soil surveys.

Soil survey reports are published to make the soil maps and the accompanying text available for public use. The text includes a description of each map unit, a classification of the soils, and soil interpretations for various uses. The original purpose was primarily agricultural, but other uses such as land valuation and engineering uses of soils have since been added. The original maps were line drawings, but most soil maps are now published on an aerial photo-mosaic base. Most of the map units are named as phases of soil series.

Soil map units are classified into land use capability classes, subclasses, and units for farm planning purposes. A map colored to indicate the soil classes is useful for identifying parts of a farm that are best suited for particular uses or that have special management needs. Each soil map unit can be interpreted for crop adaptations and predicted yields, land value, drainage needs, woodland sites, wildlife habitats, recreational sites, engineering qualities for construction, and other purposes.

Soil surveys provide a sound basis for land management for farms and ranches. Precision agriculture is a modern method of management that subdivides fields into small management units and applies variable rates of seed, fertilizer, pesticides, and soil amendments. Soil surveys are also used as a guide for mechanizing cultivation in developing countries, for identifying areas of nutritional deficiencies and toxicities for animals and people, for proper disposal of wastes, and for general land development.

Soil survey reports (texts and maps) are valuable as a scientific basis for land use planning. Such plans should be made on a sustainable basis so that the land will be productive in the future as well as at present. The concepts of *prime* and *unique* agricultural lands were developed to preserve productive soil for essential food, feed, and fiber crops. Such areas are delimited only with the help of a soil survey. Town and county planning and zoning agencies are now using soil surveys as a basis for their ordinances and orders. Construction projects that result in major soil disturbances now require environmental impact statements. Soil factors are essential considerations in the preparation of environmental impact statements.

QUESTIONS

1. What are the principal purposes of a soil survey?
2. How is soil related to land?
3. What is the difference between soil classification and soil survey interpretation?
4. Describe the contents of any recent soil survey report.
5. Select two contrasting soil map unit interpretations and explain their significance to natural resource experts in these respective subjects.
6. What is meant by *precision agriculture* and *sustainable agriculture?*
7. Explain the use of soil surveys as a scientific basis for rational planning and zoning.

REFERENCES

BALDWIN, M., C. E. KELLOGG, and J. THORP, 1938. Soil Classification, p. 979–1001 in *Soils and Men,* 1938 USDA Yearbook of Agriculture. U.S. Govt. Print. Off., Washington, D.C.

BASINSKI, J. J., 1959. The Russian approach to soil classification and its recent development. *J. Soil Sci.* 10:14–26.

BONGIOVANNI, R. and J. LOWENBERG-DEBOER, 2000. Economics of variable rate lime in Indiana. *Precision Agriculture* 2:55–70.

BREGMAN, J. I., 1999. *Environmental Impact Statements,* 2nd ed. Lewis Publishers, Boca Raton, FL., 248 p.

DAVIDSON, D. A., 1992. *The Evaluation of Land Resources.* Longman Scientific & Technical, Essex, England. 198 p.

ELLSWORTH, J. W., 2001. *Dividing Cornfields into Soil Management Units for Nitrogen Fertilization.* Ph. D. Thesis, Iowa State University, Ames, IA., 87 p.

FAO, 1993. *Guidelines for Land-use Planning.* FAO Development Series 1, Food and Agriculture Organization of the United Nations, Rome, Italy, 96 p.

FAO, 1995. *Planning for Sustainable Use of Land Resources.* FAO Land and Water Bulletin 2, Food and Agriculture Organization of the United Nations, Rome, Italy. 60 p.

Gardner, D. R., 1998. *The National Cooperative Soil Survey of the United States.* USDA Natural Resources Conservation Service Historical Notes No. 7, 270 p.

GOLD, A. J., T. SAIPING, P. V. AUGUST, and W. R. WRIGHT, 1989. Using soil surveys to delineate stratified drift deposits for groundwater protection. *J. Soil Water Cons.* 44:232–234.

JOHNSON, W. M., 1975. Classification and mapping of prime and unique farmlands. In *Recommendations on Prime Lands.* Prepared at the Seminar on Retention of Prime Lands, USDA, July 16–17, p. 189–198.

KELLOGG, C. E., 1933. A method for the classification of rural lands for assessment in western North Dakota. *Journal of Land Public Utility Econ.* 9:10–14.

KLINE, C. K., D. A. G. GREEN, R. L. DONAHUE, and B. A. STOUT, 1969. *Agricultural Mechanization in Equatorial Africa.* Inst. Int. Agr., Mich. State Univ., Contract afr-459, 633 p.

KLINGEBIEL, A. A., 1977. Soil survey methodology—Use of Landsat for determining soil potential. In *Soil Resource Inventories.* Proc. of a workshop held at Cornell University, Ithaca, NY, April 4–7, p. 101–105.

KRESKE, D. L., 1996. *Environmental Impact Statements: A Practical Guide for Agencies, Citizens, and Consultants.* John Wiley & Sons, NY,. 480 p.

KUBOTA, J., 1975. The poisoned cattle of Willow Creek. *Soil Conservation* 40(9):18–21.

LONG, D. S., R. E. ENGEL, and G. R. CARLSON, 2000. Method for precision nitrogen management in spring wheat: II. Implementation. *Precision Agriculture* 2:25–38.

MARBUT, C. F., 1935. Soils of the United States. In O. E. Baker (ed.), *Atlas of American Agriculture,* Part III. USDA, U. S. Govt. Printing Office, Washington, D.C.

MCKEAGUE, J. A., and P. C. STOBBE, 1978. *History of Soil Survey in Canada, 1914–1975.* Agriculture Canada Historical Series 11, Ottawa, Canada.

NACHTERGAELE, F. O., O. SPAARGAREN, J. A. DECKERS, and B. AHRENS, 2000. New developments in soil classification world reference base for soil resources. *Geoderma* 96:345–357.

NATIONAL RESEARCH COUNCIL, 1986. *Soil Conservation: Assessing the National Resources Inventory,* Vol. 1. National Academy Press, Washington, D.C., 114 p. Data Source: *1982 National Resources Inventory.* USDA Soil Conservation Service.

ORGANISATION FOR ECONOMIC CO-OPERATION AND DEVELOPMENT, 1999. *Environmental Requirements for Industrial Permitting. Vol. 1 Approaches and Instruments.* OECD Publications, Paris, 99 p.

PRESCOTT, J. A., and R. L. PENDLETON, 1966. *Laterite and Lateritic Soils.* Commonwealth Bureau of Soil Science, Technical Comm. No. 47, 51 p.

SOIL CONSERVATION SERVICE, 1967. *Subdivision Regulations for Canfield, Ohio,* p. 20–35.

SOIL CONSERVATION SERVICE, 1971. *Two Thirds of Our Land: A National Inventory.* Program Aid No. 984, USDA, Washington, D.C.

SOIL CONSERVATION SERVICE, 1974. *Soil Survey of Cottle County, Texas.* In cooperation with the Tex. Agr. Expt. Sta.

SOIL CONSERVATION SERVICE, 1994. *From the Surface Down: An Introduction to Soil Surveys for Agronomic Use.* National Employee Development Staff, 1991, revised 1994, Lincoln, NE.

SOIL SURVEY STAFF, 1993. *Soil Survey Manual.* Agriculture Handbook No. 18, Soil Conservation Service.

SOIL SURVEY STAFF, 1999. *Soil Taxonomy: A Basic System of Soil Classification for Making and Interpreting Soil Surveys,* 2nd ed. USDA Agric. Handbook 436, U.S. Govt. Print. Off., Washington, D.C., 869 p.

STOMBAUGH, T. S., and S. SHEARER, 2000. Equipment technologies for precision agriculture. *J. Soil Water Cons.* 55:6–11.

STRUBEN, G. R., and M. E. LILLY, 1999. *Soil Survey of Pike County, Illinois.* USDA Natural Resources Conservation Service in cooperation with the Illinois Agricultural Experiment Station, 305 p. plus maps.

SYLVESTER-BRADLEY, R., E. LORD, D. L. SPARKES, R. K. SCOTT, J. J. WILTSHIRE, and J. ORSON, 1999. An analysis of the potential of precision farming in northern Europe. *Soil Use and Management* 15:1–8.

VAN ALPHEN, B. J., and J. J. STOORVOGEL, 2000. A methodology for precision nitrogen fertilization in high-input farming systems. *Precision Agriculture* 2:319–332.

VINCENT, G., 1977. Land use control by law. *Successful Farming* Oct., p. A6.

WHITNEY, M., 1900. *Field Operations of the Division of Soils,* 1899. In USDA Rep. 64, Washington, D.C.

CHAPTER

8

CROPPING SYSTEMS

Cropping systems—that is, the crops grown and the techniques used to grow them—should be chosen to provide the degree of protection that the land needs to replace the cover previously provided by trees, grass, or other plants in the natural vegetation. The usual result of a poor cropping system is accelerated soil loss that reduces soil productivity and increases environmental pollution.

Many cropping systems leave much of the soil surface exposed during part of the year. The impact of water or wind on the soil at such times can cause high rates of erosion and serious damage. The damage in many fields could be greatly reduced by adjusting the cropping systems to minimize exposure of the soil to the most erosive winds and rainstorms. Cropping systems should be designed to protect both present and future productivity by providing adequate vegetative cover to control erosion.

8–1 PLANT COVER

The first requirement for a crop is that it be adapted to the environment where it is to be grown. Tropical fruits need warm, frost-free climates. Warm-season crops such as corn and soybeans can evade cold winters if summer moisture is available, whereas earlier maturing crops such as small grains are grown where the summers are dry, or the growing season is short. Every crop has its own limited range of climatic adaptation.

Soil factors are also important for producing plant growth. A wet, puddled soil is good for paddy rice, but bad for most other crops. Blueberries, strawberries, and other iron-loving crops do best on acid soils, but most other crops produce maximum growth in the slightly acid to nearly neutral range.

Poor crop growth can be a disaster to the soil and the environment as well as to the grower. Poor growth exposes the soil to erosion by raindrop splash, runoff, and wind. The eroded soil is deposited elsewhere, often burying plants, filling reservoirs, eutrophying streams, and causing other damage to the environment. The remaining soil almost always has less productive potential than the original soil had.

8–1.1 Amount of Plant Cover Needed

The amount of plant cover needed to protect a soil depends on the erodibility of the soil and the intensity of erosive forces. Loose soil on steep slopes needs permanent vegetation to intercept raindrops and to limit the amount and velocity of runoff. Vegetation may be needed to deflect high wind velocities even if the land is immune to water erosion. Usually the need for cover is much greater at some seasons than at others. The soil-loss equations discussed in Chapter 6 provide a means of estimating the adequacy of the cover produced by various cropping systems in particular situations.

Environmental concerns add another reason to maintain a good vegetative cover. Growing plants reduce both the leaching of plant nutrients and the loss of soil by erosion.

8–1.2 Types of Crops

Many kinds of crops are grown in the world. Only a few can be included here, so this discussion will be limited to groups or types of crops that cover large areas of land. The full meaning of the term *crops* includes any plants or parts of plants grown for agricultural production. Most crops are included within the collective domains of agronomy, horticulture, and forestry. Usually some kind of management or culture is needed to encourage crop growth so that a good yield can be harvested. The harvest may come after a few weeks or months, or it may be delayed for years. Annual crops must be planted every year, but certain perennial crops may be planted once and harvested many times.

For discussion purposes, crops will be divided into row crops, small-grain crops, cultivated forage crops, and tree crops. Some characteristics of each group will be considered in this section; management factors will be discussed in later sections.

Row Crops. Many crops have traditionally been planted in rows about 40-in. (1-m) apart so they can be cultivated. Often these are a farmer's most profitable crops and therefore the ones that receive the most attention. Corn, cotton, potatoes, sorghum, soybeans, sugar beets, sugarcane, and sunflowers are examples of field crops that are usually grown in rows. Truck crops, such as most vegetables and small fruits, are also grown in rows. The rows facilitate tillage to control weeds, spraying or dusting of pesticides, application of supplemental fertilizer as side-dressing after the crop is established, and finally the harvesting of the crop.

Row crops frequently create problems for soil conservationists. An unprotected area between the rows, as shown in Figure 8–1, may be exposed to several erosive rainstorms. Cultivation keeps the soil loose and erodible. Rills form easily where the rows guide water down a slope. More sheet, rill, and gully erosion occur under row crops than under close-growing crops in a similar environment. The frequency of growing row crops in a rotation is often limited by the erosion hazard.

Small Grains. Rice, wheat, barley, oats, and rye are known as small grains or cereal crops. These crops are widely adapted and are used to produce bread, cereals, and other

Figure 8–1 The soil between these young soybean rows was puddled by raindrops and crusted when it dried. (Courtesy F. R. Troeh.)

foods for people and animals. Rice is the main staple of many areas with warm climates and will grow in wet conditions that exclude most other crops. Rye, wheat, barley, and oats produce most of their growth during cool seasons. They are often the only cash crops grown in areas where cool climates cause short growing seasons and where dry summers in warmer areas limit crop production to the cooler part of the year. Summer fallow helps to extend their range into still drier climates. In more humid areas, the small grains may be grown in rotation with row crops and forage crops.

Small grains are usually drilled in rows about 6- to 10-in. (15- to 25-cm) apart or else broadcast. Their fast early growth and relatively close plant spacing provide much more erosion control than row crops, but not as much protection as most forage crops.

Forage Crops. Crops grown to be fed to livestock as pasture or hay are known as *forage* or *fodder crops.* Many grasses and legumes are included either singly or in combination. Some are annuals, some are biennials, and some are perennials. Some are native plants, some are introduced from other parts of the world, and some have been developed by intensive plant breeding.

Forage crops are maintained as permanent cover on areas that are not cultivated because of soil or climatic limitations or some other reason. They are grown elsewhere as part of the crop rotation. They may occupy the land for only a few months in some fields, for a year in others, and for several years in succession in still others.

Forage crops are grown with close spacing between plants except in areas too dry to support dense vegetation. Maximum cover is attained by planting crops that grow rapidly, by planting the forage crop along with a companion crop, or by maintaining established stands for a long time.

TABLE 8–1 NUTRIENTS CONTAINED IN TYPICAL YIELDS OF SEVERAL CROPS

		Nutrients (lb/ac)					
	Yield (tons/ac)	N	P	K	Ca	Mg	S
Forage crops							
Alfalfa hay	4	230	15	176	118	25	22
Red clover hay	2.5	104	11	80	61	17	7
Smooth bromegrass hay	2.5	99	12	94	17	11	11
Sweetclover hay	5	219	24	135	144	39	41
Timothy hay	2	61	10	86	18	4	5
Row crops							
Corn grain	4	113	23	25	2	9	10
Grain sorghum (milo)	3	48	18	22	2	7	6
Potatoes	9	60	9	86	2	5	4
Soybeans	1.5	188	19	66	10	11	7
Sugar beets	20	77	4	80	40	12	9
Small grains							
Barley	1.5	55	10	13	2	4	4
Oats	1.5	57	10	12	2	4	6
Rice	1.5	36	10	13	2	4	1
Rye	1	38	6	9	1	2	3
Wheat	1.5	62	11	12	2	4	5

Source: Calculated from percentage compositions in M. H. Jurgens, *Animal Feeding and Nutrition,* 7th ed., 1993. Kendall Hunt, Dubuque, Iowa, 580 p.

Forage crops are often considered to be "soil-building crops." This concept is true in some ways but false in others. Improved soil fertility is often inferred because many forage crops are legumes, and nitrogen-fixing *Rhizobium* bacteria grow on their roots. Crops such as alfalfa or sweetclover have been used extensively to replenish the nitrogen supply of the soil before growing a corn or wheat crop. A large tonnage of plant material should be disked or plowed into the soil when a positive fertilizer effect is desired. Leaving only roots and stubble results in little or no net addition of nitrogen to the soil and causes a significant removal of all other nutrients. As shown in Table 8–1, forage crops remove larger amounts of many nutrients than most other crops remove. This soil-depleting effect can be reduced by feeding the forage to livestock and returning manure to the land.

The close spacing of forage plants coupled with the improved soil structure and permeability they produce provide good protection against erosion. Forage crops are therefore regarded as soil conserving and are useful where the erosion hazard is too great for other crops.

Tree Crops. Trees are grown for many purposes. They produce many valuable products, including wood, paper, fruits, nuts, medicines, and Christmas trees. They have many ornamental uses, and they are used to help restore fertility to tropical soils. Trees also help purify the air by assimilating large quantities of CO_2, lesser amounts of sulfur compounds such as SO_2, and some nitrogen gases such as ammonia (NH_3). Some trees are grown in pure stands, some are mixed with other species of trees, and some grow along with various shrubs, grasses, and herbaceous plants. Growing trees along with grass or field crops is called *agroforestry*. In its broadest sense, agroforestry includes settings that range from an occasional tree in a lawn or along the border of a field to strips of trees that help control erosion and produce valuable wood and/or food products.

Tree crops are suited to a wide range of environments. Apple trees grow where the soil freezes in the winter, oranges and other citrus fruit that are sensitive to frost are grown in warmer climates, and coconuts require tropical conditions. Trees grow in some of the warmest and wettest climates on earth and prevail far into the cold climates, but give way to smaller plants in drier climates.

Trees, combined with undergrowth and a litter layer, provide strong protection against erosion. Raindrops seldom strike the soil surface beneath trees because they are gently absorbed by the litter layer after having been intercepted several times by the leaves, branches, and undergrowth. The protected soil surface retains a porous structure and the infiltration rate is much faster than it would be if a crust could form. Increased infiltration reduces the runoff volume. Runoff velocity is slowed as the water trickles through the litter layer. Most soil loss from such settings is either by solution erosion in the percolating water or by sudden, rare, intense rains that produce enough concentrated runoff to wash away the cover and cut a gully. Even then, areas outside the main flow of runoff may be undamaged.

Not all trees provide as much protection as was outlined in the preceding paragraph. Exposed soil in a clean-cultivated orchard, for example, is subject to erosion. Raindrop impact can be damaging even under the trees because drops falling off tree branches are often large. Erosion control requires cover to intercept these drops again nearer the soil surface.

8–1.3 Plant Population and Row Spacing

A bluegrass pasture normally has millions of plants per acre (0.4 ha), while a few hundred trees may cover a nearby acre. Most other plant populations fall somewhere between these extremes. The number of plants is, of course, likely to be inversely related to their size.

Increasing the population of any given type of plant per acre can improve both crop yields and erosion control. Newer varieties of some crops have been developed to be grown at higher populations than were formerly common. For example, top corn yields are likely to require 20,000 to 28,000 plants per acre (50,000 to 70,000 per hectare) instead of the old standard of about 12,000 plants per acre (30,000 per hectare). The older varieties do not produce well at the higher densities because crowding causes many barren stalks and reduces grain production.

Increased plant population in row crops is achieved by using narrower rows, closer spacing in the row, or both. Spacing plants equally in all directions would minimize crowding and maximize protection against erosion, but cultivation and harvesting practices often require that certain crops be grown in rows.

Field crops such as corn, cotton, potatoes, sorghum, soybeans, and sugar beets were planted in rows 40-in. (1-m) apart so that a horse could walk between the rows for cultivation. The 40-in. spacing survived in many places long after tractors replaced horses. Narrower rows of 20- or 30-in. (50- or 75-cm) spacing have increased in popularity in recent decades where tractors have become the rule and herbicides have eliminated much of the cultivation. Some variable spacings are used with wider intervals for the wheel tracks than for other rows.

Narrow rows are often elected because they usually increase crop yields. The plants are better spaced to absorb sunlight, and the more complete crop canopy keeps the soil cooler so less water is lost by evaporation. Erosion control is also improved. Crops in 20-in. (50-cm) row spacing, for example, grow together and cover the space between the rows at an earlier date than those planted in 40-in. (1-m) rows. Where the rows are narrow, fewer raindrops are able to strike the bare soil and cause erosion, more water infiltrates because the soil surface is less likely to crust, and less water accumulates in any one place to become erosive runoff.

8–1.4 Soil Fertility and Fertilizers

Dramatic reductions in erosion and sedimentation can often be achieved through proper fertilization. Well-fertilized crops grow more vigorously and protect the soil much more effectively than weak ones. Fertilizers are therefore as important for conserving soil as they are for increasing yield.

A good soil fertility program combined with the best crop varieties and close plant spacing can reduce soil loss to less than half as much as would occur under the same crop with low fertility and wide row spacing. Higher fertility produces a better stand and larger plants at all stages of growth, including the critical early period when the soil is most exposed to erosion.

The proper amount and timing of fertilizer applications depend on soil, crop, and weather. The fertility status of the soil may be estimated if one knows the nature of the soil and its cropping and fertilizer history. Rather than rely on such an estimate, however, it is usually better to test the soil at least once every four years. The cost of such tests is easily regained through more accurate fertilization. Excess fertilizer is expensive and can contribute to water pollution; too little fertilizer results in reduced yield. Variable fertility can be managed with modern fertilizer equipment combined with GPS (global positioning system) that make it possible to adjust fertilizer rates as needed to better optimize fertilizer use for different parts of a field. This system, known as *precision agriculture* (Stombaugh and Shearer, 2000), produces high yields while saving fertilizer and reducing pollution.

Any available sources of organic fertilizer should not be overlooked in a fertility program. Organic fertilizers such as manures and sewage sludge are likely to not only provide plant nutrients but also increase soil organic matter and reduce erosion. For example, Nyakataw et al. (2000) found that applications of poultry manure promoted good emergence, vigorous growth, and high yields of cotton and helped control erosion in Alabama. The return of plant residues serves a similar purpose and may be especially important in developing countries where little fertilizer is used. Kayuki and Wortmann (2001) suggest that plants from hedgerows and other areas can provide green manure for Ugandan soils and others that contain little organic matter. A combination of organic and mineral fertilizers is often best; the amount of organic fertilizer can be adjusted to supply enough of one nutrient and mineral fertilizers used to add whatever other nutrients are still needed.

Each crop has its own fertility needs and its own pattern of response to variations in the availability of essential nutrients. A few examples will illustrate this point. Legumes such as beans, peas, and clovers normally do not need nitrogen fertilizer because they obtain nitrogen from their symbiotic relationship with *Rhizobium* bacteria. They do need adequate supplies of all other nutrients, and many legumes, especially those used as forage crops, need enough lime to keep the soil pH near neutral. Most nonlegume crops, on the other hand, respond dramatically to nitrogen. Large amounts of nitrogen fertilizer are used for members of the grass family, such as corn, small grains, and the forage grasses.

The tonnage produced and the part of the plant harvested influence the fertilizer needs of both the current crop and the next crop to be grown. As was shown in Table 8–1, a crop of 5 tons of sweetclover hay removes much larger quantities of nutrients than a 1.5-ton crop of barley or oats. The large hay crop demands a good supply of nutrients. Also, the next crop may require extra amounts of potassium fertilizer because of the large amount of K^+ removed in the hay.

Weather influences fertilizer use in several ways, including the choice of crop, the rate of crop growth, the loss of nutrients by leaching and erosion, the availability of nutrients that remain in the soil, and the ability of the soil to support a truck loaded with fertilizer. Even an adapted crop may suffer if the weather is cooler than usual in the spring or warmer than usual in the summer.

Weather that is cool enough to retard growth is often wet enough for poor drainage to further limit growth on wet soils, or for leaching to deplete the supply of available nutrients in drier soils. Intermittent wetness leads to the loss of much available nitrogen by denitrification. Warm, dry weather also can limit nitrogen availability as capillary movement and evaporation at the soil surface combine to concentrate nitrates in a surface crust where there is very little root activity. A good rain can wash this nitrogen back down into the root zone and cause the crop to turn green again.

Wet conditions can seriously hamper potassium absorption by plants. The absorption process requires energy because potassium is usually more concentrated inside the root than in the soil solution. Excessive wetness reduces aeration and slows the oxidation processes that provide energy for nutrient absorption and root growth. Extra potassium is therefore needed in wet years and in poorly drained soils.

8–1.5 Seasonal Changes in Plant Cover

Seasonal variations occur in both the amount of plant cover produced by a crop and in the amount of protection the soil needs against erosive forces. A storm that causes disastrous erosion when there is little plant cover on the ground might cause very little damage later in the season when plants are larger and offer more protection.

The most hazardous periods occur when the soil is exposed by tillage before a new crop is planted and during the early growth stages while the plants are too small to adequately protect the soil, as shown in Figure 8–2. The erosion hazard diminishes as the percentage of bare ground remaining between plants decreases. Such changes are considered as components of the cropping factors in the soil-loss equations discussed in Chapter 6.

Seasonal changes are most significant where annual crops are grown because the soil often has little cover during a cold or a dry season between crops. Some soils suffer excessive erosion, for example, when the land is plowed after harvest in the fall and left exposed until a spring crop is planted. The exposure time is even longer where a year of summer fallow is used. A period of bare soil may sometimes help control weeds, insects, and plant diseases, but it greatly increases the erosion hazard. Usually, there are cropping and tillage alternatives that provide cover most of the time in the form of either growing plants or plant residues. For example, one of the advantages of winter wheat over spring wheat is the cover that a fall-planted crop provides during the winter and early spring. The same is true of other small grains in areas where either a winter or a spring variety may be planted. Similarly, fall-planted chickpeas are replacing some of the spring-planted varieties in the Mediterranean

Figure 8–2 The erosion hazard is greatest while plants are small and diminishes as the growing plants cover the soil more completely. (Courtesy F. R. Troeh.)

region (Singh et al., 1997). The fall plantings of chickpeas average 70% better yield, give more consistent results, and provide winter and early spring cover that the spring plantings cannot provide. Tillage practices that leave plant residues on the surface can also help hold the soil and conserve moisture for the next crop, as discussed in Chapter 9.

Perennial crops also have seasonal changes that should not be overlooked. The cover in a hay field is greatly reduced for a time after the hay is cut. Even a pasture may be grazed heavily enough to permit storms to cause erosion. Overgrazing is especially serious when it occurs along with trampling damage during a wet season, or while the vegetation is not growing because of drought or cold weather.

8–2 MANAGING MONOCULTURES

Cropping systems that always grow the same crop on the same land are known as *monocultures.* The advantages and disadvantages of monocultures and crop rotations have been long debated without either system being eliminated. Each system is important in its own time and place. Monocultures will be discussed in this section and crop rotations in Section 8–3.

A monoculture permits a farmer to specialize in a crop and to manage the land for the benefit of that crop. Usually the chosen crop is highly profitable and well suited to the soil and climate.

Sometimes the land on a farm justifies two or more monocultures rather than a rotation. For example, a farmer in a humid region might grow a row crop continuously on level land, hay on sloping land, and pasture or trees on steep land. Another farmer in a drier climate might use large areas for rangeland, grow wheat on the most productive nonirrigated land, have an orchard on permeable sloping irrigated land, and perhaps use a crop rotation on the flatter irrigated terraces. Wet bottomland might be used for pasture. With these systems, each crop is grown where it fits best rather than over the whole farm in a rotation.

8–2.1 Annual Cash Crops

Arguments against monocultures usually center on the effects of growing annual crops such as wheat, corn, or cotton year after year on the same land. Farmers often grow these crops on as much land as possible as a source of cash income. Monocultures put the most profitable crops on the best land every year rather than alternating them with other crops.

It was long assumed that certain monocultures, such as continuous corn, would ruin land. Indeed, yields from such systems soon dropped to low levels when they were tried without fertilizers. But later it was found that well-fertilized continuous corn could produce high yields, though probably not as high as that grown in rotation (Crookston and Kurle, 1989). The yield results satisfied some people, but others argued that the soil structure was deteriorating and the rate of erosion was excessive. The argument continues even now, partly because each side has situations that illustrate its point, and partly because some people worry more than others about erosion and soil structure.

Corn as a monoculture is now considered acceptable on certain soils, but other soils are damaged by such treatment. The damage is either excessive soil erosion and sedimentation on rolling land or breakdown of soil structure and poor drainage on level land. Users of monocultures and other intensive-use cropping systems should be alert to such problems. Sometimes an intensive rotation is worse than a monoculture. For example, Van Doren et al. (1984) found erosion from a corn-soybean rotation to be 45% more than that from continuous corn. They concluded that soybeans predisposed the soil to water erosion. It should be noted that a corn-soybean rotation consists of two row crops and that both are grown as annual cash crops. As such, it lacks the regenerative features of the older corn-oats-meadow rotations and is described by some as continuous row crop rather than as a rotation.

Several things can be done to reduce the detrimental effects that may occur with intensive cropping systems. First, such systems should be limited to suitable soils. Next, crops should be planted in narrow rows and provided with all the lime, organic and mineral fertilizers, improved drainage, and other factors the crop needs to grow well. Then, cover should be provided during the off-season by leaving crop residues on the surface to protect against erosion while there is no crop growing and to return as much organic matter to the soil as possible. Or, where feasible, growing a cover crop during the off-season is an excellent way to control erosion and add organic matter to the soil.

8–2.2 Summer Fallow Systems

Large areas of semi-arid land are fallowed every second or third year, or on a variable schedule depending on weather and other factors. Much of this land produces wheat or perhaps barley or rye whenever it is cropped, so the system is a monoculture that sometimes has long intervals between crops. The purpose of the fallow is to accumulate water in the soil to increase the supply for the next crop, as discussed in Chapter 13 (Section 13–8), or sometimes to combat noxious weeds.

Weeds and volunteer plants must be controlled if water is to be conserved during the summer fallow period. Traditional summer fallow was often called *black fallow* or *clean fallow* because tillage with plow and disk implements buried all of the crop residues. Sometimes the residues were burned to make it easier to plow the field. The result was exposure to erosion by both water and wind during the 15- to 18-month period between crops. One response was to substitute a large disk for the plow. This was called *trashy fallow* because it left some residues on the surface, at least for the first few months, but excessive erosion was still common. Chisel plows have also been used for this purpose, but any tillage that penetrates very deep causes water to be lost by evaporation from deep cracks. Newer implements that kill weeds with minimal soil disturbance by undercutting at shallow depths with wide sweeps or rod weeders are discussed in Chapter 9 (Section 9–3). Another alternative, known as *chemical fallow,* is to use herbicides to kill the weeds.

Many people question the value of fallow for increasing production. Crop yields are often higher than they would be with annual cropping, but there is no yield at all during the fallow year. Nielsen (2001) discusses the use of short-season annual legumes in rotation with winter wheat in areas of the Great Plains that have traditionally used a wheat–fallow system. He suggests that early-planted chickpeas, field peas, or lentils can mature and be harvested in time for summer rains to replenish the soil water so winter wheat can be planted in September. This system provides a crop every year, nitrogen fixation by the legumes, and improved control of weeds, insects, and plant diseases.

Fallow systems are also questioned on the basis of long-term sustainability. Organic matter is lost both by decomposition during the long fallow period and by erosion, and the crop residues returned are not nearly enough to replace it. Consequently, both the fertility and the physical condition of the soil deteriorate, especially under intensive tillage systems. Schillinger et al. (1999) found that a no-till system helps slow the decomposition and reduce erosion losses, but there still are not enough crop residues to offset long fallow periods. They suggest that a no-till system may conserve enough water to make annual cropping possible in many areas and that this would be more sustainable than fallow systems.

8–2.3 Forage Crop Monocultures

Close-growing perennial grasses and legumes are often used for hay or pasture on land that is too steep for row crops and small grains. These forage crops are often combinations of grasses and legumes, so they are not strictly monocultures. However, the same crop remains on the land year after year, and it is convenient to group these mixed forages with the monocultures.

Good management of land in forage crops can increase profit and reduce soil loss at the same time. Three main factors are involved: soil fertility practices, good grazing or harvesting management, and occasional reseeding where it is feasible and needed. Some pastures in humid regions also need to be clipped (mowed) once or twice a year to keep weeds, trees, and brush from invading the pasture.

Fertilizer and lime can be used not only to increase forage production but also to help control the composition of mixed forages. Liming to near neutral pH combined with adequate phosphorus, potassium, and any other deficient nutrients will help maintain legumes in the stand. Nitrogen applications favor the grasses and are often omitted where the grasses might crowd out the legumes.

Managing the livestock in a pasture and the haying equipment in a hay field are the most important means of controlling erosion on these lands. Enough plant growth should be left on the land to protect the soil and to maintain vigorous plants. The plants should be allowed to reseed periodically, especially on land that is never tilled. Late-fall and early-spring grazing should be avoided or limited to avoid weakening the plants.

Systems of rotation grazing usually result in more forage production and better utilization than uncontrolled grazing. Large pastures may be managed like rangeland as discussed in Chapter 12. Smaller, more intensively managed pastures are sometimes subdivided into daily pastures by means of a movable electric fence. The livestock are concentrated in a small pasture so they eat all the forage in one day. This system causes livestock to eat forage that they would leave untouched if they had more room to roam. Rotation grazing thereby achieves rapid and complete forage utilization and a long regrowth period. These are normal characteristics of a well-managed hay field.

Much land in forage crops is too shallow, stony, rough, wet, or otherwise limited to ever be tilled. Some land, however, is used to produce hay or pasture because of a steep slope, dry climate, or some other factor that does not prevent occasional tillage. It is sometimes worthwhile to till such land and reseed it to improve the forage composition. Often the legume component is increased by reseeding, and the dominant grasses may be replaced with more desirable species. A small-grain companion crop may be grown with the new seeding. Erosion is limited because the sod from the previous forage crop helps hold the soil until the new crop becomes established. The fast-growing grain crop also helps control erosion, and the harvested grain helps defray the costs. Alternatively, a range or pasture may be renovated without plowing. An herbicide or perhaps shallow tillage with a disk may be used to kill part or all of the old vegetation before reseeding.

8–2.4 Tree Crops

Forest management is discussed in Chapter 12, but crops such as orchards and Christmas trees are included here. Agroforestry, where trees and cultivated land are intermixed, is discussed in Section 8–5.2. Tree crops provide income and erosion control in many places where other crops are less desirable. The scale of tree-crop enterprises ranges from a backyard tree or two to large orchards covering many acres. Some enterprises consist entirely of tree crops, whereas others use them only in odd corners, on steep slopes, or in other problem areas. Tree crops are versatile and could be useful to many people who do not presently grow them.

Tree crops occupy land for years at a time. The harvest usually extends over a period of years either as an annual fruit harvest or by selective tree-cutting. The crop may be extended indefinitely by planting new trees when old ones are removed.

Properly managed tree crops provide excellent erosion control, even on steep land, if they cover the soil with a litter layer. Trees in an orchard usually need a grass or legume cover crop between the trees. The needs of both the tree crop and the cover crop must be considered when applying lime, fertilizer, and irrigation water. A hay crop is sometimes removed from between fruit trees before the fruit is harvested. The cover crop helps protect the soil from erosion and from the traffic that passes through the orchard. The drastic amount of erosion that can occur where there is no cover crop shows in some orchards where each tree stands on an island of soil surrounded by eroded areas.

Some crop rotations in the tropics use trees as soil-improving crops in the slash-and-burn system of shifting cultivation (Section 20–3). This makes agriculture possible in places where soil fertility is difficult to maintain and currently sustains some 300 million people (Kleinman et al., 1995). The trees improve both the soil structure (through the effects of their root systems) and the soil fertility (through their ashes after they are burnt) that deteriorate under annual crops. The system works where there is enough land so that one to three years of annual crops can be rotated with 10 to 20 years of forest. Unfortunately, increasing population pressure is making the forest period too short in many areas. The forest no longer has enough time to reestablish itself and renew the soil before it is burned again. The increased burning that results adds a large amount of CO_2 and other greenhouse gases to the atmosphere, thus contributing to global warming. Commercial fertilizers can be used to help maintain the soil fertility, but they do little for the soil's physical condition and its need for erosion control.

8–3 CROP ROTATIONS

Two or more crops alternating on the same land constitute a crop rotation. A rotation may control erosion, plant diseases, and other problems on land whereas a monoculture of the most profitable crop might be disastrous. The soil-conserving crops not only protect the soil while they are growing but also have a carryover effect that reduces erosion while the next crop is growing. The soil loss from a field of cotton, for example, might be only half as much following a forage crop as it would be if the cotton followed cotton. Rotations can also break insect and disease cycles, control persistent weeds resulting from monocultures, and avoid allelopathic toxins such as those that reduce the yield of corn following corn.

A crop rotation often provides more continuous cover than is possible when the same annual crop is grown year after year. For example, growing a warm-season row crop two years in succession often leaves the soil without any crop during the winter. However, a cool-season forage crop might be grown to conserve and improve the soil during that time if the winter is not too severe. Some forage crops are plowed under as green manure before planting a row crop; others are left for one or more years of hay or pasture in the rotation.

A crop rotation may also help by providing a good supply of plant nutrients through decomposition of various types of organic matter. Sanchez et al. (2001) found that a mixture of organic residues decomposed faster and released nitrogen faster than the residues from a corn monoculture.

Crop rotations necessarily differ from one part of the world to another in response to climatic conditions, crops grown, soil properties, and the kind and severity of erosion problems. The crops and management suited to tropical areas with wet-dry seasons, for example, differ from those adapted to the warm-cold seasons of temperate regions. But both of these climates have a season when the soil may be unprotected because the weather is unfavorable to the primary crops of the area. Rotations that provide cover during these dry or cold seasons help control erosion.

TABLE 8–2 FIELD PLAN FOR A CORN-CORN-OATS-MEADOW CROP ROTATION

	Years				
	1	2	3	4	5[a]
Field 1	C	C	O	M	C
Field 2	C	O	M	C	C
Field 3	O	M	C	C	O
Field 4	M	C	C	O	M

[a]The fifth year is the same as the first. Succeeding years are repetitions of the second and later years.

8–3.1 Planning a Rotation

The sequence of crops in many rotations is fixed and repetitive and can be projected as many years into the future as desired. Such rotations are designed to produce nearly constant amounts of each crop each year as a basis for planning and managing crop production. Planning such a rotation begins by dividing the land into as many parts as there are years in the rotation. A four-year rotation, for example, requires four fields or groups of fields of fairly equal productive capacity.

As an example of a fixed rotation, consider a farmer using a corn-corn-oats-meadow rotation (CCOM). Six fields on this farm are to be rotated. These fields contain 54, 48, 45, 18, 18, and 15 ac, respectively. The three small fields are equivalent to one field of 51 ac, so the rotation can be applied on the basis of four fields containing 54, 48, 45, and 51 ac. Each field can be assigned to a specific crop each year, as shown in Table 8–2. With this arrangement, there are always two fields in corn, one in oats, and one in meadow.

A farmer using a three-year rotation on the example farm could match each small field with one of the larger fields to form three pairs totaling 69, 66, and 63 ac, respectively. Or, two 99-ac field groups could be formed for a two-year rotation.

If two or more rotations are used on the same farm, each should have its own set of fields with balanced productive capacities. The total production of the farm is automatically balanced when the individual rotations are balanced.

Fixed rotations allow productivity and erosion control to be adjusted to the capability and needs of the land, but such scheduling is not always possible. Variable weather, for example, may require that plans be adjusted sometimes. Rotations can still be beneficial under variable conditions, but their effects and benefits are less predictable than those of fixed rotations.

8–3.2 Companion Crops

A small-grain crop and a forage crop are often planted together as companion crops. The small-grain crop grows rapidly and is harvested within a few months. The forage crop is often alfalfa or a clover and/or a perennial pasture grass. It becomes well established about the time the small grain is harvested. The forage crop is used for hay or pasture for one or more years (and/or for green manure as discussed in Section 8–3.4).

The small-grain member of companion crops has sometimes been called a "nurse crop." This erroneously implies that the tiny grass and legume plants need protection from the elements. Actually, the forage crop planted alone becomes established more rapidly and often produces a better stand. Companion crops compete for water, plant nutrients, and sunlight, causing the forage crop seedlings to develop slowly. It is usually wise to plant less small-grain seed per acre when there is a companion crop than would be used for maximum grain yield.

Fertilizer can be used to help control the growth of companion crops. Nitrogen would favor the grain crop and should be used sparingly. Most other fertilizer elements and lime should be applied at rates determined by soil tests to encourage growth of the forage crop. Grass-type forages can be fertilized with nitrogen after the grain crop has been harvested.

Companion crops help control soil erosion and sedimentation. The small grain starts fast and provides reasonable protection within a short time. It covers the soil during the establishment period for the forage crop. Harvesting the small grain leaves both its residue and the growing forage crop to protect the soil. Later, the close-growing forage crop provides excellent protection, and its soil-conserving effects carry over into the succeeding crop.

8–3.3 Cover Crops

Cover crops protect the soil by filling gaps in either time or space when the other crops would leave the ground bare. Cover crops in rotations are grown during cold or dry seasons that are unfavorable to the cash crops. Cover crops in orchards are grown between and beneath the trees. Some of them are harvested, but their main purpose is erosion control. Dry climates limit their use because they use soil water that is needed for the main crop. However, in humid climates, the use of soil water by cover crops is helpful for reducing leaching. Wyland et al. (1996) found a 65 to 70% reduction in nitrate leaching when cover crops replaced winter fallow on a vegetable farm.

Hardy plants are needed to withstand the cold or dry conditions when cover crops are grown in rotations. Some of these are slow-starting plants that can be seeded as companion crops along with the preceding crop or interseeded between the rows before a row crop gets too large. Abdin et al. (1998) interseeded several different cover crops into corn when it was 6- to 12-in. (15- to 30-cm) tall. Vetch, several clovers, and black medic provided erosion protection and organic matter without reducing corn yield, but crimson clover was too competitive and stressed the corn.

Fast-growing plants that can become established rapidly can be used as cover crops by planting them after the previous crop is harvested. Fast-growing types include Austrian winter peas, rye, oats, and ryegrass. Timing is critical to allow the fast-growing types to grow enough between harvest and the cold or dry weather so they will protect the soil.

Small grains such as rye or wheat are good choices for cover crops when long-lasting residues are needed for erosion control; hairy vetch, sweet clover, red clover, crimson clover, or some other legume may be chosen when a nitrogen contribution is considered particularly important. Several studies have shown higher yields of corn or other main crops when a cover crop was grown during the winter, but corn yields can also be reduced when a cover crop uses too much of the available water. This problem is sometimes averted by using a cover crop that winterkills or by using a herbicide to kill the cover crop before the critical period.

Cover crop competition can provide assistance for weed control. Fisk et al. (2001) found that using an annual medic or a clover planted as a cover crop in no-till corn suppressed weed growth by as much as 80%. The weed control and nitrogen fixation provided by legume cover crops may be especially significant for organic farmers who are avoiding the use of pesticides and mineral fertilizers. Hively and Cox (2001) found that interseeding annual medics or clovers in a soybean crop resulted in higher yields of the following corn crop.

The fertilizer needs of cover crops must be considered along with those of the other crops. Otherwise, yield decreases may occur, especially if a nonlegume is used without adding nitrogen. However, a cover crop used for green manure (discussed in the next section) can protect nitrates and other available nutrients from leaching and other losses and release them when the next crop needs them (Meisinger et al., 1990). Herbicides are sometimes used to kill or stop the growth of cover crops so that a no-till crop can be planted. Such a system gives maximum erosion protection for growing row crops.

Perennial forage crops such as alfalfa and/or grasses are often used as cover crops in orchards and vineyards. A mixture including one or more grass species produces the thickest stand and provides the best erosion control. Legumes in the mixture help to reduce the nitrogen fertilizer requirement of the crops.

A vigorous cover crop helps control erosion in an orchard or vineyard, but it may interfere with harvesting the fruit. Two alternatives are available: the forage crop may be harvested and removed, or it may be beaten down into a mat before the fruit is harvested. Another problem in dry climates is that the cover crop competes with the fruit crop for water (Gómez et al., 2001). This water usage may be partially offset by the higher infiltration rate and reduced runoff effects of the cover crop.

8–3.4 Green Manure

Plowing or disking a growth of forage into the soil as green manure benefits a soil's organic-matter content, structure, and permeability. Either the growth produced by a cover crop or the last growth of a forage crop may be used in this manner. A crop destined for failure caused by unfavorable weather in a marginal cropping zone also may be used for green manure. The more growth there is, the better, but young succulent growth contains a higher concentration of plant nutrients and decomposes faster than older material.

Green manure easily adds more tons of organic matter per acre to the soil than would likely be added in manure or other organic materials spread on the soil. The fresh organic matter decomposes readily and the soil microbes multiply. This microbial activity produces cementing agents that have a strong positive effect on soil structure. Soil permeability and aeration are often increased by new root channels, by tillage channels that are held open by the plant residues, and by improved soil structure.

A legume cover crop can fix significant amounts of nitrogen that will be available for a succeeding crop if it is worked into the soil as a green manure. Sainju and Singh (2001) recommended hairy vetch for such use in Georgia because its residues decomposed rapidly enough to provide nearly all the nitrogen needed when the vetch was killed a few weeks before planting a corn or tomato crop. The decomposition process is also good for soil structure because active microbes produce exudates that stabilize soil aggregates.

Green manure crops are often followed by impressive crop yields attained with little or no fertilizer. Part of this effect can be attributed to the nitrogen fixed when a legume is used for green manure. But the effect also occurs with nonlegume green manures and when nutrients other than nitrogen are needed. The release of nutrients from the decomposition of the green manure is the key to this enhanced fertility. Nutrients that are normally released slowly from the soil can often be made available fairly rapidly by decomposition. The green manure crop thus serves to accumulate available nutrients for release to the next crop.

8–3.5 Crop Residue Utilization

Crop residues are much too valuable to be ignored. Sometimes they are used to feed livestock, either by allowing the livestock to graze in the field or by hauling the straw, stalks, or other residues to them. Other uses are sought by researchers seeking ways to make plastic or an energy source such as alcohol from the residues. But, to a soil conservationist, crop residue utilization means using residues either as mulch to protect the soil or as raw material for soil organic matter. Using the residues to protect the soil is important enough to claim priority. Enough residues should be left to control erosion and maintain satisfactory soil physical conditions.

The soil protection afforded by crop residues is roughly proportional to the percentage of the soil surface that they cover. Residues that have been plowed under no longer provide soil cover. Leaving residues on the soil surface is an easy way to conserve soil and water.

Different crops produce different kinds and amounts of residues. The stubble and straw from a small-grain crop usually provide good erosion protection if they are simply left in place; neither wind nor rain is able to exert much erosive power on the soil surface. The coarser residues left from a corn crop may need to be beaten down or chopped into smaller pieces to cover the soil between the rows.

Any tillage performed on the residues will affect their value as a mulch. Plowing will cover most of the residues and leave little or no soil protection. Disking covers about half of the residues but still leaves considerable protection. Chisel-type implements often leave about 80% of the residues on the surface. The practice of leaving significant amounts of residues on the surface during periods between harvest and planting, as shown in Figure 8–3, is known as *stubble mulching.*

Negative feelings toward stubble mulching and other crop-residue utilization practices arise when residues plug tillage implements. Many older implements plugged easily, but newer models have been greatly improved with straw cutters and large open frameworks so they can work through the residues without plugging. The old practice of burning wheat stubble or other crop residues has been mostly eliminated in the United States and many other countries by using these machines and adding fertilizers that help decompose residues. Aulakh et al. (2001) found that the same principle worked in Southeast Asia with a wheat-rice rotation; wheat residues supplemented with nitrogen fertilizer and disked into the soil were good for both the soil and the succeeding rice crop. Using the residues in this way instead of burning them also reduced air pollution.

Plant residues mixed with soil decompose at various rates depending on the nature of the residues, the supply of available nutrients (especially nitrogen), and the soil temperature, moisture, and aeration. Half or more of the residues will decompose during the first few months if conditions favor microbial activity. The remainder is more resistant and decomposes more slowly. The last remaining organic material is resistant enough to be considered humus. Humus is finely divided, and plant parts are no longer identifiable. Probably less than 20% of the original residue weight remains, and most of that is microbial residues.

Figure 8–3 Stubble mulching protects the soil with both crop residues and clods. (Courtesy Washington State University.)

The time required to convert residues into humus ranges from a few months in a well-drained tropical soil to many years in a cold climate or centuries in a swamp or bog.

Soil humus decomposes much more slowly than does fresh plant residue, but it does decompose and release plant nutrients. Humus decomposition needs to be offset by new humus formed from residues to maintain the organic-matter content of the soil. The microbes that carry out decomposition produce exudates that help hold soil aggregates together and stabilize soil structure. Decomposing organic materials thus contribute to both soil fertility and to a desirable soil physical condition. Organic materials that are not in the process of decomposing contribute relatively little to soil fertility and soil structure but may increase permeability and reduce erosion and sedimentation.

Decomposition is often slowed by an inadequate supply of nitrogen. Many crop residues such as straw, stalks, and leaves are high in carbon but low in nitrogen. Microbes decomposing these residues use nitrogen from the soil. A deficiency of available nitrogen at this time slows both decomposition and plant growth. Under such conditions, crops respond dramatically to nitrogen fertilizer. Enough fertilizer should be applied to meet the needs of both the microbes and the crop.

Crop residues provide shade that keeps a mulched soil cooler than a bare soil. The temperature difference on a hot day can reach 10 to 20°C (18 to 36°F) at a depth of 1 cm (0.4 in.), although average differences are only 2 or 3°C (3 to 5°F). The cooling effect of the mulch is beneficial in tropical climates and during hot summers. It is detrimental, however, in the spring when a wet soil needs to dry out and warm up before it can be tilled and a crop can be planted. Mulching in temperate regions is therefore most desirable on well-drained soils. Fortunately, the well-drained soils usually coincide with the sloping areas that most need mulching for erosion control.

Crop residues may be even more important for their contribution to soil organic matter in tropical climates than in temperate climates (Buerkert et al., 2000). The kaolinite and oxide clays that are prevalent in tropical soils contribute relatively little cation exchange capacity, so that provided by organic matter may be critical. The nitrogen, phosphorus, and sulfur contents of the organic matter are likewise very important nutrient reservoirs in a tropical setting.

8–4 MULTIPLE CROPPING

Multiple cropping includes *sequential cropping* (two or more crops a year in sequence) and *intercropping* (two or more crops intermixed on a field at the same time). Multiple cropping is related to crop rotations in that the same land is used to produce more than one crop.

One objective of multiple cropping is to increase the total production from the land. Work done with intercrops in Costa Rica and El Salvador showed yields of individual crops ranging from 30 to 107% of monoculture yields (Table 8–3). The intercropping systems were considerably more profitable than the monocultures, as the additional crops more than offset the reduced production of the primary crop. Even so, it is sometimes advantageous to delay planting a secondary crop so that it will not compete too strongly with a primary crop. Hesterman et al. (1992) found in Michigan that frost seeding either alfalfa or red clover into established wheat (seeding in late fall after a hard frost) or interseeding these legumes with oats at planting time had no effect on the grain yield. In addition, the legumes provided significant amounts of nitrogen for the next corn crop. The system increased corn yields by 4 to 62% when moisture was adequate, but decreased corn yields by 3 to 27% when precipitation was below normal.

Soil conservation is an important bonus from multiple cropping. Two or more crops grown together cover the ground better and/or longer than a single crop. More raindrops are intercepted, runoff is slowed, and erosion is markedly reduced. Erosion data are scarce, but the difference is often obvious without measurement, especially in comparison to a row-crop monoculture.

TABLE 8–3 RELATIVE YIELDS OF INTERCROPPED, SHORT-STATURED CROPS EXPRESSED AS PERCENTAGES OF MONOCULTURE YIELDS

	Fertility level		
Crop	Low	Medium	High
		Percent of monoculture yield	
Rice (January planting)	32	38	46
Rice (May planting)	44	30	29
Beans (bush-type)	87	91	78
Beans (climbing-type)	56	64	78
Soybeans (January planting)	107	70	84
Soybeans (May planting)	54	52	50
Sweet potatoes	93	92	86

Source: Soil Science Department, North Carolina State University, 1974.

Sequential Cropping. Sequential cropping is like a crop rotation all in one year. Many tropical areas have favorable growing conditions the year around that permit two or even three crops to mature, one after the other. Many such areas have wet and dry seasons, so they might grow, for example, rice in the wet season and cassava in the dry season. Sequential cropping in warm climates can often increase annual production by using a longer growing season than that of any one crop (Radulovich, 2000).

Sequential cropping in which only two crops are grown in one year is also known as *double cropping*. Double cropping extends into the warmer parts of temperate regions such as southeastern United States, where soybeans are sometimes grown in the summer after a winter wheat crop has been harvested (Blevins et al., 1994). Double cropping in temperate regions is facilitated by using no-till planting (Chapter 9) to save time. Such a system provides continuous ground cover that is excellent for erosion control.

Intercropping. Intercropping is closely related to crop rotations involving companion crops and to monocultures such as orchards with a cover crop. The difference is that intercropping overlaps the main growth phases of two or more marketable crops. The crops involved include a variety of annual crops and tropical tree crops. At least one crop is planted in rows that are wide enough for another crop to be grown in between.

An example of a combination of intercropping and sequential cropping used in some experiments in Peru is shown in Figure 8–4. Jeranyama et al. (2000) call this type of system *relay intercropping* and report that planting cowpeas or sunnhemp between the rows of 28-day-old corn caused no reduction in corn yields with fertilizer rates normally used by farmers in Zimbabwe, but did reduce yields at higher fertilizer rates. Residues from the intercropped legumes improved the yields of the next year's corn crop by 8 to 27%.

Intercropping is most common in the humid tropics. Many such areas have small holdings that are intensively used, mostly to produce food for families and local markets. Intercropping produces a more varied and nutritious diet than monocultures or ordinary crop rotations would provide. Intercropping is similarly used in gardens anywhere in the world, and some studies have shown potential for combinations such as wheat and peas (Murray and Swensen, 1985), or corn or sorghum with oats or rye (Helsel and Wedin, 1981) in temperate regions.

Many tropical crops are intercropped in various combinations. Multiple benefits are obtained, for example, by growing bananas along with coffee. The banana plants quickly grow above the coffee plants and produce an abundance of leaves that protect the coffee from sun and wind, cover the soil, and control erosion (Constantinesco, 1976).

Figure 8–5 shows intercropped upland rice and banana plants. The annual rice crop provides soil cover while the banana plants are young.

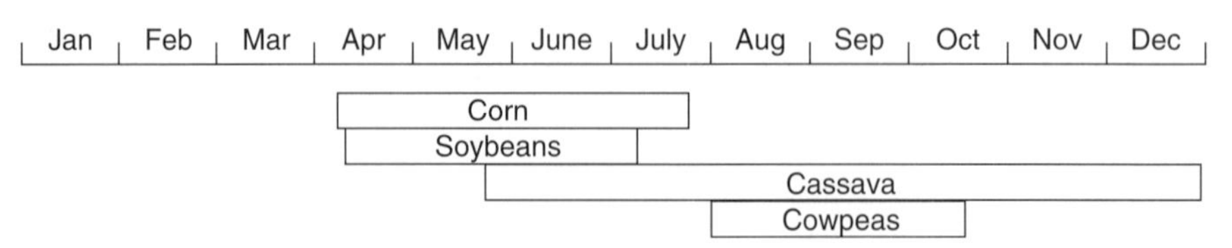

Figure 8–4 A multiple cropping schedule used in some fertility and row-spacing experiments in Peru. (After Soil Science Department, North Carolina State University, 1974.)

Figure 8–5 Bananas and rice being intercropped in The Gambia, West Africa. (Courtesy Roy L. Donahue.)

Alley Cropping. A promising variation of intercropping known as *alley cropping* has been tried in some tropical countries. In Nigeria, Kang and Ghuman (1991) planted crops in alleys 2- and 4-m (80- and 160-in.) wide between hedgerows produced by planting two woody legume species, *Leucaena leucocephala* and *Gliricidia sepium.* Trimmings from the hedgerows were used as cover and green manure to protect and fertilize the soil on slopes of about 7%. Corn and cowpeas were grown in the alleys. Corn yields were highest from the 2-m (80-in.) alleys (Figure 8–6), whereas the cowpeas did

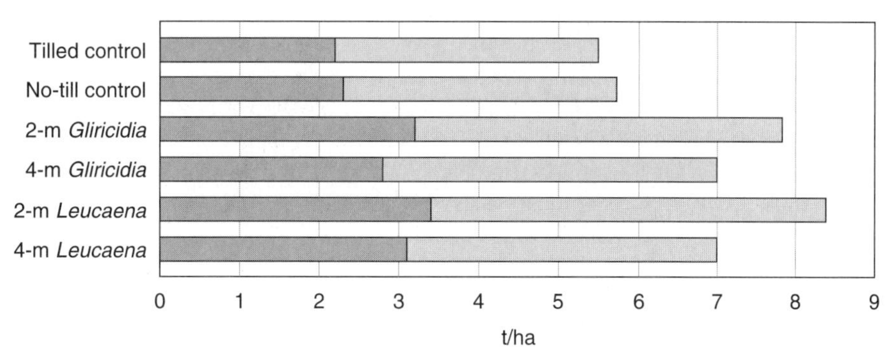

Figure 8–6 Corn yields in Nigeria as influenced by alley cropping. Hedgerows were planted with two woody legume species with 2- and 4-m alleys between them. The left part of each bar represents the grain yield and the remainder represents the stover yield. The yields were highest with alleys 2-m wide even though this took the most land out of cultivation. (Source: Based on data from Kang and Ghuman, 1991.)

best in the 4-m (160-in.) alleys—probably because shading from the hedgerows caused some reduction in cowpea growth. The tilled control plots lost much more runoff and soil than any of the other treatments. Soil analyses made after six years of cropping showed higher organic matter and nutrient contents in the alley-cropped plots than in the tilled plots.

Alley cropping is a form of agroforestry. Another form will be discussed in the section on buffer strip cropping (Section 8–5.2).

8–5 STRIP CROPPING

Strip cropping resembles a widely spaced version of intercropping. It divides a field into long, narrow bands that cross the path of the erosive force of water or wind. The strips with more vegetative cover slow runoff, reduce wind velocity, and catch soil eroded from the more exposed strips. Thus, the average soil loss may be reduced to as little as one-fourth of what it would be without the strips (see Section 6–2.5).

Some applications of strip cropping have permanent vegetation in designated protective strips and use the remaining area for either a crop rotation or a monoculture. More often, the protective vegetation is one of the crops in a rotation that shifts annually from one strip to the next. Such systems combine the favorable effects of crop rotations with contour tillage (Chapter 9, Section 9–7), and often include the use of cover crops, green manures, and crop residues. Such a combination of practices is very effective for reducing erosion.

Strip cropping is an inexpensive means of reducing erosion and is usually very effective. Nevertheless, it is used less now than it once was. The decline resulted partly from many farmers changing from crop rotations to monocultures and using larger implements. Other factors limiting its use are the susceptibility of the long exposed borders to disease and insect attacks and to the desiccating effect of hot, dry winds in semiarid climates.

8–5.1 Contour Strip Cropping

Contour strip cropping is one of the most effective means of controlling water erosion while growing crops in a rotation. It works well where the slopes are long and smooth, as shown in Figure 8–7. Variable slope gradients and rolling topography make it less practical to use contour practices of any kind.

One of the most important design factors of contour strip cropping is the width of the strips. Row crop strips must be limited in width to avoid excessive runoff and erosion. Forage crop strips must be wide enough to at least afford adequate protection and capacity to filter sediment from the runoff water. Recommended strip-width limits are given in Table 8–4. Rainfall intensity influences the maximum slope length that can be protected in this manner. These limits assume a deep soil of average erodibility where the relief (elevation difference) is about 15 ft (4.5 m) where the rainfall factor (as shown in Figure 6–2) is 250, or 25 ft (8 m) where the rainfall factor is 150. The relief may be determined from a contour map, measured with a surveying instrument, or calculated if the slope length and gradient are known. For example, a 5% slope 300 ft long equals 15 ft of relief. Terraces may be used to divide the slopes into segments where the relief is excessive.

Figure 8–7 Contour strip cropping divides the slope length into short segments to control erosion. (Courtesy F. R. Troeh.)

TABLE 8–4 RECOMMENDED LIMITS FOR STRIP WIDTHS FOR CONTOUR STRIP CROPPING

	Forage crop strip minimum width		Row crop strip maximum width	
Slope (%)	ft	m	ft	m
2	25	8	120	36
5	30	9	100	30
8	40	12	90	27
12	50	15	80	24
18	100	24	50	15

The strip widths chosen for a field should be exact multiples of the width of the row crop equipment to be used. Thus, a farmer planting, cultivating, and harvesting six 30-in. (75-cm) rows at a time would use a multiple of 15 ft (4.5 m) for strip widths. On a 4% slope, for example, the forage strips would be 30-ft (9-m) wide with 90 ft (27 m) of row crop. These widths would fit a five-year rotation including three years of row crop, one year of small grain, and one year of hay. Each year in the rotation is assigned a 30-ft (9-m) strip and the three years of row crop together total 90 ft (27 m). The small-grain strip would be included in the layout as shown in Figure 8–8 but is omitted from the calculations because its soil loss will be near the average of that for the other crops.

Contour strip cropping on slopes up to 2 or 3% may not always include forage strips. Instead, the small-grain strip may be relied on to provide the needed protection. Such strips should be about three times as wide as a hay strip would need to be and should include a winter-cover green-manure crop.

The row crop may be left out of a strip-cropping system on very steep slopes. Strips of small grain may then be grown in rotation with a forage crop. The small-grain strips can be about as wide as row crop strips on a slope that is only half as steep.

Contour strips necessarily deviate from the exact contour where slopes are variable, but the slope along a strip boundary usually should be less than 2%. Small deviations are

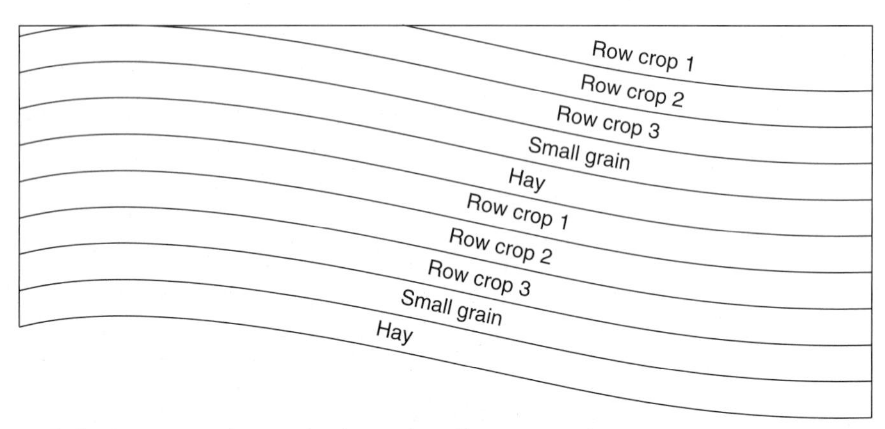

Figure 8–8 A contour strip cropping layout for a five-year rotation.

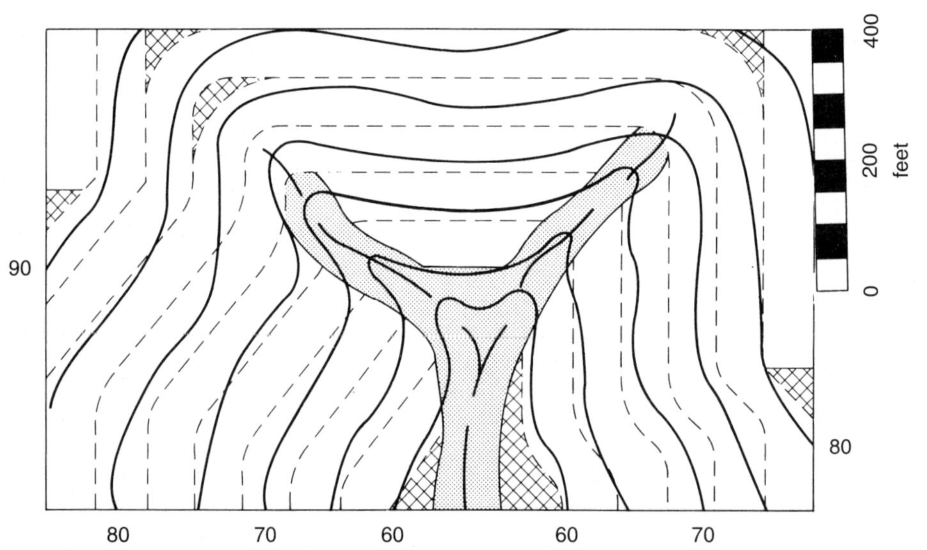

Figure 8–9 A contour strip cropping design for a small field. The solid lines are contour lines with elevations shown in feet above an arbitrary base level. The dashed lines are strip boundaries. The filler areas are crosshatched. The stippled area is grassed waterway.

acceptable if the flow of water along the rows is slow enough to be nonerosive and the flow is toward a swale rather than a ridge. An accumulation of water on a ridge would flow down the side of the ridge and probably erode a gully. Such flow can be avoided by making the contour strips slightly straighter than the true contour lines. This effect will usually result if an initial guideline is laid out along a contour line near the top of the slope. Strip boundaries are kept parallel to the initial guideline as long as the deviations from contour lines are small. Small filler areas of permanent vegetation can be used to avoid excessive deviations, as shown in Figure 8–9.

Grassed waterways are needed where enough water accumulates to cause erosion in the swales of contour-stripped fields. The waterways catch silt and raise the elevations of the swales rather than allowing gullies to form.

Contour strip cropping slows the velocity, although it may have little effect on the amount of runoff from a field. Slower velocity causes soil lost from the row crop strips to be caught in the forage strips and grassed waterways. The effectiveness declines, however, as water accumulates on the lower part of a long slope and becomes too much for the forage strips to control. Slopes longer than a few hundred feet often need a diversion terrace to remove excess water from the middle of the field. (See the relief criteria in the forepart of this section.)

Contour strip cropping usually increases the number of point rows and odd corners in a field. Point rows occur where the strips meet field boundaries at angles such as those around the edges of the field and at the waterway in Figure 8–9. Point rows reduce operational efficiency—more turning is required and small irregular areas may not be cropped.

Odd-shaped areas like the upper corners in Figure 8–9 may be planted to permanent vegetation or, since these are relatively flat, they can be planted to the same crop as the adjoining strips. Another possibility is to relocate the field boundaries to better match strip boundaries and thus minimize point rows and odd corners. The reverse has already been done on the right side of Figure 8–9, where strip boundaries parallel field boundaries that nearly follow contour lines.

The vegetation in filler areas may be harvested as hay along with an adjoining strip. Another good alternative is to manage filler areas and odd corners for wildlife purposes. Appropriate grass, shrub, and tree plantings can provide both food and cover for animals and birds. Some such areas are used as sites for bee colonies or for growing Christmas trees. Any of these uses are preferable to annual crops in such locations. Crops are seldom profitable on small irregular areas and do not provide the erosion control that results from permanent vegetation.

8–5.2 Buffer Strip Cropping

Buffer strip cropping is designed to work on rolling topography with irregular slope gradients that make contour strip cropping impractical. It lengthens the filler areas into continuous buffer strips that separate the crop strips. The crop strips are uniform in width, but the buffer strips have variable width to allow for slope irregularities. The buffer strips are positioned to include any rocky areas or other problem spots that occur in the field, as shown in Figure 8–10.

Buffer strips are planted to permanent vegetation to slow the runoff and to catch sediment eroded from the next-higher crop strips. Perennial forage crops that can be used for hay or pasture are often grown on buffer strips. Planting trees and shrubs on them makes them into *contour tree buffer strips* that are a form of *agroforestry* (Countryman and Murrow, 2000). Agroforestry may provide good long-term profits from areas planted to trees that produce valuable wood products such as oak or walnut. A grass understory may be needed for erosion control, at least until the trees are large enough to produce a continuous litter layer. *Alley cropping,* as discussed in Section 8–4, is a form of agroforestry that uses lines of leguminous trees and shrubs planted on the contour in tropical countries.

The effectiveness of buffer strips for reducing erosion depends on the nature of the protective vegetation, the crop and buffer strip widths, and the topography, soil, and climate of the area. The P factor for contour strip cropping can be used in calculating soil loss (Chapter 6) where the crop and buffer strip widths meet the requirements of Table 8–4.

Sloping land with buffer strip cropping gradually forms terraces as sediment settles out of the runoff water and is deposited at the bottom of the cropped strips and in the upper

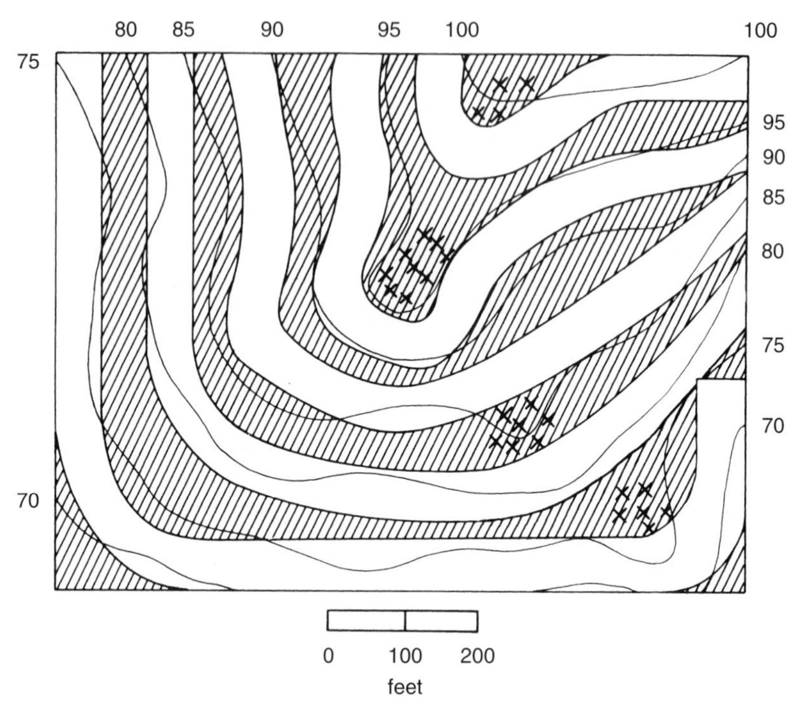

Figure 8–10 A buffer strip cropping design on a contour map. Elevations are shown in feet above an arbitrary base level. The X's represent uncroppable rocky areas.

part of the buffer strips. Poudel et al. (1999) noted that the fertility and productivity were considerably higher on the lower part of each terrace where sediment was deposited. Thapa et al. (2000) found that buffer strips combined with ridge tillage worked exceptionally well for conserving soil and producing high yields on steep Oxisols in the Philippines.

8–5.3 Field Strip Cropping

Field strip cropping is sometimes used on land that is too rolling even for buffer strip cropping. It consists of straight rectangular strips laid out parallel to one side of a field. These strips cross the general slope of the area but do not follow contour lines. The system is managed like contour strip cropping but is less effective where the deviations from the contour are too large. Grassed waterways are usually needed in low areas because water flows along the sloping rows and accumulates in the swales.

8–5.4 Wind Strip Cropping

Strip cropping designed to control wind erosion crosses the path of the prevailing winds rather than following the contour. Wind strip cropping is laid out in straight lines like field strip cropping, but the crops are different. Wind strip cropping is used in semiarid areas such as the Great Plains of the United States and the Canadian Prairies.

The soil-conserving vegetation in wind strip cropping needs to be dense enough to catch saltating sand particles and prevent them from jumping again. The protective strips need to be wide enough to keep saltating particles from jumping completely across a strip. Several feet would suffice, but cropping considerations make widths of 150 to 250 ft (50 to 80 m) more practical. Because crops shift yearly from one strip to the next, each strip should be wide enough to be cropped conveniently. The chosen width is normally a multiple of the width of equipment used in farming operations. The maximum width is limited by the increased saltation resulting from avalanching where exposed areas are too wide.

Close-growing vegetation such as bluegrass can stop saltation even when the grass is only a few centimeters tall. Bunchgrasses, however, are common in climates where wind erosion is most likely to be a problem. Bunchgrass vegetation will protect open areas between clumps of grass if it is at least as tall as the width of the open areas. Tall vegetation gives a windbreak effect in addition to its ability to catch and hold saltating sand particles.

Wind strip cropping is often used with a wheat-fallow rotation. The wheat strips control saltation quite effectively after the wheat is 4- to 6-in. (10- to 15-cm) tall. The wheat stubble continues to protect the soil after the wheat is harvested and can even provide some protection during the fallow year if it is left standing on the surface.

Strips that catch soil particles in the summer can also catch drifting snow during winter. Holding snow on the fields contributes to the soil moisture supply and increases productivity.

8–5.5 Barrier Strips

Narrow strips consisting of a few rows of small grains, grasses, or other crops can provide significant protection from wind erosion. These barrier strips must be spaced fairly close together to compensate for their narrowness. Even so, they occupy much less land than do the protective strips in wind strip cropping.

Fryrear (1963) found that two rows of either sudangrass or grain sorghum made effective barriers for Texas conditions, because they grew when protection was most needed. He recommended a 23-ft (7-m) spacing for sudangrass strips or 13 ft (4 m) for grain sorghum to protect against winds up to 40 mi/hr (65 km/hr). The highest-cut stubble afforded winter protection even after harvest.

Hagen et al. (1972) also recommended two-row barrier strips on the basis of their work in Kansas. Single-row barriers used the land most efficiently but sometimes broke and failed when wind speeds exceeded 30 mi/hr (50 km/hr). They found that winter wheat barriers 4-in. (10-cm) tall were 20% effective for trapping soil particles from a 30 mi/hr wind. Sudangrass barriers 1-ft (30-cm) tall were 60% effective under the same conditions. They calculated that a two-row rye barrier 8-in. (20-cm) tall could protect a strip about 100-ft (32-m) wide against a 30-mi/hr wind.

8–5.6 Border Strips

A border strip of grass or other close-growing vegetation can help keep soil from being carried into streams and ponds, living areas, or other sites that need protection. Border strips may be used to control movement of soil by wind after the manner of wind strip cropping, or they may be designed to restrict water transport. Their main purpose is often to control air and water pollution; reducing erosion may be secondary.

Lush grass growing between a field and a body of water is very effective for catching sediment and reducing eutrophication (Chapter 17). The grass filters soil particles from the runoff water and absorbs dissolved nutrients from the water. The low nutrient content in the purified water limits the growth of algae and other plants in ponds and streams. The water, nutrients, and fertile soil caught in the border strip help produce a lush growth that catches even more sediment and nutrients.

The dimensions of border strips vary with the situation but should usually be at least as wide as strips serving a similar purpose in fields. Most strips controlling water purity should be at least 30-ft (10-m) wide and must be wider where the water flow is large or where the slope gradient exceeds 1 or 2%. Wind-erosion control usually requires wider borders to keep the air clean—often 150- to 300-ft (50- to 100-m) across—with exact dimensions depending on topography and wind direction and velocity. These areas are often large enough to be used for pasture or hay production.

Robinson et al. (1996) measured the effectiveness of a bromegrass border strip for catching silty sediment on 7 and 12% slopes in Iowa. A 60-ft wide strip of Fayette silt loam soil above the strips was fallowed so that soil losses of 12 ton/ac (26.5 mt/ha) on the 7% slope and 29 ton/ac (65 mt/ha) on the 12% slope occurred during the year-long trial. They found that the first 10 ft (3 m) of the border strip caught 70% of the sediment on the 7% slope and 80% of that on the 12% slope, and that a 30-ft (9-m) wide grass strip filtered 85% of the sediment from the runoff on both slopes. The particles remaining in the water beyond 30 ft of filtration were very small and difficult to remove. Daniels and Gilliam (1996) similarly showed sediment removals of 60 to 90% by grass border strips 20-ft (6-m) wide in North Carolina. They also measured approximately 50% reductions in the total N and P carried in the runoff.

8–6 EVALUATING CROPPING SYSTEMS

A satisfactory cropping system must meet several standards. Economics, erosion control, pest control, physical and chemical effects on the soil, and environmental concerns are all important. Evaluation of all these factors is seldom more than a reasonable estimate based on extrapolation of long-term trends.

Economic considerations require a cropping system that produces a profit for the user. Adequate crop yields must be attainable on a long-term basis. Some economic aspects of soil conservation are discussed in Chapter 18, but much of the economics of cropping systems is beyond the scope of this book.

8–6.1 Cropping Systems and Soil Loss

The soil-loss prediction equations discussed in Chapter 6 are useful for selecting appropriate cropping systems. Several possibilities can be analyzed and their results predicted before a choice is made. The more intensive cropping systems usually give larger profits but are likely to result in larger soil losses. Continuous row crops and other intensive cropping systems are therefore commonly preferred where conditions are favorable, but they should be avoided where they would cause excessive erosion.

Erosion naturally varies according to soil and topographic conditions. A decision must therefore be made regarding which cropping systems are suitable for specific conditions. The most erodible part of the field (not just the average) needs to be checked before the system can be rated as satisfactory. The hazards of averaging can be illustrated by an example wherein the annual soil loss from a 40-ac (16-ha) field averages 2 tons/ac (4.5 mt/ha) from most of the field, but 40 tons from one acre (36 mt from 0.4 ha). The field average is 3 tons/ac (7 mt/ha) and is less than the specified tolerable rate for most soils. The system nevertheless would rapidly ruin the one acre of highly erodible land. This one acre should be taken out of cultivation or farmed with a less erosive system.

Adjusting land use and cropping systems to the most erodible land is usually impractical. Rather, these calculations help identify problem areas that need different treatment than the rest of the field. A cropping system suited to the 39 good acres in the preceding example could be used if field boundaries were changed to place the erodible acre in an adjoining pasture or woodland.

Changing field boundaries solves some problems but does not work where the erodible land is surrounded by land suitable for intensive use. Terracing can help where the erosion hazard results from a steep slope. Another approach is to use a dual cropping system. For example, if the best land is suitable for continuous row crops, the part needing protection might have a rotation of row crops, small grain, and hay. The grain and hay crops must be handled separately, but the row crop will cover the entire field some years. Such systems permit different parts of the field to be used in accordance with their potential and their need for protection as indicated by the soil-loss prediction equations.

8–6.2 Maintaining Productive Potential

Maintaining the productive potential of soil over the long term is a fundamental purpose of soil conservation. Evaluation of the productive potential, however, is difficult. Crop yields depend on weather and management as well as soil potential. New crop varieties, coupled with improved management and increased use of fertilizer, have often produced larger yields even while the soil was becoming shallower and harder to work. Thus, the crop yields can be increasing while the productive potential is decreasing.

Some items that influence productive potential are much easier to measure than the potential itself. Soil depth is one such item. The total soil thickness is important and the thickness of the A horizon is also likely to influence soil productivity. The importance of A-horizon thickness depends on the nature of the subsoil—subsoil that is too dense or otherwise unsuitable for root development increases the importance of topsoil thickness.

Another negative effect of erosion on soil productivity occurs through reduced soil fertility. Erosion has sometimes been called "the great robber" because small mineral and organic particles are carried away along with their associated fertility, while the less-fertile coarse particles are left behind. The largest differences occur when fertilizers are left on the soil surface and are eroded away with the soil. Additional fertilizer to compensate for the nutrient losses may maintain nearly equal yields, but the costs are considerable. The profit obtained from crops grown on eroding land therefore declines, even if the yield is held constant by fertilization.

Cropped soils generally contain less organic matter than similar uncropped soils because (1) most cropping systems return less organic matter to the soil after a crop is harvested, (2) tillage increases the rate of organic matter decomposition, and (3) erosion carries away significant amounts of organic matter. Smaller amounts of soil organic matter mean there is less cation exchange capacity to store cation nutrients and less organic matter to release plant nutrients as it decomposes. Also, soil structure is generally less stable with lower organic matter, so the pore space and permeability decrease and the soil crusts more readily. It is therefore worthwhile to adjust the cropping system in ways that will maintain as much soil organic matter as possible. Fertilization to maximize crop yield will also help maximize crop residues. Adding a cover crop that is used for green manure is very beneficial. Reduced tillage that leaves more residue on the soil surface helps by both slowing decomposition and reducing erosion losses.

The removal of fine particles by erosion makes many soils gradually become more sandy, gravelly, or stony. The increased percentage of coarse material makes these soils more droughty and lowers the productive potential of many of them.

8–6.3 Maintaining Soil Structure

Tillage, erosion, and reduced organic-matter content weaken the structure of intensively cropped soils. Destruction of soil structure into a puddled condition is favorable for paddy rice production but is undesirable for other crops.

Significance of Soil Structure. As soil structure weakens, the soil tilth becomes poorer, the likelihood of crusting increases, and the soil permeability decreases. Soil *tilth* refers to how easily the soil can be tilled and cropped. Tilth is most important in soils high in clay because clay soils with poor tilth form hard clods when they dry, especially if they are tilled when wet. Poor tilth increases the power required to till the soil and sometimes increases the number of tillage operations required to prepare a seedbed.

Stable structure within the various soil horizons is important for maintaining permeability in all but the very coarse-textured soils. Permeability depends on pore space between the soil peds. New pores are formed as roots and other living things force their way through the soil. Shrinking and swelling caused by moisture and temperature changes help form new aggregates and soil peds. But, tillage breaks aggregates and peds and blocks pores if the soil structure is weak. Permeability then declines, aeration and pore space are reduced, and the soil becomes less favorable for root growth.

Structural stability depends on soil texture, soil organic matter, and soil chemistry. Soil humus and clay particles, especially those with high ion-exchange capacities, can bond together into aggregates. Such bonding is strengthened by the presence of divalent cations like Ca^{2+} and trivalent cations like Al^{3+}. It is weakened in alkaline soils where sodium and other monovalent ions are relatively abundant on the cation exchange sites. Humus associated with clay is somewhat protected from decomposition, so humus contents are usually correlated with clay contents in soils of uniform climate, drainage, and vegetation.

Crusts form where raindrops beat on the soil surface and break down its structure. Structural stability and degree of exposure to raindrops determine how long it takes to loosen particles from broken soil aggregates and plug soil pores. Crust formation often reduces the water infiltration rate to a fraction of its initial value. Runoff, erosion, sedimentation, and pollution increase. After the rain, the surface hardens into a crust that may

Figure 8–11 Soil structure differences resulting from a corn–wheat–hay rotation (A) and two years of corn–soybeans (B). (Courtesy Maryland Agricultural Experiment Station.)

markedly reduce seedling emergence and crop stand. Crust formation and reduced seedling emergence are the most readily observed symptoms of weak soil structure.

Strength of Soil Structure. The terms *weak*, *moderate*, and *strong* are often used to indicate the distinctness of soil structure, although it is difficult to measure the strength quantitatively. *Aggregate stability* is a related property that can be evaluated as described in Note 8–1. Such procedures can be used to measure changes produced by different cropping systems and management practices. The soil in one field might be compared with that from a nearby field or with the less-disturbed soil in a fencerow. Such comparisons often reveal large differences in aggregate stability accompanied by observable differences in porosity and crusting, as shown in Figure 8–11.

NOTE 8–1
MEASURING STRUCTURAL STABILITY

Several methods are used to evaluate the structural stability of soils. The two outlined here involve inexpensive equipment and can be used to compare differences resulting from contrasting treatments such as cropland versus pastureland, or fields versus fencerows.

The percentage of water-stable aggregates can be determined by placing a weighed soil sample on a sieve with 0.01-in. (0.25-mm) openings. The sieve is then dipped 50 times (or some other standard number) in a container of water. The material remaining on the sieve is dried, weighed, and corrected for sand content to calculate the percentage of water-stable aggregates.

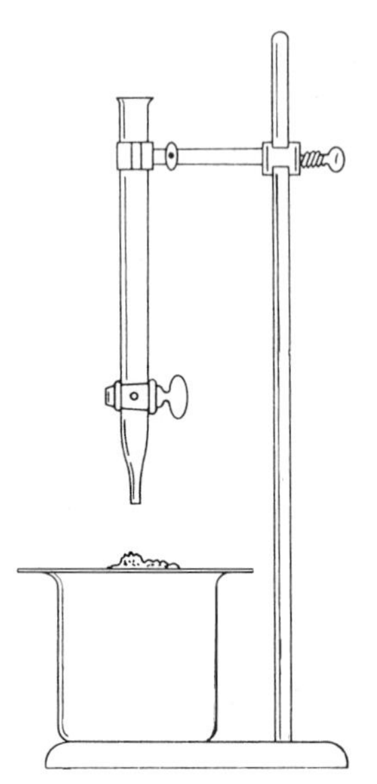

Figure 8–12 The volume of water required to wash a soil sample through a screen is an indication of structural stability.

The structural stability of soil clods can be compared by dripping water onto them from a buret as shown in Figure 8–12. The volume of water required to wash the soil through the screen and into the beaker is an indication of structural stability.

Effect of Crops on Soil Structure. The fine fibrous root system produced by a dense growth of grass helps form stable aggregates in soil. Deep taproots such as those of alfalfa help open channels in the lower soil horizons. These improvements in soil structure are important reasons why an alfalfa-grass mixture is widely regarded as a soil-improving crop. Most forage crops have similar effects but to different degrees.

Row crops such as sugar beets, cotton, corn, and beans usually weaken the soil structure and contribute to erosion and sedimentation. Narrow rows and high yields cause these crops to cover the soil faster and permit their root systems to penetrate more uniformly through the entire soil volume, thus reducing, but not eliminating, the soil deterioration. Each combination of soil, cropping system, and management practices has an equilibrium of organic matter (Fenton et al., 1999), structural stability, and of many other properties as well. As time passes, the soil properties shift toward equilibrium. The rate of change is roughly proportional to the difference between the present condition and the equilibrium condition.

Small-grain crops produce more cover and generally stronger soil structure than row crops, but less cover and weaker structure than forage crops. Tree crops are much like row crops in their effect on soil structure, unless there is a cover crop or other vegetation between the trees.

Most crops can be related to one or another of the groups already discussed. A good first approximation is that the effect of a crop on soil structure is closely related to how thoroughly its roots permeate the soil. The effect of companion crops on soil structure is often more favorable than that of either crop grown by itself. The effect of a crop rotation is an integration of the effects of all crops in the rotation. Some properties such as soil-aggregate stability change fast enough to exhibit noticeable differences from one crop to another during a rotation.

Weakened soil structure is sometimes a reason to modify an intensive cropping system. Low permeability resulting from weak soil structure causes runoff and erosion problems on slopes and wetness and sedimentation problems on flat lands. The remedy is usually to shift toward a system that includes more close-growing vegetation and/or reduces the amount of tillage. After a period of adjustment, a change to no-till may significantly increase the organic-matter content and structural stability of a soil that has been degraded by cropping with conventional tillage (Bayer et al., 2001). Inclusion of more cover crops and green manure in the cropping system also helps.

8–6.4 Environmental Effects of Cropping Systems

Most cropping systems accelerate erosion by wind and water in comparison to the rates under native vegetation. Eroded soil contaminates air with dust particles and water with sediment. The sheer mass of the eroded soil and the chemical and biological entities carried with it are significant. For example, soil carries plant nutrients that contribute to eutrophication (Chapter 17).

The early emphasis of the soil-conservation program was to protect land from excessive soil loss. Environmental concerns have made water and air pollution strong additional reasons to control erosion. Pollution caused by soil erosion extends far beyond the eroding area, making it a public as well as a private concern. Several states have responded by passing laws that require the use of soil-conserving practices under certain conditions, as discussed in Chapter 19.

Environmental considerations cause some cropping system changes that protect air and water rather than soil. Border strips of close-growing vegetation to keep sediment out of streams and ponds are a good example of adjustments that reduce water pollution.

Most soil-conserving practices benefit both the land where erosion is reduced and the environment where the eroded soil would have gone. Cropping practices such as rotations, cover crops, and strip cropping reduce air and water pollution and protect soil productivity. Mechanical practices such as terracing and conservation tillage also protect both soil productivity and the environment. Recognizing the environmental values of such practices increases the incentive to apply them.

Reducing erosion and its polluting effects to minimal values would be relatively simple if land were not needed for growing crops. However, the world's population needs to be fed, clothed, and housed. Cropping systems, therefore, need to be carefully designed to meet needs without causing excessive erosion and pollution.

SUMMARY

Native vegetation normally provides enough cover to control erosion, but many cropping systems leave the soil surface exposed periodically. Poor crop growth can allow disastrous erosion to ruin the land and pollute both air and water. Properly adapted crops and well-planned cropping systems limit erosion to acceptable rates and help reduce environmental pollution.

Most crops can be classified into four groups. *Row crops* usually produce the most profit and the most exposure to erosion. *Small-grain crops* give more protection because the plants grow fast and are usually close together. *Forage crops* produce still thicker cover and provide excellent erosion control. *Tree crops* with undergrowth and a litter layer on the soil surface almost eliminate erosion.

Practices that increase crop yields usually reduce soil erosion. Crops in narrower rows cover the soil sooner and protect it better. High soil fertility also increases both yield and soil cover. The use of organic and mineral fertilizers to supply deficient nutrients also helps conserve soil, but excessive fertilization can cause pollution.

Monocultures permit farmers to specialize and grow each crop where it is best suited. Fertilizers and lime may make it possible to maintain yields under continuous cash crops, but the soil structure may deteriorate and erosion, sedimentation, and pollution increase. Summer fallow systems increase erosion and reduce soil organic-matter contents in comparison to annual cropping. Land that is too steep for row crops and grain crops may be used for hay, pasture, or tree crops.

Crop rotations often provide more continuous soil cover than monocultures and help control erosion, plant diseases, and insects. *Companion crops* and *cover crops* help to keep the soil covered almost continuously, and sometimes provide material that is worked into the soil as green manure. This contributes to soil organic matter, soil structure, and permeability. Some crops produce carryover effects that reduce erosion while the next crop is growing. Crop residues can be used for soil cover when no crop is growing.

Multiple cropping includes *sequential cropping* and *intercropping*. It is used in many gardens and in tropical areas to increase variety and total amount of production. Its contributions to soil conservation are a bonus.

Contour strip cropping is effective for controlling water erosion; *buffer strip cropping* and *field strip cropping* are useful where the land includes small uncroppable areas or the topography is rolling. *Wind strip cropping* is laid out across the prevailing wind to catch saltating particles. *Barrier strips* consisting of one, two, or a few rows of tall, close-growing crops are also effective for reducing wind erosion if the exposed area between them is not too wide. *Border strips* are single strips of close-growing vegetation that are used to reduce pollution by filtering soil particles out of air and water.

The soil-loss prediction equations are useful for selecting appropriate cropping systems and for identifying areas that need special treatment. Appropriate cropping systems maintain adequate soil depth, fertility, and water-holding capacity to conserve the productive potential of the soil. Stable soil structure is needed to resist crusting and maintain adequate permeability. Forage crops usually have favorable effects on soil structure, whereas most row crops cause soil structure to deteriorate. The effect of any crop on soil structure is related to how thoroughly its roots penetrate the soil.

Cropping systems need to protect the environment as well as the soil. Pollution resulting from soil erosion contaminates air and water far beyond the eroding area. Well-designed cropping systems produce food and fiber for the world's population without causing excessive erosion and pollution.

QUESTIONS

1. How can a forage crop be both a "soil-building crop" and a "soil-depleting crop"?
2. What effects do rows and row spacings have on soil and water conservation?
3. Why do many farmers grow crops as monocultures rather than use crop rotations?
4. Under what conditions would the use of a companion crop for the establishment of a forage crop increase erosion? When would it reduce erosion?
5. Why do many farmers prefer to plow crop residues under rather than leave them on the soil surface?
6. Explain the differences between crop rotations, sequential cropping, and intercropping.
7. List five different types of strip cropping and distinguish them from one another.
8. How can the use of fertilizer: (a) reduce water pollution? (b) increase water pollution?

REFERENCES

ABDIN, O., B. E. COULMAN, D. CLOUTIER, M. A. FARIS, X. ZHOU, and D. L. SMITH, 1998. Yield and yield components of corn interseeded with cover crops. *Agron. J.* 90:63–68.

AULAKH, M. S., T. S. KHERA, J. W. DORAN, and K. F. BRONSON, 2001. Managing crop residue with green manure, urea, and tillage in a rice-wheat rotation. *Soil Sci. Soc. Am. J.* 65:820–827.

BAYER, C., L. MARTIN-NETO, J. MIELNICZUK, C. N. PILLON, and L. SANGOI, 2001. Changes in soil organic matter fractions under subtropical no-till cropping systems. *Soil Sci. Soc. Am. J.* 65:1473–1478.

BLEVINS, R. L., W. W. FRYE, M. G. WAGGER, and D. D. TYLER, 1994. Residue Management Strategies for the Southeast. In J. L. Hatfield and B. A. Stewart (eds.), *Crops Residue Management.* Lewis Pub., Boca Raton, FL, 220 p.

BUERKERT, A., A. BATIONO, and K. DOSSA, 2000. Mechanisms of residue mulch-induced cereal growth increases in West Africa. *Soil Sci. Soc. Am. J.* 64:346–458.

CONSTANTINESCO, I., 1976. *Soil Conservation for Developing Countries.* Soils Bull. 30, FAO, Rome, 92 p.

COUNTRYMAN, D. W., and J. C. MURROW, 2000. Economic analysis of contour tree buffer strips using present net value. *J. Soil Water Cons.* 55:152–160.

CROOKSTON, R. K., and J. E. KURLE, 1989. Corn residue effect on the yield of corn and soybean grown in rotation. *Agron. J.* 82:229–232.

DANIELS, R. B., and J. W. GILLIAM, 1996. Sediment and chemical load reduction by grass and riparian filters. *Soil Sci. Soc. Am. J.* 60:246–251.

FENTON, T. E., J. R. BROWN, and M. J. MAUSBACH, 1999. Effects of long-term cropping on organic matter content of soils: Implications for soil quality. Ch. 6 in R. Lal (ed.), *Soil Quality and Soil Erosion.* CRC Press, Boca Raton, FL, 329 p.

FISK, J. W., O. B. HESTERMAN, A. SHRESTHA, J. J. KELLS, R. R. HARWOOD, J. M. SQUIRE, and C. C. SHEAFFER, 2001. Weed suppression by annual legume cover crops in no-tillage corn. *Agron. J.* 93:319–325.

FRYREAR, D. W., 1963. Annual crops as wind barriers. *Trans. Am. Soc. Agr. Engr.* 6:340–342, 352.

GÓMEZ, J. A., J. V. GIRÁLDEZ, and E. FERERES, 2001. Analysis of infiltration and runoff in an olive orchard under no-till. *Soil Sci. Soc. Am. J.* 65:291–299.

HAGEN, L. J., E. L. SKIDMORE, and J. D. DICKERSON, 1972. Designing narrow strip barrier systems to control wind erosion. *J. Soil Water Cons.* 27:269–272.

HELSEL, Z. R., and W. F. WEDIN, 1981. Harvested dry matter from single and double-cropping systems. *Agron. J.* 73:895–900.

HESTERMAN, O. B., T. S. GRIFFIN, P. T. WILLIAMS, G. H. HARRIS, and D. R. CHRISTENSON, 1992. Forage legume—small grain intercrops: Nitrogen production and response of subsequent corn. *J. Prod. Agric.* 5:340–348.

HIVELY, W. D., and W. J. COX, 2001. Interseeding cover crops into soybean and subsequent corn yields. *Agron. J.* 93:308–313.

JERANYAMA, P., O. B. HESTERMAN, S. R. WADDINGTON, and R. R. HARWOOD, 2000. Relay-intercropping of sunnhemp and cowpea into a smallholder maize system in Zimbabwe. *Agron J.* 92:239–244.

KANG, B. T., and B. S. GHUMAN, 1991. Alley cropping as a sustainable system. In W. C. Moldenhauer, N. W. Hudson, T. C. Sheng, and San-Wei Lee (eds.), *Development of Conservation Farming on Hillslopes.* Soil and Water Cons. Soc., Ankeny, IA, 332 p.

KAYUKI, K. C., and C. S. WORTMANN, 2001. Plant materials for soil fertility management in subhumid tropical areas. *Agron. J.* 93:929–935.

KLEINMAN, P. J. A., D. PIMENTEL, and R. B. BRYANT, 1995. The ecological sustainability of slash-and-burn agriculture. *Agric. Ecosystems Environ.* 52:235–249.

MEISINGER, J. J., P. R. SHIPLEY, and A. M. DECKER, 1990. Using winter cover crops to recycle nitrogen and reduce leaching. In J. P. Mueller and M. G. Wagger (eds.), *Conservation Tillage for Agriculture in the 1990's.* Proc. of the 1990 Southern Region Conservation Tillage Conference, Raleigh, North Carolina, July 16–17, NCSU Special Bulletin 90–1, 107 p.

MURRAY, G. A., and J. B. SWENSEN, 1985. Seed yield of Austrian winter field peas intercropped with winter cereals. *Agron. J.* 77:913–916.

NIELSEN, D. C., 2001. Production functions for chickpea, field pea, and lentil in the central Great Plains. *Agron. J.* 93:563–569.

NYAKATAW, E. Z., K. C. REDDY, and D. A. MAYS, 2000. Tillage, cover cropping, and poultry litter effects on cotton: II. Growth and yield parameters. *Agron. J.* 92:1000–1007.

PAPENDICK, R. I., P. A. SANCHEZ, and G. B. TRIPLETT (eds.), 1976. *Multiple Cropping.* Spec. Publ. 27. American Society of Agronomy, Crop Science Society of America, and Soil Science Society of America, Madison, WI, 378 p.

POUDEL, D. D., D. J. MIDMORE, and L. T. WEST, 1999. Erosion and productivity of vegetable systems on sloping volcanic ash-derived Philippine Soils. *Soil Sci. Soc. Am. J.* 63:1366–1376.

RADULOVICH, R., 2000. Sequential cropping as a function of water in a seasonal tropical region. *Agron. J.* 92:860–867.

ROBINSON, C. A., M. GHAFFARZADEH, and R. M. CRUSE, 1996. Vegetative filter strip effects on sediment concentration in cropland runoff. *J. Soil Water Cons.* 50:227–230.

SAINJU, U. M., and B. P. SINGH, 2001. Tillage, cover crop, and kill-planting date effects on corn yield and soil nitrogen. *Agron. J.* 93:878–886.

SANCHEZ, J. E., T. C. WILLSON, K. KIZILKAYA, E. PARKER, and R. R. HARWOOD, 2001. Enhancing the mineralizable nitrogen pool through substrate diversity in long-term cropping systems. *Soil Sci. Soc. Am. J.* 65:1442–1447.

SCHILLINGER, W. F., R. J. COOK, and R. I. PAPENDICK, 1999. Increased dryland cropping intensity with no-till barley. *Agron. J.* 91:744–752.

SINGH, K. B., R. S. MALHOTRA, M. C. SAXENA, and G. BEJIGA, 1997. Superiority of winter sowing over traditional spring sowing of chickpea in the Mediterranean region. *Agron. J.* 89:112–118.

SOIL SCIENCE DEPARTMENT, 1974. *Agronomic-Economic Research on Tropical Soils.* Annual Report for 1974, North Carolina State University, Raleigh, NC, 230 p.

STOMBAUGH, T. S., and S. SHEARER, 2000. Equipment technologies for precision agriculture. *J. Soil Water Cons.* 55:6–11.

THAPA, B. B., D. P. GARRITY, D. K. CASSEL, and A. R. MERCADO, 2000. Contour grass strips and tillage affect corn production on Philippine steepland Oxisols. *Agron. J.* 92:98–105.

VAN DOREN, D. M., JR., W. C. MOLDENHAUER, and G. B. TRIPLETT, Jr., 1984. Influence of long-term tillage and crop rotation on water erosion. *Soil Sci. Soc. Am. J.* 48:636–640.

WYLAND, L. J., L. E. JACKSON, W. E. CHANEY, K. KLONSKY, S. T. KOIKE, and B. KIMPLE, 1996. Winter cover crops in a vegetable cropping system: Impacts on nitrate leaching, soil water, crop yield, pests and management costs. *Agric. Ecosystems Environ.* 59:1–17.

CHAPTER

9

TILLAGE PRACTICES FOR CONSERVATION

Tillage is "the mechanical manipulation of the soil . . . to modify soil conditions and/or manage crop residues and/or weeds and/or incorporate chemicals for crop production" (Anonymous, 1997). Tillage was probably the first and most important innovation that our primitive ancestors developed in their attempt to grow crops for a sufficient and stable food supply. The amount of tillage gradually increased as new implements were invented and animal power was used to pull them. Still more tillage was used on larger areas after tractors were invented. The tillage implements became bigger and more rugged while the tractors became larger and more powerful. Over time, though, tillage has come to be seen as a mixed blessing—necessary at times, but capable of causing considerable damage to the soil and to the environment. Current trends are to use much less tillage than has been customary. Optional tillage operations are likely to be omitted, and chemicals are often substituted for tillage that would otherwise be needed.

Dense, perennial vegetation, such as grass and trees, provides excellent protection against erosion. When native vegetation is destroyed in preparing land for cultivated crop production, the protection provided by perennial cover is lost. Often the land is exposed to destructive action of water and wind. Other means must be found, then, to protect the land. Every effort must be made to ensure that cultivated crops and their residues provide as much protection as possible (Moldenhauer et al., 1983). Some tillage practices bury residues and expose the soil to the erosive forces of water and wind for extended periods, but others curtail exposure by leaving crop residues on the land surface, and some help by making the soil more permeable and/or by giving it a rough surface that holds water and traps loose soil particles.

9–1 OBJECTIVES AND EFFECTS OF TILLAGE

Three major objectives of cultivation are to prepare seed and root beds, to control weeds, and to establish surface soil conditions that favor water infiltration and erosion control. Undesired effects also occur, including soil compaction, accelerated decomposition of organic matter, and increased susceptibility to erosion.

9–1.1 Preparation of Seed and Root Bed

Surface soil condition should favor effective seed placement, germination, and early emergence, and should permit unrestricted root growth and plant development. Cultivation to ensure these conditions ranges from simply stirring with the planter or grain drill to conducting many operations with a variety of implements. Often the soil is loosened and then partially recompacted to achieve desired seedbed conditions. In the process, clods are broken down into small enough sizes to provide good seed-soil contact after planting.

9–1.2 Control of Weeds

Weeds compete with crop plants for moisture, nutrients, space, and light. Tillage buries weed seeds and kills seedling and mature weeds. Chemical herbicides control many, but not all, weeds. Where herbicides are used, some tillage may be necessary to prevent resistant weeds from developing and competing with crops.

Tillage in excess of that needed to control weeds does not benefit crops on most friable medium- and coarse-textured soils. On these soils, weed control by scraping the soil surface or by applying chemicals is generally as effective as tillage. Crop growth does benefit, in the short term, from more tillage than is needed for weed control on some soils that have low permeability and low organic-matter contents.

9–1.3 Soil and Water Conservation

Tillage that stirred the soil after every rain to form a dry soil mulch (dust mulch) was recommended for moisture conservation in the early years of the twentieth century. Field tests failed to show increased moisture storage or improved crop growth, however, and because the smooth, dry surface increased susceptibility to wind erosion, the practice lost favor and was abandoned.

Cultivation to break up a crusted soil surface and to produce a rough, cloddy soil condition increases infiltration and reduces erosion, but the effect is usually short-lived because rain and wind soon smooth and compact the surface.

Crop residues on the soil surface reduce both water and wind erosion. Accordingly, tillage that leaves straw and stubble on the soil surface, instead of turning it under, has much to recommend it. This usually means minimizing the amount of tillage and seldom or never using plows that invert the soil. Most of the tillage that is used is likely to be delayed until near the time to plant the next crop.

Contour tillage that produces ridges across sloping land reduces runoff and thus helps to conserve both soil and water.

9–1.4 Tillage and Soil Movement

Tillage has both direct and indirect effects on soil erosion. Indirect effects relating to soil erodibility are related to soil organic matter contents and surface cover and will be discussed in the next section. The direct effects involve soil movement by tillage implements. These effects range from negligible for most hand tools and some narrow-tined implements to shifting a whole plow layer, forming a backfurrow with increased soil thickness where plowing begins and a dead furrow left empty where plowing ends.

Some farmers have used their plows to offset erosion by beginning at the top of the hill or slope and going back and forth across the slope with a two-way plow, so the plow always throws the soil upslope. This approach is most often used on terraced land as a means of rebuilding the terrace ridges each time the field is plowed.

Most often, tillage causes a net downslope movement that has sometimes been called *tillage erosion.* Gerontidis et al. (2001) found that a moldboard plow set 40-cm (16-in.) deep on a 22% slope threw soil as far as 97-cm (38-in.) downslope when the plow was pulled straight down the slope, and as far as 69-cm (27-in.) downslope when it was pulled along the contour. Reducing the plow depth to 20 cm (8 in.) reduced the soil movement by about 75%. They found that chisel plows can also cause major downslope soil movement, especially when set deep. Lindstrom et al. (2000) made similar measurements showing that moldboard plows, chisel plows, and disks all caused downslope soil movement in the Palouse area of Washington state. They developed a prediction equation for such soil movement and showed that tillage movement away from convex hilltops sometimes exceeds the T value for tolerable soil loss. Van Oost et al. (2000) point out that the average distance of soil movement depends greatly on the slope gradient, and that this causes a net loss of soil from convex hilltops and a net gain of soil on concave footslopes.

9–1.5 Effects of Tillage on Soil Properties

Tillage generally aerates the soil and accelerates the decomposition of soil organic matter. Additional organic matter is often lost by the sorting action that results from increased erosion. The significance of these losses in organic matter varies greatly with the soil texture, structure, and initial organic-matter content. Generally, they cause a deterioration in soil structure, an increased tendency to form crusts, reductions in soil permeability, and some loss in nutrient storage capacity. For example, Elliott and Efetha (1999) compared plots that were managed with no-till to others with conventional tillage for 11 years. They found significantly higher organic carbon contents, larger and stronger soil aggregates, and higher infiltration rates in the no-till plots. These changes are the reverse of those commonly observed during the first few years after a field of sod is tilled and cropped. No-till allows the soil to return to a condition more nearly like its virgin state, even though it is being cropped. Power et al. (1998) showed that the increased soil organic matter resulting from returning crop residues to the soil had residual effects that increased corn grain yields in Nebraska by as much as 16%.

It usually takes several years for a soil to recover from the effects of long-term tillage (VandenBygaart et al., 1999), partly because it takes time for the earthworm population to rebuild and for the earthworms and plant roots to produce numerous macropores for water transmission. Even occasional tillage can disrupt these macropores.

Tillage also influences soil crusting, runoff, and erodibility through its effects on surface cover. Implements such as a moldboard plow that bury the residues leave the soil surface exposed to raindrop impact that tends to produce a crust and leaves the surface exposed to either runoff or wind erosion. The timing of such tillage is also important. For example, fall plowing for a spring-planted crop leaves the soil exposed for a much longer time than spring plowing for a similar crop. This difference is large enough to affect the *C* factors for the soil loss equations, as shown in Table 6–7. Tillage implements that leave crop residues on the soil surface (see Section 9–3) leave the soil in better condition to resist crusting.

9–2 TYPES OF TILLAGE IMPLEMENTS

Hand tools have been used to cultivate land and plant crops since the beginning of arable agriculture. They are still the primary tools used in many parts of the developing world and are the predominant implements used in most family gardens in developed countries. Hand tools and power equipment generally perform the same tasks, but power equipment works faster and often is used to stir or mix the soil deeper and to bury more vegetative cover and plant residues. Hoes, such as those shown in Figure 9–1, are used to "plow" the land, to prepare a seedbed, to open the furrow or hole for seed placement, to weed the growing crops, and even to harvest some crops. Hoes (A) and (B) are made in developing countries and are effective agricultural tools where capital is scarce and labor is plentiful and cheap. Many individuals in developed countries also use hoes. A hoe with a long handle that allows the user to work from an erect position is used by most home gardeners in Europe and North America to accomplish the same tasks that the farmers in the developing world perform. These hoes are also used in many commercial orchards and vegetable gardens.

Brochures from agricultural machinery companies are replete with pictures and descriptive names of many specific tillage implements, but there are relatively few distinctly different tillage tools. There are a variety of types of plows, several kinds of disk implements, a number of different tine tools, and some other miscellaneous machines.

9–2.1 Plows

The original animal-drawn plow was a sharpened spike drawn through the soil. It did not invert the soil and it killed few weeds. In Europe, its soil-penetrating part was modified and a soil-turning part, originally made of wood, was added. It became the *moldboard plow.*

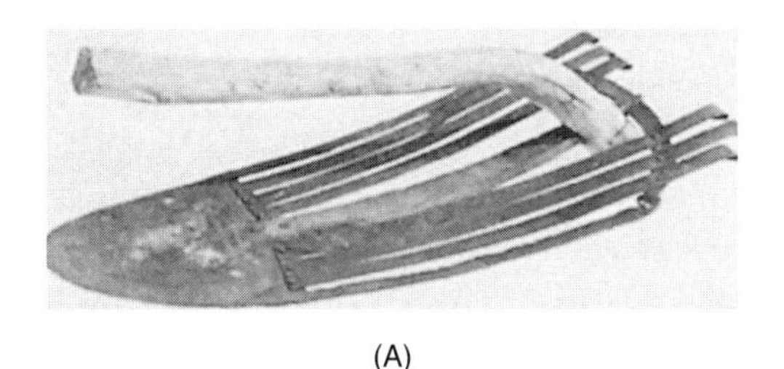

(A)

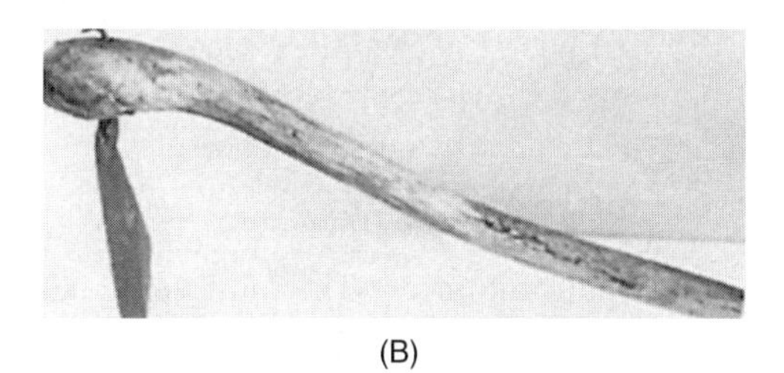

(B)

Figure 9–1 Hoes are the traditional tillage tools where hand labor is used for agriculture. The large hoe (A) is used to "plow" the land. Its blade is approximately 12-in. (30-cm) wide by 24-to 30-in. (60- to 75-cm) long. The small hoe (B) is a general purpose tool called a *lalanya* in Nigeria. It is used to control weeds, rebuild ridges, and harvest crops. The blade is roughly triangular with a cutting edge about 6-in. (15-cm) wide. (Courtesy J. A. Hobbs.)

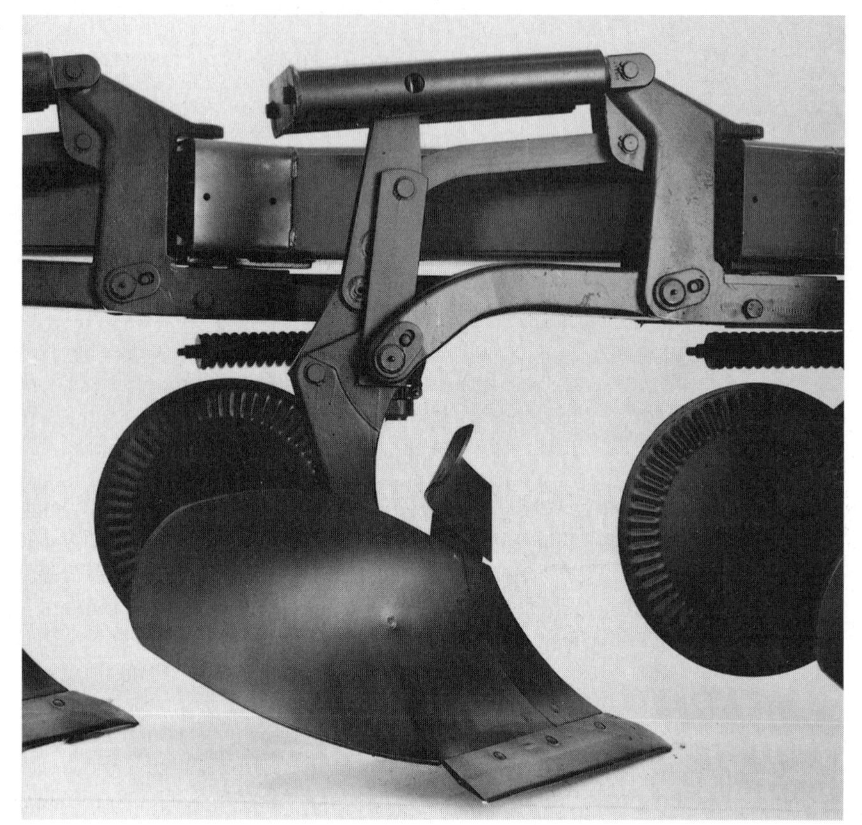

Figure 9–2 A single moldboard plow unit. The sharpened plowshare on the bottom cuts the furrow slice free from the soil. The curved moldboard, above the share, slides the furrow slice into an inverted position. The small, curved trash board guides the standing stubble and other trash beneath the furrow slice as it is inverted. The flat disk in front of the share is a coulter that cuts through crop residue and trash and partially cuts the side of the furrow slice. Not shown, because it is hidden behind the moldboard, is the landside—a flat, vertical plate—that presses against the side of the furrow and prevents the plow from sliding sidewise as it is drawn through the soil. These units can be joined together along a heavy beam to make anything from a two-bottom plow to a 10-bottom plow (or even larger). (Courtesy Deere and Company, Moline, Illinois.)

Later, the wooden moldboard was changed to iron and then to steel to enhance its ability to turn the furrows cleanly and use less power. Moldboard plows (Figure 9–2) lift and turn all the soil in the plow layer, depositing the inverted soil in the space that was occupied by the adjacent furrow. The soil is left rough and cloddy, but not ridged. Plowing requires much more power than most other types of tillage.

Repeated plowing in the same pattern causes a noticeable accumulation of soil in the *backfurrow* area, where the first furrow falls on top of unplowed soil, and a loss of soil from the *deadfurrow* area where the last furrow is not filled. Some farmers, therefore, vary their plowing pattern from one year to the next so the backfurrows and deadfurrows will fall in different places each time. Some have used this lateral soil movement to build terraces by plowing back and forth along the terrace, always moving soil toward the ridgecrest.

Figure 9–3 A four-bottom rollover (two-way) moldboard plow. The plowshare or cutting edge, the curved moldboard, and the landside are visible on the units in the air. The "two-way" feature makes it possible to turn all furrows in one direction, an excellent feature in a soil conservation program. The soil behind the plow has been inverted and crumbled. (Courtesy Massey Ferguson, Inc., Des Moines, Iowa.)

Farmers plowing land for surface irrigation prefer a two-way plow (Figure 9–3) that can move soil either to the left or to the right. They can then plow back and forth across a field without leaving a deadfurrow in the middle. This is the same feature that can allow farmers on sloping land to always move soil upslope as they plow.

Disk plows with very large sloping disks were developed for use on stony land and on soils that did not scour cleanly off moldboards. Disk plows, like moldboard plows, bury nearly all of the plant residues.

A *lister plow* loosens and moves soil into ridges on each side of the working unit. It uses either small back-to-back moldboards or disks to excavate the furrow and build the ridges. Vegetation under the ridges is buried, but the soil there is not cultivated.

9–2.2 Disk Cultivators

Disks are tillage implements with a number of saucer-shaped, steel components mounted on one or more axles. Two sets (gangs) of disks are usually joined to make a single unit; sidedraft is eliminated by facing the disks of one gang in the opposite direction to that of the disks in the other gang. The angle between the axles and the line of travel is adjustable; increasing the angle causes the disks to cut deeper and mix the soil more thoroughly. A *tandem disk,* also called a *double disk,* (Figure 9–4) has two gangs of disks mounted in front and two mounted behind. The first set turns the soil out from the center and the following set moves it back. These disks are typically used for light or secondary tillage with a penetration depth of about 3 in. (8 cm).

Figure 9–4 A tandem disk. The front disks turn the furrows out from the center, the rear disks toward the center. Such implements produce a relatively smooth surface and cover part of the crop residues. This unit is equipped with a "floating" central small disk that cuts the ridge left by the front disk gangs. (Courtesy Deere and Company, Moline, Illinois.)

A *one-way disk plow* consists of a series of large disks that are 20 in. (50 cm) or more in diameter, all saucered in the same direction. It replaced the moldboard plow as the primary tillage tool on vast areas of the Great Plains in the 1930s and 1940s. It requires much less power than the moldboard plow and it covers land more quickly. It is used less frequently now because it buries too much crop residue and leaves the surface relatively smooth and open to erosive forces.

An *offset disk,* shown in Figure 9–5, has disks mounted in tandem, single gangs that move soil in opposite directions. Disks are larger than those on the tandem disks previously described—large enough that offset disks are sometimes used for primary tillage.

9–2.3 Tine Cultivators

Some primary and much secondary tillage is performed by implements with points or blades mounted on a frame or on shanks fastened to the frame. The simplest is the *drag harrow.* Older models had rectangular frames with crossbars carrying fixed spikes about 8-in. (20-cm) long. Several frames are mounted side-by-side across the direction of travel so that units many feet wide can be pulled across a field in a single pass. More recent models have spring spikes mounted on a tool bar or on frames attached to the tool bar (Figure 9–6). This tool may be used after plowing and/or disking to break clods, smooth the soil surface, and

Figure 9–5 An offset disk. This heavy-framed, large implement has replaced the moldboard plow and disk plow as an initial tillage tool on many farms. (Courtesy Wil-Rich Manufacturing Co., Wahpeton, North Dakota.)

Figure 9–6 A spring harrow mounted behind a planter. The spring tines can be set at different angles to accommodate the purpose for which the implement is to be used—smoothing the soil surface, spreading straw, etc. (Courtesy Bourgault Industries Inc., Minot, North Dakota.)

Figure 9–7 A chisel plow equipped with narrow points. Shanks are mounted on the frame in tiers so that any bunched crop residue will flow through the machine easily. (Courtesy Wil-Rich Manufacturing Co., Wahpeton, North Dakota.)

uproot and kill small weeds. It was formerly used one, two, or even three times for the final preparation of a seedbed, but is less commonly used now that people are looking for ways to reduce tillage.

A *chisel plow* is a ripping implement that can be used with narrow points (2 in. or 5 cm), wider points (3 in. or 7.5 cm), or with attached blades up to 12- in. (30-cm) wide. The implement has several tiers of shanks on the frame, as shown in Figure 9–7. It usually penetrates 8- to 12-in. (20- to 30-cm) deep when narrow points are used. Wide blades are set to overlap 2 to 4 in. (5 to 10 cm) at the depth needed to cut weed roots.

A rugged *heavy-duty chisel,* shown in Figure 9–8, is used for soil ripping and is designed to penetrate 12- to 16-in. (30- to 40-cm) deep.

A *subsoiler* is a still more ruggedly-built chisel implement designed to penetrate 24 in. (60 cm) or more.

Sweep and *blade cultivators* have individual blades that are wider than 20 in. (50 cm); some are more than 5-ft (1.5-m) wide. Most have V-shaped blades mounted on very rugged shanks at the point of the V (see Figure 9–9), but some have straight blades with shanks at each end. Several of these cultivators can be pulled behind a single tractor. They operate at a shallow depth (about 2 in. or 5 cm) to cut weed roots.

9–2.4 Miscellaneous Cultivators

A *rotary tiller* has a series of knives or blades mounted on a rotating shaft set transverse to the direction of travel. It is powered by its own engine or by a power takeoff from the

Figure 9–8 In-Line Ripper. This implement has stout shanks and tips that are strong enough to penetrate 12- to 16-in. (30- to 40-cm) deep. The flex coulters that run ahead of each chisel cut crop residue to reduce the likelihood that bunched straw or stover will hang up on the chisel shanks. (Courtesy Thurston Manufacturing Co., Thurston, Nebraska.)

Figure 9–9 A Noble™ blade implement with rugged construction and large clearance between adjacent shanks and between blades and frame. This model has nine 6-ft (1.8-m) "V" blades for a total working width of 49.5 ft (15.0 m). The frame is flexible so that each blade will conform to the land surface and work at a uniform depth. (Courtesy Cereal Implements [Vicon], Portage la Prairie, Manitoba.)

tractor. The shaft and blades rotate at a high speed. The blades incorporate all surface residue into the soil and pulverize all large clods. Small sizes of these tillers are used to "plow" home gardens.

The *rotary hoe* has many rimless wheels mounted about 4-in. (10-cm) apart on an axle located transverse to the direction of travel. The spokes are slightly curved and are pointed. They penetrate easily into the soil, but they lift and throw the soil, crop residues, and small weeds as they exit. The rotary hoe is used to break soil crusts, smooth the soil surface, and kill small weeds.

A *treader,* really a rotary hoe pulled backwards, has spokes curved so that as they enter the soil they press residues and surface soil into the soil. The spokes leave the soil cleanly without disturbance. This tool tends to pack the soil rather than loosen it. It is sometimes used in stubble mulching to anchor loose straw in the soil. Figure 9–10 shows the spoke shapes of rotary hoes and treaders.

A *skew treader* has its axle mounted at an angle rather than perpendicular to the direction of forward movement. Two treaders can be hooked up in tandem (like a tandem disk) by joining their axles with a long chain or bar on one side and a short chain or bar on the other side. Skew treaders are sometimes used in tandem with other implements such as chisel plows, as shown in Figure 9–11.

The *rod weeder* (Figure 9–12) has a backward-rotating bar or rod that runs below the ground surface. It cuts off and lifts weeds and firms the soil under the rod. It cannot be used successfully in soils that contain stones or large pebbles. The bar needs to be equipped with cultivator points or teeth (Miller rod) if it is used on uncultivated land.

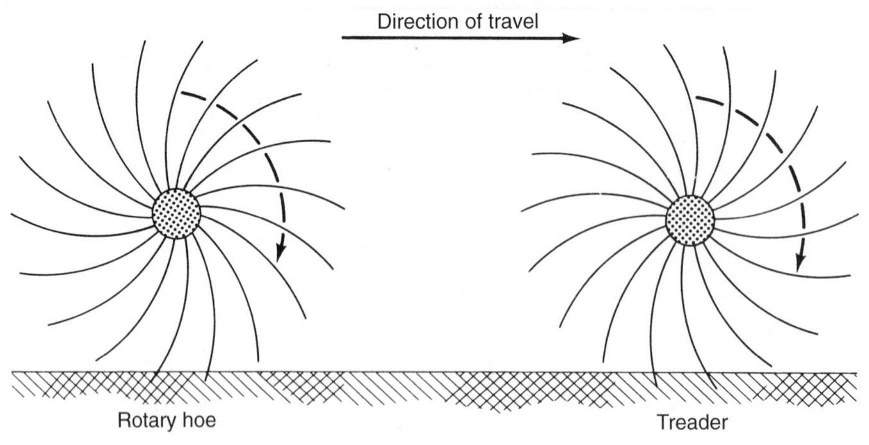

Figure 9–10 Diagram of spokes of the rotary hoe and treader wheels. The angle of entry into and out of the soil determines whether the curved tines lift the soil or press it down.

Figure 9–11 Skew treaders attached behind a gang of chisel plows. The treaders help to smooth the seedbed and to anchor the loose straw in the soil. A large amount of residue remains on the ground surface after this combined operation. (Courtesy Sunflower Manufacturing, Beloit, Kansas.)

Figure 9–12 A rod weeder. Note the overlapping rods with drive mechanisms in the centers and two bearing surfaces toward the ends of each rod. The three sections of the frame are jointed so that the bars may operate below the ground at a uniform depth. (Courtesy Hans Kok, Kansas State University, Manhattan, Kansas.)

9–3 TILLAGE, CROP RESIDUE, AND SOIL PROPERTIES

Tillage affects the nature and distribution of surface crop residues, surface soil roughness, both cloddiness and ridging, and soil-infiltration rate and permeability. Because the various tillage tools affect residue and soil differently, they affect soil and water conservation differently also. Tillage generally buries at least part of the existing crop residue. Nearly all crop residue is covered by moldboard and disk plows that invert the furrow and by rotary tillers that mix the cultivated layers. A lister, which is seldom used now, covers nearly all crop residue even though it inverts only half of the surface soil. Implements that cultivate below the surface leave 75 to 90% of the residue on top; the wider sweeps and blades leave 85 to 90%. Other implements bury intermediate amounts of residue, as indicated in Table 9–1.

Plows leave the soil surface cloddy and rough. Listers prepare a very ridged, non-cloddy surface. Rotary tillers make the surface smooth, loose, and fluffy unless they are used when the soil is moist. Disk implements produce a smoother and less cloddy surface than the plow. Chisels, especially when worked deep, leave the surface cloddy and rough. Tine implements with smaller blades usually leave the surface somewhat cloddy and rough, but wide-blade cultivators leave a smoother and less cloddy surface.

Effects of tillage on residue cover, cloddiness, and ridging depend on speed, depth of operation, and type of implement (Woodruff and Chepil, 1958). Higher speeds make larger

TABLE 9–1 EFFECT OF A SINGLE OPERATION WITH VARIOUS TILLAGE IMPLEMENTS ON CROP RESIDUE LEFT ON THE SOIL SURFACE

Implement	Proportion of original residue remaining on surface (%)
Moldboard plow (5- to 7-in. [13- to 18-cm] deep)	10
Moldboard plow (8 in. [20 cm] or deeper)	0
Chisel (twisted points)	50
Chisel (straight points)	75
Tandem disk (regular blades)	50
Tandem disk (large blades)	30
One-way disk (24- to 25-in. [60- to 63-cm] blades)	50
Chisel	75
Field cultivator (16- to 18-in. [40- to 45-cm] sweeps)	80
Skew treader	90
Blades (36 in. [90 cm] or wider)	90
Sweeps (24 to 36 in. [60 to 90 cm])	85
Rod weeder (plain)	90
Rod weeder (with small shovels or sweeps)	85
Harrow	90
Fertilizer applicator (injection)	90
No-till slot planter	90
Row crop planter	80
Furrow drill	80
Conventional grain drill (disk openers)	90

Source: Modified from the *Journal of Agronomic Education,* Volume 15, p. 23–26, 1986 (Thien) by permission of the American Society of Agronomy.

ridges, bury more trash, loosen stubble from the soil, and break down aggregates and clods more completely. A speed of about 2 mi/hr (3 km/hr) produces the most cloddy surface, but produces little ridging except with the lister. The best compromise speed for a cloddy, ridged surface appears to be about 4 mi/hr (6 km/hr). Tillage 6 in. (15 cm) or more deep, regardless of implement used in seedbed preparation, produces more nonerodible clods of greater stability than shallower tillage (2 to 4 in. or 5 to 10 cm).

Increasing the angle of attack (the angle between a horizontal line through the axis of an individual disk and the direction of machine travel) causes disks to cut deeper, buries more residue, and makes the clods finer and the ridges smaller. The angle of attack seems to be less important during second and subsequent operations.

Tillage operations that leave more crop residue on the soil surface maintain higher infiltration rates. Plowed land initially has a very high infiltration rate and permeability, but these decline rapidly, especially in soils that crust easily, because of little protective cover.

9–4 FLAT VERSUS RIDGED TILLAGE AND PLANTING

Farmers use systems of seedbed preparation and planting that experience has shown are best for their local conditions. Land for small grain and forage grass and legume seeding most often is prepared and seeded flat with no attempt to ridge the soil.

Row crops are grown on either flat or ridged fields. In dry soils, the crop may be seeded on the flat in the bottom of a furrow between two ridges. Mechanical cultivation builds new ridges over the crop rows, burying seedling weeds in the crop row and cutting off weeds between the rows. It also mounds soil around the base of the plants. This promotes the growth and development of adventitious (brace) roots.

Row crops also may be seeded on top of ridges, with subsequent cultivation maintaining the ridges. This is a common practice where the land is prepared, seeded, and cultivated by hand, especially where the soil tends to become waterlogged during rainy periods.

A modification of the ridgetop method, developed in the 1960s and known as *ridge-till,* has been used on some wet soils in the midwestern United States as a means of allowing the planted area to dry out and warm up sooner in the spring. At planting time, ridge scrapers (mounted ahead of the planter) push some soil and the dead residue left from the previous crop off the top of the ridges. Seed is planted in the residue-free, moist soil on the lowered ridge. Crop residue accumulations in the area between the ridges help to reduce erosion and control weeds. Ridges are rebuilt as the crop is cultivated for weed control.

An experimental seedbed has been developed for use in some mechanized, humid, or irrigated regions. Flat beds wide enough to accommodate two crop rows are constructed above the general soil level and are separated by furrows that provide better surface drainage and early-season aeration than does regular flat planting. Specially designed implements are needed to build these beds. Growing crops must be cultivated carefully to ensure that the beds are maintained as long as necessary. The furrows are used for irrigation on some land. This works well if the subsoil has good lateral permeability, but alternate-row irrigation is less effective on slowly permeable soils than is every-row irrigation.

Bargar et al. (1999) suggest that growing crops on ridges can result in less nitrate leaching from soils. They showed on two ridged Iowa soils that most of the water infiltrated in the furrows. Lateral water movement provided adequate water for plants growing on the ridges, but any leaching that took place was beneath the furrows. Fertilizer placed in the soil in the ridges would be carried to the plant roots rather than being leached from the soil.

9–5 CONSERVATION TILLAGE

Tillage systems that were developed in Europe and transferred to other countries when Europeans migrated usually involved the use of animal-drawn moldboard plows and tine cultivators. Systems using these implements uniformly buried all crop residues.

In much of Europe, with its gentle rains, soil erosion was not a very serious problem. However, the smooth, unprotected surfaces produced with these implements in southern and eastern Africa, in the Americas, and in Australia caused rapid and very serious soil degradation. It is now known that this type of tillage will produce extensive soil losses wherever and whenever climate and topography are conducive to erosion.

As individuals and research agencies searched for practical and economical ways to reduce erosion, it became clear that leaving crop residues on the land surfaces, rather than burying them, was very effective. As a result, several tillage methods were devised, given a variety of names, and recommended for farmer adoption. Most farmers were slow to change their tillage practices until recently, but now changes can be seen in many parts of the world.

The overall term applied to these newer techniques is *conservation tillage.* It is defined as "any tillage or planting system in which at least 30% of the soil surface is covered by plant residue after planting to reduce erosion by water, or, where soil erosion by wind is the primary concern, at least 1000 lb/ac (1120 kg/ha) of flat small-grain-residue equivalent is on the surface during the critical erosion period" (Schertz, 1988).

Crop residues have values other than soil protection. For example, they can be used as livestock feed, as a source of energy, as raw material for manufacturing building products, and so on. There is a tendency to forget their role in maintaining soil fertility and tilth. The farmer must ensure that sufficient residue for productivity maintenance and erosion control is available before any diversion is allowed to other uses (Larson, 1979).

9–5.1 Stubble Mulch Tillage

In 1938, Duley and Russell initiated a tillage concept that purposely retained crop residues on the soil surface for erosion control (Allen and Fenster, 1986). Their original tillage tool had two 22-in. (55-cm) sweeps mounted on a modified 42-in. (105-cm) corn cultivator frame. This practice came to be known as *stubble mulch tillage* or, more disparagingly, as *trashy fallow.* Its aim is to keep crop residue on the surface to protect both crops and soils from damage by water and wind erosion. This crop residue cover also serves to trap snow in areas where winter precipitation falls as snow.

The method was tested, recommended, and used to some extent in almost all climatic regions in the world where commercial agriculture is practiced. It caught on first in the drier

Figure 9–13 Undercutting wheat stubble with a four-blade implement (22.5-ft or 6.9-m wide) leaves abundant stubble standing on the soil surface. (Courtesy Noble Cultivators Limited.)

areas where wind erosion was a serious problem (Figure 9–13). Research workers, conservation specialists, and a few practicing farmers adapted it for water-erosion control in more humid regions, but it was adopted there very slowly.

Field Experience with Stubble Mulch Tillage. The first experience a farmer had with stubble mulch tillage was nearly always bad. Tillage tools plugged up with straw; weed control was often inadequate; planting, particularly with disk-type grain drills, was not always successful; and crop yields were often reduced. Zingg and Whitfield (1957) reviewed the results of much of the early research with this practice in the western half of the United States. They found that it was best adapted to semiarid regions. In more humid areas, it invariably reduced crop yields. However, the humid-area farmers who persevered and found ways to make the practice work for them were very impressed with the way it reduced erosion.

Causes of Lowered Crop Yields. Poorer crop growth with residues left on the surface appears to have been caused by less satisfactory weed control, poorer seedbed quality, delayed growth due to lower spring soil temperature, and reduced supply of available nitrogen.

Some weeds, cut off but not buried by tillage, survived unless dry conditions desiccated them. The more humid the area, the greater was the chance of rain shortly after cultivation, and the greater the chance for weed survival.

Implements other than disks and sweep machines raked up and bunched residues. If trapped residue accumulated, it lifted the implement off the surface. Weeds were not controlled and a poor seedbed resulted. Poor planter operation in heavy residues also caused poor stands with many skips and misses.

Surface mulch insulated the soil from temperature changes. Soil stayed warm longer in the fall, but warming was delayed in the spring. The mulched plots were 2 to 4°F (1 to

2°C) cooler than unmulched plots at 2 to 3 in. (5 to 7 cm) below the surface. Soil temperature below 85°F (30°C) reduces early corn growth and subsequent yields. Above 90°F (32°C), lowered temperatures are beneficial (Allamaras et al., 1964, Larson et al., 1960; van Wijk et al., 1959). Studies in the central Great Plains showed that soil temperature in stubble-mulched land in the early spring was as much as 7°F (4°C) lower than that in soil without mulch cover. Temperatures cooler than 50°F (10°C) were noted in mulched plots as long as 40 days after winter wheat regrowth started. Field and growth chamber studies show that spring regrowth of winter wheat is seriously reduced by temperatures below 50°F that last for as short a period as 18 days.

The top 36 in. (90 cm) of soil under stubble mulch tillage contained less nitrate nitrogen than similar depths under plowed land. Applications of extra nitrogen fertilizer did not always bring yields up to those on plowed plots. It should be noted that dryland crop yields are not always reduced by lower available nitrogen content because the soil may contain enough nitrogen to produce the amount of crop that the soil moisture is capable of supporting.

Some scientists warned that insects and crop diseases can be a problem where residues are not buried, but there was little evidence of this in the winter wheat or corn belt areas of the United States. Some insects were more troublesome on mulch tilled land in spring wheat areas.

Implements for Stubble Mulch Tillage. No single cultivation implement meets all the needs of stubble mulch tillage; none works equally well under all conditions to retain surface cover or to produce cloddiness and roughness. Tools with the right features for the local soil conditions must be selected (Allen and Fenster, 1986). Undercutting tools such as sweep and blade machines and rod weeders were highly recommended, but other implements were also used.

9–5.2 Reduced Tillage

In the late 1940s, scientists and some farmers in the summer-fallow area of the United States used 2,4-D, a selective herbicide, to reduce the need for tillage and to reduce tillage costs in the summer fallow operation. This was not successful because, although this chemical kills most broad-leafed weeds, most grass-family crops and weeds were immune. They survived to compete with the crops.

Increased concern over soil deterioration because of declining soil organic-matter content, soil structure degeneration, and increasing subsoil compaction developed in the early 1960s. Farmers and scientists devised ways to reduce tillage operations, especially for row crop production. The main reductions were made in the preplant period. No chemical herbicides were involved at the outset.

Reduced Preplant Tillage. Reduced tillage systems dispensed with moldboard plowing or with one or more subsequent disking or harrowing operations. Crops were produced successfully where land was prepared and seeded in one or two passes across the field, as long as weeds were controlled. Methods were given a variety of names, among them *plow-plant* and *wheel-track-plant.* A once-over tillage and planting system was called *till-plant.* Many farmers developed their own equipment, often pulling several implements in tandem behind a tractor. A commercial till-plant machine was manufactured for a brief period in the 1970s. Less traffic caused less soil compaction. The looser condition between the rows permitted more rapid infiltration, so less water ran off and erosion was reduced. Satisfactory crop stands were obtained, and yields approximated those from conventional tillage. Weeds were often less of a problem in the plow-plant and wheel-track-plant systems, but they became a severe problem in the till-plant system before wide-spectrum herbicides were available.

The partial success of these systems emphasized that there are two zones in a row crop field: the seed zone in the row area, and the moisture storage zone between the rows. Better infiltration and less runoff and erosion occur on plowed land when the latter zone is left loose, but some packing of the seed zone is usually necessary to get uniform planting and good seed-soil contact.

While results with corn and soybeans were satisfactory with these techniques, sorghum, a crop that requires a warmer seedbed and therefore a later planting date, did not do as well. Weed stands often were dense and hard to control.

Despite some success, these techniques never became really popular until effective chemical herbicides were developed. Many farmers were reluctant to use these techniques because planting rate was reduced to the speed of preparing the seedbed. Farmers with large acreages to seed to row crops want to plant rapidly.

Another approach, known as *strip tillage,* loosened a strip of soil about 4-in. (10-cm) wide, applied fertilizer and/or pesticides, planted the seed, and firmed the soil with a press wheel, all in a single pass. The soil between strips was left undisturbed and covered with residues.

Reduced Pre- and Postplanting Tillage. Additional incentive to reduce tillage came in the early 1970s with the drastic increase in cost of tractor fuel and the possibility of severely curtailed supplies for agriculture. Increasing numbers of farmers reduced cultivation or omitted it entirely (Figure 9–14). Those using *no-till* systems relied entirely on chemicals for weed control.

For successful operation of any reduced tillage system, crop residues must be handled carefully from harvest through planting time. Straw choppers and spreaders are essential on combine harvesters, especially on wide combines. Combines can now be equipped with "air reels" so that heads of small grains, sorghum, and similar crops can be harvested with a minimum of stalks. This leaves a taller standing stubble for soil protection and for moisture conservation. Smaller amounts of well-distributed straw left behind the combine make subsequent tillage and planting easier.

Planting equipment also has been modified to work through heavy residue (Hatfield and Stewart, 1994). Row crop planters have been developed that do a good job of seeding through trash. Grain drills changed more slowly, but heavy disk and shoe drills (Figure 9–15) have been modified to work through limited amounts of residue. New no-till air seeders (see Section 9–5.3) can be used for seeding through heavy residue.

Choosing the Appropriate Implement. Conservation tillage has to be flexible. In some cases, all possible residue must be kept on the surface; in others, part of the residue may need to be incorporated. Choosing the appropriate tillage implements to handle crop residues for optimum erosion control and crop production requires a knowledge of the quantity of residues needed to control erosion, the amount of residue available, and the

Figure 9–14 Row crops can be seeded in untilled soil with a no-till planter. This machine is adapted to plant on former crop ridges in order to maintain a ridged condition for erosion control. Attachments for this and other makes and models are available to apply either dry or liquid herbicides in addition to dry fertilizer. (Courtesy Fleischer Manufacturing Co., Columbus, Nebraska.)

Figure 9–15 A heavy double-disk grain drill designed to seed small grain through reasonable quantities of crop residue. It can be purchased with a variety of furrow openers. (Courtesy Deere and Company, Moline, Illinois.)

proportion of residue incorporated by each tillage operation (Table 9–1). The amount of residues needed to control erosion depends partly on the type of erosive force. Interception of raindrops is a major factor in controlling water erosion and is usually deemed adequate if the residues still cover at least 30% of the soil surface after planting. Residue cover for wind erosion control is usually evaluated in terms of an equivalent amount of flattened wheat straw (see Figure 9–16) and using the wind erosion equation discussed in Chapter 6, Section 6–5.

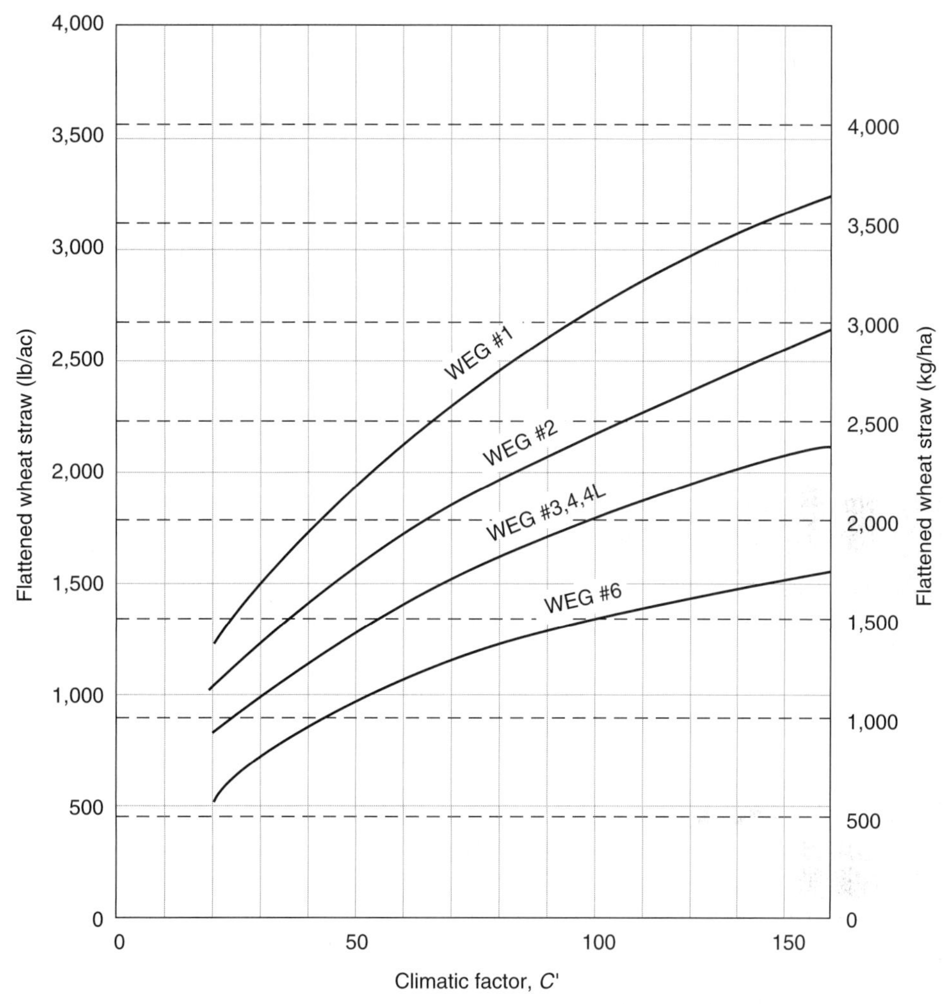

Figure 9–16 Graph showing the relationship between soil erodibility (WEG) (see Table 6–15), climatic factor C' (Section 6–5.1), and the amount of flattened wheat stubble needed to keep soil loss by wind erosion below 5 tons/ac-yr (11 mt/ha-yr) (see Figure 6–6 for converting amounts of other types of residues to equivalent amounts of flattened wheat straw). (Modified from *Crop Residue Management Systems,* p. 17–33, 1978, published by the American Society of Agronomy [Skidmore and Siddoway] by permission of the American Society of Agronomy.)

A conservation tillage implement must have at least 18 in. (45 cm) of vertical clearance between the implement frame and the soil surface. The horizontal distance between adjacent shanks must be sufficient to allow bunched residue to pass through and clear the machine. Narrow-bladed implements need several banks of shanks, one behind the other, with tines offset so that crop residue can pass through. Rolling coulters may be needed in front of each shank, especially when working through fresh, heavy residues.

Adjusting the Individual Machine. The adjustment of individual machines is very important. The "lift" of a blade affects the amount of clod disintegration and the size of ridges left by tillage. The shape of chisel points affects ridging and cloddiness produced. Implements should be adjustable for depth of cultivation, tilt of working surface, and, in disk implements, angle of attack. Disk concavity affects the amount of residue buried, efficiency of weed kill, and cloddiness and roughness produced.

Significantly more crop residue may now remain on the surface because of more effective herbicides and improvements in tillage tools and planters.

Cost and Energy Requirements of Reduced Tillage. Tillage reductions can save considerably on equipment repair and maintenance costs and cut fuel bills in half (Unger et al., 1977). This does not guarantee less farming expense, though, because the cost of buying and applying chemicals can be high. Overall costs have been reduced by judicious selection of the most effective and economical combinations of tillage and chemical treatments.

Reduced Tillage and Soil Compaction. Plowing temporarily loosens the plow layer but compacts the plowsole zone. During the first several years under reduced tillage, the top 6 in. (15 cm) of soil is often denser than it is when plowed regularly. Increased surface residues reduce surface sealing and so may maintain or improve infiltration rate.

Reduced Tillage and Runoff and Erosion Losses. Reduced tillage generally reduces runoff and erosion. Burwell and Kramer (1983) report that conservation tillage reduced runoff to 85% and soil losses to 42% of those experienced on conventionally tilled plots.

Investigations of the way reduced tillage affects wind erosion have been conducted both in the drylands of the Great Plains and on sandy soils in more humid regions. Black and Power (1965) found that chemically fallowed soil had a higher percentage of nonerodible clods in the surface soil and more surface crop residues than even stubble-mulched fallow soil. Potential erodibility of the chemical fallow soil is, therefore, even lower than that of stubble-mulched land.

Adapting Reduced Tillage Systems to Soils. Experience with reduced tillage shows that soils respond differently to various chemical-tillage combinations. For example, a seedbed preparation system was developed for sorghum in a wheat–sorghum–fallow sequence on the hardland (silt loam) soils in a 21- to 26-in. (530- to 660-mm) rainfall area in western Kansas. Two specific tillage operations and one (or two) herbicide treatment(s) controlled weeds, reduced wind erosion, and increased yields above those obtained with conventional tillage. This system did not work on sandy soils in the area or on hardland soils in slightly drier areas farther west.

Galloway et al. (1977) and Doster et al. (1991) evaluated the usefulness of various forms of tillage deeper than 6 in. (15 cm) (plowing and chiseling) versus less than 6 in. (disking, till-plant, ridge-plant, and no-till) on individual soils. They rated reduced, shallow tillage poor on muck soils. Reduced tillage, especially no-till, rated high on sloping, well-drained, permeable, coarse- and medium-textured soils. Moldboard plowing, as an initial tillage technique, rated best on poorly drained, slowly permeable soils with less than 4% organic matter and was quite satisfactory on somewhat poorly drained, permeable soils with 2 to 4% organic matter.

It takes time and effort to find the best system of reduced tillage for a particular soil, but the possibilities of reduced costs and improved erosion control make the extra effort worthwhile.

9–5.3 No-Till

The ultimate in reduced tillage is *no-till,* where the only implements used are a drill or planter, a sprayer to apply pesticides, and a harvester to bring in the crop. The major benefits of no-till systems are lower fuel and machinery costs, considerably less soil compaction, and better runoff and erosion control. Long-term benefits usually include an increase in soil organic matter and improvements in soil structure. Wildlife, too, may benefit as the residues left on the surface provide various species with food and habitat (Uri, 1998). An intangible benefit is that the farm operator has more time to think, plan, and supervise farm operations.

Several studies have been made to evaluate the effects of no-till on soil organic matter, structure, density, infiltration rate, and permeability. As indicated in Section 9–1.5, no-till at least partly reverses the negative effects of tillage on these soil properties, but it takes some time to show improvement. Initially, without the loosening effects of tillage, the soil becomes denser and less permeable. It typically takes about five years for earthworm populations to recover and to perforate the soil with their burrows. Improvements in soil organic-matter contents and soil structure are long-term effects and depend on soil characteristics and climate as well as the amount of residues returned to the soil each year. Gale et al. (2000) used radioactive tracers to distinguish between root residues and surface residues. They concluded that most of the improvement in soil organic matter under no-till is related to slower decomposition of root residues, but that earthworms and other fauna may feed on surface residues and deposit casts in the soil.

The major problems of no-till cropping are planting into soil that has had no tillage ahead of the planter and achieving adequate weed control. A cooler soil temperature is likely to delay spring planting and early growth but may become an advantage later in the season. No-till is more likely to be a good choice on well-drained soils than on wet soils.

Planters and drills designed for use on tilled land commonly plug up with residues and/or ride over the residues if they are used with heavy residues on the surface. Seeds are then planted at the wrong depth or even left exposed on the surface, and a poor stand results. No-till row crop planters are designed to cut through residues and work without plugging. Most of them not only plant and cover the seed but also apply a band of fertilizer, and they are often equipped to apply herbicides as well (see Figure 9–17). Similarly, new no-till air seeders (Figure 9–18) will seed cereal grains satisfactorily through heavy residue.

Weed control has always been one of the major reasons for tilling the soil, but it doesn't always work. Stirring the soil can bury weed seeds that were on the surface, but it can also bring dormant weed seeds from previous years back to the surface. Many weeds can be killed by timely tillage with an appropriate implement, but some are tenacious and others grow close to the crop plants where they are protected from any cultivator. Consequently, herbicides have been developed to kill weeds in a variety of crops and situations. Conventional farmers may use a combination of tillage and herbicides to kill weeds, but no-till farmers rely on herbicides. Without any tillage, they may need to apply more and/or different herbicides, and they will probably have to experiment to learn what works best.

Figure 9–17 A no-till row crop planter. The planter is equipped to open the furrow, push crop residue to the side, fertilize, and apply herbicides. This type of implement will work equally well in untilled wheat stubble or in clean-tilled land. (Courtesy Fleischer Manufacturing, Inc., Columbus, Nebraska.)

Herbicide results are not precisely predictable. Soil properties such as soil reaction, organic-matter content, clay content, and drainage characteristics, the amount of crop residue on the soil surface, and the climate all affect the activity and longevity of the chemicals. The choice of herbicides must be coordinated with the crop to be grown, since some herbicides will kill or stunt some crops. One reason that genetically modified crops (GMOs) have become popular is that they are able to tolerate some of the most effective herbicides (Broome et al., 2000). The resulting weed-free fields are impressive.

Tillage practices also influence the impact of other pests. For example, nematodes can move from one place to another in small amounts of soil that cling to a tillage implement. Eliminating tillage can therefore slow their invasion of new areas. Some plant diseases are similarly affected. For example, spotted wilt of peanuts is a serious problem in the southeastern United States. A study by Johnson et al. (2001) showed that the peanuts in a peanuts–cotton rotation had 42% less spotted wilt where reduced or no-till cropping was practiced as compared to cropping with conventional tillage, and that crop yields during their 5-year study were sustained with any of these tillage systems.

Options for fertilizer placement are reduced with no-tillage, but this is not as serious as was originally feared. Some fertilizer can be banded in or on the soil by no-till planters and additional fertilizer can be broadcast on the surface either before or after planting. Plants can absorb even immobile phosphorus satisfactorily from surface-applied fertilizer if the residue cover keeps the surface soil moist and if roots are not pruned by postplanting cultivation. Yield reductions have been reported where long-term no-till soils become dry on the surface (Vyn and Janovicek, 2001) and plant nutrients are depleted in the root zone.

Figure 9–18 A no-till grain drill with rugged, high-clearance furrow openers and a combined air seeder/fertilizer applicator. This unit comes in widths from 27 to 57 ft (8.2 to 17.4 m) and is capable of seeding through dense residues on untilled small grain or row crop fields. An herbicide application attachment is available. (Courtesy CNH Global N.V., Saskatoon, Saskatchewan, Canada.)

TABLE 9–2 EFFECT OF TILLAGE SYSTEM ON RUNOFF AND SOIL LOSS AT BILOXI, MISSISSIPPI, 1970, 1971, 1972

Tillage and crop system	Runoff (%)	Soil loss (tons/ac-yr)
No-till, soybean–wheat doublecrop	23	0.8
No-till, corn after soybeans	33	2.3
No-till, soybeans after corn	24[a]	0.6[a]
No-till, continuous soybeans	23	1.1
Conventional tillage, continuous soybeans	29	7.8

[a]Only two-year average results.

Source: From *Transactions of the American Society of Agricultural Engineers,* Volume 18, p. 918–920, 1975 (McGregor et al.).

Placement of a band of fertilizer in the soil, for example 2 in. (5 cm) to one side and 2 in. below the seed, can be very helpful in such situations.

No-till has variable effects on runoff, but it nearly always reduces erosion losses, sometimes dramatically. A Mississippi study showed that conventional tillage caused significantly more erosion than no-till in all cropping systems (Table 9–2). The no-till method of seedbed preparation was compared to conventional tillage including moldboard plowing on sandy soils in northwestern Ohio. The no-till technique left 2 to 3 tons/ac (4 to 7 mt/ha)

TABLE 9–3 EFFECT OF CHEMICAL AND MECHANICAL WEED CONTROL ON RUNOFF, SOIL EROSION, AND CORN YIELD ON MEXICO SILT LOAM AT THE MIDWEST CLAYPAN EXPERIMENT FARM NEAR MCCREDIE, MISSOURI, FOUR-YEAR AVERAGE, 1966–69

	Precipitation		Runoff			Erosion		Corn yield	
Treatment	(in.)	(cm)	(in.)	(cm)	(%)	(tons/ac-yr)	(mt/ha-yr)	(bu/ac)	(q/ha)
Chemical	40	101.6	9.5	24	23.7	10.5	23.5	105	66
Mechanical	40	101.6	8.7	22	21.8	6.7	15.0	113	71

Source: Modified from the *Journal of Soil and Water Conservation,* Volume 28: 174–176, 1973 (Whitaker et al.).

of corn residue on the soil surface, while the surface with conventional tillage was nearly bare. The conventionally treated area lost 130 tons/ac (291 mt/ha) of soil during one severe windstorm, whereas the no-till area lost only 2 tons/ac (4 mt/ha). Corn yields over the two years of the study averaged 68 bu/ac (4265 kg/ha) on the conventionally treated land and 93 bu/ac (5833 kg/ha) on the no-till area (Schmidt and Triplett, 1967).

Soil temperature is invariably lower under a mulch cover (Larson et al., 1960). Lower temperatures slow germination and early crop growth, so a later planting date may be necessary. These effects may or may not reduce the final yield. Tillage is most likely to improve yields on fine-textured soils that tend to stay wet and cold in the spring (Vyn et al., 1998). Whitaker et al. (1973) found that when Mexico silt loam (a clay pan soil) was cropped no-till, runoff and erosion increased significantly and crop yields decreased significantly (Table 9–3). The mechanical treatment was apparently better because tillage broke a pronounced crust on this soil and permitted better infiltration of rainwater. The ridge-till method discussed in Section 9–4 is often useful for growing row crops in wet soils.

Many experiments have compared crop yields with no-till to those with a wide variety of tillage systems. The results have been variable. Reduced yields often occur during the first few years of no-till. This may be considered an adjustment period, both because the earthworm population and macropore network must be rebuilt and because the farmer must learn how to manage a new system. After an adjustment period (often about five years, more or less), the no-till yields on soils with adequate drainage are generally as high as or higher than those on tilled soils. Profits are even more favorable because of savings on fuel and equipment costs.

9–5.4 Current Use and Future of Conservation Tillage

Since the Soil Conservation Service first reported on "minimum tillage" in 1963, the acreage has been steadily increasing and the technology reportedly is being adopted more quickly than any other soil and crop management innovation in U.S. agricultural history. Schertz (1988) suggests that between 63 and 82% of cropland will be conservation tilled by 2010. Some question the accuracy of these statistics and of the prediction. They claim that implement use may change as predicted, but that many fields cultivated with disks and undercutting tools have far too little crop residue remaining on the soil surface during the late seedbed preparation period to reduce soil losses appreciably or to qualify as conservation tillage (Dickey et al., 1987). A survey in Nebraska showed that 55% of the farmers believed they were using conservation tillage. Field inspections showed that less than 5% of

TABLE 9–4 PERCENT OF CULTIVATED LAND PREPARED FOR SEEDING BY VARIOUS TILLAGE SYSTEMS IN THE UNITED STATES, 1989–1996

	Year							
Type of tillage system	1989	1990	1991	1992	1993	1994	1995	1996
Conservation tillage								
No-till	5.1	6.0	7.3	9.9	12.5	13.7	14.7	14.8
Ridge-till	1.0	1.1	1.1	1.2	1.2	1.3	1.2	1.2
Mulch-till	19.6	19.0	19.7	20.2	21.1	20.0	19.6	19.8
Total[a]	25.6	26.1	28.1	31.4	34.9	35.0	35.5	35.8
Other tillage								
15–30% cover	25.3	25.3	25.7	25.9	26.3	25.8	25.2	25.8
< 15% cover	49.1	48.7	46.1	42.7	38.8	39.3	39.3	38.4
Total[a]	74.4	73.9	71.9	68.6	65.1	65.0	64.5	64.2

[a]Columns may not add to exact totals due to rounding.

Source: Courtesy USDA-ERS, 1996. (Bull and Sandretto.)

fields had 30% surface cover immediately after planting. Shelton et al. (1995) state that no-till was the only tillage system they tested that regularly provided residue-cover levels at planting time significantly greater than 30%. Two operations of any implement or combination of implements regularly reduced cover to below 30%.

Bull and Sandretto (1996), USDA-ERS, showed how methods of residue management changed in the United States during the period from 1989 to 1996. During that period, the percentage of cultivated acreage in the United States that was conservation tilled increased from 25.6 to 35.8%, as shown in Table 9–4. Nearly all of this increase was due to increases in no-till. The largest decrease in the other tillage designation was in the category "< 15% cover."

Farmers are concerned about soil losses from erosion and about rising tillage costs. Farmers in the United States also know that they must reduce erosion losses from their fields in order to participate in the benefits of the 1985 and subsequent Food Security Acts. All these factors will maintain interest in residue management and will cause increased use of conservation tillage. It is a demanding technique requiring new equipment and more sophisticated managerial ability than conventional tillage does. Extension and conservation personnel must plan and execute outstanding promotional and guidance programs to ensure the benefits of increased use.

9–6 DEEP TILLAGE

Many U.S. experiment stations have conducted studies that compared the effects of different depths of plowing, subsoiling, and even dynamiting on crop growth and yield. The results of early work showed that plowing deeper than 7 to 8 in. (18 to 20 cm) had no merit unless the soil had a root-inhibiting zone near plow depth. Soils containing pans at or below normal plowing depth that seriously restrict root penetration sometimes benefited from plowing 12 to 20 in. (30 to 50 cm) (Fehrenbacher et al., 1958). This condition is most commonly a problem in warm climates where winter frost does not penetrate deep enough to

Figure 9–19 Deep chiseling a field leaves clods on the surface and deep chisel marks in the soil. (Courtesy Allis-Chalmers Corporation, Milwaukee, Wisconsin.)

loosen the soil in a plow pan. Deep chiseling would be a likely alternative to deep moldboard plowing on such soils.

Subsequent work on certain compact high-clay soils in the Mississippi Valley has shown that soybeans and cotton grown there benefit from deep tillage (Wesley et al., 2001). Deep tillage needed to be done in the fall when the soil was dry so it would shatter large soil blocks and needed to be repeated about once every three years.

9–6.1 Deep Chiseling and Subsoiling

Deep chiseling is the practice of opening the soil 12- to 16-in. (30- to 40-cm) deep with a heavy-duty chisel or chisel plow that penetrates and rips the soil without inverting the ripped layer (Figure 9–19). *Subsoiling* involves ripping the soil 20 in. (50 cm) or more without inversion. Power requirements for deep chiseling and subsoiling are high. Chiseling the soil every 22 in. (55 cm) to a depth of 16 in. (40 cm), or subsoiling 22- to 24-in. (55-to-60 cm) deep every 44 in. (110 cm), takes more power than plowing 6-in. (15-cm) deep.

Water moves readily into chisel slots and cracks as long as they remain open at the soil surface. Water infiltration is limited by the small pores at the surface when slots are covered over or filled with soil. Subsurface cracks that are not connected to the surface can drain excess water from saturated soil, but they cannot help water infiltrate. Even tillage slots that are filled artificially with crop residues (*vertical mulching*) do not increase infiltration after the surface of the slots is covered by tillage or natural forces. Deep chiseling usually has no direct effect on the control of runoff and erosion except immediately after the tillage operation, and on soils with swelling clays (especially Vertisols), where the chisel marks may reopen as soil shrinkage cracks develop in dry weather (White, 1986).

Most investigations show little or no yield response to deep chiseling. Where pans are found at plow depth, or slightly below, chiseling between old crop rows as an initial step in seedbed preparation increases infiltration and reduces runoff and erosion (Moldenhauer et al., 1983). Shattering such pans allows roots to penetrate better into the subsoil also. Campbell et al. (1974) found that chiseling to a depth of 15 in. (38 cm) loosened Norfolk sandy loam and Varina sandy loam (Ultisols) enough to improve root penetration. The larger root systems collected enough extra moisture to permit the growing crop to evade damage during short drought periods, but they did not increase crop yields in years of adequate and timely rainfall or in years of long droughts.

9–6.2 Deep Plowing

The methods of deep tillage mentioned in the previous section involve stirring the soil or cracking it open. *Deep plowing* inverts the cultivated layer by using very large disk plows with disks up to 3 ft (90 cm) or more in diameter or with large moldboard plows. Some of the latter can turn a furrow slice 6-ft (1.8-m) deep.

Deep plowing has been employed successfully where a productive soil has been buried by unproductive erosion sediments or where an infertile or a highly erodible surface layer is underlain by more productive or less erodible material. This technique was first used successfully in California where deep layers of mixed and rocky material from eroding hilly land were deposited on productive bottomland soils. Deep plowing turned under the layer of sediment and returned the old topsoil to the surface. The buried soil had to have a high productive capacity in order to make this practice economical. Large areas of bottomland covered by relatively sterile, sandy, flood deposits also have been restored to something like their original productivity by deep plowing.

Deep plowing has also been successful in the southern Great Plains of the United States, where highly erodible sandy surface soils overlie finer-textured subsoils. Thousands of acres of sandy, highly erodible soils were deep-plowed in Kansas, Oklahoma, and Texas in the late 1940s and especially in the 1950s. Results of studies showed that many of these soils were made less erodible and more productive. Clay and organic matter contents of surface soils were increased; soil cloddiness was improved. Soil erodibility was reduced and crop yields increased. After five to six years, however, clay and organic matter had returned to levels close to those found prior to deep plowing, nonerodible aggregates were halfway back to pre-plowing quantities, and erodibility had increased dramatically. Crop yields also were lowered quickly. These degradations occurred even where subsequent wind erosion was not a factor.

The following precautions should be taken in deep plowing a sandy soil.

1. Plow only soils that contain less than 10% clay in the surface, and more than 20% in the subsoil within 12 to 16 in. (30 to 40 cm) of the surface. Do not deep plow soils that have more than 40% clay in the subsoil.
2. Plow deep enough to turn up at least 1 in. (2.5 cm) of finer-textured subsoil for each 2 in. (5 cm) of sandy surface soil present.
3. Plow in large, solid blocks to minimize the proportion of the deeply ridged, plowed land area that will be covered by sand drifting in from nonplowed areas.
4. Fields with small areas of soils that have less than 20% clay in the subsoil can be deep plowed if the extent of the sandy subsoil phase is less than 10% of the field area.

5. Furrows must be turned completely over so that the subsoil is on top rather than mixed with the topsoil. Moldboard plows invert best with minimal mixing; if a disk plow is used, the disks must have a diameter at least twice the expected depth of plowing.
6. Supplementary erosion-control practices and other good management techniques must be employed to prevent subsequent damage to the deep-plowed sandy land.

Spodosols with highly acid subsoils may also need to be deep plowed to incorporate liming materials so that deep-rooted crops like alfalfa can be grown on them. Carter and Richards (2000) reported that liming the normal plow layer was not enough on such soils on Canada's Prince Edward Island.

9–7 CONTOUR CULTIVATION

Contour cultivation (or *contouring*) is tillage and planting of crops along contour lines rather than parallel to field boundaries. It is used in humid and moist subhumid regions mainly to reduce soil erosion. It is used in semiarid and drier portions of subhumid regions primarily to increase soil moisture by reducing runoff losses. Its soil conservation features are discussed in the following pages; its moisture conservation potential is discussed in Chapter 13.

Contour cultivation is an ancient practice. In countries with a long history of cultivation, it is rare to find sloping cultivated lands in a good state of productivity that have not been contour cultivated. In the United States, contouring was used first in the Southeast. Very few farmers employed the practice until the 1940s. Adoption has been slow and resistance to acceptance is still very strong, even though contouring is an inexpensive practice to initiate.

Contour ridges produced by tillage, planting, and crop rows, such as those in Figure 9–20, form barriers that slow or stop downhill water movement. Effectiveness of ridges in trapping water and reducing soil loss decreases as slope gradient increases. Larger ridges are more effective than small ones. Where large contoured furrows are blocked (dammed) at intervals along their length, forming basins, rain is held where it falls and runoff and soil loss are further reduced. This practice is rather common in parts of Africa and Asia, where it is called *tie-ridging*. It is seldom used where agriculture is mechanized because the ties make the land too rough to be comfortably cultivated.

9–7.1 Contour Cultivation, Erosion, and Crop Yields

An early summary of results from contour studies from various state and federal sources, compiled and mimeographed for the Soil Conservation Service by J. H. Stallings, showed 25 to 76% reductions in water erosion resulting from contour cultivation compared to farming up- and downhill. Smith (1946) quotes experimental results that showed contouring reduced soil erosion losses by 40 to 80%. A study near Ottawa, Canada, on Rideau clay soil showed that contour cultivation of land in corn reduced erosion from 7.8 to 2.6 tons/ac-yr (17.5 to 5.8 mt/ha-yr) (Ripley et al., 1961).

Two points should be kept in mind when viewing contouring results from small plots. Rows in plots are short and have little row grade to carry water to a low spot, whereas contoured fields usually have some row gradient, and water does move behind the ridges.

Figure 9–20 Contour cultivation on 1 to 3% slopes in Kansas. The ridges reduce soil erosion by acting as barriers to the flow of water. (Courtesy USDA Natural Resources Conservation Service.)

Farmers usually farm parallel to a field boundary rather than directly up- and downhill. Thus, farm field erosion is likely to be less severe, and the erosion reduction a farmer can achieve by changing to contour cultivation is usually less than that obtained in experimental plots.

Studies show that farming across the slope increases crop yields. Out of more than 600 sets of comparisons compiled by Stallings, there were only seven where the contour-treated land failed to outyield noncontoured land. Average corn yields increased about 10%, average wheat yields about 29%, soybeans 11%, and sorghum about 28%. Reasons for these increases include more available moisture, a less eroded, more productive soil, less washing out of seeds, and less burying of seedlings.

9–7.2 Drawbacks of Contour Cultivation

In spite of contouring's potential benefits, much land is still farmed parallel to the fenceline. The most common reasons given for continuing to "farm with the fence" is that working a field on the contour is inconvenient, causes many short rows and extra turning, and increases labor and machinery time and cost. These objections become stronger as farm equipment gets bigger. But contouring produces longer rows as well as shorter ones, and tractors are known to operate more efficiently on the level. Costs in time and inconvenience of contouring may well be more than offset by reduced fuel costs and the benefits of erosion control and higher yields.

9–7.3 When and Where to Use Contour Cultivation

Contouring is most efficient on gentle and short slopes. Intense rainstorms on steeper slopes cause water to accumulate behind the ridges until it breaks over, rushes downhill, and forms rills and gullies. Erosion becomes progressively more severe on longer slopes until more erosion actually occurs in the gullies on the contoured land than in the rills between crop rows on noncontoured land. Limits on slope gradient and length therefore have been set beyond which contouring alone is not recommended. Terracing (Chapter 10) is the only convenient way to reduce long slope lengths and should be used where long slopes must be cropped without enough residues to provide adequate protection.

Experience with no-till and other reduced tillage systems that leave the soil surface well protected with crop residue show that abundant residue cover allows field lengths far in excess of those normally permitted on soils without residue cover.

9–7.4 Farming a Contoured Field

Wherever possible, upper and lower field boundaries should be changed to follow the contour. Natural drainageways should be prepared to handle extra water that will be guided onto them by tillage marks and crop rows. Waterways that are to serve as access roadways for implements will need to be extra-wide, as regular travel up and down the center of a waterway damages the grass cover.

Water flowing onto the field from higher land should be intercepted by a diversion (Section 10–1.7) built along the upper field boundary and carried away from the field. A new waterway may be needed or an existing one widened and reshaped to accommodate the diverted water.

A few fields have simple, uniform slopes that make it possible to farm the whole field parallel to a single guideline, but most fields have complex slopes, and several guidelines are needed to lay out a contouring plan, as shown in Figure 9–21. Terrace ridges are the best guidelines, but if a field is to be contour-cultivated without terraces, the guidelines must be laid out and marked permanently. It is seldom advisable to have these guidelines exactly on the contour. They need a slight grade so tillage marks and crop rows will direct water toward waterways. Grades between 0.1 and 0.2% can be used on permeable soils and between 0.2 and 0.6% on less permeable ones (Carter and Carreker, 1969; Harris and Watson, 1971).

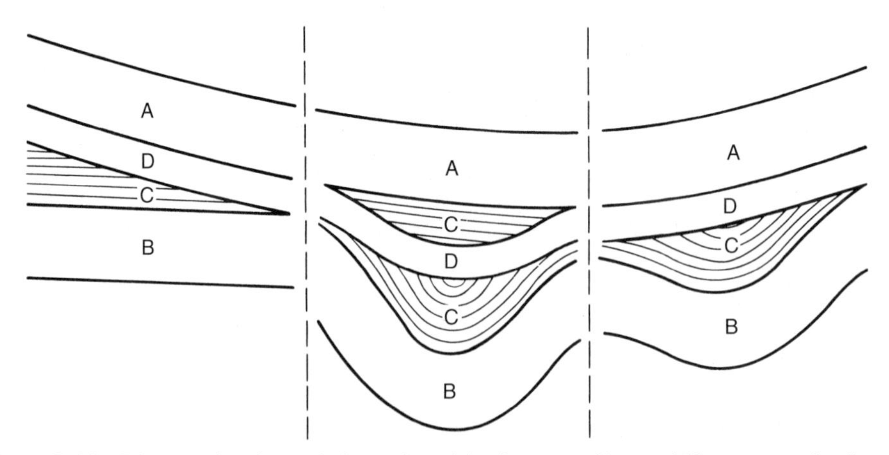

Figure 9–21 Diagram showing techniques for cultivating nonuniform-width contour strips for most efficient field management and erosion control. The areas are planted in the sequence A, B, C, D as explained in Note 9–1. Area D is the turn-row strip. (Courtesy USDA Natural Resources Conservation Service.)

The top guideline ordinarily is laid out about one terrace interval below the upper side of the field. Other guidelines are established wherever needed to keep row gradients within permissible limits. Guidelines should be laid out by engineers or conservation workers who have the necessary equipment and experience. This service is free to all conservation district cooperators in the United States (Chapter 19).

The farmer should mark the guidelines by plowing a furrow or by other suitable means before the layout crew leaves the field. Permanent stakes must be set in fence rows or elsewhere to mark the ends of guidelines that do not coincide with terraces.

All cultivating and planting operations are started along a guideline. Slopes remain the same along all tillage marks or plant rows where the guidelines are parallel but change with distance above or below nonparallel lines. Two situations arise:

1. The distance between an adjacent pair of guidelines is greater (the land is flatter) in the center of the field than at the edges. In this case, successive rows parallel to and below a guideline carry water toward the field edges along progressively steeper gradients. Rows above and parallel to a guideline move water toward the field center.
2. The width between guidelines is narrower (the land is steeper) in the center and rows above a guideline direct water toward the field boundaries; those below a guideline direct water toward the center.

All cultivation and planting operations should be performed so that each furrow leads excess water to the sides of fields or to vegetated areas established to carry the water safely to the foot of the slope. Such rigid water control is seldom practical because of the inconvenience of farming many areas of short (point) rows. Therefore, systems are developed that keep the number of furrows that lead water away from field borders and other vegetated areas to a practical minimum, but that also keep the number of point rows within bounds (Note 9–1).

NOTE 9–1
FARMING CONTOURED FIELDS

Fieldwork between parallel contour guidelines is commonly performed halfway up and halfway down from adjacent guidelines. It is advisable to work the areas between adjacent nonparallel guidelines (on irregular slopes) in the following sequence:

1. Cultivate or plant full-length rows about halfway down at the narrowest width from the upper guideline (area A in Figure 9–21).
2. Cultivate or plant the same number of rounds up from the lower guidelines (area B), leaving an unworked strip wide enough to turn at the narrowest point.
3. Fill the intervening spaces with short rows (area C) according to field conditions as shown in Figure 9–21, leaving a uniform turnrow strip across the entire field (area D) that is wide enough for the

equipment being used. Usually the area of short rows is below the turn-row strip. Field work here is done so that the ridges and plant rows grade down toward the field edges or to a waterway. Short rows should be worked out both above and below the turn-row strip where there are more short rows in C than full-length rows in B.

4. Cultivate or plant the turn-row area (area D). This area can be seeded to the crop planted in the field, to another annual crop, or to a perennial grass or grass-legume mixture.

It is especially important to follow these instructions when planting a contoured field to row crops; the sequence is less critical when planting small grains. Water will flow toward the center of the field in some strips in area A of each section of Figure 9–21. Maximum permissible slope lengths have, therefore, been established and terraces are recommended to help control runoff where the slopes are too long.

9–7.5 Contour Furrows for Rangeland and Pastureland

Overgrazed or otherwise depleted rangeland may suffer from erosion. Contour furrows and pits have been recommended as ways of reducing runoff, increasing plant growth, and reducing erosion. Contour furrows are not likely to be effective on sandy or rocky soils, on claypan areas, or on steep slopes. Best results are obtained if the range includes grass species that spread by rhizomes or by stolons. Bunchgrasses that only spread by seed or by tillering are slow to respond.

Design and Construction of Range Furrows. *Rangeland furrows* should be laid out on the exact contour for best results. Key contour lines are laid out as on cropland, about one terrace interval apart. The furrowing implement is worked halfway up and halfway down from the contour guidelines, with the odd-shaped lands filled in afterward.

Furrows are most commonly constructed with small shovels mounted on tractor tool bars. Sturdy equipment is needed to withstand the heavy pull through the sod and the fast speed required to scatter the turned sod. The tool bar should be capable of fast lift so that furrows can be stopped short of rills, gullies, and roadways. Furrows may be from 3- to 12-in. (7.5- to 30-cm) wide and from 3- to 6-in. (7.5- to 15-cm) deep. Furrows that expose subsoil may reduce or prevent improvement and spread of vegetation. Furrows should be broken at intervals by lifting the shovels out of the ground in order to reduce the danger of water concentrating at breakover sites where the furrows are not on the exact contour.

Range Pitting. *Range pitting* involves working the soil surface with specially prepared, large disk implements that have alternate disks removed. The remaining disks are either mounted eccentrically or have one-third of the circumference cut away so that each disk bites into the soil only part of the time. Range pitting is both less expensive and less effective than furrowing. Contour lines to guide the pitting operation should be laid out in the usual way at about normal terrace intervals.

9–8 EMERGENCY WIND-EROSION CONTROL

Wind erosion starts in some fields in spite of precautions taken to reduce the erosion hazard. Some soils are naturally very erodible, but the predominant factor that sets the stage for soil drifting is the lack of adequate vegetative cover. Drought, winter killing, insect depredations, and diseases affect the amount of top growth produced and the protective cover on the land. Whatever the cause of poor cover, strong winds striking bare soil can cause drifting. Once movement starts, it usually spreads across the field, pasture, or range, and may also spread to neighboring areas. Emergency tillage is used to stabilize the drifting soil as quickly as possible and to prevent the spread of erosion.

The Natural Resources Conservation Service reports that about 6 million ac (2.4 million ha) receive emergency tillage each year in the 10 U.S. Great Plains states. This figure may decline as tillage methods to leave surface residues are used more effectively, but there will always be some land in a vulnerable state.

There are two general methods of emergency soil-drifting control. One is to spread straw, manure, and other vegetative mulches, asphalt and similar sprays, and even very coarse sand and fine gravel. This method may serve well to control soil drifting around a construction project, but there is seldom enough vegetative mulching material available for large agricultural fields. The nonvegetative organic mulches and the coarse aggregates are too expensive for widespread agricultural use.

The second method is to roughen the surface of the soil by tillage to reduce wind velocity at the soil surface and to trap flying soil particles.

9–8.1 Emergency Tillage for Wind-Erosion Control

Implements that have been used in emergencies to control soil drifting include rotary hoes, disks, cultivators, chisels, sweep and blade machines, and rod weeders. The main requirement for emergency control on fallow or other bare fields is to roughen the soil surface rapidly over a large area (Figure 9–22). A strip of soil is roughened across the direction of the wind starting at the windward side of the field. The implement is then moved downwind to where soil is being picked up again by the wind and another pass is made parallel to the first. Successive passes are made across the field at intervals narrow enough to trap all saltating soil grains in the roughened soil. If wind velocity is likely to increase, the space between tilled strips should be narrowed enough to meet the expected increase in erosivity.

Spacing the passes across the field rather than tilling the whole field accomplishes three things. It speeds up the emergency operation, permitting a much greater area to be controlled; it reduces the cost of the operation; and it leaves uncultivated strips that can be tilled in a second emergency action if the clods turned up by the first tillage break down under the wind.

9–8.2 Emergency Tillage and Crop Damage

Emergency tillage destroys some plants if there is a crop on the land. A speedy decision, therefore, has to be made when erosion starts on a cropped field. Should emergency tillage be performed, or should erosion be permitted to continue in the hope that soil movement

Figure 9–22 Emergency tillage to control soil drifting in a wheat field in Kansas. A chisel with wide spacing was used to produce ridges and furrows across the path of the wind where the cover provided by the wheat crop was inadequate. A large part of the crop was not damaged by the chisel operation. (Courtesy USDA Natural Resources Conservation Service.)

will stop before much damage is done? Knowing how much a crop is damaged by tillage is helpful in making this decision.

A study near Dodge City in southwestern Kansas measured the effects of emergency tillage on wheat yields. The yields of wheat after summer fallow and of wheat after wheat are shown in Table 9–5. Yield decreases closely paralleled the area damaged by tillage on wheat after wheat. Fallow wheat yield decreases were proportionately much less than the area of crop damaged by tillage. As wind erosion was never allowed to occur on the experimental field, this trial measured only the damage done by the tillage operations; it did not measure the benefits brought about by erosion control.

Lyles and Tatarko (1982) studied the effect of emergency tillage on yields of fallow winter wheat on a sandy loam and on a silty clay soil during the period from 1977 to 1981. Tillage was accomplished with chisels 2-in. (5-cm) wide. Point spacings ranged from 60 in. (152 cm) to complete coverage. Tillage was performed in mid-March each year. Yields were reduced significantly only one year (1980) and on only one site (the sandy loam soil). In all other years and at both sites wheat yields were not significantly reduced by any tillage operation. Wind erosion was not a problem in any of the five years so the study measured the damage done by the tillage, but it did not measure any benefits that might have evolved from controlling soil drifting.

TABLE 9–5 EFFECT OF EMERGENCY TILLAGE ON WHEAT YIELDS IN SOUTHWESTERN KANSAS

	Wheat after summer fallow			Continuous wheat		
	Yield		Reduction	Yield		Reduction
Treatment	(bu/ac)	(q/ha)	(%)	(bu/ac)	(q/ha)	(%)
No emergency tillage	25.8	17.3	-	18.6	12.5	-
List every 20 ft	24.5	16.4	5	15.6	10.5	16
List every 10 ft	21.6	14.5	16	14.2	9.5	24
Chisel every 3 ft	24.3	16.3	6	14.6	9.8	22

Source: Modified from 1959 Annual Report of Southwest Kansas Experimental Field, Minneola, Kansas.

The results of both studies indicate that emergency tillage rarely causes excessive damage to wheat. Even when plants are uprooted, those remaining grow well and should compensate for tillage damage.

9–8.3 Implement Operation

Woodruff et al. (1957) studied the effect of speed and depth of tillage, spacing of points on a variable-spacing machine such as a chisel, and the type of points (narrow, heavy-duty, and shovels) on the effectiveness of the operation and on power requirements. Results can be summarized as follows.

1. Speeds in excess of 4 mi/hr (6 km/hr) give the most effective immediate results on compact soils. A larger area can be covered more quickly and at lower costs with these high speeds. Slightly slower speeds, 2.25 to 4 mi/hr (3.5 to 6 km/hr), are better for longer-term effectiveness.
2. Close spacing (27 in. or 70 cm) of chisel points is more effective for erosion control. Chisel spacings of 3 to 5 ft (1.0 to 1.5 m) may permit control with less crop damage if soil drifting is not too intense.
3. Narrow-pointed implements work best for compacted soils, and shovel points are best for looser, medium-textured soils. Very loose soils, such as sands and loamy sands, usually cannot be prevented from drifting by tillage except by deep listing. Even listing works only for brief periods.

SUMMARY

The three generally accepted objectives of tillage are to prepare a seed and root bed, to control weeds, and to prepare the soil to absorb water and resist erosion. Many tillage implements are used in soil management and crop production. Some of these make the soil more erodible under certain conditions; some generally reduce erodibility. The tools that reduce erosion the most are those that increase soil cloddiness, produce surface roughness, and leave crop residue on the soil surface. Planting through a trashy surface and leaving the soil in a ridged condition reduces erosion. Current trends are to use less tillage and leave residues on the soil surface.

Conservation tillage is a term that is applied to a variety of systems that leave the soil surface trashy, cloddy, and ridged. Water and wind erosion are reduced by tillage systems that leave more crop residue on the soil surface and make the soil more permeable. Good examples of conservation tillage are stubble mulch tillage, reduced tillage, and no-till.

Stubble mulch tillage is a method of farming with undercutting implements and associated equipment so that crop residues remain on the soil surface at seeding time to control erosion until the new crop provides its own protection. Stubble mulch tillage has been widely accepted in dryland regions but not generally adopted in humid areas.

Reduced tillage dispenses with some or all of the preplant and postplant tillage operations. Tillage operations were first omitted to reduce soil damage by implement traffic but now are omitted mainly to reduce operating costs and to conserve soil. Systems of reduced tillage generally require the use of chemical weed control.

No-till uses no tillage for seedbed preparation or weed control except for the action of the planter. The planter may be fitted with a chisel or disc to open a furrow for the seed, an extra hopper to carry fertilizer that can be banded near the seed row, and often with another hopper for chemicals to control weeds and other pests.

Powerful modern tractors and huge implements have provided the potential for very deep tillage, but tillage deeper than 7 in. (18 cm) has seldom proved beneficial or economical. *Deep chiseling* or *subsoiling* may help where there is a subsoil condition that interferes with root penetration or with nutrient or moisture uptake, especially in years when the rainfall is insufficient for normal crop growth and development. *Deep plowing* has been used to bury infertile surface material and bring up productive soil from below.

Contour cultivation reduces erosion on gentle slopes. Contouring combined with terraces has wide adaptability and great potential for erosion control. Contour cultivation on cropland and contour furrows on rangeland help conserve water and soil.

Emergency tillage with a chisel implement reduces active wind erosion by bringing clods to the surface and forming ridges that trap the drifting soil. The implement must be pulled across the direction of the wind in passes close enough together to trap all abrasive material. Emergency tillage can be used on both fallowed land and land in crop.

QUESTIONS

1. Is any tillage implement (for example, the plow) really indispensable now? Explain.
2. Under what specific conditions would a one-way disk plow be a useful instrument on a dryland farm?
3. What can a subsistence farmer in western Africa do to conserve soil while growing grain sorghum, when little if any crop residue remains on the land at the beginning of the rains when seedbed preparation starts? (What the farmer has not removed for his own use is harvested by nomadic cattle herds or by termites.)
4. Why is no-till cropping becoming increasingly popular?
5. State the soil conditions that would be necessary for deep tillage with a 26-in. (65-cm) subsoiler to be a profitable practice.

6. Diagram the proper layout of contour guidelines on a real or hypothetical field and describe how to prepare the seedbed and plant a row crop on the field.
7. Tell what kind of implement is best and how it should be used for emergency tillage to control wind erosion.

REFERENCES

ALLAMARAS, R. R., W. C. BURROWS, and W. E. LARSON, 1964. Early growth of corn as affected by soil temperature. *Soil Sci. Soc. Am. Proc.* 28:271–275.

ALLEN, R. R., and C. R. FENSTER, 1986. Stubble mulch equipment for soil and water conservation in the Great Plains. *J. Soil Water Cons.* 41:11–16.

ANONYMOUS, 1997. *Glossary of Soil Science Terms—1996.* Soil Science Society of America, Madison, WI, 138p.

BARGAR, B., J. B. SWAN, and D. JAYNES, 1999. Soil water recharge under uncropped ridges and furrows. *Soil Sci. Soc. Am. J.* 63:1290–1299.

BLACK, A. L., and J. F. POWER, 1965. Effect of chemical and mulch fallow methods on moisture storage, wheat yields, and soil erodibility. *Soil Sci. Soc. Am. Proc.* 29:465–468.

BROOME, M. L., G. B. TRIPLETT, JR., and C. E. WATSON, JR., 2000. Vegetation control for no-tillage corn planted into warm-season perennial species. *Agron. J.* 92:1248–1255.

BULL, L., and C. SANDRETTO, 1996. *Crop Residue Management and Tillage System Trends.* Econ. Res. Serv. Bull. No. 930, USDA, Washington, D.C., 27 p.

BURWELL, R. E., and L. A. KRAMER, 1983. Long-term annual runoff and soil loss from conventional and conservation tillage of corn. *J. Soil Water Cons.* 38:315–319.

CAMPBELL, R. B., D. C. REICOSKY, and C. W. DOTY, 1974. Physical properties and tillage of Paleudults in the southeastern Coastal Plains. *J. Soil Water Cons.* 29:220–224.

CARTER, C. E., and J. R. CARREKER, 1969. Controlling water erosion with graded rows. *Trans. Am. Soc. Agric. Eng.* 12:677–680.

CARTER, M. R., and J. E. RICHARDS, 2000. Soil and alfalfa response after amelioration of subsoil acidity in a fine sandy loam Podzol in Prince Edward Island. *Can. J. Soil Sci.* 80:607–615.

DICKEY, E. C., P. J. JASA, B. J. DOLESH, L. A. BROWN, and S. K. ROCKWELL, 1987. Conservation tillage: Perceived and actual use. *J. Soil Water Cons.* 42:431–434.

DOSTER, D. H., S. D. PARSONS, D. R. GRIFFITH, G. C. STEINHARDT, D. B. MANGEL, R. L. NIELSEN, and E. P. CHRISTMAS, 1991. *Influence of Production Practices on Yield Estimates for Corn, Soybean and Wheat.* ID-152, Purdue Univ. Coop. Ext. Ser., West Lafayette, IN.

ELLIOTT, J. A., and A. A. EFETHA, 1999. Influence of tillage and cropping system on soil organic matter, structure and infiltration in a rolling landscape. *Can. J. Soil Sci.* 79:457–463.

FEHRENBACHER, J. B., J. P. VAVRA, and A. L. LANG, 1958. Deep tillage and deep fertilization experiments on a claypan soil. *Soil Sci. Soc. Am. Proc.* 22:553–557.

GALE, W. J., C. A. CAMBARDELLA, and T. B. BAILEY, 2000. Surface residue- and root-derived carbon in stable and unstable aggregates. *Soil Sci. Soc. Am. J.* 64:196–201.

GALLOWAY, H. M., D. R. GRIFFITH, and J. V. MANNERING, 1977. *Adaptability of Various Tillage-Planting Systems to Indiana Soils.* Purdue Univ. Coop. Ext. Serv. Bull. A.V. 210.

GERONTIDIS, D. V. S., C. KOSMAS, B. DETSIS, M. MARATHIANOU, T. ZAFIRIOUS, and M. TSARA, 2001. The effect of moldboard plow on tillage erosion along a hillslope. *J. Soil Water Cons.* 56:147–152.

Harris, W. S., and W. S. Watson, Jr., 1971. Graded rows for the control of rill erosion. *Trans. Am. Soc. Agric. Eng.* 14:577–581.

Hatfield, J. L., and B. A. Stewart (eds.), 1994. *Crop Residue Management.* Advances in Soil Science, CRC Press, Inc., Boca Raton, FL.

Johnson, W. C., III, T. B. Brenneman, S. H. Baker, A. W. Johnson, D. R. Sumner, and B. G. Mullinix, Jr., 2001. Tillage and pest management considerations in a peanut-cotton rotation in the southeastern Coastal Plain. *Agron. J.* 93:570–576.

Larson, W. E., 1979. Crop residues: Energy production or erosion control? *J. Soil Water Cons.* 34:74–76.

Larson, W. E., W. C. Burrows, and W. O. Willis, 1960. Soil temperature, soil moisture, and corn growth as influenced by mulches of crop residue. Trans. *7th Int. Cong. Soil Sci.,* Vol. 1, Madison, WI, p. 629–637.

Lindstrom, M. J., J. A. Schumacher, and T. E. Schumacher, 2000. TEP: A Tillage Erosion Prediction model to calculate soil translocation rates from tillage. *J. Soil Water Cons.* 55:105–108.

Lyles, L., and J. Tatarko, 1982. Emergency tillage to control wind erosion. *J. Soil Water Cons.* 37:344–347.

McGregor, K. C., J. D. Greer, and G. E. Gurley, 1975. Erosion control with no-till cropping practice. *Trans. Am. Soc. Agric. Eng.* 18:918–920.

Moldenhauer, W. C., G. W. Langdale, W. Frye, D. K. McCool, R. I. Papendick, D. E. Smyka, and D. W. Fryrear, 1983. Conservation tillage for erosion control. *J. Soil Water Cons.* 38:144–151.

Power, J. F., P. T. Koerner, J. W. Doran, and W. W. Wilhelm, 1998. Residual effects of crop residues on grain production and selected soil properties. *Soil Sci. Soc. Am. J.* 62:1393–1397.

Ripley, P. O., W. Kabbfleisch, S. J. Bourget, and D. J. Cooper, 1961. *Soil Erosion by Water.* Can. Dept. Agric. Publ. 1083, Ottawa, Canada.

Schertz, D. L., 1988. Conservation tillage: An analysis of acreage projections in the United States. *J. Soil Water Cons.* 43:256–258.

Schmidt, B. L., and G. B. Triplett, Jr., 1967. Controlling wind erosion. *Ohio Rep. Res. Devel.* 52:35–37.

Shelton, D. P., E. C. Dickey, S. D. Kachman, and K. T. Fairbanks, 1995. Corn residue cover on the soil surface after planting for various tillage and planting systems. *J. Soil Water Cons.* 50:399–404.

Skidmore, E. L., and F. H. Siddoway, 1978. Crop residue requirements to control wind erosion. In *Crop Residue Management Systems.* American Society of Agronomy, Madison, WI, p. 17–33.

Smith, D. D., 1946. The effect of contour planting on crop yield and erosion losses in Missouri. *J. Am. Soc. Agron.* 38:810–819.

Thien, S. J., 1986. Residue management: A computer program about conservation tillage decisions. *J. Agron. Educ.* 15:23–26.

Unger, P. W., A. F. Wiese, and D. R. Allen, 1977. Conservation tillage in the southern plains. *J. Soil Water Cons.* 32:43–48.

Uri, N. D., 1998. Trends in the use of conservation tillage in U.S. agriculture. *Soil Use and Management* 14:111–116.

VandenBygaart, A. J., R. Protz, and A. D. Tomlin, 1999. Changes in pore structure in a no-till chronosequence of silt loam soils, southern Ontario. *Can. J. Soil Sci.* 79:149–160.

Van Oost, K., G. Govers, W. Van Muysen, and T. A. Quine, 2000. Modeling translocation and dispersion of soil constituents by tillage on sloping land. *Soil Sci. Soc. Am. J.* 64:1733–1739.

Van Wijk, W. R., W. E. Larson, and W. C. Burrows, 1959. Soil temperature and the early growth of corn from mulched and unmulched soil. *Soil Sci. Soc. Am. Proc.* 23:428–434.

VYN, T. J., G. OPOKU, and C. J. SWANTON, 1998. Residue management and minimum tillage systems for soybean following wheat. *Agron. J.* 90:131–138.

VYN, T. J., and K. J. JANOVICEK, 2001. Potassium placement and tillage system effects on corn response following long-term no till. *Agron. J.* 93:487–495.

WESLEY, R. A., C. D. ELMORE, and S. R. SPURLOCK, 2001. Deep tillage and crop rotation effects on cotton, soybean, and grain sorghum on clayey soils. *Agron. J.* 93:170–178.

WHITAKER, F. D., H. G. HEINEMANN, and W. H. WISCHMEIER, 1973. Chemical weed controls affect runoff, erosion, and corn yields. *J. Soil Water Cons.* 28:174–176.

WHITE, E. M., 1986. Longevity and effect of tillage-formed soil surface cracks on water infiltration. *J. Soil Water Cons.* 41:344–347.

WOODRUFF, N. P., and W. S. CHEPIL, 1958. Influence of one-way-disk and subsurface-sweep tillage on factors affecting wind erosion. *Trans. Am. Soc. Agric. Eng.* 1:81–85.

WOODRUFF, N. P., W. S. CHEPIL, and R. D. LYNCH, 1957. *Emergency Chiselling to Control Wind Erosion.* Kans. Agr. Exp. Sta. Tech. Bull. 90, Manhattan, KS.

ZINGG, A. W., and C. J. WHITFIELD, 1957. *Stubble Mulch Farming in the Western States.* USDA Tech. Bull. 1166.

CHAPTER

10

CONSERVATION STRUCTURES

Conservation structures are expensive enough that they are used only where a specific need exists, generally where the erosive effects of water or wind are most concentrated. The cropping systems and tillage practices discussed in the two preceding chapters are normally chosen first where they are applicable because they are less expensive and often are less obtrusive than structures. Nothing else needs to be done to control water and wind erosion where these practices work successfully; where they are not sufficient, one or more special conservation structures may be needed. Structures are also built to control runoff water and erosion from roadways and other building projects and, sometimes, to restrain nature's rampaging waters and strong winds. A wide variety of structures are used, including terraces, diversions, waterway and gully-control devices, dams, streambank protectors, flood control structures, and wind barriers.

10–1 TERRACES AND DIVERSIONS

Terraces have been used in many parts of the world to reduce erosion from cultivated soils. *Bench terraces,* the oldest type, were built where the supply of good, level, agricultural land was limited and where population pressure forced cultivation up steep slopes.

The first terraces used in the United States had small, narrow, steep-sided ridges that could not be cultivated with conventional farm implements. Water frequently overtopped them during intense storms, and weeds were a real problem.

Priestly H. Mangum of Wake Forest, North Carolina, designed the first really "farmable" terrace in 1885. It was wide enough to be cultivated, seeded, and harvested with ordinary machinery. The "*Mangum,*" or *broad-based,* combination ridge and channel terrace is now widely used.

A *graded terrace,* sometimes called a *diversion terrace,* has a graded channel that collects runoff and slowly leads it to a vegetated area or specially prepared outlet. A *level terrace,* sometimes called a *detention terrace,* is built with a level channel (see Chapter 13,

Section 13–5.3) that holds water behind the ridge until it is absorbed by the soil. Its primary purpose is to conserve water, but it also is an effective means of conserving soil.

Terraces cannot be used effectively on sandy areas, on stony land, or on shallow soils (over bedrock or fine-textured, impermeable subsoils). They are not practical on fields with complex topography and become too expensive for mechanized agriculture on slopes that exceed 8 to 12%, but are sometimes used on very steep slopes where crops are grown with hand labor and animal traction. Terraces are also used for landscaping sloping areas around buildings.

10–1.1 Bench Terraces

Early *bench terraces* were constructed by carrying soil from the uphill side of a strip to the lower side so that a level step or bench was formed. The nearly vertical back slopes below the terraces were stabilized by vegetation or by neatly fitted stonework.

Even more labor was required where erosion had left only a shallow, stony soil. Stones were gathered and carefully fitted into walls across the slope. Soil that had been deposited in the valley below was then loaded into baskets, carried back up the hill, and placed behind the stone walls.

Some early bench terraces are still being used successfully as shown in Figure 10–1. Sandor and Eash (1995) determined by radiocarbon dating that bench terraces now being farmed in the Colca Valley in Peru were built at least 1500 years ago. Bench terraces are still being built where rapidly increasing population and a dwindling food supply force cultivation of ever-steeper slopes. The high cost of this kind of terrace limits its use in commercial agriculture, except in some cases for irrigation.

Figure 10–1 Bench terraces on a steep mountain slope in the Punjab, India. (Courtesy Dr. G. S. Sekhon.)

Figure 10–2 A graded terrace carrying excess water from the terrace interval to a grassed waterway. (Courtesy USDA Natural Resources Conservation Service.)

10–1.2 Graded Terraces

Graded terraces, such as the one shown in Figure 10–2, intercept runoff and carry it to a protected outlet. In some cases runoff is not reduced, but erosion is invariably less because of shortened slope length, slowed runoff velocity, and trapped sediments. Terraces may empty onto pastures or wooded areas, or into natural waterways. Specially shaped waterways of the type described in Chapter 11, Section 11–3.2, often must be prepared.

Terraces are most effective when they are supported by contouring, but many farmers farm parallel to the field boundaries even if they have to cross their broad-based terraces (steep-backslope terraces cannot be crossed). Contour cultivation is more difficult with large equipment, but crossing terraces always causes more rapid terrace deterioration.

Advantages and Disadvantages. Graded terraces shorten effective slope length and intercept runoff water that might otherwise form erosive streams. Soil erosion, particularly rill and gully formation, is reduced, and soil productivity maintained. Localized sheet erosion may occur, but most of the sediment is deposited in the terrace channels.

Broad-based terraces actually *increase* land slope. Vertical fall from one terrace to the next is increased by the height of the ridge crest above the channel bottom, and this fall occurs along a flow distance that is less than the horizontal interval between terraces because the area from the channel to the terrace ridge has a reverse slope. With properly spaced terraces on land with a 4% slope, the average slope from the terrace crest to the center of the terrace channel immediately below is increased to between 5 and 6%.

Figure 10–3 A newly constructed steep backslope terrace in Iowa. This terrace is straight and parallel to the fence line. The near end of the channel is blocked and the water drains through a white, vertical pipe at the bottom of the channel into an underground pipeline. (Courtesy F. R. Troeh.)

Seeds, seedlings, and plants growing between the terraces are less likely to be damaged by erosion, but sections of the channel may stay wet enough to cause reduced crop yields resulting from delays in cultivation, planting, and harvesting. Terrace construction seriously reduces the productive capacity of channel areas on shallow soils and on soils with less productive or impermeable subsoils.

10–1.3 Steep-Backslope Terraces

The *steep-backslope* (also called *grass-backed* or *narrow-based*) *terrace* was developed to permit parallel-bordered strips to be cultivated safely on relatively steep land. Their ridges are usually 10- to 13-ft (3- to 4-m) wide and have relatively flat front slopes and narrow, steeply sloping (2 units horizontally to 1 unit vertically, or steeper) backslopes that are protected by a permanent grass cover. Often they are built straight and parallel to a field boundary rather than following the contour. A newly constructed steep-backslope terrace is shown in Figure 10–3.

Advantages and Disadvantages of Steep-Backslope Terraces. The major advantage of parallel steep-backslope terraces is that they are easy to farm (no point rows). They reduce the gradient of the cropped area between terraces by the amount of excavation below the terrace. This effect is maximized in "push-up terraces" built entirely from below, usually by a bulldozer. Field gradient becomes even lower as terraces are farmed and soil washes into the terrace channel. This type of terrace can therefore be used on land too steep to farm with broad-based terraces (Wittmuss, 1973).

The major disadvantage of steep-backslope terraces is the untillable backslopes. These steep slopes reduce cultivated area, confine all up- and down-field equipment movement to field edges, and can contribute to accidents. Insects and other pests increase on uncultivated steep slopes; rodents burrow through the ridges and open channels that allow ponded water to escape and possibly erode a channel through the terrace ridge.

10–1.4 Contour Bunds and Fanya Juu Terraces

Contour bunds are essentially small versions of graded terraces (Morgan, 1995). Their ridges are typically 5- to 7-ft (1.5- to 2-m) wide and may deviate somewhat from the contour. They are used to catch the runoff water from fields in some developing countries and often are built by hand. Because they are relatively small, they are typically placed at relatively narrow intervals, often between 30 and 65 ft (10 to 20 m).

Fanya juu terraces are built by farmers in countries such as Ethiopia and Kenya by digging a trench across the slope and using the soil to form a ridge on the uphill side of the ditch (Morgan, 1995). This ridge is a small embankment that is stabilized by planting grass or other permanent vegetation on it. Water ponds above the ridge and silt settles out in the ponded area; any water that overtops the ridge is carried away by the trench. When overtopping becomes excessive, the farmer adds soil to the ridge to create more water storage capacity. This increases the ridge height and gradually decreases the slope gradient of the cropped area.

10–1.5 Terrace Design

Terraces are usually designed to handle the runoff from a 10-year-frequency storm (American Society of Agricultural Engineers, 1989). The USDA National Resources Conservation Service (NRCS) standards call for a minimum of 8 ft^2 (0.75 m^2) of channel cross section on slopes that are less than 5%, 7 ft^2 (0.65 m^2) on 5 to 8% slopes, and 6 ft^2 (0.55 m^2) on slopes that are steeper than 8%. These standards are often exceeded because channels must be at least 12- to 18-in. (30- to 45-cm) deep, and broad enough to accommodate farm equipment. Steep backslope terraces usually have larger cross sections and deeper channels. Larger channel capacity must be provided where field operations angle across terraces rather than running parallel to them.

The front slopes of steep-backslope terraces and both slopes of broad-based terraces must be wide enough to accommodate the equipment that will be used in the field, certainly not less than 15 ft (4.5 m). The flatter the front and back slopes, the easier they are to farm but the more expensive they are to build. Front and back slopes steeper than 10:1 should not be cropped; those steeper than 4:1 must be seeded to perennial vegetation.

Terrace channel gradients should be sufficient to drain off water and prevent overtopping the ridge, but not great enough to cause channel erosion. Gradients may be as great as 0.7% at the upper end of a terrace where runoff volume is small, but they should be reduced in the middle of a terrace span. Average gradient for the total terrace length should not exceed 0.4%. The minimum gradient to move water along the channel without ponding in microdepressions is about 0.1% on permeable soils and about 0.2% on less permeable ones. To prevent erosion of an unprotected terrace channel, the gradient must be kept below 0.7%. Where steeper gradients are required to carry off large volumes of runoff, the channels should be seeded to perennial grasses. Systems with parallel terraces are more convenient to farm but will have variable-gradient channels. These are more expensive to build because they require larger and deeper cuts and fills during construction.

Graded terraces should not carry water more than 2000 ft (600 m) because of the danger of water buildup and overtopping in intense storms, nor longer than 1200 ft (375 m) on already gullied land. Where the distance to be protected is greater, terraces need to be split with an outlet provided for each segment. However, the distance between outlets can be doubled by grading segments so they flow in opposite directions.

Terrace Spacing. The interval between terraces must make the land farmable as well as control erosion. More permeable soils, less intense rainfall, more erosion-resisting crops, and more surface residue permit wider terrace spacing. Terrace spacing is commonly defined in terms of the vertical interval (VI). NRCS field workers calculate appropriate vertical intervals with the following equation.

$$VI = xS + y$$

where x = a rainfall factor

S = slope, percent

y = a soil and cropping factor

The values of x in the United States range from 0.4 to 0.8 ft (0.12 to 0.24 m), as shown in Figure 10–4. Values of y range from 1.0 ft (0.3 m) for a very erodible soil and a row crop

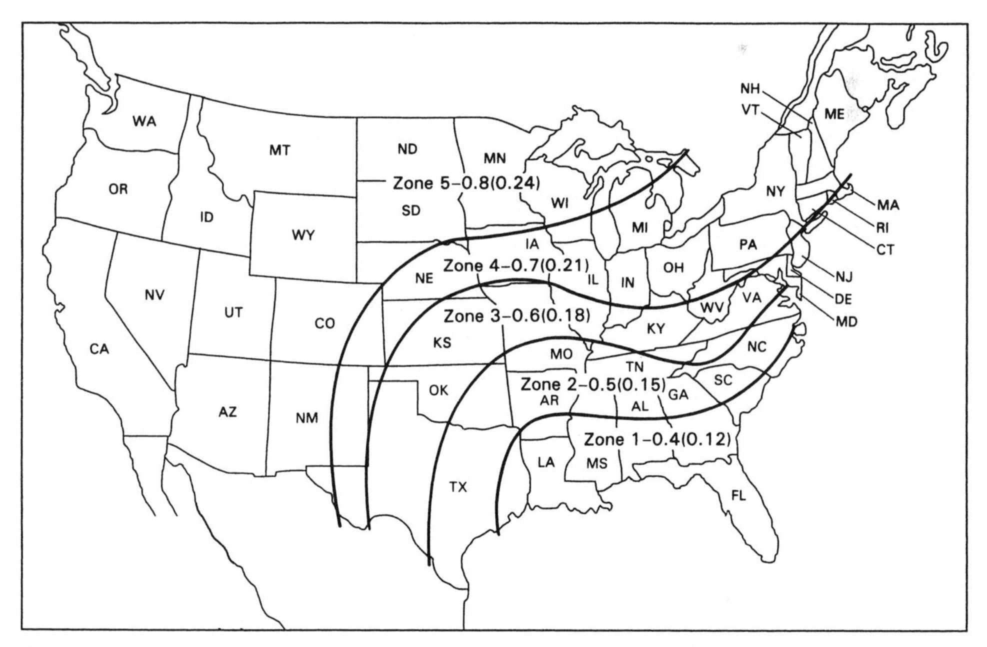

Figure 10–4 Distribution of the rainfall factor, x, for the vertical-interval equation in mainland United States. Figures in parentheses are values of x when metric units are used. (Modified from ASAE Standards, Standard S268.3, 1989, American Society of Agricultural Engineers.)

produced with conventional tillage to 4.0 ft (1.2 m) for an erosion-resisting soil combined with no-till production (a minimum of 1.5 tons of residue/ac after planting). The higher values of y should be used to calculate intervals only when residue and cropping conditions will be met every year. In the northeastern United States ($x = 0.7$ ft), the vertical interval on a 7% slope for a permeable soil (K about 0.3) producing a row crop with conventional tillage ($y = 2.0$ ft) is

$$VI = (0.7 \text{ ft} \times 7) + 2.0 \text{ ft} = 6.9 \text{ ft}$$

or

$$VI = (0.21 \text{ m} \times 7) + 0.6 \text{ m} = 2.1 \text{ m}$$

Other empirical formulae for determining vertical interval have been worked out in other countries (Bensalem, 1977). Most of these give VI values similar to those derived from the NRCS equation. The Israeli equation gives larger values for low gradients.

The horizontal interval (HI) between terraces can be calculated by means of the formula

$$HI = (xS + y) \times 100/S = \frac{VI}{S} \times 100$$

For the former example, the horizontal interval is:

$$HI = \frac{6.9 \text{ ft}}{7} \times 100 = 98.6 \text{ ft (30m)}$$

HI is actually the *horizontal* interval, not the distance between terraces measured *across the surface* of the land. The two measurements are similar, however, if the slope is not very steep. For example, the distance across the land surface between terrace lines is 98.8 ft rather than the 98.6 ft calculated for the 7% slope mentioned previously. Calculated spacings should be adjusted to the nearest multiple of the width of the tillage and planting equipment that will be used on the terraced area.

Calculated terrace spacings are often too narrow for convenient farming with modern large-scale agricultural implements. Where wider spacings are used, other measures may have to be employed to bring predicted soil loss down to tolerable levels. The NRCS does not recommend intervals narrower than 100 ft (30 m). The horizontal interval on hand-cultivated land can be considerably narrower than on mechanized farms.

Terrace Layout. Laying out a terrace system involves choosing appropriate sites for the terraces, selecting suitable channel gradients, and "shooting" the terrace lines. Some terrace systems are designed on a 50- to 100-ft topographic survey grid. The following technique is effective, quicker, and cheaper.

The conservation technician evaluates the water regime of the field from observations, soil surveys, and other information and selects waterway locations (natural draws or sites for new construction). Locations for needed vegetated waterways should be selected, staked, shaped, and seeded, and tile lines to remove runoff, if they are to be used, should be located and laid before terraces are constructed. When waterways and terraces are designed and constructed at the same time, terrace outflow must be kept off the waterways by *berms* (earth ridges) until waterways are vegetated and can handle the runoff safely.

The high point in the field is located first, and the average gradient of the top sector of the slope (to a point about 100- to 150-ft [30- to 45-m] downhill) is ascertained with an engineering level. The recommended *VI* is calculated and the point where the terrace should enter the waterway is determined (*VI* downhill from the high point). The terrace channel center is staked back from the waterway site using an appropriate gradient, increasing the gradient for the first 50 ft (15 m) to about 0.4% and, if the waterway has not already been shaped and vegetated, adding 6 in. (15 cm) of extra drop to account for the depth to the center of the finished waterway. Stakes should be placed along the proposed terrace channel about every 100 ft (30 m) or less (stakes 30- to 50-ft or 9- to 15-m apart) where relatively sharp curves or gullies are encountered. After the line is completed, it should be viewed and very crooked portions restaked. Adjustments should be kept to a minimum so that cuts and fills will not need to be too deep. Second and subsequent terraces are staked by calculating new *VI*s for each and repeating the procedure.

Parallel Terraces. The modern trend in the U.S. Corn Belt is toward parallel terraces with tile outlets. They are built straight across the landscape wherever possible, or with long-radius curves where necessary. The spacing between parallel terraces is adjusted to make it a multiple of the width of the farmer's tillage and planting equipment. Avoiding point rows and odd widths eliminates most of the difficulty of farming terraced land, especially where the terraces parallel a field boundary as well as each other. Such terraces cost more and generally require deeper soils than contour terraces do because of the extra cutting and filling required.

A guide terrace with calculated gradient is staked out in an appropriate place near midslope. Other terraces, parallel to the guide, are laid out above and below at correct intervals. Two or more guide terraces may be needed in some fields with complex slopes.

Construction. The field to be terraced should be cleared of trash. Preliminary land smoothing may be required before parallel terraces are built; dead furrows should be filled, and small ridges leveled before construction is started. Grassed waterways or underground outlets should be provided where the terrace lines cross swales or gullies. Soil productivity of the terrace channels is improved if topsoil is removed, stockpiled, and later spread over the terrace channel as construction is completed. (This increases construction costs so much that it is seldom done.)

Conventional terraces can be built with bulldozers, motor patrol graders, carryall scrapers, elevating-grader terracers, moldboard plows, disk tillers with 24 in. (60 cm) or larger disks, and with hand tools and baskets, headpans, or other carrying devices. Contractors using large-scale, earth-moving equipment build most of the terraces in countries with a commercial agriculture, where labor is expensive. Bulldozers are usually most economical for moving soil short distances; graders do a better job of smoothing and packing the terrace (see Figure 10–5). Carryall scrapers are needed for terraces involving much cutting and filling.

Publications available from state extension services and from the NRCS describe the building and maintenance of terraces. Blakely et al. (1957) describe methods for building terraces with farm implements, but these implements are not adapted for building steep-backslope terraces. Figure 10–6 shows a terrace being built by hand in a developing country.

After the terraces are built, the gradients of the completed channels and the height of the settled ridge top above the channel need to be checked to ensure that they meet specifications. High spots in the channels and low spots in the terrace ridges must be corrected before heavy rains fall. Terrace ridge tops must be level, particularly across swales or gully areas.

(A)

(B)

Figure 10–5 Large equipment constructing a terrace in Kansas. (A) A bulldozer pushing up a terrace ridge. (B) A motor patrol grader smoothing and packing the terrace. (Courtesy USDA Natural Resources Conservation Service.)

Figure 10–6 Farmers in central India building terraces with village-made hand hoes and head baskets. The state ministries of agriculture lay out and supervise the construction. (Courtesy Roy L. Donahue.)

Terrace ridges wear down and channels fill with sediment as land is farmed, so maintenance procedures must be undertaken on a regular basis if the system is to remain effective. Blakely et al. (1957) described cultivation techniques with ordinary farm equipment that can be used to maintain terraces. Powell (1989) and the American Society of Agricultural Engineers (1989) also give terrace maintenance hints. In general, tillage should move soil uphill between terraces and from the terrace channel toward the ridge on the front slopes.

10–1.6 Farming Terraced Fields

Seedbed preparation, planting, cultivating, and harvesting operations should be conducted parallel to terraces to minimize water and soil movement between terraces and to reduce tillage damage to the terrace ridges. Operations parallel to a terrace get a little more off contour as distance from the terrace increases, causing tillage marks and rows to have a gradient. Water flow toward the terrace outlet is acceptable if it is not too swift, but flow toward the upper end of the terrace is not desirable. Proper placement of the short rows assures that all water flows toward the waterway or the lower end of the terrace channel. Figure 10–7 is a guide to planning cultivation patterns on terraced land. A plan for field operations is given in Note 10–1.

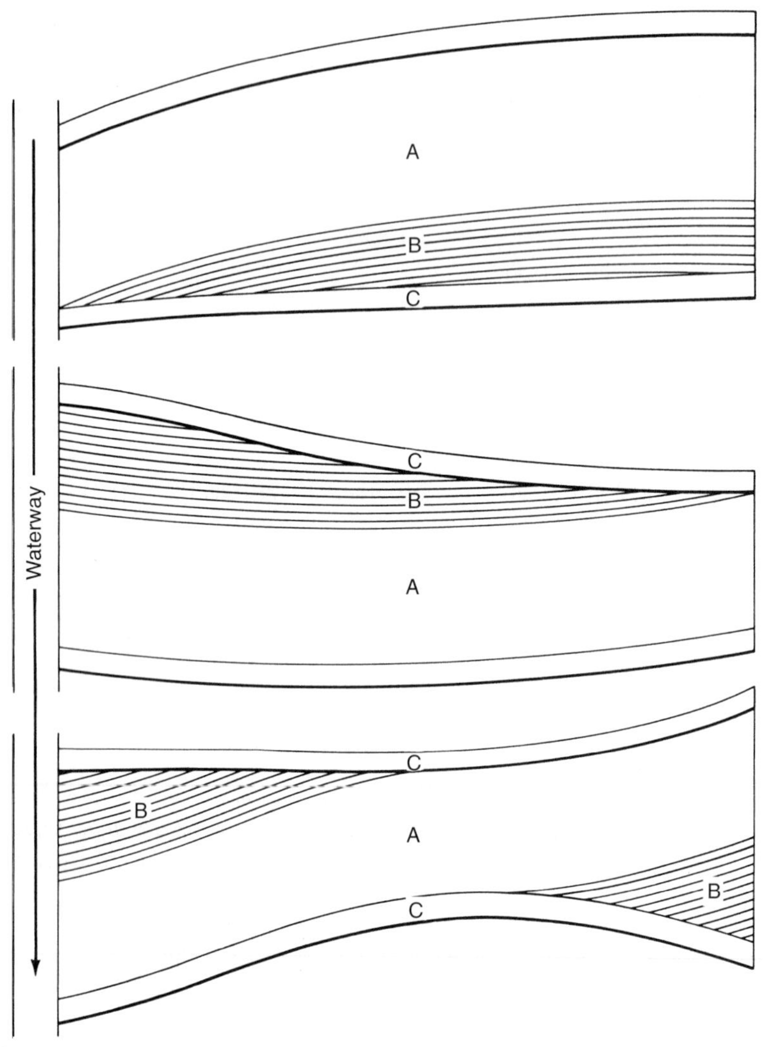

Figure 10–7 Guide to planning cultivation patterns on terraced land. In the diagram, the heavy lines represent terrace ridges and the lighter lines terrace channels. Unshaded sections represent long-row areas; shaded sections are short-row areas to be worked parallel to the shaded lines.

NOTE 10–1
FARMING TERRACED LAND

The plan proposed in Note 9–1 for farming contoured fields may be followed between terraces, or terrace intervals can be farmed so that furrows always guide water to the waterway or to the terrace channels. Where the terrace interval is narrowing in the direction that water flows in the channel, long rows or furrows *parallel to the upper terrace ridge* will all carry excess water toward the waterway, as in the upper example in Figure 10–7. The short section-B rows also will guide the water toward the waterway, but will release it into the terrace channel. Where the terrace interval is widening, all furrows *parallel to the lower terrace* will carry excess water toward the waterway, as in the middle example of Figure 10–7. The lower example illustrates a complex situation where the terrace interval is wide at both ends and narrow in the middle. The long rows will carry water all the way to the waterway if they are parallel to the upper terrace (as the interval gets narrower) and to the lower terrace (as the interval widens). The short rows should be parallel to the long rows in their section of the field so they will guide water to the waterway or to the terrace channel.

Terrace intervals can be farmed in the field by working the (A)-area long rows first, then filling in the (B)-area short rows. The front slopes of the terrace ridges (C) are used for turn areas and are worked last. Small grain and forage crops are often harvested as if there were no terraces, but with row crops the last rows planted (C) are harvested before the short rows (B).

10–1.7 Diversions

A *diversion* is a ditch or channel with an accompanying ridge on the downhill side. It is like a single large terrace designed to intercept runoff water and carry it away at a nonerosive velocity. Diversions are usually constructed at the top or the foot of steep slopes or on property lines to protect productive soil from erosion or inundation. Some diversions are used to prevent runoff water from entering gullies, to move it away from road ditches (Figure 10–8), and to direct it away from home sites and farmsteads. They can also direct water to ponds, to water-spreading sites, and to special planting areas. In many erosion-control plans, a diversion is the first structure that intercepts runoff. It must be well sited and carefully laid out and constructed. Many states have laws to control activities that change the flow path and the volume of water. These laws must be kept in mind when designing diversions.

Specifications. A diversion designed to protect a cultivated area should have sufficient capacity to handle a 10-year-frequency storm. Diversions for protection of houses, farmsteads, or expensive engineering works will need enough capacity to control up to a 50-year storm.

Diversions are usually vegetated. They need to permit more rapid water movement than graded terraces so sediment will not settle out, but not such rapid movement that channel erosion will occur. Permissible flow velocity depends in part on vegetative cover density. Gradients are often as high as 0.5% and sometimes as high as 0.8%.

Diversions commonly have flat bottoms 3- to 20-ft (1- to 6-m) wide, and straight front and back slopes (trapezoidal shape). The sides often have slopes of about 4:1, although they may range from 2:1 (for grassed diversions) to 6:1 (for regularly cultivated diversions). Channel depth ranges from 15 to 40 in. (38 to 100 cm). Actual size depends on the volume of water to be handled. Diversions, particularly those with steep side slopes, are difficult to construct with farm implements, so the work is generally done by conservation contractors.

Figure 10–8 A diversion ditch in Nigeria designed to carry water from a highway ditch safely away from the right-of-way. Spreading the water on naturally vegetated sites dissipates its energy. (Courtesy J. A. Hobbs.)

10–1.8 Outlets for Terraces and Diversions

A protected outlet is needed wherever runoff is discharged from terraces or diversions. Vegetated waterways, such as those described in Chapter 11, Section 11–3.2, usually are the cheapest and most effective man-made structures. There are conditions that reduce the effectiveness of vegetation so much that special structures are necessary to prevent the extra water from making a gully. Gully control structures may be used, as described in the next section, but underground outlets (tile lines) are often preferred in these situations.

Tile Outlets. Tile lines can serve as *underground outlets* to take runoff water from low points in terrace or diversion channels or from behind other forms of earthen embankments and carry it below ground through a pipeline to a safe discharge point. Underground outlets are used when the possible waterway sites are too steep for nonerosive water movement, the rate of peak runoff must be reduced for flood control, or the cost of installing a system of underground pipe is less than that of forming the grassed waterway and taking the area out of crop production. Underground outlets are not new, but they were not widely used until the concept of detention storage in the collecting system was introduced in the early 1960s.

There are four major parts to an underground water outlet system (Haan et al., 1994). The *tile-inlet tube* (riser) is made of plastic, metal, or concrete. It rises from a pipeline below the soil surface and has holes or slots at intervals above ground level. An *orifice plate,* positioned between the base of the inlet tube and the horizontal conducting pipe, regulates the rate of downward water movement and prevents water pressure from building up in the pipe and flooding lower-lying terrace channels. The *conducting pipe,* also of plastic, metal, or concrete, carries water from one or more inlet tubes down to the final outlet. The system *outlet* is usually located in a natural waterway with enough protection to handle the runoff water without excessive erosion.

Advantages and Disadvantages. A major advantage of an underground outlet is that less soil is lost from the field. Sediment settles out in the stored-water areas, and the water carries relatively little material to the final outlet. Low sections in terrace channels in the vicinity of former gullies tend to fill up with the sediment washed down from the land above the terrace, leveling the field. Peak runoff rates from terrace systems or diversions are reduced because of the incorporated detention storage. Where underground outlets replace grassed waterways, more land is available for cultivated-crop production. High land values make this an important advantage.

The major disadvantage is that underground outlets are costly to install and the design and construction are much more critical than for a simple grassed waterway. Crops in the detention areas may be damaged by ponding during and after excessive storms, or when the riser inlet holes or the orifice plate become plugged with trash. The system and lower-lying fields can be damaged if excess rainfall overtops the ridges.

Specifications. Detention storage is usually designed to hold rain from a 10-year-frequency storm. The underground outlets should dispose of the accumulated runoff in less than 48 hours. The actual sizes of the tile-inlet holes, orifice plates, conducting pipe, and outlet must be designed to meet these criteria. The slope of the conduit pipe influences speed of discharge and must be considered also.

The tile-inlet riser should be at least 3- in. (8-cm) higher than the adjacent terrace or diversion ridge and should be equipped with a removable cap to prevent entry of debris and to permit cleaning the orifice plate. The terrace or diversion ridge must be level in the vicinity of the major storage area to minimize the chance of overtopping. Figure 10–9 shows a terrace system in Kansas with tile inlets.

Developing an appropriate design for terrace outlets is an engineering procedure outside the scope of this book. Beasley et al. (1984) describe methods to use for arriving at satisfactory design criteria. Farmers in the United States can obtain planning assistance from NRCS technicians.

Construction. Construction of an underground outlet system is similar to the installation of tile drain lines (Chapter 14). The conduit pipe, with inlets for the surface risers, is laid in a trench with a proper grade, and the trench is filled and compacted all the way to the surface. The diversion or the terraces are then constructed with the ridges well compacted, especially over the trench area. The conduit is exposed again in the center of each channel area, and each riser with its orifice plate is installed. The soil is again packed in place and the system is ready for use.

10–2 GULLY-CONTROL STRUCTURES

Gullies may be dry much of the time, but they carry large, erosive streams of water at other times. The banks are often steep or even undercut so they may collapse easily under a load. Control measures tend to be expensive, but lack of control will allow the gully and its

Figure 10–9 A terrace system in Kansas with tile inlets. Water drains through the black riser pipes in the terrace channels. This area received 9 in. (230 mm) of rain in 3 hours the day before the photo was taken. (Courtesy USDA Natural Resources Conservation Service.)

hazards to continue to grow. Smoothing and revegetating the area as a grassed waterway (Chapter 11, Section 11–3.1) is helpful, but usually needs to be supplemented with erosion control structures. Temporary structures that control erosion until vegetation is well established suffice in some situations; others require permanent structures made of stone or concrete.

10–2.1 Broken Rock for Gully Control

Stone and broken rock have long been used for reducing erosion in waterways and gullies. *Riprap,* a covering of loose stones on the soil surface, has been widely employed for this purpose. The surface needs to be smooth before rock is applied, so that water cannot flow beneath the stones. Ideally, the materials used range from coarse sand or fine gravel next to the soil, increase in size by layers, and are covered by stones that are too large for the stream to move. The stones formerly were sorted and placed by hand, but now are more likely to be dumped over the surface and smoothed by machine. More rock is required this way, but the amount of labor is markedly reduced. The thickness of the protective stone layers is dictated by the nature of the site and the velocity of the runoff water.

Rock Barriers. A *barrier* (or a series of barriers) is often needed to reduce water's erosive power in steeply sloping waterways and in many gullies. Piles of rocks can serve to accomplish this end, but a range of sizes must be used with rocks on the downstream side and the upper surface that are too large and carefully laid to be moved by the runoff water. Smaller stones, gravel, and even sand are used in the interior to reduce water flow through the structure. Wire netting can serve as a fence to prevent the breakdown of a rock dam, or it may be used to form a basket (*gabion*). The netting is laid all the way across the channel and loose stones are placed on its upstream half. The remaining netting is folded over the rocks and is wired to the edge of the netting on the ground. These rock barriers are flexible enough to maintain contact with the soil even if the ground settles under them.

A stepped condition develops where several barriers are installed in a waterway or gully. Erosion is reduced on the flatter slopes between barriers, and vegetation has a chance to grow there (Chapter 11, Section 11–3.1).

10–2.2 Brush, Log, or Timber Barriers

Temporary wooden structures have been used to slow runoff and trap sediment in waterways and gullies where labor was more plentiful than money. Wood and other natural materials are still recommended and used regularly in many developing countries, even though the failure rate is fairly high (Morgan, 1995).

Brush and Log Barriers. Posts are used to anchor brush barriers in place. They should be driven vertically into the soil at 3-ft (1-m) intervals in two parallel rows about 18- in. (45-cm) apart across the waterway or gully bed. Loose branches, small trees, or logs are packed tightly between the rows of posts, making sure that the bottom members make firm contact along the sides and bottom surface. Ends of the brush piles should be dug into the channel walls and soil should be packed tightly around them. Logs must be dug into the bottom as well as the sides of the channel. Materials are wedged and piled tightly between the upright posts and held in place by cross pieces spiked or wired to the posts. The barrier must be impermeable enough to prevent water jetting through and undermining the structure. A large flat notch should be left in the middle of the barrier so that overflow water will be guided onto an apron and will not wash away the banks (Heede, 1976).

Timber Barriers. Large-dimension lumber or timber, or a series of thick posts held closely together, can be used. Posts may be driven vertically into the soil to support the barrier, or the lumber itself may be driven vertically into the ground and reinforced with horizontal timbers. In the latter case, more than half the length of the lumber should be imbedded below the gully floor. A rectangular notch must be used as a spillway. Larger or thicker pieces of wood and treated or termite-resistant timbers last longer.

10–2.3 Brick Barriers

Brick barriers can be used to stop gully erosion. Fired construction bricks (expensive), and sand-cement or soil-cement blocks can be used. Sun-dried clay bricks (easily weathered) and laterite blocks cut from soil and desiccated are available in many developing countries.

A poured concrete or layered rock foundation is necessary for any barrier built with bricks or blocks. The ends of the barrier must penetrate the gully walls and be sealed with tightly packed soil material (not topsoil) between the barrier ends and the excavated gully walls. Water must not be permitted to seep around the ends.

A straight wall of bricks and mortar is relatively weak; it must be buttressed on the downstream side for strength. Walls that arch upstream are also strong. This design passes water force from the center to the outer ends of the wall, where it butts into the soil. The gully walls must have very strong compression resistance where these barriers impinge.

All barriers must have a notch or low area in the top center that is large enough to permit passage of the largest expected flood. The gully floor immediately below the notch must be protected so that the falling water cannot erode the gully surface and undercut the barrier. Loose stones are suitable for protection if they are large enough to resist movement by floodwater. Smaller stones must be anchored in some way on the gully floor.

10–2.4 Drop Structures

The barriers discussed previously are built to control erosion temporarily until a permanent vegetative cover is established. Sometimes, however, permanent structures are needed. A *drop structure* is such a structure It is a small dam with a built-in spillway and is used to stabilize steep waterways or to level waterways in cultivated fields so they will be less erodible and thus may not need to be planted to perennial vegetation. Drop structures can also serve as inlets and outlets for highway culverts, pipelines, and ditches. They can lower the elevation of irrigation canals in sloping areas so less fill is required in their construction (Novak et al., 2001). They also trap sediment. They can handle large volumes of runoff water and are effective where falls are less than 8 ft (2.5 m).

A drop structure can be built of timber, sand-cement blocks, rubble masonry, poured concrete, or corrugated metal. Figure 10–10 shows the main structural features. The cutoff (main) wall is constructed across the gully or channel. Cutoff walls extend into the banks of the channel to anchor the structure and ensure that water will not cut around it. The main wall has a notch that serves as a spillway. A horizontal apron or *stilling basin* is provided

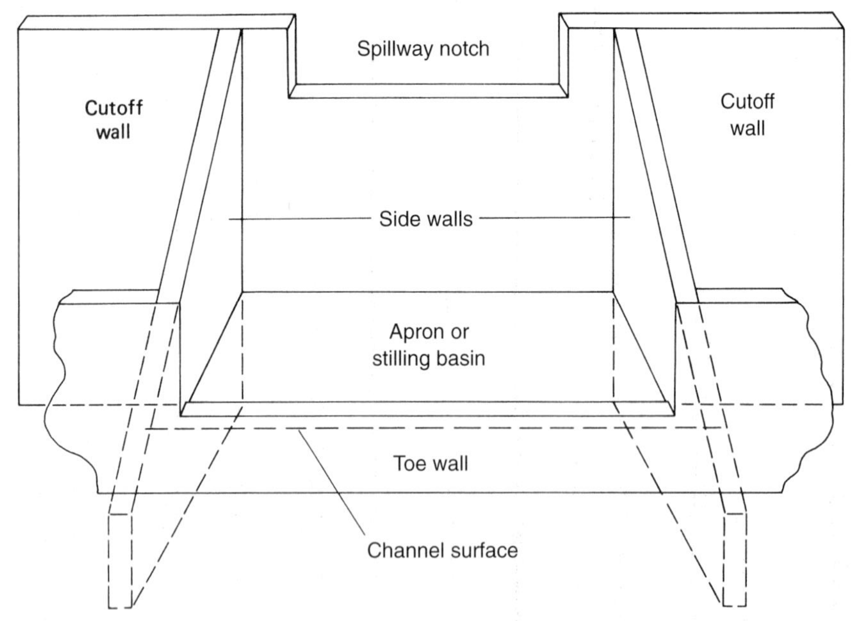

Figure 10–10 Diagram of a concrete drop structure. The side walls join the structure's cutoff wall and extend to the toe wall, as indicated by the solid lines, if the waterway below the structure is parallel to the cutoff wall. Where the waterway below the structure is perpendicular to the cutoff wall, the side walls extend down the waterway, as indicated by the dashed lines.

below the spillway to absorb the energy of the falling water. The apron front is reinforced and stabilized by a toe wall that projects into the soil at the end of the apron. Side walls extending from the main wall to the end of the apron or beyond serve as buttresses for the main wall and help to control bank erosion. Structures with long main walls may need additional buttress supports between the side walls on the downstream face.

Drop structures used as inlets must focus a shallow flow of water into a concentrated stream to maximize the capacity of a pipeline or culvert. Some are built in the shape of a box, often with water entering from three sides rather than from only one direction. Their capacity can be further increased by adding a vertical vane to prevent the water from flowing in circles. Becker and Foster (1993) describe a drop structure with a semicircular inlet placed at the head of a gully. The semicircular wall forms a longer spillway and a higher capacity than that of the straight-wall drop structure illustrated in Figure 10–10.

Failure of drop structures usually results from water washing soil away at the structure-soil interfaces, so the cutoff walls and the toe wall must be long enough to protect against water undercutting the apron. Soil material must be packed very tightly against these elements to prevent water damage. The notch size required can be calculated from the weir formulas in Figure 15–6, with due allowance for the slower flow that occurs if the drop structure has a flat sill that the water crosses rather than the sharp crest assumed for the weir equations (see Chapter 15, Section 15–4.1). Flow rates into a box inlet or a semicircular inlet can be predicted from the combined length of the three sides of the open box (or the length around the semicircular inlet) and the depth of water flowing over the lip unless the inlet is flooded. A flooded inlet indicates that the flow is regulated by an orifice plate or by the capacity of the outlet pipe or ditch (Haan et al., 1994).

10–2.5 Chutes

A *chute* is a spillway that collects runoff water at one elevation and carries it down a slope to a lower elevation (Figure 10–11). Prefabricated chutes may be used as alternatives to drop structures where terrace outlets spill into revegetated gullies (Yadav and Bhushan, 1994). Chutes for short slopes and for small amounts of runoff (1 to 3 ft^3/sec or 0.03 to 0.09 m^3/sec) may be built without forms by excavating the desired shapes in the surface of the ground. Mixed concrete is poured onto the prepared surface and smoothed in place. Wire mesh reinforcing is required in areas where large temperature variations are expected.

Long chutes usually have some provision to aerate the water as a means of avoiding shock waves in the flow and to make it easier to dissipate flow energy at the bottom (Novak et al., 2001). Some form of roughness or jump in the floor near the top of the chute can serve for aeration. The lower part of the chute often has a reduced gradient that helps slow the flow, but the outlet still requires protection.

Flat-slab limestone and other rocks were used to form chute spillways in the early days of conservation work. Rocks were placed on the channel and the joints were filled with a sand-cement mortar. These structures were reasonably satisfactory, but they frequently cracked at the stone edges and deteriorated with time, especially where freezing and thawing occurred. Their main advantages were that less cement was needed and the chute surface was usually rough enough to reduce flow velocities.

The shapes of the inlet and outlet sections of chutes are important. Water attains very high velocities on the smooth surfaces, and energy-dissipating structures must be provided

Figure 10–11 A concrete chute designed to carry runoff water safely down a slope to a road ditch at the bottom. (Courtesy F. R. Troeh.)

at the outlet. These usually consist of boxlike devices that hold pools of water and overflow at ground level. Boxes partly filled with stones that are too large to be moved by the expected water velocities serve well as energy dissipators.

Formed sidewalls and cutoff walls are needed where larger amounts of water must be carried or where the drop is more than 6 ft (1.8 m). Formed wing walls may be required to channel runoff into the chute.

10–3 ROAD DITCHES AND CULVERTS

Large numbers of ditches, culverts, and other structures are used in road construction, not only for highways but also for less-traveled roads. Proper design and placement of these structures is important for the stability and usefulness of the road and also for drainage and erosion control on the adjoining land. They are normally designed by trained engineers and built by contractors, but some installations on private land or in remote areas may not have much supervision. Often they are inconspicuous except when mistakes are made that cause problems for travelers or adjacent landowners.

10–3.1 Road Ditches

Road ditches are built along with the road bed. The ditch area often serves as a source of fill material for the road bed. Most road ditches have ample capacity for the runoff water they need to carry, but they need to be properly graded and vegetated to control erosion. Low points need to be filled or drained by a culvert or other structure. Pipelines, chutes, or drop structures may be needed where a ditch carries a large volume of water down a steep slope.

The ditch banks, consisting of the road fill on the inside and a bank cut into the landscape on the outside, need to be properly shaped and vegetated (see Chapter 11, Section 11–1.2). Maximum slopes on these surfaces depend partly on the nature of the material and partly on whether there will be any traffic on them, perhaps to mow the grass or for other maintenance. Most materials should not have slopes steeper than 2:1 (2 units horizontal to 1 vertical) to avoid slumping, and some materials that are high in clay need to be flatter than this. Loess banks are unique; a dry loess bank is stable with a nearly vertical surface, but a wet loess slope may slump on a 2:1 or even a 3:1 gradient. A thick cover of deep-rooted vegetation helps to stabilize road banks. Stone riprap is used occasionally to protect an especially vulnerable surface.

10–3.2 Culverts

Culverts allow a stream of water to pass under a road. They need to be large enough to carry the runoff during a wet season, but excess size will add greatly to the cost. Many culverts are round metal pipes made with corrugations for added strength; some have an oval cross section or a three-lobed shape with a broad base that allows the stream flowing through them to be wider and shallower than that of an equivalent round culvert. The length is also important; they need to pass through the road fill but not extend beyond it. Usually they are provided with flared ends or concrete walls at each end to prevent erosion at the inlet and outlet.

A culvert that is placed too high will cause water to pond along one side of the road and possibly back up into an adjacent field; one that is placed too low may trap sediment and lose capacity if its outlet is restricted. A culvert that is placed too low and has an unrestricted outlet may be the beginning point of a gully in the road ditch or in a stream from an adjacent field.

Water must flow through a culvert, not alongside it. The fill material next to the culvert must therefore be compacted tightly against the pipe. The most difficult part is the area beneath the pipe, and this is the place most likely to have external water flow. Collars are sometimes installed around the pipe to serve as cutoff walls that reduce seepage.

10–3.3 Other Roadway Structures

Various other water-control structures are also used in roadways, especially for major highways. The chute shown in Figure 10–11 is one such structure. The small screened openings in the median of a divided highway are structures that serve as inlets to pipelines that carry water away from such areas. Various types of temporary check dams or barriers are often used in newly graded areas that will ultimately be protected by vegetative cover. Of course, there are also bridges, overpasses, and other major structures that must be individually designed by competent engineers.

10–4 EARTHEN DAMS

Earthen dams are used to trap sediment, to stabilize drainage ways and reduce erosion, to store excess water temporarily to reduce flood damage, and to store water for livestock, irrigation, household, recreation, or municipal use. Dams for water storage are discussed in Chapter 13, Section 13–5.7.

Earthen dams are divided into three classes based on the severity of the damage that present or possible future development might suffer if the dam failed and the impounded water rushed downstream.

Class A dams are located in strictly rural or agricultural areas. Failure would cause water to inundate farm fields and could damage local (township or county) roads. Farm buildings, but not homes, may lie in the path of the floodwaters.

Class B dams are located in predominantly rural and agricultural areas. Failure could cause damage to isolated homes, as well as to farm buildings, main (state and federal) highways, branch railroad lines, and public utilities.

Class C dams are located where failure would cause serious damage to homes, industrial and commercial buildings, utilities, main highways, and main railroad lines.

Most dams constructed for individual farmers are Class A dams. Most dams designed by the NRCS in the small watershed program (PL 566) also belong to Class A, but some are Class B, and a few near urban centers are Class C structures.

Soil excavated from the reservoir area is commonly used to build dams. This provides the cheapest source of fill material and also enlarges the storage capacity of the reservoir, but some precautions are needed. Topsoil is likely to be too high in organic matter and may not have enough clay to be used in the core of a dam; subsoil material is likely to be more suitable if it is not too stony. A trench may need to be dug to remove unsuitable soil from the site of the dam. Fill material is then placed in layers, and each layer is compacted to increase its stability and reduce its permeability. The outer surfaces of the dam may be loamy soil, but the core needs to be as compact and impermeable as possible, especially if the dam forms a pond of significant size. Large dams usually need a drain to intercept seepage water; otherwise there will be a seep spot at the toe of the dam.

The outer face of the dam and the inner face above the waterline are normally protected by perennial grass cover. A strip just above and below the waterline may need protection from wave action, often by covering it with a layer of gravel or stones.

The *spillway* is a critical part of any dam-building project. It must have enough capacity to prevent floodwater from overtopping the main dam, because overtopping can quickly cut through a dam and cause it to fail. The cheapest spillway is usually a grassed waterway placed alongside the dam in undisturbed soil on a gentle gradient that can carry the runoff water without forming a gully. A concrete or rock-lined channel should be used if the spillway gradient is too steep or if it must be built in fill material. An alternative where the runoff volume is not too large is to install a *pipe outlet* with a *drop inlet.* Some reservoirs have both a pipe outlet and an emergency spillway.

10–4.1 Soil-Saving Dams (Sediment-Storage Dams)

A *soil-saving dam* is a simple earthen barrier placed where it will intercept and trap waterborne sediments in what is often called a *silt pond.* It has a notch or box as a principal spillway that passes the water off slowly so that most sediment will settle out. Sufficient *freeboard* (height above the spillway) must be provided so floodwater will never overtop the dam. The pool of water behind the dam is usually relatively shallow and serves mainly as a *settling basin* rather than as a place for water storage, fishing, or swimming. It often has a tile drain outlet that reduces its size or even allows it to dry out when there is little precipitation.

Figure 10–12 A Class B dam constructed in a gully to stop erosion and to provide a waterfront for recreation. (Courtesy USDA Natural Resources Conservation Service.)

The basins formed behind soil-saving dams are typically not very large. They commonly fill up with sediment within a few years and serve as a reminder of how much erosion is occurring on the fields above them. Some farmers have modified their cropping and tillage practices after seeing how fast their (or their neighbor's) pond filled with sediment. When filled, they may be drained and converted into fertile cropland, but any erosion that continues to occur will then flow on down the stream.

10–4.2 Grade-Stabilization Dams

Grade-stabilization dams are used to prevent gullies from eating back into fields (Figure 10–12), to stabilize or raise gully channel floors, or to drop water from terraces, waterways, or diversions to stream channels at lower elevations. The size and cost of the structure or set of structures required increases rapidly as a gully grows, so speedy response to gully stabilization needs is important. Gullies generally flow on fairly flat gradients because the nonvegetated surfaces and deep, narrow channels are conducive to high flow velocities or reduced slope gradients or both (Manning's formula, Note 4–2, can be used to evaluate these effects). Low slope gradients cause many gullies to become deeper as they eat back into landscapes.

The most effective way to stop a gully is to place the dam at the site of the overfall, but some dams are placed downstream because the site at the gully head is not suitable. Several small dams rather than one large one should be placed in a long gully. The need for land shaping above a dam should be considered. High vertical walls should be pushed in, shaped, fertilized, and revegetated, especially where further sloughing is expected. The water pool will fill with sediment in time.

10–4.3 Farm Ponds

Farm ponds range in size from a fraction of an acre up to a few acres (up to 1 or 2 ha) and have many uses. They provide reserve supplies of water for livestock (Section 13–5.7), fire protection, irrigation, and many other needs. They also provide recreational areas for fishing, swimming, bird watching, and winter ice skating. Most of them are made by simply building a small dam across a small drainageway, but some are more costly installations that involve excavating into a hillside or even a flat area and/or building a long dam across a gently sloping area.

The size of the drainage area above a pond is important. If it is too small, there may not be enough runoff to fill the pond. If it is too large, it will require excessive spillway capacity to protect the dam. An alternative is to place the pond on a small tributary to a larger stream and use a diversion from the larger stream to fill the pond. The types of vegetation needed to protect various parts of the dam and pond banks are discussed in Chapter 11, Section 11–3.4.

10–4.4 Flood Control Dams

Flood control in the United States comes under the jurisdiction of two government agencies. The U.S. Army Corps of Engineers is responsible for flood control along navigable rivers. USDA-NRCS is responsible for flood control on small watersheds in the upper reaches of rivers and streams. The Small Watershed Law (566-1947) provided for watershed districts to be set up under state laws. Federal funds are funneled through these districts to design and build flood control structures. Most of the small flood control dams in the United States have been designed, constructed, and paid for under this watershed program.

Most small flood control dams serve two main purposes: flood control and grade stabilization. As the reservoirs behind dams always trap part or all of the solid materials carried in by the streams that feed them, sediment storage is designed into each reservoir to provide a life expectancy of about 50 years.

Flood control dams are built with enough capacity to temporarily store the runoff from a 10- to 50-year storm. Floodwater passes from the storage pool through the principal spillway for several days. This spillway is usually a pipe through the dam. Excess runoff from unusually intense storms passes immediately over an emergency spillway—usually a grassed waterway.

Some flood control dams in arid regions seldom contain any water but have large capacities to control flash floods. Figure 10–13 shows one of several structures built to protect Las Cruces, New Mexico. These structures impound large volumes of runoff water and release it at a controlled rate. They have handled all runoff since they were built with no flood damage to the city.

10–4.5 Design and Construction of Dams

A wide variety of detailed technical information is needed to select a suitable site and to plan, justify, and build an earthen dam. It should be undertaken only by well qualified and

Figure 10–13 Flood control dam protecting Las Cruces, New Mexico. This is one of several similar structures built to impound runoff water from heavy rainstorms in the hills above the city. This structure is about 0.5-mi (1-km) long and has a road along the top. (Courtesy F. R. Troeh.)

experienced people (Beasley et al., 1984). NRCS Area Engineers may plan small stock water dams, but plans for larger dams in the Small Watershed Program must be developed by the NRCS State Engineer. Dams are defined as "larger" if the product of the height of the dam crest above the spillway floor (ft) times the dam capacity (ac-ft) exceeds 500, or more simply, if the watershed area is greater than 25 to 100 ac (10 to 40 ha), depending on rainfall.

10–5 STREAMBANK PROTECTION

Bank erosion in perennial streams is often severe. Not only is soil lost, but new channels meander across bottomlands and wash away fields, structures, and even buildings. In severe cases, banks may erode inland at a rate of 175 to 250 ft/yr (50 to 75 m/yr).

Vegetative and mechanical means of protecting streambanks have been developed. Vegetative control measures are discussed in Chapter 11. The nonvegetative techniques can be divided into those that divert the faster-flowing water away from the bank and those that protect the erodible bank with mechanical covers.

In the United States, it is necessary to obtain a permit before the direction or volume of flow in a perennial stream can be changed. The permit will usually be issued if inspection by representatives of the U.S. Army Corps of Engineers (and of the state water resources protection agency, where necessary) shows that the planned work will not damage land across the channel or downstream.

10–5.1 Current Deflectors

There are two main types of dikes that are used to deflect faster-flowing water away from eroding stream banks: regular dikes and vane dikes.

Regular dikes consist of loose stone or rock piles that extend above the stream surface and are built out into the water at a slight angle downstream. They may extend 3 ft (1 m) or so above the low-flow water level or may go to full bank height. They are usually shorter in height farther from the bank. The stones and rock pieces used must not move in flood currents. Dikes can also be made of poles driven vertically into the streambed. Regular dikes are probably the most reliable streambank protectors, but they are expensive.

Vane dikes have no bank contact. They are placed far enough upstream to deflect the current away from the bank needing protection. A small, well-placed vane dike may protect a site that would require a much larger regular dike, but a shift in the upstream channel can negate the effectiveness of a vane dike more readily than that of a regular dike.

Groynes resemble short dikes used to define the channel location of a wide braided river (Przedwojski et al., 1995). They are set in groups extending either at right angles from the river bank or with an upstream inclination. The space between groynes is relatively narrow and tends to fill with sediment, thus establishing a new bank along the ends of the groynes.

Controlling river flow with dikes or groynes sometimes causes it to erode its bed excessively. *Sills* placed across the river channel may be needed to hold sediment and prevent bed erosion. Traditional sills are made of concrete or large stones anchored in trenches across the channel floor. A much less expensive alternative made from plastic bags filled with 0.25 to 0.30 m^3 (9 to 10.5 ft^3) of sand has proven to be effective and reasonably durable (Przedwojski et al., 1995).

10–5.2 Bank Protectors

Mechanical covers to protect eroding streambanks include toe protectors, hardpoints, and revetments.

Toe Protectors. The *toe* of a streambank is the part where the gradient of the relatively steeply sloping bank changes abruptly near the water line. *Toe protectors* cover the toe and part of the more steeply sloping bank with stones or other protective material (Figure 10–14). The material can be either hand placed or dumped and spread by machine. Hand placing saves material but requires much more labor. Usually the toe is protected intermittently so it covers the more seriously eroding sites along the bank. The distance between areas of protection depends on the severity of the erosion hazard.

Hardpoints. *Hardpoint* protection usually consists of stones dropped over the edge of the bank onto the toe of the slope. No attempt is made to place or spread the pieces. In general, the height of the protection up the bank slope is limited, but the protected spots along the bank are closer together than in toe protection.

Revetments. A *revetment* is a protective cover such as a retaining wall, or a covering of protective material on a slope. One of the longest-lasting types of revetments,

Figure 10–14 A toe protector to stabilize a streambank. The rock covers the bank from the low water level to the height of normal flood stage. (Courtesy Kansas City District, Corps of Engineers, U.S. Army, Kansas City, Missouri.)

shown in Figure 10–15, is a vertical stone wall shielding the land from the stream. A variety of materials are used, including stones, concrete, or most any other durable material that can form a continuous cover and be fastened securely enough to stay firmly in place.

A fence consisting of a double row of posts driven vertically into the ground, cabled together, and filled with protective material such as stones or logs can be used as a temporary revetment. Riprap of graded gravel and stones can also be used. Broken concrete from old highways or building foundations can serve effectively, if slabs are laid flat on the sloping area needing protection. Gabions of smaller stones also may be used.

A *windrow revetment* is a pile of stones and rocks placed on the surface of the soil at the upper edge of the bank, as shown in Figure 10–16. As the bank erodes, the rocks drop down and lodge near the toe, thus forming a line of protection along the bank. A windrow revetment contains from 1.5 to 5 tons of rock per linear ft (4 to 15 mt/m) of bank.

10–6 FLOOD CONTROL

Rampaging flood waters are a dramatic example of nature's power. There is always a threat that it will breach the levees that form its banks and flood the adjacent lowlands. The flow is slower as it crosses a lowland, but it may still carry considerable power and the water level may rise quickly. The slower flow cannot carry as much sediment as the surging river

Figure 10–15 Rock wall revetment protecting a cultivated field from floods of the Urubamba River near Machu Picchu, Peru. (Courtesy F. R. Troeh.)

Figure 10–16 A windrow revetment. The stones are piled in a windrow near the edge of the streambank. As the stream erodes the bank, the stones fall and protect the toe of the slope from further erosion. (Courtesy Kansas City District, Corps of Engineers, U.S. Army, Kansas City, Missouri.)

and therefore causes sediment to be deposited on the lowland and inside any buildings that the water enters. Cleanup costs involve repairing the damage done by the water and removal of sediment from areas where it is unwanted.

Flood costs would be much lower if people did not put buildings and structures on floodplains, but they do. These level areas of rich soil are attractive sites when they are dry. The fact that a nearby river or stream floods the area occasionally is easy to overlook. A weather forecast of an impending storm along with a prediction of a flood higher than the levees can be a wakeup call. The situation may call for emergency action to save lives and protect property. Teams assemble to fill sandbags and place them on top of existing levees and hope they will hold long enough for the flood to pass.

Sandbag levees are porous, but their leakage is a minor concern if they can hold the river in place. The sandbags are usually removed after the water recedes, and a decision must then be made. Is it time to raise the levee? The problem gets worse with time because the levees prevent sediment from being deposited on the floodplain and cause some of it to be deposited in the river channel instead. The protection encourages people to build on the floodplain, but the hazard becomes more and more serious. After many cycles of raising the levees and the river raising its bed, the river may be as much as 50 ft (15 m) above the level of the adjacent land. This is the situation for a long distance along the Huang He (Yellow) River in China, for a considerable distance along the Loire River in France, and also along any other major rivers of the world where levees have been used for a long time to contain the water. The Mississipi River in the United States has continuous levees for 380 mi (611 km) downstream from Pine Bluff, Arkansas; the levees are about 30-ft (9-m) high along much of this distance.

When a major levee fails next to a populated area, the resulting flood can cause hundreds of deaths and hundreds of millions of dollars of property damage in a very short time.

10–6.1 Levees and Dikes

Levees are used to keep water off land that is subject to flooding but would normally be dry most of the time. Dikes are similarly used to keep water off land that was previously under water most or all of the time. Both are built like long earthen dams. Depending on height, they may be as wide as 100 ft (30 m) or more at the base. The sides slope, so the top is narrower than the base, but the structure is wide enough to carry traffic along the top, often with a paved or gravel road. The dry surfaces are protected with an erosion-resistant grass such as Bermuda grass. The side next to the water is protected as a riverbank, often surfaced with stones, concrete, or some other type of revetment. Concrete slabs linked together to form a continuous mat form some of the best covers for the waterline and the slope below it. The linked slabs hold each other in place, and the covering is flexible enough to accommodate any settling that may occur.

The best-known dikes in the world are those that protect the polders in The Netherlands (Chapter 2, Section 2–4.4). These were built over a period of several centuries to reclaim wetland and use it for agriculture to support a dense population. The dike system holds out both river water and sea water. Pumps are used to lower the water table enough so that land below sea level can be farmed. Cities, towns, and farmsteads in the area are generally built on raised areas called *terps* where the land has been built up above sea level.

10–6.2 Floodways

Floodways are used to protect valuable property (urban usually) from flooding and erosion damage by providing an alternate channel that keeps the floodwater away from the protected area. These vary in size and complexity depending on the nature of the hazard and the value of the property. The city of Winnipeg in western Canada has suffered over the years from flooding caused by rapid snow melt and ice jams in the Assiniboine River and the Red River of the North. A series of levees was constructed to protect houses in low lying areas and a floodway was built after the very severe 1950 flood. The *floodway* consists of a very wide, shallow channel with broad, low levees on each side. It starts at the Red River south of the city's built-up area. When flood conditions develop, two steel gates are raised from the bed of the river, just north (downstream) of the entrance to the floodway. This forms a submerged dam that diverts excess water into the floodway, which carries it to the east and north and dumps it back into the river 29.4 mi (47.3 km) downstream.

High river water levels have not caused serious damage in Winnipeg since the structure was completed. In the spring of 1997, the Red River crested higher south of the city than it has in any year since 1826. Tremendous flood damage was inflicted on cities, towns, and farms south of Winnipeg, both in the United States and in Canada. But inside the floodway-protected area of Winnipeg, the river crest was about 6 ft (1.8 m) lower than the crest of the 1950 flood. The floodway carried the excess river flow around the city. A few houses in Winnipeg proper were flooded, but the damage was limited.

10–6.3 Large Reservoirs

A significant part of the storage capacity of reservoirs behind large dams is used for flood control. In the United States, these reservoirs are mostly built and managed by the U.S. Army Corps of Engineers. The structure in New Mexico shown in Figure 10–13 makes a reservoir that was built only for flood control. It is allowed to drain completely, so that its total storage capacity is available when a sudden flood occurs.

Most of the reservoirs in more humid regions are multipurpose. They are never allowed to drain completely because part of their capacity is intended for recreation, navigation, electric power production, irrigation, or other uses, and only a fraction of the reservoir capacity is available for flood control. The construction details covered elsewhere (e.g., Novak et al., 2001) are outside the scope of this book, but the principle guiding their use is to allow upstream flood waters to fill reservoirs during wet periods. The water is released from the reservoirs as a relatively uniform flow throughout the year, thus avoiding both the high peaks and the low flows of the unregulated river system. This practice protects more land when many medium-sized reservoirs are built in upstream locations rather than (or in addition to) a few large reservoirs in downstream locations.

The Mississippi River in the United States is an example of a large river with many flood control structures. It has levees along much of its length and reservoirs in its upper parts. Use of the reservoirs to regulate flow is favorable for navigation as well as for flood control. Navigation by large barges used to carry freight is also facilitated by a canal that cuts across the delta in Louisiana and by controlled flow through 27 locks and dams placed at strategic locations to make the river navigable between St. Louis, Missouri, and Minneapolis, Minnesota (Petterchak, 1999).

The locks provide a channel with gates at both ends so the water level can be raised and lowered to carry river traffic around the dams (McCartney et al., 1998).

Unfortunately, uniform flow is not good for some of the river's natural ecosystems. Occasional floods and other variations in flow allow the system to renew itself and support a wide variety of plant and animal life. Decisions regarding dams and reservoirs require that priorities be assigned and choices made that will not satisfy everyone.

10–7 WIND EROSION-CONTROL STRUCTURES

Soils that are extremely susceptible to wind erosion should be returned to perennial vegetation, with special protection provided while the vegetation is being established, or they should be protected permanently by vegetative barriers. Occasionally an overwhelming need for food requires that highly erodible soils be used indefinitely for cultivated crop production. In some cases, animal or human traffic may damage or destroy vegetative cover. These areas often need special protection. Artificial wind barriers may be used to protect the cultivated soil or to assist in reestablishing perennial cover. The barriers control wind erosion for a combined windward-leeward distance of about ten times their height. For the most effective protection, the barrier must be placed perpendicular to the direction of the erosive winds.

10–7.1 Woven Mat Barriers

Mats of woven plant materials are erected as barriers to erosive winds in many areas where land scarcity forces the use of erodible soils. Grasses, reeds, and crop stalks are sometimes woven into mats that are later erected at sites where they are needed. In other places, the barriers are constructed at the site and cannot be moved without being dismantled.

Figure 10–17 shows barriers of reeds used to protect tomatoes along the Mediterranean Sea about 40 mi (65 km) west of Algiers. The principal barriers run perpendicular to the sea coast at intervals of about 35 to 50 ft (10 to 15 m). The secondary barriers, perpendicular to the primary ones, are spaced about 30-in. (75-cm) apart between tomato rows. These barriers control soil movement and also reduce desiccation damage from the hot, dry winds.

10–7.2 Snow Fence and Other Wooden Barriers

Temporary fences are used in parts of the United States and in other countries to stop snow from drifting onto highways, airport runways, and other critical areas. A snow fence consists of slats of wood about 0.25-in. (0.65-cm) thick, 1.5-in. (4-cm) wide, and 48-in. (122-cm) long fastened together by wire so that the slats are spaced about 1-in. (2.5-cm) apart (approximate porosity 40%). The fencing is usually mounted on steel posts driven into the ground. It can be dismantled, stored, and used again. This commercially available fencing is often used where temporary soil protection is needed, such as for stabilizing active sand dunes or recently denuded, erodible land (Figure 5–4).

Taller barriers made of vertically mounted boards can also be used, but they are very expensive and are suitable only for high value crops or in critical situations where an attempt is being made to revegetate eroding sites. A more permanent solution, use of field shelterbelts, is discussed in Chapter 5.

Figure 10–17 Woven-reed wind barriers in Algeria. The primary barriers are about 6-ft (2-m) tall, the secondary ones about 30- to 40-in. (75-cm to 1-m) tall. (Courtesy Mrs. A. W. Zingg.)

SUMMARY

Crop and soil management practices do not always provide enough erosion control on cultivated land. Runoff and soil loss from land in native vegetation also may be excessive. Special conservation structures are needed to reduce these losses to tolerable levels.

Bench terraces are a very old means of controlling runoff and erosion and are still widely used in areas with large populations and limited areas of arable land. *Graded terraces* are common in humid areas with a commercial-type agriculture, where machinery is available and labor costs are high. A graded terrace consists of a channel with a ridge on the downhill side. The channel slopes gently toward a protected outlet. Graded terraces reduce erosion because they shorten slope lengths, reduce runoff velocity, and trap sediment. *Level terraces* are used in drier climates to hold water until it infiltrates. *Parallel* and *steep-backslope terraces* are modifications that make farming terraced fields easier. Terraces are built by bulldozers, road graders, carryalls, farm machinery, or with hand tools. They need maintenance to retain adequate channel capacity and must be repaired when storms cause major damage.

Diversions direct runoff water away from gullies, around cultivated fields and farmsteads, and to ponds and storage reservoirs. The diversion channel is usually a high-capacity grassed waterway and is seldom cropped.

Diversions and graded terraces must have protected outlets into which the accumulated runoff water is released. *Grassed waterways* are common outlets, but *underground pipe outlets* are becoming more popular because they release land for cultivated crop production.

Erosion sometimes occurs in the beds of vegetated waterways as well as in gullies on cultivated fields and on grassed areas and timberlands. Structures needed to control this erosion may be either temporary or long term. Temporary structures are often used to control erosion in waterways and gullies while perennial vegetation or other permanent control is being established. Temporary structures include rock, brush, log, timber, and brick barriers in channels. They must make firm and intimate contact with the floor and walls of the waterway or gully so that water cannot wash under or around the structure.

Concrete drop structures and *chutes* are permanent control devices used where it seems impractical to control erosion by vegetative means. Simple concrete chutes may be used for relatively small water volume and short water drops, but reinforced concrete is needed where the volume is large or the drop great. The most critical parts of drop structures and chutes are the *inlets,* the *outlets,* and the *aprons* onto which the water falls and from which it flows onto the soil at a lower level.

Roads require large numbers of *ditches, culverts,* and other water control structures to drain the right-of-way and carry streams and runoff water that come to them from adjoining land.

Several types of *storage dams* are used in soil and water conservation activities. Dams may be built to trap sediments, to stabilize gullies or channel grades, to make farm ponds, or to store runoff temporarily in order to reduce flooding, soil loss, and sedimentation.

Current deflectors and *bank protectors* have been used with varying degrees of success in attempts to control streambank erosion. No attempt to deflect a current can be undertaken in the United States without first obtaining permission from the U.S. Army Corps of Engineers or the state water control agency or both.

Floods can be devastating and are difficult to control. *Levees* and *dikes* are used to keep water from flooding low areas, but the situation often becomes more hazardous with time. *Floodways* are used in some places to provide an alternate channel that carries high river flows around an area so it will not need high levees. Large rivers are regulated by *lock and dam systems* for purposes of navigation as well as flood control.

Wind-erosion-control structures are used to stabilize erodible soils that either are or have been cropped or that have been denuded in other ways. *Woven-mat barriers* and *snow fences* have been used successfully to reduce wind velocities at ground level.

QUESTIONS

1. What are the differences in purpose and in structure of level terraces, graded terraces, and diversions?
2. Describe the important elements involved in the construction of simple barriers in waterways and gullies. Why is each important?
3. What special water-control structures along or across the right-of-way must be installed when a new highway is built?
4. What are the major differences in specifications for dams built to (a) trap sediment, (b) stabilize a gully that is cutting back into a field, (c) store flood water? Explain why these differences exist.
5. Describe the major conditions that dictate which technique (such as toe protectors, windrow revetments, or protective fences) will be chosen for reducing streambank erosion. Explain why each of these conditions is important in making the correct choice.
6. What structures are used to reduce flooding by (a) reducing peak flow in a river, (b) keeping a high peak flow from covering a floodplain, (c) preventing sea water from entering a polder?
7. Under what conditions would you expect to find artificial wind barriers being used in a country with highly-developed commercial agriculture?

REFERENCES

AMERICAN SOCIETY OF AGRICULTURAL ENGINEERS, 1989. *ASAE Standards: Design, Layout, Construction, and Maintenance of Terrace Systems.* Publ. S.268.3. St. Joseph, MI, p. 500–504.

BEASLEY, R. P., J. M. GREGORY, and T. R. MCCARTY, 1984. *Erosion and Sediment Pollution Control,* 2nd ed. Iowa State University Press, Ames, IA, 354 p.

BECKER, S. M., and G. R. FOSTER, 1993. Hydraulics of semicircular-inlet drop structures. *Trans. Am. Soc. Agric. Engr.* 36:1131–1139.

BENSALEM, B., 1977. Examples of soil and water conservation practices in North African countries, Algeria, Morocco, and Tunisia. In *Soil Conservation and Management in Developing Countries.* Soils Bull. 33, Paper 10. FAO, Rome, p. 151–160.

BLAKELY, B. D., J. J. COYLE, and J. G. STEELE, 1957. Erosion on cultivated land. In *Soil,* Yearbook of Agriculture. USDA, Washington, D.C., p. 290–307.

HAAN, C. T., B. J. BARFIELD, and J. C. HAYES, 1994. *Design Hydrology and Sedimentology for Small Catchments.* Academic Press, San Diego, 588 p.

HEEDE, B. H., 1976. *Gully Development and Control.* USDA For. Serv. Res. Paper RM-169.

JAIN, S. C., 2001. *Open-Channel Flow.* John Wiley & Sons, New York, 328 p.

MCCARTNEY, B. L., J. GEORGE, B. K. LEE, M. LINDGREN, and F. NEILSON, 1998. *Inland Navigation: Locks, Dams, and Channels.* ASCE Manual No. 94, American Society of Civil Engineers, Reston, VA, 375 p.

MORGAN, R. P. C., 1995. *Soil Erosion and Conservation,* 2nd ed. Longman Group, Essex, England, 198 p.

NOVAK, P., A. I. B. MOFFAT, and C. NALLURI, 2001. *Hydraulic Structures,* 3rd ed. Spon Press, London, 666 p.

PETTERCHAK, J., 1999. *Taming the Upper Mississippi: My Turn at Watch, 1935–1999.* Legacy Press, Rochester, IL, 208 p.

POWELL, G. M., 1989. *Maintaining Terraces.* Kansas State Univ. Coop. Ext. Serv. C709.

PRZEDWOJSKI, B., R. Blazejewski, and K. W. Pilarczyk, 1995. *River Training Techniques: Fundamentals, Design and Applications.* A. A. Balkema, Rotterdam, 625 p.

SANDOR, J. A., and N. S. EASH, 1995. Ancient agricultural soils in the Andes of southern Peru. *Soil Sci. Soc. Am. J.* 59:170–179.

WITTMUS, H., 1973. Construction requirements and cost analysis of grassed backslope terrace systems. *Trans. Am. Soc. Agric. Eng.* 16:970–972.

YADAV, R. C., and L. S. BHUSHAN, 1994. Prefabricated drop spillways for outlets of reclaimed gullied lands. *Agric. Water Manage.* 26:227–237.

CHAPTER

11

Vegetating Drastically Disturbed Areas

Many drastically disturbed areas result from human activities such as building construction, road construction, or mining operations; others result from severe localized impact of water or wind erosion. These areas are often small but conspicuous and are commonly accompanied by concentrated pollution problems. Barren areas are sources of muddy water and dusty air. The sediment from such areas can turn a pond or lake into a mud flat and contribute plant nutrients that nourish the growth of unwanted algae in the remaining open water (Chapter 17). Thick deposits of sediment can smother plants, cover productive soil, and damage structures and machines.

Erosion control structures (Chapter 10) are needed to control erosion on some sites, and all drastically disturbed areas need vegetation for stability. Compaction and fertility problems are common and must be addressed or very little vegetation will grow on them and much of the rainwater will run off rather than infiltrate. Wind erosion may be a serious problem during dry periods. Too often, these areas are left untreated and allowed to erode while the construction, mining, or erosion process goes on.

Timely action can be highly beneficial to minimize the erosion and pollution problems associated with drastically disturbed areas, but such action is often given a low priority. Vegetation established while construction, mining, or other work is in progress is likely to be torn up by continuing activities. Even so, establishment of a quick, temporary cover is often worthwhile. Herzog et al. (2000) determined that both home buyers and realtors perceived that the value of green (vegetated) building sites exceeded that of similar brown (barren) sites more than enough to pay for the cost of seeding temporary vegetation. Such vegetation can markedly reduce erosion, sedimentation, and pollution problems. The increased value and a more pleasant work environment are significant on-site benefits. There are also important public benefits from the off-site effects of reduced sedimentation and pollution.

11–1 CONSTRUCTION SITES

Constructing either a building or a road commonly begins with excavation and includes reshaping the nearby landscape. Bulldozers and other heavy equipment are used to move soil and other materials from one place to another. Subsoil or even the underlying parent material may be exposed where material is removed, and low places are often filled with whatever material is available. The result is a surface with the desired shape formed of a heterogeneous mixture of earth materials. Filled areas may contain low-density pockets that will settle unevenly, but much of the surface is compacted by human and vehicular traffic during the construction process. Little or no vegetation will grow on it unless deliberate action is taken to establish it. Rainfall often cuts rills in such surfaces and deposits sediment in nearby low areas. The water in a nearby stream may be clouded and eutrophied by the runoff.

Extensive reshaping of landscapes makes it worthwhile to strip off the topsoil and stockpile it during the early stages of construction. Water control structures are installed as needed, and the required earth moving is done on the subsoil and underlying material to shape the area before the topsoil is replaced. Steep banks may be formed into bench terraces for better stability on some sites.

An annual grass such as Italian ryegrass may be planted for a temporary cover to hold the soil during the construction process or until it is the right season of the year to plant permanent cover. Annual grasses usually become established more rapidly than perennials and are easily eliminated so they do not compete with the permanent cover after it is established. Mulches are used as an alternative on many sites. For example, straw may be spread across an area for immediate protection from rainfall. It may be held in place by running over it with a packer or a light disc implement set for shallow penetration.

11–1.1 Vegetating Building Sites

Building construction is usually near completion before a permanent cover is established on the grounds around a home, public building, factory, or other structure. Any final shaping needed for the landscape is then performed and topsoil added where needed. The usual minimum thickness of topsoil is 2 in. (5 cm) where the area is to be sodded or 4 in. (10 cm) where grass will be seeded. Holes will be dug to provide an adequate root zone of loose soil where trees or bushes are to be planted. Fertilizer (and lime or gypsum where needed) should be applied as indicated by a soil test and the area may then be seeded or sodded for a permanent vegetative cover appropriate to the setting. Some irrigation may be needed during the establishment period even in rainfed areas. Specific recommendations may be obtained from local seed sources or advisory agencies such as the Extension Service.

Large construction projects may expose so much bare soil that provisions must be made to catch the sediment produced by runoff from the area. A *debris basin* placed where runoff leaves the site may be used for this purpose, as shown in Figure 11–1. Such basins are usually removed when the project is completed and permanent vegetation is established.

Some contractors do not bother to stockpile the topsoil from their projects, and even if they do, it may not be good material for replacement. Stockpiles often are wet, causing anaerobic processes to occur. Moving material with a high clay content can result in excessive compaction that is difficult to remediate. Replacement topsoil may be brought in,

Figure 11–1 Sediment from this construction project in West Virginia would pollute the nearby stream if it were not caught in this debris basin. (Courtesy USDA Natural Resources Conservation Service.)

but this adds to the expense and is not always readily available. An alternative is to use the local topsoil mixed with sand or clay to improve its texture and to add enough (often 5 to 20% by volume) organic matter to loosen the material. Suitable organic materials include various plant residues, manure, compost, and peat. If the organic material has a wide C:N ratio (for example straw, sawdust, or most tree leaves), enough nitrogen fertilizer should also be incorporated to reduce the C:N ratio to about 20:1. Preparation of such mixtures has become a business in some areas (McCoy, 1998).

Cool-season grasses such as Kentucky bluegrass or creeping bentgrass are used in cool temperate climates such as the northern United States. They are best planted early enough in the fall to become established before winter. This will give them an early start the next spring and permit them to produce a dense cover. Creeping red fescue may be used similarly to provide cover in shaded areas.

Warmer climates require warm-season grasses such as Bermudagrass, buffalograss, or bahiagrass. Bermudagrass is established by sod, sprigs, or cuttings placed early enough in the warm season to become well established before cool weather arrives. Buffalograss and bahiagrass may be seeded in the spring (buffalograss seed needs to be treated to break dormancy before it is planted).

The grounds around buildings may be beautified by planting adapted trees, shrubs, and flower gardens. Local sources should be consulted for recommended species and planting times.

11–1.2 Vegetating Road Construction Sites

Modern highways involve major amounts of earth moving to produce a roadbed that will facilitate the flow of traffic. Road gradients are reduced by cutting through high places and filling low places. Steep banks are often produced and subsoil and substratum materials are exposed wherever there are large cuts and fills. Overpasses and underpasses create steep banks even where highways are built on relatively level topography. The maximum slope of such banks depends greatly on the stability of the material and the climate. For example, loess banks in subhumid or drier environments may be nearly vertical, but most other materials must have sloping surfaces. Slumps in a roadbank occur where the bank was too steep for the material to be stable.

Bridges, culverts, or other waterway structures are needed wherever a road crosses or blocks a stream of any size. Outlets are required to drain the median strip of a divided highway every place where water would accumulate. Usually these are inconspicuous openings that serve as inlets for buried pipelines leading to a nearby ditch or stream. Each entrance and outlet should be covered by a screen or a set of bars to keep animals and debris from entering the pipeline.

A great deal of work goes into preparing a good roadbed. It must have a solid foundation and adequate drainage, be composed of suitable material, and be well compacted to give it strength. The gravel and other materials used to make the roadbed are not favorable for plant growth, so a layer of topsoil needs to be used to cover the roadbanks as well as the associated cuts and fills where unfavorable material is exposed. Seriously compacted materials may be loosened with a chisel implement before being covered with topsoil. Fertilizer and any needed soil amendments are also applied at this time. All of this is usually done late in the project, and vegetation should be established shortly thereafter.

Vegetation for the disturbed area along a highway should be carefully chosen. It needs to be adapted to the area and easy to establish. The cover it produces should be erosion-resistant and attractive, but it should not grow tall enough to obstruct vision. Both native and introduced species of perennial grasses and legumes are used. For example, a low-growing recumbent plant like crown vetch is suitable in the northern United States, and sericea lespedeza may be used in warmer climates, but a tall-growing biennial plant like sweetclover is not appropriate for roadsides. Many of the same species that would be used for a lawn, pasture, or other perennial cover are suitable. Often a mixture of plants is used, including native wildflowers and other decorative plants. Including native flowers and grasses in roadside areas is a good way to preserve these plant species and display their natural beauty. The use of local seed sources when they are available assures that the plants are adapted to the local climate. Taller plants like tall grasses, shrubs, and trees may be used in the outer areas if the right-of-way is wide enough.

Deep cuts and fills produce large areas of steep banks that are difficult to stabilize and vegetate. In humid regions, a slurry of seed, fertilizer, and cellulose mulch can be blown onto a steep bank, as shown in Figure 11–2. This technique will not work in arid regions unless irrigation water can be applied when needed during the establishment period, perhaps by a truck-mounted sprayer.

Long slopes may need mechanical protection along with vegetative cover. For example, a deep cut in loess may be easier to stabilize if it is shaped into narrow terraces with

Figure 11–2 Seeding a roadcut by blowing a slurry onto the area. (Courtesy USDA Natural Resources Conservation Service.)

level surfaces and near-vertical sides. Deep-rooted vegetation such as alfalfa can add to the stability of such terraces. Another technique is to cover the surface with a mulch, or as suggested by Agassi (1997), a polypropylene sheet called a *geomembrane,* such as those used by engineers to separate materials in construction projects. A geomembrane can protect the surface, reduce runoff, and provide excellent erosion control for months or years. Seedlings may be planted through holes in the geomembrane.

Roadsides in mountainous areas include many steep slopes that need special treatment. One technique that was used successfully on a cut slope in a stony, coarse, sandy soil near Lake Tahoe in the Sierra Nevada Mountains was to place *wattles* (bundles of woody brush) along contour lines across the slope with a 3-ft (0.9-m) vertical interval between the lines (Figure 11–3). The wattles were staked in place and the area was smoothed by placing soil above them. The slope was then seeded with a mixture of orchardgrass, big bluegrass, crested wheatgrass, pubescent wheatgrass, intermediate wheatgrass, and cicer milkvetch. Big sagebrush, penstemon, squawcarpet, pinemat manzanita, and bitterbrush were planted separately among the grasses.

Vegetation can usually be established readily in a humid region if the soil texture, pH, and fertility are either suitable or corrected. If necessary, a mulch may be used to control erosion until the season is right for seeding and during the establishment period. Establishing vegetation is more difficult in arid and semiarid regions unless irrigation water can be provided, so special techniques may be required in such places. Selection of drought-tolerant species is helpful, but most seedlings are less drought-tolerant than established plants. A transplanting technique reported by Hodder (1970) may be helpful in difficult circumstances. In a 10-in. (250-mm) rainfall zone in Montana, plastic tubes about 2.5 in. (6 cm) in diameter and 2-ft (60-cm) long were filled with fine soil in the bottom and coarse

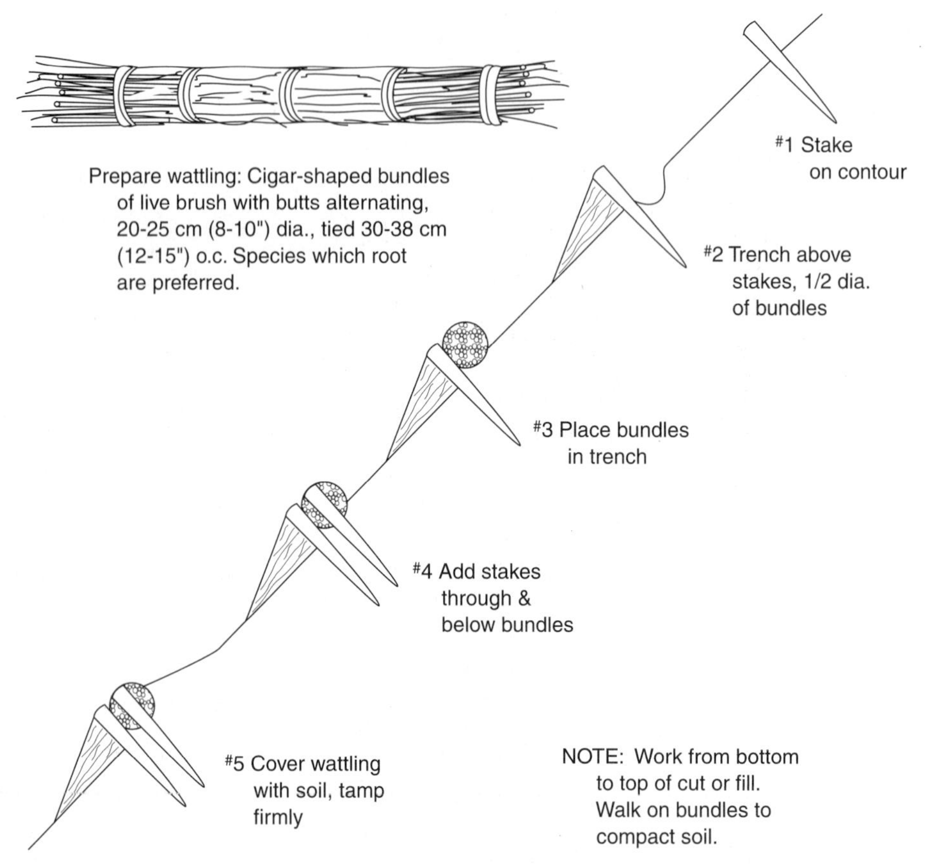

Figure 11–3 A schematic diagram of a wattling installation near Lake Tahoe in the Sierra Nevada Mountains. Adapted from a study conducted for CALTRAN and sponsored by the Federal Highway Administration. (Leiser et al., 1974, as cited in U.S. Environmental Protection Agency, 1975, *Methods of Quickly Vegetating Soils of Low Productivity, Construction Activities.* EPA-440/975-006.)

soil on top. Seeds of shrubs such as caragana were planted in the tubes and were kept in a greenhouse until the plants were well established in the tubes. They were then taken to construction sites and inserted into holes made by a power-driven auger. Survival was reported as satisfactory.

Hodder (1970) also described another technique for planting shrubs on construction sites in arid areas by using a plastic sheet to induce water condensation. This technique involves digging a small pit about 10 in. (25 cm) in diameter and 4-in. (10-cm) deep for each shrub. The shrub is planted in a small mound about 2-in. (5-cm) high in the center of the pit. A sheet of clear plastic film with a slit for the shrub stem covers the entire pit, as shown in Figure 11–4. Moisture condenses on the underside of the plastic and soaks into the soil around the plant roots.

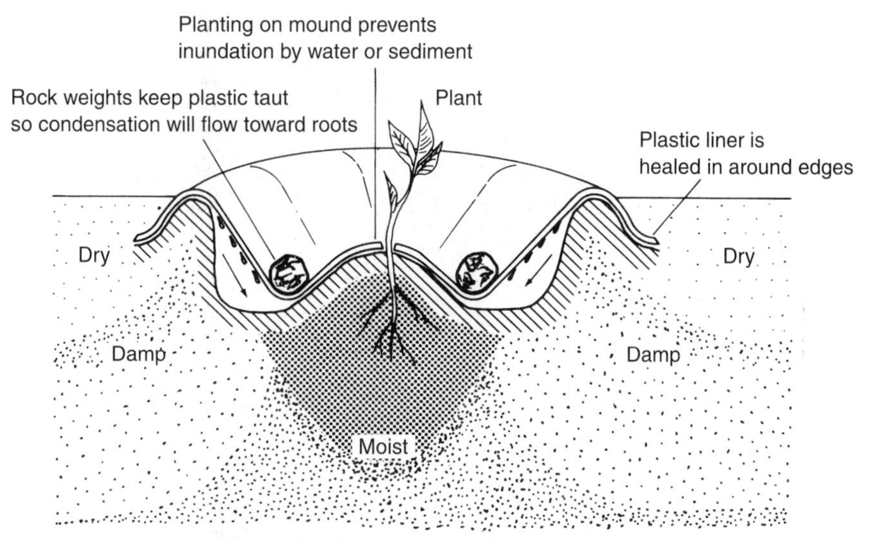

Figure 11–4 A plastic sheet covering a small pit can concentrate soil moisture in the root zone of a developing plant. (Courtesy Richard L. Hodder, Montana State University.)

11–2 MINED AREAS AND MINE SPOILS

Mining operations can be divided into two main types: *surface mines* and *shaft mines.* Surface mining is sometimes called *strip mining* because it begins by removing whatever soil and rock overlie the deposit that is to be mined, usually from a strip across the area, as shown in Figure 11–5. The mining typically proceeds from one strip to another across the area, with spoil material from each subsequent strip being deposited on the mined area that preceded it. After they were mined, many of the older mines were abandoned and left as large areas covered by steep-banked mounds of mixed materials. More environmentally friendly operations now sort the materials by layers and replace them at approximately the same depths. Toxic materials are buried in the bottom layers and fertile soil is retained to cover the new surface. The final topography may be smoother than the original; terraces may be formed to produce nearly level land even in hilly areas.

It is cheaper to mine shallow deposits by surface mining, but removal of the overburden becomes too expensive when it is more than about 150-ft (45-m) thick. Access to deeper layers of coal, ore, or other material to be mined is obtained by digging a vertical shaft or a sloping shaft from the surface above the deposit or through a drift from a nearby hillside. Shaft mines disturb much less of the land surface because they are mostly underground. However, they create another hazard that may strike the area decades later. The ceilings of the old mine caverns and tunnels may eventually collapse when the timbers that were placed to support them become rotten. Longgood (1972) states that two million acres had collapsed in this way and caused many buildings and structures on the surface to fall, break, or crack. Such damage was reported in 28 states and was most concentrated in the area of old coal mines in Pennsylvania.

Figure 11–5 A small surface (strip) coal mine in Pennsylvania. (Courtesy EPA-DOCUMERICA.)

Large amounts of associated materials are moved or mined along with the mineral ore, coal, or gravel extracted from a mine. Much of this material is sorted out at the site and deposited near the entrance of an underground mine or within the mined area and the nearby vicinity of a surface mine. For example, about 40% of all the material ever removed from U.S. coal mines was discarded as mine waste that now covers more than 177,000 ac (71,000 ha) of land (Neufeld, 1990). Much of the waste material from a strip mine is *overburden* that has been pushed aside to reach the deposit that is to be mined; the rock material adjacent to the deposit that is removed during the mining process and immediately sorted out is called *waste rock;* the finely-ground material that is removed during the processing of the coal or ore is called *tailings. Mine spoils* includes all of these solid waste materials left when the area is mined. These materials are often deposited in heaps at the end of some kind of conveyor.

Spoil heaps commonly have steep, unstable slopes, and much of the material is compacted. Most mine spoils are low in fertility and some are stony and droughty, while other materials that are high in clay are likely to form large anaerobic blocks. Some spoils, especially those from coal mines, are sterile because of extreme acidity that is often accompanied by toxic concentrations of copper, iron, aluminum, and/or manganese, while others in arid regions may be highly saline or sodic (saline and sodic conditions are discussed in Chapter 15). Such conditions need to be identified and treated.

Some mine spoils have been smoothed and revegetated, but a large proportion were mined before land reclamation was mandated and remain as the mining operation left them—bare or sparsely vegetated, gullied, and unsightly. Such areas are major sources of runoff and sediment that pollutes surface waters. Some of them include small ponds of clear water that is too acid for plant life and fish to grow in it.

11–2.1 Extremely Acid Mine Spoils

Spoils from coal mines and those from ores that are mined as sulfides (e.g., many copper and zinc deposits) often produce large amounts of sulfuric acid when they are oxidized (Note 11–1). As an essential plant component, sulfur is present in the plant materials that accumulate in wet areas. Such sulfur is converted into pyrite (FeS_2) when the organic materials are transformed into coal.

NOTE 11–1
OXIDATION OF PYRITE

Pyrite forms gold-colored crystals that are commonly known as *fool's gold.* Such crystals are found in coal and often also in overlying layers that are stripped off when the coal is mined. They can also occur in some other settings such as peat deposits. No harm is done as long as the material is buried because pyrite and other sulfides are essentially insoluble under anaerobic conditions. But, when coal is mined or a peat bog is drained, the pyrite is likely to be oxidized (other sulfides may similarly be oxidized, but pyrite is used here as the example):

$$2\ FeS_2 + 7\ O_2 + 2\ H_2O \rightarrow 2\ FeSO_4 + 2\ H_2SO_4$$

A second reaction takes place when the *ferrous* sulfate is oxidized to *ferric* sulfate:

$$4\ FeSO_4 + O_2 + 2\ H_2SO_4 \rightarrow 2\ Fe_2(SO_4)_3 + 2\ H_2O$$

The ferric sulfate then hydrolyzes to form colloidal ferric hydroxide (called *yellow-boy* by miners) and sulfuric acid:

$$Fe_2(SO_4)_3 + 6\ H_2O \rightarrow 2\ Fe(OH)_3 + 3\ H_2SO_4$$

The combined result of these three reactions produces two molecules of sulfuric acid for each molecule of pyrite that is oxidized. The ferric hydroxide that accompanies it is inert because of insolubility.

The production of sulfuric acid in mine spoils is a microbial process that can continue for many years (as long as a supply of pyrite remains exposed to oxygen). The microbial activity ceases when the pH drops to 2 or 3 and becomes so acidic that neither the sulfur bacteria (*Thiobacillus* sp.) nor any other living thing can tolerate it. The acidity can be neutralized with lime, but the neutralization may last for only a few months because the bacteria are reactivated and soon produce more sulfuric acid. Any seepage water that drains from

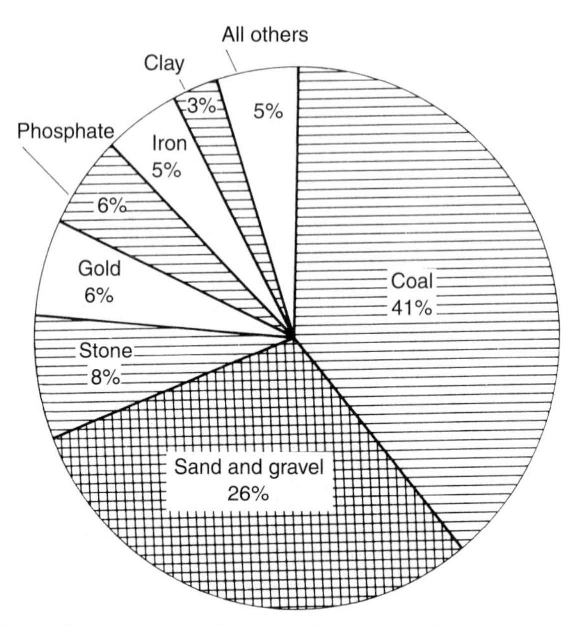

Figure 11–6 Land area disturbed by mining for various materials, as percentage of the total land area disturbed by mining in the United States. (Courtesy EPA-DOCUMERICA.)

such spoils is also likely to have a pH between 2 and 3 and to sterilize the land surface it crosses. Such seepage water may also acidify a nearby stream, pond, or lake and kill all the fish and other organisms that would normally inhabit the water.

The best way to avoid acidification by the oxidation of sulfides is to bury such spoils deep enough to keep them anaerobic. This should be done as a part of the mining operation by replacing materials in the same approximate level where they originally were. If this is not done, the only way to prevent the acid seepage from killing plants and animals is to apply lime repeatedly to the affected soil and water bodies. This alternative is both expensive and inconvenient.

11–2.2 Areas Disturbed by Mining

Mined areas are distributed throughout the United States and all other nations. Data from the EPA indicate that in 1974 the United States had about 3.63 million ac (1.47 million ha) of land that had been severely disturbed by mining operations, or about 0.16% of the total land area in the 50 states. As indicated in Figure 11–6, about two-thirds of this area was mined for either coal or sand and gravel. The materials mined from the remaining parts include stone, clay, industrial metals such as iron, copper, zinc, or uranium, precious metals such as gold and silver, fertilizer materials such as phosphate or potash, and so forth.

Sand and gravel pits are mostly located in alluvial deposits on valley floors or in terraces, although some occur on uplands as a result of glacial deposits. Such materials are used for building roads, making concrete, and many other purposes, and are therefore

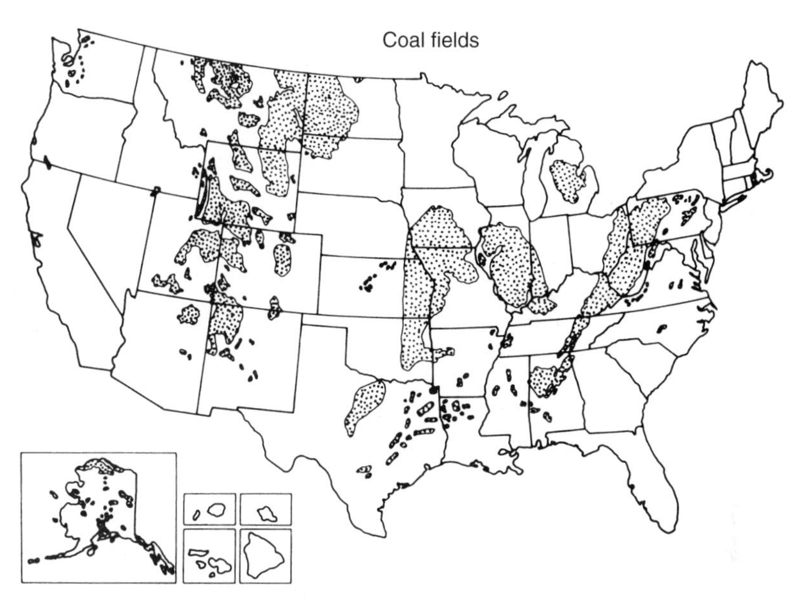

Figure 11–7 At least 28 of the 50 states have significant coal fields. Erosion, sedimentation, and toxic drainage waters from the spoils are common problems in many areas. (Courtesy U.S. Department of Agriculture.)

widely sought. They occur in multitudes of sites, most of which are small in area. Many farmers have an old gravel pit or a borrow pit for topsoil or fill material someplace on their land. Some of these areas still have enough soil to be smoothed and vegetated for use as pastureland, woodland, or even cropland and may be incorporated into the farming operation. Many other borrow areas, however, have simply been abandoned. Such areas may have steep slopes and even vertical surfaces that represent physical hazards to people and animals.

Coal deposits are more concentrated than sand and gravel, but are still widely distributed, as shown in Figure 11–7. About 69% of the abandoned coal spoils are in Appalachia, especially in Pennsylvania, Ohio, West Virginia, Kentucky, and Tennessee. Another 29% are in the Midwest, especially in Illinois, Iowa, Missouri, Kansas, and Oklahoma. Only about 2% of them are in the far western states, partly because the coal deposits there have been less exploited than those in the eastern states.

Metal deposits and ores tend to occur in areas of past tectonic activity and are therefore commonly associated with mountains and foothills. Chemicals such as phosphates, potash, or other salts occur mostly as *evaporite deposits* where the climate is or once was arid, so they occur mostly in the western states. Some deposits of soluble salts are mined by solution processes that create very little surface disturbance.

11–2.3 Surface Mining Legislation

A nationwide program to regulate surface coal mining was created in 1977 when the U.S. Congress passed Public Law 95-87, the Surface Mining Control and Reclamation Act.

Individual states also passed their own laws on the subject that often set stricter standards than those in the federal law. The federal law includes provisions to protect prime farmland that is mined by requiring that a restoration process begin within 10 years after a surface mining operation has ceased. This process must include removing and replacing key soil horizons separately to produce a root zone as good and as deep as the soil was before it was mined and to return the area to soil conditions that will produce crop yields equivalent to those for similar soil in the surrounding area. A performance bond that must be posted before the project is allowed to begin is returned only after these standards are met (Plotkin, 1987; U.S. Congress, 1977).

The surface mining law assigned its administration to the Office of Surface Mining in the U.S. Geological Service. This agency manages an Abandoned Mine Reclamation Fund that collects $0.35 per ton of underground coal mined and $0.10 per ton of lignite and uses the money to pay for reclamation of mined land. From its inception in January 30, 1978, to September 30, 2000, this fund had collected more than 5.8 billion dollars and appropriated more than 4.3 billion dollars (Web site, USGS Office of Surface Mining).

The surface mining law also requires that runoff from the site be controlled during the mining operation to keep sediment, toxic ions, and excessively acid water out of nearby streams. A sediment pond is usually needed to meet this requirement. Sediment will settle in the pond, and lime may be added if needed to neutralize acidity and cause heavy metals to precipitate. Excess water may then be allowed to leave the area.

11–2.4 Vegetating Mine Spoils

Shaft mines affect much smaller areas than strip mines. The principal problems associated with shaft mines are compaction that results from heavy equipment and traffic in the area and large spoil piles that commonly are deposited on a nearby land surface. Adding organic materials and ripping the area with a tined subsoiler (Section 9–6.1) can help remedy the compaction, but cannot solve the problem of the spoil piles. Spreading the spoil to reduce its depth and slope often is not a viable alternative because it would cover much more area. Trucking the material away is expensive and merely transfers the problem elsewhere unless the material is useful, perhaps as fill for building a highway or some other construction project. Replacing the material in the underground mine is appealing from the environmental perspective and has been shown to be feasible but expensive (Fairhurst et al., 1975).

The topography of many mine sites must be smoothed before it can be stabilized, especially where the spoil is left in piles and where surface mining has produced a vertical surface known as a *highwall* at the border of the excavated area. Many highwalls remained in place when old mines were abandoned, but current laws require that they be smoothed to make the area safer and more useful. The difference is illustrated in Figure 11–8. This Montana mine was in a mountainous area that had naturally steep slopes, so a 33% slope was satisfactory. Lesser slopes may be required in other areas. Some sites need to be terraced. The final topography should blend in with the rest of the landscape and provide both erosion control and adequate surface drainage (Hutchison and Ellison, 1992).

Any kind of mine spoils may be buried in a deep fill, but the surface needs to be covered with soil or similar rooting medium. Stockpiling and replacing the original soil can serve this purpose where conditions are favorable, but conditions are not always favorable. The original soil may be shallow, excessively stony, very sandy or clayey, strongly acidic or alkaline, and so forth. Usually the topsoil and subsoil are mixed. Heavy traffic leaves much

Figure 11–8 The highwall labelled A in this photo of a mine in Montana resulted from strip mining that was done before Montana law required that ponds be filled and slopes be smoothed. The area labelled B was mined later and reduced to a 3:1 (33%) gradient to meet state and federal requirements. (Courtesy USDI Bureau of Mines.)

of the replaced soil so compacted that it needs to be loosened. Considerable effort is required to meet the requirements for making the area as productive as it was before it was mined.

Daniels and Amos (1985) recommend that restored mine areas be covered by at least a meter (3.3 ft) of non-compacted soil material. The amount of stockpiled soil is often inadequate to provide this much cover, especially in mountainous areas where the natural soil may be thin and stony or rocky. In the Appalachian coal fields of the eastern United States, the soil material is often supplemented with rock material that was shattered by blasting during the mining process. Daniels and Amos (1985) call this material *mine soil* after it has been placed on a reclaimed surface and exposed to natural weathering processes. The nature of the material depends greatly on the type of rock in the area. It is likely to be chemically fertile, but the lack of organic matter makes it subject to compaction and crusting problems.

Some spoil materials can be made more suitable for use as rooting media by adding organic matter and loosening compacted layers with a chisel implement. Materials such as plant residues, manure, compost, and peat were suggested in Section 11–1 for remedying such problems on construction sites and may be used similarly on mine spoils if they are available. Another material that has been successfully used on mine spoils is sewage sludge. For example, Jorba and Andrés (2000) found that mixing either 7 or 15% sludge into mine spoils greatly facilitated the establishment of vegetation as compared to either untreated spoils or sludge alone. They concluded that the sludge needs to be mixed with the spoils

Figure 11–9 This grain sorghum was seeded on raw coal mine spoil in Wyoming and grew as tall as 2 ft (60 cm) even though no irrigation water was applied. The average annual precipitation at the site is 14 in. (355 mm). (Courtesy USDI Bureau of Mines.)

rather than placed on top. However, when Malik and Scullion (1998) examined a site where sewage sludge had been applied to mine spoils nine years before, they found that the treatment effects had almost disappeared. They indicated that long-term improvements in soil stability and fertility require repeated applications of sewage sludge.

Vegetation should be established after the topography has been shaped and the area is covered with soil or other usable growth media. The type and species of vegetation must be chosen to fit the climate, soil conditions, and intended use of the area. It will generally resemble that of the surrounding area. Sites that were used as cropland before they were mined may be restored as cropland afterwards, as in Figure 11–9, but sites that were previously covered with grass or trees are likely to be restored to their former use. Various forage grasses and legumes, shrubs, vines, or trees for lumber or pulp may be planted. If the timing and conditions are favorable, such vegetation may be seeded or planted as soon as the area is ready. Sometimes a fast-growing annual cereal grain is either seeded first as a temporary cover or along with the perennial vegetation as a companion crop.

11–3 AREAS OF HIGH EROSION HAZARD

Erosion processes are often concentrated on specific sites. These are often small areas that suffer severe localized erosion. Gullies are formed and streams erode their banks. Steep banks collapse in landslides or soil slips, usually when the soil is wet. Sediment from any of these processes can cover lowlands, fill ponds and lakes, and cause pollution problems.

Wind, too, can be a destructive agent that removes soil from a blowout area, deposits sandy materials in unstable dunes, and clouds the air with dust. Soil particles carried by the wind give it an abrasive action. The sandblasting effect of particles in the wind stream can etch the stems and leaves of plants, wear away fenceposts, and strip the paint from buildings and machinery.

Severely eroded areas are usually raw and unprotected. They continue to erode and expand until special measures are taken to reclaim or at least stabilize them. Vegetative cover is needed, but the site usually needs preparation before plants can be established. The landscape generally must be reshaped, and mechanical structures may be required to protect the vegetation, especially during the establishment period. These measures are likely to be expensive, especially if calculated on an area basis. Fortunately, the eroding areas are often small, and controlling the erosion automatically benefits other areas by eliminating a source of sediment and pollution.

11–3.1 Reclaiming Gullied Land

Gullies are scars on a landscape that are too large to be crossed and smoothed by normal farming operations. They range in size from a short channel that is only about 3-ft (1-m) deep to large systems that are well on their way to becoming valleys, as shown in Figure 11–10. Reclaiming a gully is expensive and becomes even more so for a large gully such as this one.

Reclamation of gullied land involves controlling the water that flows into the gully, reshaping the landscape, and establishing vegetative cover. All too often, trash such as stones, dead trees, and old machinery has been discarded in the gully. All such materials

Figure 11–10 Part of a gully near Lumpkin, Georgia. Gullying in this area was neglected for more than 100 years. The gully became several miles long and 50- to 200-ft (15- to 60-m) deep, affecting an area of about 100,000 ac (40,000 ha). (Courtesy USDA Natural Resources Conservation Service.)

must be removed before the area can be reclaimed. After that, the process may be divided into five steps:

1. Gullies grow upslope, toward any stream of water flowing into them (Section 4–1.3). Therefore, the first step is to divert the water away from the head of the gully. This is most often accomplished by building a diversion terrace or a diversion ditch across the slope above the head of the gully. The diversion carries the water to a safe disposal area such as a dense forest, a thick pasture, or a grassed waterway. The diversion may be removed after the gully is stabilized. Alternatively, a set of permanent terraces may be constructed in the field above the gully head.

2. The head and sides of the gully need to be smoothed to slopes no steeper than the angle of repose of the soil. The angle of repose for most soils is between a 2:1 (50% or 26.5°) and a 3:1 (33% or 18.5°) slope. The gully may also be straightened at this time.

3. The channel slope of the waterway should be smoothed to a slope no steeper than 0.5%. A series of drops should be installed if the overall gradient is steeper than 0.5%. The drops may be permanent concrete or rock masonry structures such as those described in Section 10–2, or they may be temporary check dams made of loose rock, lumber, logs, or brush laid between posts. Such structures must extend at least 6 in. (15 cm) into undisturbed soil on the bottom and sides of the graded gully. The lowest part is in the center so the water will flow through the structure and onto an apron installed on the downslope side of the check dam.

4. A grassed waterway (See Section 11–3.2) is established in the area that will carry water. It needs to be wide enough to carry the largest flow that is likely to occur in the area even after a significant amount of siltation has occurred in the waterway. The rest of the area may be vegetated with whatever grasses, legumes, shrubs, and trees the decision maker may desire (Figure 11–11). Some of it may be used as cropland if it has been graded to blend into an adjacent field.

5. The entire area usually needs to be fenced to exclude domestic animals until the vegetation has become well established. Fencing may also be needed to prevent wild animals such as deer from browsing on young trees and shrubs.

The channel gradient of many reclaimed gullies is reduced by placing a structure either above the head end of the gully or below the foot end, or both. The grassed waterway between the structures may then be flat enough to not require any intermediate structures such as those described in item 3 above. The structure at the head end may be either a flume such as that shown in the foreground of Figure 11–11 (B) or a drop inlet like that shown in Figure 10–10. A drop structure at the lower end of a grassed waterway is shown in Figure 11–12.

11–3.2 Grassed Waterways

It is better and much cheaper to establish a grassed waterway before serious gully erosion occurs than it is to wait until serious damage has been done. *Grassed waterways,* also known as *grass waterways* or *vegetated waterways,* are either natural or constructed drainageways that are vegetated with a thick grass cover. The grass needs to be tough enough to resist the erosive effect of runoff water and to tolerate the period of submergence that occurs during runoff.

(A)

(B)

Figure 11–11 The large, active gully in Alabama shown in the upper photo (A) was transformed into the grassed waterway shown in the lower photo (B) by reshaping the area and installing a flume at each end to serve as a drop structure. (Courtesy USDA Natural Resources Conservation Service.)

Figure 11–12 A concrete drop structure that serves as an outlet for a grassed waterway. The round hole visible in the lower left corner of the structure is the outlet for a tile drain that reduces the time the waterway stays wet. (Courtesy USDA Natural Resources Conservation Service.)

The decision to install a grassed waterway is often made because gullying has begun and it is obvious that some kind of control measures are needed. If gullies are present, they must be smoothed as described in the preceding section. Some land shaping may be required to straighten and shape the waterway even if there is no gully. Grassed waterways may also be installed as outlets for terraces or diversions that concentrate water in areas that otherwise would not carry enough water to cause erosion. A tile drainage line may also be installed to reduce the time the soil will stay waterlogged between runoff events. Such tile lines are normally placed on one side of the waterway to minimize the danger of the fill over the tile being eroded away by runoff water.

The maximum gradient of the waterway channel depends on the amount of water to be carried, the vegetation to be used, and the erodibility of the soil material. Maximum permissible water velocities range from about 3 to 8 ft/sec (0.9 to 2.4 m/sec), depending on the grass and soil conditions. Waterways that carry large flows of runoff should generally have channel gradients of no more than 0.5%. The size and shape (Figure 11–13) of the waterway are designed to carry the maximum rate of runoff that is likely to occur from at least a 10-year storm, or often that from a 25-year storm. Most waterways are built oversize to allow them to accumulate some sediment and still carry whatever runoff is likely to occur.

Some grassed waterways are established during a season of the year when little runoff is likely to occur at the site, but this is an especially favorable circumstance. Usually a diversion ditch (Chapter 10, Section 10–1.7) is needed to prevent water from entering the waterway during the establishment period. Following land shaping, a seedbed is prepared, fertilizer and soil amendments are applied as needed according to a soil test, and the waterway is seeded or planted. A mulch may be applied if needed to provide early erosion control. A wide variety of materials are used as mulch, including straw, shredded wood, paper,

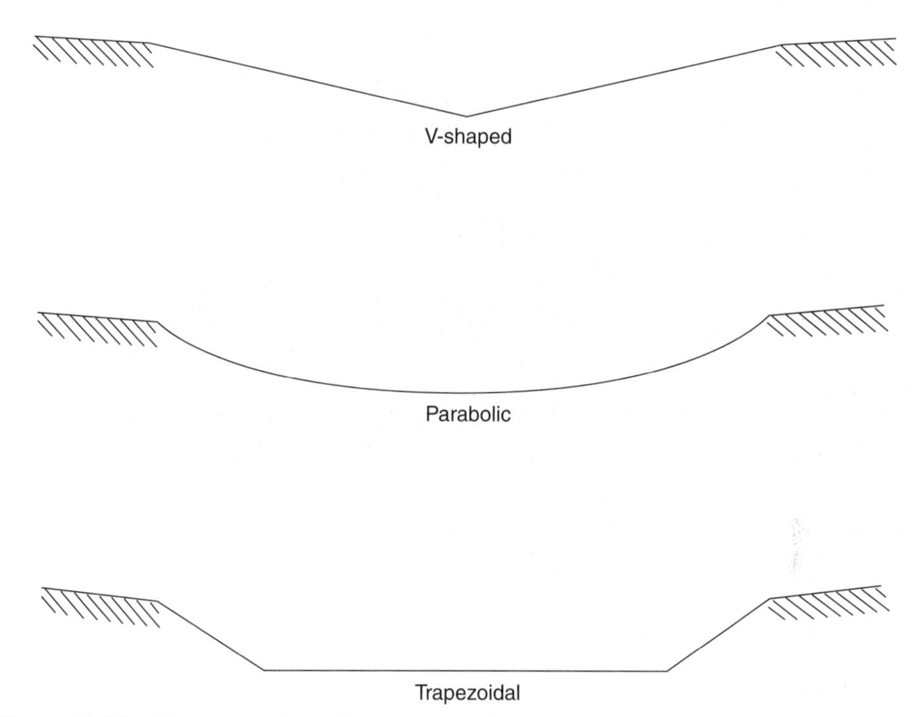

Figure 11–13 The cross sections of most vegetated waterways are V-shaped, parabolic, or trapezoidal. The side slopes are limited either by the stability of the soil or by the desire to have them crossable by machinery. (Courtesy USDA Natural Resources Conservation Service.)

or plastic strips, etc. Some mulches are pressed into the soil by running over them with a disk implement at a shallow setting. Others are held in place by staking netting over them. Ideally, the mulch should protect the soil long enough for grass to become established and grow up through it; the mulch should then decompose. Following are descriptions of eight grasses that are used extensively for grassed waterways in the United States.

Bermudagrass is a tough, vigorous, sod-forming, perennial grass that is well suited for vegetating waterways in the southern, central, and southwestern United States. It grows best on fertile, fine-textured soils and tolerates both wet and dry conditions as well as acidic, alkaline, or saline soils. It propagates by aboveground stolons, underground rhizomes, and by seed. It has a tendency to spread and can become a serious weed in cultivated fields.

Italian ryegrass is a cool-season annual bunchgrass that is adapted to the far western states (west of the Cascade Mountains) and as a winter-season grass in the South. It is especially useful as a temporary cover on disturbed soils because it becomes established rapidly and can stabilize the soil until the proper season for establishing perennial vegetation.

Kentucky bluegrass is a cool-season, sod-forming, long-lived perennial grass that spreads by underground rhizomes and by seed. It grows in both humid and irrigated areas in most of the United States except the far southern part along the Gulf Coast. It can survive at a soil pH of 6.0 or above, but grows best where the pH is near 8.0.

Figure 11–14 This V-shaped grassed waterway in North Carolina was vegetated with tall fescue in preparation for terracing the field. It provides a safe channel to carry runoff water from the terraces to a natural stream. (Courtesy USDA Natural Resources Conservation Service.)

Redtop is a cool-season perennial grass that grows well in northern humid areas on poorly drained soils as acidic as pH 4.0, even if the soil fertility is low. It reproduces by underground rhizomes.

Reed canarygrass is noted for its ability to grow on poorly drained soils and even in standing water. It will grow throughout the United States on soils ranging from acid (pH 5.0) to alkaline (pH 8.0). In dry sites, it is a cool-season bunchgrass, but in wet conditions it forms a sod-like surface mass of erosion-resisting adventitious roots.

Smooth bromegrass is a widely used, cool-season, sod-forming perennial grass that grows best in soils with high fertility. Two distinct types are identified: northern smooth bromegrass that is adapted to western Canada and the northern Great Plains, and southern smooth bromegrass that is adapted to the U.S. Corn Belt and the central Great Plains.

Tall fescue (Figure 11–14) is a cool-season, perennial bunchgrass that grows vigorously and forms a sod-like cover when seeded heavily. It grows in humid and irrigated areas of the United States except on the lower Gulf Coast. It grows best on fine-textured soils and can tolerate low soil fertility and a soil pH ranging from 5.0 to 8.0.

Western wheatgrass is a cool-season, sod-forming, perennial grass that grows from Wisconsin south to Texas and west to the Pacific coast. It is drought-resistant and tolerates high soil alkalinity and even sodic conditions.

Grassed waterways need to be well managed and maintained. They should be mowed and fertilized regularly to maintain a vigorous stand of grass. Both animal and vehicular traffic should be kept off them when they are wet enough to be damaged. Tillage equipment must be raised during crossings so it will not tear up the grass.

Many grassed waterways catch enough sediment from nearby fields to gradually fill their channels. Such waterways must be cleaned and reshaped periodically to keep the water flowing in the waterway rather than alongside where it might cut a gully. Some neglected waterways have become sites for two gullies—one on each side of the grassed strip.

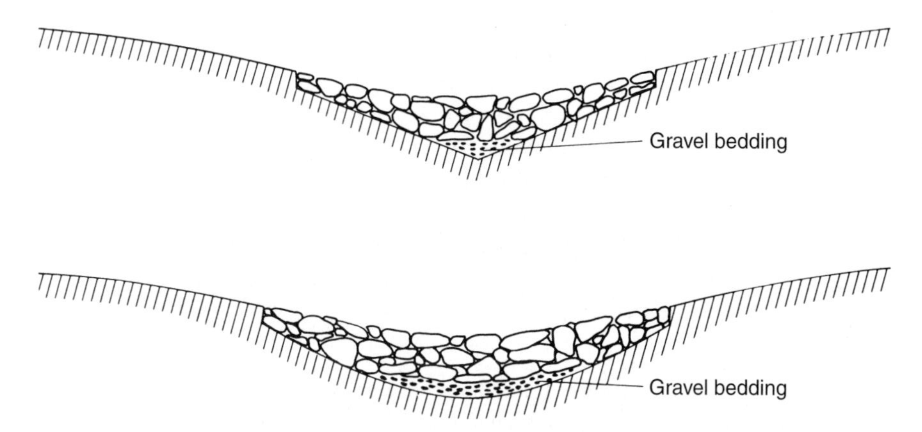

Figure 11–15 Stone linings may be used for waterways constructed where the soil is wet too much of the time or where the gradient is too steep for grass cover to protect the channel adequately. Stone linings work well with either the V-shape (upper drawing) or the parabolic shape (lower drawing). A blanket of gravel beneath the stones helps keep the wet soil from coming up between the stones. (Courtesy USDA Natural Resources Conservation Service.)

Some waterways that carry water almost constantly are too wet for most grasses. A stone lining installed as shown in Figure 11–15 in the wet part of the channel is a good practice to use in such sites. A stone lining may also be used for waterways with gradients that are too steep for grass to give them enough protection.

11–3.3 Streambank Stabilization

Streams naturally erode their banks and shift their channels as part of the geologic erosion processes (See Note 11–2), but the rate is extremely variable. A stream may be stable for many years and then shift abruptly to a new location as a result of a single large runoff event. Human activities can influence this process by making the banks either more stable or less stable. Also, activities upstream from a site can influence the rate of runoff and therefore alter the erosivity of the stream.

NOTE 11–2
MEANDERING STREAMS

Stream channels naturally meander, as may be seen by observing the path of any major river and many smaller streams. The stream tends to erode the downstream part of its outer bank, where the current strikes with the most force. Part of the sediment is carried a short distance and deposited in a sandbar where the water strikes the other bank and is slowed as it turns back the other way. The whole system shifts downstream gradually during normal flow of the stream.

> Larger flows that occur during flood stage can cause more dramatic changes. The stream may cut across a narrow neck of land, leaving the old channel as an oxbow lake, or it may shift to an entirely different channel (usually one that it has used before).
>
> Over long periods of time, the stream meanders across its entire bottomland area. When conditions change enough that it cuts to a lower level and no longer crosses part of the old bottomland, that part becomes a stream terrace. Some streams gradually shift sideways, leaving a broad plain on the side they came from as they erode into the hills on the other side.

People usually prefer for a stream to remain where it is rather than for it to erode its banks and perhaps move out from under a bridge or undercut the foundation of a building. Even where there are no structures, a shifting stream can erode away part of a farmer's field or cause trees to fall into the water.

Work on navigable streams in the United States is the responsibility of the U.S. Army Corps of Engineers. Erosion problems along the banks of large streams should therefore be referred to them. Adjacent landowners do not have the authority and usually do not have the resources to handle such problems themselves. Landowners may, however, stabilize the banks of smaller streams. Guidance and assistance are available from the hydraulic engineers and agronomists of the Natural Resources Conservation Service.

Raw soil banks and the exposed roots of trees or shrubs indicate that the stream is eroding its bank at a rapid rate. Stabilization work usually begins by fencing the area to keep out livestock and then clearing the stream channel of fallen trees or shrubs and any other debris that may be present. Trees and shrubs adjacent to the bank may also have to be removed. The bank may then be graded to reduce its slope to no steeper than 1.5:1 (67% or 34°) so it can be stabilized either by a mechanical structure or by vegetative control measures. Some banks need a combination of mechanical and vegetative control measures. Mechanical structures such as dikes, toe or hardpoint protectors, revetment, and riprap are discussed in Section 10–5.

Vegetative cover on a streambank needs to be erosion resistant and water tolerant. Grasses such as those used for grassed waterways may suffice in some places, but appropriate bushes and trees are usually needed (Figure 11–16). Willows are often used because they thrive in wet places, grow readily from cuttings, and form a dense, tangled root system. Other species that are used include cottonwood, river birch, sycamore, red maple, and forsythia (Kohnke and Boller, 1989). Helpful techniques for establishing such vegetation are illustrated in Figure 11–17.

11–3.4 Pondbank Stabilization

Farm ponds range in size from a fraction of an acre to a few acres (from tenths of a hectare to 1 ha or more), partly because they are used for various purposes. Some are used to supply water for livestock or irrigation; others are used to control erosion, as where an earthen dam is built across the lower end of a gully. Some ponds are used for swimming, fishing, bird watching, or hunting.

Figure 11–16 This streambank in Virginia was stabilized by planting woody cuttings. The shrubs next to the water are forsythia; the trees are willow and cottonwood. (Courtesy Northern Virginia Soil and Water Conservation District.)

Damming a gully is likely to be cheaper than smoothing it and vegetating the banks. Also, this usually will produce a pond that is less expensive than an excavated or embankment pond of the same volume. However, the gully may not be in the ideal location for a pond, and it is likely to result in a long, narrow body of water with steep banks. Most farm ponds are therefore formed by excavation at a site where the watershed above them is the right size to fill the pond with water and not overload its capacity, usually between 25 and 100 ac (10 to 40 ha). Streams that drain a larger area may be used to fill such a pond, but it is usually best to build the pond on a small tributary stream and divert the right amount of water into it from the larger stream. Larger ponds, like the one shown in Figure 11–18, are sometimes built to control runoff as part of a watershed project.

A pond has several distinct areas that can benefit from protective vegetation. If the pond is filled by a diversion from a nearby stream, the diversion channel may be a grassed waterway unless the flow is so erosive that it requires a lined ditch. Likewise, the spillway that carries water away when the pond overflows may be a grassed waterway unless conditions are erosive enough to require a rock or concrete structure. The dry parts of the earthen dam may be protected by almost any kind of perennial vegetation that provides good ground cover.

The area along the waterline of a pond is often the most difficult part to keep vegetated, especially if the water level fluctuates very much. Most pond banks have relatively steep slopes, both because they are built in a small valley and because extensive areas of shallow water are undesirable. Shallow water can produce unwanted aquatic vegetation and

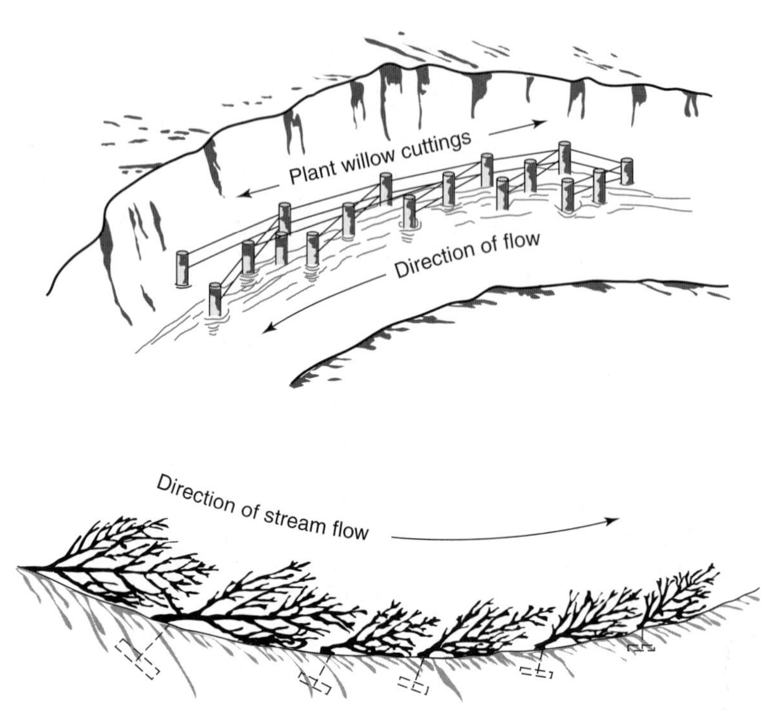

Figure 11–17 The upper drawing shows a jetty made of willow poles driven into the bank above and below the water line and wired together. More willow cuttings should be planted above the jetty so the whole assembly will grow into a dense thicket and stabilize the bank. The lower drawing shows a tree revetment used to stabilize a streambank. The tree butts are placed next to the bank and point upstream; "deadman" logs at least 8 in. (20 cm) in diameter and 40-in. (1-m) long are buried in the bank and tied to the trees by cables at least 3/8 in. (1 cm) in diameter. (Drawings courtesy USDA Natural Resources Conservation Service.)

may serve as a breeding area for mosquitoes and other insects. Water flow velocities in a pond are so slow that they are non-erosive, but wave action can be a problem if the water surface is exposed to strong winds. The wave energy can be dissipated by anchoring bales of straw on the bank at and just below the water line or by floating logs anchored near the shore until permanent vegetation is well established on the bank. The vegetation used on the banks is often the same as that used for a grassed waterway.

11–3.5 Landslides, Soil Slips, and Mudflows

Erosion sometimes occurs suddenly as mass movement of an entire body of soil and/or rock material. Depending on the material and the nature of the event, such movement is called a landslide, rock slide, debris slide, soil slip, mudflow, or debris flow (Dikau et al., 1996). A *landslide* involves a considerable mass of material, often including rock, soil, and vegetation, that moves suddenly to the bottom of a steep slope, much as an avalanche sends a mass of snow hurtling down a slope. A *soil slip* is generally a much smaller mass and typically moves only a short distance. A *mudflow* is mostly soil material that flows slowly as a soupy

Figure 11–18 This large pond is part of the West Willow Creek Watershed project in Minnesota. It is designed to store 200 ac-ft (250,000 m^3) of water from a 2000-ac (800-ha) drainage area. Its banks are stabilized with adapted vegetation. (Courtesy USDA Natural Resources Conservation Service.)

mass that can carry stones and vegetation along with it. A *debris flow* is mostly weathered rock fragments with enough fine material to make it move like a mudflow when saturated with water. The fine particles are surrounded by water and they in turn surround the stony material.

Water serves as a lubricant for mass movement, but gravity is the driving force, and an earthquake or other source of vibration may serve as a trigger. Most mass movements occur on steep slopes, but some saturated soils can produce a mudflow on a gentle slope. Some soil materials are thixotropic when wet; that is, they are stable or somewhat elastic until a certain stress point is reached, but then they flow as a viscous liquid (Coussot, 1997). Such movement can be triggered by shaking, perhaps by a minor earthquake, blasting, or even by heavy traffic passing nearby.

Most large mass movements are in mountainous areas, or at least in areas of steep hills. For example, an irrigation district in southern Idaho was concerned about how to raise money to enlarge its reservoir until a large landslide formed a dam that blocked the valley above them, and thus created a new reservoir. The landslide came down the slope with such speed that the fill rose higher on the other side than it was on the side it came from. Fortunately, no one was in the way at the time. Other landslides have buried people without warning. For example, an earthquake on the night of August 17–18, 1959, near the west entrance to Yellowstone Park caused a large landslide that covered a campsite and killed

Figure 11–19 This lake formed in the Madison River Canyon of west Yellowstone as a result of the earthquake of August 17–18, 1959. The landslide bared the central part of the ridge in the background of this photo and produced a dam about 200-ft (60-m) high. (Courtesy F. R. Troeh.)

28 men, women, and children, 19 of whom were never found. It filled the Madison River Canyon to a depth of about 200 ft (60 m) and produced the lake shown in Figure 11–19.

Smaller mass movements often occur in steep slopes produced by construction projects, especially roadcuts. For example, a survey of counties in Iowa (a state generally considered to be nearly level) found that 80% of the 60 counties that responded had experienced problems with landslides or soil slips (Chu, 2001). Most of these mass movements were on slopes between 3:1 and 1:1 (33 to 100% or 18 to 45°). They occurred in a variety of materials, including 28% in undifferentiated fill, 24% in glacial till, 21% in loess, 13% in alluvium, 7% in shale bedrock, and 7% in other materials. Probable causes included heavy rainfall −28%, high water table −22%, too steep design −21%, maintenance or construction work −14%, loading at the crest of the slope −5%, and other −10%.

Landslides and other mass movements in remote areas are likely to be left as they are. It may take a long time, but nature will gradually weather exposed rock surfaces and cover the area with vegetation. Landslides that damage a building, block a road, or even leave a raw surface in a roadbank or near someone's home are another matter. Many of these are induced by construction projects, and remedial action is required. Preferably, sites where such movement is likely to occur should be identified and appropriate steps taken to prevent it (Campbell, Bernknopf, and Soller, 1998).

The most appropriate remedial action depends on what caused the mass movement or makes a future occurrence likely in a particular location. For example, the gradient of a steep bank may need to be reduced. Campbell, Bernknopf, and Soller (1998) examined various landslides in the area around San Francisco and found that materials from hard igneous or metamorphic rocks were generally stable up to about 40° (84% slope gradient), materials from moderately consolidated sandstones and shales were stable up to about 35° (70%), and those from unconsolidated surficial deposits were stable up to about 30° (58%). The latter condition would apply to most road fills and many roadcuts, so these should usually have slopes less steep than 30°, probably about 3:1 (33%) or flatter unless the material is exceptionally stable.

Many sites can be protected from mass movement by providing drainage to keep excess water from accumulating in the area. The source of the water needs to be identified so that it can be diverted away from the problem area. A grassed waterway placed above the area to be protected may serve to intercept and carry away surface water. An interceptor tile line may be needed if underground seepage water causes the problem.

Sometimes a significant difference can be made by a prolific vegetative cover that transpires enough water to keep the soil from becoming waterlogged. Deep-rooted vegetation is especially helpful, both for transpiring water and for anchoring the soil. Of course the loss of vegetative cover by fire, overgrazing, or other cause has the reverse effect. Campbell, Bernknopf, and Soller (1998) found that a brush fire increased the susceptibility of an area to landslides for a period of months or years until the vegetative cover was reestablished.

A series of narrow terraces, a layer of stones, or various structural support measures have been used to prevent landslides in some critical areas. Structures are expensive, so they are likely to be used only where they protect valuable property.

11–4 SAND DUNES

Sand dunes are produced by strong winds blowing across an area that serves as a sand source. The source may be a sandy beach along the coast of an ocean or lake, or it may be an area of open land such as a desert with little vegetation. Broad outwash plains served as source areas for both sand dunes and loess deposits during glacial times. An unusual example occurs at White Sands, New Mexico, (Figure 11–20) where white gypsum sand has been deposited in dunes as much as 60-ft (18-m) deep. The gypsum crystals formed on alkali flats upwind from the dunes. An area of 144,458 ac (58,460 ha) was set aside as White Sands National Monument in 1933, but most of it is now part of the White Sands Missile Range.

Most desert sand dunes occur in the bands on both sides of the Tropic of Cancer (23.5° N latitude) and the Tropic of Capricorn (23.5° S latitude). In the United States, sand dunes are most common in areas adjacent to the coasts along the Pacific Ocean, Atlantic Ocean, Gulf of Mexico, and Great Lakes. A few significant areas occur in the interior, including White Sands, New Mexico, the Sand Hills region of Nebraska, and parts of southern Idaho. It is estimated that about 11% of the world land area, about 3.58 billion ac (1.45 billion ha), has conditions conducive to the formation of sand dunes.

Sand dunes drift downwind, covering vegetation and low-lying objects along the way, unless they are stabilized by vegetative cover. Shelterbelts and mechanical barriers that reduce wind velocity are helpful but are seldom adequate by themselves. Woven mat barriers, snow fences, and other wooden barriers that can be used for this purpose are described in Section 10–7. They generally need to be supplemented by vegetative cover on the sandy area.

Figure 11–20 The eastern side of the gypsum sand dunes of White Sands, New Mexico. The gypsum deposit here is about 20-ft (6-m) thick. The upper surface is a series of sand dunes. Scattered shrubs form most of the sparse vegetation on the dunes. (Courtesy F. R. Troeh.)

11–4.1 Stabilizing Coastal Sand Dunes

Sand dunes are common near coastlines because oceans and some large lakes produce sandy beaches with little or no vegetation, and the temperature variations between land and water cause winds to blow across the beaches (Klijn, 1990). The sand supply on beaches is constantly renewed by the erosive action of the water on exposed rocky areas and steep banks. Salt spray from the ocean creates saline and alkaline conditions that make it harder to establish vegetation on the dunes.

Shifting sands can be stabilized by planting adapted salt-tolerant vegetation. The work should start at the high waterline of the ocean so there will not be any drifting sand coming from the upwind direction to cover the newly established plants. The work can then move progressively inland. Recommended plants for various coastal areas of the United States are shown in Table 11–1. Marramgrass and Baltic marramgrass are regarded as two of the best grasses for stabilizing dunes along the Atlantic coast of Europe (Dieckhoff, 1992). They are easily established from locally-gathered cuttings and are resistant to drifting sand accumulations up to one meter thick per year.

Grass should be planted rather than seeded for stabilizing dunes because strong winds along the coasts blow the seed away and sandblast young seedlings. A complete fertilizer should be broadcast over the plantings as soon as growth starts in the spring. Brush cuttings or other mulch is sometimes laid over the plantings in critical areas to prevent them from blowing out. Adapted shrubs and trees should be interplanted among the grasses to further stabilize the sand.

Coastal dunes provide vital protection to keep the sea from flooding the polders in The Netherlands. Various vegetative and mechanical approaches have been tried to maintain the dunes, and it has been concluded that the beach and dune system needs to be considered together as a dynamic natural combination. The sand is mobile, but it establishes a

TABLE 11–1 RECOMMENDED PLANTS FOR STABILIZING COASTAL SAND DUNES

Atlantic Coast Maine to N. Carolina	Atlantic and Gulf Coasts N. Carolina to Louisiana	Gulf Coast Texas vicinity	Pacific Coast
	Grasses and legumes		
American beachgrass	sea-oatsgrass	seacoast bluestem	tall fescue
European beachgrass	panicgrass	weeping lovegrass	red fescue
Volga wildrye	broomsedge bluestem	veldtgrass	hairy vetch
	trailing wildbean		beachpea
	Shrubs and trees		
beach plum	sand pine	(same as Atlantic	Scotch broom
bayberry	Virginia pine	and Gulf Coasts)	Monterey pine
Scotch pine	loblolly pine		shore pine
mugho pine	sweetgum		
Austrian pine	willows (local species)		
pitch pine			

steady state that is reasonably stable. Stabilizing eroding areas by adding more sand is sometimes better than building structures and planting vegetation (Van Bohemen and Meesters, 1992). The local geography must be studied to determine whether the sand should be added to the front or the back of the dunes or to the beach area. Adding sand to shallow offshore areas is also a possibility where the ocean currents would carry it to the beaches. The wind will pick it up from the beach and carry it to the dune area.

Strange as it may seem, it is sometimes possible to stabilize dunes too well. The stabilized dunes may deprive a downwind area of the supply of sand that it needs to balance its own losses. The southern tip of Africa is one place where this has happened. Sand driven eastward on land by the wind and underwater by the ocean currents crosses several large bays near Port Elizabeth (McLachlan and Burns, 1992). A dune area near the city was stabilized early in the 20th century by spreading city refuse on the dunes and vegetating them. This cut off the sand supply for beaches near the city and erosion exposed the underlying rock, thus damaging the recreational value of the beaches. The city has elected to build expensive erosion control structures to hold the remaining sand on its beaches because bringing in replacement sand would be a continual expense and destabilizing the upwind area is unacceptable for housing that has developed there.

11–4.2 Stabilizing Great Lakes Dunes

The Great Lakes region has more than 530,000 ac (215,000 ha) of sand dunes. Some of the early practical but scholarly work on sand dune stabilization was done around the Great Lakes by Michigan State University (Sanford, 1916). It was confirmed by field tests that the proper place to start control was on the windward side of dunes. A typical planting of vegetation to stabilize sand dunes consisted of several rows each of:

1. European beachgrass, American beachgrass, and/or Volga wildrye at the windward edge of the shifting sands.
2. Beachpea, wild lupine, sandgrass, wildrye, or a combination of these herbaceous plants.

3. Wild rose, ground hemlock, wax myrtle (bayberry), sweetgale, wild red cherry, Virginia-creeper, redosier dogwood, snowberry, or a combination of these shrubs.
4. Willow cuttings from any local willow species.
5. A mixture of adapted fast-growing hardwoods and slower-growing conifers. Recommended hardwoods include cottonwood, white poplar, trembling aspen, white birch, red oak, sassafras, silver maple, and black locust. Approved conifers include jack pine, Scotch pine, white spruce, hemlock, and white cedar.

11–4.3 Stabilizing Inland Dunes

Sand dunes in the interior of the United States have been stabilized by two contrasting techniques, one with irrigation water and one without irrigation. The nonirrigation system of stabilization consists of first establishing drought-resistant annual plants such as grain sorghum, sudangrass, or rye. Adapted grasses such as sand bluestem, side-oats grama, Indiangrass, switchgrass, and Canada wildrye are seeded in the residues of the annual plants just prior to the next rainy season.

When irrigation water is available, sand dunes can be leveled to slope gradients less than 5%, fertilized according to soil test, and planted to alfalfa, wheat, corn, Irish potatoes, or other adapted field or forage crops. Crop residues should be left in place on harvested areas to protect the area until the next growing season. No-till planting is advantageous on such areas to avoid leaving the soil bare while the new crop is becoming established. Irrigation water is needed on a timely basis to keep the soil surface from becoming dry enough to erode easily by wind. Sprinkler irrigation is the most common type on "blowsand" areas because surface irrigation streams can erode the sand and are easily disrupted by even a small amount of drifting sand, including sand from an adjacent nonirrigated area. Center-pivot irrigation (Section 15–5.3) is the dominant type in the Sand Hills region of Nebraska, an area of about 20,000 square miles (52,000 sq km).

11–4.4 Dune Leveling

Crests of sand dunes sometimes need to be leveled to reduce the slope gradient or to make the deposit thinner and permit deep plant roots to reach the underlying finer-textured soil. Also, work on European coastal dunes has indicated that heights above 40 to 50 ft (12 to 15 m) cause excessive wind turbulence (Barrère, 1992).

Modern earth-moving equipment may be used to take the tops off dunes that are too high where labor is expensive, immediate results are required, or the downwind area must be protected from drifting sand. Where such limitations do not exist, wind deflectors and barriers placed so the wind can blow beneath them have been used to direct the wind so it will sweep the sharp crests off of active dunes. Snow fence has sometimes been used for this purpose. As the wind lowers the crest, the fence is lowered progressively so its lower edge is always a short distance above the surface. When the crest is reduced enough, the snow fence or other barrier is lowered to the surface and used to protect the area from wind erosion while vegetation is being established.

11–4.5 Managing Stabilized Dunes

Good management is essential to prevent deterioration after sand dunes have been stabilized by perennial vegetation. Sand may start blowing again when the vegetation is destroyed along recreation trails for off-road vehicles. Even a foot path or animal trail may be enough to destroy the protective vegetation. Overgrazing and fires are common causes of damage to vegetation. Homesites or commercial buildings located on stabilized sand dunes can easily kill enough vegetation to permit the sand to become mobile again. For example, an area of stabilized dunes 12- to 15-m (40- to 50-ft) high and 230-km (140-mi.) long was established in the 19th century to protect the forest along the west coast of France (Barrère, 1992). It served its purpose well for several decades, but was damaged by large storms in 1912, 1917, and 1926 and by military activities during World War II. More recently, the vegetation has suffered from excessive tourist and other traffic. Damaged areas have been eroded by the sea, by runoff water cutting gullies, and by wind-produced blowouts. The blowing sand is invading the forest and damaging dwellings in the area. Both mechanical and vegetative remedies have been proposed to restabilize the dunes, and it is emphasized that the vegetation must be protected from traffic of all kinds.

Persons managing stabilized sand dunes must be alert to all of the preceding hazards. Prompt action must be taken whenever the vegetation is damaged and the soil exposed to the wind. Traffic should be diverted from such areas and steps taken to revegetate them as soon as possible. Mechanical wind barriers may be used either temporarily or permanently in critical areas.

11–5 DISTURBED ALPINE AND ARCTIC SITES

Alpine sites are areas of high elevations named for the Alps mountains in southern Europe, but the term is used for other mountainous areas where the growing season is only about 60 days, winds are strong, and precipitation comes mostly as snow. Soils are usually thin, rocky, and infertile. *Arctic sites* occur at high latitudes where similar cold temperatures and short growing seasons prevail and the soils are shallow because they are underlain by permafrost. This includes Alaska, northern Canada, Greenland, and much of Siberia. Antarctica would also be included if it were not covered by glacial ice. Areas that are cold most of the year may nevertheless produce rapid growth during the summer because the days are long and the temperatures are favorable at that time. The annual precipitation may not be very much, but little moisture is lost during the cold weather, so the water supply is likely to be adequate or even equivalent to a humid condition. For example, central Alaska is classified as humid even though its annual precipitation is only about 4 to 10 in. (100 to 250 mm).

The soil and vegetative cover are fragile in both alpine and arctic sites. Fortunately, human activity is rare in such areas, so most of them remain in their natural state. Any disturbance that does occur may have long-lasting effects. Tracks from a single vehicle crossing an open area in Alaska, for example, are likely to remain visible for several decades. Disturbed alpine sites that require rehabilitation include areas overgrazed by sheep or goats, trails for hiking and recreation vehicles, areas where roads, pipelines, powerlines, or reservoirs have been built, and sites of mining and mineral exploration.

Figure 11–21 The route of the Alaskan pipeline from Prudhoe Bay on the Arctic coast to Valdez on the southern coast. (Courtesy Alaskan Pipeline Service Company.)

The native vegetation is sparse and includes relatively few species. Reseeding a disturbed area usually takes 3 to 5 years for the vegetation to become well established. Common native species are best for seeding. For example, in alpine sites in Colorado these include tufted hairgrass, alpine bluegrass, alpine timothy, spike trisetum, slender wheatgrass, and sedges (Brown and Johnson, 1981).

Arguments arose after oil was discovered in 1968 at Prudhoe Bay on the Arctic coast of Alaska. How could the oil be transported to market? Arctic ice makes the area unsuitable for transporting the oil by ships, so a pipeline across Alaska to the port of Valdez on the southern coast was proposed (Figure 11–21). Environmentalists successfully argued during the 1970s that a buried pipeline carrying oil warm enough to flow would melt the permafrost. The pipeline would then collapse and pollute the environment. The permafrost would not melt if the line was supported above the ground, but such a line could obstruct animal migration. The 789-mile (1270-km) pipeline was finally built on tall supports.

The Alaskan pipeline could not be built without disturbing soil and vegetation along its route. The areas disturbed by building the pipeline and its service road were stabilized by planting adapted cold-tolerant plants, including Garrison variety of creeping foxtail, white Dutch clover, alsike clover, and Nugget or Merion varieties of Kentucky Bluegrass.

SUMMARY

Drastically disturbed areas are usually localized spots that are exposed to rapid erosion as a result of either natural conditions or human activities. They are especially common around construction sites and mining operations. They are sources of muddy water, dust-laden air, and other pollution problems. Some of these sites need erosion control structures; virtually all of them need a good vegetative cover. Many of them have been compacted and need to be loosened with a chisel implement before vegetation will grow on them.

Construction sites produce both cuts and fills that may be highly erodible. Annual grasses are often suitable for temporary cover, and a debris basin is sometimes needed to catch runoff during the construction period. It is usually best to stockpile the topsoil so it can be replaced on the surface when construction is complete. The permanent cover should be attractive as well as erosion-resistant and can include many native and introduced species that are adapted to the climate and soil conditions.

Surface mines disturb more area than shaft mines, but both produce mine spoils that can cause environmental problems. Materials excavated from coal mines and some ore mines contain sulfides that produce sulfuric acid when oxidized. Such materials need to be buried deep enough to prevent oxidation. Restoration of mined areas is now legally required, and an Abandoned Mine Reclamation Fund has been established to reclaim sites damaged in past years. The topography needs to be smoothed, covered with good soil, and revegetated. Nitrogen and phosphorus fertilizers are commonly needed. Sewage sludge and other organic materials have proven useful for making it easier to establish vegetation on mine spoils.

Gullied land needs prompt attention. Gullies will usually continue to enlarge until they are stopped by diverting the water, smoothing the area, and establishing erosion-resistant vegetative cover. Mechanical drop structures are needed if the channel gradient is excessive. Grassed waterways are used to stabilize reclaimed gullies and in areas where water flow could cut a gully if they were left unprotected. Several tough sod-forming perennial grasses are useful for this purpose.

Streams naturally meander and erode their banks, but this process can be moderated by protecting the banks with well-placed structures and vigorous water-tolerant vegetation such as willows. Pondbanks may also need protection from the erosive action of waves.

Landslides and other forms of mass movement of soil and rock material typically occur on steep slopes when the material is very wet. Protection therefore involves either reducing the slope gradient or diverting enough water to avoid excessive wetness. Vegetative cover can help by transpiring water as well as anchoring the soil with its roots.

Sand dunes are most common along coastlines and in desert areas. They drift gradually downwind as long as a lack of vegetative cover exposes them to the wind. Irrigation can turn some sandy soils into productive cropland; the rest need permanent vegetative cover. Sands are droughty, and vegetation is usually best established by planting rather than seeding. The process should begin on the upwind side. Once established, the vegetation needs to be protected from excessive grazing and traffic.

Alpine and Arctic soils and their vegetative covers are very fragile. Any areas that are disrupted by road, pipeline, or powerline construction, mining, or other human activities need to be revegetated with some of the relatively few perennial plant species that can endure the long cold period and short growing season.

QUESTIONS

1. What are the most common problems that must be overcome when vegetating a construction site?
2. How can sediment be kept out of a nearby stream while construction is in progress on a site?
3. Why do mine spoils often produce strong acidity? What should be done about it?
4. Why do highwalls remain in many old mining sites but not in more recent sites where mining has been completed?

5. What are the five steps suggested for reclaiming a gully?
6. What should be done if the channel gradient at a site is too steep for a grassed waterway to provide enough protection?
7. What can be done to stabilize an eroding streambank? Who should do it?
8. Can landslides be prevented? If so, how?
9. Are sand dunes best stabilized by structures or by vegetation? Why?
10. Why was the Alaskan oil pipeline built high above the ground rather than buried?

REFERENCES

AGASSI, M., 1997. Stabilizing steep slopes with geomembranes. *Soil Tech.* 10:225–234.

BARRÈRE, P., 1992. Dynamics and management of the coastal dunes of the Landes, Gascony, France. In R. W. G. Carter, T. G. F. Curtis, and M. J. Sheehy-Skeffington (eds.), *Coastal Dunes: Geomorphology, Ecology and Management for Conservation.* Proc. Third European Dune Congress, Galway, Ireland, June 17–21, 1992. A. A. Balkema, Rotterdam, 533 p.

BROWN, R. W., and R. S. JOHNSON, 1981. Reclaiming disturbed alpine lands. *West. Wildlands,* 7(3):38–42.

CAMPBELL, R. H., R. L. BERNKNOPF, and D. R. SOLLER, 1998. *Mapping Time-Dependent Changes in Soil-Slip-Debris-Flow Probability.* USGS I-2586. 16 p. plus map.

CARLSON, C. L., and J. H. SWISHER, 1987. *Innovative Approaches to Mined Land Reclamation.* Published for Coal Extraction and Utilization Research Center, Southern Illinois University at Carbondale. Southern Illinois University Press, Carbondale, IL, 752 p.

CHU, S.-J. N., 2001. *Landslide and Slope Stability of Iowa.* Iowa State University thesis. 89 p.

COUSSOT, P., 1997. *Mudflow Rheology and Dynamics.* A. A. Balkema, Rotterdam, 255 p.

DANIELS, W. L., and D. F. AMOS, 1985. Generating productive topsoil substitutes from hard rock overburden in the southern Appalachians. In American Society for Surface Mining and Reclamation, 1984 National Meeting, *Symposium on the Reclamation of Lands Disturbed by Surface Mining: A Cornerstone for Communication and Understanding.* Science Reviews Ltd., Middlesex, England, p. 37–57.

DIECKHOFF, M. S., 1992. Propagating dune grasses by cultivation for dune conservation purposes. In R. W. G. Carter, T. G. F. Curtis, and M. J. Sheehy-Skeffington (eds.), *Coastal Dunes: Geomorphology, Ecology and Management for Conservation,* Proc. Third European Dune Congress, Galway, Ireland, 17–21 June 1992. A. A. Balkema, Rotterdam, 533 p.

DIKAU, R., D. BRUNSDEN, L. SCHROTT, and M.-L. IBSEN, 1996. Introduction. In R. Dikau, D. Brunsden, L. Schrott, and M.-L. Ibsen, (eds.), *Landslide Recognition: Identification, Movement and Causes,* John Wiley & Sons, Chichester, England. 251 p.

ENVIRONMENTAL PROTECTION AGENCY, 1975. *Methods of Quickly Vegetating Soils of Low Productivity, Construction Activities.* EPA-440/9-75-006, p. 149–166.

FAIRHURST, C., J. P. BRENNAN, T. V. FALKIE, L. A. GARFIELD, J. E. GILLEY, R. E. GRAY, W. N. HEINE, D. R. MANEVAL, J. D. MCATEER, and W. N. POUNDSTONE, 1975. *Underground Disposal of Coal Mine Wastes.* National Academy of Sciences, Washington, D.C., 172 p.

HERZOG, M., J. HARBOR, K. MCCLINTOCK, J. LAW, and K. BENNETT, 2000. Are green lots worth more than brown lots? An economic incentive for erosion control on residential developments. *J. Soil Water Cons.* 55:43–49.

HODDER, R. L., 1970. *Roadside Dry-Land Planting Research in Montana.* Montana State Univ., Bozeman, MT.

HUTCHISON, I. P. G., and R. D. ELLISON (eds.), 1992. *Mine Waste Management.* Lewis Pub., Boca Raton, LA, 654 p.

JORBA, M., and P. ANDRÉS, 2000. Effects of sewage sludge on the establishment of the herbaceous ground cover after soil restoration. *J. Soil Water Cons.* 55:322–327.

KLIJN, J. A., 1990. Dune forming factors in a geographical context. In T. W. Bakker, P. D. Jungerius, and J. A. Klijn (eds.), *Dunes of the European Coasts: Geomorphology-Hydrology-Soils.* Catena Suppl. 18, Catena Verlag, Cremlingen, Germany, 227 p.

KOHNKE, R. E., and A. K. BOLLER, 1989. Soil bioengineering for streambank protection. *J. Soil Water Cons.* 44:286–287.

LEISER, A. T., B. L. KAY, J. PAUL, J. J. NUSSBAUM, and W. THORNHILL, 1974. *Revegetation of Disturbed Soils in the Tahoe Region.* Final Contract Report, California Department of Transportation R. T. A. 12945–191210.

LONGGOOD, W. F., 1972. *The Darkening Land.* Simon and Schuster, New York, 572 p.

MALIK, A., and J. SCULLION, 1998. Soil development on restored opencast coal sites with particular reference to organic matter and aggregate stability. *Soil Use and Management* 14:234–239.

MCCOY, E. L., 1998. Sand and organic amendment influences on soil physical properties related to turf establishment. *Agron. J.* 90:411–419.

MCLACHLAN, A., and M. BURNS, 1992. Headland bypass dunes on the South African coast: 100 years of (mis)management. In R. W. G. Carter, T. G. F. Curtis, and M. J. Sheehy-Skeffington (eds.), *Coastal Dunes: Geomorphology, Ecology and Management for Conservation,* Proc. Third European Dune Congress, Galway, Ireland, 17–21 June 1992. A. A. Balkema, Rotterdam, 533 p.

NEUFELD, R. D., 1990. Coal refuse to energy. In EPA/600/9-90/039 *The Environmental Challenge of the 1990s,* p. 445–453.

PLOTKIN, S. E., 1987. The Office of Technology Assessment's review of prime farmland reclamation under the 1977 Surface Mining Control and Reclamation Act. In C. L. Carlson and J. H. Swisher (eds.), *Innovative Approaches to Mined Land Reclamation.* Southern Illinois Univ. Press, Carbondale and Edwardsville, IL, p. 31–58.

POWER, J. F., R. E. RIES, and F. M. SANDOVAL, 1978. Reclamation of coal-mined land in the Great Plains. *J. Soil Water Cons.* 33:69–74.

SANFORD, F. H., 1916. *Michigan's Shifting Sands: Their Control and Better Utilization.* Mich. Agr. Expt. Sta. Spec. Bull. 79, 31 p.

SUTTON, P., and J. P. VIMMERSTEDT, 1973. *Treat Stripmine Spoils with Sewage Sludge.* Ohio Rep. 58:121–123. Ohio Agricultural Research and Development Center, Wooster, Ohio.

U. S. CONGRESS, 1977. *Surface Mining Control and Reclamation Act of 1977.* Public Law 95–87, Washington, D. C.

VAN BOHEMEN, H. D., and H. J. N. MEESTERS, 1992. Ecological engineering and coastal defence. In R. W. G. Carter, T. G. F. Curtis, and M. J. Sheehy-Skeffington (eds.), *Coastal Dunes: Geomorphology, Ecology and Management for Conservation,* Proc. Third European Dune Congress, Galway, Ireland, 17–21 June 1992. A. A. Balkema, Rotterdam, 533 p.

CHAPTER

12

Pastureland, Rangeland, and Forestland Management

Erosion rates are normally low on well-managed pastureland, rangeland, and forestland because good managers maintain an adequate cover of pasture grasses and legumes, range grasses, or forest trees and shrubs on their land. These lands have close-growing perennial vegetation that intercepts rain drops, thus absorbing impact energy and decreasing soil dispersion and splash and sheet erosion. The velocity of surface water flow is reduced by plant stems and residues. The result is less soil erosion and sediment, clean water flowing slowly along the soil surface, and more infiltration of water into deep soil horizons. Vegetation also stabilizes soil against wind erosion by the plant roots anchoring the soil and by aboveground growth and residues decreasing the wind velocity at the soil-atmosphere interface. Furthermore, the accumulation of organic materials in trees, grasses, and plant residues is credited with reducing global warming by sequestering carbon that would otherwise accumulate as carbon dioxide in the atmosphere.

Of course, the favorable effects of good vegetative cover are negated when the vegetative cover is removed by excessive grazing or torn up by machines or other human activity. In fact, grazing and forest lands are commonly so steep, shallow, sandy, or otherwise limited in productivity that their cover is fragile and easily damaged. Damaged cover may leave the land susceptible to rapid runoff and severe erosion by either water or wind.

The effectiveness of vegetation in increasing infiltration of rainfall, and thereby decreasing runoff and erosion, is illustrated in Figure 12–1. Soils used for well-managed pastures and meadows have more rapid infiltration than similar soils used for clean-tilled crops. However, heavy grazing can cause bare areas in pastures that, like bare areas in cropland, have reduced infiltration rates and increased runoff and erosion.

12–1 PASTURELAND, RANGELAND, AND FORESTLAND

Pastureland usually refers to an intensively managed grazing area in a humid climate or to similar irrigated areas in an arid or semiarid climate. Pastureland supports forage such as native or improved grasses or a mixture of grasses and legumes. Pasture may be either con-

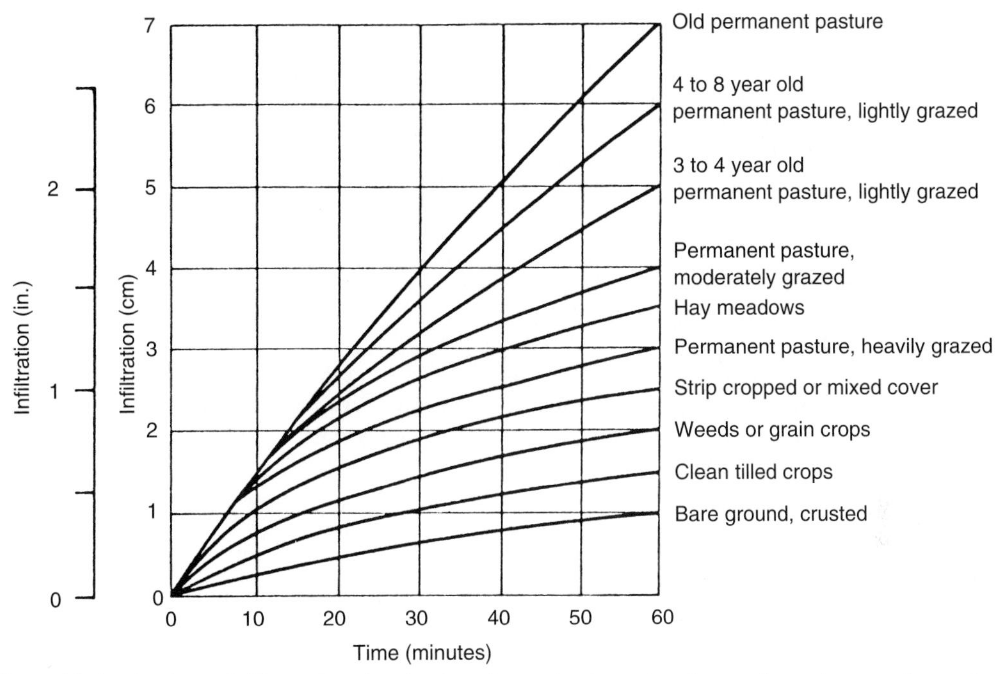

Figure 12–1 A schematic illustration showing how increasing intensity of use reduces the rate of water infiltration into soil. (Courtesy USDA Agricultural Research Service, Hart, 1974.)

tinuous or rotated periodically with other crops. Some pastures occupy areas that suffered severe erosion during a period of cropping and were converted to pastureland to reduce the rate of erosion. About 9% of U.S. pastures are irrigated during normal years.

Rangeland refers to unplowed areas where native grasses, forbs, shrubs, and trees are used for forage. Some ranges may be fertilized and reseeded to native or exotic grasses, but they are seldom or never plowed.

Forestland is an area with growing trees or of soil capable of supporting trees. A tree is any woody plant that is at least 15-ft (4.5-m) high supported by a single stem. Table 12–1 shows the area of pastureland, rangeland, and forestland in the United States.

The general distribution of pastureland, rangeland, and forestland can be inferred from the map of the native vegetation of the 48 conterminous states shown in Figure 12–2. Most of the pastures are in areas that originally grew either forests or tall grasses. Present rangeland for domestic livestock includes most of the areas shown as shortgrass, mesquite-grass, sagebrush, creosote bush, and arid woodland. A large percentage of the forestland of southeastern United States is also used as rangeland. Most present-day forestlands occupy parts of the areas identified in the map as having had forest as their native vegetation, especially those parts that are mountainous or otherwise unsuitable for cropland. Much of the forest has been harvested one or more times and either replanted or allowed to regenerate

TABLE 12–1 PASTURELAND, RANGELAND, AND FORESTLAND IN THE 50 UNITED STATES

Land use	Total area (million acres)	Total area (million ha)
Pastureland	133.3	53.9
Rangeland	405.9	164.3
Forestland	393.8	159.4
Total pasture-, range-, & forestland	933.0	377.6

Source: National Research Council, 1986a.

naturally. Some extensive areas of forestland in eastern United States occupy areas that were once cropped but later abandoned or replanted to trees.

12–2 PASTURELAND MANAGEMENT

Pastureland management involves the production of forage and its utilization by livestock. High production requires an agronomic approach whereby the manager uses pasture species suited to the soil and climate of the area and applies whatever agronomic practices are needed to make them productive. In the humid part of the United States, important pasture grass species include bluestem, bermudagrass, lovegrass, annual ryegrass, dallisgrass, johnsongrass, and rhodesgrass (Figure 12–3). Lime and fertilizers should be applied according to soil test recommendations. Tillage operations can be used when needed to improve the quality of the pasture. All of these activities need to be coordinated with the livestock specialist to optimize livestock production. Stocking rates that are too low limit production by not utilizing enough of the forage; stocking rates that are too high also limit production because overgrazed plants lack the size and vigor needed to produce well and the livestock do not gain well without adequate forage. Optimum usage depends on the plant species, but the simple rule of using half of the plant growth is a good beginning point.

On average, pastures in the United States are well managed for erosion control. Soil erosion by water averages 1.4 tons/ac (3 mt/ha) annually, and erosion by wind is near zero. However, streambank trampling and erosion can be a problem where a stream flows through a pasture. Such areas usually should be fenced to keep livestock away from the streambanks. Leaving an ungrazed filter strip alongside the stream is a good way to reduce water pollution by animal droppings and runoff from the pasture. On irrigated pastures, this should be coupled with careful control of the water to limit runoff to a small amount (Tate et al., 2000).

12–2.1 Fertility Management of Pastures

Soil tests can identify needs for fertilizers and soil amendments. Lime and other soil amendments applied when needed help make essential plant nutrients more available, avoid toxicities, and ensure that plants can utilize nutrients efficiently. On acidic soils, the application of lime will reduce acidity, lower the exchangeable aluminum, and make soil

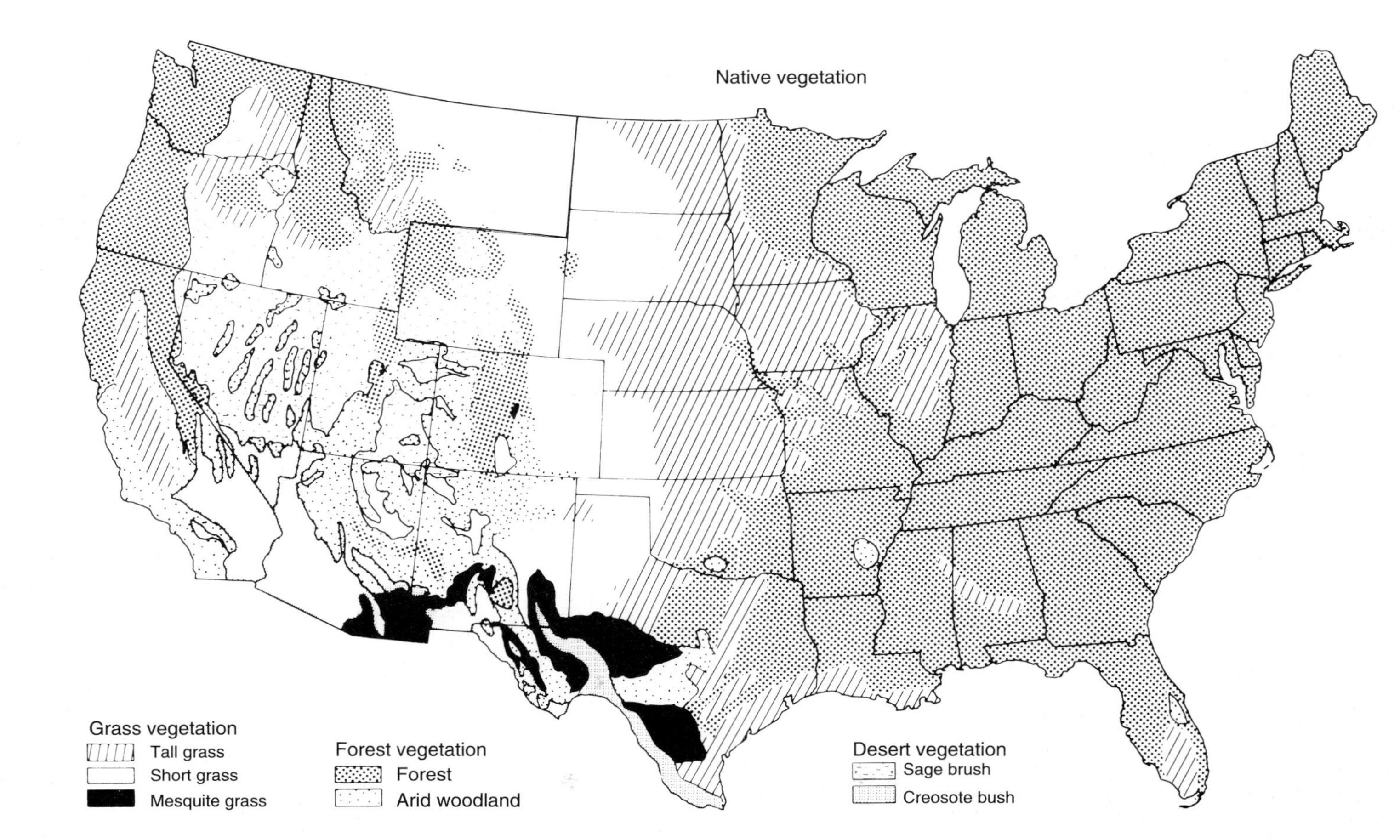

Figure 12–2 Native vegetation in the conterminous 48 states. (Courtesy Raphael Zon, USDA Forest Service, and H. L. Shantz, USDA.)

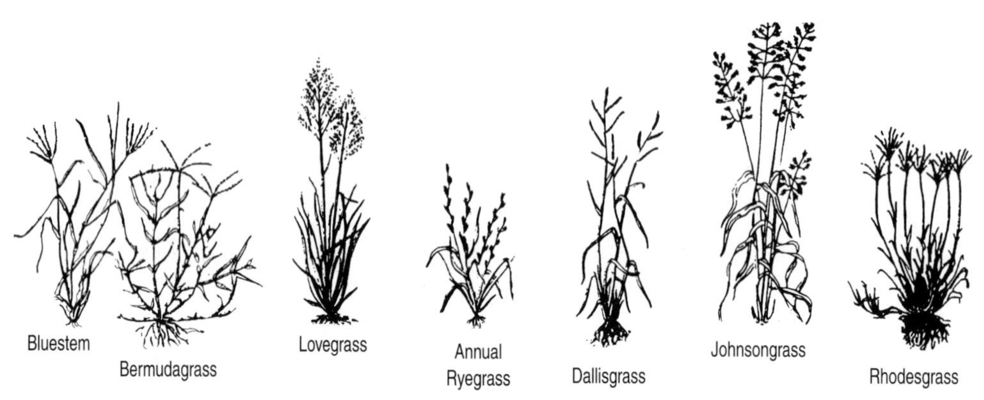

Figure 12–3 Some important pasture grasses in the humid U. S. environment. (Courtesy Instructional Materials Services, Texas A&M University.)

phosphorus more readily available. Excessive alkalinity caused by high sodium content may be corrected with gypsum.

Nitrogen is the nutrient most likely to limit the growth of pasture grasses. However, excessive applications cause luxury consumption and lead to a large increase in top growth but not in root growth. High rates of nitrogen fertilizers favor grasses in a legume-grass mixture and decrease atmospheric nitrogen fixation by *Rhizobium* bacteria. Even moderate rates of N fertilizer on sandy soil may leach into the water table and increase the nitrate (NO_3) level (see Chapter 17).

Phosphorus is often the second most critical nutrient in pasture production and, indirectly, in controlling soil erosion. Phosphorus fertilizer is most efficiently applied by mixing or injecting it into the soil before or at the time of seeding. As much as a three-year supply of phosphorus can often be applied at this time because very little will leach from the soil. Phosphorus fertilizer applied later as topdressing moves downward into the root zone very slowly.

Potassium is required on many pastures, especially in humid regions and on some sandy soils even in semiarid areas. Plants will absorb more potassium than they need (luxury consumption) when it is applied in surplus. Fortunately, much of the excess potassium is contained in plant residues and manures that return to the soil.

Low available soil magnesium or surplus plant absorption of K^+ or NH_4^+ may cause a decrease in plant uptake of magnesium. When forages contain less than 0.20% magnesium (Mg) and cattle blood serum is less than 1.5 mg Mg^{2+}/100 ml, grass tetany (hypomagnesemia) usually results. This nutritional disease is often fatal to cattle.

Manure dropped in pastures by grazing animals is a relatively efficient means of returning some of the nutrients consumed by the animals (Franzluebbers et al., 2001). Additional nutrients can be supplied by spreading manure from barns and feedlots, other organic sources such as sewage sludge, or mineral fertilizers. The amounts and times of application, however, must be reasonable to comply with pasture plant requirements as well as to ensure environmental integrity (Chapter 16).

12–2.2 Grazing Management of Pastures

Good grazing management is a very important part of pasture management, but it is often neglected. Excessive grazing or grazing too early in the season can severely damage the most desirable forage species in a pasture and cause serious soil compaction. Moderate grazing is generally sustainable, but heavy grazing is not. Wienhold et al. (2001) reported that a North Dakota pasture under moderate grazing produced a 10-yr average of 4260 kg of forage per hectare and the animals gained an average of 120 kg per hectare annually, whereas a similar pasture under heavy grazing produced an average of only 2030 kg of forage per hectare and the animals gained an average of 42 kg per hectare annually. The heavily grazed pasture suffered from soil compaction and a decrease in the quality of the forage as the species composition shifted toward the least palatable grasses.

The principal grazing management systems are continuous, rotational, and deferred grazing. These same systems are also used on rangeland.

Continuous grazing means placing livestock on a pasture and allowing them to graze there throughout the grazing season. When the stocking rate is low to moderate, this system is suitable for such grass species as bermudagrass, Kentucky bluegrass, pangolagrass, perennial ryegrass, and tall fescue. There is a tendency, however, to graze the pasture with the same stocking rate throughout the grazing season with the result that it is underused during the spring flush of growth and overgrazed during dry weather and dormant periods. Overgrazing causes pasture productivity to decrease and soil erosion and sediment yield to increase.

Rotational grazing means placing fences so the livestock can graze only part of the pasture at a time. This grazing system is best adapted to grasses such as smooth bromegrass and intermediate wheatgrass. Such grasses should not be grazed closer to the ground than about 4 in. (10 cm). This system can produce more total pasturage per unit of land and better utilization of forage than that resulting from continuous grazing. The pasture grows more vigorously when it is not being grazed, and restricting the animals to a smaller area at any one time causes them to graze forage that might otherwise be wasted. Also, different subdivisions can be planted to grasses or grass mixtures with different seasons of growth. For example, the grazing season can be extended by using a cool-season grass such as tall fescue or crested wheatgrass in one pasture subdivision for spring and fall grazing. Another subdivision could be seeded to a warm-season grass such as bermudagrass or switchgrass for midsummer grazing. Surplus switchgrass can be harvested for biomass as an energy source (Sanderson et al., 1999). Alternate areas such as grassed waterways, harvested cropland, or mountain pastures are sometimes included in the rotation at appropriate times. Some pastures are rotated on a daily basis by means of movable electric fences. Daily pastures with many animals in a small area cause the livestock to utilize the vegetation more completely rather than choosing only the most succulent growth. This technique reduces the bloat hazard of grazing legumes such as alfalfa.

Deferred grazing entails delaying the start of grazing beyond the normal beginning of the grazing period. This delay permits the grass species that are likely to be damaged or crowded out by grazing to become more vigorous and in some cases to produce seed. Sometimes grazing is deferred until after a fall freeze. Such "frosted pasturage" is often ideal when other forage is scarce or expensive. Deferred grazing and rotational grazing work well in combination. The livestock can graze in one pasture while another is being deferred. Different pastures can be deferred each year in a rotation sequence that maintains productive vegetation in all of them.

12–2.3 Pasture Renovation

Most perennial pastures deteriorate at times because of overgrazing, soil compaction, drought, insects, diseases, a decrease in percentage of legumes, or a depletion of one or more essential plant nutrients. On sloping pastures, deterioration is usually accompanied by soil erosion and sediment transport that pollutes surface waters. The solution is pasture renovation followed by good grazing management.

Pastures should be renovated when their productivity declines to between 50 and 75% of their potential. A small-grain crop is sometimes grown before reseeding pasture if the soil is suitable, but pasture is reseeded immediately where the erosion hazard is high. Either way, a suitable seedbed must be prepared, preferably with crop residues left as a protective mulch.

Renovation of a pasture is usually accomplished by chiseling or disking on the contour, sometimes applying herbicides to kill existing vegetation, adding lime and fertilizers as indicated by soil tests, and then reseeding with the pasture mixture recommended for that specific soil series and the intended season of use. Livestock must be kept off the renovated pasture until the new vegetation is well established.

A newer technique to reduce soil erosion during pasture renovation includes heavy grazing of the existing forage followed by the use of a contact herbicide and then seeding the pasture mixture directly into the dead sod.

In the southern United States, cool-season legumes such as crimson clover may be seeded into a warm-season grass pasture. In the northern United States, legumes such as white clover, alfalfa, birdsfoot trefoil, and crownvetch can be seeded in this manner.

12–3 RANGELAND MANAGEMENT

The native vegetation on rangeland is predominantly grasses, grass-like plants, forbs, or shrubs suitable for grazing or browsing (Figure 12–4). Rangeland includes natural grasslands, savannas, shrublands, open woodlands, deserts, tundra, mountain (alpine) meadows, coastal marshes, and wet meadows (Wright and Siddoway, 1982). Nearly half of the land on Earth can be classed as rangeland (Menke and Bradford, 1992). Most of it is either unsuitable or of low quality for use as tilled cropland because it includes steep areas, shallow and/or stony soils, or dry and/or cold climates.

People use rangeland for raising livestock. Much of it is less productive than it could be because the native vegetation has been replaced with inferior species where the land has been overgrazed. Large areas of rangeland that once were shortgrass prairies in the western United States are now dominated by sagebrush and an introduced annual grass known as cheatgrass. The name "cheatgrass" tells how its value compares to the native perennial bunchgrasses that it replaced.

The range manager should take an ecological approach, using mostly native plants adapted to the soil and climate, and managing kinds and numbers of livestock to encourage optimum growth of the most desirable forage species. The control of soil erosion and sedimentation is of immediate concern to the range manager because water and wind erosion reduce the forage produced for the range livestock. Sedimentation cannot be ignored be-

Figure 12–4 Rangeland in eastern Texas. The dominant vegetation in this photo is little bluestem grass, a species that is well adapted to both humid and semiarid rangelands. (Courtesy USDA Natural Resources Conservation Service.)

cause it reduces the storage capacity of watering ponds after each heavy rain. Since the range manager usually cannot directly control erosion, the best recourse is to control it indirectly by limiting the rate of stocking so that adequate vegetative cover is always present. Vegetative cover damaged by overgrazing is likely to require years to recover even if livestock are removed from the area. Greenwood et al. (1998) tested the soils of Australian rangeland grazed by sheep and found that seven months was not enough time but that soils that had not been grazed for two-and-a-half years had recovered their normal density and permeability. The length of time required for recovery would be expected to vary considerably according to how much or how little remains of the desirable plant species.

On privately owned rangelands in the United States, wind erosion is estimated to average 1.5 tons/ac-yr (3.4 mt/ha-yr) and water erosion about 1.4 tons/ac-yr (3.1 mt/ha-yr) (National Research Council, 1986b). The beneficial effects of proper grazing management can be seen in Figure 12–5.

In addition to wind and water erosion, some rangelands have steep slopes that are susceptible to landslides. For example, periods of prolonged rainfall make landslides a major problem on pastoral hillslopes in New Zealand (Hawley, 1991). Investigation showed that overgrazing makes the area more susceptible to landslides. The weakened vegetation has fewer deep roots to serve as anchors, and the soil is more likely to become saturated because sparse vegetation uses less water.

12–3.1 Grazing Management of Rangelands

Although rangelands include much of the open pine and hardwood lands of humid regions, the principal areas are in the arid and semiarid regions. Rangelands in these drier regions

Figure 12–5 Deteriorated, private rangeland on alkaline soils in Texas (left) where a sparse growth of buffalograss has replaced the bluestems, and high productive bluestem grasses on rangeland managed by the U.S. Forest Service (right) in this 32-in. (800-mm) precipitation zone. (Courtesy USDA Natural Resources Conservation Service.)

include tall grass prairies, short grass plains, and semidesert areas supporting sparse bunchgrasses, forbs (broadleaved, herbaceous plants), and woody shrubs.

The following range management practices increase the amount and efficiency of use of range forage and protect the soil against wind and water erosion, as shown in Figure 12–6.

1. Seed improved grass cultivars.
2. Delay spring grazing until the grasses have a good start and the soil is dry enough to avoid trampling damage.
3. Graze the range simultaneously with cattle and sheep.
4. Adjust the stocking rate to the growth rate of grass as affected by season and precipitation.
5. Leave about 50% of all range forage for reserve and residue. In the words of the best ranchers, "graze half and leave half."
6. Integrate range grazing with the grazing of irrigated pastures.
7. Practice rotational and deferred grazing. This usually requires building more cross fences, establishing more water facilities, as in Figure 12–7, and adding more salt boxes.
8. Move salt periodically to undergrazed areas away from the water, so cattle will graze the range as uniformly as possible. Cattle need both salt and water, but not together. Beef cattle can walk a mile or more for a drink, and they need only one drink every second day.
9. Clear brush from ranges to encourage the growth of desirable forage grasses.

(A)

(B)

Figure 12–6 Stocking rates must be adjusted to the rate of growth of the grass, according to the season and to the precipitation, as shown by these semiarid rangelands in California. (A) Overgrazed rangeland with benchlike slumping caused by cattle trails and weakened grasses. (B) Well-managed rangeland with abundant grass and stable soils. (Courtesy Agricultural Experiment Station, University of California.)

Figure 12–7 Good grazing management includes establishing enough fences and water facilities to permit rotational and deferred grazing. A cowboy noticed wild iris growing here and realized it meant wetness. This small pond, dug with a bulldozer, provides livestock water throughout the grazing season. (Courtesy U.S. Department of Agriculture.)

Proper grazing management will limit wind and water erosion to acceptable levels (T-values), except in areas where conditions are too severe for treatment to be feasible. One such area is the Sonoran Basin, a desert area in southern California, western Arizona, and southern Nevada. Mean annual precipitation is 2 to 4 in. (50 to 100 mm). Although water erosion is negligible, wind erosion is severe (National Research Council, 1986b).

12–3.2 Reseeding Rangeland

Reseeding is a less favorable option for rangeland than it is for pastures because of the generally lower productivity and relatively low intensity of use on rangeland. Also, many areas of rangeland are too steep, too stony, or too shallow to be tilled for reseeding purposes. Nevertheless, the vegetative cover on some of the areas with favorable soil has been so seriously damaged by overgrazing that reseeding is the best option.

Often it is best to reseed rangeland with grasses and legumes that are native to the area. However, there are several problems that may occur. Seed is not always available commercially, and gathering it from native plants is expensive. Many of these species produce only low yields of seed and the germination rate is also likely to be low. Consequently, an introduced species may be chosen. In western United States, this is usually crested wheatgrass (Broersma et al., 2000) because it grows well from seed, its seed is readily available, it has a wide adaptation, it yields well, and livestock grazing on it produce good weight gains. Crested wheatgrass is a cool-season grass that was introduced from the steppe region of Russia in 1898 and again in 1906 (Wienhold et al., 2001).

Rangeland areas in western United States have dry summers, so reseeding is done when there is more rainfall, usually in late fall or early spring. Tillage, fire, or chemicals are used to eliminate undesirable vegetation. Often the area is disked to prepare a seedbed, but drills are now available that can seed into untilled soil. Minimizing soil disturbance and leaving residues on the surface are desirable practices for limiting erosion during the establishment period. Fertilizer should be applied according to local recommendations and soil tests.

Livestock should be kept away from reseeded rangeland until the new vegetation is well established. After that, normal good management is needed to maintain the vegetation. It will usually take several years to repay the cost of reseeding.

12–3.3 Burning Rangeland Grasses

Much native grassland in western United States was burned during the settlement period to control sagebrush that had increased dramatically when the land was overgrazed. This led to invasion by annual grasses such as cheatgrass. Livestock will graze on cheatgrass, but its productivity is variable from year to year and seldom is as high as that of the native grasses. Some of these areas are now dominated by less palatable annual species such as medusahead. Such areas are well protected from erosion because the medusahead forms a thick cover and its residues accumulate on the ground—livestock avoid eating it.

In the Great Plains, annual burning of the little bluestem and big bluestem range increases the quality of forage for livestock. Reasons cited in favor of burning include weed control, insect control, earlier growth, more total growth, and improved forage quality. Opponents of range burning claim it increases air pollution and soil erosion and is poor aesthetically.

Hyde and Owensby (1970) cite results of a 20-year period of range burning in the 30-in. (760-mm) annual precipitation belt in Kansas. They concluded that burning a bluestem range when the bluestem grasses start to grow, usually about May 1, results in the greatest gain of yearling beef cattle, 104 lb/ac (117 kg/ha) versus 94 lb/ac (105 kg/ha) for the unburned range. This is true despite the fact that the unburned range produced the most total forage. The reason probably lies in the composition of the forage. Burning favors the bluestem grasses over other species that increase or invade with overgrazing and fertilization (Owensby and Smith, 1979). Environmental damage from range burning must be weighed against the 10 lb/ac additional gain in beef each year.

Large areas of flat, wet grassland in the Pampas of Argentina are commonly burned in the winter. The results are highly favorable; the paspalum grass that is dominant there recovers quickly after the fire and sometimes produces more than twice as much forage as that on the unburned range (Sakalauska et al., 2001). The increased production appears to be related to increased water use and a lowering of the high water table that prevails in the Pampas. Plants in the burned area produce more roots, tiller more, and produce higher quality forage than those in the unburned area. There is, however, some invasion by "opportunist" grass species that probably produce lower quality forage.

Figure 12–8 A USAID specialist in Somalia (eastern Africa) helping a Somalian raise tree seedlings to control gullies, improve the environment, and produce firewood. (Courtesy USAID.)

12–4 FORESTLAND MANAGEMENT

Trees sequester more organic carbon and release more molecular oxygen per unit area than other plants. Trees are therefore the most effective vegetation to improve the environment by neutralizing the "greenhouse effect" and reducing global warming (Brown, 1993). However, it would take millions of acres of new forest trees annually to absorb and store the carbon dioxide produced by the combustion of all the fossil fuels (coal, petroleum, and natural gas) now being used for heat, transportation, production of electricity, etc. Tree plantings (Figure 12–8) are favorable for environmental protection because they fix carbon, slow runoff, and reduce erosion. Conversely, rapid clearing of forestlands is considered ecologically damaging because these environmental benefits are lost. Forest fires also cause great damage by converting sequestered carbon into carbon dioxide, clouding the atmosphere with soot, and exposing the soil to erosion.

Much of the erosion on forestland is associated with the roads built to provide access to the area and to remove logs and other products. Grace (2000) reported that up to 90% of the sediment from forested areas comes from the roads, especially from sloping roadbanks. Temporary roads built by a bulldozer are commonly used for logging operations and then left without maintenance. Gullies often form either in or alongside such roadways.

Some roads are needed in forests, but they should be carefully placed, either on ridges where there is little water accumulation, in valleys where the slope is gentle, or on a gentle gradient with adequate water controls where they must cross a hillside. The roadbanks and other disturbed areas should be vegetated and generally need to be mulched until a vegetative cover is established (see Section 11–1.2 in Chapter 11).

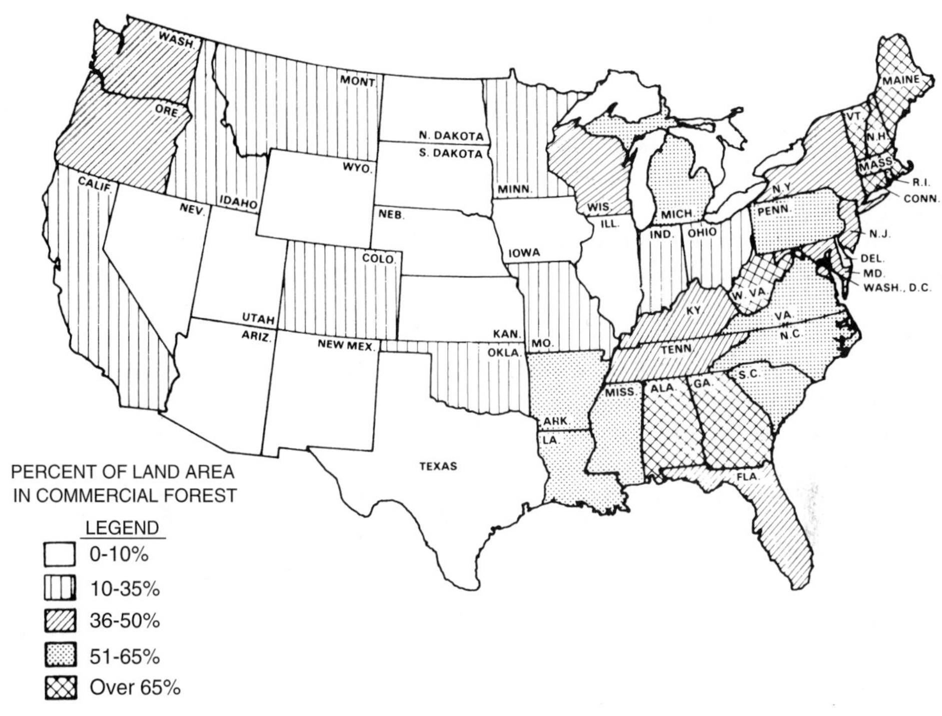

Figure 12–9 Commercial forestland in the conterminous United States. (Courtesy USDA Forest Service.)

Many people in developing countries depend on wood for fuel. They often spend hours collecting trimmings and fallen branches from trees and brush in the forests and along roads. Many of them cook with stoves that are inefficient, so there is a large potential for saving fuel by introducing more efficient stoves (Orians and Millar, 1992).

Forests grow mostly in humid regions; however, the term "forestlands" includes many semiarid areas. The definition of forestland requires only 10% of the area to be occupied by trees. Total area in the United States so classified is 393.8 million acres (159.4 million hectares) distributed as shown in Figure 12–9.

12–4.1 Forests and Watersheds

Well-managed forests are unsurpassed as vegetative cover for watersheds. Leaves, branches, and the organic litter layer on the forest floor break the velocity of falling raindrops. Rain may slowly infiltrate through these layers or be held by the spongy leaf litter and fine roots. Some water moves down branches and trunks and infiltrates into the soil along root channels. The net result is that a properly managed forest has more infiltration and less runoff, erosion, and sedimentation than it would have as cropland. Furthermore, slow, deep seepage gives rise to many springs and to a three- to five-day delay in flood crests.

Forest trees are excellent protectors of the watershed but only after a uniform cover of at least 2 in. (5 cm) of leaves and/or needles has accumulated and remains on the soil surface throughout the year. It normally takes at least four or five years for a new tree planting to produce enough litter layer to minimize soil erosion. After that, the litter layer needs to be preserved to protect the forest soil from erosion. Much forestland has both steep slopes and enough intense rainfall to quickly cut gullies where the litter layer is damaged (Elliot et al., 1999).

America had abundant forest when it was colonized by Europeans. Since then, large areas have been cleared for other uses, especially for agricultural cropland. Some of this land, especially land with acidic soils in the northeastern United States, has since been reforested, either by replanting trees or by abandonment and natural regrowth. Attitudes and government policies regarding forestland have changed over time. Hamdar (2000) divides U.S. public land policy into four periods: acquisition, disposition, reservation, and management, explaining that land was acquired in the early years of the nation, then disposition occurred as land was made available to individuals (e.g., by homesteading), and after that, areas were reserved for national forests, parks, and monuments. Such transfers are now mostly past, and the land is now considered to be in the management stage, with attention being given to beneficial use of forestland to produce wood products, provide wildlife habitat and recreation, and to protect watersheds from excessive runoff and erosion.

12–4.2 Harvesting Methods and Erosion Potentials

Harvesting forest trees always makes the soil less productive. Heavy logging machinery compacts the soil, thus increasing its bulk density and soil strength, decreasing water infiltration, and decreasing growth of the next generation of trees by 5 to 15%. Furthermore, from 5 to 10% of the forest area must be dedicated to haul roads, thereby decreasing the area of tree growth and causing erosion on exposed sloping soils. Between the roads, skid trails lay bare up to 10% of the forest tract being logged. Such trails can cause serious erosion, soil compaction, and loss of soil productivity (Startsev and McNabb, 2000). The loss of productivity is likely to last for decades, but may be reduced by limiting the amount of traffic in the forest, lifting logs rather than dragging them, and by using wide tires and low air pressure to reduce compaction.

A large amount of "slash" is left behind when trees are harvested. *Slash* includes the branches that are trimmed from the trees, plus the brush and small trees that are cut to provide access to the trees being harvested. In developing countries, much of the slash may be picked up and used for fuel. In some places, it is piled in rows that provide habitat for small animals and birds. Often it is burned in place after the logs have been hauled from the area. The burning of slash in the forest is detrimental to soil productivity. Hot slash fires make some soils *hydrophobic* (resistant to wetting), resulting in less infiltration and more runoff and erosion. Slash burns expose the darker soil surface, which absorbs more heat from the sun. Tree seedlings on the bared soil, therefore, may be killed by excessive heat. Controlled burning reduces the hazard of wildfire, but it also pollutes the atmosphere, removes wildlife habitat, and increases the erosion hazard (Orians and Millar, 1992). Ash from slash burns may be a beneficial fertilizer in reasonable amounts, but large concentrations of ash could increase surface soil pH enough to reduce availability of P, Fe, Mn, Cu, and Zn (Ballard, 1988; Froehlich, 1988; and Megahan, 1988).

Studies by the Southeastern Watershed Research Laboratory at Tifton, Georgia, have shown that streamside trees effectively trap sediment and nutrients that could contaminate surface waters. Trees filtered out an average of 19.4 tons of sediment/ac-yr (43.5 mt/ha-yr) that was deposited outside the stream. The streamside trees also absorbed substantial amounts of N, P, K, and Mg (Lowrance and Leonard, 1988). Care should be taken to preserve these environmental benefits when harvesting trees near a stream.

The amounts of bare soil and sediment produced by a variety of pine timber-harvesting methods were compared in eastern Texas. The traditional method is to clear cut the marketable timber, kill hardwoods by knocking them down with a bulldozer, pile up the slash and burn it, and then replant pine. For comparison, clearcutting was followed by a large, sharp chopper-roller. The chopper exposed about one-fourth as much bare soil and resulted in less than a tenth of the sediment yield of the bulldozed areas (DeHaven, 1983).

There are four principal methods of harvesting trees: *selection, shelterwood, seed-tree,* and *clearcutting.* Each method has its own advantages and effects on erosion potential and the general environment as follows.

Selection. The selection method consists of removing individual trees or small groups of trees as they mature. The result is a stand of trees with all ages intermixed. It is adapted to tree species that will reproduce satisfactorily under severe competition for soil moisture, nutrients, and light. Less tolerant tree species need larger openings in the forest canopy to favor their reproduction. The only significant erosion potential resulting from the selection method of harvest is that caused by heavy traffic to remove the harvested trees.

Shelterwood. The shelterwood system removes all mature trees in a series of harvests several years apart. Heavy-seeded species, such as oaks, usually reproduce well under the partial shade resulting from this system. The resulting forest stands are nearly even-aged, yet there is always an overstory to shelter the site. Very little erosion and sedimentation result from the shelterwood method.

Seed Tree. The seed-tree method leaves only enough trees to bear seed for natural regeneration. The seed trees may be harvested after a new, even-aged forest is established. It is applicable to trees such as the southern pines (loblolly, longleaf, shortleaf, and slash), which have light seed that can be borne by the wind. The potential for soil erosion and sediment pollution is great for a few years after harvest.

A version of the seed-tree method of harvest was used in Europe for centuries in a system known as *coppice with standards* (Orians and Millar, 1992). In this system, a few trees of selected species are chosen as "standards" and allowed to grow to large size so they can ultimately be made into large timbers. The remaining trees are "coppiced," that is, a thick stand of trees is encouraged, and when the trees reach a moderate size, they are cut back to tall stumps. The trimmings are used for cooking, fence posts, and small construction projects. The coppiced trees are allowed to regrow, and the process is repeated several times, whenever the new branches reach adequate size. This system nearly died out in the 20th century as coal replaced much of the wood used for cooking. There is current interest in reviving it, because the coppiced woodlands produce a wide variety of wildlife as well as trees.

Clearcutting. Clearcutting removes all the trees from the logged area. The purpose is to clear the area to establish a new, even-aged stand of a valuable, fast-growing species that will not reproduce satisfactorily in the shade of other trees. It is often used for Douglas fir in the Pacific Northwest, for certain species of pine, and for black cherry in the Allegheny Mountains. The clearcut area may consist of patches, strips, or an entire watershed. It has the advantage of permitting efficient use of high-lead, skyline, balloon, or helicopter logging

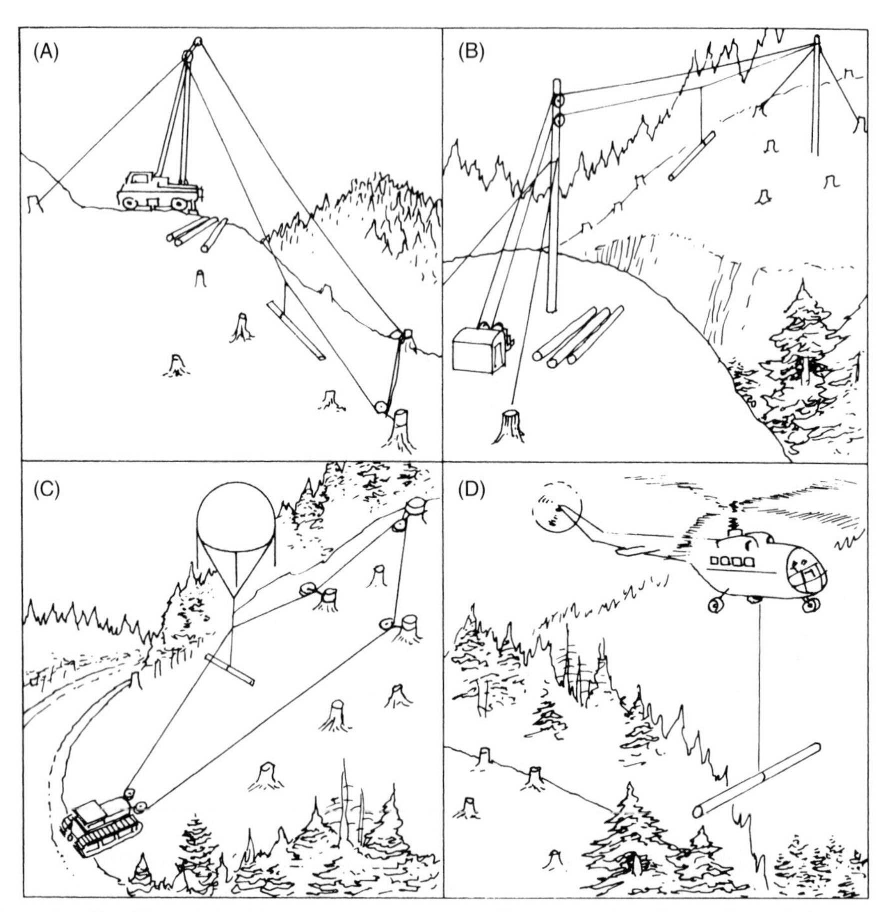

Figure 12–10 The principal log transport systems used to remove logs from a forest being harvested are (A) high-lead cable, (B) skyline cable, (C) balloon cable, (D) helicopter, and (not shown) tractor logging.

systems. Clearcutting is strongly opposed by many environmentalists because it causes a severe erosion hazard until a new stand of trees is well established. This hazard is especially great when clearcutting is combined with systems that create skid trails, such as tractor logging.

12–4.3 Log Transport Systems and Soil Disturbance

The principal log transport systems include logging by tractor, highlead cable, skyline cable, balloon cable, and helicopter, as shown in Figure 12–10. In general, the systems that cause the least soil disturbance are the most expensive. Typical soil disturbance in the northwestern United States caused by each of the log transport systems, except helicopter, is presented in Table 12–2. Tractor logging causes more bare soil and more soil compaction than any other log transport system. Logging by balloon or helicopter causes the least soil disturbance. Balloon logging causes 83% less bare soil and 94% less compacted soil than tractor logging. Soil disturbance by the helicopter system would be about the same as that for balloon logging.

TABLE 12–2 EFFECTS OF LOG TRANSPORT SYSTEMS ON SOIL DISTURBANCE

	Soil disturbance[a]	
Log transport system	Logged watershed with bare soil (%)	Logged watershed with compacted soil (%)
Tractor	35.1	26.4
High lead	14.8	9.1
Skyline cable	12.1	3.4
Balloon cable	6.0	1.7

[a]Compacted soil averaged about 40 to 50% higher in bulk density than uncompacted soil. Actual values of the 0–2 in. (0–5 cm) depth of the latter were between 0.6 and 0.7 g/cm^3.

Source: Rice et al., 1972.

Tractor Logging. Tractor logging is the cheapest and therefore the most popular system of moving logs to the log yard for loading, especially on slopes of less than about 30%. Soil disturbance by tractor logging, however, is greater than for any other system because the tractors and the logs they drag scrape the litter layer off the forest floor along the skid trails. Less damaging systems such as skyline, balloon, or helicopter transport are more likely to be used where slopes greater than 30% prevail, and on very fragile soils. The bulk density of soil on skidtrails made by tractor logging was studied in the northwestern United States by Froehlich et al. (1985). In skid trails on granitic soil, the 2-in. (5-cm) depth was the only soil layer whose bulk density returned to normal in 23 years. At all other depths on granitic soil and at all depths on volcanic soil, the bulk density remained higher than normal. Dense soil layers restrict plant root growth.

High-Lead Cable. The high-lead system of logging uses a mobile spar and yarder with mounted engine and winches to drag logs toward a loading yard. Only the front ends of the logs are lifted to clear obstacles or to reduce soil disturbance. The logs make skid trails as they are moved to the yarding area. Each skid trail can become a gully.

Skyline Cable. As early as 1915, skyline cable logging was tried as a method adaptable to remote areas, steep slopes, and unstable soils, where road building creates excessive erosion from landslides and exposed cuts and fills. When operated skillfully, skyline cable logging does not produce skid trails because the entire log is lifted in transport. Erosion is, therefore, minimal except in the yarding areas and traffic lanes.

Balloon Cable. This system uses a balloon to lift and transport logs. It is well adapted to steep slopes (up to 90%) and to shallow or fragile soils, where only helicopter logging or skyline logging may compete. The system is also adapted to selective logging, where the minimum harvest is about 1000 ft^3/ac (70 m^3/ha).

Balloon logging causes soil disturbance only at the yarding areas, from which trucks haul the logs to the mill. Yarding areas can be as far as 3000-ft (900-m) apart, but they must be downhill from the logged areas and, therefore, may be a hazard to streams. Balloon logging is more expensive than all other logging systems except helicopter logging.

Helicopter Logging. Logging by large helicopter so minimizes erosion that it is the apparent answer to the fondest dreams of concerned environmentalists. Logging by helicopter requires fewer access roads and, therefore, results in minimized sediment pollution of streams. It is the most versatile system of moving logs from where they are cut to a yarding area for truck loading and hauling, but it costs more per board foot of lumber. A weakness in the helicopter system is the need to enter the forest on the ground to replant, thin trees, take out the commercial thinning (poles), and control fire.

12–4.4 Reforestation

Planting or seeding a new stand of trees on land that was once forested is called *reforestation* not only where the previous stand was recently cut or burned but also where the land has been used for cropland, mining, or other purposes for some period of time. The term is sometimes extended to include tree plantings in areas that have never before been forested, although this is more properly called *forestation.* Natural regrowth is commonly called *regeneration,* although regeneration and reforestation are sometimes used interchangeably for either natural regrowth or for areas where trees have been planted or seeded.

Several nurseries have been established in various parts of the country to provide trees suitable for planting. For example, the Lucky Peak Nursery near Boise, Idaho, was established in 1959 to supply tree and shrub species for replanting forest in six western states.

Tree plantings often involve uniform spacing of trees that are all the same species. The spacing is chosen to optimize the tree growth on the site, according to the soil, climate, and tree species. The resulting pattern of straight rows of trees with uniform spacing is readily identifiable, as shown in Figure 12–11, because natural regrowth has variable spacing and usually involves several species.

Figure 12–11 This 10-year-old slash pine plantation in Oklahoma was established on abandoned, eroded cropland. These trees and their thick needle layer have been fully protecting the area against erosion and sedimentation for at least five years. (Courtesy USDA Natural Resources Conservation Service.)

TABLE 12–3 TREE AND SHRUB SPECIES BEING GROWN AT THE LUCKY PEAK NURSERY (IDAHO) FOR SOIL STABILIZATION PURPOSES AND THEIR ECOLOGICAL SITE ADAPTATION

Species	Ecological site adaptation		
	Dry sites (Mountain brush)	Intermediate sites (Ponderosa pine)	Moist sites (Douglas-fir)
Alder, Thinleaf			X
Barberry, Creeping		X	X
Bearberry			X
Bitterbrush, Antelope	X	X	X
Bladdersenna	X	X	
Ceanothus, Deerbrush		X	X
Ceanothus, Snowbrush	X	X	X
Ceanothus, Squawcarpet		X	X
Ceanothus, Wedgeleaf	X		
Cherry, Bessey	X	X	X
Cherry, Bitter	X	X	
Chokecherry, Black	X	X	
Cinquefoil, Bush		X	X
Dogwood, Redosier		X	X
Elder, Blueberry	X	X	
Eriogonum, Sulfur	X	X	
Honeysuckle, Tatarian	X	X	
Juniper, Common		X	
Locust, Black	X		
Oceanspray, Bush	X	X	
Penstemon, Bush		X	
Pine, Lodgepole		X	X
Pine, Ponderosa		X	X
Rose, Woods	X	X	X
Sagebrush, Big	X		
Sagebrush, Mountain			X
Serviceberry, Saskatoon	X	X	
Snowberry, Common		X	X
Snowberry, Mountain	X	X	
Spirea, Douglas	X	X	
Sumac, Rocky Mountain	X		
Virginsbower, Western	X	X	
Willow, Scouler		X	X
Wormwood, Oldman	X	X	

Source: U.S. EPA, 1975. *Methods of Quickly Vegetating Soils of Low Productivity, Construction Activities.* EPA Publ. 440/9-75-006.

The selection of tree species that are adapted to the soil is important in reforestation, whether done primarily to grow commercial forests or to reduce soil erosion and sedimentation. Guidance is given by several agencies, but the reports and maps of the National Cooperative Soil Survey are the most site-specific. The soil interpretations for woodland use contained in soil survey reports are based on 22,000 plots of trees planted or managed on representative soil series throughout the United States, wherever climate and soils favor tree growth. For example, as shown in Table 12–3, certain species such as thinleaf alder, redosier

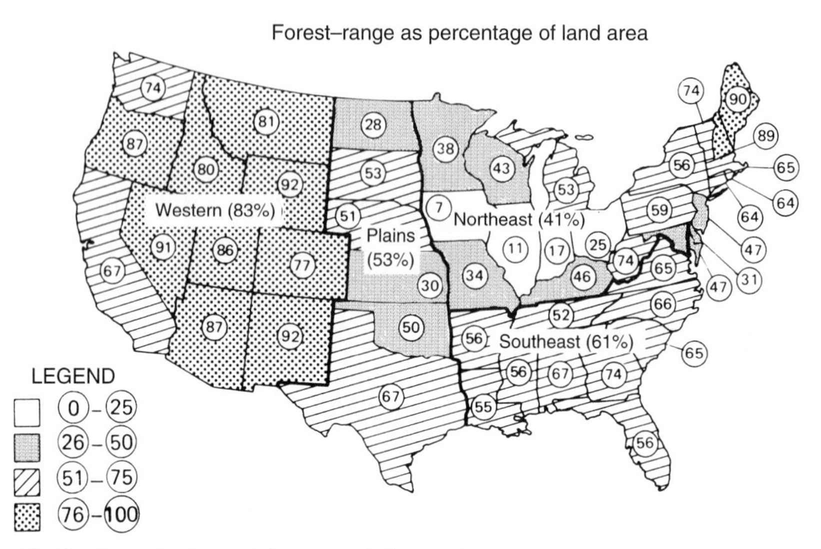

Figure 12–12 Forestland grazed by domestic livestock as a percentage of the total forestland area by states and regions in the United States. (From U.S. Department of Agriculture, 1974.)

dogwood, lodgepole pine, and ponderosa pine are suited for plantings on moist sites in Idaho, whereas none of these species is suitable on nearby dry sites. On the other hand, bitter cherry, blueberry elder, black locust, and bush oceanspray are among the plants that are suitable for the dry sites but not for the moist sites.

12–4.5 Grazing Forestlands

Foresters generally have negative opinions regarding livestock grazing in woodlands because the results are nearly always detrimental to tree reproduction and growth. Animals trample and compact the soil (especially fine-textured soil) and injure shallow roots. Nevertheless, about half of all forestlands are grazed by domestic livestock that browse on tree seedlings and sprouts. The percentage varies from 41% in the northeastern to 83% in the western United States, as shown in Figure 12–12.

Grazing of nonfarm forestlands is usually not as destructive as grazing of farm woodlands because of fewer cattle per unit area of forest, more coarse-textured and stony soils, less soil compaction by trampling, and more conifers that are not browsed as much as hardwoods. In addition, the pinelands of the southern and western states include many open forests with areas of forage grasses. The grasses, forbs, and hardwood seedlings and sprouts make fair to good forage during the months of April through July. Supplemental forage can be supplied by establishing improved pasture on all fire lanes and forest roads. When livestock graze these fire lanes, forest roads, and the forestlands in general, they reduce the fire hazard by removing vegetation that is flammable when dry.

Grazing of forestlands always increases the hazard of water erosion. Table 12–4 indicates that the average grazed forest in the United States erodes at the rate of 2.3 tons/ac-

TABLE 12–4 EROSION BY WATER AND WIND ON GRAZED AND UNGRAZED FORESTLANDS IN THE UNITED STATES

Forest management land use	Water erosion		Wind erosion	
	tons/ac-yr	mt/ha-yr	tons/ac-yr	mt/ha-yr
Grazed	2.3	5.2	0.1	0.2
Not grazed	0.7	1.6	0.0	0.0

Source: National Research Council, 1986a.

yr (5.2 mt/ha-yr). Forests not grazed erode at less than one-third of this rate. Wind erosion in forests is negligible.

12–4.6 Farm and Home Woodlands

Small areas of woodland are common on farms and large home plots in many places, both in the United States and in other nations. Some of these areas were deliberately planted to trees for erosion control, wood production, wildlife habitat, or recreational use, but most such land comes from areas that were never cleared because the land is too steep, stony and rocky, or otherwise unfavorable for cultivation. Some of it simply occurs in small areas that didn't seem worthwhile to clear. It may be used for gathering wood, observation of wildlife, or other recreational purposes, and some of it is grazed by livestock, but much of it receives little management.

Woodland offers much-needed erosion control on many sites that have steep slopes or shallow soils and on bottomland soils that are subject to flooding. Maintaining adequate vegetative cover on such land is important because of the high erosivity of these sites. It is usually possible to harvest wood products from them by selective cutting, but care should be exercised to avoid leaving any areas unprotected.

12–4.7 Fertilizing Forestlands

The use of chemical fertilizers on forests has been increasing along with the value of forest products. Currently, the fertilizer nutrient most commonly applied on forest is nitrogen. Its use is centered in the Douglas fir region of the Pacific Northwest and in the southern pine region. In addition, young stands of commercial redwood in northern California and the western hemlock-sitka spruce forests along the coasts of Alaska, Washington, and Oregon are judged likely to respond economically to applications of fertilizer.

Douglas-fir has responded to nitrogen fertilization with about 30% faster growth during a five- to seven-year period. Trees as old as 300 years have shown growth acceleration.

Southern pines on well-drained soils respond well to fertilization. Nitrogen alone is expected to enhance their growth by about 5% a year. Another area of present and predicted future response is in the flatwoods coastal plains where both nitrogen and phosphorus give increased growth of pines.

Helicopters are preferred for use in fertilizing forests. They are environmentally safer than fixed-wing aircraft because they can fly slower, spread fertilizer more accurately on the land intended, and avoid streams and lakes.

Guidelines for environmentally safe application of fertilizers to forests include the following (Environmental Protection Agency, 1973):

1. Fertilize only when a soil test indicates that benefits are expected to be economical.
2. Fertilize at rates that do not exceed the adsorption capacity of the soil and the uptake capability of timber stands.
3. Frequent fertilization at low rates is environmentally safer than infrequent application at high rates.
4. Do not fertilize water courses; leave buffer strips between streams and fertilized areas.
5. Apply fertilizers when wind drift is minimal.
6. Avoid fertilization just prior to periods of anticipated heavy rainfall.
7. Coarse-pelleted fertilizers are environmentally safer than fine pellets or dusts; liquid fertilizers have the greatest fugitive loss potential and the greatest water-pollution hazard.

Besides the use of fertilizers, two other fertility management techniques have been researched: growing legumes in the forest and using a symbiotic fungus. Fifty-two understory legumes were sown on forest sites in the lower Coastal Plains of Virginia, North Carolina, and South Carolina, and the Piedmont of Virginia and North Carolina. The objective was to find legumes that would fix atmospheric nitrogen (N_2) to enrich the soils and enhance the growth of accompanied forest trees. The eight perennial legumes that were successfully established were tick clover, false anil indigo, bicolor lespedeza, sericea lespedeza, virgata lespedeza, birdsfoot trefoil, and big trefoil. Sericea lespedeza grew the best (Jorgensen and Craig, 1983).

A symbiotic mycorrhizal fungus, *Pisolithus tinctorius,* has been used successfully to inoculate trees and shrubs in the nursery. Woody plants, so inoculated, have greater survivability when planted in adverse environments such as dry soils, low-nutrient soils, and mine spoils. Common plants that are benefited by mycorrhizae include pines, birches, beeches, hemlock, oaks, and spruces (Figure 12–13). Plants inoculated with mycorrhizae survived and grew more than 20% better than those not inoculated (Cordell et al., 1987).

SUMMARY

Pasture, range, and forest provide excellent erosion control where they are well managed to maintain good vegetative cover and high productivity. Pastures usually are intensively managed grazing lands in humid regions or are irrigated. They are seeded to improved grass plus legume cultivars, and are limed (if acidic) and fertilized. Ranges normally have low-intensity use and are in semiarid and arid regions or are a secondary use of land such as open forest in humid or semiarid regions. Native species are used for grazing, although some exotic species have been used to reseed the range. Forests, by definition, must have 10% of the area covered by trees, and a tree is a woody plant at least 15-ft (4.5-m) high supported by a single stem.

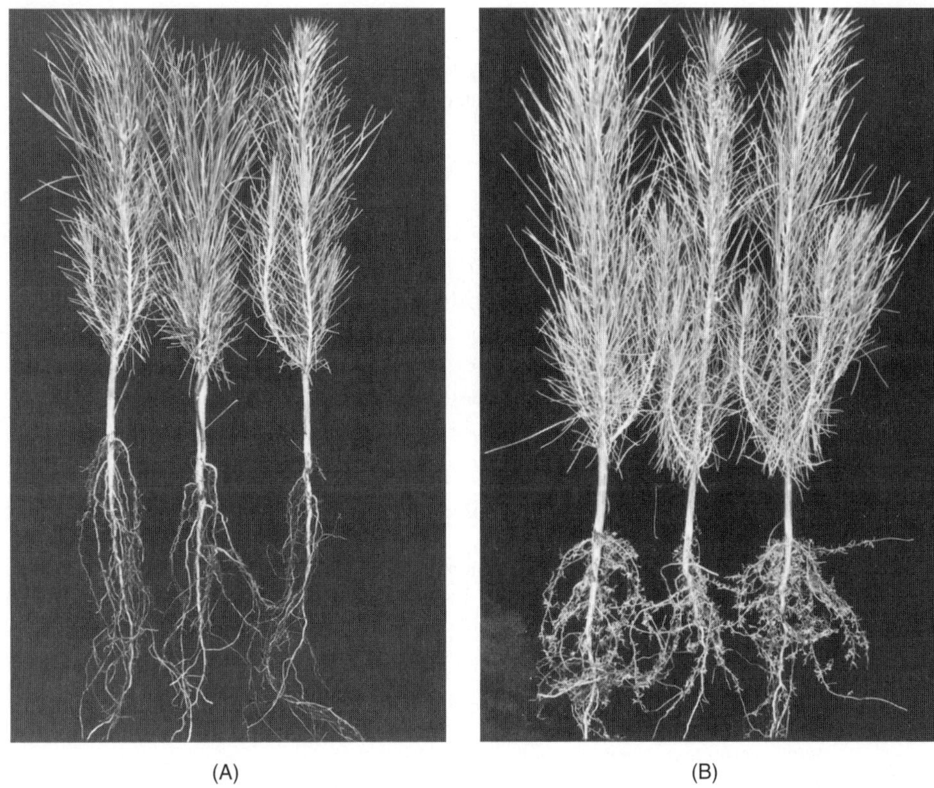

Figure 12–13 One-year-old loblolly pine seedlings. (A) Only naturally occurring mycorrhizal fungus present. (B) Ectomycorrhizal fungus, *Pisolithus tinctorius,* has been added. (Note fungal hyphae on roots.) (Courtesy USDA Forest Service.)

There are many erosion-resisting pasture grasses, pasture legumes, and range grasses, and each one has its own area of adaptation. Nitrogen fertilizer makes grasses more productive but may cause them to crowd out legumes. Pastures should be renovated when their productivity declines to 50 to 75% of its potential. Rotational and deferred grazing are good management techniques to improve pasture and range productivity.

Much rangeland has been damaged by overgrazing and is less productive than it would be if it were stocked with only enough livestock to graze half of the forage it produces. Some rangeland can be reseeded, but it takes several years to repay the costs. Certain types of rangeland benefit from burning, but this can lead to invasion by annuals and other less desirable species.

Forest cover usually keeps erosion to very low levels, except when roads are built and trees are logged. Trees are harvested by the selection, shelterwood, seed-tree, or clearcutting methods. Log removal is by tractor, high-lead cable, skyline cable, balloon cable, or helicopter. About half of all forestland is grazed by livestock; the percentage is highest in the south and west where there are open areas with grass vegetation in the forest. Reforestation and forest fertilization are increasing now along with the value of forest products.

QUESTIONS

1. Define: pastureland, rangeland, forestland.
2. Compare and contrast wind-and-water erosion potentials on pasturelands, rangelands, and forestlands.
3. Why is pastureland more likely to be fertilized than rangeland?
4. What fertilizer nutrients are most likely to be applied to pastureland, rangeland, and forestland?
5. An agronomic approach is recommended for pasture management and an ecological approach for rangeland management. Explain the difference and the reasons for it.
6. Why is grazing management the key to successful range management?
7. Describe a timber harvesting method and a log transport system that minimize soil erosion and sediment yield. Why aren't these two methods used universally?
8. Should woodlands be grazed by livestock? Explain.

REFERENCES

BALLARD, T. M., 1988. Soil degradation associated with forest site preparation. In *Degradation of Forested Land: Forest Soils at Risk.* Proc. 10th B.C. Soil Science Workshop, Feb. 1986. Land Mgmt. Rep. 56. Ministry of Forests, British Columbia, Canada, p. 74–81.

BROERSMA, K., M. KRZIC, D. J. THOMPSON, AND A. A. BOMKE, 2000. Soil and vegetation of ungrazed crested wheatgrass and native rangelands. *Can. J. Soil Sci.* 80:411–417.

BROWN, S., 1993. Tropical forests and the global carbon cycle: The need for sustainable land-use patterns. *Agric. Ecosystems Environ.* 46:31–44.

CORDELL, C. E., M. E. FARLEY, J. E. OWEN, AND D. H. MARX, 1987. A beneficial fungus for mine land reclamation. In C. L. Carlson, and J. H. Swisher (eds.), *Innovative Approaches to Mined Land Reclamation.* Southern Illinois Univ. Press, Carbondale and Edwardsville, IL, p. 499–524.

DEHAVEN, M. G., 1983. *Assessment of Stormflow and Water Quality from Undisturbed and Site Prepared Forest Land in East Texas.* Tech. Rep. 122. Texas Water Resources Institute, Texas A&M University, College Station, TX, 125 p.

ELLIOT, W. J., D. PAGE-DUMROESE, AND P. R. ROBICHAUD, 1999. The effects of forest management on erosion and soil productivity. Ch. 12 in R. Lal (ed.), *Soil Quality and Soil Erosion.* CRC Press, Boca Raton, FL, 329 p.

ENVIRONMENTAL PROTECTION AGENCY, 1973. *Process, Procedures, and Methods to Control Pollution from Silvicultural Activities.* Publ. EPA 430/9–73-010, 91 p.

FRANZLUEBBERS, A. J., J. A. STUEDEMANN, AND S. R. WILKINSON, 2001. Bermudagrass management in the southern Piedmont USA: I. Soil and surface residue carbon and sulfur. *Soil Sci. Soc. Am. J.* 65:834–841.

FROEHLICH, H. A., 1988. Causes and effects of soil degradation due to timber harvesting. In *Degradation of Forested Land: Forest Soils at Risk.* Proc. 10th B.C. Soil Science Workshop, Feb. 1986. Land Mgmt. Rep. 56. Ministry of Forests, British Columbia, Canada, p. 3–12.

FROEHLICH, H. A., D. W. R. MILES, AND R. W. ROBERTS, 1985. Soil bulk density recovery on compacted skid trails in central Idaho. *Soil Sci. Soc. Am. J.* 49:1015–1017.

GRACE, J. M. III, 2000. Forest road sideslopes and soil conservation techniques. *J. Soil Water Cons.* 55:96–101.

Greenwood, K. L., D. A. MacLeod, J. M. Scott, and K. J. Hutchinson, 1998. Changes to soil physical properties after grazing exclusion. *Soil Use and Management* 14:19–24.

Hamdar, B., 2000. A review of land utilization in the U.S.: Agricultural and forestry land use competition. *J. Sust. Agr.* 17(1):71–87.

Hart, R. H., 1974. Crop selection and management. In *Factors Involved in Land Application of Agricultural and Municipal Wastes.* USDA, Beltsville, MD. p. 178–200.

Hawley, J. G., 1991. Hill country erosion research. In W. C. Moldenhauer, N. W. Hudson, T. C. Shen, and S. W. Lee (eds.), *Development of Conservation Farming on Hillslopes.* Soil and Water Conservation Society, Ankeny, IA, 332 p.

Hyde, R. M., and C. E. Owensby, 1970. *Burning Bluestem Range.* Publ. L-277, Kansas State Univ., Manhattan, Kansas.

Jorgensen, J. R., and J. R. Craig, 1983. *Legumes in Forestry: Results of Adaptability Trials in the Southeast.* Southeast. For. Exp. Stn. Res. Pap. SE-237.

Lowrance, R., and R. A. Leonard, 1988. Streamflow nutrient dynamics on coastal plain watersheds. *J. Environ. Qual.* 17:734–740.

Megahan, W. F., 1988. Roads and forest site productivity. In *Degradation of Forested Land: Forest Soils at Risk.* Proc. 10th B.C. Soil Science Workshop, Feb. 1986. Land Mgmt. Rep. 56. Ministry of Forests, British Columbia, Canada, p. 54–65.

Menke, J., and G. E. Bradford, 1992. Rangelands. *Agric. Ecosystems Environ.* 42:141–163.

National Research Council, 1986a. *Soil Conservation: Assessing the National Resources Inventory.* Vol. 1, p. 7. Data are for 1982.

National Research Council, 1986b. *Soil Conservation: Assessing the National Resources Inventory.* Vol. 2, p. 163–203.

Orians, G. H., and C. I. Millar, 1992. Forest lands. *Agric. Ecosystems Environ.* 42:125–140.

Owensby, C. E., and E. F. Smith, 1979. Fertilizing and burning Flint Hills bluestem range. *J. Rge. Mgmt.* 32:254–258.

Rice, R. M., J. S. Rothacher, and W. F. Megahan, 1972. *Erosional Consequences of Timber Harvest: An Appraisal.* Proc. Symp. on Watershed in Transition, Ft. Collins, Colorado, June 19–22, p. 321–329.

Sakalauska, K. M., J. L. Costa, P. Laterra, L. Hidalgo, and L. A. N. Aguirrezabal, 2001. Effects of burning on soil-water content and water use in a *Paspalum quadrifarium* grassland. *Agric. Water Mgmt.* 50:97–108.

Sanderson, M. A., J. C. Read, and R. L. Reed, 1999. Harvest management of switchgrass for biomass feedstock and forage production. *Agron. J.* 91:5–10.

Startsev, A. D., and D. H. McNabb, 2000. Effects of skidding on forest soil infiltration in west-central Alberta. *Can. J. Soil Sci.* 80:617–624.

Tate, K. W., G. A. Nader, D. J. Lewis, E. R. Atwill, and J. M. Connor, 2000. Evaluation of buffers to improve the quality of runoff from irrigated pastures. *J. Soil Water Cons.* 55:473–477.

U.S. Department of Agriculture, 1974. *Land Use Planning through the United States Department of Agriculture,* 45 p.

Wienhold, B. J., J. R. Hendrickson, and J. F. Karn, 2001. Pasture management influences on soil properties in the northern Great Plains. *J. Soil Water Cons.* 56:27–31.

Wright, J. R., and F. H. Siddoway, 1982. Determinants of soil loss tolerance for rangelands. In *Spec. Publ. 45,* American Society of Agronomy and Soil Science Society of America, Madison, WI, p. 67–74.

CHAPTER

13

WATER CONSERVATION

Water is a constituent of all living things, from the smallest bacteria to the largest animal and the tallest tree. It is also found in inanimate objects such as rocks and minerals. It is all around us—in the air, in rivers and streams, in ponds and lakes, in seas and oceans, even piled high in solid form on the Earth's surface in polar regions and in many high mountain areas. It is also stored in the ground under our feet.

Water is the most abundant liquid on Earth, and it is the only common substance that exists naturally in all three states of matter: solid, liquid, and gas. It cycles readily from one state to another and from one place to another. It is almost universally present, but often not in the right amount for the optimum growth of plants and animals or to meet human needs. A shortage of water is the most common limiting factor for plant growth, but an excess of water is also a common limiting factor.

Average annual precipitation ranges from almost zero in the deserts of Africa and central Asia to more than 650 in. (16,500 mm) at one location in Hawaii. Even in areas where abundant rainfall is the rule, precipitation fluctuates widely from year to year. Thus, many countries suffer from periodic or regular water shortages. According to the World Bank, 80 countries with 40% of the world's population presently have water shortages that could cripple agriculture and industry. Even now, control of water is an issue that underlies strife between countries in several parts of the world.

Besides being vital to life and the principal component of living things, water serves many other functions in the world. Its participation in the erosional processes that shape landscapes is discussed in Chapter 3, and the problems caused by excessive erosion are a major topic throughout this book. Water also acts as a weathering agent that participates in the gradual decomposition of minerals, thus releasing nutrients that help sustain life. It serves as a medium in which living microbes decompose the residues of life, thus preventing them from accumulating to the point of tying up too much carbon and other necessities of life. It acts as a transporting system that carries both solid particles and solutes from the land to the sea. It reduces temperature variations through a buffer system based on its high heat capacity, mobility, and ability to change states. Other substances may supplement its action, but none can replace water.

Fortunately, there are some remedies that can be used when the wrong amount of water is present for growing plants. Removal of excess water by soil drainage is discussed in Chapter 14, and water additions by irrigation are the topic in Chapter 15. The remaining alternative, water conservation, is the subject of this chapter.

What can be done to overcome or ameliorate water deficiencies? Actually a lot can be done, and a lot is being done. Individuals can become conscious of the need to conserve water and can stop wasting it in their homes and in their gardens. Industries can find ways to use less water in their operations and can make greater efforts to stop polluting it. Farm irrigators can certainly use water more efficiently, and rainfed farming operations can be made more effective in trapping the rain that falls on the land and in using the water that is stored in the soil for crop production.

Far more water is used to grow agricultural crops than for any other purpose. Consequently, agriculturalists always hope that some new weapon, some panacea, will be discovered that will easily and cheaply increase moisture for crop growth. Any number of exotic avenues have been examined—weather modification, chemicals that will make soils more permeable to water, or that might reduce evaporation and/or transpiration, and many more. Farmers have learned that the old methods, or modifications of them, often work best. These include proper and timely tillage, effective and efficient use of crop residues, selection of adapted and high-yielding crop varieties planted on time and at proper rates, land forming, irrigation where feasible, and—most important of all—control of weeds and volunteer plants over the whole land-preparation and crop-growing period. These are the tools for conserving water; they work when they are used properly, but the search for better ways to perfect their effectiveness must continue. That search begins with an understanding of water cycles, weather, and related phenomena.

13–1 THE WATER CYCLE

Water evaporates and condenses on a grand scale; most of the condensation falls as rain, but some falls as snow, sleet, or hail. The solid forms melt and the liquid forms can freeze, vaporize, or flow to another location. Water is highly mobile in both liquid and vapor forms, and even the solid form can be mobile in the form of a glacier or avalanche. Nevertheless, much water is locked in place in rocks and ice, and only a small fraction of the water present cycles from one state to another during a year's time. By far the largest amount (97% of all water on Earth) is held in the seas and oceans that cover 71% of the surface. This water circulates within the oceans, but only a tiny fraction of it escapes annually by evaporation. Even so, this small fraction of the total is the principal source of fresh water that sustains life and performs all the other functions alluded to in the introduction to this chapter.

13–1.1 Precipitation Patterns

Weather may seem erratic, but it follows broad climatic patterns that relate to the nature of the Earth. Rainfall is abundant in certain areas year after year, while other areas are consistently arid. Much of this is controlled by the way air circulates around the globe. Air rises

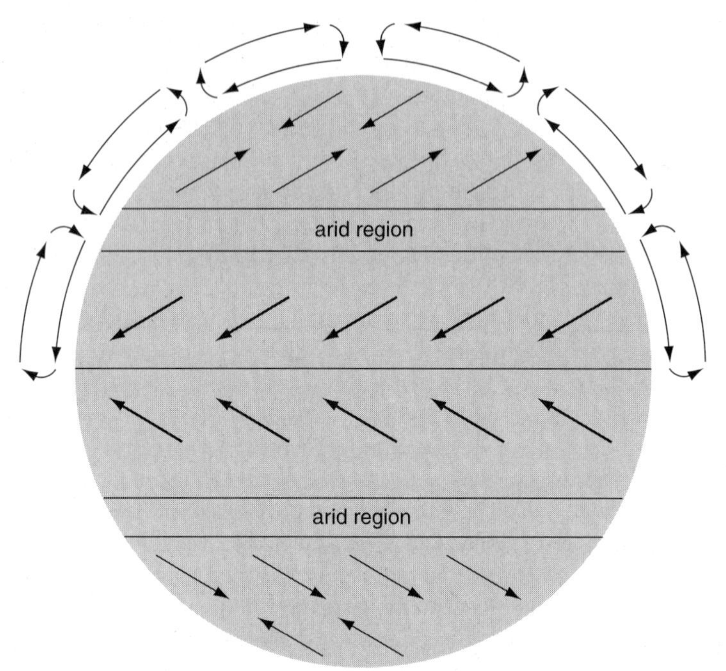

Figure 13–1 An idealized diagram of the global air circulation on Earth. The three cells shown for the Northern hemisphere also occur in the Southern hemisphere as a mirror image. They result from air rising at the warm zone near the equator and sinking at the cold zone around each pole, with a counter-flowing zone in the middle resulting from the size of the Earth. The rotation of the Earth adds an east-to-west component to the equatorial and polar winds and a west-to-east component to the winds in the temperate zone. Arid regions occur where the air has a strong downward movement because the air is compressed and warmed as it descends. These zones shift north and south with the summer and winter seasons and are somewhat disrupted by mountain ranges.

in equatorial regions where the direct rays of the sun cause warm temperatures. It descends at the poles where the temperatures are cold. High-altitude winds, therefore, transport air from the equator to the polar areas, while surface winds carry a net air movement from the poles to the equator. That might constitute the main pattern if the Earth were smaller, but in actuality, the movement breaks up into three cells each in the Northern and Southern hemisphere. The middle cell rotates counter to the other two and makes the prevailing winds in temperate regions blow in the opposite direction from those of the rest of the world (Figure 13–1).

The zones between 20° and 30° north and south latitude are predominantly arid because the air has a strong downward movement in these zones. Most of the world's deserts are located in or near these zones. They become very hot in the summers because the sun moves nearly overhead at that time and there is little water to provide a cooling effect.

The idealized zones shown in Figure 13–1 shift north and south with the summer and winter seasons. This shift results from the tilt of the Earth's axis that causes the noonday sun to appear directly overhead (and therefore hottest) in a band that moves up to 23.5° north of the equator in June and 23.5° south of the equator in December. The equatorial region, therefore, includes a band near the equator that is humid all year and bands north and south of it that are humid in summer but dry when the other hemisphere has the sun overhead. The reverse occurs in Mediterranean climates where the summers are dry and most of the precipitation occurs during the cooler parts of the year.

The whole picture is further complicated by topographic features of the Earth that divert the air movement. The upward movement causes precipitation where air cools as it rises to go over a mountain. The other side of the mountain casts a rain shadow where the air tends to be dry because it warms again as it descends. Ocean currents add their effect by producing warm and cold zones in the water. The resulting increase or decrease in evaporation can have a considerable effect on the amount of precipitation occurring in the areas downwind from these currents.

13–1.2 Water Returns to the Oceans

Much of the precipitation that occurs on land comes from water that evaporated from the oceans. Thus, the atmosphere provides a large net transfer of water from the oceans to the land. Runoff plus small amounts of seepage and modest amounts of water from melting icebergs provide an equal movement of water (37,000 km^3 annually according to Bloom [1998]) from the land to the oceans, so there is an overall balance. The return trip to the ocean may take any amount of time from a few seconds for water to flow down the side of a shoreline rock to millions of years for water that gets locked into buried rock structures. In general, water that enters a small mountain stream and follows the drainage system into a larger stream and then a major river should take several days or a few weeks to reach the ocean. However, some of the water that is held in a lake or reservoir can remain there for months or years.

Much of the water from precipitation soaks into the soil through pores and cracks. Macropores such as cracks, worm channels, and other visible openings are very important because water moves through them much faster than it can move through the soil mass. Such passages permit water to reach subsoil layers without having to saturate the topsoil first. Soil permeability typically decreases with time as soil particles absorb water and expand enough to close the soil cracks. The rate of infiltration depends on either the rate of supply (rainfall, snowmelt, and runon minus runoff) or the soil permeability, whichever is limiting. The depth of water penetration depends on the infiltration rate, on how long water remains available at the soil surface (much longer in concave areas that receive runon and hold water than in convex areas that lose it by runoff), and on the storage capacity of the soil material.

Most soil water returns to the atmosphere by evaporation and transpiration by growing plants. It may then fall again as precipitation over either land or water. A small part of the water absorbed by plants becomes a part of plant tissue and may be held there for weeks, months, or years. Another portion of the water that soaks into the soil penetrates to an underlying saturated zone known as the *water table.* Water tables may be undetectable in some extremely arid locations, but generally they occur at depths ranging from a few to many feet. Water table depths may fluctuate during the year, rising during a season of surplus precipitation and sinking during drier months. Some of this groundwater escapes through plant transpiration when the water table is high enough for access by plant roots and by evaporation where it reaches or approaches the land surface, as in a bog.

The surface of a water table normally has some slope; in fact, the topography of the water table typically resembles a subdued form of the land surface (Bloom, 1998). Underground water, therefore, moves slowly toward an escape zone that may be a nearby stream, a spring, or a bog. This slow underground flow contributes to the consistency of stream flow by supplying water at times when surface runoff has ceased. This stabilizing effect is limited, of course, by the underground storage capacity, so the flow rate declines when the supply is depleted during a long dry period. Dry summers cause most streams in arid regions to dry up, although subsurface flow usually continues in the streambed.

Sand and gravel layers such as those typically found in valley floors and beneath terrace soils can hold or transmit a lot of water. Sand layers that have been cemented into sandstone often are still quite permeable. Massive rocks have very low permeability and storage capacity, but some rock layers have large passageways. Some basalt rock formations enclose large tubes that can conduct water for long distances, for example the 120-mile (200-km) passage discussed in Section 14–2. Seismic shifts in the Earth's crust produce joints in otherwise massive rock layers, and water may move through these joints. Solution processes enlarge such passageways in soluble rocks such as limestone, sometimes into large cavern systems with underground streams flowing through them.

Some groundwater comes from recent rainfall, but some of it has been there for time periods ranging up to millions of years. The deeper it is and the farther it has moved from its entry point, the longer it has probably been there, and the longer the replenishment time will be if it is removed. This stored groundwater is the supply that is tapped when one digs or drills a well. The long-term value of the well depends more on how fast the water can be replaced by fresh supplies than on how much water is stored in the rock layers. Even an underground reservoir as large as the Ogallala aquifer (Section 15–3.4) can be depleted by excessive use. Water quality is also a factor, as varying amounts of soluble salts are dissolved while the water seeps through soil and rocks.

13–2 WHAT IS DROUGHT?

The terms *dryland* and *drought* are associated, but are not synonymous. A dryland area is one where the major factor limiting plant growth is shortage of water. A specific average annual rainfall cannot serve to distinguish between drylands and humid areas because other factors such as rainfall distribution and reliability, humidity, temperature, rate of evaporation, and soil characteristics all influence whether a given amount of rainfall will be sufficient for, or will limit, plant growth. Droughts of varying severity and duration are common in drylands, but they also occur in more humid areas.

13–2.1 Definition of Drought

Attempts have been made to define drought in terms of precipitation, but temperature and wind also affect the amount of water plants need. The evapotranspiration concepts of Thornthwaite (1948) (discussed in Section 6–5.1) provide a basis for describing sufficiency or deficiency of rainfall in an area for normal or average conditions, but not for specific periods.

Climatic conditions alone cannot define drought either. Soil properties and crop and soil management practices influence water uptake by plants. Crops may survive a dry period without serious damage if the soil initially contains an abundance of stored water within the root zone.

In terms of crop production, *drought* occurs when lack of water reduces growth and final yield of the staple crops of a region. In terms of wheat production, drought is relatively common in the Great Plains area of the United States, but is rare in the Corn Belt. Drought occurs more frequently in the Corn Belt if it is assessed in terms of corn growth.

There are two types of drought—atmospheric drought and soil drought. Atmospheric drought occurs when some combination of low humidity, high temperature, and high wind velocity causes high transpiration rates; soil drought occurs when the water supply is diminished by low soil moisture resulting from low precipitation, low soil permeability, and/or low soil storage capacity. Both types cause plant stress. Severity and duration of the stress determine the amount of plant damage. Plant stress increases in both severity and extent as a dry period continues—first on soils with limited stored water, but eventually on all soils. Short periods of stress reduce plant growth; extended stress increases the damage and eventually kills plants.

13–2.2 Effect of Drought on Plants, Animals, and People

A plant's ability to survive dry conditions depends on the severity of the drought and on plant characteristics. Some plants are in a race with drought. The annuals that have very short growing periods (desert ephemerals) *escape* drought by germinating, growing, and producing flowers and seeds during a very brief life span, usually maturing if there is enough moisture to germinate the seed. Some plants *evade* drought by having large absorbing root systems, low transpiration rates, or mechanisms for reducing leaf area or closing stomata. A third group *endures* drought by means of massive water-storage organs (cacti) or by shedding leaves and becoming dormant (mesquite).

Crops for agriculture must come from the drought-evading group. The drought escapers have too limited production; the drought endurers grow too slowly. Actually, a cultivated crop's ability to survive drought is less important than its ability to produce some grain or forage even under drought stress. Cereal grains are preeminent dryland crops despite their relatively rapid collapse under severe drought stress.

Society also suffers when drought kills plants and the domestic and wild animals that feed on plants. People could migrate to escape a drought, but distance to less droughty areas often is excessive, and fear of the unknown and attachment to home make flight unwelcome. Furthermore, droughts come and go without advance warning. The best time to leave would be at the beginning of a drought, but people are more likely to leave after the drought has depleted their reserves and possibly is nearly over.

Inhabitants of most large, developed countries can survive a drought because food and finances from outside the drought area can be made available. In many developing countries with dense populations and poor transport systems, there is no quick way for people to seek or get outside help. Massive famines have occurred regularly in the developing world as, for example, in the Sahel region of western and central Africa in the 1970s, and in Ethiopia and the Sudan in East Africa in the 1980s.

Even the soils of an area show the effects of drought. The decrease in organic matter when there is little plant material to return to the soil can lead to a deterioration in soil structure and an increase in bulk density that make the soil less suitable for plant growth (Albaladejo et al., 1998).

13–3 COMBATING DROUGHT

Combating drought is one of the primary requirements of successful dryland farming and can be important in more humid areas. Every effort must be made to conserve rainfall, store it in the soil, and use it wisely. Much of the problem is the uncertainty resulting from irregular rainfall (Rockström et al., 1999). An area that produces a good crop one year may have a crop failure another year. Techniques that improve the consistency of production are very important in areas that are subject to drought.

Crops need to be adapted to the area. Many experiment stations show that the best-adapted varieties of crops commonly outyield poorer ones by 54 to 194% or more (Roozeboom et al., 1996). Yet the best and poorest varieties use nearly equal amounts of water. Farming and ranching success is not assured even by efficient water use. Several successive drought years may reduce current production and income below that necessary to meet even minimum living requirements and farm expenses. Dryland farmers must accumulate reserves during favorable years.

Livestock owners face additional difficulties. Water supplies for livestock often are depleted during dry periods. Deeper wells may solve this problem, or additional ponds and dugouts that store runoff water and snowmelt may help. Reserve feed is more difficult and often more expensive to accumulate and store, especially in hotter areas. One possibility in the higher latitudes of temperate regions where summer fallow is common is to plant a cereal crop suitable for use as animal forage on a field that has been fallowed. The crop can then be fed to the livestock if forage is needed, or it can be harvested for grain if enough other forage is available.

Often it is necessary to reduce the number of livestock during times of drought. In these areas, the best livestock venture is one that can cut back quickly when a drought occurs and build up speedily when more normal weather returns.

13–4 WHAT HAPPENS TO RAINFALL?

Briggs and Shantz (1914) estimated that a wheat crop at Akron, Colorado, used about 17% of the 16.5 in. (419 mm) of annual precipitation; more recent evapotranspiration studies show values up to about 40% in cool climates, but considerably less than that in warm climates. What happens to the rest of the rainfall? Why isn't more used by the crop? Answers to these questions will suggest ways to conserve water by using wisely the rain that falls both in dryland and in humid areas.

Precipitation (P) is lost as it moves from the air to the soil because it is intercepted (I), or because it runs off (R). If the rainwater passes through the soil surface (infiltrates), it is stored in the soil to be used by crops unless it evaporates from the soil surface (E), is transpired by weeds or volunteer plants (T), or percolates below root depth (D). The amount of water available for storage in the soil and for use by a crop (A) may be expressed as

$$A = P - (I \pm R + E + T + D)$$

Three of the five avenues of moisture loss—runoff, evaporation, and transpiration by weeds—cause water losses that can be significantly reduced. Water loss by deep percolation is important in many humid areas and on very coarse-textured soils in dryland regions. Water conservation must include management methods that:

1. decrease runoff
2. reduce evaporation
3. reduce deep percolation
4. prevent unnecessary loss from storage

13–4.1 Interception

Vegetation intercepts and holds some rain during each storm. The amount is variable, but an amount such as 0.1 in. (2 or 3 mm) is sometimes assumed as a representative value. Some intercepted water may reach the soil by flowing down plant stems, and a small amount may be absorbed by the plant tissue, but most of it evaporates into the air. Evaporation raises the humidity and cools the plants and the air around them and may thereby temporarily reduce the amount of water transpired by the plants.

13–4.2 Runoff

Some of the rain that penetrates the vegetative canopy runs off the land instead of soaking into the soil. Runoff from individual sites ranges from zero on highly permeable, vegetated, and level soils to more than 75% of the rain on impermeable, poorly vegetated, steeply sloping sites. Runoff from watersheds (including both surface runoff and subsurface flow) in the continental United States ranges from 57% in the Middle Branch of the Westfield River watershed in Massachusetts to 2.4% for the Smoky Hill River watershed in Kansas. Runoff is affected by rainfall intensity, soil properties, land configuration, and vegetative cover.

Coarse-textured and well-aggregated soils have high infiltration rates; fine-textured, poorly aggregated soils have low infiltration and high runoff rates. Infiltration rates are usually higher at the beginning of a storm, when soils are dry, but drop off quickly as soils become wet.

Increasing slope gradient increases the amount and velocity of runoff; surface soil depressions hold water and permit local water, and even runon from adjacent areas, to be absorbed. Vegetation, both living plants and dead crop residues, reduces the number of raindrops that hit the soil directly, thus reducing soil crusting and maintaining higher infiltration rates.

Frozen soil markedly increases the runoff from both snowmelt and rainfall during the early spring season in temperate areas. Standing vegetation and/or a rough soil surface will hold more snow in place and slow the runoff, but there still may be considerable loss. Techniques for reducing runoff losses are discussed in Section 13–5.

13–4.3 Evaporation

A saturated soil loses water by evaporation as fast as a free water surface. When surface soil is dry, evaporation is reduced to the rate that water vapor moves upward through the dry layer. Losses by evaporation often exceed 50% of the annual rainfall. Dark-colored soils absorb more heat than light-colored soils, so they are hotter and lose more water by evaporation. South- and west-facing slopes in the Northern hemisphere (north- and west-facing slopes in the Southern hemisphere) are warmer and have higher evaporation rates than north- and east-facing slopes. Drylands with low relative humidities and high wind velocities lose a larger proportion of rainfall by evaporation than do humid regions. Vegetation, either living or dead, reduces evaporation by insulating against heat and deflecting wind away from the soil.

13–4.4 Transpiration

Transpiration by a crop is difficult to measure or predict accurately, but estimates range from about 2.7 in. (69 mm) to produce a crop of dryland wheat to 16.5 in. (420 mm) for corn in a humid region. Transpiration by crop plants is consumption rather than a loss of water because it is being used to grow the crop. Transpiration by weeds or volunteer grains is a loss whether it occurs on fallowed fields, on land being prepared for immediate seeding, or in fields producing a crop. These losses are more completely under the control of the farmer than almost any other type of moisture loss.

13–4.5 Deep Percolation

Water that percolates below plant roots is lost to those plants. Some plant roots can penetrate as deep as 30 ft (9 m), but usual root penetration is no more than 50 to 60 in. (125 to 150 cm) for annual spring-seeded cereals, 60 to 70 in. (150 to 180 cm) for winter cereals, and 100 in. (250 cm) or so for some fibrous-rooted perennial grasses. The limited rainfall in dryland areas seldom penetrates below root depth in medium- and fine-textured soils, even when the land is summer fallowed; in humid regions, considerable water is lost for plant growth by deep percolation.

13–4.6 Storage

Rainfall not lost by one of the foregoing processes is stored in the soil and available for use by crops. Stored water is the key to plant survival in dryland areas. It is the cushion that helps to level the excesses and deficiencies of rainfall. Rains rarely occur daily, so between rains plants must use stored water. The more arid the region, the longer the duration of these dry periods, and the longer plants have to depend on stored water. Water-conservation activities, water-storage techniques, and efficient use of stored water all help to maintain adequate stored water supplies.

13–5 DECREASING RUNOFF LOSSES

Water that infiltrates into the soil cannot run off the surface. Therefore, runoff is reduced by any practice that increases a soil's infiltration rate or maintains it at a high level, and by any practice that causes water to stand on the land longer.

Leaving vegetation and crop residue on the surface rather than burying it, applying mulches, and using soil-conserving cropping systems increase infiltration and reduce runoff. Contour cultivation, level terraces, and water-spreading devices help hold rainwater on the soil longer and give it time to infiltrate. Land forming that leads water to prepared absorption sites will reduce runoff from the whole field, even though some water evaporates as it crosses the contributing area to reach the storage area.

Runoff is a loss to upland farmers, but people downstream may benefit from it. Impounded water may be used for irrigation. Where upland farmers use water conservation practices, inflow into storage reservoirs is reduced. This is already posing problems in some areas east of the Rocky Mountains in the United States where reservoirs have been constructed to trap and store excess river flow for irrigation.

13–5.1 Use of Plants and Surface Crop Residues

The most practical way to maintain high infiltration rates is to keep vegetation and vegetative residues on top of the soil. Plant material intercepts raindrops and reduces the energy that is released by raindrops hitting the ground, and thus reduces crust formation, runoff, and soil erosion. Ramos et al. (2000) found that ten minutes of rainfall on bare soil reduced the infiltration rates of ten representative soils in Spain to 1 to 7 mm/hr, but that protecting the soil surface with mulch maintained infiltration rates from 50 to 200 times as fast as those on the bare soils. See Section 8–6.3 (Chapter 8) for a discussion of the effect of crops and crop rotations on soil structure and permeability.

Vegetation and crop residue are also useful in preventing snow from blowing off fields in the higher latitudes. Stubble is most useful for trapping snow when it is standing tall, not flattened. Use of chemical herbicides or of undercutting implements leaves most of the stubble standing. Black and Siddoway (1971) recommended the use of tall wheatgrass strips across cultivated fields to control erosion and trap snow. Double rows of tall wheatgrass spaced 40- to 60-ft (12- to 18-m) apart are especially effective in trapping snow. In a trial at Swift Current, Canada, nearly an extra inch of water was saved in the shelter of tall wheatgrass barriers compared to that in the open plain (Steppuhn and Waddington, 1996). Snyder et al. (1980) found that extra moisture from trapped snow increased crop yields. They estimated that net returns from these increases varied in value from $5.00 to $13.00/ac depending on the region.

Three ways to get the greatest possible moisture conservation benefits from vegetation on cropland are to:

1. Use crop rotations, cropping systems, and crop management practices that keep the soil well covered for as long as possible each year.
2. Leave as much of the crop residue on the soil surface as practical, particularly between crops.
3. Apply mulches to fields that would otherwise be bare and unprotected.

Figure 13–2 Contour cultivation on broad, gentle slopes in the central Great Plains in a crop system of alternate wheat and fallow. (Courtesy USDA Natural Resources Conservation Service.)

A fourth method that may be used where these practices are inadequate is to plant vegetative barriers on the contour to intercept the runoff. Sharma et al. (1999) described the use of such barriers in India and reported reductions in runoff of 28 to 97%. They recommend the use of locally adapted native grass species planted in lines with a 1-m (40-in.) vertical interval (about a 33-m or 110-ft horizontal spacing on their 3% slope). These grass barriers also trapped sediment and nearly eliminated soil loss from the area. They were well accepted by farmers, because the grass provided much-needed forage for their livestock.

13–5.2 Contour Cultivation

Contour cultivation, as shown in Figure 13–2, produces miniature furrows and ridges across the slope. These furrows trap rainwater and give it more time to infiltrate. More water is conserved by contouring on gentle slopes than on steep slopes because channel capacity decreases as the slopes become steeper. Large ridges, such as those produced by a lister, trap more water than the small ridges thrown up by a disk, chisel implement, or cultivator.

Deep ripping would be expected to increase infiltration and reduce runoff, especially if the ripping is done on the contour. Few benefits, however, have been measured from use of this practice. This may be because the slot that permits water entry also allows evaporation to occur. An exception to this arises in the case of ripping for conservation of snow melt. If ripping is used in conjunction with practices to retain snow on fields, and if the slots are closed after melt is completed, extra water will often be saved. Pikul and Aase (1998) recommended contour ripping in the fall as a means of increasing infiltration into frozen soil during the winter and spring. They used a single subsoiler shank to rip the soil to a depth of 1 ft. (0.3 m) on 20-ft (6-m) intervals on a sandy loam soil with 5% slope used to grow spring wheat in Montana. The rip lines provided storage volume as well as openings for infiltration, and resulted in 32 mm more water stored in the soil profile for a distance of 5 ft (1.5 m) downslope from the rip line.

Tillage furrows are seldom exactly on the contour, so some water moves to low spots. A *damming* or *basin lister* was developed to prevent movement in lister furrows. This implement reduces lateral water movement, but the dams make tractor travel so rough and uncomfortable that few mechanized farmers use the machine. With larger tractors, and especially with larger, wider tires, there is renewed interest in this practice in the southern Great Plains (Baumhardt et al., 1992). A similar system called *tie-ridging* has been successful in countries where hand tillage is used.

A study on Rideau clay soil showed that contour cultivation of a corn field reduced runoff from 1.51 to 0.63 in. (38 to 16 mm) or from 8.7 to 3.6% of seasonal rainfall (Ripley et al., 1961). A classic study by Fisher and Burnett (1953) on a 0.5% slope at Spur, Texas, showed that farming cotton up and down the slope permitted 2.75 in. (70 mm) of runoff compared to 1.95 in. (50 mm) from contour-cultivated land. Water loss was reduced from 13.7% of the 20-in. annual rainfall to 9.7%.

When less water runs off, more is stored in the soil, so crop yields are generally higher on contoured plots. At Spur, Texas, lint-cotton yields were 25% higher on contoured plots (146 versus 117 lb/ac or 167 vs. 131 kg/ha).

13–5.3 Terracing

Terraces that are designed to conserve runoff must hold the water on the soil surface until it infiltrates. These are called *detention terraces* or *level terraces* because the channel must be level to be effective, not graded like the diversion terraces used in humid regions to dispose of water. There are two major types of level terraces: level ridge-type terraces and conservation-bench terraces.

Level Ridge-Type Terraces. These are combination ridge and channel structures that resemble the broad-based terraces described in Chapter 10, except that the channels are level. Runoff water ponds in the channels, as shown in Figure 13–3. These terraces help to reduce water erosion and they improve water conservation.

Figure 13–3 A level-terraced field with water ponded in the terrace channels after a heavy rain. (Courtesy USDA Natural Resources Conservation Service.)

Level terraces were built on the experiment stations at Spur, Texas, and Goodwell, Oklahoma, in 1926. A few terraces had partly closed ends and periodic blocks along the channel so that water could not escape. This design allowed excessive rainwater to run off, but held all the water from less-intense storms, reducing runoff losses and improving water-conservation efficiency. At Spur, terraces supported by contour cultivation increased lint-cotton yield by 29% over contouring alone.

Level terraces usually are designed for 10-year-frequency storms. Channel capacity is generally larger than for graded terraces on comparable slopes. Terrace spacing is determined with the vertical and horizontal interval equations used for regular terraces (Chapter 10). Under no circumstances should the horizontal spacing exceed 400 ft (120 m).

Level terraces should be constructed only on deep, permeable soils that have large water-storage capacities. Land slope should be gentle, not exceeding 5%. Terrace length may be up to 3000 ft (925 m); it can be even longer if the terrace channel is blocked at intervals. Blocking not only enhances water storage, it guards against excessive damage if the structure is overtopped in intense storms. Excess runoff must be discharged safely. Outlets may be natural waterways, grassed areas, shaped and vegetated waterways, or underground conduits.

Three general designs of level ridge-type terraces are shown in Figure 13–4. *Normal ridge-and-channel terraces* can be used anywhere that conditions suit level terraces. *Steep-backslope terraces* can be used only on deep loess or other permeable soils and must be provided with outlets. They can be used on slopes up to 20%, but the structures on steep slopes are so massive and close together that construction costs are very high. *Flat-channel terraces* are used to spread the water over a broad area so it will infiltrate more uniformly. They can be used where regular level terraces are suited, but with two additional requirements: all soil exposed in the cut area must be permeable, and the slope must not exceed 4%. Fields should have quite uniform slopes. The largest permissible slope variation along each terrace site is one-fifth of the average slope for that particular terrace. The smaller the slope variation, the easier

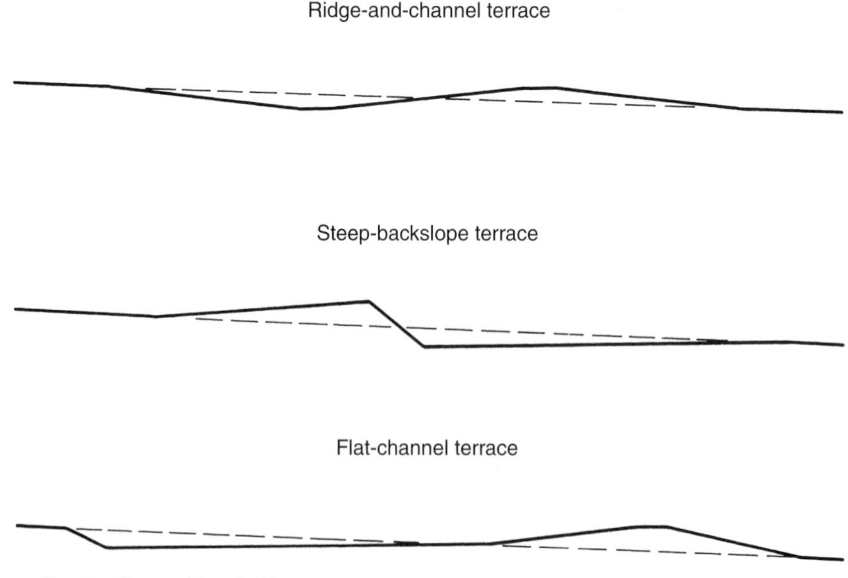

Figure 13–4 Types of level ridge-type terraces.

Figure 13–5 Diagram of a conservation-bench (Zingg) terrace.

it is to design and build the terrace system. The width of the flat channel is 100 ft (30 m) on a 0.5% slope with a ridge height of 20 in. (0.5 m); narrower channels are used on steeper slopes.

Conservation-Bench (Zingg) Terraces. Zingg and Hauser (1959) described an experimental system consisting of a series of level benches with steep, narrow ridges. The terraces were separated from each other by unleveled, runoff-contributing areas (Figure 13–5). A set was built on a land slope of 1% to 1.8% at the ARS Field Station at Bushland, Texas, where the proportion of contributing area to leveled area was 2:1. Benches occupied 13 ac (5.25 ha) in a 40-ac (16-ha) field. They varied in length from 1100 to 1400 ft (335 to 425 m) and in width from 80 to 145 ft (24 to 44 m). Maximum cut-and-fill depths were about 12 in. (30 cm). Four other sets were constructed in Colorado, Kansas, South Dakota, and Texas. Contributing area to benched area ranged from 1:1 to 3:1. Contributing areas were used for cropping systems common in the area, including summer fallow. The benched areas were seeded each year to an adapted crop, often grain sorghum, that can tolerate standing water for short periods.

Under conditions favorable for runoff production, water losses were markedly reduced, and yields on leveled areas were nearly doubled (Hauser, 1968, and accompanying articles by Black, Cox, and Mickelson in the same publication). These terraces were not totally successful. Crop yields on the excavated sections of the benches were lower than on the built-up areas, but this changed over time with good management. Water ponding in more humid regions and in wetter-than-normal years in drier areas caused occasional crop damage.

Conservation-bench terraces have not been built on many production farms. A farmer in Kansas started to replace regular level terraces with conservation benches in 1969; the system was completed within 10 years. The lips on the downhill side of his benches have outlets so that excessive collected water can be released, but he claims that no runoff water leaves his farm. Crop yields on the benches are as high each year as yields on untreated land that is fallowed every second year. It is strange that his neighbors have not taken up this practice.

13–5.4 Land Imprinting

Land imprinting is a mechanical method of land forming that generates microwatershed and microcatchment areas on degraded land. It has been used successfully to revegetate arid, barren (desertified) soils. Imprints accelerate the revegetation process by improving seed germination and seedling growth.

Infiltration can be impeded in soils with fine, smooth surfaces because the initial rainfall seals a thin surface soil layer (Dixon, 1995). This seal prevents soil air from escaping to the atmosphere. The small pressure buildup caused by the entrapped air slows infiltration or may even prevent additional water from gaining entrance to the soil. Infiltration continues where air can escape through associated macropores. Dixon and his coworkers designed an implement, the rolling land imprinter, that produces small drainage beds and open diamond-shaped slots for water entrance and storage. Each slot has an area of about 1 ft^2 (900 cm^2) and can hold from 3 to 5 qts (3 to 5 l) of water. The slots trap the water that runs off the microwatersheds and hold it until it infiltrates. The presence of macropores on the ridges and in the diamond-shaped catchments promotes higher rates of infiltration. This imprinter, with a drop seeder mounted above the imprinter drum, was used extensively and very successfully to reestablish native vegetation on severely denuded land in the Sonoran and Mojave Deserts of southern Arizona and southern California (Dixon, 1995).

13–5.5 Use of Chemical Wetting Agents

Chemical wetting agents (surfactants) were first suggested for increasing water infiltration into dryland soils in the early 1950s. Several materials were developed and tested, but none was practical on normal dryland soils. More recently, wetting agents have been recommended as additives for use in sprinkler irrigation systems. Some farmers use these materials, but there is no proof that they increase water intake rates on normal irrigated soils. Letey (1975) demonstrated that surfactants can increase the infiltration rate in water-repellent soils in California, but he does not recommend their use on normal soils (Note 13–1).

NOTE 13–1
EFFECT OF SURFACTANTS ON INFILTRATION

The major force moving water into and through soils during rainfall is capillary attraction. Capillary attraction is equal to $2y \times \cos \theta / r$ (twice the product of the surface tension of water and the cosine of the wetting angle between water and the solid it contacts, divided by the radius of curvature). A surfactant (wetting agent) reduces both y and θ. Effective wetting agents decrease surface tension (y) in laboratory systems by more than 50%. If the capillary attraction is to be increased (to increase infiltration), θ must be decreased enough so that $\cos \theta$ is increased more than the surface tension is decreased. The wetting angle between normal soil and water is generally 10° or less. Since $\cos 10° = 0.985$, it is impossible to decrease this angle enough to increase $\cos \theta$ by 50%. Thus, it is theoretically impossible for wetting agents to increase infiltration in normal soils.

Nonwetting soils have wetting angles much greater than those for normal soils; the angle may exceed 90°. Because the cosine of an angle greater than 90° has a negative value, these soils are truly water repelling. Capillary force will be greater and infiltration will be increased if this angle can be reduced sufficiently by adding a surfactant.

13–5.6 Improving Soil Structure

Runoff is inversely related to the infiltration rate and permeability of the soil. Soil texture and structure determine the rate of water movement into and through soil. Changing soil texture is impractical, but soil structure can be altered by management practices including the use of sod crops, crop and animal residues and mulches, reduced tillage, and the addition of chemical soil conditioners.

Cropping Systems and Soil Structure. The crops raised and the practices used in their culture affect soil structure. In dryland agriculture, where moisture conservation is most important, there are only a few adapted crops and practical cropping systems. Small-grain crops are prevalent in dryland cropping systems; these systems do not include sod crops in rotations because it is difficult to get good stands and they take a long time to establish.

Crop and Animal Residues and Organic Mulches. Crop residues incorporated into the soil have a beneficial effect on soil structure. Leaving residues on or adding an organic mulch to the soil surface helps maintain a better and more stable structure. Annual applications up to 25 t/ac (55 mt/ha) of manure or sewage sludge may be used to improve soil structure and increase soil productivity if the supplement contains no harmful chemicals. Applications to dryland soils often upset soil moisture relations and may reduce crop yields temporarily.

Tillage. The immediate effect of tillage on soil permeability is often beneficial. A recently tilled field is much more permeable than firm, untilled land unless the untilled area is protected by a dense vegetative cover. Postplanting cultivation of fine-textured, low-organic-matter soils improves permeability and reduces runoff, but the effect is temporary.

The long-term effect of tillage usually is reduced permeability and increased decomposition of organic matter. The more tillage a soil receives, the denser it becomes, the lower its infiltration rate, the slower its permeability, and the greater the runoff. This explains, in part, the beneficial effect of no-till crop-production methods on row crop yields on dense soils that tend to form surface crusts (Freese et al., 1995). Similarly, Bonfil et al. (1999) found that no-till management improved infiltration and water storage on soils in the 237-mm precipitation zone of southern Israel. They generally obtained higher wheat yields every year with no-till than those produced in alternate years with a conventional wheat–fallow system. Leaving the straw on the surface as a mulch increases infiltration, reduces evaporation, and avoids the soil loss that usually occurs with summer fallow.

Chemical Conditioners. Chemicals have been added to soils to improve soil structure and permeability and to reduce runoff. If the exchangeable sodium in sodic soils is replaced by calcium (by application of gypsum, as discussed in Section 15–7.3), the colloid will flocculate and desirable structure will regenerate, improving soil permeability.

Several synthetic organic chemicals have been promoted as soil conditioners. Hydrolyzed polyacrylonitrile (HPAN) was developed and promoted as Krilium™ by Monsanto Chemical Company in 1950. Several research articles on Krilium and other synthetic soil conditioners appeared in the June 1952 issue of *Soil Science.* Since that time, claims for many other materials with soil-conditioning properties have been made. Some materials improve soil structure or increase aggregate stability, but all have been economically impractical. Duley (1956) reported that applications of 0.5 to 2 t HPAN/ac (1.1 to 4.4 mt/ha) increased the size of aggregates and also increased infiltration rate. He also showed that applications of 2.5 t straw/ac (5.6 mt/ha) had about twice as much effect on infiltration as HPAN. The chemical then sold for $1.00 to $1.60/lb.

A new technique using polyacrylamide (PAM) on irrigated soils to increase infiltration was developed more recently. PAM applications as small as 0.6 lb/ac (0.7 kg/ha), added

Figure 13–6 A farm pond in Kansas that combines catfish production with water storage for irrigation. (Courtesy USDA Natural Resources Conservation Service.)

to the water in the irrigated furrows, increased infiltration rates by as much as 30% and reduced runoff greatly (Sojka and Lentz, 1994; Trout et al., 1995). See Chapter 15, Section 15–5.1, for more details. Unfortunately, applications of these low rates of PAM to nonirrigated soils do not increase infiltration or reduce runoff.

13–5.7 Water-Storage Structures

Methods for reducing runoff losses described so far all involve water storage in the soil. Aboveground storage in ponds, lakes, and reservoirs to meet irrigation, livestock, industrial, and other needs is also common.

Farm Ponds and Dugouts. Trapping surface runoff behind small earthen dams is a common way to provide water for livestock in areas where suitable well water is scarce or unavailable, or where additional watering sites are needed to ensure better use of pasture or range. Stock-water dams are usually sited on small watercourses that drain 25- to 100-ac (10- to 40-ha) grassed watersheds. They are relatively simple structures with a grassed spillway alongside the dam and about 3-ft (1-m) lower than the crest of the dam (Figure 13–6). See Section 11–3.4 for a discussion of the use of vegetation to control erosion around farm ponds. *Dugouts* may be used to store water on more level land, especially in higher latitudes where snow accumulates over winter and melts during a short spring thaw. Dugouts should be large and deep to hold enough water for the entire summer season (Figure 13–7).

Sites for both dams and dugouts must be selected carefully. Bottoms and sides of reservoirs and the dam embankment must be impermeable to retain stored water. Construction material for dams is usually obtained from the site of the reservoir. This minimizes hauling distance and makes the reservoir deeper—a factor that reduces the proportion of stored water lost by evaporation.

Figure 13–7 A dugout excavated in glacial till to impound water for livestock. A pump is located in the far corner. (Courtesy Canada-Manitoba Soil Survey.)

If impermeable soil is not available at the site, bentonite, a swelling clay, may be purchased and spread over the floor of the reservoir at a rate of 1 to 3 lb/ft^2 (5 to 15 kg/m^2) and mixed with the top 6 in. (15 cm) of soil. This adds materially to the cost of the structure. The bed of the reservoir and the dam itself should be thoroughly compacted during construction using a sheepsfoot packer or by herding cattle or other animals in the basin to help seal the surface.

An alternative is to place plastic film, butyl rubber, or a thin asphalt layer over the entire reservoir bed. The film or layer must be applied very carefully to ensure complete coverage with no breaks, and the film must be protected with 6 in. (15 cm) or more of fine (not cloddy or stony) soil. It is absolutely essential that a site so treated be fenced to exclude livestock.

Small dams and dugouts can be constructed by individual farmers, but many will want to have the work done by a competent conservation contractor with large earth-moving equipment. Construction specifications for both dams and dugouts should be obtained from the Natural Resources Conservation Service or from a competent engineering consultant.

Ponds should be fenced to prevent bank trampling and to reduce eutrophication. A pipe should be installed through the dam to carry water to a trough or tank below the dam. Dugouts must also be fenced to keep livestock away from the steeply sloping sides. Animals may be permitted access to the water by way of the ramp used by the earth-moving equipment during construction, but it is better to keep them out and pump water to a tank above the dugout. A reservoir providing water for home use must be fenced, and animals should be kept out of the entire watershed to reduce the possibility of contamination.

Municipal Water-Supply Reservoirs. Many small- to medium-sized dams have been built to trap streamflow (runoff) for use in municipal water systems. More of these reservoirs will be required as population increases.

Large Irrigation-Water Reservoirs. Flow in the major rivers of most countries is seasonally greater than the needs of downstream users. Where irrigation water can

Figure 13–8 Lovewell Dam and Reservoir in north-central Kansas was built by the Bureau of Reclamation as an irrigation storage reservoir. The irrigation water discharge outlet and canal are in the foreground beside the spillway. Water was being discharged from the spillway when the picture was taken. (Courtesy USDA Natural Resources Conservation Service.)

be used effectively, seasonal surpluses can be stored for use during the growing season. Many reservoirs, such as the one shown in Figure 13–8, have been built in the western half of the United States and in dryland areas of other countries. Because more surplus water flows in the rivers and streams of humid areas than in most dryland rivers, storage for use during the short, dry periods in humid regions may be a reasonable project of the future.

Organic mercury compounds can build up in reservoir water and in large boggy areas where dense stands of large trees are submerged. Mercury levels in the waters of some reservoirs, constructed for hydroelectric-power generation in northern Canada, have increased dramatically. Fish from these reservoirs now contain so much mercury that they pose a health threat to humans that consume them.

It is essential to recognize that water stored behind dams submerges valley land, often the best and most productive soils in the area. Gains from the use of this stored water must be discounted by the production losses from inundated land.

13–6 REDUCING EVAPORATION LOSSES

Evaporation losses from soils probably amount to 30 to 50% of the annual precipitation in subhumid areas and may exceed 70% in dryland regions. Reducing these losses, especially in drylands, would significantly increase the water supply for crop production.

The earliest recommended method for reducing evaporation from soil was to till the surface into a *dust mulch.* The idea was to disrupt capillary rise, but evaporation from the tilled soil usually left no net gain in water storage, and the practice produced an extremely erodible soil surface. Using vegetative mulches and other mulching materials, and forcing water to percolate deeper, are now the preferred methods to reduce evaporation from soil.

A hand-tillage variation of the dust mulch appears to be more successful for conserving water. A special hoe called the *hilaire* is used for this purpose in semiarid West Africa (Payne, 1999). The hilaire has a blade that is moved in a horizontal plane 4 to 5 cm (1.6 to 2 in.) below the soil surface to cut soil capillaries and the roots of weeds. This type of hand tillage performed immediately after each significant rainfall reduced evaporation losses by as much as 47 mm (1.8 in.) for the season and increased millet yields about 70%. The careful timing and the relatively shallow depth of tillage probably improved the results as compared to the use of a disc implement for dust mulching.

13–6.1 Vegetative Mulches

Mulches of crop residue, forest litter, sawdust, and wood chips not only reduce runoff (as discussed in Section 13–5.1), they also reduce evaporation by reducing soil temperature and wind velocity near the soil. A German investigator first demonstrated the effects of mulches on evaporation. In a one-month study with constantly moist soil, he found that 2 in. (5 cm) of fir needles reduced evaporation 89.2%, from 2.26 in. (5.739 cm) from bare soil to 0.24 in. (0.621 cm) from a mulched soil (Eser, 1884).

More recent studies show that evaporation losses from moist, unmulched soils are much more rapid than those from mulched soils. Over time, losses from the unmulched soil become slower as the soil dries out, while those from the mulched soil continue until it, too, becomes dry if additional moisture is not supplied. Mulches reduce water loss most in years and in regions in which rain wets the soil frequently (Brun et al., 1986).

13–6.2 Nonvegetative Mulches

Nonvegetative mulches also reduce evaporation losses. Stones and gravel on surface soil increase infiltration and reduce water loss. Valentin and Casenave (1992) showed that increasing amounts of small-sized stones increased infiltration, whereas increasing quantities of larger stones (> 2.9 cm), especially when they were embedded in the crusted surface layer of soils, decreased infiltration. Fairbourn and Gardner (1975) coated natural soil aggregates with water-repellent materials, applied the treated aggregates as mulches, and found they decreased the moisture lost by evaporation.

Black polyethylene film has been used in irrigated areas and in humid parts of the United States to produce high-value vegetable crops. Increases in vegetable yields resulted from weed control and from higher soil temperature (which favored early growth) as well as from reduced evaporation.

In an Iowa study, corn was grown to maturity on plots covered with black plastic film that prevented evaporation but also stopped rainfall from entering the soil. Corn on the plastic-covered plot, with only the water that was in the soil at seeding time, produced a yield nearly as high as the adjacent unmulched plot that received normal rainfall supplemented by several irrigations (121 to 129 bu/ac or 75.9 to 80.9 q/ha) (Shaw, 1959).

In an Arizona study, soils sprayed with asphalt immediately after cotton and sorghum seeding held more moisture in the seed zone and had improved germination and early growth. Good results were obtained on some soils, but the surface had to be smooth to economize on the asphalt used. If soil shrinkage occurred, the film cracked and was useless.

Tar-paper caps placed on ridges between sorghum rows increased yields from 38 to 240% in western Kansas. Yield increases were credited to three effects of the caps:

1. Evaporation was reduced because a smaller surface area was exposed to sun and wind.
2. Rainfall was concentrated onto smaller areas so it percolated deeper into the soil and was less subject to evaporative forces.
3. The cover maintained faster water absorption by preventing raindrops from compacting a large part of the soil surface.

13–6.3 Forcing Deeper Water Penetration

Water close to the soil surface is more subject to evaporation than water deeper in the profile, so deeper percolation should save water. The Kansas study using tar-paper caps (previous section) caused deeper percolation because water was concentrated on a smaller area of soil. Fairbourn and Gardner (1974) used this technique in Akron, Colorado. The soil surface was formed into beds 6.6-ft (2-m) wide. Each bed had two 29.5-in. (75-cm) sloping (3:1) contributing areas leading down from each side onto a 20-in. (50-cm), flat, collecting area. The flat area had a 6-in. (15-cm) deep vertically mulched slot running down the middle with a row of sorghum planted 6 in. on each side of the slot. The slopes were treated with hydrophobic material to further reduce water penetration on the sloping shoulders. Check dams were constructed at intervals across the flat furrows to prevent water movement in the furrow. The treated plots saved 41% more rainfall and produced from 37 to 150% more sorghum grain than did regular flat-planted, unmulched land.

A similar system designed to make better use of light rainfall in China is described by Li et al. (2001). They made *contour ridges* 60-cm wide with 60-cm flat areas between them, as shown in Figure 13–9. The ridges were covered with sheet plastic, and corn was

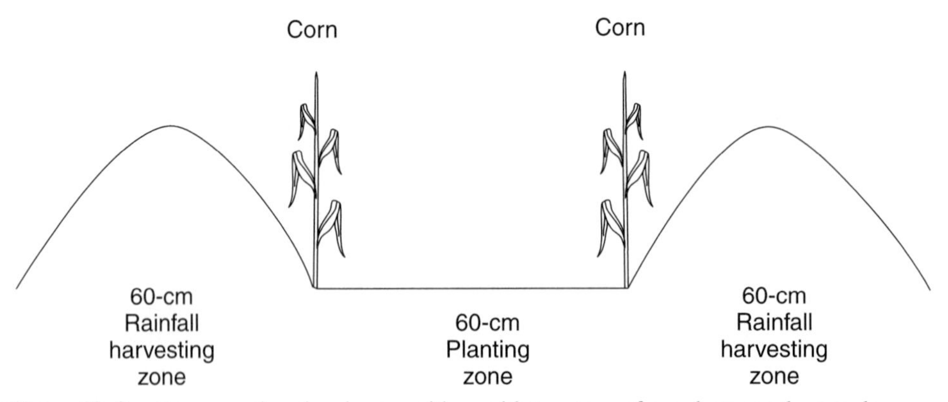

Figure 13–9 A cross section showing two ridges with two rows of corn between them as described by Li et al. (2001). The ridges are covered with sheet plastic so rainwater will flow from the ridges into the soil where the corn is growing.

planted in the flat area next to each ridge. Even a light rain produced enough runoff from the ridges to be stored in the soil for plant use. Their corn yields were increased by 10 to 143% by this system, combined with a pebble, gravel, or sand mulch in comparison to control areas without these treatments. They considered the system to be economical in China in spite of the high labor input required.

Dixon's rolling land imprinter (Section 13–5.4) is effective because the land forming that results from its use concentrates rain from the microwatersheds onto the microcatchment areas. This extra moisture penetrates deeper, and less is lost by evaporation.

13–6.4 Reducing Losses from Reservoirs

Much of the runoff water caught in reservoirs is lost by evaporation. Annual evaporation from open-water surfaces in agricultural regions ranges from about 13.8 in. (350 mm) in north-central Canada (54° N latitude) to more than 78 in. (2000 mm) in northern Nigeria (13° N). Hot desert areas suffer even larger losses. Evaporation rates are related directly to temperature and wind velocity, and inversely to relative humidity. Losses from small reservoirs often exceed livestock and household use.

Methods for reducing evaporation losses from reservoirs are constantly being sought. For example, a field shelterbelt planted around a reservoir site reduces evaporation losses by reducing wind velocity over the water. Trees beautify the site and make a useful recreational area and wildlife sanctuary. They also trap snow for reservoir recharge. Trees should be at least 100 ft (30 m) from the water in order to prevent them from taking water from the reservoir and transpiring it.

Materials such as hexadecanol and other long-chain fatty alcohols form monomolecular films on water surfaces, reducing evaporation about 20%. Gentle winds blow the films toward the water's edge and leave the water surface less protected. They are not very practical because they are biodegradable and lose effectiveness with time. Nicholaichuk (1978) described rafts made of inexpensive, lightweight concrete that are as effective as monomolecular films and much more durable.

13–7 REDUCING DEEP PERCOLATION LOSSES

Deep percolation losses are of little significance in nonirrigated dryland areas because the limited rainfall can usually be trapped in the root zone, except possibly in sands or in sites that receive runon water. Percolation losses are likely in humid and irrigated areas.

Deep percolation is a common avenue of water loss in humid regions, but this loss does not seriously reduce the amount of water available to plants unless the soil has a low water-holding capacity. Most deep-, medium-, and fine-textured soils hold 6 to 10 in. (150 to 250 mm) of available water before water moves below the root zone. This is usually sufficient to last the crop through a short dry period. Sandy soils lose large amounts of water through percolation because some of them hold less than 2 in. (50 mm) of available water in the root zone. Reducing deep percolation losses in sands could increase water available to crops and thereby increase crop yields.

Erickson et al. (1968) demonstrated that horizontal asphalt barriers 22 to 24 in. (55 to 60 cm) below the soil surface in Bridgman fine sand (unclassified) and Grayling loamy sand (Typic Udipsamment) appreciably increased the water available to plants. Vegetable crop yields were increased by 35 to 40% on the fine sand, but only small increases occurred on the loamy sand soil. A trial in Taiwan on a sandy soil showed paddy rice yields were increased 10- to 14-fold with barriers placed at depths of 8 to 24 in. (20 to 60 cm) (Erickson, 1972). Material costs have increased greatly since 1968, but there are situations where the technique may still be usefully employed.

Somani and Sharma (1985) achieved some reduction in seepage losses by compacting subsurface layers of sandy soils in a dryland area of India. This practice is not likely to be practical in commercial agriculture unless it is mechanized. Not all sands can be compacted to reduce the percolation rate.

A chemical that absorbs water could prove useful for holding more water in sandy soils. A copolymer of starch and acrylonitrile named *super slurper* was reported to absorb up to 1400 times its weight of water. Preliminary reports indicated that additions of this material increased water-holding capacity of sandy soils, but its use was not economically feasible.

Deep percolation leaches nutrients and makes soils more acid. Controlled runoff (without serious erosion) in humid regions actually may be less damaging than deep percolation, particularly where fertilizers and lime are expensive and hard to obtain. In many tropical areas with monsoon rainfall, water conservation is important mainly at the beginning and the end of the rains. During the heart of the rainy season, good water management should emphasize controlled surface drainage rather than surface detention and infiltration.

13–8 STORING WATER IN SOIL

Soils in humid regions are usually wet to field capacity beyond rooting depth at least once each year. Water stored over winter in North America's Corn Belt soils makes high corn and soybean yields possible even though normal rainfall is not sufficient to meet crop short-term needs during dry periods in July and August. It can be equally important in sections of the Intermountain Valleys and the Pacific Northwest. Winter storage would be important in dryland areas also, but it is limited by meager precipitation. Crop production under these conditions may be possible if water can be accumulated in the soil, either by concentrating the supply with water harvesting techniques or by using summer fallow to store water from one year to add to the next year's supply.

13–8.1 Water Harvesting

Water harvesting is a general term for several methods of collecting water and concentrating it in a small enough area to grow plants that require more water than the normal rainfall of the area provides. Installations that gather water and use it for irrigation are discussed in Chapter 15, Section 15–3.3. Smaller installations where water is stored only in the soil are discussed here.

The use of *catchment structures* to concentrate runoff and trap it in prepared sites is an ancient art in sections of the Negev and Sinai Deserts that are too dry for normal crop production (Medina, 1976). These early catchment structures and channels, often covered with flat stone flakes, diverted runoff from winter rainstorms down hillsides and onto prepared, permeable sites in the valleys. The valley sites were crisscrossed with stone fences that trapped soil and held water until it moistened the soil to considerable depths. The moistened sites were then used to produce food for the relatively large population.

Runoff concentration was also accomplished by constructing individual *microwatersheds,* each delivering runoff toward its center where a single fruit tree or a small patch of vegetables or vine crops was located. Contributing land, as much as 20 to 25 times the area of the cultivated site, was compacted and kept free of vegetation (Medina, 1976). Israeli farmers use this technique to grow crops in areas that receive less than 8 in. (200 mm) of annual rainfall.

Ojasvi et al. (1999) describe *micro-catchments* used in a 264-mm rainfall zone in India to grow jujube trees as an agroforestry fruit crop. Each tree was planted in the center of a conical depression with a 1-m (40-in.) radius; the outer part of the cone was 15-cm (6-in.) higher than the center. They tried various mulches on the soil surface and found that either stone or marble waste from building construction was a very effective mulch for improving water storage in the micro-catchments and increasing the establishment and growth rates of the jujube trees.

Another method using *trenches* to increase infiltration in a heavy clay soil in Jordan is described by Abu-Zreig et al. (2000). Trenches 80-cm (32-in.) deep and 1-m (40-in.) wide were dug across the slope between tree rows in an olive orchard. The trenches were filled with sand and fractured rock that were available locally. Runoff water was caught in the trenches and seeped gradually into the adjoining soil, thus significantly increasing the amount of water stored without any interference with traffic through the orchard.

Lagoon-leveling has been studied as a means of improving water distribution. A lagoon (pot-hole, slough, or swale) is a small depression that collects runoff water from the surrounding land and remains wet or ponded for extended periods. Lagoons in pastures provide temporary water for livestock, but the wet spots in cultivated fields are a nuisance and rarely produce a good crop. Lagoon water can be spread more widely by moving soil from above normal water level to the center of the depression. Water is absorbed more quickly when spread over more surface. This improves both farming convenience and crop yields.

Several lagoons were leveled on the ARS station at Akron, Colorado. One 2-ac (0.8-ha) lagoon was leveled to 4 ac (1.6 ha) by backsloping the lagoon walls to a 5% slope and using the excavated material to fill the center of the depressed area. Maximum cuts were about 24 in. (60 cm); maximum fills were 20 in. (50 cm). Cost of leveling was considerably less than the price of productive land. Annual forage yields are as good as, or superior to, those on unleveled upland sites. The lagoons are no longer a hindrance to field operations (Mickelson and Greb, 1970).

Polyethylene film and similar impermeable materials have been used to waterproof soils and cause 100% runoff to catchment sites. The extra water makes crop production possible every year on half the land instead of alternating fallow and crop every other year on all the land. The cost of the film and its short life in windy climates make the practice uneconomical.

Mehdizadeh et al. (1978) sprayed microwatersheds in Iran with an impermeable asphalt formulation and used the runoff to supplement rainfall for tree production. The contributing area was 1.5 times the cultivated area. During five years, the collecting areas received 19.2 in. (487 mm) of extra runoff, or 35% of the total rainfall. This is an attractive alternative to providing irrigation water for tree growth.

Recharging underground aquifers by spreading runoff water on soil is an important practice where water is available and where the surface conditions and the underlying geological formations are suitable, as discussed in Chapter 15 in Section 15–3.4.

13–8.2 Summer Fallow

Summer fallow is the practice of leaving a field without any crop during a growing season. The major objective of summer fallowing is to store water in the soil for subsequent crop use. Weed control, release of extra nitrogen for the use of a succeeding crop, and spreading the workload of seedbed preparation over a longer period may be reasons for using summer fallow in some areas.

The process is not very efficient, but some water can be stored in the soil during a growing season if plant growth is prevented. This water will supplement rainfall in the next growing season. This must be distinguished from the *fallow* under several years of forest or grass regrowth used for soil recuperation in shifting cultivation.

Summer fallow has been used in various parts of the world for several centuries. It was used extensively in England during and after the Roman occupation. It was first used in North America in the early 1800s by the Selkirk settlers in the area north of Winnipeg, Canada, but it was not widely used in dryland regions until the 1880s, shortly after the Great Plains area of the United States was first settled. The area of fallowed land west of the Mississippi River in the United States averages a little less than 35 million ac (14 million ha). Summer fallow is also common in western Canada, over vast areas of the former USSR, in Australia, and in many of the wheat-growing areas in southern South America. Recent studies in a semiarid area in southern Africa show that fallow may improve water-use efficiency as well as increase crop yields (Jones and Sinclair, 1989).

Efficiency of Water Storage by Fallow. A large number of federally supported agricultural research stations and several state stations in the United States and Canada studied moisture storage in fallow-crop and continuous-crop systems from the early 1900s until the late 1940s. The average amount of rainfall stored in the summer rainfall area during a 15- to 21-month fallow period ranged from 1.5 in. (38 mm) (6%) at Dalhart, Texas, to 6.17 in. (157 mm) (24.6%) at Sheridan, Wyoming. During the same period, moisture stored between harvest and planting in continuous-crop systems ranged from 0.76 in. (19 mm) (10%) at Dalhart to 3.88 in. (99 mm) (40%) at Sheridan (Mathews and Army, 1960). Fallow storage efficiency at Swift Current in western Canada was nearly 27% (Doughty et al., 1949). These results explain why summer fallow is used less commonly in hot drylands. More recent studies quoted by Smika and Unger (1986) show increasing efficiency of water storage in the central Great Plains with improved cultivation methods (Table 13–1).

Water-storage efficiency of summer fallow in the winter-rainfall area of the United States is usually greater than in the Plains area. More than 60% of the precipitation is stored during the early winter period when temperature is low and precipitation is relatively abundant. There is often a net loss of soil moisture during the relatively dry summer season.

Soil permeability also affects the efficiency of water storage in fallowed land. Sandy soils are permeable and absorb rainfall readily. Percolating water goes deeper into a sandy soil than it does into a finer-textured soil, and less is lost by evaporation. Sandier soils store rainfall more efficiently only if they have sufficient storage capacity to hold the percolating water in the root zone. The best soils for efficient water storage are those with sandy surface layers overlying medium-textured subsoils.

Efficient storage of moisture in fallowed land depends on successful weed control and on maintaining crop residues on the soil surface to retard evaporation, as discussed in

TABLE 13–1 EFFECT OF IMPROVING TECHNOLOGY ON WATER STORAGE AND WHEAT YIELDS ON SUMMER FALLOW IN THE CENTRAL GREAT PLAINS

Years	Tillage during fallow[a]	Fallow water storage			Wheat yield	
		inches	mm	Percent of precipitation	bu/ac	q/ha
1916–30	Maximum-tillage; plow, harrow (dust mulch)	4.0	100	19	15.9	10.7
1931–45	Conventional tillage; shallow disk, rod weeder	4.6	116	24	17.2	11.5
1946–60	Improved conventional tillage; began stubble mulch in 1957	5.4	137	27	25.7	17.2
1961–75	Stubble mulch; began minimum tillage with herbicides in 1969	6.2	157	33	32.1	21.5
1976–90	Projected estimate; minimum tillage; began no tillage in 1983	7.2	183	40	40.0	26.8

[a]Based on 14-month fallow period, mid-July to mid-September of the next year.

From *Advances in Soil Science,* Volume 5, p. 111–138, 1986 (Smika and Unger) by permission of Springer-Verlag, New York, and the authors.

Section 13–6. Weeds must be killed with chemicals or with a minimum amount of shallow tillage. Extra tillage dissipates moisture, increases costs, buries residues, and does not increase crop yields. Horizontal blade implements such as those discussed in Section 9–5.1 minimize soil mixing as they cut the roots of weeds below the surface. These implements bury less residue and cause less water to evaporate than tillage with a disc implement.

Improved water conservation methods and more drought-resistant crop varieties help make annual cropping more attractive as a means of avoiding the soil and water losses that are associated with summer fallow. Wheat and other small grains are likely to have higher yields when they follow fallow, but annual cropping gives a crop every year and is likely to produce more during the two-year period. An alternative in areas where summer rainfall is significant is to alternate wheat with a summer crop such as corn or sorghum. Farahani et al. (1998) found that a three-year rotation of wheat, corn or sorghum, and fallow made better use of rainfall in the Great Plains than the prevailing wheat-fallow system.

Frequency of fallow in a cropping system should be governed by soil moisture and by climatic conditions at the time. A dryland soil should be fallowed when its moisture content just before seeding is insufficient to assure a reasonable yield. Medium- and fine-textured soils in dryland regions are generally fallowed once in two to four years. Government restrictions on crop production in the United States have forced farmers to cut back on the area seeded to the major cash crops and caused more land to be summer fallowed than necessary in many areas.

Crop Response to Summer Fallow. Crop yields on summer fallow in dryland areas are higher than yields on land that produces a crop each year. The amount of increase depends on how favorable the conditions are for extra moisture storage by fallowing. Average wheat yield increases until the early 1950s ranged from 0.2 bu/ac (13 kg/ha) (4.4%) at Big Spring, Texas, to 10.7 bu/ac (7.2 q/ha) (120%) at Colby, Kansas, in the Great

Plains, to 30.2 bu/ac (20.3 q/ha) (153%) at Moscow, Idaho, in the winter-rainfall area of the intermountain valleys.

Other adapted crops respond to summer fallow about the same as wheat does. Percentage increases of oats and barley yields are as great as those of wheat in the northern Great Plains. Percentage increases of sorghum yields are as great as, or greater than, those of wheat in the southern Great Plains. Corn is not well adapted to most dryland areas; its response to fallow is small and erratic. Yield increases indicate, as does water-storage efficiency, that summer fallow is most useful in dryland areas with cool temperatures where evaporation losses during storage are relatively small.

13–8.3 Summer Fallow and Saline Seeps

Considerable areas of *saline seeps* are developing in the northern Great Plains. Summer fallow is one of several factors causing saline seeps, as discussed in Section 14–3.7. The likelihood of deep percolation causing water losses from fallowed soils increases as the amount of moisture storage increases. Stored water may move downward in the soil beyond the reach of crop roots during a summer fallow season and be lost to crop production. Percolating water carrying dissolved salts from the soil may resurface farther down the slope and provide plants there with extra water and nutrients. The salts in excess of those used by plants accumulate in the soil and in time may produce saline seeps.

One obvious method of reducing the magnitude of saline seeps is to restrict fallowing to areas where the water that can be saved in the soil will not exceed the amount that can be stored in the root zone. This would mean a shift from alternate fallow and crop to less frequent occurrence of fallow or to continuous cropping when water losses are reduced. A technique for developing a more flexible method of combining crops and fallow is described in Section 13–9.1.

13–9 EFFICIENT USE OF STORED SOIL WATER

Water storage in soils is not an efficient process, but stored water is extremely important to a subsequent crop. The soil profile in most dryland areas is usually dry to rooting depth at harvest time. Moisture recharge starts at the surface of the soil, with rain forcing the moisture deeper as water accumulates. The amount of stored water, therefore, can be determined by assessing how deep the soil has been moistened. Depth of moist soil multiplied by the available water-holding capacity of the soil gives the amount of available stored water.

Stored soil water must be husbanded carefully if it is to serve the crop and the farmer well. There are several ways to reduce unnecessary loss of stored soil water:

1. Plant a crop only when there is enough stored water for a reasonable chance of successful production.
2. Grow efficient crops.
3. Plant and cultivate with timeliness and precision.
4. Plant at the proper seeding rate.
5. Control weeds and volunteer crop plants.
6. Use windbreaks to reduce transpiration.

13–9.1 Predicting Successful Crop Production

Water stored in a soil is wasted if a planted crop fails to mature and produce a harvest. A technique for predicting cropping success could help reduce this unnecessary water loss. For example, the depth of moist soil at planting time has been used to estimate the likelihood of successful crop production and to make the decision on abandoning a winter wheat crop. Early abandonment (in April) permits saving some of the stored soil moisture for the next crop.

Brown et al. (1990) and Brown and Carlson (1990) studied the relationship between water used (stored water at seeding time plus growing season rainfall) and crop yield. They developed yield equations for the cereal grains and safflower grown in Alberta, Saskatchewan, Montana, and North Dakota. One such equation for winter wheat in northern Montana is

$$Y = 5.8\,(ET - 3.9)$$

where Y is yield in bu/ac, and ET is the sum of the initial available soil moisture and the growing season rainfall in inches. The figure 3.9 is the *initial yield point* (*IYP*), the amount of moisture that is required to produce the vegetative portion of the crop with no grain yield. Every inch of water used above *IYP* produces an additional 5.8 bu/ac of winter wheat. Estimates of the amount of soil moisture that must be available for a profitable crop can be based on the likelihood of rainfall in the area.

On the basis of their results, Brown et al. (1990) recommended a flexible cropping system—seeding a crop or fallowing the field based on the amount of stored water available at seeding time plus expected growing season precipitation.

To translate the quantity of soil moisture needed for a reasonable yield into depth of moist soil, it is necessary to know the available water-holding capacity of the soil. Values for various textures are given in Table 13–2. If it is established that 4 in. (10 cm) of available soil moisture is required to produce a crop under normal rainfall conditions, then it can be determined from the table that an average silt loam soil must be wet 2-ft (60-cm) deep in order to store this much water.

A soil moisture probe, soil sampling probe, or even a metal rod with a handle on one end can be used to determine moisture depth by pushing it into the soil (without turning) until it meets strong resistance. Probes readily penetrate moist, stone-free soil, but it is difficult to force them into dry soil. Using either a soil sampling probe or a rod with a short piece of auger on its tip will permit a sample of soil to be removed to check texture.

TABLE 13–2 APPROXIMATE AVAILABLE WATER-HOLDING CAPACITIES OF VARIOUS SOIL TEXTURES

Soil textural class	Plant-available water[a]	
	in. per foot of soil	cm per cm of soil
Sand	0.6	0.05
Loamy sand	1.0	0.08
Sandy loam	1.5	0.125
Loam, silt loam, or silt	2.0	0.17
Clay loam, silty clay loam, or sandy clay loam	2.1	0.175
Clay, silty clay, or sandy clay	2.0	0.17

[a]Actual measurements commonly vary as much as ± 25% from these values, depending on soil organic matter, soil structure, type of clay, etc.

When the soil is too dry to support a successful crop, the land should be set aside and kept free of weeds (i.e., fallowed) so that additional water can be stored for producing a crop during the next growing season. Computer programs have been developed to estimate yield potential based on plant-available soil moisture at seeding time, growing-season precipitation, and management factors, such as crop species and variety, crop rotation, potential weed and insect problems, soil fertility, and planting date.

Wise postponement of planting and judicious use of abandonment permit farmers to obtain the advantages of summer fallow without specifically setting aside large areas of land for this purpose every year. It gives them more flexibility.

13–9.2 Growing Efficient Crops

Crop plants generally use between 15 and 30% of the rainfall, mostly for transpiration. A small portion of the water becomes part of the plant tissue, but much of this is lost if the plant material is dried. A comparison of the amount of water transpired to the amount of dry matter produced is called the *transpiration ratio.* Some early transpiration ratios published by Briggs and Shantz (1914) showed average values of 298 for sorghum, 368 for corn, 481 for wheat, and over 1000 for some grasses. This might suggest that sorghum and corn should use less moisture and be better adapted than wheat to regions where moisture is in short supply, especially since the corn and sorghum roots are capable of reaching deeper than those of wheat. But, when these crops are grown under similar conditions in a semiarid area, each one uses all of the available water at a rate that depends on the crop. The corn runs out of water first, and then the sorghum goes dormant (but may revive if moisture conditions improve). Wheat and other small grains are grown in such areas because they grow during cool weather and thereby use water more efficiently and produce some grain even under relatively dry conditions. Even so, some farmers grow corn on part of their land as a fodder crop in semiarid areas. The corn stover they produce probably contains more nutrients than stover from mature corn because there has been no translocation to the grain.

In less arid areas, selecting crop varieties based on the number of days required to reach maturity permits partial control of the amount of soil water that is used and how much remains in the soil at harvest time. It also determines how much time will be available for water recharge between harvest and the next seeding.

Farmers in many developing countries use mixtures of varieties and even of species, seeding drought-tolerant and water-loving types together, so that the chance of crop failure is reduced in drier years without losing productive potential in more humid periods. Such mixtures are not used in commercial agriculture.

13–9.3 Timeliness of Operations

There is an optimum time for most crop-production operations—controlling weeds, planting crops, and so forth. Delayed cultivation or delayed spraying to control weeds means moisture loss. For example, seeding winter wheat too early causes the crop to grow luxuriantly and use excess stored water in the fall; if too late, the crop will grow slowly and may fail to establish properly before winter. In humid regions, it is usually possible to plant within a few days of the optimum date; in drier areas, planting often must wait for a "planting" rain to replenish surface soil moisture to ensure germination and early growth. There may be only one "planting" rain or none close to the optimum planting date.

Land must be ready for planting—weed free and in good shape to receive seed—when seeding time arrives. When the rain comes, seeding should begin as soon as the surface dries off, without wasting time or moisture. If the soil needs cultivating when the rain is received, moisture will be lost by tillage and the soil may become too dry for planting. If so, the moisture is lost, and planting must wait for another rain.

Because of the uncertainty of rainfall at planting time in dryland areas, there is psychological pressure on the farmer to plant winter wheat when a good rain comes two or three weeks before the optimum planting date. Early planting often causes very vigorous early growth and excessive use of stored water. Judicious pasturing of the wheat in the fall and early spring can reduce excessive depletion of moisture reserves. The pastured wheat still protects the soil from wind erosion and often yields more than an unpastured crop. Still, early planting is a risky practice. Early planting can also expose the crop to certain diseases and to attack by the egg-laying Hessian fly. This insect can devastate a wheat crop and, to a lesser extent, a barley or rye crop, so planting should await the announcement of the fly-free date unless a resistant variety is used.

Timeliness is also important in tropical areas with wet-dry seasons. Delayed planting at the end of the monsoon may allow too much water to be lost from the soil by evaporation and leave too little for the dry-season crop.

13–9.4 Rate of Seeding

The amount of water available for crop production dictates the number of plants that can be supported. Population density must be restricted where moisture is limited, or all plants will suffer moisture stress.

Corn, soybeans, and other crops that do not tiller must be planted at the population dictated by the climate, soil fertility, and other factors. Widely-spaced corn plants (low populations) may produce bigger ears under favorable conditions, or more than one ear per plant in some varieties, but attaining the full yield potential for that set of ecological conditions requires an adequate plant population. Wheat, other small grains, and sorghum often counteract low seeding rates by tillering profusely under favorable conditions. Thus, wheat yield may not be greatly different whether seeded at rates of 30 or 75 lb/ac (35 or 85 kg/ha) because the thin crop will tiller to fill the space. But if too much seed is used, the plants compete for water and space so that no individual plant grows well. In extreme cases, water is insufficient to bring the crop to maturity. A thinner stand under the same low-moisture regime may survive and produce a harvestable crop.

Studies were conducted in the drylands of North America to discover if planting row crops in widely-spaced rows (80 in., or 2 m, as opposed to the then standard 40-in. rows) would reduce water use enough to increase the growth of the following wheat crop. Moisture use was reduced by this technique, and the yield of the succeeding wheat crop was increased. The yield of the widely-spaced row crop, however, was reduced, and the total production per unit area for the two-year sequence was always less on the widely-spaced plots. For example, at Colby, Kansas, wheat yields averaged 3.9 bu/ac (2.6 q/ha) more on the widely-spaced corn land, but the wide-row corn yielded 5.1 bu/ac (3.4 q/ha) less. Use of wide-spacing of row crops did not ensure against wheat failure either. Only once in 31 years was wheat grain harvested from the wide-spaced plots when the crop was destroyed by dry conditions on the narrow-row plots (Kuska and Mathews, 1956).

13–9.5 Controlling Weeds and Volunteer Plants

All plants absorb and transpire water. Weeds and other unwanted plants use moisture from summer fallow and from cropped fields. One Russian thistle plant uses about as much water as three sorghum plants, and two wild sunflowers use as much as five corn plants. Widely-spaced Russian thistle plants (100 ft^2 or 9 m^2 per plant) in western Canada used nearly 500,000 lb water/ac (560,000 kg/ha). This is enough to produce 8 bu/ac (4 to 4.5 q/ha) of wheat.

Weeds and volunteer plants of the wrong crops must be prevented if maximum conservation of available water is to be achieved and maximum crop yields obtained. Transpiration by weeds and volunteer crop plants often is the largest single cause of stored-water loss. It is especially costly under dryland conditions, where every drop of water is precious. The control of this loss is almost completely in the hands of the farmer.

13–9.6 Windbreaks and Field Shelterbelts

Barriers that are perpendicular to the wind reduce wind velocity and affect air temperature near the ground. Potential transpiration close to the barrier is reduced. Reductions in transpiration of 10% or more are obtained for as far as 15 to 20 times the height of the barrier to leeward. Greatest reductions are at night, with the maximum, about 30%, being found at a point twice the windbreak height to leeward. Greatest reduction at midday, about 20%, is attained about eight times the windbreak height to leeward (Woodruff et al., 1959).

Changes in transpiration are sufficient at times to prevent desiccation of plants in the shelter of the windbreaks and to reduce the effect of drought on yield. Unfortunately, reductions in crop transpiration are offset in part by the use of water by the trees, reducing crop yields next to the windbreak unless the tree roots are severely pruned.

13–9.7 Antitranspirants

Long-chain fatty alcohols and some long-chain fatty acids are known to reduce transpiration by plants. Wilting of transplanted tree seedlings and other plants has been reduced by spraying leaves with octadecanol and hexadecanol. Trials with field crops also show reduced transpiration, but plant growth is reduced also. Similar applications to soils rather than to plants have no significant effect on water use (Peters and Roberts, 1963).

Fuehring (1973) reported reductions in water use and significant increases in crop growth from small applications of phenylmercuric acetate (PMA), atrazine, and Folicote to sorghum just before the boot stage. Brengle (1968) showed that PMA reduced water use when applied to spring wheat at heading or flowering, but plant growth also was reduced. Interest in antitranspirants remains high, and other materials may be found that decrease transpiration without decreasing yield. It seems unlikely, however, that any will prove economically useful.

SUMMARY

Water is important for many reasons, including being essential for life, causing erosion, and functioning in weathering processes and in the decomposition of organic materials. It is almost universally present, yet the supply is commonly either deficient or in excess. Deficiencies may be overcome by irrigation or by water conservation.

The oceans contain about 97% of Earth's water. Evaporation carries a small part of the ocean water over the land where it condenses as rainfall, snow, sleet, or hail. It flows and seeps back to the ocean, but along the way it supplies the needs of living things and performs all its other functions.

Dry periods of varying length cause problems for crops, livestock, and people all over the world. Problems in the driest climates are aggravated by irregularity of precipitation. Without irrigation, the only way to meet the challenge of dry weather and drought is to conserve and use wisely the water that is received. Good water management prevents damage from nearly all water shortages in humid regions. The use of all practical conservation measures prevents damage and hardship in only the least-severe droughts in arid areas.

Water from precipitation is lost by plant interception, runoff, evaporation, transpiration, and deep percolation. The largest losses in humid areas are from runoff and deep percolation. Evaporation and unnecessary transpiration are most serious in dryland regions.

The principal ways to reduce runoff are to increase the soil's infiltration rate, to use vegetative barriers, terraces, contour tillage, or pits to hold water on the soil surface longer so that it can infiltrate, and to trap runoff in dugouts, ponds, and reservoirs.

Evaporation losses are hard to control. Some reduction results from the use of no-till along with crop residues and other mulches. Soil bedding, designed to concentrate water in the crop rows, has increased yields, but the technique is too complicated and too costly for common field use. Use of chemical monolayer and other surface evaporation barriers on reservoirs are not economically feasible.

Crops in dryland areas generally use all the water that is available each year, so crops are grown that can produce some yield in spite of water deficiencies. Control of weeds and volunteer crop plants reduces unnecessary transpiration loss. Windbreaks have some effect over a limited area. Antitranspirants have not reduced water loss economically in the field.

Water harvesting and summer fallow are often useful practices where the precipitation is not sufficient for profitable annual crop production. Water harvesting uses several methods to concentrate water on a small area where plants are grown. Water storage by summer fallow is generally low but can be increased by better weed-control and residue-management practices. Fallow works best where temperatures are not too high, especially during the season of precipitation and crop growth. It has been used successfully for wheat and other small grains and for sorghum production, but water conservation often makes it unnecessary. Increased moisture-storage efficiency is the cause of some saline seeps in the northern Great Plains and elsewhere. Reducing the proportion of fallow in the cropping systems may prevent further buildup of saline seeps, and may help to reclaim ones already developed.

The precious water that has been stored in the soil must be guarded zealously. Great care must be taken to grow the right crops in the right way so that water is used efficiently and crops are grown successfully.

QUESTIONS

1. Describe the five avenues of water loss that affect the efficiency of the storage of precipitation in soil and the factors that influence each.
2. Describe the water-control structures that are useful in dryland areas.
3. What are the principal mechanisms by which surface crop residues increase the amount of soil water available for crop use in dryland areas? What is the relative effectiveness of each mechanism?

4. List the approaches that have been used to reduce water losses from soils, and explain how each one works.
5. What is meant by water harvesting? Describe two types of water-harvesting systems.
6. How efficient is moisture storage by summer fallowing? Where is summer fallow recommended for regular use?

REFERENCES

ABU-ZREIG, M., M. ATTOM, and N. HAMASHA, 2000. Rainfall harvesting using sand ditches in Jordan. *Agric. Water Mgmt.* 46:183–192.

ALBALADEJO, J., M. MARTINEZ-MENA, A. ROLDAN, and V. CASTILLO, 1998. Soil degradation and desertification induced by vegetation removal in a semiarid environment. *Soil Use and Management* 14:1–5.

BAUMHARDT, R. L., C. W. WENDT, and J. W. KEELING, 1992. Chisel tillage, furrow diking, and surface crust effects on infiltration. *Soil Sci. Soc. Amer. J.* 56:1286–1291.

BLACK, A. L., and F. H. SIDDOWAY, 1971. Tall wheatgrass barriers for soil erosion control and water conservation. *J. Soil Water Cons.* 26:104–111.

BLOOM, A. L., 1998. *Geomorphology: A Systematic Analysis of Late Cenozoic Landforms,* 3rd ed. Prentice Hall, Englewood Cliffs, NJ, 482 p.

BONFIL, D. J., I. MUFRADI, S. KLITMAN, and S. ASIDO, 1999. Wheat grain yield and soil profile water distribution in a no-till arid environment. *Agron. J.* 91:368–373.

BRENGLE, K. G., 1968. Effect of phenylmercuric acetate on growth and water use by spring wheat. *Agron. J.* 60:246–247.

BRIGGS, L. J., and H. L. SHANTZ, 1914. Relative water requirements of plants. *J. Agric. Research* 3:1–63.

BROWN, P. L., A. L. BLACK, C. M. SMITH, J. W. ENZ, and J. M. CAPRIO, 1990. *Soil Water Guidelines and Precipitation Probabilities for Barley, Spring Wheat, and Winter Wheat in Flexible Cropping Systems—Montana and North Dakota.* Montana Coop. Ext. Bull. 356, Bozeman, MT, 30 p.

BROWN, P. L., and G. R. CARLSON, 1990. *Grain Yields Related to Stored Soil Water and Growing Season Rainfall.* Montana AES Spec. Rept. 35, Bozeman, MT, 22 p.

BRUN, L. J., J. W. ENZ, J. K. LARSEN, and C. FANNING, 1986. Springtime evaporation from bare and stubble-covered soil. *J. Soil Water Cons.* 41:120–122.

DIXON, R. M., 1995. Water infiltration control at the soil surface: Theory and practice. *J. Soil Water Cons.* 50:450–453.

DOUGHTY, J. L., W. J. STAPLE, J. J. LEHANE, F. G. WARDER, and F. BISAL, 1949. *Soil Moisture, Wind Erosion, and Fertility on some Canadian Prairie Soils.* Can. Dept. Agric. Publ. 819, Tech. Bull. 71, Ottawa, Canada.

DULEY, F. L., 1956. The effect of a synthetic soil conditioner (HPAN) on intake, runoff, and erosion. *Soil Sci. Soc. Amer. Proc.* 20:420–422.

ERICKSON, A. E., 1972. Improving the water properties of sand soil. In D. Hillel (ed.), *Optimizing the Soil Physical Environment Toward Greater Crop Yields.* Academic Press, New York, p. 35–41.

ERICKSON, A. E., C. M. HANSEN, and A. J. M. SMUCKER, 1968. The influence of subsurface asphalt barriers on the water properties and the productivity of sand soils. *Trans. 9th Intern. Cong. Soil Sci.*, Adelaide, Australia, Vol. 1, p. 331–337.

ESER, C., 1884. Investigations on the influence of the physical and chemical properties of the soil on the evaporation potential. *Forsch. Gebeite Agric. Phys.* 7:1–124.

Fairbourn, M. L., and H. R. Gardner, 1974. Field use of microwatershed with vertical mulch. *Agron. J.* 66:741–744.

Fairbourn, M. L., and H. R. Gardner, 1975. Water-repellant soil clods and pellets as mulch. *Agron. J.* 67:377–380.

Farahani, H. J., G. A. Peterson, D. G. Westfall, L. A. Sherrod, and L. R. Ahuja, 1998. Soil water storage in dryland cropping systems: The significance of cropping intensification. *Soil Sci. Soc. Am. J.* 62:984–991.

Fisher, C. E., and E. Burnett, 1953. *Conservation and Utilization of Soil Moisture.* Texas Agric. Exp. Sta. Bull. 767, College Station, TX.

Freese, R. C., D. K. Cassel, and H. P. Denton, 1995. Infiltration in a Piedmont soil under three tillage systems. *J. Soil Water Cons.* 48:214–218.

Fuehring, H. D., 1973. Effect of antitranspirants on yield of grain sorghum under limited irrigation. *Agron. J.* 65:348–351.

Hauser, V. L., 1968. Conservation bench terraces in Texas. *Trans. Am. Soc. Agric. Eng.* 11:385–386, 392.

Jones, M. J., and J. Sinclair, 1989. Effects of bare fallowing, previous crop and time of ploughing on soil moisture conservation in Botswana. *Trop. Agric.* 66:54–60.

Kuska, J. B., and O. R. Mathews, 1956. *Dryland Crop-Rotations and Tillage Experiments at the Colby (Kansas) Branch Experiment Station.* USDA Circ. 979.

Letey, J., 1975. The use of nonionic surfactants on soils. In *Soil Conditioners.* Spec. Publ. 7, Soil Sci. Soc. Amer., Madison, WI, p. 145–154.

Li, X.-Y., J.-D. Gong, Q.-Z. Gao, and F.-R. Li, 2001. Incorporation of ridge and furrow method of rainfall harvesting with mulching for crop production under semiarid conditions. *Agric. Water Mgmt.* 50:173–183.

Mathews, O. R., and T. J. Army, 1960. Moisture storage on fallowed wheatland in the Great Plains. *Soil Sci. Soc. Amer. Proc.* 24:414–418.

Medina, J., 1976. Harvesting surface runoff and ephemeral streamflow in arid zones. In *Conservation in Arid and Semi-Arid Zones.* Conservation Guide 3. FAO, Rome, p. 61–73.

Mehdizadeh, P., A. Kowsar, E. Vaziri, and L. Boersma, 1978. Water harvesting for afforestation: I. Efficiency and life span of asphalt cover. *Soil Sci. Soc. Amer. J.* 42:644–649.

Mickelson, R. H., and B. W. Greb, 1970. Lagoon levelling to permit annual cropping in semiarid areas. *J. Soil Water Cons.* 25:13–16.

Nicholaichuk, W., 1978. Evaporation control on farm-size reservoirs. *J. Soil Water Cons.* 33:185–188.

Ojasvi, P. R., R. K. Goyal, and J. P. Gupta, 1999. The micro-catchment water harvesting technique for the plantation of jujube (*Zizyphus mauritiana*) in an agroforestry system under arid conditions. *Agric. Water Mgmt.* 41:139–147.

Payne, W. A., 1999. Shallow tillage with a traditional West African hoe to conserve soil water. *Soil Sci. Soc. Am. J.* 63:972–976.

Penman, H. L., 1948. Natural evaporation from open water, bare soil, and grass. *Proc. Roy. Soc., Ser. A.* 193:120–145.

Peters, D. B., and W. J. Roberts, 1963. Use of octa-hexadecanol as a transpiration suppressant. *Agron. J.* 55:79.

Pikul, J. L., Jr., and J. K. Aase, 1998. Fall contour ripping increases water infiltration into frozen soil. *Soil Sci. Soc. Am. J.* 62:1017–1024.

Ramos, M. C., S. Nacci, and I. Pla, 2000. Soil sealing and its influence on erosion rates for some soils in the Mediterranean area. *Soil Sci.* 165:398–403.

RIPLEY, P. O., W. KABBFLEISCH, S. J. BOURGET, and D. J. COOPER, 1961. *Soil Erosion by Water.* Can. Dept. Agric. Publ. 1083, Ottawa, Canada.

ROCKSTRÖM, J., J. BARRON, J. BROUWER, S. GALLE, and A. DE ROUW, 1999. On-farm spatial and temporal variability of soil and water in pearl millet cultivation. *Soil Sci. Soc. Am. J.* 63:1308–1319.

ROOZEBOOM, K., D. JARDINE, G. WILDE, P. EVANS, K. KELLEY, K. KOFOID, A. SCHLEGEL, M. WITT, M. CLAASEN, W. B. GORDON, W. HEER, K. JANSSEN, B. MARSH, and V. MARTIN, 1996. *1996 Kansas Performance Tests with Grain and Forage Sorghum Hybrids.* Kansas Agr. Exp. Sta. Rep. Prog. 775, Manhattan, Kansas.

SHARMA, K. D., N. L. JOSHI, H. P. SINGH, D. N. BOHRA, A. K. KA, and P. K. JOSHI, 1999. Study on the performance of contour vegetative barriers in an arid region using numerical models. *Agric. Water Mgmt.* 41:41–56.

SHAW, R. H., 1959. Water use from plastic-covered and uncovered corn plots. *Agron. J.* 51:172–173.

SMIKA, D. E., and P. W. UNGER, 1986. Effect of surface residues on soil water storage. In B. A. Stewart (ed.), *Advances in Soil Science* Springer-Verlag, New York, N.Y. 5:111–138.

SNYDER, J. R., M. D. SKOLD, and W. W. WILLIS, 1980. Economics of snow management for agriculture in the Great Plains. *J. Soil Water Cons.* 35:21–24.

SOJKA, R. E., and R. D. LENTZ, 1994. Time for another look at soil conditioners. *Soil Sci.* 158:233–234.

SOMANI, L. L., and S. P. SHARMA, 1985. Improving moisture conservation and nutrient utilization by wheat through subsurface compaction of a sandy soil. *Trans. Indian Soc. Desert Technology.* 10:82–87.

STEPPUHN, H., and J. WADDINGTON, 1996. Conserving water and increasing alfalfa production using a tall wheatgrass windbreak system. *J. Soil Water Cons.* 51:439–445.

THORNTHWAITE, C. W., 1948. An approach toward a rational classification of climate. *Geog. Rev.* 38:55–94.

TROUT, T. J., R. E. SOJKA, and R. D. LENTZ, 1995. Polyacrylamide effect on furrow erosion and infiltration. *Trans. Amer. Soc. Agric. Eng.* 38:761–766.

VALENTIN, C., and A. CASENAVE, 1992. Infiltration into sealed soils as influenced by gravel cover. *Soil Sci. Soc. Amer. J.* 56:1667–1673.

WOODRUFF, N. P., R. A. READ, and W. S. CHEPIL, 1959. *Influence of a Field Windbreak on Summer Wind Movement and Air Temperature.* Kans. Agric. Exp. Sta. Tech. Bull. 100, Manhattan, Kansas.

ZINGG, A. W., and V. L. HAUSER, 1959. Terrace benching to save potential runoff for semiarid land. *Agron. J.* 51:289–292.

CHAPTER

14

Soil Drainage

The terms *poorly drained* and *well drained* tell whether potential crop growth is or is not limited by excess water. Well-drained soils have no excess-water limitations, and moderately well-drained soils have only minor ones. Somewhat poorly drained soils can be cropped but have high water tables long enough to cause problems and delays in planting, tillage, or other practices. Poorly drained soils usually cannot be cropped without artificial drainage, and very poorly drained soils are saturated with water most of the time.

Artificial drainage is appropriate in some places but inappropriate elsewhere. Some soils are naturally well drained, and the wet conditions of others may benefit the most appropriate use of the land. Wetlands in the United States are protected under the "swampbuster" provisions of the 1985 Food Security Act, so those that have not already been drained will be used mostly for wildlife and recreation rather than cropland or pasture. Some drained land becomes so acidic from the oxidation of sulfides (Section 14–3.4) or so alkaline from sodic conditions (Section 14–3.5) that nothing will grow on it. Artificial drainage should be provided only where soil conditions are suitable and the land use will benefit.

The 1985 Food Security Act and its subsequent amendments include provisions intended to preserve remaining wetlands. As a result of this act, *wetlands* are defined not only on the basis of a water table being present but also as soils that have water-loving vegetation growing on them and that show evidence of wetness in their profiles (Section 14–3). Drainage systems can be maintained on land that has already been drained, but they generally cannot be improved or extended into new areas. The law also provides for restoring limited areas of drained land to its previous wet condition. Of course, these restrictions on drainage are unpopular with many of the farmers, road builders, land developers, and others who are prevented from installing projects that would have been unrestricted before 1985, but the value of retaining wetlands for wildlife, recreation, and hydrologic benefits is considered paramount under the new law. Provisions are included for exceptions to be granted, often with the requirement that an alternate area be returned to wetland to replace an area where drainage is proposed. These provisions affect large areas of land. For example, it has been estimated that 20 to 30% of the area of the Corn Belt states of Ohio, Indiana, Illinois, and Iowa was originally wetland and that 90 to 99% of the wetland in these states has been drained, mostly for agricultural purposes (Lant et al., 1995; Mathias and Moyle, 1992).

Artificial drainage for land made wet by human actions is usually less debatable than that applied to naturally wet areas. Construction projects sometimes create wet areas by blocking the natural escape routes of water. Such situations often are remedied either by installing tile lines or by filling the low area with soil. On irrigated cropland, a large amount of wetness results from irrigation. Some excess water must be added to avoid accumulating excess salts in irrigated land. Even more excess water is added simply because it is impossible to apply exactly the amount needed in all parts of an irrigated field. Excess irrigation water causes water tables to rise, often to the point that drainage is needed. Lesaffre et al. (1992) estimate that about half of the irrigated land in arid and semiarid areas needs improved drainage. They estimate that drainage systems have been installed in about 150 million ha (370 million ac) of land throughout the world.

Drainage systems that remove water from land also remove solutes. This is helpful when the solutes are salts that need to be removed for soil reclamation (Section 15–7), but salts, including nitrates and other plant nutrients, and dissolved pesticides become pollutants in the streams, lakes, and seas that receive the water. For instance, De Vos et al. (2000) reported nitrate leaching of 10 to 25 kg of N/ha (9 to 20 lb/ac) annually from tile-drained land in the Polders of The Netherlands. Pitts (2000) notes that nitrates from tile drains in the upper midwest are major contributors to high nitrate levels that pass through the Mississippi River and cause algal growth and oxygen depletion in the Gulf of Mexico. He suggests that the water quality could be much improved and large amounts of nutrients retained in the fields if control devices were added to the tile drainage systems to reduce or eliminate the outflow when no crop is growing on the land. Pollution would be reduced and the farmers could save money on fertilizer. During a dry spring, the crop would also benefit from the additional water retained in the soil.

14–1 VALUE OF UNDRAINED WETLANDS

Laws protecting wetlands are a relatively recent development that represents a significant change in the dominant attitude toward their value. Roadbuilders tend to think of wetlands as problem areas to be crossed, and agriculturalists have been accustomed to thinking of the value these lands would have for cropland if they were drained. For a long time, few people were concerned about their value for purposes other than agriculture. More recently, those interested in wildlife and environmental issues have pointed out that wetlands are an important part of the landscape, and that their wet condition is valuable. The fact that large areas have already been drained makes the remaining wetlands more important.

The potential of detrimental effects from excess drainage is well illustrated by an example from India described by Varshney (1992). A large pond and adjoining swampy area were drained in 1986 at the insistence of the local populace because they were annoyed by mosquitoes and other pests that lived there. They recanted the next year, however, because draining the swamp also drained the nearby paddy rice fields and seriously affected the livelihood of many of the same people.

Wetlands have many functions, some of which will be considered here. They are vital to many species of plants and animals, they serve as recreational areas, and they are significant to the hydrology of the area. They also offer protection against some forms of water pollution.

Preserving genetic diversity in plants and animals must now be considered an integral part of the environmental impact of proposed projects. The results are sometimes ridiculed when a new highway has to be rerouted to avoid destroying the habitat of an obscure plant or animal, but the alternative of continuing to ignore undesirable effects is no longer acceptable morally or legally. Water-loving plants include a wide variety of grasses, forbs, ferns, bushes, and trees that cannot survive if the soil is drained. Fish thrive in the water associated with wetlands along with amphibians and other aquatic animals. Waterfowl depend on wetlands for food and resting areas. Mathias and Moyle (1992) estimate that 35% of rare and endangered animal species depend on wetlands in some way.

Recreational uses of wetlands, such as nature studies and hunting, depend on the wildlife present there. The species that occupy these areas may not be seen elsewhere and would disappear entirely if the last of the wetlands were drained. Boating and fishing are recreational uses of open-water areas associated with wetlands.

The hydrologic value of wetlands is less obvious, but also important. Runoff water would escape during a much shorter period of time if there were no wetland reservoirs to absorb large quantities of water and release it gradually. Thus, they serve to reduce flooding problems during wet periods and to help maintain the flow of streams during dry periods. Also, many of them serve as important recharge areas for underground aquifers that supply water for many uses.

The slow passage of water through a wetland provides time for natural processes to remove pollutants such as nitrates and other plant nutrients. They can either be used by the growing plants or absorbed by exchange processes. If there is excess nitrate nitrogen, denitrification can eliminate it in the reducing conditions of a wetland. The wetland filters and purifies water before allowing it to seep back into a stream.

One option worth exploring when cropland is too wet for one crop is to grow another crop that better tolerates wet conditions. Varshney (1992) provides a partial classification of crops as follows:

- Highly tolerant crops (able to produce 80 to 100% of normal yield with a water-table depth of 0.5 m)—sugarcane, potatoes, broadbeans, rice, plums, strawberries, and several grasses
- Medium tolerant crops (able to produce 60 to 80% of normal yield with a water-table depth of 0.5 m)—sugar beets, wheat, barley, oats, peas, cotton, citrus, bananas, apples, pears, and blackberries
- Sensitive crops (produce less than 60% of normal yield with a water-table depth of 0.5 m)—corn, pears, cherries, raspberries, date palms, and olives

14–2 OCCURRENCE OF WETLANDS

Water accumulates in low places even in arid regions. Wetness occurs wherever the water supplied by precipitation, overland flow, irrigation, and seepage into an area exceeds the amount of water lost by outward flow, seepage, evapotranspiration, and extraction for human projects. For example, the valley of the Dead Sea and Death Valley both accumulate salt water at their lowest points because these areas are below sea level and the water has no other place to go. The same is true of basins at higher levels such as the area around Great Salt Lake in Utah. The water level rises until there is enough wet area for all incoming water to escape by evapotranspiration. Salt is left behind when the water evaporates. Irrigation systems aggravate the wetness problems in arid areas through deep percolation of excess irrigation water applied to the land and by large seepage losses from many canals and ditches. Large irrigation projects generally need drainage systems to remove excess water.

Seep spots occur in many footslopes and hillsides where water flows above impermeable strata. The water may infiltrate on a nearby level hilltop in a humid region, or it may enter the soil and rock layers in hills or mountains, pass through several kilometers of rock, and reach the surface again in an arid valley. Such water can be under enough pressure in its aquifer to produce an *artesian well* (Figure 14–1). The water flow through aquifers varies from slight seepage to voluminous springs. At the largest springs in the world, near Thousand Springs, Idaho, water flows from basalt cliffs in such volume that Idaho Power Company uses it to produce electricity (Figure 14–2). Much of this water is believed to enter the basalt layers 120-mi (200-km) away where Big Lost River, Little Lost River, and some smaller streams disappear into openings in the rock.

Not surprisingly, wetlands are more abundant in humid regions than in arid ones. Much of the cropland in humid areas could not be cropped without artificial drainage. Any broad, flat area that is underlain by a layer with low permeability can have wet soil when precipitation exceeds evapotranspiration. Floodplains and deltas typically have wet soils because the streams that brought the sediment that formed these nearly level areas also supply water to them. Wet soils are also common on the uplands of the glaciated areas of north-central and northeastern United States and in Europe. Some areas are wet for only part of the year and may have a water deficit at other times. The limitations imposed by wetness depend on how wet, how long, and when. Many crops in temperate regions are unaffected by wetness during winter, but wetness in the spring may delay planting, and wetness in the fall often interferes with harvest.

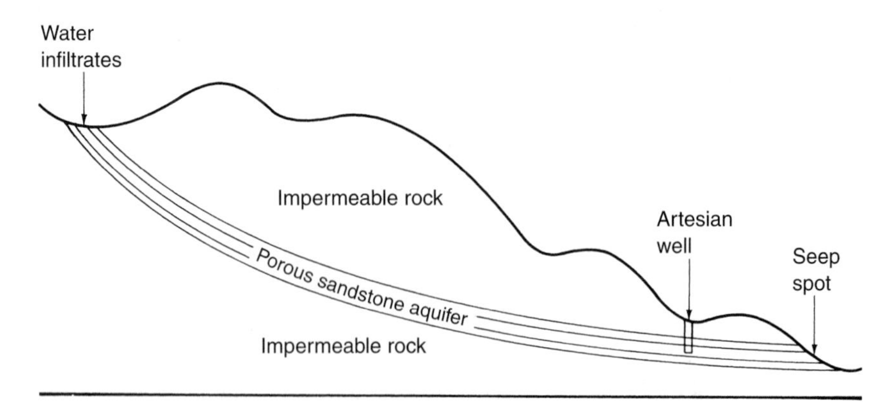

Figure 14–1 A cross section of a landscape showing an aquifer carrying water to an artesian well and a seep spot.

(A)

(B)

Figure 14–2 Electric power is produced from the springs emerging from the canyon wall near Thousand Springs, Idaho. (A) The springs as they once appeared. (B) The power plant. (Courtesy S. Z. Thayer.)

14–3 CHARACTERISTICS OF WET SOILS

Poorly and very poorly drained soils produce sedges, reeds, and other water-loving vegetation. The vegetation stays greener and includes plants that are not usually found on better-drained soils. Crops grown on such soils may be markedly better or poorer than on nearby soils, even where artificial drainage is provided. Several distinctive soil characteristics result from wetness. Features that help identify wet areas include the vegetation, soil-color patterns, organic-matter content, clay type and amount, and soil pH and related chemical factors.

14–3.1 Colors of Wet Soils

A wet soil has characteristic colors and color patterns that persist even during dry periods. Most poorly and very poorly drained soils have relatively thick, dark (often black) surface colors because of high organic-matter contents. The subsoil colors are normally grayer and duller than those common in drier soils, although rust mottles occur in soil horizons that sometimes receive oxygen. Mottles show how much the water table fluctuates. The permanently wet part has a bluish-gray color and is often called "blue clay" (even though its texture ranges from sand to clay). White colors result from salt accumulations on the surface of arid-region soils where water moves upward from a water table.

14–3.2 Organic Matter in Wet Soils

Soil organic-matter contents increase progressively from the driest to the wettest soils of an area. The driest soils usually have the least plant growth to produce organic matter and suffer the largest losses of organic matter by erosion and oxidation. The wet soils accumulate organic matter, not only because they receive organic matter eroded from dry soils above them, but also because decomposition is slowed and organic matter preserved by oxygen deficiencies.

High organic-matter contents contribute to high fertility and favorable soil structure in many wet soils. Drainage improves soil aeration, however, and causes the organic-matter content to decline gradually. Subsequent cultivation of the land causes the fertility and structural conditions to move toward the norms for cropped soils of the area.

Organic-matter losses are most serious in peat and muck soils. These soils usually subside considerably by shrinkage and compaction when they are first drained. Early subsidence is often large enough and irregular enough to disrupt the drainage system. Therefore, it is advisable to install a temporary initial drainage system, and to replace it after four or five years with a more permanent system (Lesaffre et al., 1992). Subsequent losses by wind erosion and oxidation cause many of these organic soils to subside at rates of 1 to 2 in. (2 to 5 cm) per year when they are drained and cropped. Organic soils disappear in a few decades or, at most, a few centuries of cropping. These soils are highly productive while they last, but they cannot be cropped without being gradually lost. The rate of loss can be minimized by keeping the water table as high as possible at all times and saturating the soil when there are no crops.

14–3.3 Clay in Wet Soils

Erosion and weathering increase the clay contents of many wet soils compared to those of their drier neighbors. Weathering progresses faster and forms more clay where the soil stays moist. Erosion takes fine particles from sloping soils and deposits them on flat areas, making the sloping soils coarser and the flat soils finer textured. Erosional sorting is most effective in materials that contain a wide range of particle sizes (are well graded). It has little effect on materials that are already well-sorted, such as many loess deposits.

Wetness also influences the type of clay in mature soils. Most soils of warm climates are dominated by kaolinite and oxide clays that form as advanced weathering removes silica. The wet soils, however, retain their silica and enough bases to form smectite clay instead. The smectite clay contributes to higher cation-exchange capacities and fertility in areas where the drier soils have very low fertility.

14–3.4 Reducing Conditions in Wet Soils

A few wet soils have a moving water table that carries dissolved oxygen. Most, however, have stagnant water in their saturated zones. Decomposing organic matter uses up the oxygen, produces reducing conditions, and slows microbial activity in such zones. Some microbes reduce nitrates to N_2, N_2O, or NO gases where oxygen is deficient. Others reduce ferric iron to ferrous iron and, thus, produce bluish-gray colors. Rust mottles are formed where air enters and oxidizes the ferrous iron back to the ferric state. Tubular concretions with a high iron content sometimes form along root channels where oxygen enters and interacts with the soil and chemicals released from the roots.

Certain wet soil and rock materials, especially those associated with coal deposits, contain iron pyrite (FeS_2), also known as "fool's gold." Like most sulfides, pyrite has very low solubility and causes no problem in a wet soil, even where it is present in the nodular concretions known as "cat-clay." Drainage, however, ruins such a soil by causing the sulfide to oxidize to sulfate:

$$2\,FeS_2 + 7\,O_2 + 2\,H_2O \xrightarrow[\text{bacteria}]{\text{Sulfur}} 2\,FeSO_4 + 2\,H_2SO_4$$

Sulfuric acid formed by this and subsequent reactions (see Note 11–1) acidifies the soil to a pH between 2 and 3. The reaction stops when nothing will grow, not even the sulfur bacteria. Lesaffre et al. (1992) indicate that there are about 12.6 million ha (31 million ac) of these acid sulfate soils in the world.

Neutralization of sulfuric acid produced by oxidation of sulfides requires so much lime that it is usually better to avoid draining such soils. Material containing sulfides is sometimes drained, however, when someone fails to recognize cat-clay (a costly oversight), or when coal-mine spoil is left in a heap (an old practice that is now illegal in most states). Acidity from coal-mine spoil is especially objectionable because it endangers nearby bodies of water.

14–3.5 Alkalinity in Wet Soils

Low-lying soils in arid regions may have excess water in their subsoils at least part of the year. These soils accumulate soluble salts when water moves upward from a saturated zone. Such salts commonly include carbonates that cause alkaline reactions. Sheikh (1992) considers salt accumulation to be the most significant threat to the sustainability of irrigation projects. Lesaffre et al. (1992) indicate that there are more than 300 million ha (750 million ac) of salt-affected soils in the world. Another 1 to 1.5 million ha in irrigated areas are added to that total each year.

The salinity of the groundwater has a significant effect on how many years it takes for a soil to become saline. Arid-region groundwater contains salt leached from soil and rock layers. Evaporation and plant use of water leave salt behind and increase its concentration in the groundwater. Flat, low-lying valleys in arid regions are likely to have considerable salinity in their groundwater, even if it is many feet below the land surface. Water tables deeper than 6 or 7 ft (2 m) will not affect the soil above them, but salinity makes them undesirable as water sources for most uses.

Irrigation causes the level of the water table to rise. Also, irrigation is likely to add more salt to the system and further salinize the groundwater. A temporary reversal is possible, however, if the irrigation water is low in salt. The largest contiguous irrigated area in the world, 13 million ha (32 million ac) in the Indus Plain of Pakistan, represents such a situation. The native groundwater is strongly saline, but it occurs at considerable depth. Irrigation has raised the water table, but the added water is less saline and therefore less dense, so it has not mixed much with the native groundwater (Sheikh, 1992). Under these conditions, it is possible to pump water from the upper layers and reuse it for irrigation. Such pumping must be done with care, however, because excessive withdrawal causes "upconing" of the more saline native groundwater. Even though most of the irrigation on the Indus Plain has been instituted since 1960, 39% of the soils are now affected by salinity.

14–3.6 Types of Alkaline Soils

The amount of water moving up and down through the soil is important and depends greatly on the depth of the saturated zone and the nature of the soil material. Types of soils influenced by water tables are shown in Figure 14–3. The water table may be regional, but in many places it is a temporary, perched water table or even a saturated zone with only a few yards (meters) of lateral continuity.

A layer of soil just above the water table, known as the *capillary fringe,* is saturated with water by capillary rise. The capillary fringe may be 6- or 8-in. (15- or 20-cm) thick where the soil pores are small but is thinner where larger pores are present. The soil to a height of 12- to 16-in. (30- to 40-cm) above the water table is usually poorly aerated and preserves enough organic matter to be black. Salts do not accumulate here because they diffuse back down through the moist soil.

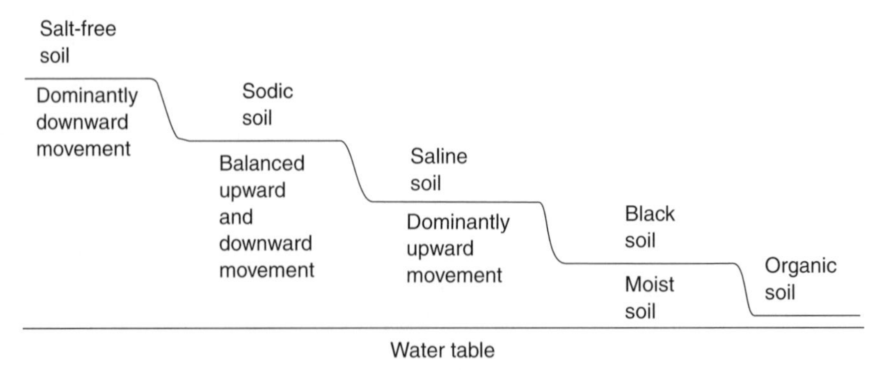

Figure 14–3 Soils developing in uniform material above a water table in an arid region.

Soils with white salt crusts occur where water moves upward to a dry surface. The water table usually occurs at a depth of 20 to 40 in. (50 to 100 cm), depending on the size of the soil pores. The white crust disappears when the soil is wet but reappears as the soil dries. These soils, known as *saline soils,* develop high osmotic concentrations when dry. Lowering the water table makes them less suitable for plant growth unless they are leached to reduce their salt contents. This process, known as *soil reclamation,* is discussed in Chapter 15, Section 15–7.

Sodic soils form where upward and downward movement of water are nearly balanced. Their salt contents are not high, but Na^+ constitutes more than 15% of their exchangeable cations. This condition develops where the climate is arid and the water table is at a depth of 3 to 5 ft (100 to 150 cm). The sodium component increases because its salts are highly soluble and easily carried back to the surface. Other factors, such as the salt content of the water table and the sodium content and rate of weathering of the parent material, also influence the formation of sodic soils. A sodic condition causes dispersed clay, shrinkage and cracking when dry, and very low permeability when wet. High sodium concentrations accompanied by carbonate anions cause the soil pH to rise above 8.5. The organic-matter contents are normally low, and what little there is, dissolves at the high pH and moves with the soil water. Sodic soils are called *black alkali* because a thin, black, organic coating forms as water evaporates at the soil surface. Water seeping from a sodic soil into a drainage ditch or other outlet appears brown and oily with dissolved organic matter. Little or nothing grows on such soils, and they are very difficult to reclaim (Section 15–7.3).

Saline, sodic, and salt-free soils are often intermixed on alluvial bottomlands that have spots and lenses of gravel, sand, silt, and clay. The variations in texture and/or small differences in elevation cause variations in capillary rise that produce the soil differences. Salt-free soils occur, for example, where a gravel lens interrupts the capillary rise.

14–3.7 Saline Seeps

Saline seep is a term applied to a soil-salinization process that has been accelerated by dryland farming activities that conserve water or alter its subsurface movement. Saline seeps can be distinguished from other saline soils "by their recent and local origin, saturated root zone profile, shallow water table, and sensitivity to precipitation and cropping systems" (Brown et al., 1982). They occur throughout the Great Plains in the United States, the Prairie Provinces in Canada, and in drylands of other countries. First identified and noted extensively in the early 1940s, they have spread dramatically and now occupy more than 2 million ac (810,000 ha) in the northern Great Plains alone.

Initially it was thought that summer fallow was the essential cause of saline seeps. By 1982, however, extensive research in the northern Great Plains and in the Prairie Provinces showed that the following conditions may contribute to the water accumulation that causes this problem:

1. Summer fallow
2. High precipitation periods
3. Poor surface drainage
4. Snow accumulation

5. Gravelly and sandy soils
6. Drainageways
7. Constructed ponds and dugouts that leak
8. Artesian water flows
9. Roadbeds across natural drainage ways
10. Crop failure

The causes of saline seeps must be identified before control measures can be initiated successfully. Because there is always a seepage origin for these soils, one of the necessary ingredients for control and reclamation is the removal of the source of seepage water. Plant use is a good way to remove water either by changing from a summer-fallow system to annual cropping, or by choosing a crop that uses more water. Otherwise, a drainage system will probably be needed. Since the waters contain salts, disposal of the drainage is often a problem.

Research and farmer experience have shown that much, but not all, of this land can be reclaimed by using (a) deep-rooted perennials, such as alfalfa, (b) flexible, cultivated crop/fallow systems, and (c) surface drainage of wet areas. Depending on the characteristics of the specific site, a combination of these measures may be necessary to reclaim saline seeps on farmland. Cultivated crops (and summer fallow) managed in a flexible planting system can reduce water buildup. A discussion of such a flexible system is provided in Chapter 13, Section 13–9.1.

Wind barriers, such as rows of tall wheatgrass, may be used to trap and distribute snow more evenly over the fields and to reduce evaporation between the barriers. The saved moisture, often in excess of 1 in. (2.5 cm), makes it possible to grow crops more frequently and to fallow the land less often, thus reducing the amount of groundwater recharge.

After the high water tables in the recharge and seep areas have been lowered or eliminated, steps can be taken toward reclamation. Depending on the salinity of the seep area, annual or perennial crops can be selected that will tolerate the prevailing salt content. The salinity will be reduced gradually by crop removal and a minimum of leaching, but it is a long, slow process. The levels of the groundwater tables in the problem areas must be monitored continually and, when necessary, steps taken to control them.

14–4 LIMITATIONS RESULTING FROM WETNESS

The government land-office maps for the original survey of some of the flatter parts of the U.S. Corn Belt labeled whole counties as "unfit for agricultural use" because of poorly drained soils. Rolling land where excess water could escape was considered best. Erosion on the hills and drainage on the flats have since reversed the ratings. The flat land now produces higher yields than the hills.

The surveyors undoubtedly realized the physical problems of wet soils—especially the lack of support for animals and vehicles, and the stickiness of soils with high clay contents. Mosquitoes and flies also would have caught their attention. These factors are still important, along with other physical and chemical limitations.

14–4.1 Physical Limitations of Wet Soils

Wetness has a strong influence on soil strength. Water weakens the bonds between clay particles and makes it easier for them to shift under a load. A wet clay soil becomes very weak, especially for supporting a load concentrated on a small area.

Sand behaves quite differently than clay, as a sandy beach will illustrate. Dry sand is easily blown into dunes and has little strength to support moving objects. Moistening the sand increases its strength. Sand castles with vertical walls can be built with moist sand because the water films bind the particles together. Vehicles travel easily on the moist sand a few meters inland from the water line but are likely to get stuck if they stray into the dry sand above or the wet sand below the water line. Neither the dry sand nor the saturated sand has air–water interfaces to produce the surface tension that holds sand particles in place. Also, water provides buoyancy that makes it easier to move particles. Upward-flowing water can add enough buoyancy to separate the sand particles and create a quicksand condition.

Most soils contain enough clay to provide binding strength to support a load when dry, but neither clay nor sand provides much strength under saturated conditions. Also, thick films of water make clay particles sticky, causing soil to adhere to feet, wheels, and other objects.

Traffic across a wet soil causes soil damage like that shown in Figure 14–4. Soil compacted by feet or wheels is often converted into a puddled mass that dries to a smooth, hard surface and later resists penetration by plant roots. Animal and vehicular traffic should not be permitted on wet soils for the sake of both soil and traffic.

Figure 14–4 Cattle trampled this pasture when it was wet, killed about half of the grass, and produced a rough surface. (Courtesy F. R. Troeh.)

Wetness often delays tilling and planting for several days. The more clay the soil contains, the longer the delay. Soil wetness in temperate climates causes low soil temperatures in the spring (Note 14–1). Seed planted too early will rot and not germinate. Even a few wet spots in a field create a problem. The farmer must either wait for them to dry before working the field or work around them and come back later.

NOTE 14–1
EFFECT OF WETNESS ON SOIL TEMPERATURE

Wetness lowers the temperature of soil in two ways—by evaporation and by the heat capacity of the water. Each gram of water evaporating at 20°C absorbs 585 cal of heat (539 cal at 100°C). More water evaporates and, therefore, more heat is withdrawn from a wet soil than from a dry one. Cooling by evaporation remains significant until the soil surface becomes dry.

The heat capacity of water does not actually cool the soil, but it does increase the amount of heat required to warm the soil. The heat capacity of water is 1.0 cal/g whereas that of dry soil is about 0.2 cal/g. A comparison of a dry soil and a wet soil will illustrate the effect of heat capacity:

	Soil near wilting point	Saturated soil
Percent H_2O by weight	20%	60%
Heat capacity per gram of solid:		
From solids	0.2 cal	0.2 cal
From water	0.2 cal	0.6 cal
Total heat capacity	0.4 cal	0.8 cal

Heat applied to the saturated soil in this example would raise its temperature only half as fast as it would that of the dry soil, even without allowing for evaporation.

The actual temperature difference between wet and dry soils is reduced because cool soil absorbs heat faster and loses heat more slowly than warm soil. Wesseling (1974) cites data for temperature differences of 2° to 4°C (3.5° to 7°F) between drained and undrained sandy soils and from 0.5° to 1°C (1° to 2°F) in clay soils. The drained clay soils retained more water and showed less difference than the sandy soils.

A rain that produces runoff often makes ponds in low spots even after a crop has been planted. Sometimes the crop is drowned, and the wet spot becomes a weed patch unless it is replanted. Heavy rain may cause another problem when a crop has been planted but has not yet germinated. The pounding raindrops may puddle the surface and form a crust that may dry so hard it prevents seedlings from emerging.

14–4.2 Chemical Effects of Wetness

Most chemical effects of wetness are caused by oxygen deficiencies resulting from poor aeration. Plant roots need oxygen for respiration to provide energy for growth, nutrient absorption, and other life processes. Small amounts of oxygen may be obtained from water, as in solution culture, but this source requires frequent replenishment. Rice and some other water-loving plants absorb oxygen above the water level and transport it inside the plant to the roots. But most plants rely on oxygen absorbed from soil air. Their roots will not grow into saturated soil or even into soil that has isolated pockets of air in its larger pores. Soil microbes use most of the oxygen from such pockets, so what remains is mostly nitrogen with an elevated concentration of carbon dioxide.

Researchers have shown that root growth of a wide variety of plants needs minimum oxygen diffusion rates between 5 and 25×10^{-8} g of O_2/cm^2 per minute. Patt et al. (1966) reported that citrus trees need 8 to 10% of aeration porosity by volume in the 10- to 30-in. (25- to 75-cm) layer to produce an adequate root density. The effective depth of many soils is limited by poor aeration rather than by soil strength.

Certain plant-nutrient deficiencies can be attributed to soil wetness. Two of these will be discussed here—potassium and nitrogen. Several others could be added, especially if the effects of pH changes produced by draining wet soils, as discussed in Section 14–3, were considered.

Potassium absorption by plant roots is slowed by poor aeration. Plants normally use energy from root respiration to absorb potassium and make it much more concentrated inside the roots than in the soil water. Potassium is said to be physiologically unavailable when a deficiency is caused by a shortage of oxygen.

Nitrogen deficiencies are most likely to occur in soils that are alternately saturated and unsaturated for a few days at a time, like the soil shown in Figure 14–5. Nitrifying bacteria convert ammonium ions to nitrates during the periods when oxygen is available. Other microbes reduce the nitrates to gaseous N_2, N_2O, and NO during the saturated periods. This denitrification process can use up large quantities of available nitrogen.

Another chemical problem results where metal pipelines cross both wet and dry soils. Redox potential differences cause an electrolytic action that ionizes metal, creating pits in the pipe that may become holes in one-fourth to one-half of its normal life expectancy. Saline soils make the problem worse because dissolved salts increase the electrical conductivity of the water.

14–5 WATER REMOVED BY DRAINAGE

Soil water is commonly classified as unavailable, available, and gravitational, according to how tightly it is held in the soil. Drainage systems remove only gravitational water, and not even all of that. The line between available and gravitational water is called field capacity and is commonly taken to be one-third bar (0.33 atmosphere, 0.34 kg/cm^2, or 0.033 megapascals). This line is gradational and varies from values as low as one-tenth bar for very sandy soils to as high as one-half bar for clay soils. These fractional differences in soil-moisture tension near field capacity influence the amount of available water as much as do

Figure 14–5 This corn is pale green as a result of nitrogen deficiency caused by denitrification in this intermittently wet area. The cracked and curled soil surface is a result of ponding. (Courtesy F. R. Troeh.)

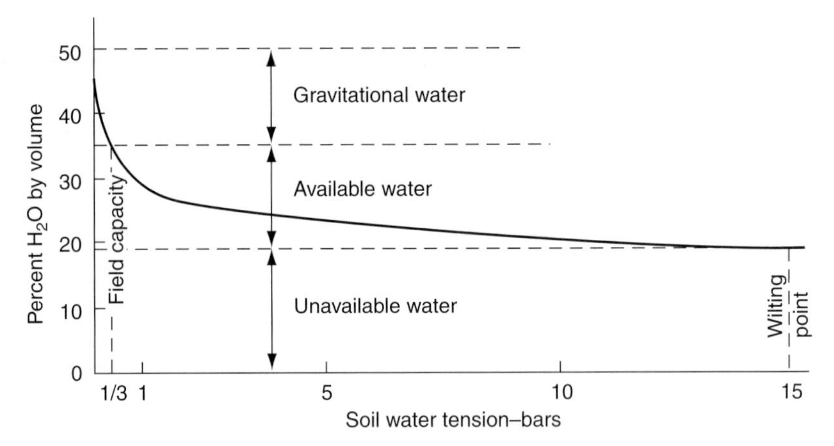

Figure 14–6 A soil-moisture-retention curve showing water relationships in a representative loamy soil with 50% pore space.

variations of several bars near the wilting point, as illustrated by the soil-moisture-retention curve in Figure 14–6.

Field capacity is defined as the amount of water retained by the soil when downward movement into dry soil below nearly ceases. The time required varies from about one day for sandy soils to three or four days for soils with low permeability. The dry soil below is important because the soil-moisture tension of a loamy soil at field capacity

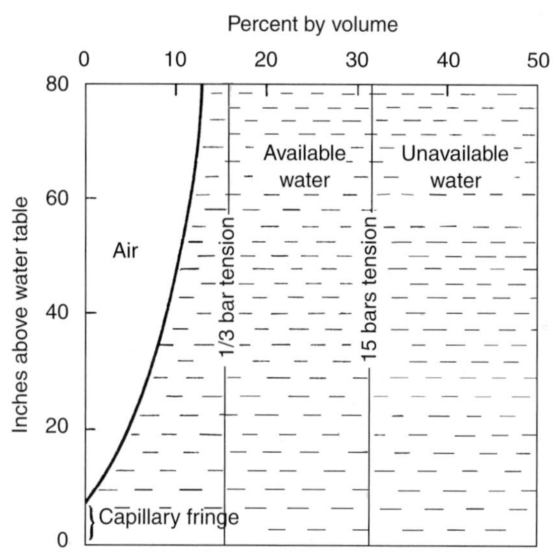

Figure 14–7 Air and water held by a soil like that of Figure 14–6 above a water table at 80 in. (2 m).

exerts as much force on the water as about 10 ft (3 m) of gravitational head. The corresponding value for a sandy soil is 3 to 6 ft (1 or 2 m), and for clay soils it can be as much as 16 ft (5 m).

Soils with water tables obviously do not have dry soil below to exert soil-moisture tension. A subsurface drainage system functions by gravitational force and would theoretically have to lower the water table to a depth of 11 ft (3.4 m) to produce a tension of one-third bar at the soil surface. Few drains are placed that deep, so less tension is created and more water is left in the soil. Figure 14–7 shows the amount of air space resulting from a drain at a depth of 80 in. (2 m) in a soil with water relations like those illustrated in Figure 14–6.

The water removed by a drain is not necessarily the water that is closest to the drain. Rather, *preferential flow* lines carry much of the water, or at least carry some water from the soil surface to a drain tile in a relatively short period of time. Kung et al. (2000) found that marker ions or molecules sometimes reached a tile line in New York at a depth of 3 ft (0.9 m) in as little as 13 minutes. Similarly, Jaynes et al. (2001) found markers reaching a tile line in Iowa at a depth of 4 ft (1.2 m) in as little as 15 minutes with an application of irrigation water as small as 1 mm. The water must have been following *macropores* such as earthworm channels or soil cracks to move through the soil so quickly. Preferential flow such as this can contribute to water pollution by carrying surface-applied pesticides to the drains without the pesticides interacting with the soil. Shipitalo and Gibbs (2000) found that animal manures injected into the soil also can be transported to tile lines in a similar manner, especially in no-till fields where earthworm channels are most abundant. They recommended that tile-drained fields be tilled to disrupt flow channels before the application of animal manures or that the tile drains be blocked for a long enough time to allow the contaminated water to soak back into the soil.

14–6 SURFACE VERSUS SUBSURFACE DRAINAGE

Most areas that need artificial surface drainage are either nearly level or depressional. The wetness may be caused by either low soil permeability, or heavy rainfall and/or runon. Most of the water comes from precipitation on either the wetland or nearby higher areas. *Surface drains* remove excess water that has not had enough time to infiltrate. The soil moisture conditions are improved by reducing the amount of water entering the soil, rather than by taking it out after it gets there.

Subsurface drainage is used where there is a high water table. The water may have come from precipitation on the land, surface flow, and seepage from nearby land, or it may have seeped a long distance through an aquifer. Irrigation water also contributes to the need for subsurface drainage in irrigated areas. Whatever the source, excess water is already in the soil and must be removed from below.

14–7 METHODS OF REMOVING WATER

Ditches are often used for surface drainage and tile for subsurface drainage, but there are exceptions. Ditches can be used for subsurface drainage and tile lines can have surface inlets. Several other drainage methods also are available, including land smoothing, wells, bedding systems, and mole drains. Drainage methods are chosen according to the amount of water to be removed, soil characteristics, cost and convenience factors, availability of equipment and materials, and personal preferences.

14–7.1 Land Smoothing

Filling low places in a field is called *land smoothing* or, in irrigated areas, *land leveling.* Leveling improves both the irrigation and the drainage of irrigated land. Actually, the leveled land usually has a gentle slope that is as uniform as possible. Elimination of high and low areas in a field is costly, but leveling makes it possible to greatly improve the uniformity of water application by surface irrigation.

Land smoothing for drainage purposes can improve both surface and subsurface drainage. Eliminating depressions lets water flow across the area without being trapped. Subsurface drainage is improved by more uniform infiltration and by the soil surface of the former low area being higher above the water table.

Land smoothing a field like that shown in Figure 14–8 may be the only drainage practice needed, or it may be used to make a ditch or tile drainage system function better. An important advantage of land smoothing is that it usually requires no maintenance.

14–7.2 Drainage Ditches

Cato discussed Roman methods of farm drainage by ditches as an established practice in 200 B.C. (King, 1931). The Egyptians, Babylonians, and perhaps others practiced drainage centuries before Cato's time. Ditches offer high capacity for either surface or subsurface drainage, even where there is little slope. Installation costs for a ditch drain are usually less

Figure 14–8 Land smoothing to allow water to escape would have saved the crop in this depression in an Iowa soybean field. (Courtesy F. R. Troeh.)

Figure 14–9 This 20-ft (6-m) deep ditch in Idaho serves as an overflow for irrigation water if the canal above gets too full, as an interceptor for seepage water, and as an outlet for drainage systems in the valley. (Courtesy F. R. Troeh.)

than those for a covered drain installed at the same depth. Ditches often require more maintenance than do covered drains to remove sediment and unwanted vegetation that inhibit the water flow, but it is easy to see when maintenance is needed. Ditches such as that shown in Figure 14–9 are often used for main drains and as outlets for tile and other covered drains. The principal disadvantage of ditches is that their channels and banks occupy land that could be cropped if a covered drain were used. In addition, ditches can be hazardous obstacles to the movement of people, animals, and machines. Ditches are often placed along field boundaries to minimize these disadvantages.

Variations in ditch gradient are much less serious than variable gradients in covered drains. Ditches, therefore, are preferred for the initial drainage of unstable land, such as a peat bog that may settle unevenly. Covered drains may be installed after settling has ceased. Many ditches are nearly flat because there is little elevation difference between the wetland and the outlet. The steepest gradient used should produce a nonerosive velocity of no more than 0.06 to 0.24 ft/sec (0.2 to 0.8 m/s), depending on the erodibility of the soil.

Ditches are usually widely spaced—hundreds of yards (meters) apart where possible—to minimize the obstacle problem and the amount of land removed from crop production. Ditches for surface drainage should be designed to carry the maximum anticipated flow when filled to no more than 80% of their depth. Their capacities can be calculated by Manning's formula, as explained in Note 4–2.

Ditches for subsurface drainage usually are deeper and spaced farther apart than tile drains would be under similar conditions. Ditches 6- to 10-ft (2- to 3-m) deep provide adequate drainage in many soils even though they are spaced 300 to 650 ft (100 to 200 m) or more apart. Soils that need more drainage may have tile lines that empty into ditches one-fourth mile (400 m) or more apart. These ditches, too, need to be relatively deep to serve as outlets for the tile.

Steep ditch banks minimize the amount of soil to be moved and the land area occupied, but a safety margin should be allowed, and maintenance requirements need to be considered. Donnan and Schwab (1974) indicate that common side slopes in clay soils range from 0.5:1 (1 ft horizontal to 2 ft vertical) to 1.5:1. Coarser-textured soils need side slopes between 1:1 and 2:1, and some very sandy soils require side slopes of 3:1. Cleaning by dragline requires side slopes no steeper than 1:1, grazing by livestock requires slopes of 2:1 or flatter, and banks to be mowed or crossed by machinery should be no steeper than 3:1 (Soil Conservation Service, 1973).

Vegetation, as shown in Figure 14–10, is both helpful and detrimental in drainage ditches. Vegetation helps stabilize ditchbanks against slumping and erosion, and may be of value to wildlife. Water lilies, cattails, and other water plants, however, slow the flow, raise the water level, and reduce the effectiveness of the drain. After all, the effectiveness of subsurface drainage depends on the water level rather than the ditch depth.

14–7.3 Tile Drainage

Tile lines are commonly used when a drainage system must extend into the interior of a field to achieve a reasonably uniform depth to the water table. They are especially common for subsurface drainage, but some have surface inlets to drain depressions or low areas above terraces. This last use has developed in recent years as a means of making terraces as straight and as parallel to one another as possible (Chapter 10) and to eliminate the grassed waterways that were used previously.

Tile drainage has been practiced for hundreds of years. Horseshoe tile were invented in France in about the fourteenth or fifteenth century, then forgotten, and later reinvented in England in the seventeenth or eighteenth century (King, 1931). They were made by hand from baked clay with a U-shaped cross section and were placed with the open side down in the bottom of a trench, sometimes with a flat clay pallet beneath them. A version of horseshoe tile using pallets with raised sides that interlock with the half-round top por-

Figure 14–10 Vegetation stabilizes the banks of this drainage ditch, but the vegetation growing in the water slows the flow and raises the water level. (Courtesy F. R. Troeh.)

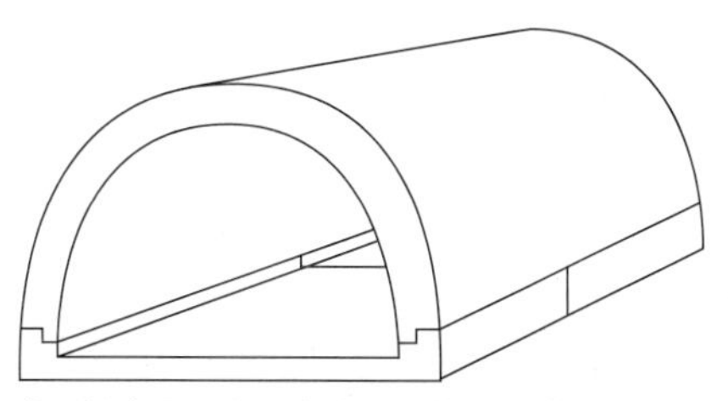

Figure 14–11 Horseshoe-tile design currently in use in southern China. A typical size is 90-mm (3.5-in.) wide outside and 60-mm (2.4-in.) inside by 330-mm (13-in.) long. The top and bottom joints are staggered to help keep the tiles aligned. (Redrawn from Zhaoyi and Rongkai, 1992.)

tion (Figure 14–11) is currently in use in southern China (Zhaoyi and Rongkai, 1992). At least 20% of the 33 million ha (82 million ac) of cropland in southern China need drainage to avoid low yields resulting from waterlogging.

The practice of tiling was brought to the United States from Scotland in 1835 (Wooten and Jones, 1955). A tile-making machine was invented in England in 1841, but tile continued to be placed in hand-dug trenches until a steam-powered trenching machine was marketed about 1883. Tile are now placed by wheel-type or bucket-ladder-type trenching machines, most of which can dig as deep as 6 ft (1.8 m). Cuts deeper than 6 ft require special, large trenchers, backhoes, or draglines and are considerably more expensive.

Tile drainage permits normal equipment and livestock traffic across the field. A good system should function for hundreds of years with little maintenance (some sewer tile thousands of years old are still serviceable). The biggest disadvantage is cost—unless, of course, something prevents tile drainage from working. Conditions that limit the use of tile include shallow soil, excessive stoniness, and soils with hydraulic conductivities below 0.2 to 0.4 in./hr (0.5 to 1 cm/h).

Types of Tile. Three types of tile are now in common use—clay, concrete, and plastic—with the plastic type now dominant for new installations in developed nations. All three types have round cross sections. Clay tile are the traditional type, made by forming moist clay into pipes 1-ft (30-cm) long and firing them until they are dry and hard. They are available with inside diameters ranging from 4 to 12 in. (10 to 30 cm). The walls are usually between 0.5- and 0.6-in. (12- to 15-mm) thick. Clay tile are brittle and will deteriorate if subjected to freezing and thawing, but are otherwise very durable.

Concrete tile are similar to clay tile except that they are usually 2- or 3-ft (60- or 90-cm) long and are available in larger sizes up to 3-ft (90-cm) diameter. Concrete tile resist freezing and thawing and are, therefore, better than clay tile for lines that are above the frost line. Concrete is subject to attack by acids, so it should not be used in extremely acidic soils.

The newest type of tile is made of PVC, polyethylene, or polypropylene plastic and began to be marketed in the 1960s. It is thin-walled and corrugated to give it strength and flexibility. As shown in Figure 14–12, it comes in rolls 200- to 300-ft (60- to 90-m) long and has slots for water to enter. Diameters range from 4 to 10 in. (10 to 25 cm) in the United States, but sizes down to 2 in. (5 cm) are used in Europe (Dierickx, 1992). Plastic resists damage by both freezing and acids, but there have been reports of rodents chewing holes in it. It is less rigid than clay or concrete and relies more on proper packing of soil around it for support to keep it from collapsing. The long lengths and easy joining of plastic tile

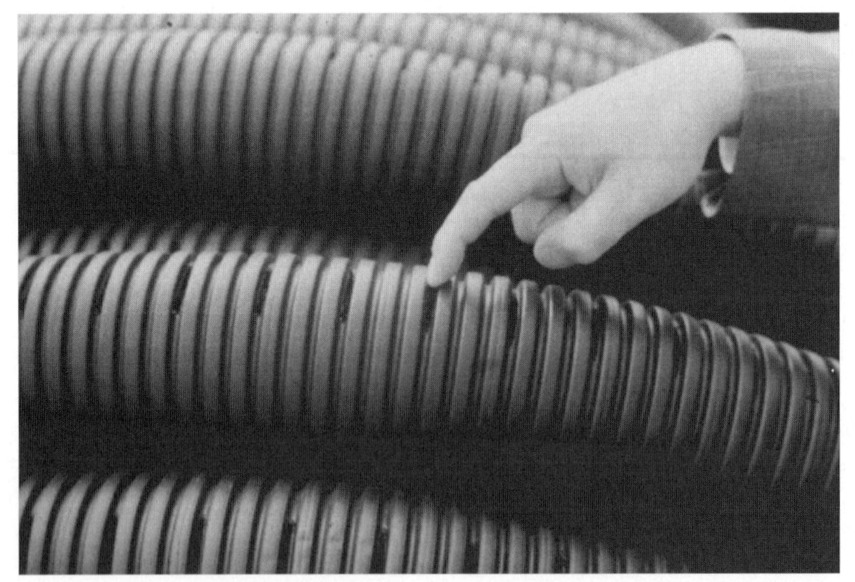

Figure 14–12 The newest type of tile is made of corrugated plastic, comes in rolls, and has holes for water to enter. (Courtesy F. R. Troeh.)

eliminate the misalignment problems that can occur with clay or concrete. The effectiveness of plastic drains is commonly enhanced by enclosing them in a drain envelope composed of any of several materials, including coarse sand, fine gravel, glass wool, fibrous peat, straw, wood chips, or plastic fibers such as carpet waste (Dierickx, 1992). The light weight of plastic drains and their ease of installation make them less expensive than clay or concrete, so they are now the dominant type used for new installations in most of the world.

Several types of pipe, including steel, aluminum, rigid plastic, and fiber are used where strength or rigidity is needed in a drainage system. Most of these are not true drainage tile because they have no holes for water to enter. A 20-ft (6-m) length of corrugated steel is commonly used at the outlet of a tile line. Rigid pipe also may be needed where a tile line passes under a road, through a very deep cut with heavy overburden, or through soils that may shrink and settle unevenly.

Depth and Spacing of Tile Lines. Hand-dug tile lines were often placed at depths of 20 to 24 in. (50 to 60 cm) and were spaced only 15- or 20-ft (5- or 6-m) apart in some fields. Most lines are now placed at depths of 3 to 5 ft (90 to 150 cm) in humid regions and about 6 or 7 ft (2 m) in arid regions. Many tiling contractors charge extra for placing tile deeper than 4 ft (120 cm). Deeper placement protects against frost, heavy vehicles, and plant roots and permits wider spacings between lines. Shallower placement (or alternatively, provision to raise the water level as described in Section 14–9.4) makes more water available to crops during dry periods. The 2-m depth for arid regions prevents capillary rise from carrying salts to the soil surface. Tile lines are now commonly placed between 30- and 120-ft (10- to 40-m) apart depending on the soil and the depth of placement.

Soil permeability is a factor in determining the depth and spacing of tile lines. Less-permeable soils need closer-spaced, shallower lines to remove the water within a reasonable time. The horizontal permeability is most important because drainage water must move farther horizontally than it does vertically. Basak (1972) found horizontal hydraulic conductivity to be 1.0 to 1.6 times as high as in the vertical direction. Variation in the properties of soil horizons also should be considered. Going deeper into a less-permeable horizon may not affect water removal, but reaching a more permeable layer may be very beneficial.

Maintaining suitable slope gradients for water flow in tile lines is more important than constant depth. Donnan and Schwab (1974) suggest a minimum gradient of 0.1% for 4-in. (10-cm) drain lines and 0.05% for 6-in. (15-cm) lines. Larger tile lines should have gradients of at least 0.05%. Line gradients steeper than 1% require that tile joints be wrapped with durable material such as tar-impregnated paper and packed tightly with soil to keep the tile properly aligned. Sewer tile with sealed joints can be used on steeper grades that do not need drainage. Breathers (vents) to allow air entrance are sometimes needed at the top of a slope, and relief wells or larger-diameter tile at the bottom to prevent pressure buildup and facilitate water flow.

Laying Tile. Tiling is greatly facilitated by trenching machines. Some older machines made a wide trench so a worker riding inside a trailing shield could place the tile by

Figure 14–13 Concrete tile being placed with tiling hooks behind a trenching machine. (Courtesy W. H. Lathrop.)

hand. Other workers used tiling hooks, as shown in Figure 14–13. Newer machines provide a tile chute that guides tile from an aboveground loading point down into place in the trench. Nearly all flexible plastic tile is placed automatically as shown in Figure 14–14. Such machines lay tile at rates up to about 130 ft (40 m) per minute.

Topographic surveys are needed for planning tile depths and gradients. The trenching machine is usually controlled by either a guideline or a laser beam. The guideline method involves setting stakes to hold a line at a specified height above the level of the tile line. The machine operator then keeps a pointer on the machine at the level of the guideline.

The laser method is an example of new technology applied to an old problem. The laser beam is aimed parallel to the desired grade line. A photo sensor on the tiling machine detects the laser beam and controls the machine so it stays on grade. The laser unit can control the depth for distances up to 1700 ft (500 m) with a vertical accuracy of 1 cm.

Plastic tile has holes for water to enter, but clay and concrete tile allow water entrance only at the joints. Spacings of 1/8 in. (2 or 3 mm) are recommended for most soils. Larger openings, such as those resulting from changes in direction, should be covered to keep soil from entering the tile.

The soil or other material next to the tile needs to be permeable to water, stable enough to stay out of the tile, and shaped to support the tile on all sides. Subsoil often does not meet these requirements. Some surface soils with stable granular structure will serve for *blinding* (surrounding and covering) the tile. Coarse sand and gravel mixtures also can be used for blinding. Various crop residues and synthetic materials also have been used for this purpose.

Figure 14–14 Flexible plastic tile being installed by a tiling machine. (Courtesy USDA Soil Conservation Service.)

14–7.4 Other Closed Drains

Closed drains are those that are covered and concealed. These include tile drains (already discussed), box drains, rock drains, and mole drains.

Box drains are made of boards assembled into boxes with triangular or rectangular cross sections and buried in a trench. They have a short life expectancy unless they are treated with wood preservative to prevent rotting.

Rock drains are constructed by placing several layers of rocks in the bottom of a trench and covering them with soil. The capacity is low and becomes even lower if soil fills the space between the rocks. Box drains and rock drains were usually homemade and were most common in pioneer times.

Mole drains are drawn rather than laid. A torpedo-shaped "mole" hooked behind a long shank is pulled through the soil to create an unlined channel about 4 in. (10 cm) in diameter. Mole drains are commonly made at depths of 20 to 30 in. (50 to 75 cm) and spaced 10- to 15-ft (3- to 5-m) apart. The close spacing and the fracturing produced by the shank help make mole drains effective. Any open slots left by the shank should be tilled shut as soon as possible to keep soil from falling into the mole channel.

The use of mole drains in land with surface irrigation has led to rapid failure caused by water flowing straight down into the drain through the soil fractured by the mole shank. Christen and Spoor (2001) describe an equipment modification that makes the mole channels last longer. The lower part of the shank is angled forward and 30° to one side to change the fracture zone from vertical to sloping. They found a significant advantage when they used the angled shank in unstable soils, such as those affected by sodium.

Mole drains are relatively inexpensive and can be made quickly when the soil is dry on top and moist below. They should be drawn more slowly in wet soil because a vacuum can cause the channel to close behind the mole. Sandy soils are generally too unstable for mole drains, but properly installed mole drains in clay subsoils probably will last from 3 to 15 years; those in organic soils may last 3 to 5 years. Mole drains have been popular for draining clay soils in England, where they are reported to last three or four years before the roof and sides of the drain collapse (Harris et al., 1993). A new set of mole drains is drawn when the old set becomes ineffective.

Jha and Koga (1995) had favorable results from mole-drain trials in waterlogged Vertisols containing 63% clay on the Bangkok Plain of Thailand. They reported improved aeration leading to a 46% increase in soybean yields on moled plots, as compared to control plots. The Bangkok Plain has 3365 km^2 of such soils.

Mole drains are often a good choice for temporary drainage of an area where the need may disappear (as for removing excess salts from a saline soil) or where a more permanent drainage system will be installed later (as for draining an organic soil that will settle unevenly at first).

14–7.5 Bedding Systems and Mounds

Bedding systems and *mounds* are sometimes used for surface drainage on land with slope gradients of less than 1%. Bedding systems are the most common type of surface drainage in Europe. Mounds are used in swampy areas in Africa. Both are used where slow permeability and moderate soil depth prevent the use of most other types of drainage. The crops are often pasture or hay, although some row crops and truck crops are grown on bedded land, and cassava is a common crop on mounds in Africa.

Bedding is also called *crowning* because the areas between drains are usually graded into a raised, convex shape. The beds can be formed by repeatedly plowing backfurrows in the beds and deadfurrows in the drains, but it is much faster to use earth-moving equipment. Grassed waterways about 20-in. (0.5-m) deep with about 8:1 side slopes are located between the beds. Row crops may be grown on the crowned areas but not in the waterways. The rows run parallel to the waterways. The crowned areas are usually about 65-ft (20-m) wide, their exact width being a multiple of the width of the equipment used.

Mounds are used in tropical Africa to grow cassava during the dry season in valleys that grow paddy rice during the wet season (Mohamoud, 1994). The mounds are steep-sided and 30 to 60 cm (1 to 2 ft) or more tall. The soil dries out during the dry season, but is too wet for cassava during its early and late stages. The mounds provide drained areas that permit the owners to plant and harvest a cassava crop in the same fields that produce paddy rice during the wet season.

14–7.6 Vertical Drainage

Both dry wells and pumped wells are used for drainage. *Dry wells,* otherwise known as *drainage wells* or *vertical drains,* act as wells in reverse. Either surface or subsurface water is permitted to run into the well and down to an adsorbing layer below. Most states have public health laws restricting or prohibiting the use of dry wells because this simple solution to a drainage problem entails significant hazards. One, the possibility that sediment may plug the pores in the adsorbing layer, can be reduced by filtering the water before it enters the well. Another peril is the pollution potential from surface water entering an aquifer. A very similar pollution potential exists in karst topography where natural sinkholes drain into cavities in limestone bedrock. In either case, runoff draining into the bedrock can carry pesticides, soluble fertilizers, manure, and any other chemicals that happen to be present. Possible remedies include plugging the well or sinkhole, diverting the water elsewhere, limiting the use of chemicals and manure in the vicinity, or installing a filter strip of permanent vegetation in the area around the well or sinkhole.

Pumped wells are relatively expensive to install and operate, but they can tap a deep gravel layer and thus reduce the water table to lower levels than other drainage methods. The pumped water may be useful for irrigation or livestock. Pumped wells also may be used to remove water from aquifers and, thus, reduce their contribution to high water tables and springs. They are likely to be the preferred option for problems caused by deep artesian sources that work their way up through the substratum and cause waterlogged soils.

Irrigation of the Indus Plain in Pakistan caused a rapid rise in the water table during the middle of the twentieth century. Pumped wells, locally known as tubewells to be placed at an average of one per square mile, were chosen as an immediate solution (Sheikh, 1992). The initial project involving the installation of 2069 wells was completed in 1963. The project was successful and the work continued rapidly, partly because the large number of wells protected individual wells from being overwhelmed by water from adjoining areas. By 1990, the Indus Plain had 12,800 wells in use for both drainage and irrigation, and 1800 wells used only for drainage because their water had too much saline to be used for irrigation. Disposal of the saline water is a major problem (Sheikh, 1992). The ideal solution is to have its own outlet all the way to the ocean, but this is expensive. Some of it can be emptied into rivers during the summer when high flow provides adequate dilution. Irrigation of salt-tolerant crops and the use of evaporation ponds are other potential solutions.

The efficiency of many of the pumped wells in Pakistan has declined considerably over time as a result of plugging of their intake screens or of the surrounding filter material by *chemical encrustation* (mostly calcium and magnesium carbonates) and biofouling (mostly caused by iron bacteria). Several ways of reducing this problem have been developed, including the use of nonferrous screens, injecting acid into the well to dissolve carbonates, and surging (Hussain et al., 1992). *Surging* uses either air or water pressure to reopen passageways around the well. It commonly needs to be repeated annually. Increased irrigation, reduced well efficiency, and an increasing number of saline wells have led to a new program of using tile drainage in parts of the Indus Plain (Sheikh, 1992). Tile drainage permits withdrawal of the nonsaline water without getting into the deeper layers of saline water.

Pumped wells can lower the water table to a depth of several meters when this is desirable, or the pumping can be halted to let the water table rise at other times. The original intent in Pakistan was to lower the water table to 10 to 15 ft (3 to 4.5 m), but 5 to 7 ft (1.5 to 2.1 m) is now generally considered adequate.

Pumped wells are also used extensively in India, Egypt, and China. Egypt began installing them because of waterlogging and salinity resulting from increased irrigation after the Aswan High Dam was built in the 1960s (Fatma, 1992). China has a large number of pumped wells in the North China Plain, an area that has monsoon rains in the summer but requires irrigation in the winter (Zhang, 1992). The water table is allowed to rise to 0.5 to 1 m (20 to 40 in.) by the end of the monsoon season so that it will benefit the winter crops, and then is lowered to at least 3 to 5 m (10 to 16 ft) so the soil reservoir can help absorb the excess water from the next monsoon season.

Another variation of well drainage is used in Norway in areas that have waterlogging caused by artesian pressure. Wells drilled 3- to 15-m (10- to 50-ft) deep into the underlying aquifer are fitted with outlet pipes at a depth of 2 to 3 m (6 to 10 ft). The artesian water rises through the wells and is carried away through the outlet pipes without any pumps being required. This system has functioned well for more than 20 years in an area where tile drainage failed because the lines were fouled with iron precipitates (Bergedalen, 1992). Oxygen in the tile lines causes the iron to precipitate, but the well systems are closed so oxygen does not enter.

14–8 RANDOM, REGULAR, AND INTERCEPTOR DRAINS

The topography of the area, source of excess water, and pattern of wetness are important factors in determining whether the design of a drainage system will be random, regular, or interceptor.

Random drains are used in rolling topography with small wet areas. The drains follow the lowest parts of the landscape connecting one wet area to the next. Either ditch or tile drainage may be used, but the layout depends on the locations of wet spots.

Regular drainage systems consist of several parallel ditches or tile lines draining broad, flat areas. A variety of patterns are possible, including two basic ones known as *parallel* and *herringbone,* as shown in Figure 14–15. Laterals may enter the main drain from one or both sides in either system. The herringbone pattern allows both the laterals and the main line to flow down the general slope of an area rather than directly across it.

Interceptor drains are placed between the source of excess water and the area needing drainage. The drain must be deep enough to intercept the main flow of water. The water being intercepted in humid regions comes from natural precipitation, whereas in arid regions it may come from an irrigation canal.

14–9 DESIGN FACTORS FOR DRAINAGE SYSTEMS

The first consideration in the design of a drainage system is what land should be drained. Areas that will benefit and other areas that might be harmed by removal of water should be identified and mapped. Soil survey maps, topographic maps, geologic maps, and hydrologic maps are all useful for planning drainage systems.

A soil survey helps identify both problems and potentials. It locates wet soils whose drainage may not be improved legally and identifies problem soils high in clay, organic matter, and/or sodium. Soils shallow to bedrock or other impermeable layers are shown on soil maps. Soil surveys also show which soils are permeable enough to drain easily and which ones have adequate natural drainage for their intended use.

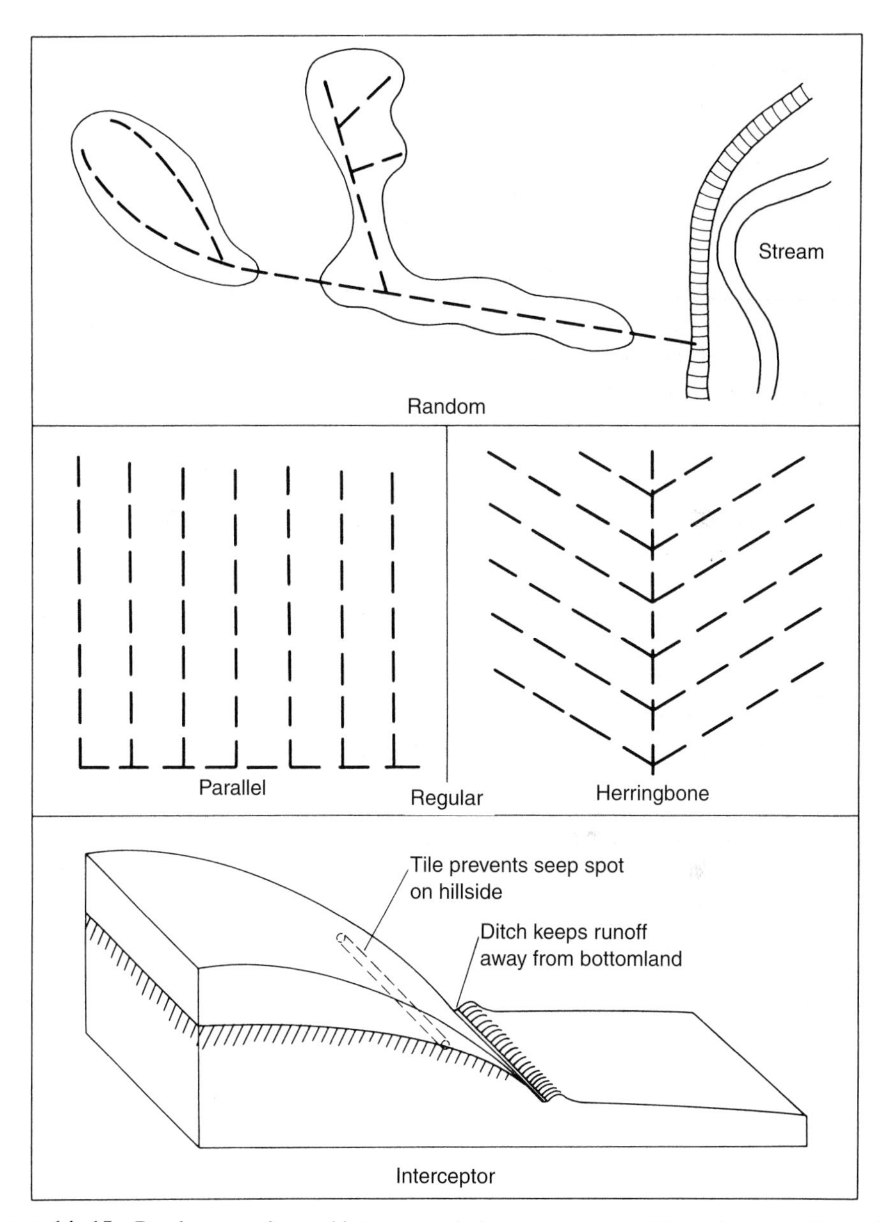

Figure 14–15 Random, regular, and interceptor drainage systems each have their own distinctive layout patterns.

Precise topographic information is needed for drainage design. Elevations of the outlet and of high and low points throughout the system must be known. Often, a full topographic survey is worthwhile. Some elevations may need to be measured even if a topographic map is already available.

Geologic maps tell what underlies the soil to a greater depth than do soil survey maps. Permeable layers that can carry water either in or out of the area are identified, and the nature and locations of impermeable rock layers are revealed.

Hydrologic maps show the source of the water, its direction of movement, and the depth of the water table at various points. They help identify places where an interceptor drain can be used instead of placing several drainage lines across the field.

All of these types of maps plus field observations may be necessary for a complex project, but many simpler projects are designed from one or two maps plus some field observations. Aerial photographs are often very helpful. For example, a sandy strip that carried water into the corner of an orchard in Idaho was much easier to see on an aerial photograph than on the ground.

14–9.1 Choosing the Type of System

Decisions regarding drainage systems are based on the information sources discussed in the preceding paragraphs and the preferences of the persons involved. Some choices become obvious as information is gathered. The Idaho orchard mentioned in the last paragraph should be drained by an interceptor system. Rolling land with several wet spots needs a random system, whereas a large, uniformly wet area calls for regularly spaced parallel drains.

Surface drainage works where water stays on the land surface long enough to be gathered and carried away. Subsurface drainage is required when the water enters the soil rapidly or from beneath. Sometimes the best system is a combination of surface and subsurface drainage.

The decision to use ditches, tile lines, wells, or some other method to remove excess water is strongly influenced by economics and convenience factors; preferences and customs are also factors. Tile lines are advantageous in fields where ditches would be in the way. Ditches, on the other hand, are often used for main outlets where the large tile required to carry the water would be too expensive. The special adaptations of other types of systems were explained in Section 14–7.

14–9.2 Layout of a Drainage System

The outlet is a critical part of a drainage system and is a natural beginning point for a design. Its location and elevation are significant to the placement of ditches and tile lines. Also, the outlet must have adequate capacity to handle the water and be legally accessible for that purpose. Some installations provide controls near the outlet to regulate the outflow so that the water table level can be managed (Section 14–9.4).

Figure 14–16 shows three different kinds of drainage systems on a farm in a humid region. Water from all three systems drains into a main ditch that belongs to, and is managed by, a drainage company. The smaller ditches and the tile lines belong to the farm owner. The bottomland in the southwest corner of Figure 14–16 is surrounded by ditches on all sides. The main ditch is about 8-ft (2.5-m) deep across this bottomland, and the two

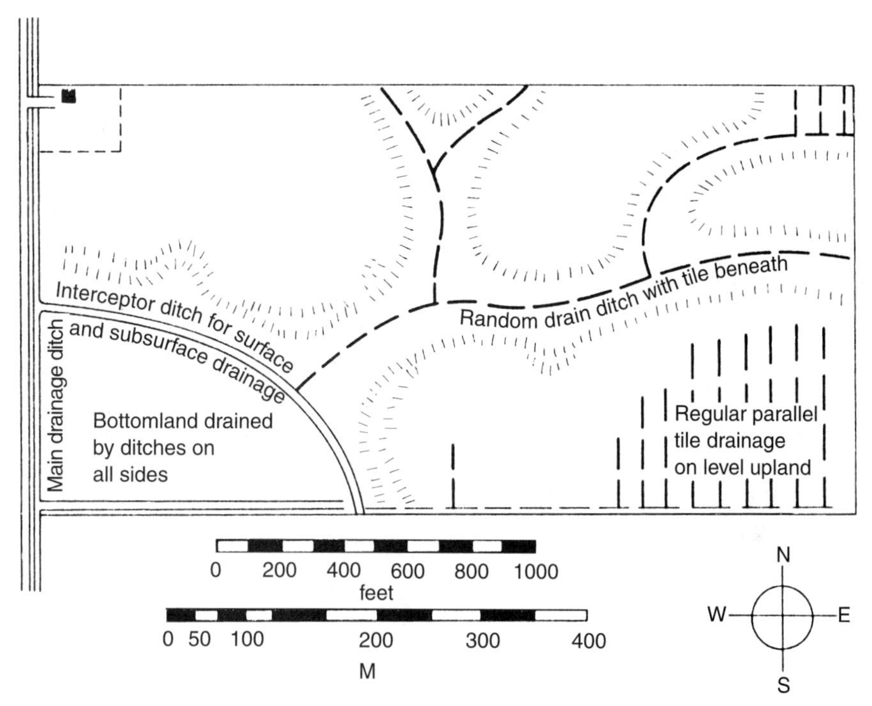

Figure 14–16 This 80-ac (32-ha) area has drainage ditches surrounding a bottomland, a random system of ditches with tile beneath following natural drainageways, and a regular parallel-tile drainage system on a level upland. The single branch tile line near bottom center of the map allows air to enter the line at the top of a slope.

farm ditches have gradients of about 0.1%, so the water flows at nonerosive velocities into the main ditch. All three ditches have side slopes of 1.5:1. This bottomland soil is sufficiently permeable that these three ditches provide adequate drainage, especially since both surface and subsurface water from the upland are intercepted by one of the ditches. The ditch capacities could have been calculated by Manning's formula as explained in Note 4–2, but this was unnecessary because of their large surplus capacity. If it had been needed, additional surface drainage for the bottomland could have been provided by a few field ditches or a bedding system, or subsurface drainage could have been supplemented by tile lines.

The level upland in the southeast corner of Figure 14–16 is drained by a parallel-tile system. The most important design factors for a regular system such as this are the depth, spacing, gradient, and size of the tile lines. These are approximately 4-ft (1.2-m) deep at the main line and as shallow as 3 ft (0.9 m) at the upper ends of the laterals. A spacing of 80 ft (25 m) between lines was chosen for this moderately permeable soil. The laterals are 4-in. (10-cm) clay tile placed on a controlled 0.2% gradient. The upper end of each lateral is closed to keep out soil and burrowing animals.

Most drainage laterals have more than adequate capacity, but the size of the main line required must be calculated. For the system shown in Figure 14–16, the *drainage coefficient* (the depth of water to be removed in 24 hours) is taken as 0.4 in./24 hr (1.0 cm/24 hr). Elsewhere, it might range from 0.1 to 1.5 in./24 hr (0.3 to 4 cm/24 hr) (Table 14–1), or even

TABLE 14–1 DRAINAGE COEFFICIENTS FOR SUBSURFACE TILE DRAINAGE[a]

	Field crops		Truck crops	
	in./24 hr	cm/24 hr	in./24 hr	cm/24 hr
Arid regions (irrigated)	0.12–0.25	0.3–0.6	0.25–0.50	0.6–1.2
Humid regions				
Mineral soils	0.25–0.50	0.6–1.2	0.50–0.75	1.2–1.8
Organic soils	0.50–0.75	1.2–1.8	0.75–1.60	1.8–4.0

[a]The values should be doubled or a known amount added where surface drainage is also carried by the tile.

Source: Based on Soil Conservation Service, 1973; Donnan and Schwab, 1974.

larger if the drainage system has surface inlets that must protect a residential area or other sensitive site from flooding. The drained area in this example is equivalent to a rectangle 500 ft by 830 ft (150 m by 250 m), so the amount of water to be removed is

$$500 \text{ ft} \times 830 \text{ ft} \times 0.4 \text{ in./24 hr} \times 1 \text{ ft/12 in.} = 13{,}833 \text{ ft}^3\text{/24 hr}$$
$$\text{or } 150 \text{ m} \times 250 \text{ m} \times 1.0 \text{ cm/24 hr} \times 1 \text{ m/100 cm} = 375 \text{ m}^3\text{/24 hr}$$

The main line in Figure 14–16 has a slope gradient of 0.1% and needs a 6-in. (15-cm) tile to transport this quantity of water (Table 14–2). An 8-in. (20-cm) tile would have been needed if the line had had surface inlets.

The gradient of the main line increases to about 1% for the last 300 ft (100 m) before it empties into the ditch and could cause a vacuum that would draw soil into the lines. This problem can be overcome (1) by installing a surface inlet at the top of the slope so air can enter the line, or (2) by allowing air to enter through a branch line in dry soil as shown in Figure 14–16. A 20-ft (6-m) length of corrugated-steel pipe protects the outlet of the main line by spilling water into the middle of the drainage ditch so it cannot start a gully by undercutting tile. Three rods inserted through holes in the end of the pipe keep animals out. Animals that crawl into a tile line are likely to die there and plug the line.

The random drains shown in Figure 14–16 are a combination of field ditches and tile. The ditches range from 20 to 40 in. (0.5 to 1 m) in depth, have bottom widths of 0 to 7 ft (0 to 2 m), and have side slopes between 3:1 and 4:1. They are crossable with farm machinery and have a cover of smooth bromegrass to control erosion. Other sod-forming grasses such as Kentucky bluegrass, tall fescue, Bermudagrass, and Dallisgrass can be used in their appropriate climatic zones. Reed canarygrass is used in cool climates and Bahiagrass in warm climates where tough sods are needed to resist gully formation. Switchgrass has become important for field ditches because it resists the triazine herbicides that have killed other grasses in many waterways.

Tile lines located under one side of the field ditches provide subsurface drainage to dry the waterway after a rain. These are mostly 4-in. (10-cm) tile, but 6-in. (15-cm) tile is used for the lower part of the main line. Locating them at the side of the ditch area reduces the likelihood of surface water eroding holes down to the tile. The upper ends of the lines are closed and the lower end is protected by a steel pipe, like the one already described for the regular tile drain. The three short lines in the northeast corner of Figure 14–16 form a regular drainage system that outlets through the random tile line.

TABLE 14–2 FLOW VELOCITIES[a] AND CARRYING CAPACITIES[b] OF WELL-ALIGNED CLAY OR CONCRETE TILE[c]

	Tile gradient				
Tile diameter	0.05%	0.1%	0.2%	0.5%	1.0%
4-in. (10-cm) tile:					
Velocity of flow					
ft/sec	[d]	0.82	1.18	1.83	2.59
m/sec	—	0.25	0.36	0.56	0.79
Capacity					
ft^3/24 hr	—	6,030	8,500	13,450	19,020
m^3/24 hr	—	171	241	381	539
5-in. (12.5-cm) tile:					
Velocity of flow					
ft/sec	0.69	0.95	1.35	2.13	3.02
m/sec	0.21	0.29	0.41	0.65	0.92
Capacity					
ft^3/24 hr	7,730	10,900	15,420	24,420	34,510
m^3/24 hr	219	309	437	692	978
6-in. (15-cm) tile:					
Velocity of flow					
ft/sec	0.75	1.08	1.54	2.43	3.41
m/sec	0.23	0.33	0.47	0.74	1.04
Capacity					
ft^3/24 hr	12,560	17,750	25,090	39,520	56,100
m^3/24 hr	356	503	711	1,120	1,590
8-in. (20-cm) tile:					
Velocity of flow					
ft/sec	0.92	1.31	1.84	2.92	4.13
m/sec	0.28	0.40	0.56	0.89	1.26
Capacity					
ft^3/24 hr	27,030	38,110	53,990	85,400	120,680
m^3/24 hr	766	1,080	1,530	2,420	3,420
10-in. (25-cm) tile:					
Velocity of flow					
ft/sec	1.08	1.51	2.13	3.41	4.79
m/sec	0.33	0.46	0.65	1.04	1.46
Capacity					
ft^3/24 hr	49,050	69,160	98,100	154,910	219,140
m^3/24 hr	1,390	1,960	2,780	4,390	6,210
12-in. (30-cm) tile:					
Velocity of flow					
ft/sec	1.21	1.71	2.43	3.84	5.41
m/sec	0.37	0.52	0.74	1.17	1.65
Capacity					
ft^3/24 hr	79,750	112,570	159,150	251,950	356,400
m^3/24 hr	2,260	3,190	4,510	7,140	10,100

[a]Velocity based on Manning's formula with slope as a decimal and tile radius in feet ($V = 86\ R^{2/3}S^{1/2}$) or meters ($V = 58.5\ R^{2/3}S^{1/2}$).

[b]Capacity = πr^2 velocity × time.

[c]The corresponding values for corrugated-plastic drain tubing are about two-thirds as large.

[d]Four-in. tile is not recommended on 0.05% slope.

14–9.3 Installing a Drainage System

The installation of a drainage system begins at the outlet and proceeds up the grade, so the water can escape without ponding. In fact, some systems have been installed without surveying by watching how fast the water flowed away while the ditch or trench was being dug. Surveying is strongly recommended, though, so the system can be planned and the installation optimized.

Junctions involve special techniques in both ditch and tile systems. A branch ditch is often given a short length of steeper gradient near the junction so that its water level will be slightly above that of the larger ditch. This prevents water from flowing up the branch ditch and depositing sediment.

Many lines have been installed by "chipping" tile (breaking out pieces to make them fit), but the use of manufactured junction tile is recommended for a faster installation and a tighter fit that will keep soil from entering. Special tile are available for junctions, corners, and size changes. These and the pipe needed for the outlet and for road crossings should be on hand when the job begins.

Installing surface inlets to tile lines involves special techniques. Inlets need to be located in low points on the land surface, although the low point can be moved some by land smoothing. Many surface inlets are installed in fencelines or enclosed in a small area to protect them from machinery. Others are tall, conspicuous standpipes with holes or slots that control water entrance. Some inlets are flat, so traffic can cross them, and are covered by gratings. These tend to catch trash that needs to be removed frequently. Another variation, the *blind inlet,* consists of a pit filled with gravel or small stones.

Any type of inlet should have openings small enough to keep out animals and trash. Surface inlets should enter a branch line about 15-ft (5-m) long rather than going directly into a main tile line. The main line will then function even if the inlet line is plugged.

Occasionally, one wants to drain an area that is too wet to enter with drainage equipment. Dynamite has been used successfully for the initial drainage of some areas with cohesive soils. This method is not satisfactory for gravel or loose dry soils. Bennett (1947) suggested that charges be placed about 20- to 30-in. (50- to 75-cm) apart and 30-in. (75-cm) deep. A few trials should be used to determine the optimum placement and size of charge.

14–9.4 Water Table Management

A drainage system that provides optimum soil-moisture conditions during wet periods or when crops need maximum rooting depth may cause the soil to be too dry at other times (Mejia et al., 2000). It may also carry away plant nutrients that are useful in the soil but become pollutants in the water. These problems can be addressed by installing controls in the drainage system to manage the water table. Usually these controls are installed near the outlet so they will affect the entire area of the drainage system; additional controls may also be needed at critical points in the system if there are large elevation differences in the drained area.

A *check structure* similar to that used for irrigation (Section 15–4.2) can be used to raise the water level in a ditch and thus raise the water table in land drained by the ditch.

Boards placed in notches in the check raise the water level when no crop is growing and can be removed one by one as a growing crop needs more rooting depth.

Tile drains can be managed by having them flow through a control box with a removable divider across the flow path. The control box may be made of poured concrete, or it may be made of concrete blocks with concrete poured into their cores to hold them together. A slot on each side of the box will hold boards that will raise the water level. Controlling the water level was found to reduce nitrate leaching and increase corn yields by 19% and soybean yields by 64% as compared to an uncontrolled drainage system in Ohio (Fisher et al., 1999).

The addition of a source of water can convert a controlled tile drainage system into a subsurface irrigation system (Section 15–5.2) that removes water during wet periods and supplies water during dry periods.

14–9.5 Maintenance of Drainage Systems

Drainage ditches need to be kept free of excess vegetation and debris. Many ditches, especially the flatter ones, accumulate sediment that must be removed once every few years. Regular maintenance is essential to keep ditches functioning properly.

Grass growing on ditch banks is usually desirable for erosion control, but reeds and cattails in the water slow the flow, raise the water level, and make the ditch less effective. Mowing, burning, use of herbicides, and hand removal are all used to limit the amount of growth in ditches, but none gives a permanent remedy.

Tile lines also need maintenance, although less frequently than ditches. A tile outlet covered by debris or by high water in an outlet ditch may cause sediment to be deposited in the line and, thus, reduce its capacity. Pressure may build up in the line and cause water to flow outward and wash out a hole. Tile falling into such holes cause misalignments that allow soil to enter and block the line. Similar problems result when a heavy load crushes one or more tile.

Tree roots can plug tile lines if they are allowed to grow nearby. Poplar trees are a special hazard because they grow horizontally and produce fine roots that can easily enter the tile lines. Furthermore, a new tree can sprout from the roots and, thus, extend the root system. Awan (1992) recommends keeping poplars completely away from tile-drained fields. Willows, eucalyptus, and silver mulberry pose moderate hazards and should be kept at least 50-ft (15-m) away from tile lines, whereas some hardwoods and fruit trees might be safe at a distance of 15 ft (4.5 m). Deep-rooted crops, such as alfalfa, also have some potential for plugging tile lines. Problem spots in tile lines produce wet spots and/or holes in the field. Either of these conditions should be remedied promptly by digging a hole to expose the tile so that the damage can be repaired. Excessive delay permits more soil to be washed into the line. Sections of the line can become so blocked by sediment that they have to be dug up, cleaned, and relaid.

14–9.5 Inadequate Drainage Systems

Sometimes a drainage system becomes inadequate. Perhaps the soil has lost permeability, or a new cropping system may be planned. More lines may then be needed to make the system function adequately again. The procedure for a ditch system may be as simple as digging new ditches between the existing ditches.

A need for additional tile lines may be noticed when the area midway between lines stays wet too long after a rain. Unfortunately, the exact location of the existing tile lines may not be obvious. If available, a map of the original system can be very helpful. Sometimes the lines can be seen faintly on an aerial photo or located by watching for tile chips or different-colored soil in the field. The fastest-drying soil and the earliest-maturing crops also help identify the approximate locations of tile lines. Precise locations can be determined with a probe and suitable plans made for enlarging the system.

SUMMARY

Wetlands occur in arid as well as in humid climates and are identified by black, rust, bluish-gray, and white soil colors. They usually have higher organic-matter contents and often have more smectite clay than neighboring soils. Reducing conditions influence their chemical behavior. Wet soils containing sulfides can become strongly acidic when drained, and those with excess sodium salts can become highly alkaline.

The limitations resulting from wetness depend on how wet the land is, when the wetness occurs, and how long it stays wet. Clay soils become sticky and too weak to support heavy loads when wet. Sandy soils are strongest when moist but not wet. Traffic can puddle wet soil so it will harden when it dries. Wet soil delays planting because of coldness as well as wetness. Lack of oxygen prevents root growth of most plants and results in denitrification and reduced availability of potassium.

Either surface or subsurface water can be removed by *open ditches, tile lines,* or various other methods. *Land smoothing* and *bedding systems* are means of surface drainage. Mole drains are the least expensive and shortest-lived means of subsurface drainage. Both *pumped wells* and *dry wells* are sometimes used for drainage.

Ditches offer high capacity and are less expensive than tile, but they occupy significant land areas and are inconvenient to have in fields. Tile lines cost more, especially in large sizes, but require less maintenance than ditches and do not restrict traffic. Tile are available in clay, concrete, and plastic and are usually placed at depths between 3 and 5 ft (90 to 150 cm) in humid regions or about 6 or 7 ft (2 m) in arid regions.

Drainage systems are designed in *random, regular,* or *interceptor* patterns. The outlet is a very important feature of a drainage system and is the beginning point for installation. A *drainage coefficient* must be assumed in calculating how much capacity is needed before minimum tile or ditch sizes and gradients can be determined. After a drainage system is installed, it needs proper maintenance to keep it functioning properly.

QUESTIONS

1. Explain how well drained, somewhat poorly drained, and poorly drained soils are identified in the field.
2. Why should some land be left undrained?
3. What causes some wet soils to become strongly acidic when drained and others to become highly alkaline?
4. What factors determine whether a drainage system should be surface or subsurface, ditch or tile, and random or regular?

5. Describe a circumstance where an interceptor drain would be useful and explain how it would be installed.
6. How does land smoothing improve drainage?
7. What special precautions are needed for locating and installing a tile drainage outlet?

REFERENCES

AWAN, M. S., 1992. Clogging of subsurface drain pipes by tree roots. In W. F. Vlotman (ed.), *Subsurface Drainage on Problematic Irrigated Soils: Sustainability and Cost Effectiveness.* Proc. 5th Int. Drainage Workshop, Feb. 8–15, 1992, Lahore, Pakistan, Vol. 1, 296 p.

BASAK, P., 1972. Soil structure and its effects on hydraulic conductivity. *Soil Sci.* 114:417–422.

BENNETT, H. H., 1947. *Elements of Soil Conservation.* McGraw-Hill, New York, 406 p.

BERGEDALEN, J., 1992. Iron contaminated groundwater and land drainage. In W. F. Vlotman (ed.), *Subsurface Drainage on Problematic Irrigated Soils: Sustainability and Cost Effectiveness.* Proc. 5th Int. Drainage Workshop, Feb. 8–15, 1992, Lahore, Pakistan, Vol. 2, 356 p.

BROWN, P. L., A. D. HALVORSON, F. H. SIDDOWAY, H. F. MAYLAND, and M. R. MILLER, 1982. *Saline-seep diagnosis, control, and reclamation.* USDA Cons. Res. Rept. No. 30, 22 p.

CHRISTEN, E. W., and G. SPOOR, 2001. Improving mole drainage channel stability in irrigated areas. *Agric. Water Mgmt.* 48:239–253.

DE VOS, J. A., D. HESTERBERG, and P. A. C. RAATS, 2000. Nitrate leaching in a tile-drained silt loam soil. *Soil Sci. Soc. Am. J.* 64:517–527.

DIERICKX, W., 1992. Research and developments in selecting drainage materials. In W. F. Vlotman (ed.), *Subsurface Drainage on Problematic Irrigated Soils: Sustainability and Cost Effectiveness.* Proc. 5th Int. Drainage Workshop, Feb. 8–15, 1992, Lahore, Pakistan, Vol. 1, 296 p.

DONNAN, W. W., and G. O. SCHWAB, 1974. Current drainage methods in the U.S.A. In *Drainage for Agriculture,* Agronomy Monograph No. 17, American Society of Agronomy, Madison, WI, p. 93–114.

FATMA, A. R. A., 1992. Application of tubewell drainage in Egypt—Implications and results. In W. F. Vlotman (ed.), *Subsurface Drainage on Problematic Irrigated Soils: Sustainability and Cost Effectiveness.* Proc. 5th Int. Drainage Workshop, Feb. 8–15, 1992, Lahore, Pakistan, Vol. 2, 356 p.

FISHER, M. J., N. R. FAUSEY, S. E. SUBLER, L. C. BROWN, and P. M. BIERMAN, 1999. Water table management, nitrogen dynamics, and yields of corn and soybean. *Soil Sci. Soc. Am. J.* 63:1786–1795.

HARRIS, G. L., K. R. HOWSE, and T. J. PEPPER, 1993. Effects of moling and cultivation on soil-water and runoff from a drained clay soil. *Agric. Water Mgmt.* 23:161–180.

HUSSAIN, M. S., I. A. SHEIKH, and A. W. NAGI, 1992. Deterioration of scarp tubewells in Pakistan and recommendations for their rehabilitation. In W. F. Vlotman (ed.), *Subsurface Drainage on Problematic Irrigated Soils: Sustainability and Cost Effectiveness.* Proc. 5th Int. Drainage Workshop, Feb. 8–15, 1992, Lahore, Pakistan, Vol. 1, 296 p.

JAYNES, D. B., S. I. AHMED, K.-J. S. KUNG, and R. S. KANWAR, 2001. Temporal dynamics of preferential flow to a subsurface drain. *Soil Sci. Soc. Am. J.* 65:1368–1376.

JHA, M. K., and K. KOGA, 1995. Mole drainage: Prospective drainage solution to Bangkok clay soils. *Agric. Water Mgmt.* 28:253–270.

KING, J. A., 1931. *Tile Drainage.* Mason City Brick and Tile Co., Mason City, IA, 108 p.

KUNG, K.-J. S., T. S. STEENHUIS, E. J. KLADIVKO, T. J. GISH, G. BUBENZER, and C. S. HELLING, 2000. Impact of preferential flow on the transport of adsorbing and nonadsorbing tracers. *Soil Sci. Soc. Am. J.* 64:1290–1296.

LANT, C. L., S. E. KRAFT, and K. R. GILLMAN, 1995. The 1990 farm bill and water quality in Corn Belt watersheds: Conserving remaining wetlands and restoring farmed wetlands. *J. Soil Water Cons.* 50:201–205.

LESAFFRE, B., J. C. FAVROT, M. PENEL, D. ZIMMER, and M. P. ARLOT, 1992. Overview of diagnosis and identification of problematic soils. In W. F. Vlotman (ed.), *Subsurface Drainage on Problematic Irrigated Soils: Sustainability and Cost Effectiveness.* Proc. 5th Int. Drainage Workshop, Feb. 8–15, 1992, Lahore, Pakistan, Vol. 1, 296 p.

MATHIAS, M. E., and P. MOYLE, 1992. Wetland and Aquatic Habitats. *Agric. Ecosystems Environ.* 42:165–176.

MEJIA, M. N., C. A. MADRAMOOTOO, and R. S. BROUGHTON, 2000. Influence of water table management on corn and soybean yields. *Agric. Water Mgmt.* 46:73–89.

MOHAMOUD, Y. M., 1994. Effect of mound height and cassava cultivar on cassava performance under a fluctuating water table. *Agric. Water Mgmt.* 26:201–211.

PATT, J., D. CARMELI, and I. SAFRIR, 1966. Influence of soil physical condition on productivity of citrus trees. *Soil Sci.* 102:82–84.

PITTS, D., 2000. Drainage management to improve water quality and to enhance agricultural production. *J. Soil Water Cons.* 55:424.

SHEIKH, A. W. F., 1992. Pakistan's experience in tubewell drainage. In W. F. Vlotman (ed.), *Subsurface Drainage on Problematic Irrigated Soils: Sustainability and Cost Effectiveness.* Proc. 5th Int. Drainage Workshop, Feb. 8–15, 1992, Lahore, Pakistan, Vol. 1, 296 p.

SHIPITALO, M. J., and F. GIBBS, 2000. Potential of earthworm burrows to transmit injected animal wastes to tile drains. *Soil Sci. Soc. Am. J.* 64:2103–2109.

SOIL CONSERVATION SERVICE, 1973. *Drainage of Agricultural Land.* Water Information Center, Inc., Port Washington, NY, 430 p.

VARSHNEY, R. S., 1992. Necessity of demarcating waterlogging and drainage limits. In W. F. Vlotman (ed.), *Subsurface Drainage on Problematic Irrigated Soils: Sustainability and Cost Effectiveness.* Proc. 5th Int. Drainage Workshop, Feb. 8–15, 1992, Lahore, Pakistan, Vol. 2, 356 p.

WESSELING, J., 1974. Crop growth and wet soils. In *Drainage for Agriculture.* Agronomy Monograph 17. American Society of Agronomy, Madison, WI, p. 7–90.

WOOTEN, H. H., and L. A. JONES, 1955. The history of our drainage enterprises. In *Water,* Yearbook of Agriculture. USDA, Washington, D.C., p. 478–491.

ZHANG, W., 1992. Tubewell drainage in semi-arid areas of North China Plain. In W. F. Vlotman (ed.), *Subsurface Drainage on Problematic Irrigated Soils: Sustainability and Cost Effectiveness.* Proc. 5th Int. Drainage Workshop, Feb. 8–15, 1992, Lahore, Pakistan, Vol. 2, 356 p.

ZHAOYI, L., and S. RONGKAI, 1992. Drainage amelioration of waterlogged soils in south China. In W. F. Vlotman (ed.), *Subsurface Drainage on Problematic Irrigated Soils: Sustainability and Cost Effectiveness.* Proc. 5th Int. Drainage Workshop, Feb. 8–15, 1992, Lahore, Pakistan, Vol. 2, 356 p.

CHAPTER

15

IRRIGATION AND RECLAMATION

Irrigation and reclamation are powerful means of increasing land productivity. Adding irrigation water can overcome drought limitations and improve both the quality and the quantity of crop production. Reclamation increases land productivity through irrigation, drainage, salt removal, or other amelioration of soils so that more and better crops can be grown on them. Both irrigation and reclamation help supply the food and fiber needs of the world's growing population.

Irrigation and reclamation are obviously significant in arid regions (Note 15–1). Most land reclamation occurs in arid or semiarid climates, but supplemental irrigation to meet special or occasional needs is used in subhumid and even humid areas. High-value specialty crops are more likely to be irrigated than others in humid areas.

NOTE 15–1
ARID CLIMATES AND ARID REGIONS

An *arid climate* is defined in *Soil* (the 1957 Yearbook of Agriculture, p. 752) as "A very dry climate like that of desert or semidesert regions where there is only enough water for widely spaced desert plants." *Arid regions* are defined as "Areas where the potential water losses by evaporation and transpiration are greater than the amount of water supplied by precipitation." Arid regions encompass the areas of arid climates plus large areas with semiarid climates.

The maximum precipitation that defines arid climates and arid regions depends on temperature and, to some extent, on the season when most of the precipitation occurs. The upper limit of precipitation for arid climates is about 10 in. (250 mm) per year in cool regions and about 20 in. (500 mm) per year in the tropics. Arid region limits are two to three times as much as those for arid climates.

The precipitation in arid climates is seasonal and erratic as well as limited in amount. Plants must either grow during short periods of favorable moisture or rely on irrigation.

Some wet soils occur even in arid climates. These regions include hilly or mountainous areas with humid conditions at their higher elevations. Water flows or seeps down to the lower elevations and saturates the soil in some low areas. Even the limited precipitation within an arid area generates runoff and seepage that cause local wet spots. These wet spots can often be used as sources of water for household or livestock needs. Some of the larger ones can be developed for irrigation water sources.

Irrigation is practiced in many countries, but Asia has more than 60% of the total irrigated land in the world. India with 146 million ac (59 million ha) of irrigated land has recently surpassed China; China is in second-place with 133 million ac (54 million ha) of irrigated land; the United States is a distant third with 55 million ac (22 million ha) (Table 15–1). Worldwide irrigated land approximately doubled between 1961 and 1999, but the increase was spotty. Japan had a decrease and several countries (such as the United States) showed modest increases, whereas India, China, the former Soviet Union, and several smaller countries had large increases during this period.

15–1 EFFECTS OF IRRIGATION

Irrigation can change low-priced grazing land into expensive cropland. New crops can be grown and drought risk can be greatly reduced for adapted crops, even in humid regions. Farming systems, including field arrangements, crops, fertilization, and field operations are adjusted to work with the irrigation system. Costs and returns both increase, and good management is needed to take advantage of the opportunities while avoiding pitfalls. The changes require many more workers to manage the land.

15–1.1 Increased Production with Irrigation

Arid-region food production can be multiplied several fold if irrigation water is available for land reclamation. Low-intensity cropland and grazing land become diversified, high-producing cropland.

Irrigation in humid regions may not change the crop to be grown, but it increases production by increasing yields. Humid region soils may lose water by runoff and deep percolation, but there are often droughty periods as well. The available water-holding capacity of the soil helps determine whether irrigation is needed. A deep, loamy soil able to store 10 or 12 in. (25 or 30 cm) of available water can support a crop through a much longer dry period than a shallow or sandy soil storing only 2 to 4 in. (5 to 10 cm).

A reliable water supply increases yields so much that about 40% of the world's harvested crops comes from the 20% of cropland that is irrigated (Oomen et al., 1990). Adequate water often improves the crop quality as well as increasing the quantity. Drought causes quality problems in various growth factors and in fruit production.

TABLE 15–1 IRRIGATED LAND IN THE MAJOR IRRIGATING COUNTRIES OF THE WORLD FOR SELECTED YEARS FROM 1961 TO 1999 IN THOUSANDS OF HECTARES

	1961	1970	1980	1990	1999
India	24,685	30,440	38,478	45,144	59,000
China, People's Republic of	30,411	38,121	45,470	47,957	53,740
United States of America	14,000	16,000	20,582	20,900	22,400
Soviet Union (former USSR)	9,400	11,100	17,200	20,800	19,921*
Pakistan	10,751	12,950	14,680	16,940	17,950
Iran, Islamic Rep. of	4,700	5,200	4,948	7,000	7,562
Mexico	3,000	3,583	4,980	5,600	6,500
Indonesia	3,900	3,900	4,301	4,410	4,815
Thailand	1,621	1,960	3,015	4,328	4,750
Turkey	1,310	1,800	2,700	3,800	4,500
Bangladesh	426	1,058	1,569	2,936	3,985
Spain	1,950	2,379	3,029	3,402	3,640
Iraq	1,250	1,480	1,750	3,525	3,525
Egypt	2,568	2,843	2,445	2,648	3,300
Vietnam	1,000	1,200	1,700	2,900	3,000
Brazil	490	796	1,600	2,700	2,900
Italy	2,400	2,400	2,400	2,711	2,698
Romania	206	731	2,301	3,109	2,673
Japan	2,940	3,415	3,055	2,846	2,659
Afghanistan	2,160	2,340	2,505	2,500	2,386
Australia	1,001	1,476	1,500	1,832	2,251
France	360	539	870	1,300	2,100
Sudan	1,480	1,625	1,800	1,946	1,950
Myanmar (Burma)	536	839	999	1,005	1,841
Chile	1,075	1,180	1,255	1,600	1,800
Saudi Arabia	343	365	600	1,600	1,620
Argentina	980	1,280	1,560	1,560	1,561
Philippines	690	826	1,219	1,550	1,550
Korea, People's Dem. Rep. of	500	500	1,120	1,420	1,460
Greece	430	730	961	1,195	1,441
South Africa	808	1,000	1,128	1,290	1,354
Morocco	875	920	1,217	1,258	1,305
Peru	1,016	1,106	1,140	1,190	1,195
Syrian Arab Republic	558	451	539	693	1,186
Korea, Republic of	1,150	1,184	1,307	1,345	1,159
Nepal	70	117	520	950	1,135
Madagascar	300	330	645	1,000	1,090
World	138,989	67,803	209,716	244,305	274,166

*The Soviet Union divided into 15 nations in 1991. Seven of these had more than a million ha of irrigated land in 1999; in thousands of ha, these were Russia, 4,600; Uzbekistan, 4,281; Ukraine, 2,434; Kazakhstan, 2,350; Turkmenistan, 1,800; Azerbaijan, 1,455; and Kyrgystan, 1,072.

Source: Food and Agriculture Organization Web site.

Figure 15–1 Excess salts (forming white crusts) and pH > 8.5 (causing dark organic coatings) inhibit corn growth on these irrigated bottomland soils in Idaho. (Courtesy F. R. Troeh.)

In addition to overcoming drought, sprinkler irrigation systems have been used to protect sensitive fruit crops from frost damage. The heat released by freezing water holds the temperature near 0°C rather than letting it drop low enough to freeze plant tissue.

15–1.2 Hazards of Irrigation

Some notable hazards occur along with the benefits associated with irrigation and reclamation. One is the economic risk from high irrigation costs, and another is the spreading of water-related diseases. A third hazard is productivity loss caused by increased erosion and by excess salts accumulating from improper use of water, as shown in Figure 15–1. Some such soils can be reclaimed, as discussed later in this chapter, but the effort is not always economical. A fourth hazard is the risk of causing water pollution, both by sediment and associated chemicals in runoff water and by chemicals leached downward into the ground water.

Irrigation expenses arise from obvious factors such as the costs of buying water or drilling a well, purchasing additional equipment, and paying for the increased labor required to apply water to the land. Several other inputs, such as fertilizer, usually need to be increased to allow the crops to yield according to their new potentials. Oomen et al. (1990) indicate that 60% of all fertilizer used is applied to irrigated land. Another expense is caused by the constant struggle to maintain the water distribution system and to keep it free of pests such as burrowing animals and weeds. Rodent holes can cause canal banks to fail suddenly with disastrous results. Weeds growing in a canal disrupt the system by slowing the flow of water. Even weeds growing on a canal bank are a problem because their seed can be carried by the water and distributed through the fields watered from the canal. One of the best solutions is to graze sheep on the canal banks. Farmers in Idaho found that the sheep controlled both the weeds and the rodents because the rodents depended on the vegetation for cover (Fiege, 1995).

Several water-related diseases that can be spread by irrigation are discussed by Oomen et al. (1990). These include malaria, bilharzia, African sleeping sickness, liver fluke infections, and several others. Malaria can be controlled by varying the water level in reservoirs by about 1 ft (0.3 m) every 7 to 10 days so that fish in the reservoir can eat the mosquito larvae. Bilharzia (also called schistosomiasis) is caused by a small worm that is parasitic in human blood and needs snails as an intermediate host. The snails breed in any vegetation that grows in the water around the edge of a reservoir and can be controlled by eliminating such vegetation. African sleeping sickness (trypanosomiasis) is transmitted by the tsetse fly and is best avoided by controlling the flies.

Salt accumulations and increases in alkalinity are common in irrigated soils. Poor quality irrigation water can bring in much larger quantities of various salts than the plants growing on the soil can possibly incorporate into their tissues. Any remaining salt must either accumulate in the soil or leach into the ground water. If the ground water becomes saline, either from downward leaching salts or from salts moving laterally from higher-lying soils, it can serve as a source of salts in soils with a high water table. Generally, the most saline soil will be found at relatively low elevations, for example at the lower end of a field with surface irrigation (Ben-Hur et al., 2001).

Erosion hazards often increase with irrigation. Because surface irrigation often cuts rills, the cumulative effects of many such irrigations can be ruinous. Deposition of sediment can also be so damaging that continued irrigation becomes impractical. A well-designed program of soil and water conservation may reduce erosion to tolerable rates, but the losses usually still exceed the natural rate.

Eroded soil may carry more plant nutrients and chemical pesticides from an irrigated field than from one where rainfed crops are grown, both because the chemicals are likely to be applied at a higher rate and because there is likely to be more runoff and erosion from the irrigated field, especially where surface irrigation is practiced. There is also likely to be more leaching and more movement of chemicals downward to the water table. For example, Butters et al. (2000) found even more leaching of atrazine than was anticipated on the basis of the amount of irrigation water applied, possibly because of rapid water flow through soil cracks or some kind of interaction with mobile soil components.

The hazards associated with irrigation have led some observers to suggest that all irrigation systems are temporary. Some irrigation systems have never produced a marketable crop, most often because of sodic soils. Others have failed after some decades of use, most commonly because of erosion and sedimentation problems. Even so, some irrigation systems have endured, and it is unfair to suggest that the failure of some means that all will fail.

Gulhati and Smith (1967) enumerate many examples of long-lasting irrigation systems. Egyptians have been practicing basin irrigation for more than 5000 years. Iranians are still using 2500-year-old *kanats* (tunnels) to supply irrigation water, as shown in Figure 15–2. Japanese paddy fields have been irrigated for more than 2500 years. Other ancient irrigation works occur in India, Pakistan, China, Peru, Central America, southwestern United States, and elsewhere. Some of these have been abandoned, but many are still in use. Some have been resurrected when modern surveyors have discovered that their new survey followed the precise course of a long-lost, sediment-filled canal.

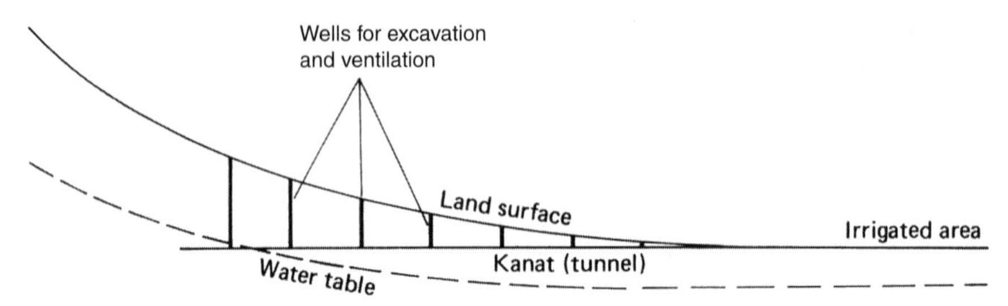

Figure 15–2 Kanats are used in several Asian countries to tap the water table under higher land and carry it out to irrigate the lower land.

15–2 SELECTING LAND FOR IRRIGATION

Several factors must be favorable for irrigation to be practical. Soil factors will be considered in this section and water in the next. Other important factors such as climate, suitable crops, roads, and markets will be assumed to be favorable.

Important soil factors that influence the profitability of irrigation include depth, texture, slope, and salt content. These factors may already be known if the land has been cropped before, either with or without irrigation. A preliminary evaluation of uncropped land can be made visually—smooth topography and good vegetative cover usually indicate good soil. Soil maps and topographic maps are very helpful for evaluating land.

The minimum depth of soil for irrigation depends on other soil properties, the crop, and the irrigator. Any soil less than 5-ft (150-cm) deep may reduce yields, and a soil depth less than 3 ft (90 cm) is often considered a limiting factor. Nevertheless, some irrigated soils are only 1-ft (30-cm) deep to hard bedrock. Such shallow soils require frequent irrigation because they cannot store much water. Deep tillage is out of the question, and the topography must be left unchanged because little or no land leveling can be done on these shallow soils. Erosion control is vital to retain what little soil there is.

Loamy soil textures throughout the solum are generally desirable, although sandy loams may be preferred for ease of tillage, faster infiltration rate, and easy harvesting of root crops. Very sandy soils have low water-holding and nutrient-storage capacities and are easily eroded by surface irrigation. Wind erosion becomes a serious factor with fine sand. Clay soils often have slow infiltration rates, waterlogging, and stickiness. A clay subsoil can cause waterlogging, even if the surface soil is favorable. Nearly pure silt, as found in some loess or alluvial deposits, is the easiest texture of all to erode by running water. Stony conditions cause tillage difficulties and reduce the water-holding capacity of the soil.

The ideal slope gradient for surface irrigation is about 0.5%. Surface irrigation is difficult on slopes flatter than 0.2% because the water stream required to irrigate them becomes too deep. Such problems can be solved by using basin, trickle, or sprinkler irrigation or by "leveling" the field into segments that have enough slope for the water to flow. Steep or irregular slopes cause more serious problems. Trickle, sprinkler, or contour irrigation offer only limited solutions. Large elevation differences make uniform water application more difficult no matter what method is used. Variable soils and slow infiltration rates are common complications.

Erosion is a potential problem wherever surface irrigation is practiced on slopes. In the United States, the usual maximum slope for surface irrigation varies from 3% in the

Gulf Coastal Plain, 6% in the central Great Plains, and 20% for pasture and hay in Colorado to 35% with good cover in the Pacific Northwest (Maletic and Hutchings, 1967). Irregularities in slope gradient and direction make irrigation more difficult than it is on uniform slopes.

Salt content may be easy to overlook, yet it may make the soil barren, as shown in Figure 15–1. Salt problems vary from minor and easily remedied to severe and not worth reclaiming. Excess salts, high percentages of sodium, and boron concentrations are serious hazards to plant growth.

15–3 WATER FOR IRRIGATION

The water supply for irrigation is usually more limiting than the area of suitable soil. Most of the available water in arid regions is already in use. In fact, most of the streams in arid regions have dry streambeds during at least part of the irrigation season (some of them dried up naturally in the summers even before any water was diverted). A dried-up stream is obviously disastrous for fish and other aquatic life. Humid regions have more water, but their water supplies also tend to run low when they are most needed. Inadequate supplies of water result in competition among irrigators, would-be irrigators, livestock operations, and other users such as cities and factories. Furthermore, streams and lakes need to have water left in them for fish, wildlife, recreational, and navigational uses. Water shortages are becoming more and more common along with population increases. Postel (1992) indicates that global water use now is more than three times as much as it was in 1950. Overuse of available water causes dropping water tables, shrinking lakes, and disappearing wetlands. Less water is left then to sustain aquatic life and for people to use. A nation needs about 1000 m^3 of water per person per year to meet modern living standards, but more than 200 million people in 26 countries now have less than that amount available (Postel, 1992). Water for irrigation is either essential or highly beneficial for agriculture in all 26 of these countries, but the water supply is not equal to the need.

15–3.1 Water Rights

Water use has both economic and legal aspects. The simplest way to allocate water is to sell it to the highest bidder. However, the interaction between water and land values requires a more complex analysis—often the water is not for sale apart from the land. Laws regulating the use of water also influence its economic value.

Two legal doctrines known as *riparian rights* and *prior appropriations* underlie water rights in the United States and many other countries. The riparian rights doctrine is based on old English law and allows the owners of land along a stream or lake to use the water (Johnson and Paschal, 1995). Riparian rights are used in the eastern part of the United States (Wisconsin, Iowa, Missouri, Arkansas, Louisiana, and all states east of these states). Nobody actually owns the water, but it can be used for domestic needs including household, livestock, garden, and lawn regardless of the effect on stream flow. Other uses are allowed if they do not diminish downstream flow or if they are within the land owner's reasonable share of the water remaining after all domestic needs are satisfied. Irrigation has a low priority for water use under riparian rights.

The prior appropriations doctrine is based on *beneficial use* of water and is used in the western United States (Minnesota and all states farther west, except that California, Oklahoma, North Dakota, and South Dakota use a combination of riparian rights and prior appropriations). A potential user must apply to the state for a permit to divert and use a certain amount of water from a certain place (Dufford, 1995). The permit is granted if the water has not been granted to someone already. The land does not have to be adjacent to the water source. A license is granted after "beneficial use" has been established. This license is renewable as long as the beneficial use is uninterrupted. It is treated as real property that can be bought and sold. Prior appropriations gives the highest priority to whatever use is established first and is therefore more favorable to irrigation, mining, and other nondomestic water use than riparian rights is. One problem with the prior appropriations doctrine is that it encourages established users to keep their use "uninterrupted," so they will not lose their water rights even if their use is inefficient. Wasteful practices are thereby protected.

Other laws can modify the riparian rights and prior appropriations doctrines. Municipalities and some organizations can obtain water by condemnation proceedings in some states. In some places, 20 years of use establishes a legal right to continue. Some recent proposals would encourage water conservation by making water rights a marketable commodity (Perala and Benson, 1995).

Many recent laws designed to maintain water quality place limitations on the sewage, chemicals, and other pollutants that can be emptied into streams or lakes. Also, water used for cooling a power or manufacturing plant must not be warm enough to damage aquatic life when it is returned to the stream. The Clean Water Act of 1987 recognizes the importance of aquatic life in the overall management of water use. Studies indicate that the proportion of aquatic organisms that have become extinct or rare (34% of fish, 65% of crayfish, and 75% of unionid mussels) is larger than that of terrestrial organisms (Karr, 1995). For example, low water flows have been particularly hard on the salmon industry of the northwestern United States and have contributed to the decline in the annual salmon catch to about 20% of what it was at its peak. Because about 88% of the diverted water is used for irrigation, there is much competition between agriculture and the salmon industry (Benson, 1995).

15–3.2 Surface Water

Water from streams and lakes is used for irrigation on both small and large scales. Irrigation began thousands of years ago with simple diversions of stream water onto nearby bottomland. Large diversions supply water to canal and ditch systems that distribute it to many users.

Bottomland is usually irrigated first because water diverted a short distance upstream can reach the land by gravity flow. Higher land requires a long canal to carry water diverted far upstream, a high dam, or a means of lifting the water. Many lifting devices have been used, including water wheels, Archimedean screws, and pumps. Pumps were once powered by people or animals but, in developed countries, they are now driven by engines or electric motors. The ancient methods are still common in developing countries.

Irrigation by diversions is limited by low stream flow during dry seasons. An upstream reservoir can make the water supply more reliable. Such reservoirs are often placed on a small tributary rather than across the main stream. A diversion from the main stream can be used to fill the reservoir without overloading it with all the flood waters and sediment carried by the main stream.

15–3.3 Harvesting Rainwater

Land surfaces are sometimes treated to decrease infiltration and make more runoff water available for irrigation and other uses. The runoff can be stored in a reservoir to supply water for households, livestock, gardens, small fields, and wildlife. Rainwater harvest has been practiced for thousands of years and can be used where average annual rainfall is as low as 2 to 3 in. (50 to 75 mm). It can also be used in more humid areas that need additional water, for example for paddy rice production. Srivastava (2001) concluded that a system of this type in eastern India should have a catchment area three times as large as the irrigated field to provide for a rice crop during the monsoon season and another crop during the dry season.

Ditches were used in ancient times to harvest rainwater from hillsides or gentle slopes where the soil permeability was slow. Newer practices include decreasing soil permeability by treatment with sodium salts or using water-repellent compounds such as asphalt, paraffin, or silicone to resist infiltration. Some projects produce runoff by using large sheets of plastic covered by a layer of gravel.

Harvested rainwater may be guided directly to a field or garden and distributed through some form of irrigation system. A simple water-spreading system can be very helpful in an arid climate if the soil is deep enough to store additional water. Systems that deliver water to a small area where it is stored in the soil are discussed in Chapter 13, Sections 13–5.3 and 13–5.4.

Storage reservoirs are needed when collected water is to be saved for later use. Evaporation can be a problem because the air is usually very dry in such areas. A deep reservoir with minimum surface area helps reduce evaporation. Still less evaporation occurs from an oversize reservoir filled with stones, gravel, and sand as shown in Figure 15–3. The sand also filters the water and thus improves its quality. A pumped well is used to remove water from such reservoirs.

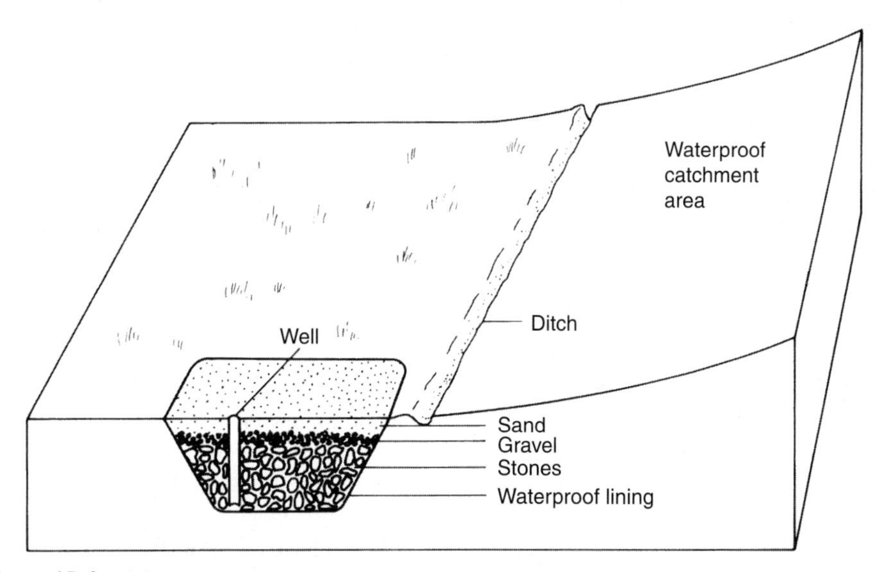

Figure 15–3 A landscape cross section showing a catchment area and storage basin for harvesting rainwater.

Small-scale rainwater harvesting systems with underground storage tanks have been used for centuries in northwestern China for domestic water. A large-scale comprehensive system to be used for irrigation was instituted in China's Gansu province in 1995, using larger tanks made of concrete and steel (Cook et al., 2000). Terraces and plastic soil covers were used on the catchment surfaces to increase the water supply. Drip irrigation and drought-tolerant wheat and corn varieties were used to maximize the crop production gained by irrigating during the critical months of May and June when water supplies are usually deficient in the area. Irrigation ceases after that because most of their rainfall comes during the months of July, August, and September.

15–3.4 Underground Water

Far more fresh water is stored underground than aboveground. Thomas and Peterson (1967) estimate that there is enough underground water to cover the land surface to a depth of about 100 ft (30 m). Much of this water is held in tiny pores or occurs at great depth. Also, much of it occurs in humid regions where it may not be needed. Still, there is much underground water that could be used for irrigation, as shown in Figure 15–4.

Good supplies of groundwater depend on porous rock layers that hold water loosely and have surface connections so they can be recharged. Most gravel and sand layers in valleys are good sources; sandstone, limestone, and some basalt flows are potential sources. Most other rocks are too dense to supply water at a significant rate.

Groundwater has several advantages over surface water:

1. The underground water reservoir minimizes supply fluctuations from wet to dry periods.
2. The water quality is relatively constant and favorable in most places.
3. A well may make long conveyor systems and the attendant rights-of-way unnecessary.
4. Groundwater may be used to irrigate land that has no riparian rights to surface water (but the prior appropriations doctrine often applies to both groundwater and surface water).

The Ogallala aquifer underlying the Great Plains states from South Dakota to Texas is an outstanding example of how underground water can be used and abused. It occupies gravel beds washed out of the Rocky Mountains during Pleistocene times. The aquifer underlies some 174,000 mi^2 (450,000 km^2) of arid to semiarid land and is typically 150- to 300-ft (45- to 90-m) thick and 50- to 300-ft (15- to 90-m) below the land surface (Opie, 2000). It was charged with more than 3 billion ac-ft (3.7 trillion m^3) of water during glacial times, but the input has been largely cut off not only by the retreat of the glaciers but also by the valleys cut since that time by the Pecos River and the Rio Grande.

Part of the area underlain by the Ogallala aquifer is the famous Dust Bowl area of the 1930s around Garden City, Kansas, and extending into the Oklahoma and Texas panhandles. By 1960, this area was beginning to flourish because thousands of wells were drilled to use Ogallala water for irrigation. The water table began to drop at rates from 6 in. (15 cm) to 2 ft (60 cm) or more per year as millions of acres were irrigated. The former Dust Bowl and adjoining areas prospered and now constitute about seven million acres of cropland and

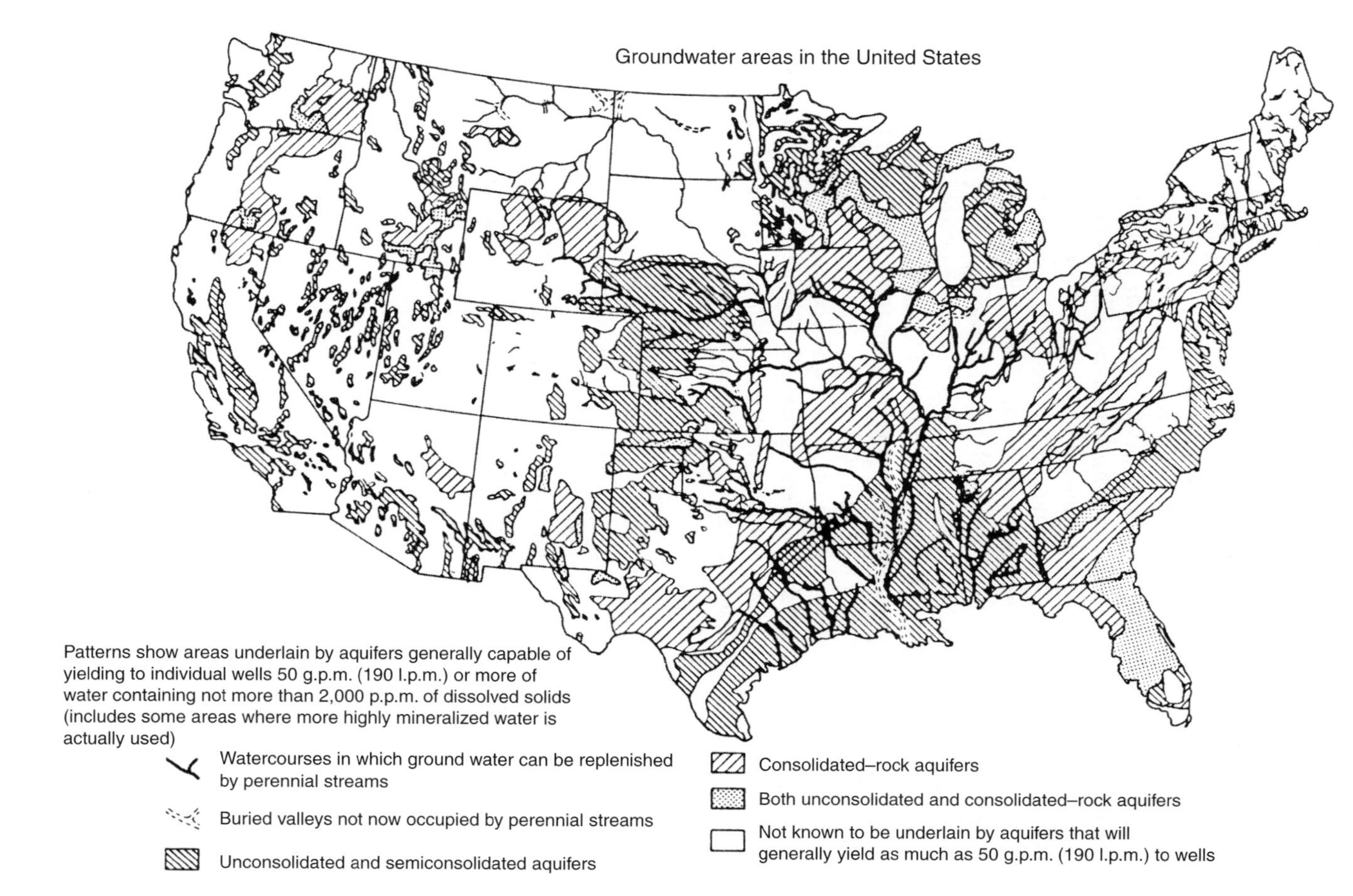

Figure 15–4 Areas in the United States where groundwater aquifers will supply individual wells with 50 gal/min (190 l/min) or more of water containing less than 2000 ppm dissolved solids. (Source: Thomas, 1955.)

large cattle feedlots that provide up to 40% of the beef consumed in the United States, but this may change soon. The farmers used about one-third of their total water reserve in 30 years from 1960 to 1990. The last third may not be extractable, so much of the Ogallala source is likely to go dry by 2020. Efforts are being made to extend its life as long as possible by using more efficient irrigation systems and monitoring water use, but the water levels are still dropping. Many wells have gone dry already. Some of them have been drilled deeper and pumped again, but that increases the expense, and the potential is limited. Already, some farmers are reverting to rainfed agriculture rather than paying the costs of pumping from wells as deep as 1000 ft (300 m). This trend will continue because the use of "fossil water" is equivalent to mining a finite resource. Long-term groundwater use must be based on recharge rates rather than on the amount present. The current recharge rate in this area is probably less than an inch of water per year (Opie, 1993).

Farther north, the Sand Hills Region of Nebraska is also underlain by the Ogallala aquifer. The Sand Hills also have been converted into a fruitful area by irrigation in spite of the droughty character of the soils there. Center-pivot irrigation (described in Section 15–5.3) is well suited to this situation because it permits an automated system to provide light irrigations at frequent enough intervals (about once a week) to produce good crops. Furthermore, liquid fertilizer can be added to the water to take care of the problem of sandy soils that have low fertility. The future outlook for the Sand Hills farmers is better than for those of the former Dust Bowl area because the sandy soils provide some local recharge for the underlying aquifer, and their current reserves represent about 67% of the remaining water in the Ogallala (Opie, 2000).

Because recharge rates limit the long-term use of wells, some projects have arranged to recharge an aquifer artificially. This may be possible if the aquifer connects to a higher area that has surplus water. The technique normally involves spreading the water over an area of porous soils that overlie part of the aquifer.

Sometimes there are two water tables under the same land—a shallow *perched water table* and a deeper *permanent water table,* as shown in Figure 15–5. The perched water

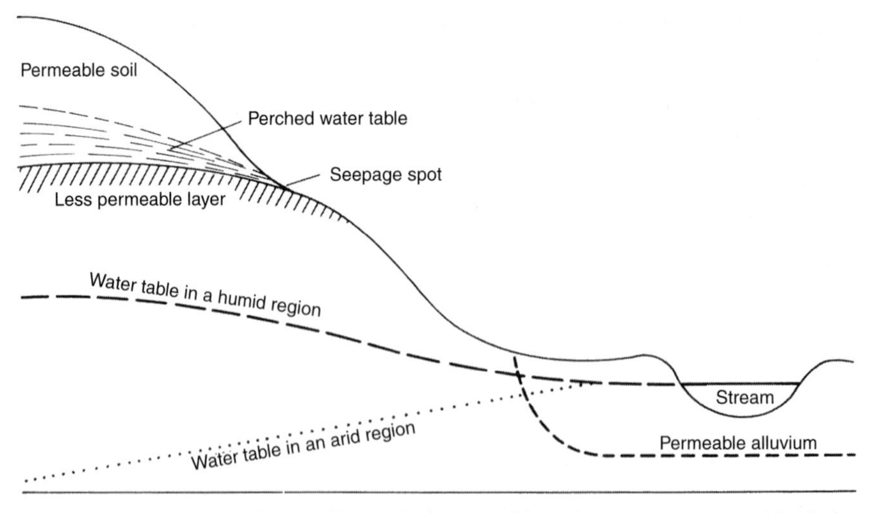

Figure 15–5 Representative forms of a perched water table and permanent water tables in humid and arid regions.

table is not a reliable water supply; it will usually disappear during dry seasons. Another factor shown in Figure 15–5 is the difference between permanent water tables in humid and in arid regions. Water percolates through humid-region soils down to a water table in the substratum. Water accumulates until the water table is high enough for water to flow toward an escape point such as a stream or a spring. In an arid region, evaporation and plant growth use all of the available water, so water rarely reaches the substratum. Water tables in arid regions originate from streams or low-lying wet soils. Consequently, the water table slopes down and away from these water sources.

Underground water is most often obtained from wells, but other possibilities such as gravel pits, rock quarries, and other excavations should not be overlooked. Some springs are large enough and reliable enough to be used for irrigation. The largest spring system in the world, Thousand Springs in southern Idaho (shown in Figure 14–2), supplies about 6000 ft^3/s (170 m^3/s), according to Fiege (1995). That would be enough water to irrigate close to 500,000 ac (200,000 ha) if that much land were conveniently placed, but the Hagerman Valley, where the springs are located, has much less irrigable land than that. However, the aquifer that supplies water to the springs is tapped by many wells on the north side of the Snake River Valley.

15–3.5 Icebergs

Glacial ice contains about three-fourths of the world's supply of fresh water. Some glacial ice melts before it reaches the ocean, but much of it breaks off and floats away as icebergs. Many icebergs are huge—miles across and hundreds of feet thick. They melt and break apart as they float in the ocean.

The idea of using icebergs as sources of fresh water for desert lands was explored in a conference at Ames, Iowa, in 1977. The water is good, but technical and economic problems of navigating the icebergs, containing the meltwater, and distributing the water are still unsolved.

15–3.6 Desalinized Water

Oceans contain 97% of the water on earth, but it is too salty for most uses (its salt content is about 3.5%, mostly NaCl). Ocean water can be desalinized by such techniques as distillation, reverse osmosis, and ion-exchange resins. These methods are useful for modest needs such as supplying water for a ship at sea or even for city water supplies, but all are too expensive for irrigation purposes. Furthermore, transportation and lifting costs could be prohibitive if the water were to be used anywhere except at low elevations near a coast.

15–3.7 Sewage Effluent

Sewage effluent is sometimes used as a source of irrigation water (see Section 16–3.2 regarding its application to land for the sake of disposal). Mexico has been using sewage from Mexico City for irrigation since 1912 (Siebe, 1998) and now has 257,000 ha of such use; Peru, Chile, and other parts of Latin America also have similar use of sewage, partly as a water source and partly as a sewage disposal system (Nava, 2001). India uses sewage effluent as a substitute for fertilizer and as a water source (Brar et al., 2000). Israel began a large-scale project to use sewage effluent for irrigation in 1972 and now uses about two-thirds of its total sewage in that way (Avnimelech, 1993). Saudi Arabia has also turned to sewage effluent to supplement its scarce supply of fresh water (Abo-Ghobar, 1993). The plant nutrients contained in the sewage are a bonus that provides a significant part of the fertility needs of the cotton and other crops being watered this way. Such nutrients are not always supplied in optimal amounts, though, nor at the most appropriate times. Drawbacks include possible health problems, reduced infiltration rates caused by the sediment and sodium content of the water, the hazard of groundwater pollution, and the possibility of accumulating heavy metals in the soil if much sludge is distributed with the sewage effluent (Brar et al., 2000). The hazards are minimized when the effluent is thoroughly filtered and applied through a trickle irrigation system.

15–3.8 Water Quality

Water falling as rain or snow is nearly pure H_2O. The initial runoff into a nearby stream is still relatively pure unless it picks up a sediment load. Water that seeps through soil and rock layers, however, dissolves various materials. Water is classified as *medium hard* if it contains more than 50 parts per million (ppm) calcium and magnesium salts, and *hard* if it contains more than 100 ppm, because such water requires much soap for cleaning purposes. These limits are far below the acceptable amount of salt for irrigation water.

Streams pick up more and more seepage water and dissolved salts as they flow into flatter areas and lower elevations. The salt content increases more rapidly in arid regions than in humid regions because the *dilution factor* is smaller. In addition, the water in arid regions is likely to have been diverted, used for irrigation (or for city or industrial purposes), and returned to the stream several times. Each cycle returns less water and more salt to the stream. The highest concentrations occur in arid regions in the lower parts of long, small rivers. Arid soils contain leachable salts, and long, small streams permit several leaching cycles to occur with little dilution.

Four factors should be considered for predicting the effect of irrigation water on soils and crops: (1) total salt concentration, (2) the proportion of Na^+ to Ca^{2+} and Mg^{2+} ions, (3) toxic ions, and (4) solid matter such as weed seeds and sediment.

High salt concentrations make it difficult for plants to absorb water. The plants therefore need a high moisture content in the soil. Additional water is needed to leach the salts from the soil, or it may become saline. The traditional *leaching requirement* has been based on salt concentrations in the irrigation and drainage waters as:

$$\text{leaching water} = \frac{\text{salt concentration in irrigation water}}{\text{salt concentration in drainage water}} \times \text{amount of irrigation water}$$

The salt concentration in drainage water for this calculation is often taken as that which would cause a 50% decrease in yield on uniformly saline soil. Some researchers have maintained that a smaller amount of leaching water will suffice if salts are allowed to precipitate outside the root zone with trickle irrigation or in the lower part of the root zone with surface irrigation. Sharma et al. (1994) point out that salt accumulation is not a problem in tropical areas where any salts that accumulate during the dry season are washed out of the soil by the monsoon rains during the following wet season.

Established plants are more tolerant of salinity than are seedlings. Some irrigators need to supplement their limited supply of fresh water with available saline water; others need to dispose of saline water in some way other than dumping it in a stream. Several workers have found success by using fresh water to establish their crops and then switching to more saline water (probably drainage water or sewage effluent) later in the season. If additional fresh water is available, the two supplies may be alternated. This procedure has proven to be more productive than the alternative of mixing the two water sources for use throughout the season (Sharma et al., 1994).

Electrical conductivity in mmho/cm has been widely used to evaluate salt concentration because it is easily, quickly, and reliably measured with a conductivity meter. Other measures of concentration can be estimated by these approximations:

$$1 \text{ mmho/cm} = 0.1 \text{ siemen/m} = 1 \text{ dS/m} \approx 640 \text{ ppm} = 0.064\% \approx 10 \text{ meq/liter} = 0.01 \text{ N}$$

Water quality classes for salt concentration as used by the U.S. Department of Agriculture (Richards, 1954) are as follows:

- 0–0.25 mmhos/cm = low salinity hazard
- 0.25–0.75 mmhos/cm = medium salinity hazard
- 0.75–2.25 mmhos/cm = high salinity hazard
- $>$ 2.25 mmhos/cm = very high salinity hazard

The *sodium hazard* of irrigation water is evaluated by the *sodium adsorption ratio* (SAR). The SAR value is calculated as:

$$\text{SAR} = \frac{Na^{+}}{\sqrt{(Ca^{2+} + Mg^{2+})/2}}$$

where the ion concentrations are expressed in milliequivalents per liter. High SAR values are hazardous because such water tends to increase the *exchangeable sodium percentages* (ESP) and produce saline-sodic and sodic soils. The U.S. Department of Agriculture (Richards, 1954) rates the sodium hazard for irrigation water with low salinity (electrical conductivity = 0.1 mmho/cm) as follows:

- SAR 0–10 = low sodium hazard
- SAR 10–18 = medium sodium hazard
- SAR 18–26 = high sodium hazard
- SAR $>$ 26 = very high sodium hazard

The sodium hazard for water with higher electrical conductivities depends greatly on the likelihood of the soil being leached by rainfall or low-salinity irrigation water at a later time. Higher SAR values can be acceptable if leaching will not occur, but lower SAR limits are needed if irrigation with saline water will be followed by leaching with low-salinity water. For example, the above values should be divided by two if water with an electrical conductivity of 2 mmho/cm is used on a soil that will later be thoroughly leached by rainfall. The tendency to produce sodic soils increases if the water contains a high proportion of bicarbonate ions because HCO_3^- reacts with Ca^{2+} and Mg^{2+} and precipitates $CaCO_3$ and $MgCO_3$. Precipitation of Ca^{2+} and Mg^{2+} increases the SAR and ESP values. The likelihood of producing a sodic problem also depends on the salinity of the soil and the clay content of the soil. Salinity helps keep clay flocculated even in the presence of a high percentage of sodium.

Borate ions are the most common toxic ions in irrigation water. Boron concentrations as low as 1 ppm may injure apples, cherries, grapes, and several other fruit and nut crops. Boron-tolerant crops such as alfalfa, asparagus, and date palms tolerate up to 3 or 4 ppm of boron (Ayars et al., 1993). Chlorides and some other ions may reach concentrations that are high enough to be detrimental to plant growth, but they seldom kill the plants. Several ions, however, can be toxic to animals and humans without injuring plant growth. These include ions of arsenic, fluorine, lithium, nitrogen, selenium, and several heavy metals. The more soluble of these ions (e.g., fluorides, nitrates, and selenates) commonly become much more concentrated in drainage water than in the original irrigation water, and sometimes cause problems in the disposal of the drainage water. For example, selenium in a drainage-water reservoir that served as a wildlife refuge caused deaths and deformities of migratory birds in California in the 1980s (Ayars et al., 1993).

Solid matter in irrigation water can cause a variety of problems or, occasionally, be helpful. Sediment can block irrigation furrows and smother young plants, or it can plug soil pores and thus reduce permeability. Plugging some pores might be good for a very porous soil, but it is detrimental to most soils. Sediment may also carry pesticides and other chemicals from the area where they were applied to other crop areas or bodies of water where they are not wanted. Soil conservation is the best solution to sediment problems.

Weed seeds are another detrimental form of solid matter. Weeds line many irrigation reservoirs and ditch systems. Irrigation ditches, therefore, may act as conveyors distributing weed seeds to fields.

15–4 DISTRIBUTING WATER

Canals and ditches are used to distribute water for surface and subsurface irrigation; pipelines are needed for sprinkler and trickle irrigation. Canal systems are operated by irrigation districts, companies, or cooperatives that sell water to individual farmers. Sometimes a government agency such as the U.S. Bureau of Reclamation builds the reservoirs, diversions, canals, and main ditches to deliver the water to users.

Canals flow on a flatter gradient than the rivers from which they were diverted. Thus, they gradually reach a position on the side of the valley that is higher than the river. They may have to cross permeable alluvial fans, go around or through hills, and cross the valleys of tributary streams. Canals need clay or concrete linings to prevent large water losses on the alluvial fans, not only to conserve water but also to prevent them from producing seepage spots below the canal. Upland drainage and reduced grazing are needed to stabilize some hillsides where landslides and hillside sediment may obstruct the canal. Often the canal must cross tributary streams in a flume, or through a large pipeline. Valleys are expensive to cross, even if the streams are dry during the irrigation season.

Water from a canal is often subdivided three or four times on its way to the fields. First, a main ditch or lateral leaves the canal and flows down the divide between two tributary streams. Smaller ditches that serve a few farms branch from the lateral. Even after the water reaches the individual users, it may be divided into still more parts to irrigate individual fields and parts of fields.

15–4.1 Measuring Water Flow

Each division point requires a metering device to control the flow of water. These may be adjustable metal gates that partially cover a submerged opening, or an adjustable opening for the water to spill over. A device to measure how much water is being delivered may be included. *Weirs* are the most common measuring devices; they are relatively inexpensive and a good installation has an accuracy of $\pm 2\%$. Three styles of weirs are shown in Figure 15–6.

Rectangular weirs are simple to build, but the flow constriction below the notch requires extra arithmetic to calculate the water flow. The *Cipoletti weir* has a trapezoidal shape with 1:4 side slopes that offset the constriction tendency and simplify the calculations. *Triangular weirs* have a wide range of capacity with good percentage accuracy at small flows as well as large, but they require a large head loss between the water levels above and below the weir.

Weirs need a pool on their upstream side so the water will not approach them fast enough to increase the flow. The head should be measured in the pool at least four times as far from the notch as the head (depth) of water flowing over the notch. Undersize or sediment-filled pools increase the discharge by as much as 10 or 15%. Weir notches need a sharp crest to cut the water cleanly and a free fall below the notch so the water escape is unrestricted. The pocket of air on the downstream side of the weir should be at atmospheric pressure; the rate of flow is increased if a vacuum forms there (Jain, 2001). The flow over a *sill* (a weir with a flat surface rather than a sharp edge) depends on how much friction the flat surface causes; often such flow is about half as fast as the same depth of flow over a sharp-crested weir.

Parshall flumes (lower part of Figure 15–6) are often used where the land is too flat to allow the head loss required for free flow over a weir. Parshall flumes speed the water flow down a narrow passage and then shoot it back up nearly as high as the inflow. A good installation should measure water flow with an accuracy of $\pm 3\%$, but a faulty one might be $\pm 10\%$ (Robinson and Humphreys, 1967). Excessive flow that submerges the drop and rise characteristics of a Parshall flume may reduce delivery by 25% or more.

The delivery equation for Parshall flumes given in Figure 15–6 is only approximate. Robinson and Humphreys (1967) list values of coefficients and exponents that vary by about $\pm 3\%$ from those given. They also describe a flume with a trapezoidal cross section instead of vertical walls. Several other devices, such as submerged orifices and current meters, can also measure water flow rates, but these are less common than weirs and flumes.

15–4.2 Water-Control Structures

Canals are usually placed on a constant nonerosive grade that requires few structures other than outlets to laterals. An emergency spillway is sometimes needed because cost factors require the canal to be built with little spare capacity. A heavy rain or damage to a ditch could require rapid diversion of a large volume of water.

Laterals and smaller ditches often flow down slopes that are steep enough to require erosion control. Concrete-lined ditches are one solution. Drop structures like the one diagrammed in Figure 10–10 are another way to keep a ditch from eroding. Often they are placed like stairsteps, with the apron of each drop structure level with the notch of the one below it, as in Figure 15–7. Large drop structures are usually made of poured concrete; smaller ones may be poured or made of concrete blocks that have their cores filled with concrete.

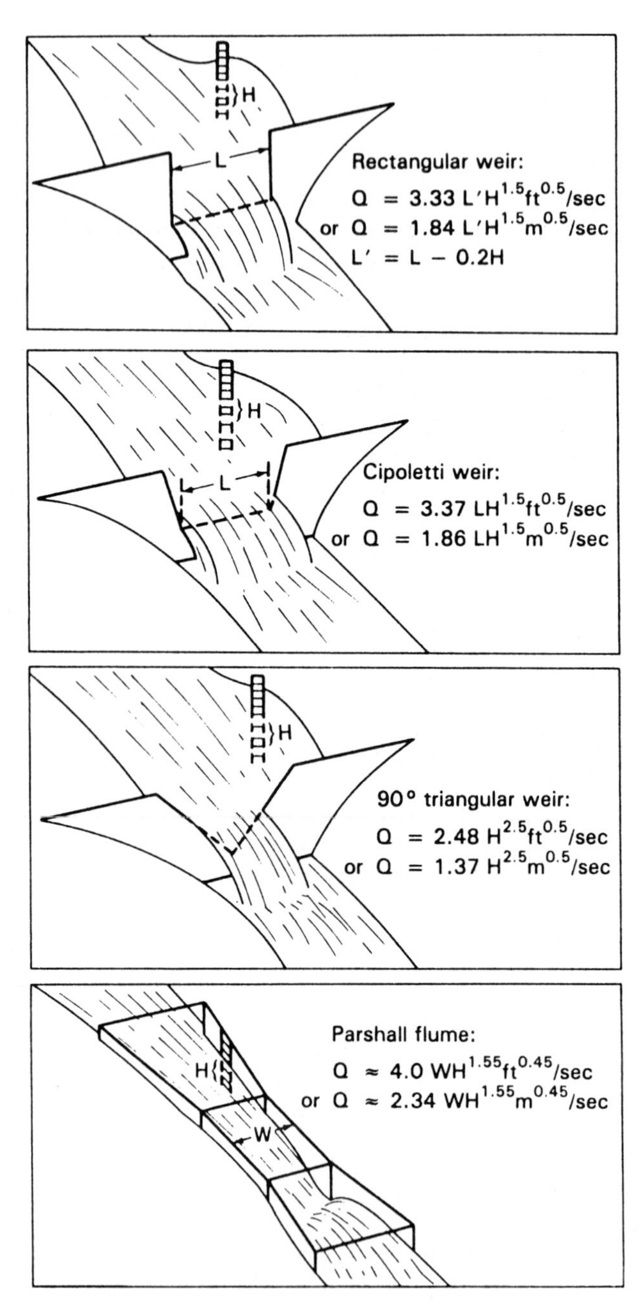

Figure 15–6 Water-measuring devices and equations for calculating their flow. The $ft^{0.5}$ or $m^{0.5}$ in the equations completes the cubic form of the units. The constants must be adjusted if units other than feet or meters are used. These weir equations assume that the head is measured where the water velocity is negligible, that the weir has a sharp crest, and that an air pocket at atmospheric pressure (not a vacuum) is present beneath the water below the weir.

Figure 15–7 A series of drop structures protecting a ditch in Idaho. (Courtesy USDA Natural Resources Conservation Service.)

Check structures are used to raise the flow of water in a ditch to irrigate the field next to it. A drop structure with boards in its notch can serve as a check. A simple check structure consists of the main wall and notch of the drop structure but without the apron and side walls.

A *division box,* such as that shown in Figure 15–8, controls the direction of water flow. Boards dropped in slots close one channel while another is opened. Division boxes take various forms to fit ditch patterns and topography. Some are made of wood, but those made from poured concrete or concrete blocks are more durable. Special concrete blocks are cast with slots to hold the control boards.

15–4.3 Distributing Water From Ditches

When water finally reaches the field to be irrigated, it must somehow be taken out of the ditch and applied to the land. Usually the water has been flowing below ground level and must be raised by blocking the ditch. Boards may be placed in a check for this purpose, or a temporary blockage such as a *sod dam* or a *canvas dam* (a canvas sheet attached to a pole) may be used. The canvas lies in the ditch upstream from the pole and is held there by water pressure. A sod dam is made of pieces of sod taken from the ditchbank with a shovel.

Turnouts (notches through the ditch bank) are the simplest way to turn water onto the field, but other methods offer better control. *Siphon tubes,* as shown in Figure 15–9, are a popular means of irrigating. They come in several sizes and their flow can be adjusted by raising or lowering the outlet end. One type has water traps on each end to hold its prime when the ditch is dry. *Spiles* (tubes permanently installed at ground level through the ditch bank) are another means of removing water from ditches. Spiles usually have adjustable outlets to control their flow.

15–4.4 Irrigation Pipelines

Metal pipelines are required to sustain the pressure of most sprinkler irrigation systems. The main lines of permanent systems are often buried, but portable lines are used for

Figure 15–8 A division box for diverting water in any of three directions. (Courtesy F. R. Troeh.)

Figure 15–9 Siphon tubes conducting water from an irrigation ditch in Idaho. (Courtesy F. R. Troeh.)

temporary installations. Either type uses portable branch lines that attach to outlets on the main lines.

Metal pipe is expensive whether it is made of steel for maximum strength or of a lightweight metal such as aluminum for portability. Systems are therefore designed to use minimum sizes and amounts of metal pipe. Irrigation engineers use flow-rate tables to calculate the required pipe sizes and friction losses for a particular system.

A buried pipeline that carries water from one high point across a low area to another high point is called an *inverted siphon.* Fills or flumes can serve the same purpose, but because they are above ground, they are barriers to traffic, and they occupy land area that could be cropped if buried pipelines were used. Concrete pipe is the least-expensive type that is generally available and that meets strength and durability requirements. Ordinary concrete pipe can withstand pressure heads up to 15 or 20 ft (5 or 6 m) of water. Higher heads require the use of reinforced concrete, steel, or other high-pressure pipe.

Carrying capacities of inverted siphons and other concrete pipelines can be calculated using the average slope from the inlet to the outlet and the data in Table 14–2. An intervening low area has no effect on carrying capacity unless air is trapped in the pipeline. Air entrapment is avoided by having only one low point in the line.

Pipelines need to be installed below frost depth to avoid damage by freezing. Most of them need concrete structures at each end for erosion protection. Some pipelines are built into a division box at one or both ends. Properly installed pipelines should last for many decades and nearly eliminate water losses by seepage and evaporation.

15–5 IRRIGATION METHODS

Irrigation methods can be divided into five main types—surface, subsurface, sprinkler, trickle, and pot irrigation—and many subtypes. *Surface irrigation* is the oldest type and still constitutes about three-fourths of all irrigation. *Subsurface irrigation* is limited in its adaptation. *Sprinkler irrigation* can be used in any climate, is the most popular method in humid regions, and is still expanding in use. *Trickle irrigation,* the newest type, and *pot irrigation* make the most efficient use of water. Sprinkler and trickle irrigation are both generally more efficient than surface and subsurface irrigation, but they also involve more investment in equipment.

15–5.1 Surface Irrigation

Surface irrigation includes both furrow and flood types. These are distinguished by whether the water flows in small channels or covers the entire surface.

Furrow Irrigation. *Furrow irrigation* uses the ridges formed by cultivation to guide water across fields of row crops. The rows can be fed with siphon tubes or spiles, or in groups from ditch turnouts. A serious erosion hazard occurs because the water flows where cultivation has loosened the soil. The maximum nonerosive stream flow can be estimated from the following equation:

$$Q\text{max} = \frac{9.5 \text{ gal/min}}{\% \text{ slope}} \text{ or } \frac{36 \text{ l/min}}{\% \text{ slope}}$$

Qmax from this equation should be decreased if the soil is known to be more erodible than average. This equation should not be used for slopes of less than 0.3% because the furrow capacity will limit the flow.

Irrigating furrows without eroding them is difficult on slope gradients that are steeper than 2%. Large streams of irrigation water would erode the soil, and small streams will flow only a short distance. Contour furrows overcome this problem because they are placed on

TABLE 15–2 SUGGESTED MAXIMUM LENGTHS IN FT (M) OF CULTIVATED FURROWS FOR DIFFERENT SOILS, SLOPES, AND DEPTHS OF WATER IN IN. (CM) TO BE APPLIED

Furrow slope (%)	Clays				Loams				Sands			
	Average depth of water applied (in.)											
	3	6	9	12	2	4	6	8	2	3	4	5
	Maximum furrow length in ft											
0.05	1000	1300	1300	1300	400	880	1300	1300	200	300	500	620
0.1	1100	1440	1540	1640	590	1100	1440	1540	300	400	620	720
0.2	1200	1540	1740	2030	720	1200	1540	1740	400	620	820	980
0.3	1300	1640	2030	2620	920	1300	1640	1970	500	720	920	1300
0.5	1300	1640	1840	2460	920	1200	1540	1740	400	620	820	980
1.0	920	1300	1640	1970	820	980	1200	1540	300	500	720	820
1.5	820	1100	1410	1640	720	920	1100	1300	260	400	620	720
2.0	720	880	1100	1300	590	820	980	1100	200	300	500	620

Furrow slope (%)	Clays				Loams				Sands			
	Average depth of water applied (cm)											
	7.5	15	22.5	30	5	10	15	20	5	7.5	10	12.5
	Maximum furrow length in m											
0.05	300	400	400	400	120	270	400	400	60	90	150	190
0.1	340	440	470	500	180	340	440	470	90	120	190	220
0.2	370	470	530	620	220	370	470	530	120	190	250	300
0.3	400	500	620	800	280	400	500	600	150	220	280	400
0.5	400	500	560	750	280	370	470	530	120	190	250	300
1.0	280	400	500	600	250	300	370	470	90	150	220	250
1.5	250	340	430	500	220	280	340	400	80	120	190	220
2.0	220	270	340	400	180	250	300	340	60	90	150	190

Source: Booher, 1974. Courtesy Food and Agriculture Organization of the United Nations.

a gradient of about 0.5% across the main slope. Extra care must be taken to be sure that water does not cross from one furrow to the next. The lower furrow could overtop and begin a chain reaction that would produce a gully.

Uniform irrigation requires that irrigation furrows be no longer than the distance that water will flow during one-fourth of the irrigation period. For example, irrigation water should reach the lower end of the rows in 2 hours out of an 8-hour irrigation period. The maximum length of irrigation rows, therefore, depends on the soil infiltration rate and erodibility, the slope, and the amount (depth) of water to be applied. Table 15–2 contains estimates of appropriate row lengths for various conditions.

A recently developed method known as *surge irrigation* has enabled some irrigators to improve irrigation uniformity and reduce soil erosion on longer rows (Burt et al., 2000). Water is applied alternately to two sets of furrows for periods of 15 to 30 minutes. Switching the water allows the wet part of the rows to partially seal as the water soaks into the soil. Then, when the water is turned into these rows again, a surge flows quickly across the partially sealed section into the dry part of the rows to continue the irrigation. Some automated systems accomplish surge irrigation by using diverters on gated pipe to switch the water so the flow alternates between two sets of furrows, one to the right and one to the left of each opening in the pipe. Kanber et al. (2001) measured reductions of 21–38% in the tailwater flowing from the ends of the rows and calculated 19–70% reductions in deep percolation losses when they compared surge irrigation to continuous flow on nearly level land with permeable soils in Turkey. Total water application was reduced by 2–22%. Sirjacobs et al. (2001) found that the sealing effect was greater on a silt loam soil than on a clay soil, and that soil loss by rill erosion was only 16% as much on the silt loam soil and 18% as much on the clay soil as it was with continuous flow irrigation on the same soils.

Another approach to improving furrow irrigation efficiency by stabilizing soil structure with soil conditioners is discussed by several authors in volume 158 of *Soil Science.* Sojka and Lentz (1994) indicate that chemical soil conditioners were used to stabilize soil for new airfields and roads in Europe during the closing years of World War II. Since that time, several different polymers have been studied for soil stabilization purposes. One of the first was announced by Monsanto Chemical Company in 1952 and marketed under the trade name of Krilium™. These materials stabilize soil structure without modifying the soil fertility, and they have been useful research tools for projects that needed to distinguish between structural effects and fertility effects. The amounts required to treat a plow layer of soil have generally made the cost of these materials too high for commercial agriculture, but a new way of using them may overcome that problem.

The new soil conditioning approach is to add a small amount of soluble soil stabilizer to the irrigation water during the early part of an irrigation period. The stabilizer helps the soil resist erosion and maintains an infiltration rate two to three or more times as fast as that of the unstabilized soil (Shainberg and Levy, 1994). The most effective materials identified for this purpose are very high molecular weight ($>10^7$ g/mole) anionic polyacrylamides (PAMs). These are formed by polymerization of acrylamide ($[-CH_2-CHCONH_2-]_x$) or related monomers (Barvenik, 1994). These PAMs are commercially available because they are used in several industries to clarify water, process foods, dewater sewage sludge, recover petroleum, and so forth.

PAM applications as low as 0.6 lb/ac (0.7 kg/ha) reduced furrow sediment loss by 80 to 99% and increased net infiltration by an average of 15% on highly erodible silt loam soils in Idaho (Lentz and Sojka, 1994; Sojka et al., 1998). Furthermore, these treatments reduced the total phosphorus, nitrate, and biochemical oxygen demand in the runoff water. The estimated cost is about $3/ac ($7/ha) per application (Trout, Sojka, and Lentz, 1995). A fresh application is needed following each cultivation, and perhaps one or two additional times during the season to maintain good erosion control.

Furrow irrigation, according to the principles outlined in the preceding paragraphs, can achieve about 60% efficiency in water use. The other 40% is lost by evaporation, by deep percolation in the upper ends of the rows and in the most permeable soil, and by runoff (waste water) from the lower end of the rows. Waste-water loss can be lessened by reducing the size of the irrigation streams as they approach the lower end of the rows. The extra water can then be used elsewhere, perhaps on a pasture that needs a short irrigation period.

Principles similar to those outlined for furrow irrigation apply to other forms of row irrigation. Small furrows known as *corrugations* are used for noncultivated grain and forage crops. Such vegetation protects the soil better than row crops, but corrugations are too small to carry large streams of water. The row lengths are therefore similar to those for furrow irrigation.

Flood Irrigation. The three main types of flood irrigation are basin irrigation, border irrigation, and wild flooding. Basin irrigation is a simple method and is probably the oldest method of all. It was practiced in Egypt more than 5000 years ago (Gulhati and Smith, 1967) and is still used for long-term flooding of paddy rice or for shorter periods for many other crops.

Basin irrigation requires a narrow ridge between 6 and 20 in. (15 to 50 cm) or more high on all sides of each area to be flooded. The entire basin should be as level as possible—certainly within a range of 2 to 4 in. (5 to 10 cm). Basins range in size from those designed to irrigate individual trees or small areas of vegetable crops to rice paddies that occupy several acres. Their maximum size may be limited by elevation changes, by the area that the available water supply can cover uniformly, or by cropping factors. A ditch or other water supply must be available for each basin. Water is turned in until the desired depth is reached, then cut back to hold a constant depth for paddy rice or shut off completely for other crops. The water in the basin is generally allowed to infiltrate completely, except that some may be drained onto a lower basin after a specified time on a few low-permeability soils.

The basins used for growing paddy rice may be small or large, but they must be extremely level. Normally the water depth is kept between 2 and 4 in. (5 to 10 cm) throughout the season for the growing rice. Much more than this may drown the rice, and much less may reduce yields. Bouman and Tuong (2001) found that rice yields were reduced by about 6% when they kept the soil saturated but not flooded and declined by 10 to 40% if the soil became noticeably dry.

Paddy rice production works best where it is grown on all the land in a solid block rather than intermixed with other crops that require aerated soil. The difference in water level at the borders not only causes water loss from the rice paddy and excess water in the adjoining fields but also is likely to cause salt accumulation as water rises from the water table in the adjoining fields. If these areas have a controlled drainage/subirrigation system (Section 15–5.2), that system and the cropping pattern should be coordinated with subsections that can be either drained or irrigated, with the entire subsection either flooded for rice or drained for other crops (Broughton, 1995).

A version of basin irrigation called *spate irrigation* that utilizes spring runoff from ephemeral streams is widely used in northern Africa and the Middle East. Tesfai and Stroosnijder (2001) describe its use in Eritrea on the coastal plain next to the Red Sea. Water from the interior highlands is diverted into deep basins and held long enough to fill the soil profile to a depth of 2 m (80 in.) or more before being released into the next basin. About 10 days after the spring runoff stops, the soil is tilled and planted to sorghum, corn, or other crops, and no further irrigation is applied until the next year. The soils in the area hold 40% or more water at field capacity. If half of this is available, it provides at least 80 cm (20 in.) of water for plant growth and is enough to produce their crop.

Border irrigation can be described as elongated basins with a gentle slope in the long direction. Water supplied at the upper end flows down the length of the border as though it were a very wide furrow, as shown in Figure 15–10. Borders range from 10- to 100-ft (3- to 30-m) wide and must be level across their width so the water will spread uniformly across

Figure 15–10 Border irrigation in California. (Courtesy F. R. Troeh.)

them. Their lengths are about the same as those of furrows on similar soils and slope gradients (Table 15–2).

Border irrigation can be used on slope gradients that are between 0.2 and 2% for cultivated crops, up to 4 or 5% for small grain or hay crops, and up to about 8% for pastures. Extensive land leveling is often required because the topography must be smoother than for furrow irrigation. Land-leveling costs are offset by the lower labor requirement for turning water into a few borders rather than into many furrows or corrugations.

The slope range of border irrigation is increased by the use of irrigation terraces. The length of the terrace surface may either slope like a border or be level like a basin.

Wild flooding is used on uneven topography to irrigate pasture or hay, and sometimes small grains. The pastures may have slope gradients as steep as 10 or 15%. Water floods across the land from ditches on the ridges. The irrigator uses a shovel to make small furrows and ridges that guide water to any areas that would otherwise remain dry. Water that accumulates in swales may be redistributed with short spreader ditches.

Wild flooding uses both water and labor inefficiently (Burt et al., 2000), but it irrigates land that cannot be managed by other methods of surface irrigation. The soil may be too shallow or stony to have its surface smoothed by land leveling, and it may not be used intensely enough to justify investment in a sprinkler system.

15–5.2 Subsurface Irrigation

Subsurface irrigation, also called *subirrigation,* is essentially a controlled drainage system (Fisher et al., 1999). Ditches are usually used, but some systems use tile lines. Water is removed during wet seasons and added during dry seasons so that the water table is always near a specified depth. That depth might be as little as 1 ft (30 cm) for shallow-rooted vegetation in a coarse sandy soil, or as great as 4 ft (120 cm) in some loamy soils. The surface soil should be dry but most of the root zone should be moist. The field can even be cultivated and irrigated at the same time.

Relatively little land is subirrigated because the required conditions are very stringent. The land surface must be quite smooth and have a slope gradient of less than 0.5%. The subsoil must be highly permeable, but it must have a shallow water table or be underlain by an impermeable layer that permits a perched water table to be maintained. Both the soil and the irrigation water must be low in salts to avoid forming saline and sodic soils. Suitable conditions for subsurface irrigation most often occur on glacial outwash plains, terraces, or deltas in humid or subhumid areas. The area may increase in the plains section of the Corn Belt in Canada and the central United States, where the land already has a tile drainage system that can also be used for subirrigation by pumping water into the lines. Mejia et al. (2000) used this type of system in eastern Ontario, Canada, to drain the land in the spring and subirrigate it in the summer. They showed yield increases of 3 to 7% for corn and 8 to 32% for soybeans during their two-year field trial.

Subirrigation has also been increasing in areas with monsoonal climates, such as the Punjab in India (Broughton, 1995). Nearly level fields need drainage during the monsoonal rains, but water is a limiting factor during the intervening dry season. Whatever salts accumulate in the soils from subirrigation are leached away during the next monsoon season. In somewhat drier areas, subirrigation should be supplemented with enough surface irrigation to remove excess salts.

15–5.3 Sprinkler Irrigation

Sprinkler irrigation is much newer than surface irrigation because the necessary pipes, pumps, and power supplies were not available long ago. Advantages such as portability, adaptability to a wide range of soil and topographic conditions with little or no land smoothing, and good control of water application have made sprinkler irrigation popular. Efficient water application may save energy, reduce leaching of nitrates and other nutrients, and help avoid erosion. Disadvantages include high equipment and operating costs, moving lines in muddy conditions, salt damage to some plants if poor-quality water is used, and disease problems with some plants.

Sprinkler irrigation is often the method chosen for supplemental irrigation in a humid or subhumid region. A portable pipe system usually costs less than it would to level the land for surface irrigation, and the system can be moved from one field to another or placed in storage when it is not needed. Some systems are designed to spray sewage effluent as a combination of irrigation and an approved method of waste disposal (Scherer et al., 1999). Most field sprinklers use a rotating sprinkler head such as the one shown in Figure 15–11. The sprinklers are usually mounted on either moving or portable lines, although they may be fixed in permanent locations for limited areas of high-value crops.

Portable Lines. Hand-moved sprinkler lines are conventional in many areas. They are used on a regular schedule throughout the growing season in arid climates, but in humid climates they are often kept in storage except during periods of drought.

Irrigation once every seven to ten days with applications of 3 to 4 in. (7 to 10 cm) of water each time is common. The application rate should be slow enough to avoid runoff. Sprinklers are available to apply water at rates between 0.1 and 0.2 in./hr (3 to 5 mm/hr), but faster rates are usually more efficient. Rates that result in irrigation sets of 8 or 12 hr are convenient for work schedules.

Figure 15–11 A sprinkler at the top of a riser on a portable line in a potato field. Water strikes the protruding arm and makes it work back and forth against a spring. This action rotates the sprinkler head and helps distribute the water. (Courtesy F. R. Troeh.)

The number of irrigation lines needed depends on the size of the field, the irrigation period and frequency, and the area irrigated by each line. A field 1/4-mi (400-m) square (40 ac or 16 ha) might be irrigated from a main line across the middle of the field, as shown in Figure 15–12. Each lateral line could have 21 sprinkler pipes, each 30-ft (9-m) long, plus a 15-ft (4.5-m) coupling to the main line. Outlets every 30 ft along the main line would irrigate an area one-eighth mi by 30 ft (0.45 ac or 0.18 ha) at any one time. Complete irrigation of the 40 ac requires 88 sets (2 sides × one-quarter mi ÷ 30 ft). An irrigation period of 8 hr (three per day) and a frequency of once every 10 days would result in 30 irrigation periods. Such a system requires three lines (88 sets ÷ 30). Four lines may be used so the soil can dry and one line can be moved while the other three lines are operating.

The system described above and illustrated in Figure 15–12 is designated as a 30-ft by 30-ft system because it uses 30-ft pipe lengths and 30 ft between lines. Other systems may use pipe lengths from 20 to 40 ft and line spacings from 30 to 60 ft.

Irrigation pipes have quick-coupling devices on each end that permit them to be uncoupled, carried to the next position, and reconnected. A sprinkler is attached to the reinforced area near the coupling. Most sprinklers spray past the next sprinkler position for complete coverage. Sprinkler irrigation is usually about 75% efficient in use of water—the other 25% is lost by evaporation and by deep percolation where the water application is heavier than average. Excess wind distorts the pattern and reduces efficiency.

Rolling Lines. One way to make sprinkler lines easier to move is to mount them on wheels. Some lines run through the hubs of large wheels. Other systems use small wheels on each side of the line. Some rolling lines have long, flexible supply lines so they

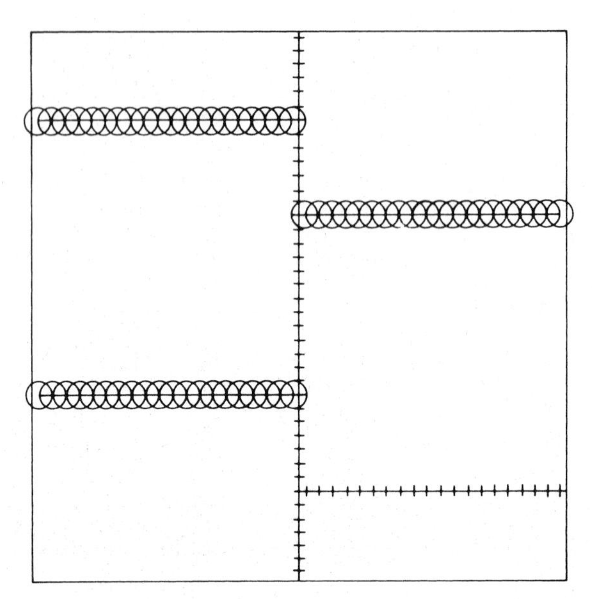

Figure 15–12 A hand-moved sprinkler irrigation system for a field 1/4-mi (400-m) square using four lines with three operating at any one time. The lines are spaced around the field so they do not have to be moved past each other.

can be motor-driven to roll during the irrigation period. Others are detached, moved, and reconnected much like a hand-moved line.

Center-Pivot Systems. Center-pivot irrigation systems are the most convenient and the most expensive movable systems. Either a well or a buried main line supplies water to the pivot point. The sprinkler line is supported at about 100-ft (30-m) intervals by two-wheeled, motor-driven towers that carry the line at a height of about 7 to 10 ft (2 or 3 m), as shown in Figure 15–13. Either hydraulic or electric power moves the towers in concentric circles at a rate proportional to their distance from the pivot point. The number and size of sprinklers vary along the line to equalize the water applied.

The original center-pivot irrigation systems sprayed water upward from the top of the line, some 7 to 10 ft (2 or 3 m) above the ground. This design works well except that it allows a significant amount of water to evaporate before reaching the ground. Some of the newer lines have drop sprinklers that extend downward from the line and spray downward, thus achieving water efficiencies up to 80% (Opie, 1993).

The most common size of center-pivot system is a line 1/4-mi (400-m) long that irrigates a 125-ac (50-ha) circle in a 160-ac (64-ha) square. Water spraying beyond the end of the line will cover a few more acres. The corners may or may not be irrigated. Some lines have an extension that irrigates the corners but trails behind when it is not needed. Another method uses a very large sprinkler known as a "big gun" at the end of the line to irrigate the corners.

The use of center-pivot systems has increased rapidly in recent years, especially in the Great Plains and the southwestern parts of the United States. Their high installation cost is offset by the convenience and labor savings of their automated operation.

Figure 15–13 A center-pivot irrigation line in Texas. (Courtesy F. R. Troeh.)

15–5.4 Microirrigation

A recent adaptation of sprinkler irrigation using small, fixed sprinklers is called *microirrigation.* The equipment used is sometimes described as *microsprinkler* if it uses sprinkler heads with moving parts or *microspray* if it sprinkles by means of a jet of water striking a fixed plate (Burt et al., 2000). These systems are either permanent installations, for example in an orchard, or left in place for an entire irrigation season for annual crops. They became popular in the 1980s, mostly as conversions of trickle irrigation systems because they offered advantages such as less stringent filtration than that required for trickle irrigation and some degree of frost protection by spraying fruit crops at critical times (Burt et al., 2000).

15–5.5 Trickle Irrigation

Trickle irrigation, also called *drip irrigation,* is a relatively new method. It supplies water to individual plants through small plastic lines. It and pot irrigation are the only methods efficient enough to deliver 90% of the irrigation water to the plant root zone. Water is supplied either continuously or so frequently that the root zone is constantly moist.

Trickle irrigation is especially suitable for watering trees or other large plants. Much of its use has been in orchards and vineyards, but it has also been used to irrigate many row crops, including various vegetables and fruits. Its advantages are greatest where areas between plants can be left dry. It has no advantage for close-growing vegetation such as lawns, pastures, or small-grain crops.

An Israeli engineer named Symcha Blass developed the idea of trickle irrigation in the 1930s (Shoji, 1977), but practical systems had to wait until plastic tubing was available.

Trickle irrigation in the United States increased from 100 ac (40 ha) in 1960 to about 1,500,000 ac (600,000 ha) in 1991, out of a worldwide total of about 1,770,000 ha. Nearly half of the trickle irrigation in the United States is in California, some of it in avocado orchards with slopes up to 50 or 60%. Erosion is not a problem because there is no runoff.

A bonus with trickle irrigation is its ability to use water with a higher salt content than surface or sprinkler methods—up to about 2500 mg/liter. The constant flow of water from the trickle emitter toward the outer edges of the plant root zone carries the salt along with it. Salt concentrations become very high in the dry areas between plants but not in the active root zone.

Trickle irrigation saves water, functions well in all but extremely coarse- or fine-textured soils, works on almost any topography without causing erosion, and requires little labor after it has been installed. The main problems are high equipment costs and plugging of the lines by sediment, salt encrustation, or algae.

Trickle irrigation normally includes a control box that filters the water, regulates its pressure, and adds fertilizers and herbicides. Chlorine may be added to eliminate algal growth. Trickle irrigation normally uses 5 to 15 lb/in.^{2} (0.4 to 1 kg/cm^2) of water pressure (Shoji, 1977) as compared to 15 to 120 lb/in.^{2} (1 to 8 kg/cm^2) for sprinkler irrigation. Some trickle controls increase the pressure periodically to flush the lines and reduce clogging.

Trickle irrigation lines have multiple branches at three or four stages to provide the many outlets required. The last stage is often a flexible plastic lateral line 0.5 to 1.25 in. (12 to 32 mm) in diameter that lies either on or below the soil surface. It applies water either through small holes in the line or through emitter nozzles. Emitter nozzles provide more constant flow and reduce plugging by leading the water through a long spiral path that slows the flow and permits a larger emission hole to be used.

Subsurface location of emitter lines and nozzles involves extra installation cost but avoids having to redo the installation each year. It also reduces evaporation losses and leads to more efficient use of fertilizer applied through the irrigation system. Martinez Hernandez et al. (1991) reported an increase in sweet corn yields from 4.3 kg/m^2 with surface emitters to 4.9 kg/m^2 with emitters placed at a depth of 30 cm. Ayars et al. (2001) found that subsurface trickle irrigation once or twice per day increased both cotton and tomato yields in the San Joaquin Valley of California as compared to yields with furrow irrigation. Use of trickle irrigation also contributed to better management of the depth and salinity of the chronically high water table in the area.

15–5.6 Pot Irrigation

Pot irrigation, also called *pitcher irrigation,* is little used in America but has long been used in parts of Africa and Asia (Bainbridge, 2001). It consists of burying an unglazed clay pot so it barely protrudes from the soil and planting seeds or small plants next to it, as shown in Figure 15–14. Water seeps out of the pot and is absorbed by the plant roots. The pot is filled periodically with water and covered with a tight-fitting lid. The rate of water seepage depends on the porosity of the pot and the soil moisture tension, and it maintains a nearly constant soil moisture content for the growing plants. The interval between fillings is normally a few days, depending on the size of the pots, the soil, the plants being grown, and the climate. The result is a restricted and very efficient use of water that produces less than

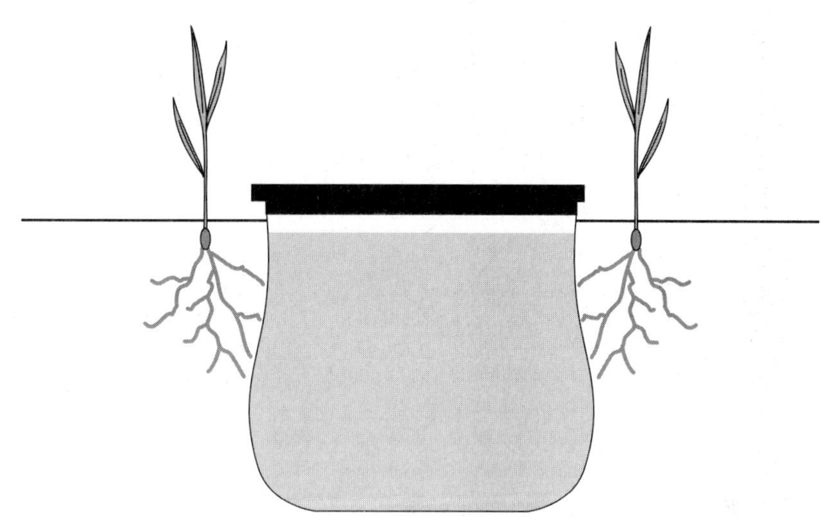

Figure 15–14 A buried clay pot can supply water to plants growing next to it.

the maximum yield attainable with other irrigation methods but uses much less water. Investment costs are low and the technology is simple, but hand labor is required to place the pots and to fill them with water. The system is adaptable to a variety of circumstances and is reported to make it possible to grow plants in saline soils or with saline water that causes failure with other types of irrigation.

15–6 IRRIGATION FREQUENCY

Trickle irrigation may be either continuous or quite frequent and pot irrigation is continuous, but the other methods normally have several days between irrigations. Longer intervals result in more extensive root systems, but the surface should not be allowed to dry out enough to curtail transpiration. The usual solution is to irrigate when about half of the available soil water has been used by the growing crop. Enough water is then applied to refill the soil profile (putting on much less would leave a dry zone that could become a root barrier). Each water application should therefore apply about half of the soil's available water-holding capacity. The proper interval between irrigations can be calculated by dividing the amount of water to be applied by the daily consumptive use (evapotranspiration) of water. The result may be as short as a few days or as long as two or three weeks.

The calculations outlined in the preceding paragraph form a good basis for planning an irrigation system, but the optimum timing and amount of irrigation also vary from week to week and season to season according to other factors such as precipitation temperature, humidity, and rate of crop growth. A more precise way to irrigate is to monitor the water content of the soil to determine when to start and stop applying water. The most popular means for doing this is to bury gypsum blocks in the soil (Peavy, 1992). Wires embedded in the blocks extend to the soil surface so that a meter can be attached to read their electrical resistance. The resistance drops when the block is wet and gradually increases as the soil and block dry out. A few sets of blocks buried at strategic depths, commonly at 1, 2, 3, and 4 ft (30, 60, 90, and 120 cm), allow an irrigator to keep track of moisture conditions in the soil. As an alternative, Zur et al. (1994) describe a depth probe with sensors at 2-in. (5-cm) depth intervals designed to monitor the depth of the wetting front in the soil.

The water potential in the leaves of growing plants can also be used as an indicator of the appropriate time to irrigate. Ayars et al. (2001) found that leaf water potentials of -1.0 MPa in tomatoes and -1.4 MPa in cotton were appropriate indicators of the time to initiate trickle irrigation. They showed that allowing this much water stress where there was a shallow water table caused tomatoes to obtain about 10% of their water requirement from the water table, and cotton obtained 40 to 60% of its water requirement from the water table in spite of the fact that the ground water was saline. Comparable management with furrow irrigation caused cotton to obtain 40% of its water requirement from the water table but caused a net addition to the ground water in the tomato field.

15–7 LAND RECLAMATION

Reclamation, in its broad sense, means modifying land to make it suitable for cropping. Vegetating mine spoils and construction sites (Chapter 11), drainage of wetlands (Chapter 14), and irrigation of arid lands are all reclamation processes. One more type, the reclamation of saline and sodic soils, will be considered in this section.

The formation of saline and sodic soils by water moving upward from a water table was discussed in Sections 14–3.5 and 14–3.6. Pessarakli (1991) indicates that the world has 400,000,000 ha (1 billion ac) of potentially arable land that is affected by excess soluble salts. Saline soils form where upward water movement is dominant; sodic soils form where upward and downward water movements are nearly balanced. Saline and sodic soils can be defined as follows:

> *Saline soils* have electrical conductivities of saturation extracts higher than 4 mmhos/cm (or higher than 2 mmhos/cm if certain salt-sensitive crops are grown). This is measured by saturating a soil sample with water, extracting the water by vacuum, and measuring the conductivity. Saline soils contain less than 15% exchangeable Na^+.
>
> *Sodic soils* have more than 15% of their cation-exchange capacity occupied by exchangeable Na^+. They are low in total salt content and electrical conductivity.
>
> *Saline-sodic soils* have electrical conductivities higher than 4 mmhos/cm and have more than 15% exchangeable Na^+.

The one difference of immense practical importance between saline-sodic and saline soils is that leaching changes saline-sodic soils into sodic soils, whereas leaching reclaims saline soils.

Both saline and saline-sodic conditions are called *white alkali* because the soluble salts form a white deposit on dry soil. The soluble salts make it more difficult for plant roots to absorb water, thus decreasing plant growth and making seriously affected areas barren. Plants vary widely in salt tolerance (see Table 15–3). Electrical conductivity of 4 mmhos/cm represents an arbitrary division point that sometimes needs to be lowered or raised according to the kind of crop being grown. The stage of growth is also important—

TABLE 15–3 SALT TOLERANCE OF CROPS DURING THEIR RAPID GROWTH PERIOD IN TERMS OF ELECTRICAL CONDUCTIVITIES OF SATURATED SOIL EXTRACTS

	Conductivity in mmhos/cm at 25°C causing yield reduction of		
	10%	25%	50%
Forage crops:			
Bermuda grass	13	16	18
Tall wheatgrass	11	15	18
Crested wheatgrass	6	11	18
Tall fescue	7	10	15
Perennial ryegrass	8	10	13
Bird's-foot trefoil	6	8	10
Beardless wildrye	4	7	11
Alfalfa	3	5	8
Orchard grass	2	4	8
Alsike clover, red clover	2	3	4
Field crops:			
Barley	12	16	18
Sugar beets, cotton	10	12	16
Wheat, safflower	7	10	14
Sorghum	6	9	12
Soybeans	6	7	9
Corn, paddy rice	5	6	7
Flax	3	4	6
Field beans	1	2	3
Vegetable crops:			
Beets	8	10	12
Spinach	6	7	8
Tomatoes, broccoli	4	6	8
Cabbage	2	4	7
Potatoes, corn, sweet potatoes	2	4	6
Lettuce, bell pepper	2	3	5
Onions	2	3	4
Carrots	1	3	4
Beans	1	2	3

Source: Bernstein, 1964.

plants are more sensitive to salinity during the germination and seedling stages than they are later.

Very few soils are as unproductive and as difficult to reclaim as sodic soils. The high Na^+ percentage causes the soil colloids to disperse and the pH to rise above 8.5 (often to about 10) when the Na^+ is not masked by a high salt concentration. The permeability of very sandy soils drops to less than 0.1 in./hr (1 or 2 mm/hr) and that of silty or clayey soils becomes negligible. A smooth, crusted, barren surface forms that is so slippery when it is wet that sodic soils are commonly called "slick spots." The organic-matter content of sodic soils is usually less than 1%, and what organic matter there is becomes soluble at the high pH. Soil water carries some of it to the soil surface where it forms a thin, black coating that is the basis for the common name "black alkali." Dissolved organic matter gives a brown, oily appearance to drainage water seeping from sodic soils.

Mace and Amrhein (2001) suggest that much of the permeability loss in sodic soils is associated with swelling of the colloids that hold exchangeable Na^+, and that such swelling reduces soil permeability even if the exchangeable Na^+ value is considerably less than 15%. They indicated that permeability losses caused by swelling can be reversed by replacing Na^+ with Ca^{2+}, whereas permeability losses caused by dispersed clay particles blocking soil pores are largely irreversible. Sumner et al. (1998) also indicate that exchangeable Na^+ levels of less than 15% often cause clay dispersion and reduced permeability, especially in soils that contain smectite clays. They suggest that any soil with physical properties adversely affected by the presence of exchangeable Na^+ should be considered sodic.

15–7.1 Reclaiming Saline Soils

Suitable irrigation water, appropriate means of application, and good drainage are required for reclaiming saline soils. Large quantities of water are needed, and it must have a low enough salt content to leach salts from the soil until the salinity is reduced enough for the desired purpose. If leaching water is not available in the required amount and quality, the area may still be used for crops that are highly salt-tolerant. Pasternak and De Malach (1987) report reasonable crop yields for selected varieties of broccoli, carrots, Chinese cabbage, lettuce, melons, and tomatoes using trickle irrigation with water that had electrical conductivities up to 8–10 mmhos/cm. Broccoli and carrots still did fairly well, even with water that had an electrical conductivity of 10–15 mmhos/cm. Good results were also obtained with onions and corn by starting them with fresh water and changing to saline water after they were well established. Other crops may be selected from Table 15–3.

Flood irrigation is usually best for leaching salts from soils because the entire soil surface must be covered, often for two, three, or more days. Methods such as furrow and trickle irrigation reclaim part of the soil at the expense of the rest by moving salt to the drier parts.

Salt-tolerant vegetation helps reclaim saline soils, especially those that are fine textured. Plant roots help keep the soil permeable, and the top growth helps control erosion. The plant growth must also tolerate enough leaching water to remove the excess salts. This is often a water depth between 0.5 and 1.5 times the depth of soil to be reclaimed.

A good drainage system is required to remove the leaching water and keep the water table from rising while a saline soil is being reclaimed. Intermittent leaching is sometimes used to give more time for drainage and to reduce the leaching-water requirement. Allowing one- or two-day "dry" intervals permits about 70% as much leaching water to dissolve the same amount of salts.

Reclamation is futile unless the land is well managed afterwards. The water table should be kept low enough to prevent the soil from becoming saline again, and enough irrigation water should be applied for drainage to remove the salts contained in the water. Section 15–3.8 explains how to calculate the amount of drainage water needed.

15–7.2 Reclaiming Saline-Sodic Soils

Saline-sodic soils need a soil amendment such as gypsum or sulfur to replace Na^+ prior to the leaching process described for saline soils. The amount of amendment can be calculated as in Note 15–2 when the required chemical data are available.

NOTE 15–2
SOIL-AMENDMENT CALCULATIONS

Problem

A saline-sodic soil contains 9 meq exchangeable Na^+/100g plus 0.5% soluble Na_2CO_3 by weight. Calculate the amounts of gypsum, sulfur, or calcium chloride needed to amend the soil to a depth of 20 in. (50 cm).

Solution

1. The weight of soil can be calculated by assuming an average bulk density of 1.3 g/cm^3 (81.12 lb/ft^3)

$$1 \text{ ac} \times 20 \text{ in.} = 43{,}560 \text{ ft}^2 \times 20/12 \text{ ft} \times 81.12 \text{ lb/ft}^3$$
$$= 5{,}889{,}312 \text{ lb} = 2945 \text{ tons}$$

or

$$1 \text{ ha} = 10^4 \text{ m}^2 = 10^8 \text{ cm}^2$$
$$10^8 \text{ cm}^2 \times 50 \text{ cm} \times 1.3 \text{ g/cm}^3 = 6.5 \times 10^9 \text{ g} = 6500 \text{ mt}$$

2. Each 100 g of soil contains
   (a) 9 meq × 23 mg/meq = 207 mg = 0.207 g of exch. Na^+
   (b) 0.005 × 100 g × (46/106) = <u>0.217</u> g of soluble Na^+
   (0.5%) (wt of 2 Na/Na_2CO_3) 0.424 g of Na/100 g of soil

3. Each acre of soil 20-in. deep (hectare 50-cm deep) contains

$$2945 \text{ tons} \times 0.424 \text{ g Na/100 g of soil} = 12.5 \text{ tons of Na/ac}$$

or

$$6.5 \times 10^9 \text{ g} \times 0.424 \text{ g Na/100 g} = 2.76 \times 10^7 \text{ g of Na/ha}$$
$$= 27.6 \text{ mt of Na/ha}$$

4. The required amounts of each amendment can be calculated from the equivalent weights of each material divided by that of Na:

(a) $CaSO_4 \cdot 2H_2O$	12.5 tons × 86/23	= 46.7 tons/ac
(172/2 = 86)	27.6 mt × 86/23	= 103 mt/ha
(b) S	12.5 tons × 16/23	= 8.7 tons/ac
(32/2 = 16)	27.6 mt × 16/23	= 19 mt/ha
(c) $CaCl_2$	12.5 tons × 55.5/23	= 30.2 tons/ac
(111/2 = 55.5)	27.6 mt × 55.5/23	= 67 mt/ha

Step 2 shows that about half of the soil amendment that this soil needs is required to replace exchangeable Na^+ and the other half reacts with the soluble Na_2CO_3. Actually, some Na^+ would remain because the process is not 100% efficient, but any of these amendment applications should reduce the sodium content of this soil to a safe level to a depth of about 20 in. (50 cm); larger or smaller applications will reclaim a soil depth proportional to the amount applied (Costa and Godz, 1998). Other materials that have been used to replace sodium include fly ash, pyrite (Tiwari et al., 1992), and cottage cheese whey (Robbins and Lehrsch, 1992). Appropriate amounts of such materials must be based on chemical analyses to determine their capacity to replace sodium.

The gypsum requirement may also be measured directly by mixing a soil sample with a saturated solution of $CaSO_4$ and determining how much Ca^{2+} is adsorbed by the soil or how much Na^+ enters the solution. The gypsum in this test reacts with both the exchangeable Na^+ and the Na_2CO_3 because Ca^{2+} is adsorbed more strongly than Na^+ and insoluble $CaCO_3$ precipitates.

Gypsum is usually the cheapest amendment for saline-sodic soils, but a large amount is required and the process is slow. Gypsum may be added to the irrigation water, but an acre-foot of irrigation water will dissolve only about 2 tons (1,000,000 l or 10 ha-cm of water will dissolve 1.4 mt) of gypsum. Mixing the gypsum into the soil helps, but reclamation still takes months or years.

The low equivalent weight of sulfur minimizes the amount required for sodium replacement. Sulfur, however, has its own limitations. Soil bacteria need months or years to oxidize the sulfur to sulfuric acid. The H^+ of the acid will react with soil lime to release Ca^{2+}, which in turn exchanges for soil-adsorbed Na^+. Only then is the Na^+ ready for leaching. Lime is common in saline-sodic soils but its presence should be verified rather than assumed.

Calcium chloride is a highly soluble, fast-acting soil amendment that can be spread on a saline-sodic soil or added to the irrigation water. Reclamation then proceeds in the same manner and at the same rate as if it were a saline soil. The main problem is high cost.

The amount of soil amendment needed to reclaim a saline-sodic soil may be greatly reduced if salty irrigation water is available. The unamended soil would become sodic if leached with pure water, but salty water can be used as a preliminary treatment. Water containing mostly calcium salts is best, but even sea water has been used. After partial reclamation, the reduced amendment requirement can be applied and nonsaline water used to finish reclaiming the soil.

The presence of growing plants facilitates the reclamation process (Ilyas et al., 1997) by increasing water flow and improving the soil chemistry. For example, alfalfa can provide root channels for water movement, lower an alkaline pH by releasing CO_2 into the soil, and reduce both Na^+ and Cl^- concentrations in the soil solution. Qadir et al. (2001) reduced soil salinity from 10 mmhos/cm to 7 mmhos/cm and reduced the exchangeable sodium percentages to about half of their beginning levels in a saline-sodic soil in Pakistan during three years of double-cropping rice and wheat, even though they irrigated with poor quality saline-sodic water. They called the process *phytoremediation* and suggested that carbon dioxide released by the growing plant roots reacted with calcium carbonate in the soil, thus providing a source of calcium to replace exchangeable sodium.

15–7.3 Reclaiming Sodic Soils

Sodic soils are difficult to reclaim because their slow permeability prevents water from carrying amendments into the soil. Mechanical mixing can help get the amendment into the soil and open up some passages for water percolation. Unfortunately, deep plowing may not provide much mixing, and any method that mixes the soil is usually quite expensive. However, sodic soils often occur as small spots mixed with more productive soils. Some such spots have been improved by mixing the soil with a backhoe. Many sodic soils have layers containing gypsum and lime concentrations underlying the solum. Mixing incorporates these free soil amendments into the soil.

TABLE 15–4 EXCHANGEABLE SODIUM PERCENTAGE (ESP) VALUES IN THE TOP 15 CM (6 IN.) OF SOIL IN WHICH SELECTED CROPS BEGAN TO SHOW YIELD REDUCTIONS (THRESHOLD ESP) AND THAT CAUSED 50% YIELD REDUCTIONS (50% ESP) ON SANDY LOAM SOILS IN INDIA

Crop	Scientific name	Threshold ESP*	50% ESP*
Sesbania	*Sesbania aculeata*	46.9	67.7
Rice	*Oryza sativa*	24.4	80.0
Wheat	*Triticum aestivum*	16.4	40.2
Pearlmillet	*Pennisetum typhoideum*	13.6	32.8
Cowpea	*Vigna unguiculata*	13.5	19.0
Linseed	*Linum usitatissinum L.*	13.3	25.0
Guar (Cluster bean)	*Cyamopsis psoralodes D.C.*	11.9	27.5
Sunflower	*Helianthus annuus L.*	11.3	56.8
Rye	*Secale cereale*	11.0	30.2
Onion	*Allium cepa*	9.8	32.5
Garlic	*Allium sativum*	9.5	37.3
Sesamum	*Sesamum indicum*	9.0	22.8
Barley	*Hordeum vulgare* (P)	8.5	22.3
Groundnut (Peanut)	*Arachis hypogaea*	8.0	29.7
Soybean	*Glycine max*	8.0	22.3
Pea	*Pisum sativum*	7.7	19.9
Gram	*Cicer arientinum*	7.7	17.7
Raya (Indian mustard)	*Brassica juncea L.*	7.6	70.1
Safflower	*Carthamus tinctorius*	7.6	17.2
Lentil	*Lens esculentum*	4.9	14.0

*ESP values indicate average soil ESP of 0–15 cm layer

Source: Gupta and Sharma, 1990.

Sodic soils are low in soluble salts and therefore require less soil amendment than saline-sodic soils. The soil in Note 15–2 would have required only half as much amendment if it had been sodic instead of saline-sodic. The reclamation process is very slow, however, and few plants will grow on sodic soil until some reclamation has been achieved. Even weeds should be encouraged because their roots open channels that improve soil permeability. Growing tolerant crops (Table 15–4) at an early stage helps reclaim the soil. Permitting the soil to dry and crack open occasionally is also helpful (sodic soils shrink and swell more than other soils of similar clay content).

Salty water is much better than pure water for the initial stages of sodic soil reclamation. Salts help flocculate the soil colloids and increase the soil permeability, often by one or two orders of magnitude. Calcium ions in the water are especially helpful because they replace exchangeable Na^+.

15–8 CONSERVATION IRRIGATION

A careless irrigator can waste large amounts of soil and water. Attitude is important because the easiest way to irrigate is often not the best way, and the best methods often require extra time and effort. One interaction is usually favorable—irrigation methods that conserve water also conserve soil.

The method of irrigation influences how much soil and water are wasted. Careful management of surface irrigation can achieve about 60% efficiency in water use, typical sprinkler irrigation systems are about 75% efficient, and trickle irrigation is about 90% efficient.

Subsurface, sprinkler, trickle, and pot irrigation should not produce any runoff and, therefore, should not cause erosion. In fact, the increased vegetative cover may reduce both wind and water erosion. Sprinkler and trickle irrigation are sometimes used on steep slopes that need both vegetative and mechanical erosion-control practices. These needs would exist if the land were cropped without irrigation.

Surface irrigation can be a significant cause of erosion because large amounts of water flow across the land. The worst case results when preirrigation is used to store water in the soil before a crop is planted. Residues from the previous crop can reduce the hazards of preirrigation. Irrigation frequency is also important, especially where row crops are cultivated between irrigations. Most of the erosion occurs during the first hour of irrigation while the soil is still loose. A few long irrigations therefore erode less soil than an equal amount of water applied in several shorter irrigations.

Long furrows, corrugations, and borders are convenient because it takes fewer rows and less work to cover the same area. But, long rows require large, erosive streams of water and make it more difficult to irrigate uniformly. Rows that are too long therefore waste both soil and water. One solution to long-row problems is to place portable gated pipe, such as that shown in Figure 15–15, across the middle of the rows. The lower half of the field is irrigated first, then the pipe is removed, and the upper half is irrigated. Tillage, planting, and harvesting operations run the full length of the rows.

Figure 15–15 Gated pipe being used in Iowa for furrow irrigation. Each opening in the pipe has an adjustable gate to control the water flow. (Courtesy USDA Natural Resources Conservation Service.)

Figure 15–16 A land plane is used to smooth land surfaces for more uniform surface irrigation. (Courtesy F. R. Troeh.)

Variations in either soil permeability or slope gradient can cause nonuniform applications that waste water. Soil erosion may also result on the steeper parts. Variable soil permeability usually cannot be corrected, but variable slope may be smoothed by land leveling. Major land leveling requires staking the field on a grid pattern, surveying the elevations of the stakes, and preparing a plan that balances the cuts and fills. Cuts and fills are then marked on the stakes, and large earth-moving equipment levels the land to the prescribed uniform slope gradient. Carryalls such as those used for building roads are used for major leveling. Lighter smoothing work is done with a land plane such as that shown in Figure 15-16.

Miller et al. (1987) found that the size of the maximum nonerosive irrigation stream increased considerably when crop residues were present, especially those grown in place. They combined the use of residues with surge irrigation (described in Section 15–5.1) to obtain maximum uniformity of water application in furrows. In surge flow, they alternately ran water in a furrow for 20 minutes, then turned it off for 20 minutes. These surges carried water to the ends of the furrows nearly as soon as continuous flow. Continuous flow was used after the water reached the end of the furrows. A more expensive but highly effective alternative for stabilizing the soil and thus reducing erosion is to add PAM to the irrigation water, as discussed in Section 15–5.1.

Ditch erosion is another irrigation hazard. Farm ditches placed on an ideal gradient of about 0.15% seldom erode much (larger ditches need to be flatter). But, ditches that must go down a steeper slope can easily cause erosion. Drop structures, concrete-lined ditches, and irrigation pipelines are all used to avoid ditch erosion. Lined ditches and pipelines are also good ways to minimize seepage losses.

SUMMARY

Irrigation is steadily increasing in extent and is widely practiced. More than 60% of the world's irrigated land is in Asia. Irrigation causes changes in both crop and soil management. Average worldwide yields with irrigation are more than twice as high as rainfed yields; crop quality may be improved as well. However, irrigation costs are high, certain diseases spread by waterborne vectors such as mosquitoes and snails may increase, and erosion and sedimentation may result.

Soil suitable for irrigation must have satisfactory depth, texture, structure, and topography. Maximum slope gradients vary from 3% to more than 35%, depending on soil, climate, crop, and type of irrigation system. *Water rights* are an important issue for irrigators and are based on either riparian rights or prior appropriations. Water for irrigation comes from surface water diverted from streams, harvested rainwater, and underground water. Some water is returned to the stream after use and then reused downstream. Falling water tables are a common problem where wells are used to tap underground water. *Water quality* depends on salt content, sodium adsorption ratio, toxic ions, and solid matter such as sediment and weed seeds. Salts in either the soil or the water can produce saline and sodic soils.

Water for small projects is distributed by ditches or pipelines. Large projects require *canal systems* that branch into *laterals* and *ditches* for individual users. *Weirs* and *flumes* measure water flow, and *drop structures, checks,* and *division boxes* are used to control it. Water flow in ditches is blocked by checks, canvas dams, or sod dams and delivered to fields through turnouts, siphon tubes, or spiles. Pipelines are used for sprinkler irrigation and for problem spots in ditch systems.

Irrigation includes surface, subsurface, sprinkler, trickle, and pot methods. *Surface* methods include furrows, corrugations, basins, borders, and wild flooding. *Subsurface* irrigation is limited to specific conditions where a controlled drainage system can be used. Most *sprinkler* irrigation uses portable lines, rolling lines, or center-pivot systems; but *microirrigation* is a new method that uses small stationary sprinklers. Trickle irrigation is a relatively new method; it and pot irrigation are the most efficient for watering individual plants. They can use water containing more salt than the other methods.

Land reclamation includes drainage of wetlands, irrigation of arid lands, and correction of saline and sodic conditions. *Salinity* can be leached from a soil, but sodium replacement requires a *soil amendment* such as gypsum or sulfur. Plants vary greatly in their tolerance of saline and sodic conditions.

Sprinkler irrigation usually applies water more uniformly than surface methods; trickle and pot irrigation supply water directly to the plant root zone for maximum efficiency. Erosion from surface irrigation is reduced by limiting lengths of irrigation runs, using less frequent but longer irrigation periods, careful land leveling, using surge irrigation, and/or by stabilizing the soil with PAM. The other methods of irrigation should not produce runoff and therefore do not cause much increased erosion.

QUESTIONS

1. Irrigation would improve crop yields in most humid regions. Why is it not used more there?
2. Irrigation competes with other uses for available water. What are some of the competing uses?

3. What difference does it make to an irrigator whether water rights are based on the riparian doctrine or on prior appropriations?
4. How is water delivered from a ditch onto a field?
5. What happens to the salt contained in irrigation water applied by surface irrigation? by trickle irrigation?
6. Why is a sodic soil worse than a saline-sodic soil?
7. What erosion problems may be caused by irrigation?

REFERENCES

ABO-GHOBAR, H. M., 1993. Influence of irrigation water quality on soil infiltration. *Irrig. Sci.* 14:15–19.

AVNIMELECH, Y., 1993. Irrigation with sewage effluents: The Israeli experience. *Environ. Sci. Technol.* 27:1278–1281.

AYARS, J. E., R. A. SCHONEMAN, F. DALE, B. MESO, and P. SHOUSE, 2001. Managing subsurface drip irrigation in the presence of shallow ground water. *Agric. Water Mgmt.* 47:243–264.

AYARS, J. E., R. B. HUTMACHER, R. A. SCHONEMAN, S. S. VAIL, and T. PFLAUM, 1993. Long term use of saline water for irrigation. *Irrig. Sci.* 14:27–34.

BAINBRIDGE, D. A., 2001. Buried pot irrigation: A little known but very efficient traditional method of irrigation. *Agric. Water Mgmt.* 48:79–88.

BARVENIK, F. W., 1994. Polyacrylamide characteristics related to soil application. *Soil Sci.* 158:235–243.

BEN-HUR, M., F. H. LI, R. KEREN, I. RAVINA, and G. SHALIT, 2001. Water and salt distribution in a field irrigated with marginal water under high water table conditions. *Soil Sci. Soc. Am. J.* 65:191–198.

BENSON, R. D., 1995. Water rights deals, water law reform: Restoring water to northwest rivers. *Illahee* 11:12–16.

BERNSTEIN, L., 1964. *Salt Tolerance of Plants.* USDA Agric. Inf. Bull. 283, 24 p.

BOOHER, L. J., 1974. *Surface Irrigation.* FAO Agricultural Development Paper No. 95, Food and Agriculture Organization of the United Nations, Rome, Italy, 160 p.

BOUMAN, B. A. M., and T. P. TUONG, 2001. Field water management to save water and increase its productivity in irrigated lowland rice. *Agric. Water Mgmt.* 49:11–30.

BRAR, M. S., S. S. MALHI, A. P. SINGH, C. L. ARORA, and K. S. GILL, 2000. Sewage water irrigation effects on some potentially toxic trace elements in soil and potato plants in northwestern India. *Can. J. Soil Sci.* 80:465–471.

BROUGHTON, R. S., 1995. Economic, production, and environmental impacts of subirrigaiton and controlled drainage. Ch. 11 in H. W. Belcher and F. M. D'Itri, (eds.), *Subirrigation and Controlled Drainage.* Lewis Pub., Boca Raton, FL, 482 p.

BURT, C. M., A. J. CLEMMENS, R. BLIENER, J. L. MERRIAM, and L. HARDY, 2000. *Selection of Irrigation Methods for Agriculture.* American Society of Civil Engineers, Reston, VA, 129 p.

BUTTERS, G. L., J. G. BENJAMIN, L. R. AHUJA, and H. RUAN, 2000. Bromide and atrazine leaching in furrow- and sprinkler-irrigated corn. *Soil Sci. Soc. Am. J.* 64:1723–1732.

COOK, S., L. GENGRUI, and W. HUILAN, 2000. Rainwater harvesting agriculture in Gansu Province, People's Republic of China. *J. Soil Water Cons.* 55:112–114.

COSTA, J. L., and P. GODZ, 1998. The effects of gypsum applied to a Natraquoll of the Flooding Pampas of Argentina. *Soil Use and Management* 14:246–247.

DUFFORD, W., 1995. Washington water law: A primer. *Illahee* 11:29–39.

FIEGE, M., 1995. Creating a hybrid landscape: Irrigated agriculture in Idaho. *Illahee* 11:60–76.

FISHER, M. J., N. R. FAUSEY, S. E. SUBLER, L. C. BROWN, and P. M. BIERMAN, 1999. Water table management, nitrogen dynamics, and yields of corn and soybean. *Soil Sci. Soc. Am. J.* 63:1786–1795.

GULHATI, N. D., and W. C. SMITH, 1967. Irrigation agriculture: An historical review. In *Irrigation of Agricultural Lands,* Agronomy Monograph 11, American Society of Agronomy, Madison, WI, p. 3–11.

GUPTA, S. K., and S. K. SHARMA, 1990. Response of crops to high exchangeable sodium percentage. *Irrig. Sci.* 11:173–179.

ILYAS, M., R. H. QURESHI, and M. A. QADIR, 1997. Chemical changes in a saline-sodic soil after gypsum application and cropping. *Soil Tech.* 10:247–260.

JAIN, S. C., 2001. *Open-Channel Flow.* John Wiley & Sons, New York, 328 p.

JOHNSON, R. W., and R. PASCHAL, 1995. The limits of prior appropriation. *Illahee* 11:40–50.

KANBER, R., H. KÖKSAL, SERMET ÖNDER, S. KAPUR, and S. SAHAN, 2001. Comparison of surge and continuous furrow methods for cotton in the Harran plain. *Agric. Water Mgmt.* 47:119–135.

KARR, J. R., 1995. Clean water is not enough. *Illahee* 11:51–59.

LENTZ, R. D., and R. E. SOJKA, 1994. Field results using polyacrylamide to manage furrow erosion and infiltration. *Soil Sci.* 158:274–282.

MACE, J. E., and C. AMRHEIN, 2001. Leaching and reclamation of a soil irrigated with moderate SAR waters. *Soil Sci. Soc. Am. J.* 65:199–204.

MALETIC, J. T., and T. B. HUTCHINGS, 1967. Selection and classification of irrigable land. In *Irrigation of Agricultural Lands.* Agronomy Monograph 11, American Society of Agronomy, Madison, WI, p. 125–173.

MARTINEZ HERNANDEZ, J. J., B. BAR-YOSEF, and U. KAFKAFI, 1991. Effect of surface and subsurface drip fertigation on sweet corn rooting, uptake, dry matter production and yield. *Irrig. Sci.* 12:153–159.

MEJIA, M. N., C. A. MADRAMOOTOO, and R. S. BROUGHTON, 2000. Influence of water table management on corn and soybean yields. *Agric. Water Mgmt.* 46:73–89.

MILLER, D. E., J. S. AARSTAD, and R. G. EVANS, 1987. Control of furrow erosion with crop residues and surge flow irrigation. *Soil Sci. Soc. Am. J.* 51:421–425.

NAVA, H., 2001. Wastewater reclamation and reuse for aquaculture in Perú. *J. Soil Water Cons.* 56:81–87.

OOMEN, J. M. V., J. DEWOLF, and W. R. JOBIN, 1990. *Health and Irrigation.* ILRI Publication 45, International Institute for Land Reclamation and Improvement, Wageningen, The Netherlands, 304 p.

OPIE, J., 2000. *Ogallala: Water for a Dry Land,* 2nd ed. Univ. of Nebraska Press, Lincoln, NE, 475 p.

PASTERNAK, D., and Y. DE MALACH, 1987. Saline water irrigation in the Negev Desert. In *Agriculture and Food Production in the Middle East.* Proceedings of a Conference on Agriculture and Food Production in the Middle East, Athens, Greece. January 21–26, 1987.

PEAVY, L., 1992. Conserving Colorado's Ogallala aquifer. *Soil Water Cons. News* 13(2):20.

PERALA, O., and R. BENSON, 1995. Water transfers: Can we get there from here? *Illahee* 11:16–18.

PESSARAKLI, M., 1991. Formation of saline and sodic soils and their reclamation. *J. Environ. Sci. Health* A26:1303–1320.

POSTEL, S., 1992. Water scarcity. *Environ. Sci. Technol.* 26:2332–2333.

QADIR, M., A. GHAFOOR, and G. MURTAZA, 2001. Use of saline-sodic waters through phytoremediation of calcareous saline-sodic soils. *Agric. Water Mgmt.* 50:197–210.

RICHARDS, L. A. (ed.), 1954. *Diagnosis and Improvement of Saline and Alkali Soils.* Agr. Handbook No. 60, U.S. Department of Agriculture, Washington, D.C., 158 p.

ROBBINS, C. W., and G. A. LEHRSCH, 1992. Effects of acidic cottage cheese whey on chemical and physical properties of a sodic soil. *Arid Soil Res. Rehab.* 6:127–134.

ROBINSON, A. R., and A. S. HUMPHREYS, 1967. Water control and measurement on the farm. In *Irrigation of Agricultural Lands,* Agronomy Monograph 11, American Society of Agronomy, Madison, WI, p. 828–864.

SCHERER, T. F., W. KRANZ, D. PFOST, H. WERNER, J. A. WRIGHT, and C. D. YONTS, 1999. *Sprinkler Irrigation Systems.* MidWest Plan Service, Iowa State University, Ames, IA, 243 p.

SHAINBERG, I., and G. J. LEVY, 1994. Organic polymers and soil sealing in cultivated soils. *Soil Sci.* 158:267–273.

SHARMA, D. P., K. V. G. K. RAO, K. N. SINGH, P. S. KUMBHARE, and R. J. OOSTERBAAN, 1994. Conjunctive use of saline and non-saline irrigation waters in semiarid regions. *Irrig. Sci.* 15:25–33.

SHOJI, K., 1977. Drip irrigation. *Sci. Am.* 237(5):62–68.

SIEBE, C., 1998. Nutrient inputs to soils and their uptake by alfalfa through long-term irrigation with untreated sewage effluent in Mexico. *Soil Use and Mgmt.* 14:119–122.

SIRJACOBS, D., I. SHAINBERG, I. RAPP, and G. J. LEVY, 2001. Flow interruption effects on intake rate and rill erosion in two soils. *Soil Sci. Soc. Am. J.* 65:828–834.

SOJKA, R. E., and R. D. LENTZ, 1994. Time for yet another look at soil conditioners. *Soil Sci.* 158:233–234.

SOJKA, R. E., R. D. LENTZ, and D. T. WESTERMANN, 1998. Water and erosion management with multiple applications of polyacrylamide in furrow irrigation. *Soil Sci. Soc. Am. J.* 62:1672–1680.

SRIVASTAVA, R. C., 2001. Methodology for design of water harvesting system for high rainfall areas. *Agric. Water Mgmt.* 47:37–53.

SUMNER, M. E., P. RENGASAMY, and R. NAIDU, 1998. Ch. 1 in Sumner, M. E., and R. Naidu, (eds.), *Sodic Soils: Distribution, Properties, Management, and Environmental Consequences.* Oxford University Press, New York, 207 p.

TESFAI, M., and L. STROOSNIJDER, 2001. The Eritrean spate irrigation system. *Agric. Water Mgmt.* 48:51–60.

THOMAS, H. E., 1955. Underground sources of our water. In *Water,* Yearbook of Agriculture, USDA, Washington, D.C., p. 62–78.

THOMAS, H. E., and D. F. PETERSON, JR., 1967. Groundwater supply and development. In *Irrigation of Agricultural Lands,* Agronomy Monograph 11, American Society of Agronomy, Madison, WI, p. 70–91.

TIWARI, K. N., D. N. SHARMA, V. K. SHARMA, and S. M. DINGAR, 1992. Evaluation of fly ash and pyrite for sodic soil rehabilitation in Uttar Pradesh, India. *Arid Soil Res. Rehab.* 6:117–126.

TROUT, T. J., R. E. SOJKA, and R. D. LENTZ, 1995. Polyacrylamide effect on furrow erosion and infiltration. *Trans. ASAE* 34:761–766.

ZUR, B., U. BEN-HANAN, A. RIMMER, and A. YARDENI, 1994. Control of irrigation amounts using velocity and position of wetting front. *Irrig. Sci.* 14:207–212.

CHAPTER

16

SOIL POLLUTION

Pollution is contamination that makes things unclean or impure. Soil can either pollute or be polluted. Eroded soil can become either a water pollutant or an air pollutant, or it can bury a fertile soil beneath unproductive sediment. Soil may be physically polluted by being covered with trash or chemically polluted by contamination with heavy metals or other toxic materials. Soil may be biologically polluted either by toxic materials or by poor management that exhausts its fertility, depletes its organic matter, and forms clods and crusts that restrict air and water movement. Most of these effects are localized, but recent research is highlighting an important cumulative worldwide effect. Soil organic matter sequesters a large amount of carbon (Sá et al., 2001). Cropping, especially with intensive tillage, tends to accelerate the decomposition of soil organic matter with a consequent release of carbon dioxide, a greenhouse gas that contributes to global warming. Cropping with reduced or no tillage reverses this trend and thus sequesters more carbon dioxide than conventional cropping does.

The rapid growth of world population in recent times (Figure 16–1) increases the value of productive soil. The pollution problem can no longer be ignored because there is no desirable new land to replace that which is "worn out." Earth's resources are a closed system of limited size that is much easier to damage than to improve. Soil pollution squanders basic resources that support life.

16–1 CONCERN ABOUT POLLUTION

There was relatively little concern about pollution until recent times because the population was only a fraction of its present size. There was room for both the people and the pollution they caused. In fact, the careless discards of bygone civilizations have yielded much of the available information about ancient people. The Earth is now so fully populated, however, that space and resource restrictions limit the standard of living in many areas. Soil pollution further decreases the standard of living.

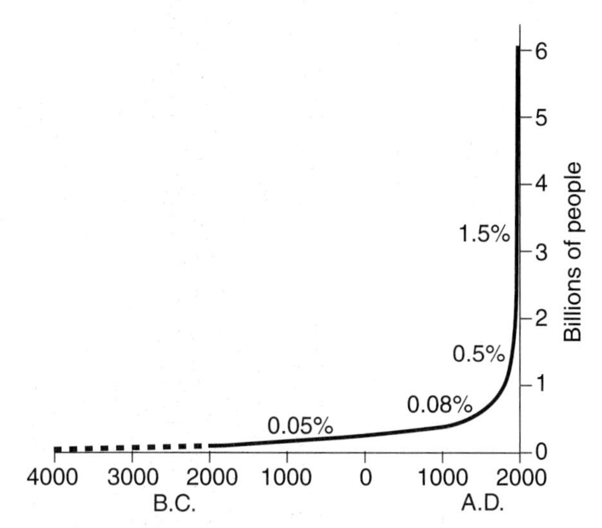

Figure 16–1 The world population growth rate was less than 0.1% per year until about 1600 A.D.; since then the growth rate has increased to about 1.5% per year. The numbers beside the population curve tell the approximate annual increase in population.

The damage caused by pollution is difficult to evaluate accurately, but it certainly adds up to many billions of dollars each year in the United States alone. Air pollution causes more than half a billion dollars of damage to agricultural crops annually (Shaw et al., 1971). Damage to buildings, machinery, and health must be added to that, as well as all of the detrimental effects of water pollution (see Chapter 17) and soil pollution. The total of all these various costs of pollution will increase as population increases unless billions of dollars are spent on pollution prevention.

Another reason for increased concern about pollution is the variety of new types of pollutants. Items made of plant materials and animal products decompose naturally and cause little pollution. The smelting of ores to produce metals has caused a growing amount of pollution for centuries. The wide variety of useful products coming from chemical industries has been accompanied by an equal variety of pollutants. Exotic materials are likely to become pollutants rather than decompose into harmless chemicals. Fortunately, the trend toward increasing chemical pollution is being slowed or reversed by considering decomposition rates when new products are developed.

Some soil pollutants are hazardous to health (Oliver, 1997). These include disease vectors, heavy metals and other toxic chemicals, and radon. Illness can result from direct consumption of soil (geophagia), from inhaling radon or dust particles such as asbestos that can cause malignancy, or by eating foods that have accumulated toxic materials from the soil.

Nuclear wastes cause much concern because of the very long lifespan of many radioactive wastes and the extreme toxicity of some materials, especially plutonium. Safe disposal of such materials is a difficult problem that has yet to be resolved on a long-term basis. Also, what should be done with nuclear power plants when they are worn out or obsolete?

The increasing concern about pollution has been accompanied by more sensitive means of detecting pollutants and measuring their concentrations. This increased sensitivity identifies pollutants that would have escaped notice before and raises the question of how much hazard such tiny amounts can cause. The long-standing requirement of zero tolerance imposed by the "Delaney clause" on any chemical that causes cancer in laboratory animals was changed in 1996. Tolerable levels are now assigned specific values rather than the minimum detectable concentrations.

16–2 SOURCES OF POLLUTANTS

A pollutant is sometimes defined as a resource out of place—much like dirt is soil out of place or a weed is a plant out of place. Most pollutants are either once-useful products that have worn out or are by-products of something useful. The problem is to eliminate the pollution without losing the useful products.

The sources of pollutants can be divided into many types, but three broad classes will be used here: (1) people-related sources, (2) industrial sources, and (3) agricultural sources. People-related wastes come from homes and offices and are proportional to the number of people involved in a particular environment. For example, the amount of sewage produced daily by a city can be estimated by multiplying 13.4 ft^3 (0.38 m^3) by the population (Thomas and Law, 1977). In addition, Americans produce an average of 3 to 5 lb of solid waste per person each day (Holmes et al., 1993). Industrial wastes are related to products and outputs rather than to the number of people. Agricultural sources are related to land areas, crop production, and livestock numbers. Each of these sources produces its own unique pollutants and needs its own kind of management to control pollution.

16–3 PEOPLE-RELATED WASTES

Population centers must provide means for disposal of solid and liquid wastes. Gaseous wastes are also produced, especially by burning solid wastes, but the resulting air pollution is outside the scope of this book. The solid and liquid wastes will be considered because they can be significant soil pollutants.

16–3.1 Solid Wastes

Metal, wood, paper, plastic, glass, and other materials (Table 16–1) are dumped into trash cans and hauled away. The disposal site was once the city dump where piles of trash accumulated and scavengers searched for usable items in the debris. Open dumps had fires that caused air pollution, were havens for rats and other pests, and created health hazards. Dumps were replaced by sanitary landfills where the trash is covered by a layer of soil the same day it is dumped. Sanitation is improved and the hazards are reduced, but satisfactory landfill sites are scarce. Percolating water may pollute the water table, and sight and odor pollution still cause people to protest the location of a landfill site near their residences.

TABLE 16–1 PERCENTAGE COMPOSITION OF SOLID WASTE FROM TWO CITIES AND TWO NEARBY COUNTIES

	City		County	
Solid Waste	Oakland, California	Phoenix, Arizona	Sacramento County, California	Pima County, Arizona
Cans and metals	7.4	7.1	6.4	8.0
Glass	10.1	8.9	9.8	9.7
Paper	38.0	42.5	26.0	39.2
Organics and yard trim	31.0	27.2	29.0	28.5
Plastics	5.0	6.2	1.3	4.8
Textiles (rags)	2.5	3.1	1.1	1.6
Wood	2.5	1.2	0.8	1.0
Tires	2.2	1.0	1.0	1.0
Other (ashes, dirt, etc.)	1.4	2.8	24.6	6.2

Source: Reproduced from *Soil for Management of Organic Wastes and Waste Waters,* Chapter 18, Fuller and Tucker, 1977, p. 472–489 by permission of the American Society of Agronomy, Crop Science Society of America, and Soil Science Society of America; data credited to Benjamin Petrucci, Sacramento, California, and W. H. Fuller, University of Arizona, Tucson.

Solid-waste recovery plants are the newest approach to solid-waste disposal. The first municipally owned plant of this type began operation at Ames, Iowa, in 1975. The Ames plant shreds the waste into small pieces and sorts it into combustible material, ferrous metal, aluminum, other metals, and reject material, as shown in Figure 16–2. The metals are sold and the combustible material is mixed with coal for electric power production. Another benefit to the city is the low sulfur content of the combustible portion of the waste.

Marketable products may cover the operating costs of solid-waste recovery plants, but probably not the depreciation costs. Such plants are increasing in number, however, because the alternatives are becoming less tenable. Recycling is likely to become more and more important with time.

16–3.2 Liquid Wastes

Sewer systems transport liquid wastes to sewage plants for treatment, as discussed in Chapter 17. The output of such plants includes a liquid material known as *sewage effluent* and a semi-solid material known as *sewage sludge.* Sludge represents only about 0.1% of the sewage volume, but that is still a large amount. The wet sludge is usually dried on a sand bed. The dried sludge is a mellow material of variable composition averaging about 4% N, 3% P, and 0.3% K (Sommers, 1977). It can be used to loosen and fertilize garden soils, or it can be applied to fields when the supply is large. Sometimes it is finely ground and mixed with effluent for distribution through a sprinkler irrigation system. Lester (1990) discusses the use of sludge and the problems associated therewith. The chief concern is the content of heavy metals in the sludge (see Section 16–7.3).

Sewage effluent normally is fairly low in heavy metals because they occur as insoluble compounds and are held in the sludge. Effluent is widely used for irrigation in Latin America (Nava, 2001), often with little pre-treatment. In some places, it is the only water

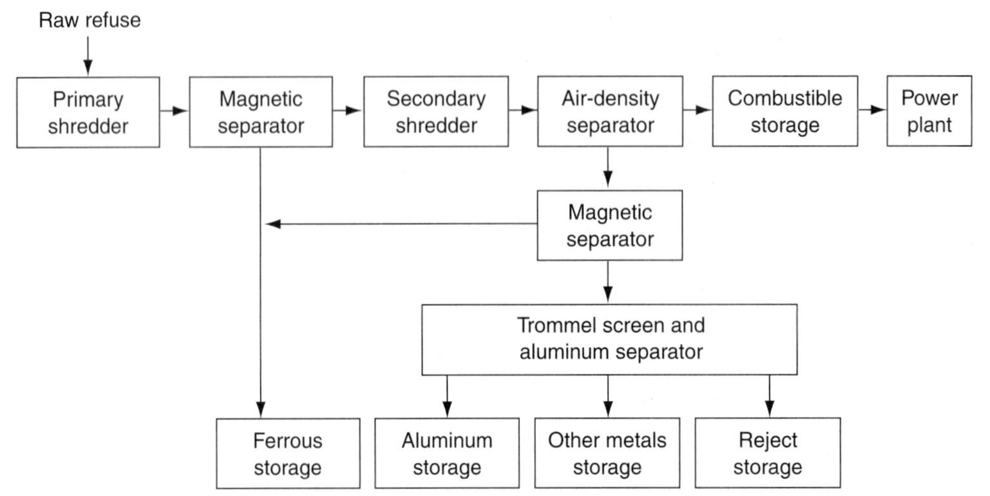

Figure 16–2 A flow diagram for the solid-waste recovery system at Ames, Iowa. (Courtesy City of Ames.)

available, and it carries a bonus in the form of plant nutrients that substitute for fertilizer. Direct application of such effluent to fields is less hazardous to health than the alternative of discharging it into a stream. Water from the stream may also be used for irrigation, but some of it is likely to enter municipal water supplies.

Tertiary treatment removes dissolved materials from sewage effluent. Relatively few sewage plants provide tertiary treatment, but the number is increasing because of pollution-control requirements. Two methods are used—chemical treatment and disposal on land by sprinkler irrigation. Chemical treatment requires large quantities of expensive resins to absorb anions and cations. Most interest in tertiary treatment is therefore directed toward disposal on land (Fuller and Warrick, 1984).

The soil to be used for tertiary sewage treatment must be permeable and well drained. High application rates are used to minimize the land area and the size of the irrigation system required. Large sprinkler nozzles are required to avoid clogging problems. These nozzles produce large droplets that are highly erosive. A forage crop should be grown on the land to protect it from erosion.

Waste disposal on land usually applies water and nutrients at rates that are much higher than needed for plant growth. The systems should therefore be managed to dispose of water and nutrients. Tile drainage may be needed to prevent waterlogging. Forage crops should be harvested to remove nutrients. Luxury consumption will make them rich in nutrients, especially nitrogen and potassium. Nitrogen loss by denitrification can be promoted by keeping the soil wet for several days and then allowing it to dry for several days. Excess phosphorus will accumulate in the soil as insoluble phosphates of calcium or iron and aluminum. Potassium is held on cation-exchange sites.

Heavy-metal-ion content of the sewage is usually the factor that limits application amounts or even precludes application of sewage sludge to soil. Many of these ions are toxic to animals and plants. For example, cadmium is toxic to animals if it is present in significant concentrations. Some heavy metals are quite toxic to plant growth. Their toxicity

Figure 16–3 Trash discarded in a gully is unsightly, may harbor mice and rats, can cause water pollution, and must be removed before the gully can be repaired. (Courtesy F. R. Troeh.)

may persist for decades because the metallic ions are held by cation-exchange sites. The *zinc equivalent* may be used to monitor the problem. It has been defined as the zinc concentration plus double the copper concentration and eight times the nickel concentration. According to Larson et al. (1975), the zinc equivalent added to agricultural soils should not exceed 5 to 10% of the soil's cation-exchange capacity. Typically, a soil with a cation-exchange capacity of 20 meq/100 g can probably accept a total of 200 to 600 tons/ac (500 to 1500 mt/ha) of average sewage sludge (a depth of about 2 to 6 in. or 5 to 15 cm) before being limited by the zinc equivalent. (See also zinc in Section 17–2.6.) High metal concentrations in sludge can be reduced through acid leaching. Tyagi et al. (1993) discuss the use of sulfur-oxidizing bacteria to acidify sludge economically in preparation for leaching.

16–3.3 Rural Home Wastes

Homes not connected to sewer systems still have to dispose of both solid and liquid wastes. Residents of many such homes use their own small version of the old city dump, often by discarding their trash in a gully or on a stony area as shown in Figure 16–3. One advantage many rural homes have is that garbage can be fed to hogs or other livestock and therefore need not be included with the other solid waste.

Septic tanks can be a very effective disposal system for liquid wastes. They provide both settling and bacterial action in the tank, followed by disposal through a tile drain field. A system that functions properly provides the equivalent of primary and secondary sewage treatment in the tank and tertiary treatment in the soil. Problems arise when the tank fills with sediment (it then needs to be pumped out) or when the soil in the drain field will not absorb all of the effluent. Soil permeability is often reduced by bacterial products clogging the pores around the drain lines. Septic tank effluent may then reach the surface and cause unpleasant odors and possible health hazards.

16–3.4 Litter

People have a distressing habit of littering wherever they go. Discarded wrappers, containers, and worn-out items fall by the wayside and clutter the landscape. These items injure the appearance of the area, and some of them stifle plant growth by cutting off air, water, and light. Items such as broken glass and sharp pieces of metal can inflict wounds on persons or animals, and some discarded poison containers can be deadly. A small percentage of the population undoubtedly produces most of the litter. The problem seems to be worst when leisure time is involved and people are relaxing.

Cities, towns, and other public bodies spend large sums cleaning up litter. Fines may be levied on litterbugs, but it seems that few are caught. A newer approach of requiring a deposit on all beverage containers has been more successful but cannot solve the whole problem. Changes in attitudes and habits are needed. One approach that has proven helpful is the "adopt-a-highway program" whereby a group or an individual agrees to clean a designated road section periodically.

16–4 INDUSTRIAL WASTES

Industries produce a wide variety of solid, liquid, and gaseous wastes. Heat often accompanies the wastes. The solid wastes include slag from smelters, gypsum from phosphate fertilizer plants, sawdust from sawmills, and all kinds of scrap pieces or remnants. The material may be wood, metal, cloth, paper, glass, rubber, plastic, etc. Reactive chemicals are important concerns, especially if they are toxic to people, animals, or plants. Some wastes are dense, and some are bulky; some are combustible, whereas others are inert and resistant to decomposition.

16–4.1 Solid Industrial Wastes

Disposal methods for solid industrial wastes depend on the nature and amount of the waste. Municipal solid-waste disposal systems accept some industrial wastes, and some large industries manage their own landfills. Combustible waste may be burned in high-temperature incinerators. This disposes of part of the waste, but much still accumulates and pollutes land near factories.

Scrap metals and some other materials can be recycled if they are separated from other wastes. Proper planning, coupled with attitude changes that result in increased recycling, can turn wastes into resources, reducing both the waste-disposal problem and the consumption of raw materials.

Some industrial wastes are organic materials that can be applied to soils. For example, large amounts of wastes are produced by paper mills and by de-inking plants that recycle used paper. Aitken et al. (1998) suggest that both paper mill sludge and de-inking sludge may safely be applied to soil and thus kept out of landfills. They found that additional nitrogen fertilizer was needed to offset the high cellulose content of the sludge. Increased crop yields have resulted, especially on acid soils, probably because of the liming effect of the sludge and the buildup of soil organic matter. Maximum lifetime applications would be limited by heavy metals such as copper after several years or decades of sludge applications.

16–4.2 Liquid Industrial Wastes

Petroleum products and water contaminated by acids, bases, or other chemicals pose difficult disposal problems. They can pollute water, soil, and air. They often are released in concentrations that overwhelm the ability of biological systems to degrade them. Extreme examples of petroleum pollution will be discussed in Section 16–7.4. Lesser spills occur at multitudes of sites where gasoline, oil, and other petroleum products are handled.

One of the most famous toxic chemical sites is the Love Canal in the community of Niagara Falls, New York. Chemicals buried over a 25-year period contaminated the soil and caused sickness, death, and the abandonment of a school and its surrounding neighborhood (Colten and Skinner, 1996). Multitudes of chemical plants have discarded chemical wastes that contaminate the soil in less well-known sites.

16–4.3 Nuclear Wastes

Atomic bombs, nuclear-powered ships, and nuclear electric plants inevitably produce nuclear wastes. These wastes are so hazardous and so long-lived that no acceptable long-term means of disposal has yet been found. The disposal problem, the remote but frightening possibility of a nuclear accident, the fearsome prospects of nuclear warfare, and the limited supply of uranium have raised strong opposition to the use of nuclear energy. Equally strong support is based on energy needs and the military potency of atomic power.

A nuclear reactor uses only a few thousand tons of uranium oxide during its lifetime and produces a comparatively small amount of radioactive wastes. But the twin hazards of long-lasting radioactivity and extreme toxicity of waste components, such as plutonium, make the storage and disposal methods very important. Corrosion-resistant steel tanks embedded in thick concrete have been used to contain the wastes. Some of these containers rest on the ocean floor, where they were dumped several years ago. They are now regarded as potential hazards to fish and other life because the wastes may remain toxic 1000 times as long as the projected lifespan of the containers.

Consideration is being given to disposal of atomic wastes in deep rock formations. The formations need to be dense and free of earthquake faults and other fractures. There must be a way to form a cavity, place a waste capsule in it, and seal the hole. Preferably, the capsule should be retrievable if the need should arise. Certain deep salt deposits appear to best fulfill the requirements, but it is difficult to be sure that nothing would ever escape from them.

Obsolete or worn-out nuclear reactors pose another problem. None has yet been dismantled and removed from its site. Removal must be accomplished in a way that protects the workers from radioactivity and leaves the site uncontaminated. Radioactive parts of the plant present a disposal problem similar to that of the spent fuel. Until the problem is solved, old power plants will have to be sealed and guarded to keep out people and animals.

16–4.4 Gaseous Wastes

Gaseous wastes initially become air pollutants, but they become soil pollutants when they are washed from the air by rainfall or otherwise deposited. The atmosphere is sometimes the preferred waste disposal site (e.g., exhausting heat through a radiator or a cooling tower). Carbon dioxide and water vapor are expelled along with heat when materials are burned. Unfortunately, burning may produce gaseous oxides of sulfur and nitrogen plus fine particles of solid matter that pollute the atmosphere. Such materials often concentrate in droplets of rain, mist, or fog and sometimes react to produce smog. The droplets become soil and water pollutants when they fall.

Oxides of sulfur and nitrogen dissolved in water produce sulfurous, sulfuric, nitrous, and nitric acids. These can acidify rainwater to a pH as low as 2. *Acid rain* corrodes metals and concrete, and virtually eliminates the growth of flowers and many other plants in coal-burning industrial centers. Pollution-control efforts to solve the acid rain problem need to be targeted at problem sites rather than applied universally. Areas where acid rain is not considered a problem commonly receive free fertilizer benefits amounting to 10 to 15 lb/ac (11 to 17 kg/ha) of available N and 15 to 20 lb/ac (17 to 22 kg/ha) of available S annually.

16–5 AGRICULTURAL WASTES

Livestock in the United States produce between 1.5 and 2 million tons of waste per year. Plant residues total hundreds of millions of tons. These materials are generally beneficial when returned to the soil, but some of them get into streams, ponds, and lakes where they cause pollution. In addition, agriculture uses many chemicals, containers, and other potential soil or water pollutants.

The task of reducing agricultural pollution is complicated by the *nonpoint* nature of the sources. Livestock wastes may be deposited wherever an animal happens to be and some may eventually reach a nearby stream. Similarly, chemicals applied to either soil or plants can be washed into streams by runoff water. Pollution-control efforts must be applied to large areas to significantly reduce the effects of nonpoint sources.

16–5.1 Livestock Wastes

The handling of livestock wastes varies widely. Dried cattle droppings are prized as fuel in India where other fuel is not affordable. However, the manure from large feedlots in the western United States accumulates in piles that constitute a disposal problem of mammoth proportions. Operations that feed tens of thousands of animals seldom have access to enough land to spread the manure as fertilizer. Smaller operations can use their manure on nearby land, but the hauling distance becomes too great for large operations. Attitude is also important—many livestock managers think of the manure as a nuisance rather than a resource. In fact, in years past, many feedlots were deliberately placed on sloping land next to a stream so the manure would be washed away.

Much manure has been carelessly handled or ignored because its fertilizer value was deemed to be too low to pay the handling costs. The composition varies with types of livestock and management (Table 16–2), but an average ton of wet cattle manure contains about 12 lb of N, 3 lb of P, and 10 lb of K (6 kg of N, 1.5 kg of P, and 5 kg of K per 1000 kg). The commercial fertilizer value of these nutrients was about $2 in 1970, $4 in 1980, and $5 in

TABLE 16–2 AVERAGE WATER AND NUTRIENT CONTENTS OF ANIMAL MANURES

		Nutrients (%)					
Animal	H_2O (%)	N	P	K	S	Ca	Mg
Dairy cattle	79	0.56	0.10	0.50	0.05	0.28	0.11
Fattening cattle	80	0.70	0.20	0.45	0.085	0.12	0.10
Hogs	75	0.50	0.14	0.38	0.135	0.57	0.08
Horse	60	0.69	0.10	0.60	0.07	0.785	0.14
Sheep	65	1.40	0.21	1.00	0.09	0.585	0.185
Broiler	25	1.70	0.81	1.25			
Hen	37	1.30	1.20	1.14			

Source: Calculated from *Soils for Management of Organic Wastes and Waste Waters,* Chapter 8, Olsen and Barber, 1977, p. 197–215 by permission of the American Society of Agronomy, Crop Science Society of America, and Soil Science Society of America; data credited to R. C. Loehr.

1990. These values approximate the cost of handling the manure. Given this equal cost option, many people choose to handle a bag of commercial fertilizer instead of a ton of manure. This reasoning ignores three significant factors: (1) cleanup costs should be charged to the livestock operation rather than to fertilizer value; (2) manure contains other nutrients in addition to N, P, and K; and (3) manure is good for the soil structure. These factors added to the value of the N, P, and K make a ton of manure worth much more than its handling cost. Furthermore, manure management would be important to control pollution even if it had no value for crop production.

Much manure has accumulated in piles behind barns until there was a convenient time to haul and spread it. Such manure usually contains straw or hay used as bedding for the animals. Bedding helps absorb the liquid excrement and the nutrients it contains and thereby increases the tonnage without diluting the nutrient content. Odors from manure piles indicate that nitrogen is volatilizing as ammonia, and sulfur as hydrogen sulfide. Volatilized nitrogen and sulfur compounds are either absorbed by soil or water in the downwind area or circulated in the atmosphere until they are brought down by rain. Airborne ammonia from cattle feedlots may add more nitrogen to nearby lakes and rivers than they receive in runoff and drainage water. Volatilization can be minimized by compacting the manure to keep it anaerobic and by hauling the manure to the field and incorporating it into the soil as soon as possible. Plowing or disking it into the soil is important because aeration accelerates volatilization losses in the field. Lauer et al. (1976) found ammonia losses ranging from 61 to 99% during 5 to 25 days after surface applications.

Manure must be left on the surface when it is applied to pasture, hay, or other permanent vegetation or to cropland in a no-till system (although some researchers have experimented with injecting manure slurries beneath the soil surface). Ammonia will volatilize, but most of the rest of the nitrogen will leach into the soil. Odor problems from such applications can be minimized by composting the manure for a few months before it is spread (Eghball and Power, 1999).

Manure piles should have a roof over them to keep out rain. A covered area in a feedlot works well because trampling by the animals compacts the manure and helps make the pile anaerobic. Piles that must be in the open should be tall and rounded to shed water. Water leaching through a manure pile removes nutrients and is likely to cause water pollution.

Lagoons are another means of handling livestock wastes. A pipeline may be used to carry the wastes from a pit to the lagoon, inasmuch as water must be added anyway to dilute them. Most lagoons are anaerobic and therefore produce some odors. Most of the solid matter eventually liquefies by anaerobic digestion; the remainder settles as sludge on the bottom of the lagoon. Much of the organic matter decomposes to CO_2 and H_2O. Half or more of the nitrogen and sulfur may volatilize when long-term retention is practiced. The effluent from a lagoon still contains much organic matter and many plant nutrients that can be applied to land by a sprinkler system or a spray truck (sometimes called a "honey wagon"). This effluent should not be emptied into a stream.

The phosphorus content of manure is usually relatively high in comparison to crop needs (Smith et al., 1998). Application rates based on phosphorus needs often must be supplemented with commercial nitrogen fertilizer, and possibly potassium. More often, the rates are calculated to supply the nitrogen requirement, and phosphorus accumulates in the soil. This is an inefficient use of phosphorus, but it usually has no detrimental effects.

Manure application rates up to about 10 tons/ac (22 mt/ha), when properly applied and incorporated into the soil, are effective for fertilizer purposes and cause little or no pollution. Commercial fertilizers can be used to meet any remaining fertility requirements. Although they represent inefficient use of the nutrients in the manure, rates up to 30 tons/ac (70 mt/ha) have been used in humid regions for at least 40 years without causing harmful buildups of nutrients or soluble salts in the soil (Sommerfeldt et al., 1973).

Manure is sometimes applied to land at rates of 50 to 100 tons/ac (110 to 220 mt/ha) to minimize the amount of land area required for disposal purposes. Such high rates can cause pollution because crops cannot use all of the available nutrients. Nitrates and other soluble salts accumulate in the soil and in drainage water. Runoff water, too, may be contaminated, especially if manure is left on the soil surface. To avoid such pollution, the manure should be used on more land at lower rates. If high rates must be used, luxury consumption and denitrification should be encouraged, as discussed for sewage applications in Section 16–3.2.

16–5.2 Plant Residues

Plant residues on and in the soil are usually an asset rather than a cause of pollution. They help to control soil erosion, retain plant nutrients, and produce soil organic matter. Nevertheless, plant residues can cause problems by accumulating as debris either on land or in water. The debris is unsightly and may block a watercourse or plug a screen. Sometimes the residues smother vegetation. Sometimes they cause odors as they rot, or produce smoke and soot as they burn.

Tumbleweeds are a common source of debris in arid regions. They accumulate in fences and ditches, and against buildings, where they may obstruct passage and increase the fire hazard. Broken tree branches and fallen trees also produce litter that can block pathways and even roads, or they may fall into a stream and accumulate in quiet water or next to a dam. The decomposing debris adds to the biological oxygen demand in the water.

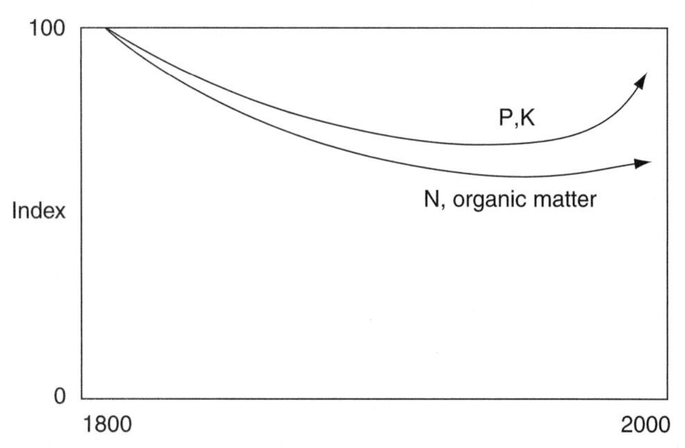

Figure 16–4 A schematic illustration of the general trends of soil fertility in the U.S. Corn Belt from 1800 to 2000. (Courtesy L. M. Thompson.)

Plant residues are so abundant that it is fortunate that they usually cause only minor pollution problems. The usual solution is simply to clean up the offending residues.

16–5.3 Fertilizers

The world's farmers use about 120 million tons of fertilizer N, P, and K annually as mineral fertilizers. About 15% of this fertilizer is used in the United States. Fertilizer application rates range from zero to several hundred pounds of nutrients per acre. High-value horticultural crops tend to receive the highest fertilizer rates, but agronomic crops such as corn, cotton, and tobacco also receive more N, P, and K than is being removed in the harvest. An illustration of the effects on soil fertility of changing cropping and fertilizer practices is shown in Figure 16–4.

The dramatic increase in fertilizer use during this century has improved both the quality and the quantity of crops produced. This trend is likely to continue to be an essential component of efforts to provide food and other needs for the world's growing population. Nevertheless, several environmental charges made against fertilizers need to be taken seriously. Three such concerns will be considered here—the effects of fertilizers on water pollution, the effect of denitrification on the ozone layer, and the effects of fertilizers on soil microbes.

Plant nutrients can reach a stream either in overland flow or in drainage water. Both overland flow and drainage water carry dissolved nutrients. In addition, overland flow carries nutrients that are attached to suspended soil particles. The nutrients attached to eroding soil particles constitute by far the largest of these losses and will be considered in Section 17–2.2.

Solution losses are proportional to the concentration of the plant nutrients in the water. Nitrate nitrogen (NO_3^-) is a prime concern because it is highly soluble and very mobile in the soil. In contrast, both ammonium and potassium ions (NH_4^+ and K^+) are held by cation exchange, and phosphates ($H_2PO_4^-$ and HPO_4^{2-}) have both low solubility in water and strong attraction to anion-exchange sites in soil.

TABLE 16–3 NITRATE NITROGEN LOSS IN TILE DRAINAGE WATER FROM VARIOUS RATES OF NITROGEN APPLICATIONS TO CONTINUOUS CORN ON FINE-TEXTURED WEBSTER SOIL IN MINNESOTA

N fertilizer rates,				
lb/ac	18	100	200	400
kg/ha	20	112	224	448
NO_3- N in tile drainage water,				
lb/ac	17	22	53	107
kg/ha	19	25	59	120

Source: Gast, et. al,1978.

Leaching losses of nitrate nitrogen are usually small but can be large under certain conditions. Table 16–3 contains data from a nearly level Minnesota soil with tile drainage. The recommended annual rate for fertilizer nitrogen in this area is about 150 lb/ac (170 kg/ha) for continuous corn. About one-fourth of that nitrogen may be lost by leaching, according to this study. Higher fertilizer rates increased the leaching loss; lower rates decreased but did not eliminate the loss. About 16 lb/ac (18 kg/ha) of nitrate nitrogen from decomposing organic matter would have been leached even without any fertilizer application.

Drainage makes cropping possible on the Webster soil represented by the data in Table 16–3. Unfortunately, drainage also makes leaching possible. The leaching of nitrates from well-drained soils in humid regions is usually slower than that indicated in Table 16–3 because the water has farther to go. Longer percolation times give deep-rooted plants time to use more nitrogen. Denitrifying microbes also consume more nitrate and reduce its concentration in the deeper soils. Nitrate leaching is therefore most likely to occur in permeable soils that are shallow to tile drains or to a gravel layer that drains the water. No leaching occurs from soils of arid regions unless they are irrigated or located in wet positions on the landscape.

Vigorous plant growth accompanied by a thriving microbial population can immobilize considerable amounts of nitrogen from the root zone and thus reduce leaching losses (Cookson et al., 2000). Nitrate leaching in deep soils can be reduced by growing deep-rooted crops. Alfalfa is one of the best crops for this purpose, especially if it can be left about three or four years to fully develop its root system. Even though it is a legume, it will absorb available nitrogen from its root zone. Entz et al. (2001) showed significant nitrate removal by alfalfa down to a depth of 9 ft (270 cm).

Fertilization can increase nitrate leaching if the nitrate concentration in solution is increased at times when leaching losses occur. The losses can be minimized by applying only as much nitrogen as necessary and by putting it on as late as possible while still meeting the needs of the crop.

The form of nitrogen fertilizer also makes a difference in leaching rates, especially when the weather is cool. Ammonium and organic forms of nitrogen must be oxidized to nitrate before much nitrogen can be leached from soils that have significant cation-exchange capacities. The nitrification process proceeds slowly at soil temperatures below 50°F (10°C) and stops when the soil freezes. The nitrification rate increases as the soil warms up in the spring, but plant growth should be using the nitrate nitrogen at that time.

Runoff losses of nitrogen fertilizer range from none to a few percent, rarely exceeding 10% of the amount applied. Dunigan et al. (1976) measured runoff losses in Louisiana on a fertilized silt loam soil on a 5% slope. Less than 3% of the applied nitrogen was lost in the runoff, except when heavy rain on the third day after fertilization carried away 9.5% of the nitrogen from one treatment. Phosphorus and potassium losses were less than 1% of the amounts applied.

Most rains have a significant wetting period before runoff begins. Soluble fertilizer is carried down into the soil during these first few minutes, so little is lost in the runoff. Exceptional losses can occur if intense rain strikes very suddenly or if the rain cannot infiltrate because of frozen or impermeable soil.

Ammonia volatilization and denitrification cause nitrogen to be lost from the soil to the atmosphere. *Ammonia volatilization* can become significant when urea from either commercial fertilizer or manure is applied and left on the soil surface, especially if the soil is alkaline. The urea reacts with water to form ammonium carbonate:

$$CO(NH_2)_2 + 2\,H_2O \longrightarrow (NH_4)_2CO_3$$

Reaction with a base under alkaline conditions can then volatilize ammonia:

$$(NH_4)_2CO_3 + 2\,NaOH \longrightarrow Na_2CO_3 + 2\,H_2O + 2\,NH_3$$

Much of the volatilized ammonia is probably carried back to the soil somewhere in rainwater, although some may be oxidized to N_2O, NO, HNO_2, or HNO_3.

Denitrification occurs when a soil contains nitrates but lacks aeration (Xu et al., 1998). The nitrates may come from fertilizer or from nitrification that occurred either in an aerated part of the soil or while the soil was dry enough for air to enter. Increased wetness or movement of nitrate ions to a wetter part of the soil causes anaerobic bacteria to take oxygen from nitrate ions and produce gaseous N_2, N_2O, and NO. The dinitrogen molecules are harmless, but the oxides may damage the ozone layer in the stratosphere. The atmosphere normally contains small quantities of these oxides. Some are formed in the atmosphere by lightning, and some rise from the soil and other sources. The concern is whether increased fertilization followed by denitrification might cause higher levels of nitrogen oxides that will help decompose too much ozone. Nitrogen oxides from automobile exhausts and other sources add to this problem.

The amounts of nitrogen oxides produced by denitrification are largest where large amounts of nitrate nitrogen occur in soils having alternate wet and dry conditions. This situation is sometimes deliberately created for waste disposal, as described in Section 16–3.2. Denitrification is generally undesirable, however, not only because of its effects on the ozone layer but also because it decreases the nitrogen supply for plant growth. The best remedy for these losses in cropped land is to drain the wet spots and thus improve aeration and reduce denitrification. Undrained wet spots usually should not be cropped and fertilized.

Van Cleemput (1971) showed that under denitrifying conditions an acid medium evolved N_2O and NO, but an alkaline medium produced N_2. Liming of soils where denitrification may occur should therefore reduce the output of nitrogen oxides. Another approach is to inhibit denitrification. Bollag and Henninger (1976) found that fungicides such as captan, maneb, and nabam inhibited denitrification and that the herbicide 2,4-D had a lesser but significant inhibitory effect.

Chemical changes resulting from fertilization can have considerable impact on the microbial population. Some effects show within a few days, as when added nitrogen speeds up the decomposition of organic materials in the soil. Long-term effects may develop over periods of years, as when the oxidation of nitrogen fertilizers containing ammonium compounds or urea causes acidification, which in turn changes the microbial population. Fortunately, such changes are normal and reversible. The microbial population is able to adjust to the changing conditions.

Banded fertilizer applications create different conditions in the bands than in the rest of the soil and can have a strong impact on the soil in the band. The salt effect may dehydrate and kill microbes in a fertilizer band. Even so, there is a zone around the band where the added nutrients cause microbes to multiply, and the total microbial population is larger than it would be without the fertilizer. Similar effects result from ammonia injection into soil. The pH in the injection zone rises to about 10 and stops all microbial activity. Microbes thrive on the margin of the zone, though, and gradually convert the ammonia to nitric acid, thereby acidifying the zone as they work their way toward its center.

Fertilizers, like most things, can be misused. Low to moderate fertilizer rates cause little or no environmental damage because the nutrients are used by growing plants. In fact, the improved plant growth often reduces pollution by reducing erosion. However, fertilizer nutrients that are not used by plants can contribute to pollution. The excess nutrients may be held by the soil for a while, but eventually they will move into either water or air.

Most fertilizer pollution studies concentrate on nitrogen. Phosphorus and potassium are hard to leach, so not much is lost. Table 16–4 shows what happened to the nitrogen that was applied to two soils on the North Carolina coastal plain. Crops used about half of the applied nitrogen. Surface runoff removed more than would be expected from the poorly drained soil. Although the runoff volume was low, its nitrate concentration was high. The unmeasured nitrogen loss would have included denitrification and any deep percolation that bypassed the tile drains.

Two indirect forms of fertilizer pollution should be mentioned here. One form is caused by their manufacture. Processing plants necessarily produce by-products, such as

TABLE 16–4 ANNUAL NITROGEN INPUTS AND REMOVALS FROM TWO FERTILIZED COASTAL PLAIN SOILS IN NORTH CAROLINA (TWO-YEAR FIELD AVERAGES)

	Field 1 (moderately well-drained soil)		Field 2 (poorly drained soil)	
	lb/ac	kg/ha	lb/ac	kg/ha
Nitrogen input—fertilizer	143	160	175	196
Nitrogen removal:				
Grain harvested	82	92	81	91
Surface runoff	20	22	26	29
Subsurface drains	23	26	14	16
Total removal measured	125	140	121	136
Nitrogen not measured	18	20	54	60

Source: Reproduced from *Journal of Environmental Quality,* Volume 4, No. 3, Gambrell, Gilliam, and Weed, 1975, p. 317–323 by permission of the American Society of Agronomy, Crop Science Society of America, and Soil Science Society of America.

gypsum and other materials, that may become pollutants. The second form results when mineral fertilizers are substituted for manure. The unused animal wastes cause much more pollution than properly handled manure used to fertilize fields.

16–5.4 Pesticides

Crop yields can be drastically reduced by the combined impact of weeds, insects, and plant diseases. Farmers and gardeners have fought pests with tillage, crop rotations, and selections of resistant crops. Some have gone out against hopeless odds to battle insects in person. The pests may be small, but they are exceedingly numerous. A single field may harbor more insects than the human population of the entire nation. In addition, the weed seeds in a field soil are generally much more abundant than the crop seeds planted in it.

Chemical warfare against agricultural pests began in the nineteenth century. Two of the earliest pesticides were Paris green (a mixture of arsenic trioxide and copper acetate used as an insecticide since about 1870), and Bordeaux mixture (a copper sulfate and lime mixture used as a fungicide since the 1880s). Bordeaux mixture was originally used to make grapes look poisonous so schoolboys would not steal them. Paris green may have been a by-product of political intrigue, as arsenic compounds were favorite poisons during the Middle Ages.

Highly toxic materials such as arsenic compounds and heavy metals such as copper present serious environmental problems. These problems are amplified by the high application rates required for these materials and the nondecomposable nature of their active elements. Furthermore, heavy metals are strongly held by cation exchange in the soil, and toxic buildups can last for decades.

The development of DDT as an insecticide in 1940 and of 2,4-D as an herbicide in 1941 opened the way for the metallic pesticides to be replaced by organic materials. The organic pesticides have several advantages: lower rates of application and fewer applications are needed, they have more selective toxicity than the metals, and they decompose into simple, harmless compounds. Their use, especially that of DDT, expanded rapidly until it was realized that they, too, can cause problems. DDT became highly controversial and was banned after the discovery that it threatened birds and other unintended victims by bioaccumulation. Metcalf (1971) cites data from Lake Michigan showing that DDT concentrations of two parts per trillion in the water increased progressively through the food chain to 99 parts per million in herring gulls.

Thousands of different chemicals are now in use for agricultural purposes. Very large quantities of many chemicals are spread over the immense areas used for producing crops and livestock. There would certainly be some problems, even under ideal conditions. Actually, weather, soil, plant, and animal interactions with chemicals often produce conditions that are far from ideal. Chemical users and their equipment may apply too much chemical in one place and not enough elsewhere. Some users are untrained or perhaps inclined to experiment. Spills occur, especially in areas where tanks are filled. Small amounts of each pesticide enter streams by runoff and groundwater by leaching (Williams et al., 1999). A pessimist could easily predict that chemical problems would be much more common and severe than they are.

Pesticides now go through an elaborate testing procedure to prove that they will work without harming the environment. Their effects on various forms of life are determined, their decomposition rates and tendencies to be adsorbed by soil components are measured, and the specific conditions under which they may be used are defined. Thousands of chemicals are screened to find a few that can be used.

There are still problems with pesticides in spite of the precautions. Some short-lived pesticides must be applied repeatedly during a single season; others last too long and damage the next year's crop. Some need to be applied at different rates on soils that have different organic-matter contents, cation-exchange capacities, or pH. Crop residues on the soil surface intercept sprays (Sadeghi et al., 1998) and alter the action of the pesticide. Wind may cause fine sprays or volatile chemicals to drift, and runoff may carry some chemicals into places where they are harmful. Many grassed waterways, for example, have been killed by atrazine washed from cornfields. Another problem is the tendency of the pests to adapt and develop resistance. New materials are needed to maintain control when the old materials become ineffective.

Pesticides, especially insecticides, applied to the soil can have a significant impact on soil animals ranging in size from earthworms to protozoa. The impact depends on the toxicity and persistence of the chemical. It is stronger if the chemical is mixed into the soil than if it remains on the surface. Fortunately, the soil animals can usually recover and repopulate the soil within a few months.

Despite the problems, modern pesticides are making possible several agricultural improvements. They have helped increase crop production during recent decades and are needed to continue that trend in the future. Herbicides save time, energy, and soil by reducing or even eliminating the cultivation of row crops.

Several biological control methods are favored as pesticide alternatives in specific circumstances. Crop breeding to develop resistance to pests is an old practice and yet is current enough to send workers all over the world to search for native plants that might carry significant genetic material. Similar searches are conducted for natural predators that can control insect pests. Other biological control methods that have been effective against certain insects include the release of large numbers of sterile male insects to interfere with normal insect breeding, and the use of synthetic sex attractant compounds to confuse the insects. Such methods are laborious because every situation must be handled separately, but they provide ways to control pests without harming the environment.

16–6 AEROSOLS

An *aerosol* is defined as a suspension of small solid or liquid particles in a gas. The most common aerosols are tiny droplets of water, as in fog, dust from windblown soil, and salt spray from ocean waves, but many other kinds of small particles are suspended in the air in variable quantities. Most aerosols are concentrated near their source and extend downwind in a plume that is gradually diluted with distance. Some, however, may circle the globe before they are entirely removed by settling, rainout, or chemical reactions.

Aerosols change from air pollutants to soil and water pollutants when they are deposited on land or water. Even inert dust particles are a nuisance when they coat otherwise clean surfaces. They not only soil the surfaces on which they settle, they also cause wear in machinery, contaminate experiments, and dull the appearance of coated objects. Some dust, however, is not inert. It may carry unwelcome microbes, chemicals, or even radioactivity. Some of the microbes could cause disease in exceptional circumstances, but this hazard is more often associated with liquid aerosols, such as those released when a person sneezes.

Transported dust particles may carry toxic chemicals, heavy metals, and radioactivity that are serious hazards, but fortunately, their concentrations are normally very low except in specific areas downwind from a nearby source. Heavy metals will be discussed in more detail in Section 16–7.3.

Polycyclic aromatic hydrocarbons (PAHs) are an example of the detrimental chemicals deposited as dust on the soil. PAHs are a large group of organic compounds with two or more interlocking ring structures. Some PAHs are mutagenic and/or carcinogenic. They are released into the atmosphere by wear from rubber tires or by burning organic materials in open fires, engines, incinerators, cigarettes, etc. They are deposited on soil and water, and can enter the food chain from there. The Environmental Protection Agency has listed 16 PAHs as priority pollutants (Menzie et al., 1992).

Radioactive particles such as those released by the Chernobyl accident in 1986 in Ukraine (part of the Soviet Union at the time) cause long-lasting concern. The effects at the time were widespread—for example, fish caught 2000 km (1200 miles) away in Britain contained high accumulations of radioactive cesium from the accident (Mason, 1990). The residents of farms and villages immediately downwind from the reactor had to be evacuated, and a significant area is still unsafe for habitation or even for cropping because the soil holds a high level of radioactivity. An estimated 8000 deaths were attributed to the accident and its aftereffects.

Additional aerosols that can cause problems are asbestos particles, soot, coal dust that can cause black lung disease, pollen that causes allergic reactions in sensitive individuals, and salt spray that can cause saline and sodic soil conditions in coastal areas.

16–7 POLLUTED SOIL

There are many different pollutants that may seriously degrade soil quality. Some of the most important soil pollutants are sediment, acids from aerosols and other sources, heavy metals, and industrial chemicals.

16–7.1 Sediment

Fertile bottomlands are built from thin layers of loamy sediment that are rich in organic matter. Such sediment is generally beneficial, but sediment deposition can also be detrimental. Some sediment has much lower fertility than the soil it covers. Other sediment is so high in clay that it is unfavorable for plant growth and for tillage operations. Sediment is frequently deposited in such thick layers that plants are smothered and killed, as shown in Figure 16–5. Similar problems can occur with sediment from wind erosion. The sedimentation problems discussed in Chapters 4 and 5 can be considered as a form of soil pollution.

16–7.2 Soil Acidification

Soil acidification is the most widespread form of soil pollution. Gradual acidification is natural in humid regions, but often it is greatly accelerated by human effects. Some of this is inherent in the growth of crops, but some comes from other sources. Some of it could be and needs to be avoided. Fortunately, acidity can generally be corrected by liming the soil.

Figure 16–5 A heavy spring rain deposited about 18 in. (45 cm) of sediment on this Iowa pasture, killing the grass and half burying the fence. (Courtesy USDA Soil Conservation Service.)

Not enough liming is done, though, because there is a cost involved, essential participants may be uninformed or disinterested, and there are physical difficulties in some places.

Natural acidification occurs as a soil is leached by rainwater that removes cations such as Ca^{2+} and Mg^{2+} from the exchange sites and replaces them with H^+. The process is generally quite slow and eventually reaches an equilibrium. Consequently, natural soils of humid regions are generally acid, but they still support acid-tolerant vegetation. Most of the favorable areas of these soils have been converted to cropland. Cropping removes some of the plant material and therefore does not return all of the nutrients that the growing plants took from the soil. Fertilizers are then added to compensate for the loss of fertility. Certain fertilizers, notably anhydrous ammonia and those containing ammonium ions or urea, form acids in the soil through biological oxidation processes. Even nitrogen fixation by legumes or free-living soil bacteria provides ammonium ions that are oxidized into acids. As a result, cropping generally accelerates the acidification process. Alternate fertilizers that carry their own neutralizing cations are available, for example calcium nitrate, but it is usually more economical to allow the acidification to occur and then to raise the soil pH periodically by adding lime. This works well on cropland and in gardens. It is more of a problem in pasture and forest, where there is no tillage to mix the lime into the soil and the economic returns may be low.

Soil acidification from sources unrelated to cropping can be a more serious problem. The external source can be fairly local, such as the leachate from a pile of mine tailings or other waste material, but there is one such source that is widespread. That source is acid rain. The local sources can be dealt with by intercepting the leachate and liming it or by liming the soil that has become too acidic. The source of the leachate often makes it evident who should pay for the remedial action. The acid rain problem is much more complex.

The acidity that causes acid rain comes from many sources. The most important input is sulfur dioxide coming from the combustion of organic materials. Oxides of nitrogen are another important contributor because they can react with water and oxygen to form nitric acid, just as sulfur dioxide can react with water and oxygen to form sulfuric acid. Burning fossil fuels, such as coal and petroleum products, contributes much of the acidity and is especially significant because it removes acid-forming materials from long-term storage and puts them into circulation. Worldwide rainfall is therefore much more acidic now than it was prior to the industrial age.

The effects of acid rain have become much more evident during the last few decades than they were before. Dead and dying forests in eastern Europe, the eastern United States and Canada, and elsewhere are attributed to a combination of the direct effects of the rain on the trees and indirect effects through soil acidification. Carter and Turnock (1993) attribute widespread damage to forest and other land in eastern Europe to the central planning practiced under Communism and the widespread use of the local high-sulfur lignite coal in those nations. In the United States, concern over air pollution problems in general and acid rain in particular led to the passage of the Clean Air Act in 1970 as a start toward controlling air pollution. Since then, significant reductions in sulfur dioxide emissions from automobile exhausts, industrial smokestacks, and electric-power generating stations have occurred, but the problem is far from under control. Similar efforts are being made in Europe and most other industrialized areas, but much less has been done in the developing nations.

Acid rain affects much more than just forests. Cropland and gardens are acidified, too, but these can be limed. Limestone and concrete are etched and metal and stone objects are corroded much faster than they would be without the added acidity. The health of humans and animal life can also be affected, especially that of fish in acidified waters. Rodhe et al. (1995) report that several countries in northern Europe are liming repeatedly approximately 11,000 lakes and streams and more than a million hectares of forest land to neutralize the effects of acid rain.

16–7.3 Toxic Elements

Almost anything can become toxic if it becomes too concentrated. However, certain things are much more likely than others to reach detrimental concentrations. Some of these are complex compounds that will be considered in the next section. Others are ions of heavy metals and other elements that will be discussed in this section.

Metals form cations that can be held on the cation-exchange sites of clay and organic matter. They can stay there and resist leaching for long periods of time—years, decades, centuries, or even longer. Normally, the dominant cations on the cation-exchange sites are calcium, magnesium, and hydrogen ions. Smaller amounts of potassium and sodium are also held on these sites along with tiny amounts of micronutrient cations like copper, iron, manganese, and zinc. These micronutrient cations and many nonnutrients are heavy metals. Any heavy metal cation can cause problems if it becomes too concentrated in the soil. Some that have been identified as most likely to cause toxicity are cadmium, copper, iron, lead, manganese, nickel, and zinc. Common sources of these and other heavy metals include a wide variety of waste materials such as mine tailings, sewage sludge, exhaust from vehicles and power plants, trash, and industrial discharges.

Copper and other heavy metals were formerly used as components of pesticides such as Bordeaux mixture, but these have been replaced by organic pesticides. Nevertheless, some old orchard sites and other areas that were treated with these pesticides still contain high metal concentrations in the soil. Arsenic and selenium should be added to the list of potentially toxic elements, even though they are transitional elements that can act either as metals or nonmetals. Aluminum, a light metal, will also be discussed.

Aluminum, Iron, and Manganese. Aluminum (Al^{3+}) toxicity is the most common form, with serious effects on soil microbes as well as most plants when the soil pH is near 5 or below (Robert, 1995). Trivalent iron (Fe^{3+}), quadrivalent manganese (Mn^{4+}), and divalent zinc (Zn^{2+}) ions become toxic to plant growth when the soil pH drops to about 4 or below (the precise value depends on the plant, the soil temperature, and other factors). These ions can seriously impair the growth of root hairs and thereby inhibit the absorption of water and nutrients and the growth of the entire plant. The detrimental effect is related to these ions rather than to the pH per se. Plants can grow very well with their roots in nutrient solution with the pH more acidic than this as long as these ions are not concentrated in the solution and the essential plant nutrients are kept available. Soil minerals automatically provide excesses of these ions when the pH drops below 5. The remedy is to apply sufficient lime to keep the soil pH out of the danger range.

It should be noted that aluminum, iron, and manganese ions become toxic to growing plants without being toxic to animal life that might eat what plant growth there is. The remaining elements to be discussed in this section can be present in concentrations that allow plants to grow well but are toxic to animals and humans that eat the plants.

Cadmium. Cadmium sulfide is used to make yellow pigments and cadmium is used as a coating to protect iron from corrosion. Rechargeable nicad batteries contain cadmium and phosphorus fertilizers contain small amounts of cadmium. Plants bioaccumulate cadmium as they absorb it from soil, and it is further bioaccumulated by animals, especially in their kidneys and livers. The bioaccumulation factor is greater than that determined for any other metal, and the absorbed cadmium remains in the body for a long time. Its half-life in body tissue is between 10 and 30 years (Muelle et al., 1992). Excess cadmium causes organ damage, cardiovascular disease, and even death.

Cadmium concentrations in soil are normally very low, but they tend to increase with long-continued applications of phosphorus fertilizer, and where the soil receives large applications of wastes such as sewage sludge or leachate from solid waste disposal sites, zinc mines, or certain chemical plants. Stigliani et al. (1993) calculated that the cadmium concentrations in agricultural soils in the Rhine Basin of Germany have been increasing by 1 to 3% annually and have doubled since 1950. They recommend a consistent liming program to keep the soil from becoming too acidic, because acidity increases the availability and plant uptake of cadmium and other heavy metals. Proposed permissible levels of Cd in soils range from 1 to 5 ppm (Piotrowska and Dudka, 1994). The limits should be somewhat proportional to the soil's cation exchange capacity. Limits used in the United Kingdom for cadmium and several other metals are shown in Table 16–5.

Lead. Lead is used in much larger amounts and has a much longer history of human use than most other metals. It is considered to be the element that constitutes the most serious environmental and health hazard of all toxic elements (Ma et al., 1995). It causes anemia, hypertension (Oliver, 1997), and lead encephalopathy—a disease of the brain and spinal cord. It is believed that many Romans died of lead poisoning caused by drinking acid

TABLE 16–5 UNITED KINGDOM LIMITS ON METAL ADDITIONS THROUGH THE APPLICATION OF SEWAGE SLUDGE TO ARABLE SOILS

	Annual limit		Soil concentration limit	
Metal	kg/ha	lb/ac	kg/ha	lb/ac
Arsenic	0.33	0.29	10	8.9
Cadmium	0.17	0.15	3.5	3.1
Copper	9.3	8.3	140	125
Nickel	2.3	2.1	35	31
Lead	33	29	550	490
Selenium	0.167	0.15	3	2.7
Zinc	18.6	16.6	280	250

Source: Based on data from Lester, 1990.

wine from lead goblets. Lead pipes were in use at the same time. Of course, the lead came from mines and was processed in smelters, both of which are likely to have contaminated workers and loaded the soil in nearby areas with lead that may remain concentrated there to the present time. More recently, lead has been used in batteries, solder, gasoline, paint, and other commonly used products. It is now banned from some of these uses, such as paint, gasoline, and solder for water pipes. Small children are particularly susceptible to lead poisoning, partly because they may eat chips of old paint that is high in lead oxide, a favorite white pigment before it was banned.

Some soils have accumulated excess lead washed from the paint of a nearby building. In fact, the "chalking" that allowed the surface of the paint to wash or rub off was once considered a positive feature that gave it a degree of self-cleaning. Also, soils near major thoroughfares have accumulated lead that was exhausted from automobiles during the decades when tetraethyl lead was used as an anti-knock ingredient in gasoline. Small children should not play in such areas because they may ingest significant amounts of lead by eating the soil. Brinkman (1994) indicates that lead concentrations above 500 ppm are considered hazardous to human health, and that concentrations that high or higher may be present in the soil of any long-settled urbanized area. Some authorities recommend soil lead concentration limits as low as 50 or 100 ppm that would give some margin of safety.

Old orchard sites are another setting where lead concentrations in soil may be high because acid lead arsenate was used as an insecticide prior to the introduction of DDT (Peryea and Creger, 1994). Even sports enthusiasts have distributed lead in soil and water through the lead shot used in ammunition and the lead weights used for fishing sinkers. Recent laws stopping their use do nothing to clean up the lead materials distributed in past years. More than 11 million tons of lead were used in the United States alone during the twentieth century, mostly for gasoline additives and as an ingredient in paint (Ma et al., 1995).

Lead phosphates are the least soluble and therefore most stable inorganic lead compounds in soils, but lead is more closely associated with the soil organic matter than with the mineral matter. Soil organic matter complex can lead and hold it for very long periods of time. Fowler et al. (1995) suggest that the mean residence time for lead can be as long as 2000 years in organic soils. Release of the complexed lead depends greatly on soil pH and increases as the pH becomes more acidic. Released lead can be absorbed by plant roots and thus enter the food chain.

Zinc, Copper, and Nickel. These three elements are micronutrients that are normally found in trace amounts in soil, plants, and animals. They become toxic to animals or humans that eat plant material from the soils, or to the plants themselves if excessive amounts are present. The source most likely to increase their soil concentration to toxic levels is sewage sludge. The zinc equivalent concept, as discussed in Section 16–3.2, has been used to estimate the maximum amount of sewage sludge that can safely be applied to a soil. Other more localized sources are mine tailings and seepage from waste deposit sites. All three of these metals are used for electroplating and other metallurgical purposes, so metallurgical plants are potential sources for polluting the soil in their nearby vicinity. Ali et al. (1992) found heavy metal concentrations in the soil 5000 ft (1500 m) from a lead and zinc smelter in Egypt that ranged up to 500 ppm Pb, 1200 ppm Zn, and 50 ppm Ni and Cd. Elevated concentrations of these metals extended more than 3 miles (5 km) from the smelter. Dust on the leaves of plants indicated that at least part of the heavy-metal distribution was airborne. Plant growth is reduced but continues in such soil because the heavy metals are held strongly by the soil solids and are much less concentrated in solution.

Misra et al. (1994) showed that when Co, Ni, and/or Zn were present in solution at concentrations near 2 ppm, they stimulated root elongation of the faba beans they were testing. But higher concentrations of these metals (4–10 ppm) reduced root elongation. Ideally, one would never allow heavy-metal contents to reach excessive levels in soils. However, something needs to be done with the sewage sludge and other waste materials that may contain heavy metals. Chemical tests can determine which sludges contain unacceptable metal concentrations, and these can be treated, for example, by acid leaching to remove much of their heavy-metal content. Even if the metal concentration does build up in a soil, it may be possible to reduce plant uptake of the heavy metals by applying lime. Raising the soil pH from acidic to near neutral can reduce the heavy-metal concentrations in solution and thereby make it possible to grow crops where the metal ions would otherwise be toxic.

Arsenic. Arsenic is a semimetallic element with chemical behavior much like that of phosphorus because it occurs directly beneath phosphorus in the periodic table. It occurs in many forms and is present in small amounts in most soils and in plant and animal tissue. Arsenic concentrations up to 20 ppm are quite common in soil, and up to 40 ppm may be considered within the normal range. However, a few soils in arid climates have accumulated much higher natural arsenic concentrations that are toxic to plants and/or animals. Contaminated soils near mining or smelting sites may have arsenic concentrations up to several thousand ppm that can cause the sites to be barren of life.

Arsenic's toxic properties have been known and used for centuries, and it is also a carcinogen (Oliver, 1997). It is most abundant in the oxidized As(V) form, but becomes many times more toxic when reduced to the As(III) form. Korte and Fernando (1991) explain that the high toxicity of As(III) is related to its ability to react with sulfhydryl groups and thereby be held in living tissue longer than other forms. Various arsenic compounds have been used as bactericides, herbicides, and insecticides. Before DDT was introduced, acid lead arsenate ($PbHAsO_4$) was the most widely used insecticide in deciduous fruit orchards (Peryea and Creger, 1994). Orchards were sprayed several times each season with high rates of arsenic compounds. The drippage from the trees loaded the soil with arsenic and lead that may continue to reduce plant growth in the worst spots for decades. Organic methyl arsenic compounds were produced and used after that time, and some of these are still in use in some places.

Arsenic substitutes for phosphorus to a small degree in phosphate rock and therefore is a likely component of phosphorus fertilizers and other materials made from phosphate rock. Arsenic from either phosphorus fertilizer or soil storage sites can be absorbed by plant roots and incorporated into plant tissue. Jiang and Singh (1994) found that using NPK fertilizer spiked with 0.3% As in the form of sodium arsenite increased the As concentration in food and fodder crops so much that they could not be used safely, even though the arsenic did not depress the yield of the ryegrass and barley.

Areas where the soil may be contaminated with arsenic include mining sites, metal-smelting sites, waste-disposal sites from chemical producers or tanneries, and others. Peryea and Creger (1994) present data showing that high arsenic concentrations gradually leach down into the subsoil. Such leaching might eventually contaminate the groundwater. The state of Washington now recommends soil remediation for soils containing more than 20 ppm As based on recent studies indicating that arsenic in food can be carcinogenic (Peryea and Creger, 1994).

Selenium. Selenium is related to sulfur, much as arsenic is related to phosphorus. A small amount of selenium is essential for human health (Oliver, 1997), but excessive amounts are toxic. Like arsenic, selenium can accumulate to toxic concentrations in some soils of arid regions. With selenium, however, the toxicity depends on the vegetation as well as the soil, and it affects only animals. Certain plants, commonly known as *locoweed,* are *selenium accumulators.* That is, they absorb selenium preferentially and thus concentrate the soil selenium. Locoweed growing on soil that is unusually high in selenium produces forage that can cause cattle to behave erratically (go "loco") in a condition known as the "blind staggers." Control techniques include efforts to eradicate the locoweed or to use fencing or other means to keep cattle away from the affected areas. Another approach is to fertilize the area heavily with sulfur, so the plants will absorb more sulfur and less selenium.

Selenium toxicity has been reported to have affected up to half of the human population in the worst-affected areas of China (Neal, 1995). A combination of circumstances caused the people to rely heavily on locally grown vegetables in an area where the soil was high in selenium. The most common symptoms were loss of hair and nails.

Other Toxic Elements. Toxicity problems are known to have occurred with several other elements, including beryllium, boron, chromium, cobalt, fluorine, mercury, molybdenum, uranium, and vanadium. However, problems with these elements in soil are rare and will not be discussed here.

16–7.4 Toxic Chemicals

Some toxicity is related to chemical compounds rather than to the elemental components of the compounds. This is especially true of organic compounds and is the reason that organic pesticides are so potent. There are large numbers of such compounds, but they can be grouped into a few classes for discussion purposes. The principal groups of organic chemicals known to have caused significant soil pollution problems are pesticides, petroleum products, and PAHs.

Pesticides. Pesticides were discussed in Section 16–5.4 and will also be considered in Chapter 17 as a factor in water pollution, so only a short section is needed here. Modern organic pesticides are highly potent and are applied at very low rates. One of the properties of an ideal pesticide is that it can be decomposed in the soil after it is no longer needed to protect the crop for which it is applied. Thus, they should not cause much soil pollution when they are properly handled and used. Indeed, most pesticide problems result from either improper handling or from atmospheric drift that allows them to get where they are not wanted. Improper handling of pesticides can create soil pollution problems that are pertinent for discussion here.

Some of the early pesticides, such as DDT, were quite persistent and therefore more likely to cause soil pollution problems. Some developing nations still use persistent pesticides because of their low cost, but most of the pesticides now in use are less persistent. Nevertheless, it is possible to overload the soil with any pesticide where a spill occurs, where a container is discarded, or perhaps where the rinse water from a sprayer is dumped. Also, it is possible for a pesticide user to miscalculate the amount of pesticide needed and to apply excessively high rates in spite of all the warnings and regulations. Such problems are not limited to the agricultural sector; significant amounts of pesticides are applied to lawns and gardens, often by users with less experience than most farm users. Excessive pesticide rates can kill the soil microbes that ordinarily would decompose the pesticide, thus making it much more persistent than normal in such sites.

Many things can happen to a pesticide in the soil. Most of them have reactive groups that can be attracted to either cation or anion exchange sites in the soil. Soil organic matter has a complex structure and many exchange sites that can interact with pesticides and hold them, so the amount of organic matter a soil contains commonly affects the activity of pesticides in the soil. In fact, some pesticides are supposed to be applied at higher rates in soils that are higher in organic matter. A pesticide molecule that is held to a soil particle is less likely to kill a pest, cause a soil pollution problem, or leach into the groundwater. It will probably remain in place until it is found by a soil organism that can decompose it. Unfortunately, the decomposition process is not always completed in a short time, and some pesticides (DDT, for example) produce decomposition products that are more toxic than the pesticides themselves. In addition to the microbial population, environmental factors such as soil temperature, water and air content, and pH can make a difference in how fast a pesticide will decompose. Sethunathan (1989) discusses the decomposition process and indicates that a supply of organic residues that provide an energy source for the soil microbes is a vital factor for rapid decomposition of pesticides.

Petroleum Products. Petroleum products contaminate the soil all too often through spills, leaks, and discards. Buried gasoline storage tanks have become a concern in recent years because they eventually rust through and leak. Even an "empty" tank usually holds enough petroleum product to pollute adjacent soil and water. Simple hydrocarbon chains would not be much of a problem for soil microbes to decompose, but oil and gasoline contain many more complex compounds that are more likely to resist decomposition. Also, many of the sites are so heavily polluted that the soil is sterilized.

Crude oil may be worse than the refined products for soil pollution. For example, the oil that is washed up on beaches from a tanker spill causes a great deal of damage to soils as well as to water and wildlife. Probably the worst impact ever of crude oil on soil was the damage caused in 1990 by the Iraqi army when it ignited more than 600 oil wells during its retreat from Kuwait (Al-Houty et al., 1993). Soot and other air pollutants, estimated at half a million tons per day, clouded the sky and settled on vegetation, soil, and water for long distances downwind. The heat of the fires seared the surface of the soils, and gushing oil created oil lakes in depressions near the oil wells. Oil from the lakes seeped into the soil to depths of as much as a meter where it did not encounter a shallower restrictive layer. In the dry climate of Kuwait, these soils are likely to remain polluted for a very long time. Oil on soil causes water repellency that may last for 50 years or more (Roy and McGill, 2000), and contaminated spots may expand through soil mixing, liquid seepage, or vapor transport.

PAHs. PAHs (polycyclic aromatic hydrocarbons) are a group of organic compounds containing two or more interlocking ring structures. PAHs have both carcinogenic and mutagenic properties. They are produced by decomposition of organic materials, especially when materials such as rubber or plastic are burned. They occur both naturally and as a result of human activities, and are found in air, water, and soil and in plant, animal, and human tissue (Menzie et al., 1992). Most humans probably consume a few micrograms of PAHs in their food every day, with those who consume a lot of meat, leafy vegetables, or unrefined grains being in the upper part of the range. Tobacco smoke also makes a significant input of PAHs.

The organic structure of PAHs is so complex and variable that they decompose very slowly in soil. Even very small inputs therefore result in detectable concentrations in normal soils. Human activities can increase PAH levels in soil from natural concentrations of a few parts per billion to hazardous levels of hundreds or even thousands of parts per billion. The highest measured concentrations come from road dust with PAH levels of a few hundred parts per million (Menzie et al., 1992).

16–8 HAZARDOUS SITES AND SOIL REMEDIATION

Some soil areas become so polluted that they constitute an intolerable hazard to human and animal life. Toxic chemicals may leach into groundwater that eventually becomes drinking water, and heavy-metal cations may be absorbed by growing plants in amounts that can be toxic to animals and humans that eat the plants. Some chemicals may vaporize from the soil and become air pollutants, or children and animals may consume toxic chemicals by ingesting soil particles.

The Love Canal site mentioned in Section 16–4.2 is one example of a site abandoned because of excessive health hazard, and there are many other less-publicized sites that have chemical contamination. Similarly, the atomic age has created a number of sites where radiation is a hazard. These and other hazardous sites need to be identified, marked, and closed off from ready access until such time as corrective measures can be taken. Furthermore, measures must be taken to reduce pollution so that the number of new hazardous sites is minimized.

16–8.1 Superfund

The U.S. Congress enacted the Comprehensive Environmental Response, Compensation, and Liability Act (CERCLA) and established Superfund in 1980 to provide a mechanism for remediating hazardous sites. This law permits legal action to be taken to require polluters to clean up the contamination they have caused. *Superfund* is a trust fund that the Environmental Protection Agency (EPA) can use to clean up sites where the polluter either cannot be identified or cannot be forced to accomplish the needed work. By 1990, the EPA had logged more than 32,000 hazardous waste sites in the United States and identified 1236 of them as the most serious in the nation (Holmes et al., 1993). Satisfactory cleanup had been verified on only 29 of the 1236 sites by 1990. This work will have to continue for many years to make up for centuries of unregulated pollution.

Many Superfund sites were caused by wastes from chemical and petroleum industries, while others were landfill sites containing chemical products that should never have been discarded there. Transportation spills and radioactive sites also contributed to the list. The largest Superfund sites are four mining areas in western Montana known as the Clark Fork sites. More than a century of mining, milling, and smelting contaminated about 50,000 ac (20,000 ha) of soil along 140 mi (225 km) of the Clark Fork River and its tributaries with arsenic and heavy metals, including cadmium, copper, lead, and zinc (Miller, 1992).

16–8.2 Soil Remediation

Several different procedures are used to remediate polluted soil, depending on the kind of pollutant, the nature of the soil, and the surrounding environment. All of these procedures are expensive, largely because even a small polluted area involves a large mass of soil. The simplest procedure of all is one that is chosen fairly often—excavation and removal of the affected soil followed by replacement and regrading as necessary. Of course, this really is an area remediation procedure, rather than soil remediation. The polluted soil is transferred to another site.

Remediation procedures that actually remove or inactivate pollutants from soil include incineration, chemical treatment, and bioremediation. Incineration is better adapted to the disposal of mostly combustible wastes, but it is sometimes used for small volumes of soil polluted with organic contaminants. Organic materials are thereby converted to mostly water and carbon dioxide, though some air pollutants such as CO, NO_2, SO_2, and Cl compounds may be released. Materials containing large amounts of heavy metals generally should not be incinerated because some of the heavy metals may be volatilized as air pollutants. Emissions of gases and of particulate matter can be minimized by the use of scrubbers, filters, and electrostatic precipitators (Bache et al., 1991). Unfortunately, the soil organic matter and soil structure are also destroyed in the incineration process.

Chemical Treatment for Soil Remediation. Heavy-metal availability and leachability is strongly influenced by soil pH. It is therefore possible to remove heavy metals from soil by acid leaching and/or to reduce their availability by liming. Liming is by far the least expensive choice, if it provides adequate treatment by itself. The main requirement then is to add enough lime and to mix it well with the soil. Acid leaching is much more expensive because it requires both acid and lime, and it usually requires removal of the soil to some kind of container for the leaching process. Furthermore, the acid leachate must be contained and properly handled for safe disposal.

A more sophisticated chemical remediation process, known as *supercritical fluid extraction* (SFE), has only been used on a small scale (5-kg samples) but could be very desirable if it can be expanded to a more useful scale without becoming too expensive (Jain, 1993). Supercritical fluids are very potent solvents for organic compounds and therefore can be used for extraction of almost any organic pollutant. Pesticide and petroleum product spills would be likely candidates for such treatment. The procedure is a closed process in which the soil is leached with the extractant (probably carbon dioxide at 31°C and 74 bars; possibly water at 374°C and 221 bars), and then the pressure is reduced and enough heat supplied to vaporize the extractant. The extracted materials precipitate, and the extractant is available for reuse. Some of the soil organic matter may have been extracted, but the soil structure suffers much less damage than that caused by incineration, and the soil can be returned to the site from which it was taken.

A temporary method of chemical remediation was proposed by Torres and De Varennes (1998). They added beads of a polyacrylate polymer to a sandy soil that had been contaminated with copper. The polymer tied up the copper ions and rapidly lowered the concentration of copper in solution. They suggested that use of the polymer would be economically feasible for the establishment of a new vineyard or orchard in areas where the soil has been contaminated by the use of copper insecticides. The copper ions will be released again as soil microbes decompose the polymer, but that should take at least a few months and thus permit the new vines or trees to become established with root systems penetrating beyond the affected zone.

Bioremediation of Soil. Microbes and plants are useful allies to remediation workers. Microbes are well known for their ability to decompose organic chemicals. Certain plants have strong tendencies to absorb large amounts of heavy metals from soil; harvesting the plant material then removes significant amounts of these metals from the soil. It may take several cropping cycles to lower the metal concentration in the soil, but this may still be much less expensive than any other extraction procedure. The harvested plant material will need some kind of disposal or treatment, but that is much more manageable than disposal or other treatment of the soil. Shrimp et al. (1993) point out that using plants in this way has several advantages. This method produces an aesthetically pleasing effect. The plant material can be sampled and tested to show the stage of remediation. Plant uptake of water and contaminants reduces leaching of contaminants into deeper soil layers. Similar use of plants to remove excess nutrients from sewage materials applied to soil is discussed in Section 16–3.2.

Several approaches using bacteria for soil remediation are explained by Ritter and Scarborough (1995). Some soil microbes have an inherent ability to biodegrade a wide variety of organic contaminants. The presence of these contaminants stimulates increased activity of such microbes. Some approaches involve inoculating the soil with microbes that are specifically adapted to decomposing the pollutants present. This approach might be useful where a soil has been contaminated by a petroleum product spill or a pesticide spill that has soaked into the soil and needs to be degraded. Other approaches involve making the soil environment more favorable for microbial activity, perhaps by supplying oxygen through air sparging or the addition of hydrogen peroxide (Figure 16–6). Supplying easily decomposable organic materials as an energy source for the microbes may also serve as a stimulant to their activity. This kind of treatment is likely to be more useful on an older or more chronic contaminant to which the soil microbes have had time to adapt. If possible, the soil is treated *in situ,* but excavation followed by treatment and replacement is sometimes necessary.

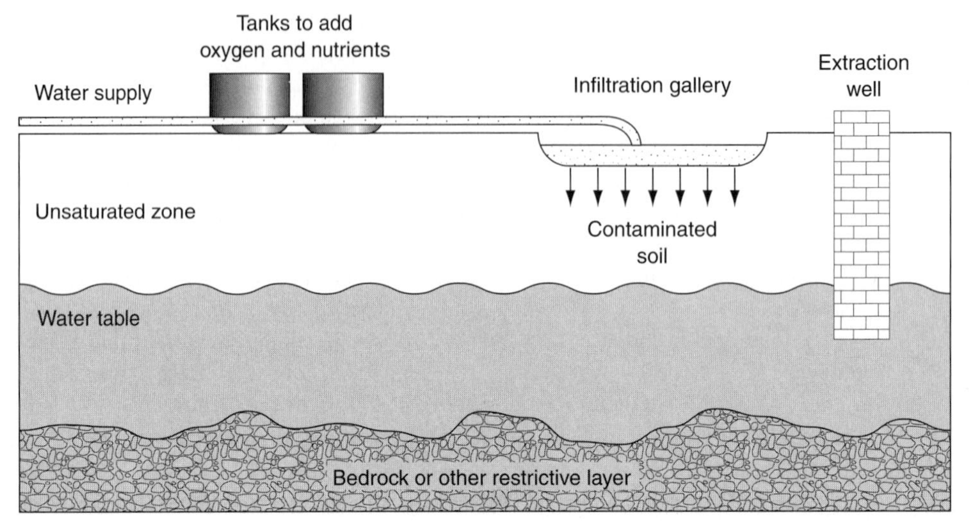

Figure 16–6 An *in situ* bioremediation setup using native soil microbes that have adapted to the contaminated environment. A similar setup using an injection well instead of the infiltration gallery can be used for bioremediation of the soil below a water table. (Based on Tursman and Cork, 1992.)

SUMMARY

Pollutants contaminate and degrade the environment. People have always caused pollution, but significant concern about pollution is a recent development related to increasing population, new types of pollutants, and increasingly sensitive detection techniques.

People-related wastes are directly related to population. Solid waste has been discarded in dumps or buried in landfills but is beginning to be viewed as a resource containing reusable materials. Sewage treatment varies from none to tertiary treatment that removes dissolved ions. Disposal by sprinkler irrigation on land is the favored method of tertiary treatment, but heavy-metal ions can be a problem.

Industrial wastes include scrap and by-product materials, some of which are highly toxic. When feasible, solid materials should be recycled rather than placed in landfills. Liquid industrial wastes include chemicals that must be removed and water that must be cooled before it can be returned to a stream or recycled in a closed system. Air pollution also pollutes soil and water because rain and snow bring sulfur dioxide, nitrogen oxides, and solid particles back to land or water. The long life and high toxicity of nuclear wastes make them the most difficult disposal problem of all.

Agricultural waste disposal is complicated by the large amounts of wastes and the large areas involved. Manure has fertilizer value that is often wasted. Losses can be minimized by keeping the manure anaerobic, protecting it from leaching, and working it into the soil as soon as applied. Manure application rates up to 10 tons/ac make effective use of its fertilizer value. Higher rates can be applied for disposal purposes, but pollution problems can arise.

Excessive fertilizer rates can raise the nutrient content of both runoff and drainage water above natural levels. Nitrate nitrogen is the nutrient most likely to be lost in large amounts in water. Nitrates may also be lost from wet soil by denitrification. Denitrification can be reduced by draining wet spots or by using certain chemicals.

Accumulations of copper and other heavy metals used in early pesticides are difficult to remove from soil. Organic pesticides developed since the 1940s are applied at lower rates than the metal-based materials and have more selective toxicity, but some, such as DDT, cause trouble by bioaccumulation. Some pests can be controlled by biological methods including crop breeding, natural predators, synthetic sex attractants, and sterile males.

Suspensions of small solid or liquid particles in the air are called *aerosols*. Radioactive dust, PAHs, and a number of other aerosols can be serious soil and water pollutants where they are deposited.

Sediment becomes a soil pollutant when unfavorable soil material is deposited on top of good soil. Soil acidification is the most widespread form of soil pollution and can be toxic to plants. Some of it comes from acid rain that has caused widespread damage to forests in eastern Europe and in the eastern United States and Canada. Several different elements, mostly heavy-metal cations, are toxic even at low soil concentrations to plant growth or to animals and humans that eat the plants. Other more-complex toxic chemicals include pesticides, petroleum products, and PAHs.

Highly polluted soils can become hazardous sites to human and animal life. *Superfund* was established in 1980 to identify the most serious sites and remediate them. The soil may be treated by incineration, chemical treatment such as acid leaching or supercritical fluid extraction, or bioremediation using plants and/or microbes to decompose or remove the pollutants.

QUESTIONS

1. What is a pollutant?
2. Why weren't people concerned about pollution centuries ago?
3. Is it economically feasible to replace sanitary landfills with recycling programs?
4. What are the most hazardous forms of industrial wastes?
5. How much value does manure have as a fertilizer? Why is this value often ignored?
6. How should manure be handled for maximum benefit and minimum pollution?
7. How can denitrification be reduced?
8. What pollution problems are caused by sediment?
9. List five toxic elements and three classes of toxic chemicals that are soil pollutants.
10. What processes are useful for remediating polluted soil?

REFERENCES

AITKEN, M. N., B. EVANS, and J. G. LEWIS, 1998. Effect of applying paper mill sludge to arable land on soil fertility and crop yields. *Soil Use and Management* 14:215–222.

AL-HOUTY, W., M. ABDAL, and S. ZAMAN, 1993. Preliminary assessment of the Gulf War on Kuwaiti desert ecosystem. *J. Environ. Sci. Health* A28:1705–1726.

ALI, E. A., Y. H. IBRAHIM, and M. M. NASRALLA, 1992. Contamination of the agricultural land due to industrial activities southern of greater Cairo. *J. Environ. Sci. Health* A27:1293–1304.

BACHE, C. A., W. H. GUTENMANN, M. RUTZKE, G. CHU, D. C. ELFVING, and D. J. LISK, 1991. Concentrations of metals in grasses in the vicinity of a municipal refuse incinerator. *Arch. Environ. Contam. Toxicol.* 20:538–542.

BOLLAG, J. M., and N. M. HENNINGER, 1976. Influence of pesticides on denitrification in soil and with an isolated bacterium. *J. Environ. Qual.* 5:15–18.

BRINKMAN, R., 1994. Lead pollution in soils in Milwaukee County, Wisconsin. *J. Environ. Sci. Health* A29:909–919.

CARTER, F. W., and D. TURNOCK, 1993. Introduction. Ch. 1 in F. W. Carter and D. Turnock (eds.), *Environmental Problems in Eastern Europe.* Routledge, London, 249 p.

COLTEN, C. E., and P. N. SKINNER, 1996. *The Road to Love Canal.* Univ. of Texas Press, Austin, 217 p.

COOKSON, W. R., J. S. ROWARTH, and K. C. CAMERON, 2000. The fate of residual nitrogen fertiliser applied to a ryegrass (*Lolium perenne* L.) seed crop. *Aust. J. Agric. Res.* 51:287–294.

DUNIGAN, E. P., R. A. PHELAN, and C. L. MONDART, JR., 1976. Surface runoff losses of fertilizer elements. *J. Environ. Qual.* 5:339–342.

EGHBALL, B., and J. F. POWER, 1999. Composted and noncomposted manure application to conventional and no-tillage systems: Corn yield and nitrogen uptake. *Agron. J.* 91:819–825.

ENTZ, M. H., W. J. BULLIED, D. A. FORSTER, Robert GULDEN, and J. K. VESSEY, 2001. Extraction of subsoil nitrogen by alfalfa, alfalfa-wheat, and perennial grass systems. *Agron. J.* 93:495–503.

FOWLER, D., R. MOURNE, and D. BRANFORD, 1995. The application of ^{210}Pb inventories in soil to measure long-term average wet deposition of pollutants in complex terrain. *Water, Air, and Soil Pollut.* 85:2113–2118.

FULLER, W. H., and A. W. WARRICK, 1984. *Soils in Waste Treatment and Utilization.* CRC Press, Boca Raton, FL, 288 p.

GAST, R. G., W. W. NELSON, and G. W. RANDALL, 1978. Nitrate accumulation in soils and loss in tile drainage following nitrogen applications to continuous corn. *J. Environ. Qual.* 7:258–261.

HOLMES, G., B. R. SINGH, and L. THEODORE, 1993. *Handbook of Environmental Management and Technology.* Wiley & Sons, New York, 651 p.

JAIN, V. K., 1993. Supercritical fluids tackle hazardous wastes. *Environ. Sci. Technol.* 27:806–808.

JIANG, Q. Q., and B. R. SINGH, 1994. Effect of different forms and sources of arsenic on crop yield and arsenic concentration. *Water, Air, and Soil Pollut.* 74:321–343.

KORTE, N. E., and Q. FERNANDO, 1991. A review of arsenic (III) in groundwater. *Crit. Rev. Environ. Control* 21:1–39.

LARSON, W. E., J. R. GILLEY, and D. R. LINDEN, 1975. Consequences of waste disposal on land. *J. Soil Water Cons.* 30:68–71.

LAUER, D. A., D. R. BOULDIN, and S. D. KLAUSNER, 1976. Ammonia volatilization from dairy manure spread on the soil surface. *J. Environ. Qual.* 5:134–141.

LESTER, J. N., 1990. Sewage and sewage sludge treatment. Ch. 3 in R. M. Harrison (ed.), *Pollution: Causes, Effects, and Control,* 2nd ed. Royal Society of Chemistry, Cambridge, U.K., 393 p.

MA, Q. Y., T. J. LOGAN, and S. J. TRAINA, 1995. Lead immobilization from aqueous solutions and contaminated soils using phosphate rocks. *Environ. Sci. Technol.* 19:1118–1126.

MASON, C. F., 1990. Biological aspects of freshwater pollution. Ch. 6 in R. M. Harrison (ed.), *Pollution: Causes, Effects, and Control,* 2nd ed. Royal Society of Chemistry, Cambridge, U.K., 393 p.

MENZIE, C. A., B. B. POTOCKI, and J. SANTODONATO, 1992. Exposure to carcinogenic PAHs in the environment. *Environ. Sci. Technol.* 26:1278–1284.

METCALF, R. L., 1971. Pesticides. *J. Soil Water Cons.* 26:57–60.

MILLER, S., 1992. Cleanup delays at the largest Superfund sites. *Environ. Sci. Technol.* 26:658–659.

MISRA, J., V. PANDEY, and N. SINGH, 1994. Effects of some heavy metals on root growth of germinating seeds of *Vicia faba. J. Environ. Sci. Health* A29:2229–2234.

MUELLE, P. W., D. C. PASCHAL, R. R. HAMMEL, S. L. KLINCEWICZ, M. L. MACNEIL, B. SPIERTO, and K. K. STEINBERG, 1992. Chronic renal effects in three studies of men and women occupationally exposed to cadmium. *Arch. Environ. Contam. Toxicol.* 23:125–136.

NAVA, H., 2001. Wastewater reclamation and reuse for aquaculture in Perú. *J. Soil Water Cons.* 56:81–87.

NEAL, R. H., 1995. Selenium. Ch. 12 in B. J. Alloway (ed.), *Heavy Metals in Soils,* 2nd ed. Blackie Academic & Professional, London, 368 p.

OLIVER, M. A., 1997. Soil and human health: A review. *Eur. J. Soil Sci.* 48:573–592.

PERYEA, F. J., and T. L. CREGER, 1994. Vertical distribution of lead and arsenic in soils contaminated with lead arsenate pesticide residues. *Water, Air, and Soil Pollut.* 18:297–306.

PIOTROWSKA, M., and S. DUDKA, 1994. Estimation of maximum permissible levels of cadmium in a light soil by using cereal plants. *Water, Air, and Soil Pollut.* 73:179–188.

RITTER, W. F., and R. W. SCARBOROUGH, 1995. A review of bioremediation of contaminated soils and groundwater. *J. Environ. Sci. Health* A30:333–357.

ROBERT, M., 1995. Aluminum toxicity: A major stress for microbes in the environment. Ch. 17, p. 227–242 in P. M. Huang, J. Berthelin, J.-M. Bollag, W. B. McGill, and A. L. Page (eds.), *Environmental Impact of Soil Component Interactions,* Vol. II, Lewis Pub., Boca Raton, FL, 283 p.

RODHE, H., P. GRENNFELT, J. WISNIEWSKI, C. ÅGREN, G. BENGTSSON, K. JOHANSSON, P. KAUPPI, V. KUCERA, L. RASMUSSEN, B. ROSSELAND, L. SCHOTTE, and G. SELLDEN, 1995. Acid reign '95—conference summary statement. *Water, Air, and Soil Pollut.* 85:1–14.

ROY, J. L., and W. B. MCGILL, 2000. Investigation into mechanisms leading to the development, spread and persistence of soil water repellency following contamination by crude oil. *Can. J. Soil Sci.* 80:595–606.

SÁ, J. C. M., C. C. CERRI, W. A. DICK, R. LAL, S. P. VENSKE FILHO, M. C. PICCOLO, and B. E. FEIGL, 2001. Organic matter dynamics and carbon sequestration rates for a tillage chronosequence in a Brazilian Oxisol. *Soil Sci. Soc. Am. J.* 65:1486–1499.

SADEGHI, A. M., A. R. ISENSEE, and D. R. SHELTON, 1998. Effect of tillage age on herbicide dissipation: A side-by-side comparison using microplots. *Soil Sci.* 163:883–890.

SETHUNATHAN, N., 1989. Biodegradation of pesticides in tropical rice ecosystems. Ch 5.2 in P. Bourdeau, J. A. Haines, W. Klein, and C. R. Krishna Murti (eds.), *Ecotoxicology and Climate.* Wiley and Sons, Chichester, U.K., 392 p.

SHAW, W. C., H. E. HEGGESTAD, and W. W. HECK, 1971. Pollution poses threat to man, farms, nature. In *A Good Life for More People,* Yearbook of Agriculture, U.S. Govt. Printing Office, Washington, D.C., p. 293–299.

SHRIMP, J. F., J. C. TRACY, L. C. DAVIS, E. LEE, W. HUANG, L. E. ERICKSON, and J. L. SCHNOOR, 1993. Beneficial effects of plants in the remediation of soil and groundwater contaminated with organic materials. *Crit. Rev. Environ. Sci. and Tech.* 23:41–77.

SMITH, K. A., A. G. CHALMERS, B. J. CHAMBERS, and P. CHRISTIE, 1998. Organic manure phosphorus accumulation, mobility and management. *Soil Use and Management* 14:154–159.

SOMMERFELDT, T. G., U. J. PITTMAN, and R. A. MILNE, 1973. Effect of feedlot manure on soil and water quality. *J. Environ. Qual.* 2:423–427.

SOMMERS, L. E., 1977. Chemical composition of sewage sludges and analysis of their potential use as fertilizers. *J. Environ. Qual.* 6:225–232.

STIGLIANI, W. M., P. R. JAFFÉ, and S. ANDERBERG, 1993. Heavy metal pollution in the Rhine basin. *Environ. Sci. Technol.* 27:786–793.

THOMAS, R., and J. P. LAW, 1977. Properties of waste waters. In *Soils for Management of Organic Wastes and Waste Waters,* Soil Science Society of America, American Society of Agronomy, and Crop Science Society of America, Madison, WI, p. 45–72.

TORRES, M. O., and A. DE VARENNES, 1998. Remediation of a sandy soil artificially contaminated with copper using a polyacrylate polymer. *Soil Use and Management* 14:106–110.

TURSMAN, J. F., and D. J. CORK, 1992. Subsurface contaminant bioremediation engineering. *Crit. Rev. Environ. Control* 22:1–26.

TYAGI, R. D., J. F. BLAIS, B. BOULANGER, and J. C. AUCLAIR, 1993. Simultaneous municipal sludge digestion and metal leaching. *J. Environ. Sci. Health* A28:1261–1379.

VAN CLEEMPUT, O., 1971. Etude de la denitrification dans le sol. *Pedologie* 21:367–376.

WILLIAMS, C. F., S. D. NELSON, and T. J. GISH, 1999. Release rate and leaching of starch-encapsulated atrazine in a calcareous soil. *Soil Sci. Soc. Am. J.* 63:425–432.

XU, C., M. J. SHAFFER, and M. AL-KAISI, 1998. Simulating the impact of management practices on nitrous oxide emissions. *Soil Sci. Soc. Am. J.* 62:736–742.

CHAPTER

17

WATER QUALITY AND POLLUTION

Civilization has many uses for high-quality water, and many ways of degrading water quality. Dirty water is obviously polluted, and even clear water often is not as clean and pure as it appears. A wide variety of solids, liquids, and gases may be dissolved or suspended in the water. Typical water contains small concentrations of dissolved cations and anions that help supply the nutrients that living organisms need. However, when the water is polluted with unusually high concentrations of dissolved or suspended materials, or even with small concentrations of certain highly toxic materials, it can be detrimental and sometimes deadly to living things. The detrimental effects of water pollutants range from barely detectable changes in growth to nearly instant death. Sometimes the effects are more hazardous to plant or animal life than to humans. Also, pollution problems such as water clouded by sediment may be aesthetically undesirable, whether they have significant health effects or not.

Water pollution affects life in many ways, not only because there are many different kinds of pollutants but also because water serves many different functions. Water is the principal component of living organisms and the basic medium for life. Most of the other components of living bodies are either dissolved or suspended in the water. Water serves as filler and as the transportation service for living cells. The temperature, pressure, pH, and ionic concentrations in the water must be within certain limits to sustain life. Pollution damage can result from interference with any part of the relationship between water and living things.

17–1 THE EARTH'S WATER SUPPLY

Water in the oceans, rivers, streams, lakes, and ponds covers more than 70% of the surface of the Earth. Water is also present in the soil and rocks that cover the remainder of the surface and is a significant component of the atmosphere. It occurs naturally in solid, liquid, and vapor phases, and cycles readily from one phase to another as well as from one location to another. The cycling process can purify water by leaving pollutants behind when it

TABLE 17–1 ESTIMATED AMOUNTS OF WATER IN VARIOUS FORMS PRESENT ON EARTH

	km^3	% of total
Water in plants and animals	1,100	0.000,1
Water vapor	13,000	0.000,9
Freshwater in lakes, ponds, rivers, and streams	100,000	0.007
Groundwater in soil and rocks	8,300,000	0.6
Water in glaciers and other ice and snow	27,500,000	2.0
Water in oceans and seas	1,350,000,000	97.4

Source: Based on data from Speidel and Agnew, 1988.

evaporates, but other parts of the cycle allow water to transport pollutants as it flows from one place to another. Some parts of the cycle, such as precipitation, occur rapidly, whereas other parts, such as glaciers and some bodies of groundwater, hold water for long epochs. Water that circulates or cycles from one place to another is generally of most interest to humans because it can be diverted and used for various purposes. Estimated amounts of the several forms of water present on Earth are given in Table 17–1.

17–1.1 Life in Water

Plants and animals require water to sustain life no matter where they live. Many of them live in aquatic environments. Aquatic life ranges from the largest living animals (whales) to the smallest living entities (viruses and bacteria). The impact of pollution on aquatic life is important, not only because it may be a life or death matter for all of these organisms, but also because they are part of the food chain for land animals and human beings. A conspicuous example of this is the way DDT in water becomes concentrated in fish and still more concentrated in birds that eat the fish, with the ultimate result being eggs with shells so thin that young birds fail to develop and hatch. Another example to be discussed later in this chapter is the accumulation of heavy metals that can make seafood toxic to human life.

Tuna, salmon, crabs, oysters, shrimp, and other seafoods originate in the water, not in the grocery store. Many forms of seafood are important economically. For example, salmon fishing, processing, and packing form the economic base for many communities in the northwestern part of the United States. However, the construction of large dams for generating electric power, irrigation, flood control, and recreation has isolated many former salmon spawning areas from the ocean and thus reduced their habitat. Urbanization, agriculture, logging, and fishing interests have taken priority over the salmon. The Columbia River salmon and steelhead catch peaked at 43 million pounds per year in the 1880s but declined to 1.2 million pounds in the 1980s (Petit, 1994). Farther south, in the Klamath Basin of California and Oregon, Sletteland (1995) reports that at least 10 of 54 salmon and sea-run trout populations are facing extinction. Efforts to reverse this trend have been made by providing fish ladders to bypass the dams and fish hatcheries to restock the streams, but these have had little long-term success. The needs of salmon and other fish are now among the many factors that are considered in the process of assigning water rights in the area, but they must compete with cities, irrigation projects, and other water users.

Aquatic plants include sphagnum moss that grows with its roots in water instead of soil, various forms of algae that are often considered nuisances because they turn the water red or fill it with green, slimy vegetative matter, and a host of sedges, reeds, and even trees such as cypress that grow with their roots underwater while their stems reach upward into the atmosphere. Over time, such plants convert open bodies of water into peat bogs, marshes, and swamps as the dead remains of many generations accumulate, because they are preserved by the water. Water pollution sometimes promotes this process by supplying plant nutrients that would otherwise be deficient, but it can also interfere with the process when the pollutants are toxic to plant growth.

The discovery of bacteria, and later of viruses, answered an old question about how certain diseases can be caused by drinking water. Learning how to purify the water was an important step forward in public health. Remembering this lesson is important because these microorganisms are still present in the environment and are still able to cause illness and death. Water supplies must therefore be monitored for the presence of harmful microbes. Such monitoring has shown that most water contains a small but significant population of living microorganisms, and that not all microbes are harmful.

17–1.2 Water Sources and Uses

Animals consume water from streams, lakes, ponds, and smaller reservoirs of accessible water. Humans add to this supply by forming artificial reservoirs to store water and by digging wells to obtain groundwater. Of course, all of these supplies would soon be depleted if nature did not replenish them through rainfall, runoff, and seepage. The replenishment is not always proportional to the need, however, so there are shortages of water in some times and places and surpluses at other times and places. Shortages and surpluses lead to the assignment of water rights and responsibilities that can be very expensive and cause much contention. Parts of this topic were covered in Chapter 14—Soil Drainage, and Chapter 15—Irrigation and Reclamation. It should be noted here that the effects of pollution on water quality add complications to the already complex matter of water rights. For example, a city's established water right is violated when pollution makes the water unfit for human consumption, even if the volume of water remains adequate. Examples of water-quality degradation include increased salinity that makes water less suitable for irrigation and many other purposes, warmer temperatures that may kill some or all of the fish species in lakes and streams, nutrient enrichment that leads to algal blooms and other unwanted vegetation growing in water, and the dumping of toxic chemicals that can kill fish and cause illness in animals and humans that eat the fish or drink the water.

Small bodies of water, such as that shown in Figure 17–1, are easily polluted because the amount of pollutant required is relatively small. However, even the oceans are polluted where oil spills from a tanker, large amounts of sewage are dumped, or some other source overloads a sector in the water. Shipping lanes may be contaminated by significant quantities of oil and debris, with only a small amount coming from each ship. Coastal waters are often more polluted than the rest of the ocean because they receive pollutants carried in by streams as well as from passing ships. Agricultural chemicals and sediment, sewage, industrial waste, and general trash enter from the land side. Furthermore, oil spills from tankers are often either near a coast or are washed toward a coast by wind and waves. Some pollutants are dense enough to settle to the bottom and destroy oyster beds and other sea life in coastal areas.

Figure 17–1 Polluted water is both a hazard and a disappointment for these children. (Courtesy EPA-DOCUMERICA.)

The situation is worse in water bodies like the Baltic Sea and the Black Sea that are surrounded by cities, villages, and farms. Increasing population reduces the fresh water inflow and increases the sewage and other pollutant inputs that reach these seas. Enell and Fejes (1995) consider the Baltic Sea to be the worst affected of any in the world, largely because of the eutrophying effect of the heavy load of sewage and fertilizer nitrogen it has received for decades. The situation is bad enough that the nations surrounding the Baltic Sea have agreed to jointly reduce both waterborne and airborne inputs of nitrogen from a total of about 1.4 million mt/yr (1.55 million tons/yr) to an estimated tolerable level of 600,000 mt/yr (660,000 tons/yr) (Enell and Fejes, 1995). Otherwise, the sea might soon be too polluted to sustain fish and other life. Birds and animals that depend on seafood would also be lost. Of course, any output of seafood may be too contaminated for safe consumption, and swimming beaches may have to be closed, even if the level of pollution is reduced considerably.

Larger water bodies, such as the Mediterranean Sea, and those that have better circulation with the ocean, such as the North Sea, have also suffered from pollution but have withstood eutrophication better than the more-isolated waters of the Baltic and similar seas.

17–1.3 Water Quality

Water quality is an important consideration, but there are difficulties in evaluating it. For one thing, it can be quite variable. For example, the water quality in a river during a spring runoff period may be considerably different than it is at other times. Another significant factor is the use of the water. Quality standards are quite different for drinking water than they are for irrigation water or for industrial use water. Irrigation water quality factors are discussed in Chapter 15, and industrial standards can be left to the industries involved. The emphasis here will therefore be on drinking water from municipal water systems.

The importance of a given factor in water quality may depend in part on the water treatment facilities that are available. The workers at a water treatment plant need to know what impurities are in the water coming into the plant so they can treat it appropriately, but household users are mainly concerned with the quality of the water after it has been treated, as it comes from their faucets.

Municipal water may be pumped from a nearby stream or from wells that reach the groundwater. Surface water tends to be softer, since water dissolves materials from soil and rocks on its way to the water table. However, the soil and rocks also provide a filtering action that removes coarse materials and most microbes from the water. The groundwater is likely to be more consistent in quantity and quality than surface water because it is less influenced by individual runoff events. Large cities often add stability to their water supplies by storing water in reservoirs for use during dry periods.

Water treatment plants are generally needed to assure water quality. The treatment details vary according to circumstances, but usually include one or more filtration systems, addition of coagulants and a period of settling to remove colloidal material, and the addition of chemicals to control microbial populations. Fluorides are often added to reduce tooth decay. Hard water may be softened by chemical treatment to raise the pH and precipitate calcium carbonate and iron compounds.

Water quality standards for drinking water (Table 17–2) are established by the Environmental Protection Agency in the United States and by similar agencies in other nations. Sampling frequency and methods along with testing methods are specified in detailed regulations along with rules for how to handle failure to meet standards.

Water users have several alternatives available if the quality of their water supply is inadequate to meet their personal needs or standards. One option, especially useful when traveling, is to purchase bottled water. At home, one may choose to install a water softener or a filter system. Filters range from a simple carbon filter that attaches to a faucet to a multi-stage reverse osmosis system that will even remove dissolved ions from water.

17–2 WATER POLLUTANTS

For discussion purposes, most water pollutants can be placed in one of four groups: sediment, heat, organic pollutants, and heavy metals. Natural conditions introduce some of each of these into water, but ill effects are mostly associated with the relatively high concentrations that most often result from human activities.

17–2.1 Sediment as a Pollutant

No other pollutant occurs in amounts comparable to eroded soil and rock particles. Streams in the United States carry more than 700 times as much eroded soil as sewage. Soil is an important factor in water pollution because of its large volume, the murkiness it produces, the plant nutrients, pesticides, and other polluting chemicals it carries, and the microbes that may be present.

TABLE 17–2 WATER-QUALITY STANDARDS FOR SELECTED CONTAMINANTS IN DRINKING WATER

	U.S.EPA	Canada	World Health Organization
Fecal coliforms	0	0	0
Asbestos, fibers/L	7 million		
Turbidity, turbidity units		5	5
pH		6.5-8.5	< 8.0
Aluminum, mg/L			0.2
Antimony, mg/L	0.006		
Arsenic, mg/L	0.05 (0.01 after January 23, 2006)		
Barium, mg/L	2		
Beryllium, mg/L	0.005		
Boron, mg/L		5.0	0.3
Cadmium, mg/L	0.005	0.005	0.003
Chloride, mg/L		250	250
Chromium, mg/L	0.1	0.05	0.05
Copper, mg/L	1.3 (goal)	1.0	2
Cyanide, mg/L	0.2		
Fluoride, mg/L	4	1.5	1.5
Lead, mg/L	0 (goal)	0.05	0.01
Mercury, mg/L	0.002	0.001	0.001
Nitrate as N, mg/L	10	10	
Nitrite as N, mg/L	1		
Selenium, mg/L	0.05		
Sodium, mg/L			200
Sulfate, mg/L		500	250
Thallium, mg/L	0.002		
Zinc, mg/L	5 (recommended)	5.0	3
DDT, mg/L		30	2
Heptachlor, mg/L	0.004		
Lindane, mg/L	0.0002	4.0	2
PCBs, mg/L	0.0005		
Total dissolved solids, mg/L		500	1000

Source: Compiled from Code of Federal Regulations, CFR 40, § 141.21–141.23, July 1, 2001, Barzilay et al. (1999); Chapman (1997).

Muddy Water. A small amount of clay and organic matter will make water cloudy, and some of this effect is natural. Even so, agricultural soil erosion contributes greatly to the opacity of streams and lakes, thereby affecting aquatic life that depends on light. For example, certain fish species do not thrive in murky water. The Missouri River was nicknamed "The Big Muddy" because it naturally carried enough sediment to make it cloudy. Soil erosion from the large areas of cropland in its drainage basin now adds greatly to its sediment load. Other human activities also contribute to muddy waters. For example, the Clearwater River in Idaho was so named because runoff from the forest lands through which it flows carried very little sediment, but mining operations made the water murky during the decades they were active in the area.

Soil particles ultimately settle out of muddy water, but continuing erosion typically replaces them and keeps the water cloudy. The settled particles accumulate as sediment on the beds of ponds, lakes, and streams. This reduces the storage capacities of ponds and lakes and makes streams wider and shallower. Water that becomes too shallow can be warmed by sunlight to considerably higher temperatures than exist naturally in deep water bodies. Section 17–2.4 will consider the detrimental effects of excess warming on organisms that live in water.

Eroding fields tend to have reduced infiltration rates and increased runoff. A certain amount of flooding is natural, but increased runoff flowing through channels filled with sediment greatly increases both the frequency and severity of flooding. Surging floodwaters can damage almost anything that lies in their path. Vegetation and structures are torn apart and dumped in inopportune places. The debris and sediment left behind by a flood cause large losses, both in crop and property damage and in cleanup costs.

Rapid sediment deposition is detrimental. Sediment is frequently deposited in such thick layers that plants are smothered and killed, as shown in Figure 16–5. Some sediment has much lower fertility than the soil it covers. Other sediment is so high in clay that it is unfavorable for plant growth and for tillage operations. Similar problems can occur with sediment from wind erosion. The sedimentation problems discussed in Chapters 4 and 5 can be considered forms of land pollution.

Sediment problems usually last longer in water than on land. Sedimentation damage on land is often temporary because new crops can be planted on the sediment, or other vegetation will grow and make the land productive again. However, sediment deposited in streams, ponds, and lakes displaces water in ways that have long-term effects. Sediment-filled reservoirs no longer serve well for flood control, electric-power production, and recreation. Sediment on deltas and riverbeds raises water levels and increases flood hazards. Sediment in canal and ditch systems requires costly clean-out operations. The progressive and recurring nature of sedimentation problems makes them particularly bothersome in bodies of water.

Eroded soil serves as a carrier that transports a wide variety of pollutants. These include almost anything that may be present in the soil. Soil transport is especially important for charged particles that are held on cation and anion exchange sites. These may be as simple as a calcium ion or as complex as an organic pesticide. Without erosion, these ions are held in the soil where they may serve a useful purpose, but when the soil particles are eroded, the associated ions go with them. Pesticides transported in this way are an obvious problem, but it is less obvious that even the plant nutrient ions carried into a body of water may have a negative effect.

17–2.2 Plant Nutrients in Sediment

Sediment carries much larger quantities of plant nutrients and other chemicals than are dissolved in the water. Several studies have shown that most of the nitrogen and more than 90% of the phosphorus moving from fields into streams is carried by sediment. These and other nutrients are present in both the organic and the mineral fractions of sediment.

Nitrogen carried by sediment is mostly organic and must be mineralized before the nitrogen will be released into the water. Mineralization provides a gradual but long-lasting input of nitrates. Nitrates in water promote plant growth both in the water and on the banks of streams and ponds. Nitrates are also a concern in drinking water because excessive amounts can cause methemoglobinemia (sometimes called "blue baby disease"). The limit on the nitrate nitrogen content of drinking water is commonly taken as 10 ppm. Excess nitrogen fertilization is a significant cause of excess nitrates in water. Erosion control coupled with reducing high nitrogen fertilizer rates to optimum levels and applying the fertilizer at the right time can reduce the nitrate content of runoff water (Shankar et al., 2000).

The phosphorus content of surface waters is important because it is usually the easiest nutrient to control adequately to prevent eutrophication. Sediment and sewage are the two principal sources of phosphorus in water. A combination of soil-conserving practices and sewage treatment, therefore, is the best way to control eutrophication. Proper fertilization can help reduce pollution by increasing plant growth enough to reduce erosion and the associated nutrient losses.

Filter strips are effective for removing both sediment and plant nutrients from water. A vigorous growth of perennial sod-forming grass between a field or feedlot and a stream or pond can trap nearly all of the sediment and even remove much of the dissolved nutrients from runoff water. A typical strip width is about 30 ft (9 m), but appropriate widths may be more or less than this depending on the volume of water that crosses it, the land slope, the amount of sediment in the runoff, and the degree of protection required for the body of water (see Chapter 8, Section 8–5.6).

Eutrophication can be a problem even in very large bodies of water. The Baltic Sea was mentioned earlier in this chapter as an extreme example of eutrophication. The Gulf of Mexico is another example. Plant nutrients carried in by the Mississippi River and other sources nourish an abundant growth of algae. Decomposition of the algae causes *hypoxia* (serious depletion of oxygen in a body of water) in a large section of the Gulf of Mexico. Water analyses trace much of the cause to tile drainage systems in the midwestern United States. Pitts (2000) suggests that controls could be installed on the outlets of drainage systems to raise the water table when drainage is not needed and thus reduce nutrient losses considerably.

17–2.3 Pesticides Carried by Sediment

Chemical analyses of water and aquatic life reveal that pesticide residues are widely distributed in streams, lakes, and oceans. Concentrations are usually low, but the occurrence is widespread, and many different pesticides are represented. Some pesticides dissolve in water, but many are adsorbed by soil colloids and carried by the sediment. Sediment gradually releases adsorbed pesticides into the water, thus maintaining a low but significant concentration for months or years.

Nicholson (1969) lists the principal sources of water pollution by pesticides as (1) runoff from land that has been treated to control pests, (2) industrial wastes from plants that either produce pesticides or use them in producing textiles, (3) accidents and carelessness in using chemicals or disposing of remnants and containers, and (4) the use of pesticides to control aquatic life. Erosion-control practices can reduce the amount of sediment carried in runoff; holding ponds and other waste-treatment practices can reduce the industrial source; and educational and licensing programs are currently aimed at reducing the careless and improper use of pesticides. Rester (1990) points out that one way to overcome problems with pesticide residues in sprayers is to catch and recycle the water that is used to rinse them when a job is done.

It should be noted, too, that pesticide use occurs on lawns and in buildings in cities, as well as in farm fields. Home use of pesticides is very common, and homeowners need to use appropriate handling techniques for pesticides. Runoff from lawns commonly drains into storm sewers that carry it directly to a stream.

Insecticides are more likely than other pesticides to be harmful to animals and humans. Herbicides are more likely to harm plants than animals. One exception was the implication of the herbicide 2,4,5-T as a cause of birth defects after it was widely used to kill trees and brush in Vietnam. Sethunathan (1989) points out that insecticides are the most-used group of pesticides in many tropical countries, whereas more herbicides are used in temperate regions. The most troublesome insecticides have been the chlorinated hydrocarbons (DDT and its relatives), because they combine high potency with long persistence and a tendency to accumulate in living things. These materials have been outlawed in most of the developed countries of the world but are still used in some of the developing nations because they are effective and inexpensive.

17–2.4 Heat as a Water Pollutant

Heat is easy to overlook as a water pollutant, but it can be quite significant. Gases such as oxygen become less soluble as water becomes warmer, and the decreased oxygen supply becomes inadequate for several forms of aquatic life. Trout and salmon are important examples of fish that need cool water. Duncan (1994) points out that salmon need temperatures of 42 to 58°F (6 to 14°C) (depending on species) for spawning and incubation and are likely to die in water temperatures above 75 to 78°F (24 to 26°C). He indicates that high water temperatures and depleted stream flows are the most significant habitat limitations for salmon in some watersheds of the western United States.

One example of indirect heat pollution has already been mentioned, where water depth is decreased to the point that the summer sunshine raises its temperature and reduces its oxygen content. The effects of irrigation and of industrial use of water will be considered here. Most water users in humid regions take only a small part of the available water and do not seriously alter stream characteristics. In contrast, the streams in arid regions where irrigation is practiced often are used so completely that little or no water is left in them during the summer season. Any water that remains flows slowly or is held in shallow pools and is subject to warming by the summer sunshine.

Much of the water that does flow in streams of arid regions is wastewater that runs off the bottom of irrigated fields after first having flowed through supply ditches and irrigation furrows. It may have been cool and clear when it was diverted from the stream, but it is warm and cloudy when it leaves the field and returns to the stream. Furthermore, it becomes polluted by plant nutrients and pesticides that were intended to benefit the crop in the field.

Industry may be the source of excessively warm water in humid regions. Factories that use large amounts of coal or other fuel to smelt and refine ore, produce metal products, refine chemicals, or employ any other heat-producing procedure must dissipate this heat in some way. Generally, heat disposal is through either water cooling or air cooling. Many factories have been placed next to rivers so they will have a large supply of water available for cooling and for other purposes. In times past, the heated water was simply returned to the stream through a large pipe. The local effects of dumping hot water can sterilize a section of the stream. Consequently, some industries have used diffusers to mix their hot water more thoroughly with the stream water. The effects then are less severe but more widely distributed. Generally, the environmental effects of heat disposal are less severe where air cooling is used instead of water cooling.

17–2.5 Organic Pollutants in Water

There are many kinds of organic pollutants that may be present in water. Most natural organic compounds are biodegradable, but some of them can be serious pollutants when their concentrations are too high. Synthetic compounds are more likely to resist decomposition—some of them indefinitely. The three largest sources of organic pollutants in water are sewage, manure, and petroleum products.

Sewage. Hundreds of cities and towns routinely dump raw sewage into streams, and about 150 of them dump it into coastal waters, bays, and lakes (Marx, 1988). Many more, including some large cities, dump raw sewage during peak flows or plant malfunctions. About half of the sewage treatment plants in the United States provide only *primary treatment,* removing materials that will either float or settle out of quiet water. The effluent enters a nearby stream still carrying dissolved materials, suspended solids, and about two-thirds of the biological oxygen demand (BOD) of the sewage.

The sewage treatment plants of most cities provide *secondary treatment* that uses bacteria to remove about 90% of the solid matter and BOD from the effluent of primary treatment. The bacteria work either in an activated sludge tank, through which air is blown to promote microbial activity, or on the surfaces of stones about 4 in. (10 cm) in diameter that fill trickling-filter tanks. The effluent from secondary treatment is usually chlorinated to kill harmful bacteria and then discharged into a stream.

The water pollution from treated effluent is much less serious than it would be without treatment, but the effluent still contains plant nutrients that cause eutrophication. Nitrogen and phosphorus are the most important nutrients because they are the most likely to be deficient enough to prevent algal growth in quiet water. Algae form an unsightly scum on water and then die. Decomposition of the algal mass depletes the water of dissolved O_2 and produces a rotten odor. Eutrophication can be avoided or controlled if the concentration of either nitrogen or phosphorus can be held low enough in the water.

The phosphorus content of streams increases dramatically below sewer outlets and commonly results in large masses of algal growth. Most of this phosphorus comes from laundry detergents and other cleaning agents, so one way to reduce such pollution is to reduce the phosphorus content of detergents. Researchers are seeking alternate ways of producing effective cleaning agents. One reason for optimism is the earlier success in changing detergents from the hard type that produced foam in streams, as shown in Figure 17–2, to the present biodegradable soft type.

Tertiary treatment is required when dissolved materials must be removed from sewage effluent. Tertiary treatment is provided by relatively few sewage plants, but the number is increasing because pollution-control requirements are becoming stricter. The dominant method of tertiary treatment has been chemical treatment, but this approach requires large quantities of expensive resins to absorb anions and cations. Interest in tertiary treatment is therefore being directed more toward disposal on land, as was discussed in Chapter 16.

Figure 17–2 Hard detergents produced masses of foam in streams and even in drinking water. The change from hard to biodegradable detergents in the United States was completed in 1965. (Courtesy USDA Office of Information.)

One concern that arises when sewage sludge (sometimes called *biosolids*) and/or effluent is applied to land is the potential of contaminating either runoff water or groundwater. Joshua et al. (1998) investigated this possibility in two Australian soils. They found that applications of 30,000 kg/ha (13.4 ton/ac) of biosolids (on a dry weight basis) caused little increase in loss of nutrients by runoff, partly because there was less runoff from treated plots than from control plots. After 1.5 years, they found zinc and copper had penetrated to a depth of 1 ft (30 cm) and nitrates to depths of about 2 ft (50 to 70 cm). They considered the pollution consequences of normal biosolid application rates (10,000 to 15,000 kg/ha, dry weight basis) to be low, almost negligible.

Sewage effluent is generally considered to be less hazardous than sewage sludge for application to soil because heavy metals tend to be held by the cation exchange sites of soil particles. Most sewage effluent is safe for irrigation use, but effluent that includes inputs from industrial sources can contain high concentrations of heavy metals. For example, Brar et al. (1999) found a buildup of several heavy metals in a soil in India that was irrigated for several years with effluent from a leather tannery. Most notably, chromium and nickel had built up to levels in the upper 1 ft (30 cm) of soil that caused increases in the concentrations of these elements in potatoes grown there. Fortunately, the buildup was less in the tubers than it was in the leaves of the potato plants, but they warned that continued use of such sewage effluent could create a health hazard.

Manure. Fecal matter may be 50% microbial in nature. Most of these microbes are engaged in decomposing the organic materials that surround them, but some of them are capable of causing diseases in animals and humans. Proper sewage treatment should eliminate the disease-causing viruses, bacteria, protozoa, worms, flukes, etc., but most animal manure does not pass through a sewage system, and not all sewage systems provide proper treatment at all times before it is emptied into a stream. Of course, the remaining pollutants are diluted when they are mixed with stream water, and natural processes work on them in the stream. The results may be satisfactory if there is enough dilution and enough time before the water is used for some critical purpose. Additional treatment should be applied if there is any doubt about the safety of the water for its intended use. For example, cities commonly filter their water and treat it with chlorine to eliminate harmful microbes.

Fecal contamination is more dangerous in drinking water than anywhere else. Many diseases, including some common ones, such as dysentery, and some very serious ones, such as typhoid fever and cholera, can be transmitted by waterborne agents. There are so many disease organisms that routine water analyses cannot check for all of them, so *Escherichia coli (E. coli)* has been selected as a marker organism for human fecal contamination. Other coliform bacteria are used as indicators of pollution from animal sources (Fish, 1992). Municipal water systems routinely test for *E. coli* and other coliforms to assure the quality of their water. *E. coli* should be completely absent from potable water, although low levels of other coliform bacteria may be tolerated.

Animal wastes from feedlots and confinement houses are large sources of water contamination. Years ago, a gentle hillside above a stream was considered to be an ideal location for a feedlot, not only because the stream was a source of water for the livestock, but also because rainfall could flush away the excess manure (Figure 17–3). Concern about the resulting water contamination arose later as the population increased. Most of the early feedlots were small. Moving the fence some distance upslope from the stream and establishing a grass filter strip in the area remedied the problem for many of them. However, the more recent trend toward tens of thousands of cattle in a single operation and corresponding numbers of hogs or chickens in confinement houses has led to severe problems in waste disposal. Too often, the wastes accumulate into mountainous heaps. Runoff and seepage from these manure piles can pollute both streams and groundwater. Spreading the manure on cropland so it can serve as fertilizer and structure enhancer is generally regarded as the best solution, but there is an odor problem, and field runoff can and does cause water contamination during some times and in some places (Table 17–3). Also, the concentration of

Figure 17–3 Many feedlots are located next to a stream because it was formerly thought to be desirable to let the stream carry the wastes away. (Courtesy USDA Office of Information.)

TABLE 17–3 AVERAGE WATER AND NUTRIENT CONTENTS OF ANIMAL MANURES

		Nutrients (%)					
Animal	H_2O (%)	N	P	K	S	Ca	Mg
Dairy cattle	79	0.56	0.10	0.50	0.05	0.28	0.11
Fattening cattle	80	0.70	0.20	0.45	0.085	0.12	0.10
Hogs	75	0.50	0.14	0.38	0.135	0.57	0.08
Horse	60	0.69	0.10	0.60	0.07	0.785	0.14
Sheep	65	1.40	0.21	1.00	0.09	0.585	0.185
Broiler	25	1.70	0.81	1.25	-	-	-
Hen	37	1.30	1.20	1.14	-	-	-

Source: Calculated from *Soils for Management of Organic Wastes and Waste Waters,* Chapter 8, Olsen and Barber, 1977, p. 197–215 by permission of the American Society of Agronomy, Crop Science Society of America, and Soil Science Society of America; data credited to R. C. Loehr.

large numbers of animals in feedlots or buildings produces large quantities of manure that must be hauled long distances in order to find enough suitable cropland where the manure can be used as fertilizer. The hauling cost often makes proper manure disposal economically unattractive.

Petroleum Products. Modern society uses tremendous quantities of petroleum to produce gasoline, oil, organic solvents, and a host of other products. Petroleum is a natural product formed in the depths of the Earth, but it is not very common in the environment of living things. Some of its components and most of the products made from it are not readily biodegradable. Many of them are detrimental to plant and animal life, partly because they are not miscible with water.

Crude oil spills from tankers, or from pipelines used to load or unload the tankers, are serious pollutants of ocean waters and nearby coastlines. The resulting oil slick is hazardous to many forms of aquatic life because it both seals out oxygen from the air above and uses up oxygen from the water as it decomposes. Birds that land on the polluted water become coated with oil that makes them unable to fly, so they may die.

Oil products such as gasoline and motor oil are shipped in large amounts and are subject to spills that pollute water and soil. They, too, can produce oil slicks and also cause air pollution and fire hazard.

Plastics are another class of petroleum products that are widely used in modern society. They are available in many types and are well suited for many purposes. Unfortunately, most of them are very resistant to biodegradation. Discarded plastic items, and pieces thereof, persist for a long time in water—usually until they are washed up on a beach someplace. Their resistance to decomposition keeps them from causing toxicity problems, but they are bulky and unsightly and can cause physical harm to animals and children.

17–2.6 Toxic Metals in Water

Almost anything can be toxic if it is sufficiently concentrated, but the concentration required to cause toxicity varies widely from one element or compound to another. Fortunately, most things are not likely to become sufficiently concentrated to cause serious

TABLE 17–4 ELEMENTS KNOWN TO HAVE TOXIC EFFECTS ON ANIMAL LIFE, SUGGESTED LIMITS ON THEIR CONCENTRATIONS IN POTABLE WATER, AND SOME OF THEIR DETRIMENTAL EFFECTS

Element	Chemical symbol	Suggested limit, ppb	Health problems known to have been caused by excessive amounts of the element
Arsenic	As	0.05	Neurological damage; stomach and intestinal disorders; bone marrow problems
Cadmium	Cd	10	Hypertension; heart, kidney, and liver damage; Itai-itai disease
Chromium	Cr	50	Toxic and carcinogenic
Copper	Cu	1000	Toxic, causes respiration problems in fish
Lead	Pb	50	Neurotoxin; attacks brain and kidneys; causes anemia, slow growth, coma, convulsions, high blood pressure
Mercury	Hg		Neurological disorders, skin disorders, organ damage, blurred vision, birth defects, Minamata disease
Nickel	Ni		Toxic, disrupts metabolism, possible carcinogen
Selenium	Se	10	Toxic, blind staggers in cattle and sheep
Silver	Ag	50	Disrupts enzyme systems in bacteria and fish
Zinc	Zn		Iron anemia, stunted growth, respiration problems in fish

Source: Compiled from various sources.

problems. Several heavy metals are the most likely elements to cause toxicity problems. These include arsenic (As), cadmium (Cd), chromium (Cr), copper (Cu), lead (Pb), manganese (Mn), mercury (Hg), nickel (Ni), selenium (Se), silver (Ag), and zinc (Zn). It should be noted that some of these elements are micronutrients, and two of them, arsenic and selenium, are transitional elements that can react either as metals or nonmetals. A summary of the toxic effects of these elements on animal life is shown in Table 17–4.

Arsenic. Toxic arsenic compounds have been known and used for centuries. They are probably best known for their present use as rat poisons and for political assassinations during the Middle Ages. More recently, organic arsenic compounds have been used in herbicides, insecticides, and wood preservatives, but these uses are declining. Arsenic behaves much like phosphorus and occurs as an impurity in phosphorus fertilizers. The oxidized form (As^{5+}) is usually harmless, but it becomes many times more toxic when reduced to the As^{3+} form.

The EPA maximum-allowable arsenic level in municipal drinking water is 0.05 mg/l (0.05 ppm or 50 ppb) (Korte and Fernando, 1991). Levels of 100 mg/l As in drinking water may cause neurological damage, and still higher levels (9000 to 10,000 mg/l) have caused severe stomach and intestinal disorders and impaired bone marrow functioning.

Cadmium. Cadmium is less abundant than some of the other heavy metals, but it is widely distributed, and it is highly toxic at relatively low concentrations. Its concentration in drinking water should be limited to 0.01 mg/l (0.01 ppm). Excess cadmium is reported to cause hypertension, cardiovascular disease, and kidney and liver damage (Rao, 1991). In Japan, cadmium was identified as the cause of Itai-itai disease that affected the bones of older women and caused severe pain. Cadmium is subject to strong bioaccumulation, and it has a very long residual time in the body after it has been ingested. Cadmium has been shown to become concentrated in the edible portions of certain plants (Salim et al., 1995). Cadmium levels that are toxic to humans may not cause any apparent ill effects on the plants, though higher concentrations are toxic to plants and to aquatic life in general. Also, persons who smoke cigarettes receive a significant additional input of cadmium from that source (Rao, 1991).

Cadmium is associated with zinc in mineral deposits and is present, along with zinc, in galvanized coatings used to protect iron from rust. It is also present as an impurity in phosphorus fertilizers. Commercial uses of cadmium include nicad rechargeable batteries, yellow coloring agents, and pottery glazes. Waste streams from industries producing any of these materials may contain significant concentrations of cadmium. The Cd^{2+} cation is strongly attracted to cation exchange sites of particulate material. Therefore, the cadmium concentration in sediment can be very high, even when its concentration in water is low. Nevertheless, it is readily exchangeable and therefore easily becomes available to growing plants.

Chromium. Chromium is used to harden steel and make it rust resistant. It is also used in some pigments and in the tanning and dye industries. Wastewaters from any industry that uses chromium should be monitored and may need to be treated to remove chromium, as well as other heavy metals, before it is allowed to enter a sewage plant or a stream. Drinking water should not contain more than 0.05 mg/l of hexavalent Cr (Rao, 1991). High concentrations of Cr^{6+} are both toxic and carcinogenic to animal life (McGrath, 1995). Concentrations of 5 mg/l or above are likely to adversely affect plant growth. The trivalent form (Cr^{3+}) is much less toxic and is more likely to precipitate in an insoluble compound than it is to cause any problems.

Copper. Copper is a micronutrient for both plants and animals, but it is also a toxic ion when it becomes too concentrated. Its toxicity has long been known and used, since copper sulfate mixed with lime and water constituted one of the earliest pesticides. Under the name of *Bordeaux mixture,* it was used to protect vineyards and other crops from insects and fungi. Excess copper in water has been reported to be toxic to fish because it interferes with their respiration processes (Alam and Maughan, 1995).

Copper in water is usually evaluated on the basis of either the soluble Cu^{2+} concentration or the total copper concentration, but neither of these is an accurate indicator of its effects. The Cu^{2+} cation is held on cation-exchange sites of soil clays and organic matter. It can be complexed by organic chelating agents, so the concentration in solution is usually low if sediments are present. The availability of exchangeable and chelated Cu^{2+} depends on the strength of the bonding and is not easily measured. Complexing with soil humic acids decreases the availability and toxicity of copper, but Goldberg (1989) says that some chelating agents greatly increase the toxicity of Cu^{2+}.

Copper, lead, and zinc are reported to be concentrated in highway runoff and to accumulate in the sediment in road ditches (Yousef and Yu, 1992). Fortunately, the sediment holds these metal cations tightly enough that removing the sediment once every 25 years or so seems to be adequate for minimizing the likelihood of these metals causing groundwater contamination. Copper concentration is also excessive in some sewage materials. Two methods have been developed for removing it and other heavy metals from such watery materials. One method is to add lime to raise the pH so the heavy metals will precipitate. The other method is to use peat to absorb the metallic cations from the water (Viraraghavan and Dronamraju, 1993).

Lead. Lead is an abundant metal that has been used for many purposes and has been one of the most common causes of heavy-metal toxicity in water. Lead toxicity problems began long ago and continue to the present day. For example, analyses of bones have indicated that people in the days of the Roman Empire suffered from lead poisoning, probably as a result of drinking wine from lead goblets. Acid from the wine would have helped dissolve enough lead to make it toxic, and the lead accumulated in the bones of the imbibers. A related effect can cause health problems today when acidic water dissolves lead from water pipes and fixtures. Lead-based solder is now banned from drinking water installations, but many older homes have copper water pipes that were assembled with solder containing about 50% lead. Even new brass or chrome plumbing fixtures may contain lead that can leach into standing water. For that reason, it is recommended that the standing water be flushed out before water is drawn for drinking or cooking. Drinking water should contain less than 0.05 mg of Pb per liter (Rao, 1991).

Lead is a neurotoxin that affects the brain, behavior, and growth (Rao, 1991). Toxic doses can damage the kidneys and cause anemia, convulsions, delirium, comas, and death. Even low levels of lead may cause high blood pressure in adults (Holmes et al., 1993). Children are at greater risk for lead poisoning than adults, not only because it can affect growth and development, but also because children are more likely to ingest paint chips and lead-contaminated soil in addition to any lead that may be present in their food and drinking water. Paint chips and paint dust are a matter of considerable concern wherever paint produced before 1978 was used. Some of it contained up to 50% "white lead" (lead carbonate and hydroxide). Recommendations are to leave such paint in place and cover it with wallpaper or some other protective surface. Removing leaded paint by sanding or burning can lead to high lead concentrations in the air. Persons who have been exposed to potentially harmful amounts of lead should have their blood checked. The World Health Organization recommends an upper limit of 20 mg of lead per 100 ml of whole blood (Wixson and Davies, 1994), and some authorities have recommended lowering this limit to 10 mg.

Lead is toxic to aquatic organisms and to animals as well as to humans. For example, cattle have been poisoned by eating plants high in lead and by ingesting chips of paint that contained lead. Waterfowl have been poisoned by spent lead shot, and swans in England died from eating lead weights used as fishing sinkers (Smith, 1992).

Millions of tons of lead have been mined, mostly as the mineral galena (PbS), and used for many purposes. Large amounts of tetraethyl lead were used in gasoline as an antiknock compound before it was largely phased out in the 1980s. Presently, car batteries are the largest source of lead entering the waste stream, but lead is so widely used that most industrial wastes contain it. Lead is no longer used in house paint, but it is still used in some other pigments, in some of the cheaper pottery glazes, in solder not intended for water pipes, in plastics, and most of all, in batteries.

Several alternatives are available for reducing the lead content of water. The best, of course, is to keep it out in the first place. That was the purpose of the change to lead-free gasoline, paint, and solder. However, it is still possible for water to be contaminated with hazardous concentrations of lead. If the lead comes from acid water passing through home plumbing, much of it can be avoided by flushing out the standing water before use. Also, the lead input can be reduced considerably by raising the pH of the water so it is no longer acid. For moderate amounts of drinking water, it is possible to remove nearly all of the lead and other heavy metals by ion exchange or with a super-fine filter such as a reverse-osmosis system. Also, lead may be removed from water by chemical means such as treating it with phosphate rock material to form insoluble lead phosphates (Ma et al., 1995). Another method uses microbes to remove the lead (Vesper et al., 1996).

Manganese. Manganese is a widely distributed element that commonly occurs as an octahedral cation in igneous rocks and soil minerals. Manganese is an essential micronutrient for plants and animals, and nutrient deficiencies are not uncommon, especially in alkaline soils. It is present mostly as Mn^{2+} in rocks and is somewhat soluble in water in that form, especially under acidic conditions. Manganese toxicity in soils can affect growing plants if the pH is below 5.

Leachate from acid soils and from solid wastes can contain enough manganese to be toxic to aquatic life. Acid rain contributes to such leaching and to the toxicity of not only manganese but also aluminum, ferric iron, and mercury (Holmes et al., 1993). Under oxidizing conditions and alkaline pH, manganese may be oxidized to less soluble forms such as Mn^{4+}, which precipitates as MnO_2 and forms dark brown coatings, nodules, or even layers on sediments or other surfaces. This mechanism has been used to remove unwanted manganese from water by adding ozone to oxidize the manganese along with iron and organic contaminants (Ellis, 1991).

Mercury. Mercury is also known as quicksilver because its elemental form is a silvery liquid at ordinary temperatures (from -39 to $+356°C$ or -38 to $+673°F$). It is used in thermometers, thermostats, light switches, batteries, mercury-vapor lights, dental fillings, paints, insecticides, and rat poisons. Industry uses it to produce chlorine and sodium hydroxide. The ability of liquid mercury to dissolve gold and silver makes it useful for miners to extract these precious metals from sand in a sluice box. Of course, some of the mercury escapes and contaminates the stream being used by the miners. Some of the mercury also escapes into the atmosphere as it is separated from the gold and silver by distillation.

Liquid mercury can be an amusing plaything, and people used their fingers to make it roll around before its toxic nature was well known. It looks harmless, but small quantities can be absorbed through the skin and cause skin disorders, internal organ damage, and gastro-intestinal problems. Once absorbed, mercury tends to accumulate in the body and is subject to bioaccumulation through the food chain.

Metallic mercury is hazardous, but the most damage is done by mercury compounds, all of which are toxic to animal life, though generally harmless to plants in the forms and concentrations normally found in nature. Methyl mercury is a highly toxic form that is very common in organic materials because microorganisms produce it from Hg^{2+} (Steinnes, 1995). It causes neurological disorders, blurred vision, numbness, birth defects, and some deaths (Rao, 1991). A well-known example of mercury poisoning was identified in Japan in the 1950s. Dozens of people died and thousands suffered serious health effects from eating fish with high mercury contents. The problem was named *Minamata disease* after the chemical plant that dumped the mercury, the bay in which it accumulated, and the city where many of the people lived. The hazards are serious enough that use of mercury pesticides has been largely discontinued, and a ban against dumping mercury in the ocean has been instituted by international agreement. The EPA limit for mercury in fish for human consumption is 1 ppm Hg (Holmes et al., 1993). Fish that exceed this limit occur in a number of places, including several lakes formed behind hydroelectric dams in Canada. Native Indians eating fish from these lakes have suffered from mercury toxicity.

Fuels such as coal, oil, natural gas, and wood contain enough mercury to cause a significant input of elemental Hg into the atmosphere. It also enters the atmosphere through natural sources such as volcanoes. Once there, it can circulate around the Earth and ultimately be oxidized and brought down by rain, thus becoming an input of mercury into water and soil. Mercury has been found in the Greenland ice cap and other isolated regions, indicating that it must have arrived there through the atmosphere. Organic materials in soil and sediments will bind methylmercury if the pH is high enough, but they tend to release it into the water under acid conditions (Hintelmann et al., 1995).

Nickel. Nickel is a micronutrient that animals need for liver metabolism, iron absorption, and enzyme activity; plants need it for urease to function and for iron absorption (McGrath, 1995). However, the needs are so small and the natural supply of nickel so abundant that deficiencies are unlikely to occur naturally. Higher concentrations of nickel are highly toxic to both plants and animals because it can substitute for other essential cations and disrupt metabolic pathways. Nickel also is suspected of being carcinogenic.

Nickel is used in making stainless steel and as a plating metal because of its corrosion resistance. It makes stainless steel more ductile and shiny as well as corrosion resistant. Kuligowski and Halperin (1992) recommend that strongly acidic foods not be cooked in stainless steel cookware because most of it contains nickel, with the content ranging from 0 to 31% Ni. Heat combined with acidity cause small amounts of nickel to dissolve.

Nickel also is used to make nicad batteries, and cleaning products such as detergents and bleach contain up to about 0.1% Ni. Petroleum products and coal contain some nickel that was inherited from the plant material from which they formed. Consequently, burning them emits Ni into the air. This Ni can contaminate exposed water and soil surfaces. Both urban and industrial wastes contain nickel that tends to accumulate in sewage sludge along with other heavy metals.

Several different approaches have been used to reduce the content of nickel and other heavy metals from polluted water. An alkaline pH makes metallic ions less soluble, but often will not remove enough of the nickel. Adding sulfides helps to make it precipitate. Activated charcoal will absorb nickel but is fairly expensive, so other materials such as fly ash or peat have been tried with some success (Mavros et al., 1993; Viraraghavan and Dronamraju, 1993). The reverse-osmosis filtration process may be used to remove nickel and other ions from moderate amounts of already clear water.

Selenium. Selenium is a semi-metallic element that occurs directly beneath sulfur in the periodic table. Like sulfur, it has several different oxidation states and can be bonded into organic matter or occur in mineral compounds. It is an essential element for animal life, and deficiencies can cause white muscle disease in livestock (Neal, 1995). It is, however, more widely known for being toxic in certain soil areas and causing "blind staggers" in cattle and sheep (see Chapter 16, Section 16–7.3). The most toxic form of selenium is Se^{6+}.

Selenium is used in electronics, glass, plastics, ceramics, pigments, and lubricants. When excess selenium occurs in sewage materials, most of it settles out and accumulates in the sludge. Dissolved Se in shallow groundwater ranges from undetectable amounts to several mg/l (Neal, 1995). The public health limit for selenium in water is 10 mg/l.

Silver. Silver is best known as a precious metal used in coins, jewelry, and silverware. Large amounts of it are used in photography and lesser amounts in mirrors, electrical contacts, and dentistry. Monovalent silver, Ag^+, is highly toxic to microbes, algae, and fish but less so to higher plants and animals. The silver ion disrupts biological processes by binding to enzymes and other organic molecules (Edwards et al., 1995). A limit of 0.05 mg of Ag/l has been set for drinking water (Rao, 1991).

Much of the silver found in wastewater comes from its extensive use in photography. The sensitivity of its chloride and bromide salts to light, and the black images they form when reduced to finely divided metallic silver, are the foundation of photographic processes. The silver salts left over after developing the photographic image are dissolved and washed away in the process. Large photo labs recapture much of this silver by converting it to insoluble forms, but most home darkrooms allow it to go down the drain into the sewer. Fortunately, the silver that escapes is present mostly in the relatively nontoxic thiosulfate form (Wang, 1992).

Zinc. Zinc is an essential micronutrient for both plants and animals. Zinc deficiencies in animals and humans cause lack of appetite, slow growth, skin lesions, and immaturity. The recommended daily intake of zinc for adult humans is about 15 mg/day (Kiekens, 1995). Zinc in plants is essential for the function of certain enzymes, and zinc deficiencies interfere with the production of carbohydrates and proteins. Zinc deficiency symptoms in plants include stunted growth, delayed maturity, little leaf rosette of trees, and pale-green stripes on the leaves of corn and related plants. Zinc toxicity can interfere with the functioning of other nutrients and can cause stunted growth of both plants and animals. Excess zinc has been identified as a cause of iron anemia in animals (Rao, 1991) and is known to cause breathing difficulties and slow growth in fish (Alam and Maughan, 1995).

Zinc toxicity is synergistic with that of copper, nickel, cadmium, and possibly other metals. The overall hazard caused by the concentrations of these metals in sewage materials is evaluated in a combined formula known as the zinc equivalent:

$$\text{Zn equivalent} = \text{Zn} + 2 \times \text{Cu} + 8 \times \text{Ni}$$

Lester (1990) recommends that the addition of sewage sludge to land be limited to a zinc equivalent of 560 kg/ha over a 30-year period. (see also Section 16–3.2) From the formula for the zinc equivalent, it can be inferred that a given concentration of copper is considered twice as hazardous as the same concentration of zinc, and that a nickel concentration is eight times as serious as that of zinc. However, the amount of zinc used is much greater than the amount of nickel, so the overall hazard of the two is more nearly equivalent than one might infer from the formula alone. Cadmium is also highly toxic and is generally associated with zinc, both in natural mineral sources and as an impurity in zinc products. Fortunately, newer methods of production can improve the separation of metals such as these.

Large quantities of zinc are used to galvanize iron and other oxidizable metals. A zinc coating, produced either by dipping or by electroplating, is highly effective for rust resistance. Zinc is also a component of brass (along with copper and several optional metals) and is used in making certain types of batteries. Zinc input into water can come from industrial wastes, mining wastes, or the effects of acid rain dissolving zinc from the coatings on galvanized metals. It has also been identified as a component of the leachate from sewage sludge and composted materials. Current efforts to control acid rain seem to be helping reduce some of these inputs of zinc and other heavy metals. For example, Stigliani et al.

(1993) reported that by 1988 the zinc input to the Rhine River had been reduced to 30% of its peak in the 1960s, largely as a result of air pollution controls reducing SO_2 emissions.

Much of the zinc and most other metallic cations can be removed from water by raising the pH, often by adding lime. Other possibilities include reverse-osmosis filtration and adsorption by peat or other ion-exchange materials. Most such processes produce either a more concentrated solution or a sludge that must then be handled as a hazardous waste.

17–3 ACIDIFICATION OF WATER

As already mentioned, acidity is a significant cause of heavy-metal concentrations in water becoming toxic to plant and animal life. Acidity problems of varying degrees of severity are widespread, though not universal. They are very common in humid regions but relatively rare in arid regions. Areas that have been identified as having serious acidity problems include the northeastern United States, eastern Canada, most of Europe, and southern China. Most of the attention has been focused on surface waters, but groundwater is also affected. Rodhe et al. (1995) indicate that the groundwater in large areas of northern Europe is so acidic that the water from wells corrodes home plumbing and may contain enough heavy metal ions to pose a health hazard.

17–3.1 Causes of Acidity in Water

Acidity has many causes; three important ones are: acid mine drainage, acid soil leachate, and acid rain.

Acid mine drainage is the most localized of the three causes under consideration here. It results mostly from the oxidation of pyrite and other sulfides present in mines and, especially in mine spoils. Mine spoils were often left in large piles inside open pit mines or near the entrances of shaft mines before regulations required better care. Large areas of such spoil have never been reclaimed. Oxidation produces sulfuric acid from the sulfides, as discussed in Chapter 11, and rainwater leaching through the spoils carries the acid into adjoining soils and streams. The resulting acidity can be so strong that it sterilizes both the water and the soil. The supply of sulfides is commonly sufficient to last for many decades, even after the mine has been abandoned, so something must be done to stop their oxidation and contain the leachate before the affected soil can be reclaimed. In addition, the soil pH must be raised to a reasonable level to stop continued acid leaching from it into the groundwater and nearby streams.

Acid soil leachate is natural in humid regions, but the acidity should normally be mild enough to cause little damage to streams and lakes. However, it does give the groundwater and the streams in humid regions a mildly acidic starting point that reduces their capacity to safely absorb acidification from other sources. Of course, the leachate from extremely acidic soils is also extremely acidic. The most acidic soils of all are those affected by sulfuric acid, either coming in as seepage from mine spoils or produced in place by the oxidation of sulfides. Such conditions are produced, for example, by the drainage of bogs that contain pyrite and other sulfides. Concentrated sources of acid rain, usually downwind from an industrial area that burns a lot of coal, are another cause of strongly acidic soils. The natural cation content of such soils can neutralize much of the acidic rainfall for some years or decades, but the leachate will ultimately become very acid if the acidic input continues long enough.

Acid rain is the most important cause of water acidification because it is widespread. Some acidity in rainfall is normal because carbon dioxide from the atmosphere reacts with water and produces carbonic acid. The resulting pH at equilibrium is about 5.6, so rainfall is generally not called acid rain unless the pH is lower than 5.6. Rainwater has very little buffer capacity, so the natural acidity resulting from carbonic acid is easily neutralized upon contact with soil or rock materials that can provide sources of basic cations.

Acid rain resulting from significant inputs of sulfur dioxide (SO_2) or oxides of nitrogen (N_2O, NO, and NO_2) has pH values as low as 3 or 4. Because the pH scale is logarithmic, a pH of 3.6, for example, is 100 times as acidic as 5.6. Acid rain is therefore a strong weathering agent that can cause rust on iron objects and corrosion on the surface of limestone and marble, whether they are present as stones in a field, as a statue carved by a great artist, or as facings on public buildings. It can also acidify soils on which it falls and streams into which it flows or seeps. The soil acidification process is gradual because most soils have considerable buffering capacity, but that same buffer capacity makes them large reservoirs of acidity if they are allowed to become excessively acidic.

The vast amount of fossil fuel that has been burned in the last century is the biggest contributor to acid rain, because of sulfur dioxide it has released into the air. Internal combustion engines have produced large amounts of both sulfur dioxide and nitrogen oxides that also contribute to acid rain. These gases circulate around the world, making acid rain an international problem. Consequently, the pH of rainfall even in remote areas is typically between 4.0 and 5.6, whereas the rainfall in eastern North America and all of Europe usually has a pH between 4.0 and 4.5 (Crathorne and Dobbs, 1990).

17–3.2 Effects of Water Acidification

Runoff from acid rain and leachate from acidic soils and mine spoils produce acidic streams, ponds, and lakes. Lowering the pH of the water affects plant and animal life in and near the affected water bodies. For example, it damages the plankton that grow in the water and that form the base of the food chain for fish and other aquatic life (Havens et al., 1993). Studies of sediment cores indicate that lakes in Scotland began showing the effects of acidification as long ago as 1850 and Scandinavian lakes began losing fish populations to acidity as early as the 1920s (Mason, 1990). Fish, salamanders, frogs, and other aquatic creatures suffer from aluminum toxicity in water with a pH below 5.0–5.5 and fail to reproduce if the pH is lower than 5.4 or thereabouts.

The acidity problem grew more widespread and serious when more and more fossil fuels were used as a result of the industrial revolution, the development of large coal-fired electric-generating plants, and the widespread use of gasoline and diesel engines. By the 1970s, there were so many acidified lakes and streams that the problem could no longer be ignored. The best way to overcome it is to reduce the input of sulfur dioxide and nitrogen oxides into the atmosphere. Laws have been passed and progress has been made toward reducing the polluting effect of motor vehicles and industrial smokestacks, but there is still much to be done to reach a long-term solution. In the meantime, the short-term solution of liming lakes and streams has been used in some places. For example, Henrikson et al. (1995) report that Norway and Sweden apply over 300,000 tons of lime per year to more than 11,000 lakes and streams to keep them producing edible fish. Liming has even been used to improve the quality of groundwater by means of heavy applications of lime to the soil in appropriate catchment areas (Norrström and Jacks, 1993). However, the large number of water bodies involved and the recurring nature of the problem make liming expensive and unlikely to be used on an adequate scale to overcome most water acidification problems. Furthermore, liming can easily raise the pH of localized areas high enough to injure life forms that grow best in somewhat acidic conditions.

17–4 GROUNDWATER CONTAMINATION

Groundwater differs from surface water in that it cycles more slowly, typically contains more dissolved ions, and is cut off from living things that require light and gaseous oxygen. The slow cycling is a variable that is very important. Some groundwater has remained in place for a much longer time than that covered by recorded history. Digging a well and pumping out such water is equivalent to mining a mineral resource—when it is gone, it will not be replaced in time for humans to use it again. Shallow groundwater, on the other hand, might be replenished by the next big rain. Intermediate stages might take months, years, decades, or centuries to replenish the water.

The overlying layers of soil and rock give groundwater some degree of protection against pollution, but these same layers combined with slow cycling make it very difficult to remedy any pollution that may occur. Furthermore, the sparsity of life forms reduces the ability of the water to eliminate organic pollutants. Thus, protecting groundwater from contamination is a much better alternative than attempting to correct damage after it is done.

Groundwater contamination occurs when soluble materials such as nitrates are leached from the overlying soil, and when surface water finds an entrance and carries both soluble and suspended contaminants with it. Both of these contamination sources are significant. Nitrates in groundwater have increased along with increasing fertilizer usage, especially during the latter half of the twentieth century. Recent years, though, have seen more efforts to fine-tune fertilizer rates to the needs of the crop, and the newest technology (sometimes called *precision agriculture*) is making it possible to adjust the rates according to the needs of different parts of a field. Similarly, there has been much concern about pesticides in groundwater. Most agricultural pesticides have been detected in very low concentrations in groundwater, and some widely used ones, such as atrazine, are present in nearly all water samples in areas where they are used. Butters et al. (2000) found that irrigation water carried atrazine deeper than expected in the Colorado soils they studied. They suggested that some mechanism such as preferential flow, chemical nonequilibrium, or interaction with a mobile phase may have facilitated the downward movement of atrazine.

Dangerous pesticide concentrations in groundwater are rare and usually are associated with a spill of some kind, but the widespread occurrence of pesticides in groundwater indicates that they do leach to some extent. Prudence suggests that their use be limited to the minimum that is really needed and that great care be exercised to avoid spills.

When surface water enters directly into an aquifer, it can contaminate the water much more rapidly than water that filters through the soil. This type of contamination is much more common than was once supposed because soils have natural flow paths such as cracks, worm holes, and channels left by decaying roots. Kung et al. (2000) used tracers to see how long it took irrigation water to reach tile drains at a depth of 3 ft (90 cm) in loamy New York soils. Even without any visible ponding of water, the tracers applied at the beginning of the irrigation period were detected in the tile drains within 1 to 2 hours, and some tracers added later in the irrigation period reached the tile drains in as little as 13 minutes Surface water contamination is especially hazardous where the underground water flows through large passageways as is common in basalt flows and limestone bedrock. Such water does not receive the filtering action that occurs when water flows through the fine pores of a sandstone aquifer.

Earthworm channels were implicated as likely avenues of preferential flow carrying animal waste slurries into tile drains in a study by Shipitalo and Gibbs (2000). They located earthworm channels that approached tile lines in an Ohio field by blowing smoke into the tile lines and observing the smoke as it emerged from earthworm burrows within about 20 in. (0.5 m) on either side of the tile lines. They then checked the water infiltration rate and found that water entered burrows that had emitted smoke about twice as fast as it entered other burrows. Another test showed that dyed water entered the tile lines through burrows that had emitted smoke but not through other burrows. They concluded that earthworm burrows and other preferential flow channels accounted for animal waste slurries that had been injected into the soil being found in the effluent from tile drains. This problem seemed to be most prominent in no-till fields, presumably because they have more earthworm activity than tilled fields.

Wells are common access points that need to be protected from the entry of pollutants. For example, farm families have become ill because their wells were contaminated when runoff from their farmlot seeped down either alongside or through the well itself and thus reached the water table. The hazard still exists when a well is abandoned without being properly closed and sealed. Contaminants may enter an aquifer through the abandoned well and be carried to another well some distance away. Drainage wells are also vulnerable entry points into an underground aquifer. These have been used in relatively level areas with potholes that can be transformed into cropland by running the excess water into a drainage well located in the pothole or some other nearby location. Some drainage wells have a grass filter strip around them and/or a sand filter over their entrance, but it is still all too easy for fertilizer and pesticide materials to reach the water table through them. There are many of these in Minnesota and northern Iowa, and efforts are now being made to close as many of the drainage wells as possible. Some of them are being replaced with drainage ditches; some areas may be allowed to revert to wetlands.

Seawater incursion is a form of water pollution that can occur in low-lying coastal areas. Normally, the water table in coastal areas is raised above sea level by infiltrating precipitation until enough pressure is built up to cause underground water to flow slowly toward the sea. An equilibrium is established, and the outward flow of water keeps the seawater at bay. However, human settlement in coastal areas often leads to wells being drilled to tap the underground water supply. Excess withdrawal of water from the wells can lower the water table below sea level and reverse the gradient. Seawater then seeps inland, and wells that formerly produced sweet water become salty. The only remedy is to limit water usage so the water table will always be higher than sea level, but this is difficult to do when population pressure continues to increase in dry climates such as the coastal areas of southern California and Israel.

Groundwater can also be contaminated by seepage from many different kinds of local sources. Buried underground storage tanks are a major concern in this regard. Many such tanks that once stored gasoline were left behind when service stations were closed. Eventually the tanks rust through, usually along their bottom seams, and whatever gasoline or other material they hold can then leak into the surrounding soil and groundwater. Seepage from landfills is another source that can carry toxic organic compounds and heavy metals.

Old disposal sites from chemical industries are another source of serious hazard. The best-known example of this type is the Love Canal site in the city of Niagara Falls, New York, where more than 200 hazardous chemicals were found in the soil and water (Holmes et al., 1993). Hooker Chemical Company used the area to dispose of chemical wastes in the 1920s and 1930s. Through a combination of deception and bad judgment, a neighborhood developed around the abandoned site and a public school was built on the site in the 1950s. The Love Canal became world famous in the 1970s when above-average rainfall raised the water table and brought enough chemicals to the surface to cause numerous illnesses. Both the school and the surrounding neighborhood had to be evacuated and abandoned (Colten and Skinner, 1996).

A congressional investigation found that the United States has millions of tons of chemical wastes distributed among thousands of sites. They contain all kinds of solvents, acids, bases, other industrial chemicals, agricultural chemicals, petroleum products, and miscellaneous waste materials. Much of this waste is enclosed in steel drums or other containers, but these are subject to gradual deterioration and eventual leakage.

SUMMARY

Earth has a limited supply of water available for constant recycling and use by living plants, animals, and people. *Pollution* degrades water and can make it toxic to living entities. Small streams and ponds are easily polluted, and even the oceans are not big enough to be immune, especially in coastal areas. Some large water bodies like the Baltic Sea are so polluted that they may lose all aquatic life if remedies are not applied soon.

Most *water pollutants* can be placed in one of four groups: eroded sediment, heat, organic pollutants, and heavy metals. Eroded soil supplies much more material than any other source of pollutants. Eroded soil material not only produces muddy water but also carries with it plant nutrients and pesticides. The cloudiness of muddy water alters the habitat, drives some aquatic species from the area, and supplies sediment that fills channels and reservoirs. Filled channels increase flooding that damages buildings and crops and often deposits sandy sediment on top of fertile soil. Nitrogen, phosphorus, and other plant nutrients carried by sediment contribute to eutrophication of water bodies. Grass filter strips are an effective way to remove most of the sediment and nutrients from runoff water.

Heat is a significant water pollutant that can result from sunlight on shallow water or from industrial use of water as a coolant. Species such as trout and salmon require cool water, partly because oxygen solubility is higher in cool water than in warm water.

Sewage, manure, and petroleum products are the leading organic pollutants in water. *Primary sewage treatment* removes solid materials and *secondary treatment* allows microbes to decompose much of the dissolved organic matter. *Tertiary treatment* is required if plant nutrients and other dissolved ions are to be removed. Manure washed from a feedlot or a manure pile can be a serious water pollutant, and there are large amounts of it. It is best used as fertilizer, but large feedlots and confinement houses may produce too much of it for the fertilizer needs of nearby land. Petroleum is used in such large amounts that there are frequent spills of crude oil, motor oil, gasoline, and other petroleum products that contaminate soil and water.

Metals that can be toxic water pollutants include arsenic, cadmium, chromium, copper, lead, manganese, mercury, nickel, selenium, silver, and zinc. Some of these are micronutrients, but they become toxic if their concentrations are too high. Several affect the nervous systems of animals and humans and cause organ damage, digestive problems, or reproductive difficulties. They are likely to interfere with enzyme systems in plants, stunt growth, and reduce seed production. Since they form cations, these metals are attracted to the cation-exchange sites of colloidal particles suspended in water and in sediment.

Acid mine drainage, acid soil leachate, and acid rain are common causes of water acidity. *Acidification* increases the solubility of metallic ions and makes them more likely to become toxic to plants and animals. Fish and many other aquatic species fail to reproduce when the water pH is below 5.4. Extensive water-liming projects have been undertaken in Norway, Sweden, and some other places to maintain aquatic life (especially fish), but the long-term solution is to reduce the inputs of sulfur dioxide and oxides of nitrogen that cause acid rain.

Groundwater contamination occurs when soluble ions such as nitrates leach from the soil and when surface water carries pollutants through wells, cracks, earthworm channels, or other openings to the groundwater. Wells near feedlots, drainage wells, and waste-disposal sites are potential locations for groundwater contamination. Slow cycling makes groundwater contamination difficult to correct.

QUESTIONS

1. Water is abundant. Why is it considered precious?
2. Why is eroded soil a serious water pollution problem?
3. What is the difference between primary, secondary, and tertiary treatment of sewage?
4. Why isn't manure always used as fertilizer?
5. What causes eutrophication? What are its effects?
6. How can heat pollute water? What does it hurt?
7. What are the principal types of organic pollutants in water? Can they be decomposed by microbes in the water?
8. List five metallic ions that can reach toxic concentrations in water. Describe the effects of one of them.
9. What causes acid rain? Where is it most likely to occur? What problems does it cause?
10. Why is groundwater contamination more serious than an equal degree of surface water contamination?

REFERENCES

ALAM, M. K., and O. E. MAUGHAN, 1995. Acute toxicity of heavy metals to common carp (*Cyprinus carpio*). *J. Environ. Sci. Health* A30:1807–1816.

BARZILAY, J. I., W. G. WEINBERG, and J. W. ELEY, 1999. *The Water We Drink: Water Quality and Its Effects on Health.* Rutgers Univ. Press, New Brunswick, NJ, 180 p.

BRAR, M. S., S. S. MALHI, A. P. SINGH, and C. L. ARORA, 1999. Effects of effluent contaminated sewage water on micronutrients and potentially toxic elements in soils and in potato plants in northwestern India. *Can. J. Soil Sci.* 79:641.

BUTTERS, G. L., J. G. BENJAMIN, L. R. AHUJA, and H. RUAN, 2000. Bromide and atrazine leaching in furrow- and sprinkler-irrigated corn. *Soil Sci. Soc. Am. J.* 64:1723–1732.

CHAPMAN, D. V., 1997. Water-quality monitoring. Ch. 6, p. 209–248 in A. K. Biswas (ed.), *Water Resources: Environmental Planning, Management, and Development.* McGraw-Hill, New York, 737 p.

COLTEN, C. E., and P. N. SKINNER, 1996. *The Road to Love Canal.* Univ. of Texas Press, Austin, 217 p.

CRATHORNE, B., and A. J. DOBBS, 1990. Chemical pollution of the aquatic environment by priority pollutants and its control. Ch. 1 in R. M. Harrison (ed.), *Pollution: Causes, Effects, and Control,* 2nd ed. Royal Society of Chemistry, Cambridge, U.K., 393 p.

DAVIS R., R. C. BROWNSON, and R. GARCIA, 1992. Family pesticide use in the home, garden, orchard, and yard. *Arch. Environ. Contam. Toxicol.* 22:260–266.

DUNCAN, A., 1994. Proposal for a Columbia Basin watershed planning council. *Illahee* 10:287–303.

EDWARDS, R., N. W. LEPP, and K. C. JONES, 1995. Other less abundant elements of potential environmental significance. Ch. 14 in B. J. Alloway (ed.), *Heavy Metals in Soils,* 2nd ed. Blackie Academic & Professional, London, 368 p.

ELLIS, K. V., 1991. Water disinfection: A review with some consideration of the requirements of the third world. *Crit. Rev. Environ. Control.* 20:341–407.

ENELL, M., and J. FEJES, 1995. The nitrogen load to the Baltic Sea—present situation, acceptable future load and suggested source reduction. *Water, Air, and Soil Pollut.* 85:877–882.

FISH, H., 1992. Freshwaters. Ch. 3 in R. M. Harrison (ed.), *Understanding Our Environment: An Introduction to Environmental Chemistry and Pollution,* 2nd ed. Royal Society of Chemistry, Cambridge, U.K., 326 p.

GOLDBERG, E. D., 1989. Aquatic transport of chemicals. Ch. 2.4 in P. Bourdeau, J. A. Haines, W. Klein, and C. R. Krishna Murti (eds.), *Ecotoxicology and Climate.* Wiley and Sons, Chichester, U.K., 392 p.

HAVENS, K. E., N. D. YAN, and W. KELLER, 1993. Lake acidification: Effects on crustacean zooplankton populations. *Environ. Sci. Technol.* 27:1621–1624.

HENRIKSON, L., A. HINDAR, and E. THÖRNELÖF, 1995. Freshwater liming. *Water, Air, and Soil Pollut.* 85:131–142.

HINTELMANN, H., P. M. WELBOURN, and R. D. EVANS, 1995. Binding of methylmercury compounds by humic and fulvic acids. *Water, Air, and Soil Pollut.* 80:1031–1034.

HOLMES, G., B. R. SINGH, and L. THEODORE, 1993. *Handbook of Environmental Management and Technology.* Wiley & Sons, New York, 651 p.

JOSHUA, W. D., D. L. MICHALK, I. H. CURTIS, M. SALT, and G. J. OSBORNE, 1998. The potential for contamination of soil and surface waters from sewage sludge (biosolids) in a sheep grazing study, Australia. *Geoderma* 84:135–156.

KIEKENS, L., 1995. Zinc. Ch. 13 in B. J. Alloway (ed.), *Heavy Metals in Soils,* 2nd ed. Blackie Academic & Professional, London, 368 p.

KORTE, N. E., and Q. FERNANDO, 1991. A review of arsenic (III) in groundwater. *Crit. Rev. Environ. Control* 21:1–39.

KULIGOWSKI, J., and K. M. HALPERIN, 1992. Stainless steel cookware as a significant source of nickel, chromium, and iron. *Arch. Environ. Contam. Toxicol.* 23:211–215.

KUNG, K.-J. S., T. S. STEENHUIS, E. J. KLADIVKO, T. J. GISH, G. BUBENZER, and C. S. HELLING, 2000. Impact of preferential flow on the transport of adsorbing and non-adsorbing tracers. *Soil Sci. Soc. Am. J.* 64:1290–1296.

LESTER, J. N., 1990. Sewage and sewage sludge treatment. Ch. 3 in R. M. Harrison (ed.), *Pollution: Causes, Effects, and Control,* 2nd ed. Royal Society of Chemistry, Cambridge, U.K., 393 p.

MA, Q. Y., T. J. LOGAN, and S. J. TRAINA, 1995. Lead immobilization from aqueous solutions and contaminated soils using phosphate rocks. *Environ. Sci. Technol.* 19:1118–1126.

MARX, W., 1988. Swamped by our own sewage. *Reader's Digest 132*(1):123–128.

MASON, C. F., 1990. Biological aspects of freshwater pollution. Ch. 6 in R. M. Harrison (ed.), *Pollution: Causes, Effects, and Control,* 2nd ed. Royal Society of Chemistry, Cambridge, U.K., 393 p.

MAVROS, P., A. I. ZOUBOULIS, and N. K. LAZARIDIS, 1993. Removal of metal ions from wastewaters. The case of nickel. *Environ. Technol.* 14:83–91.

MCGRATH, S. P., 1995. Chromium and nickel. Ch. 7 in B. J. Alloway (ed.), *Heavy Metals in Soils,* 2nd ed. Blackie Academic & Professional, London, 368 p.

NEAL, R. H., 1995. Selenium. Ch. 12 in B. J. Alloway (ed.), *Heavy Metals in Soils,* 2nd ed. Blackie Academic & Professional, London, 368 p.

NICHOLSON, H. P., 1969. Occurrence and significance of pesticide residues in water. *J. Wash. Acad. Sci. 59*(4):75–85.

NORRSTRÖM, A. C., and G. JACKS, 1993. Soil liming as a measure to improve acid groundwater. *Environ. Technol.* 14:125–134.

PETIT, B., 1994. The value of wild salmon. *Illahee* 10:262–264.

PITTS, D., 2000. Drainage management to improve water quality and to enhance agricultural production. *J. Soil Water Cons.* 55:424.

RAO, C. S., 1991. *Environmental Pollution Control Engineering.* John Wiley & Sons, New York, 431 p.

RESTER, D., 1990. Pesticide application equipment rinse water recycling. In *The Environmental Challenge of the 1990s,* EPA/600/9-90/039, p. 563–572.

RODHE, H., P. GRENNFELT, J. WISNIEWSKI, C. ÅGREN, G. BENGTSSON, K. JOHANSSON, P. KAUPPI, V. KUCERA, L. RASMUSSEN, B. ROSSELAND, L. SCHOTTE, and G. SELLDEN, 1995. Acid reign '95—conference summary statement. *Water, Air, and Soil Pollut.* 85:1–14.

SALIM, R., M. ISA, M. M. AL-SUBU, S. A. SAYRAFI, and O. SAYRAFI, 1995. Effect of irrigation with lead and cadmium on the growth and on the metal uptake of cauliflower, spinach and parsley. *J. Environ. Sci. Health* A30:831–849.

SETHUNATHAN, N., 1989. Biodegradation of pesticides in tropical rice ecosystems. Ch. 5.2 in P. Bourdeau, J. A. Haines, W. Klein, and C. R. Krishna Murti (eds.), *Ecotoxicology and Climate.* Wiley and Sons, Chichester, U.K., 392 p.

SHANKAR, B., E. A. DEVUYST, D. C. WHITE, J. B. BRADEN, and R. H. HORNBAKER, 2000. Nitrate abatement practices, farm profits, and lake water quality: A central Illinois case study. *J. Soil Water Cons.* 55:296–303.

SHIPITALO, M. J., and F. GIBBS, 2000. Potential of earthworm burrows to transmit injected animal wastes to tile drains. *Soil Sci. Soc. Am. J.* 64:2103–2109.

SLETTELAND, T. B., 1995. Fighting for water in the Klamath Basin. *Illahee* 11:99–102.

SMITH, S., 1992. Ecological and health effects of chemical pollution. Ch. 8 in R. M. Harrison (ed.), *Understanding Our Environment: An Introduction to Environmental Chemistry and Pollution,* 2nd ed. Royal Society of Chemistry, Cambridge, U.K., 326 p.

SPEIDEL, D. H., and A. F. AGNEW, 1988. The world water budget. Ch. 3 in D. H. Speidel, L. C. Ruedisili, and A. F. Agnew (eds.), *Perspectives on Water, Uses and Abuses.* Oxford Univ. Press, New York, 388 p.

STEINNES, E., 1995. Mercury. Ch. 11 in B. J. Alloway (ed.), *Heavy Metals in Soils,* 2nd ed. Blackie Academic and Professional, London, 368 p.

STIGLIANI, W. M., P. R. JAFFÉ, and S. ANDERBERG, 1993. Heavy metal pollution in the Rhine Basin. *Environ. Sci. Technol.* 27:786–793.

VESPER, S. J., R. DONOVAN-BRAND, K. P. PARIS, S. R. AL-ABED, J. A. RYAN, and W. J. DAVIS-HOOVER, 1996. Microbial removal of lead from solid media and soil. *Water, Air, and Soil Pollut.* 86:207–219.

VIRARAGHAVAN, T., and M. M. DRONAMRAJU, 1993. Removal of copper, nickel and zinc from wastewater by adsorption using peat. *J. Environ. Sci. Health* A28:1261–1276.

WANG, W., 1992. Toxicity reduction of photo processing wastewaters. *J. Environ. Sci. Health* A27:1313–1328.

WIXSON, B. G., and B. E. DAVIES, 1994. Guidelines for lead in soil. *Environ. Sci. Technol.* 28:26A–31A.

YOUSEF, Y. A., and L. L. YU, 1992. Potential contamination of groundwater from Cu, Pb, and Zn in wet detention ponds receiving highway runoff. *J. Environ. Sci. Health* A27:1033–1044.

CHAPTER

18

ECONOMICS OF SOIL AND WATER CONSERVATION

It would be much easier to conserve natural resources such as soil and water if cost were not a consideration, and it is much easier to make an economic analysis of a farming enterprise if erosion and pollution problems are ignored. Many disagreements have arisen because different people have different viewpoints regarding the allocation and use of natural resources. Some want to use them for one purpose, some for another, and others want to keep them for the future. Some people are very concerned with the moral problem of wasteful consumption, whereas others are driven by the profit motive. These viewpoints must be resolved somehow, and it cannot be done by ignoring the opposition. In particular, conservationists need to consider economics when they make recommendations, and economists need to consider the long-term, social, and environmental values of conservation practices.

The costs of soil and water conservation are usually obvious and therefore seldom overlooked. The returns are more obscure, especially where conservation merely preserves productive potential that would otherwise be lost. Lowered potential may be masked by weather variations, more fertilizer, new crop varieties, and improved crop management. It takes a perceptive person to properly balance the obscure costs of inaction against the obvious costs of acting to conserve soil and avoid pollution. Perception must be followed by decisive action if the soil and water are to be conserved.

Soil-conserving practices often provide long-term benefits in exchange for immediate costs. It is unfortunate, but not surprising, that the evident needs of the present often win over more important but vague needs of the future. Conservation practices that include short-term benefits are therefore easier to promote than those that are strictly long term. Even so, some practices such as terracing are used in spite of the long time required to recover their costs. Changes in tillage including contour tillage, reduced tillage, and no-till systems are usually profitable, but their adoption has been slowed by appearance and convenience factors and by pest problems.

Many people may benefit from a landowner's conservation practices. For example, soil that is held on a field neither muddies the water of a stream nor becomes sediment on other land. People living downstream therefore benefit from soil conservation practices in upstream areas. In a broader sense, everyone has a stake in the productive potential of the land. Such interest is represented to some degree by governmental participation in soil conservation through research, education, cost sharing, tax benefits, and legal actions. This participation is justified by the importance of productive potential as a national asset. Reduced productive potential is the most serious long-term effect of erosion. Government participation can also be justified when conservation practices are installed for the sake of reducing water pollution and other forms of environmental degradation that impact people outside the area where the practices are installed. In fact, some problems such as global warming and the ozone hole are worldwide in both causes and effects.

The value of natural resources such as clean air and water and productive soil is obscure and therefore often ignored in economic analyses. Too often, they are left out of evaluations, thus effectively being assigned a value of zero (Lingard, 1994). Difficulty in making accurate economic analyses does not justify ignoring the costs and benefits associated with environmental protection. Depletion of natural resources should be a concern of all people.

Economic emphases of soil conservation have shifted considerably during the last 30 years from the on-farm value of tons of soil to the public cost of sediment pollution. Both aspects are important, but the pollution problem was largely ignored until environmental concerns became a public issue. On-farm losses are receiving less emphasis now because they can be partly offset by using more fertilizer (Nowak, 1988). Similarly, the benefits from water conservation include not only the increased crop production potential of the water stored in the soil but also the reduced flooding and water-pollution problems in downstream areas (El-Swaify, 1991).

18–1 BENEFITS FROM SOIL AND WATER CONSERVATION

Soil and water conservation yield many kinds of benefits. Those to be considered in this section include increases in net returns from land, retention of productive potential, reductions in erosion losses and sediment damage, and environmental benefits.

18–1.1 Increases in Net Returns

Some conservation practices produce an immediate profit, some lead to a delayed profit, and some produce no monetary profit but are used for aesthetic or other nonmonetary reasons. Some practices are usually considered under the heading of good management rather than conservation because they increase profits. Fertilization according to soil tests is such a practice. Fertilizer recommendations are usually based on an economically optimum field response. Little thought is given to the reduced soil loss that results from improved plant growth. Proper fertilization is actually an important soil conservation practice. Excessive fertilization, however, should be avoided not only for economic reasons but also because the excess nutrients are likely to become water pollutants.

Equipment is now available that can apply variable rates of fertilizer, lime, or pesticides according to the needs of small areas within a field. The use of variable-rate equipment, often guided by the global positioning system (GPS), is called *site-specific farming* or *precision agriculture* (Sylvester-Bradley et al., 1999). Decisions about the rates to be applied should be based on as much information as possible, including soil maps, soil fertility test data (probably from more samples representing smaller areas than has been customary in the past), yield data from previous crops (preferably gathered during harvest by automated recording equipment), and visual observations of growth responses, nutrient deficiency symptoms, and pest problems. A computerized compilation of such data is called a *geographical information system* (GIS) and can be used to produce field maps to represent the planned application rates. The equipment and operating costs of such a system are offset by savings where excessive amounts of materials would have been applied and increased yields where inadequate amounts would have been applied (Batte, 2000). Additional benefits such as reduced erosion and reduced pollution are also likely to occur. The fertilizer application equipment required is large and expensive, so it is most likely to be used by either a large farm or a contractor who can work on several farms.

Conservation Tillage. The various forms of conservation tillage discussed in Chapter 9 are among the most cost-effective practices available for reducing soil erosion and water pollution (Richards, 1991). In fact, they often increase net returns. Most of them reduce both the time and the energy used for tillage and thereby reduce costs. Table 18–1 shows savings of 18% for machinery and 64% for fuel when slot-planted corn was compared to conventionally tilled dryland corn. There was a small increase in pesticide input, but its cost was only a fraction of the machinery and fuel savings. Similar fuel savings are indicated in data from Ohio, as shown in Figure 18–1. The Ohio data showed equal average corn yields whether conventional tillage or no-tillage planting was used (Triplett and Van Doren, 1977). There were, however, differences among soils. No-tillage yielded more than conventional tillage did on the well-drained Wooster silt loam, but the reverse was true on the less-well-drained Hoytville silty clay loam. Time savings also can be significant by either reducing the need to hire additional help or allowing the operator to farm more land.

TABLE 18–1 ENERGY REQUIRED IN DIESEL FUEL EQUIVALENT (GAL/AC) TO PRODUCE CORN IN NEBRASKA BY FOUR TYPES OF TILLAGE UNDER DRYLAND AND IRRIGATED AGRICULTURE

	Dryland				Irrigated			
Input	Conventional	Disk and plant	Tillplant	Slotplant	Conventional	Disk and plant	Tillplant	Slotplant
Machinery	2.55	2.31	2.19	2.10	6.13	5.89	5.76	5.67
Fuel	5.15	2.74	2.60	1.87	5.15	2.74	2.60	1.87
Transport grain	1.81	1.81	1.81	1.81	3.02	3.02	3.02	3.02
Fertilizer	21.67	21.67	21.67	21.67	30.22	30.22	30.22	30.22
Pesticides	1.13	1.13	1.13	1.43	1.13	1.13	1.13	1.43
Drying	8.23	8.23	8.23	8.23	13.70	13.70	13.70	13.70
Irrigation	-	-	-	-	30.87	30.87	30.87	30.87
Total (gal/ac)	40.54	37.89	37.63	37.11	90.22	87.57	87.30	86.78
Total (l/ha)[a]	379.4	354.5	352.1	347.3	844.3	819.4	816.9	812.1

[a]Gal/ac × 9.35 = l/ha.

Source: Calculated from data of Wittmus et al., 1975.

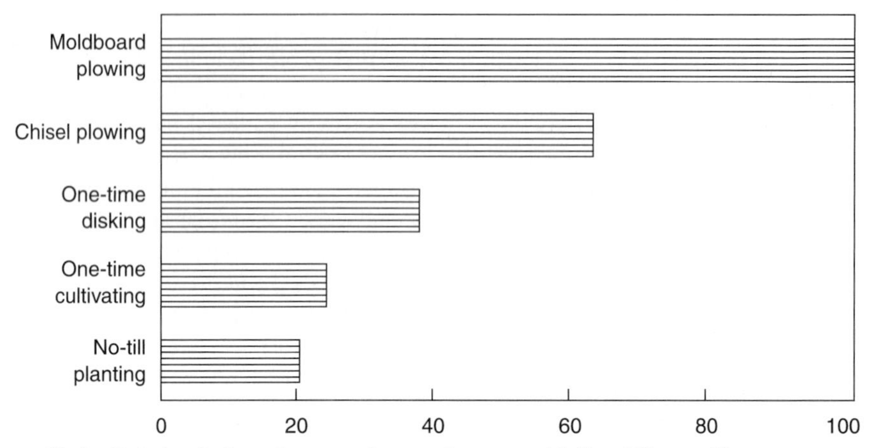

Figure 18–1 Relative fuel requirements for growing corn with five different tillage systems. (From Triplett and Van Doren, 1977.)

Weersink et al. (1992) calculated savings up to 61% in labor costs for no-till farming as compared to conventional moldboard plow systems.

Several studies have shown that changing tillage practices has no direct effect on average corn yields as long as comparable stands and adequate weed control are obtained (Van Doren et al., 1976). Minimum tillage usually shows an advantage on sloping land, where it helps conserve soil and water. Conventional tillage usually yields more than no-tillage where continuous corn is grown on land wet enough to be conducive to root disease. Wheat- and some other crop-yield responses to no-tillage have generally been less favorable than corn response has been. Reduced-tillage yields usually improve with time as the operator gains experience and the soil structure gradually improves as a result of less disruption and increased organic matter content.

Fuel and machinery savings from reduced tillage make it possible for producers to make larger profits from equal yields, even if they use more pesticides (Smart and Bradford, 1999). Rising energy costs and new types of herbicides, insecticides, and fungicides have led to a marked expansion of minimum tillage in recent years. Some farmers have calculated savings of $20 per acre with no decrease in yields when they changed to no-tillage (Bruggink, 1996). Estimates indicate that minimum tillage was used on 3.8 million acres in the United States in 1963, 34 million acres in 1974, 72 million acres in 1989, and 99 million acres in 1994 (Allen et al., 1977; Bull and Sandretto, 1996). Fuel savings of hundreds of millions of dollars per year could result from such changes, and the nation benefits from reduced petroleum consumption.

Establishing Permanent Vegetation. Permanent vegetation in the form of grassed waterways, shelterbelts, and plantings on steep slopes or other problem areas is effective for reducing erosion, but is it economical? Sometimes the answer depends on the basis chosen for comparison. A grassed waterway takes land out of cultivation and provides little or no salable product in return. It might seem to be a very uneconomical practice, unless one considers the potential cost of repairing a gully at a later date if the waterway is not established. Considering the cost of either repairing the gully or always working around it could show that a grassed waterway is the most economical choice. A similar question arises regarding shelterbelts. Can they be expected to repay the cost of their establishment and offset the lost crop production from the land they occupy? Brandle et al. (1992) found that they can. Wind protection provided more than enough yield increase in the sheltered field to pay for the costs, and government payments added to the net profit for the farmer. The shelterbelts also provided societal benefits in the form of reduced air pollution.

Water Conservation. The effect of water conservation on crop yield depends on the amount of water already in the soil, the weather, and the plant needs at the time. Water conserved before a dry period sometimes makes the difference between success and crop failure. Excess water retained during a wet period can damage the crop and reduce yields, especially in low spots. Good water control sometimes helps even in wet periods, though, by holding water on drier sloping land rather than letting it run onto level areas below.

Times when water conservation will increase yield are common in, but not limited to, semiarid climates. Subhumid and even humid climates have dry periods during which plant growth suffers. Periods of water deficit in temperate regions are common during the warmest months, as illustrated in Figure 18–2, because water requirements are highest then.

Contour tillage and other water conservation practices described in Chapter 13 might make an average of 2 in. (50 mm) more water available for plant growth on sloping lands during the growing season. This much water would typically increase yields of small grains or soybeans by 7 to 8 bu/ac (4 to 5 q/ha), of corn by 20 to 25 bu/ac (12 to 15 q/ha), or of lint cotton by 40 to 50 lb/ac (45 to 55 kg/ha).

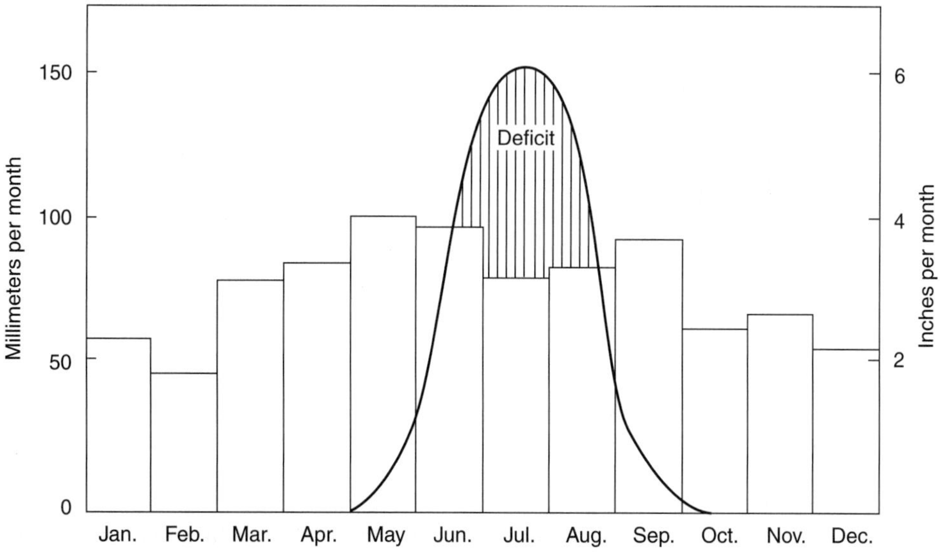

Figure 18–2 Average monthly precipitation in Illinois as compared to the water needs of corn.

Water conservation on irrigated land is also profitable, but in a different way. Any water conserved can be used to irrigate additional land and thereby increase profits. Individuals may either expand their irrigated land or reduce their water consumption.

Soil Drainage Profits. Soil drainage is another profitable practice associated with soil conservation. Land is seldom drained where it does not add to both profit and convenience. As an example, Bornstein and Fife (1973) calculated 13 and 19% annual returns on an investment in drains spaced 100 ft (30 m) and 200 ft (60 m) apart, respectively, on poorly drained soils in Vermont.

Irrigation Profits. Irrigation systems typically cost several hundred dollars per acre to install and add some tens of dollars per acre to the annual inputs of seed, fertilizer, and equipment operation, but they also increase crop yields, and profits (Opie, 1993). Crop prices that make rainfed agriculture profitable are likely to also make irrigation profitable in many places. Sanghi and Klepper (1977) concluded that exploiting groundwater for irrigation results in so much short-term profit that groundwater conservation should not be left to market forces alone. Social costs, future water needs, and the possibility of excessive erosion should also be considered.

Pest-Control Profits. Weeds, insects, and plant diseases can cause increased erosion by reducing the stand of the crop. Controlling these pests can reduce erosion and increase profits at the same time. As an example, Peters (1975) showed that use of aldrin or heptachlor to control insects on Iowa corn increased yields an average of 8 to 10 bu/ac (5 to 6 q/ha) over a 20-year period. Insecticide treatment increased yields on both good and poor soils.

Deferred Profits. Fertilizer, tillage, water management, and pest-control practices should increase profits the same year they are applied. Several years may be required to repay the investment costs of drainage, irrigation, and new machinery, but increased returns usually begin with the next harvest. The returns may be delayed longer when practices such as terracing, revegetating grazing land, replanting forests, or soil reclamation are used. These practices must therefore be considered as long-term investments.

Terracing is a very effective means of conserving soil, but it is also very expensive. To be economical, it must permit yields to be maintained on a long-term basis under more intensive use than would otherwise be feasible. Nonetheless, yields may be reduced for a few years on the areas where topsoil has been removed to build the terraces. The yield reduction can be minimized by stockpiling the topsoil and replacing it after the terraces are built, but this adds to the cost. Often it is more practical to use fertilizers, manure, green manure, and tillage practices to improve the exposed subsoil than to replace topsoil. The research of Larney et al. (2000) on four soils in southern Alberta, Canada, showed that applications of cattle manure were more effective than chemical fertilizers for restoring productivity to areas where topsoil had been removed. Similar results have been obtained by other researchers and may be considered the norm. Long-term reductions in soil loss should eventually result in higher yields than would have been obtained without the terraces.

Forage crops and trees conserve soil effectively, but they take time to grow. Good range management may require several years of little or no grazing to allow desirable grasses to recuperate from drought or mismanagement. Increased production in later years will make up for this period of decreased use. A grazing program that maintains productivity should then be established (see Chapter 12). Trees take even longer to grow and become marketable. The value is there, however, in the form of increased land value as the trees grow toward marketable size. Annual crops can sometimes be grown between tree rows while the trees are small even where the trees will be relied upon for economic value in later years (Countryman and Murrow, 2000).

Reclamation of sodic soils is another example of delayed profits. Sodic spots are often so unproductive that *any* plant growth is an improvement. Improving them so they will not only grow something but will produce a profitable crop takes time (often years) as well as investment (see Chapter 15).

Deferred profits often create a conflict between long-term and short-term values. The need for short-term income can lead to exploitation of soil and water with maximum removal and as little input as possible. Degraded vegetative cover and serious soil loss can occur under short-term exploitation of land.

18–1.2 Retaining Productive Potential

Nutrient losses and flood damage cause decreases in production for the year in which they occur, but they do not always reduce the future productivity of the land. In fact, sediment deposited by a flood may improve the soil's productive potential. However, the area that loses the soil will lose productive potential.

Reduced yield potential resulting from soil loss ranges all the way from 0 to 100%. For example, Olson and Nizeyimana (1988) found corn-yield reductions ranging from 5 to 24% on seven loess-derived soils in Illinois. The yield reduction depends greatly on the amount of erosion and on the nature of the topsoil, subsoil, and soil parent material. A 2-in. (5-cm) loss from the top of a deep, uniform soil may have little long-term impact. A 2-in. loss from another soil may cause clayey subsoil to become part of the plow layer, and productive potential may drop by 10 to 25%. An equal soil loss from a soil that is shallow to bedrock might cause a complete crop failure. The most unfortunate feature of such losses is their permanence. Future yields are decreased along with those of the year when the damage occurs. Colacicco et al. (1989) estimate that a continuation of wind and water erosion at their 1982 rates would cause U.S. yields of corn, soybean, and cotton to decline by an average of 4.6, 3.5, and 4.5% respectively during the next 100 years. They estimated that wheat yields would decline 1.6%, but that legume hay yields would decline only 0.8%. These estimated declines would have been larger had their model not assumed that fertilizer use would increase.

Soil erosion is seldom uniform across a large area of land. The value of soil conservation for preserving productive potential therefore varies for areas within a field. Severely eroded spots may be detected easily by the subsoil color and sparse vegetation. The field as a whole may produce reasonably well, but these spots will not. Calculation of costs and returns on the eroded spots will frequently show a loss from cropping that reduces the profit obtained from the rest of the field. The damaging effect is sometimes increased by sediment smothering the crop in low areas.

Productive fields that contain severely eroded spots pose management problems. Using them intensively causes more runoff, erosion, and sedimentation. Applying extra fertilizer to try to make the eroded spots produce well is a waste of inputs because these spots are less responsive and have lower potential than the rest of the field. Limiting the land use to a system suitable for the problem spots produces little income and fails to use the capability of the rest of the field. Some spots can be protected with contouring or terraces or by the use of minimum tillage. These practices are helpful because they conserve water as well as soil, and the increased water infiltration reduces the droughtiness of eroded spots. Often it is best to exclude the spots by either relocating the field boundary or going around "patches" in the field. Permanent vegetation on the eroded spots and appropriate crops on the rest maximizes the productivity of the field as a whole. The permanent vegetation may both conserve soil and be useful for wildlife or for a forage crop.

18–1.3 Reduced Erosion Losses

The value of lost soil is an obvious cost of soil erosion. Reductions in soil loss are obvious benefits of soil conservation. The value of each ton of eroded soil varies according to its clay, organic matter, and nutrient contents. Both wind and water remove fine particles while coarser particles are left behind. A ton of eroded soil is usually more fertile, and therefore more valuable, than a ton of average soil because of textural sorting combined with the effects of organic-matter accumulation, fertilization, and liming that are concentrated in the upper part of the soil.

The gross value of the plant nutrients in eroded topsoil averages about $5 per ton ($5.50 per metric ton), as shown by the calculations in Note 18–1.

NOTE 18–1
FERTILITY VALUE OF A TON OF SOIL

The fertility value of a ton of eroded soil can be calculated by using the following assumed values:

1. An average temperate-region topsoil contains about 3% organic matter, but an average ton of sediment contains more colloids, so it has about 6% organic matter.
2. The soil organic matter is about 5% N, 0.5% P, and little K apart from that held by cation exchange. The mineral matter contains little N, about as much total P as the organic matter, and about 1.5% K. (Four billion tons of eroded soil contain about the same amount of N and P and more than 10 times as much total K as the annual application of commercial fertilizer.)
3. The amounts of these nutrients that will become available annually in a temperate region can be estimated as 3% of the N, 5% of the P, and 1% of the K. (These percentages might be as high as 25% of the N in a tropical climate and as low as 1% of the N in a cold climate with the other nutrients in proportion.)

Using the above figures and 1995 prices in the United States, an average ton of eroded temperate-region soil has the following fertility value:

	Total content		Available in 1 year	
	Amount (lb/ton)	Value (1995)	Amount (lb/ton)	Value (1995)
Nitrogen	6	$1.20	0.2	$0.04
Phosphorus	1.2	0.30	0.06	0.015
Potassium	28	2.70	0.28	0.025
Other nutrients		0.80		0.02
Total value		$5.00		$0.10

A more realistic but considerably more complex value can be obtained by basing the calculations on available nutrients. This calculation requires computation of a discounted value for all nutrients that will be released in future years, as illustrated in Note 18–2. The $5 figure will be used here for simplicity and because the discount might well be offset by the value of soil components other than plant nutrients. Other items that should be added are the cost of hauling and spreading the fertilizer used to replace the lost nutrients and the cost of repairing damage done to terraces, fences, roads, and other structures by either the erosion process or the sediment it produces (Gunatilake and Vieth, 2000).

For many years, soil conservationists used an estimate of 4 billion tons of soil loss annually in the United States, with 3 billion tons of that being lost from cropped fields and 1 billion tons from all other land. As shown in Table 18–2, this value was still reasonably accurate in 1982, but it had been greatly reduced by 1992. Most of the difference can be attributed to conservation practices such as reduced tillage and to removal of land from cultivation through the Conservation Reserve Program provisions of the 1985 Food Security Act (Uri, 2000). This program is discussed in Section 18–3.3.

Applying the $5 per ton value to the 4 billion tons of soil loss annually prior to 1982 gives a nutrient-value cost of $20 billion per year that was reduced to about $15 billion per year by 1992, although the actual improvement may have been less than this because the CRP land that was removed from cultivation still had some soil loss that should be added to the amount lost from non-cropland. The amount of CRP land has been fairly stable since 1992, but the use of reduced tillage has continued to grow, so the total soil loss should show

TABLE 18–2 ESTIMATED AVERAGE ANNUAL SOIL EROSION ON CROPLAND IN THE UNITED STATES FOR 1982 AND 1992

	Cropland (million ac)	Average annual rate		Total soil loss from cropland (billion tons/yr)
		Water erosion (tons/ac)	Wind erosion (tons/ac)	
1982	420.9	4.1	3.3	3.1
1992	382.3	3.1	2.4	2.1

Source: Based on U.S. National Resources Inventory data cited by Uri (2000).

TABLE 18–3 ESTIMATED AVERAGE ANNUAL FLOOD DAMAGE IN THE UNITED STATES (EXCLUDING ALASKA AND HAWAII)

Type of damage	Annual cost in 1988 dollars (millions)	Percentage of total cost
Upstream damage (from drainages < 250,000 ac [100,000 ha]):		
Crops and pasture	$1,750	26.9
Other agricultural	780	12.0
Sediment	340	5.2
Nonagricultural	700	10.8
Indirect	350	5.4
Total upstream damage	$3,920	60.3
Downstream damage:		
Metropolitan areas	$1,550	23.8
Agricultural	850	13.1
Other	180	2.8
Total downstream damage	$2,580	39.7
Total flood damage	$6,500	100.0

Source: Calculated from several sources.

a further but less dramatic decrease after the 1992 data. Soil erosion can be slowed, but it cannot be completely stopped. Part of it is natural geologic erosion, and some accelerated erosion is unavoidable on cropped land, pastured land, and construction sites. Much land, however, is being eroded at excessive rates that could be reduced.

18–1.4 Reduced Sedimentation Damage

Sedimentation damage occurs both in fields near the source of the sediment and in downstream areas. The soil and water damage caused by floods are inseparable. Estimated amounts of various types of flood damage are shown in Table 18–3. The total flood damage in the United States is approximately $6.5 billion per year. Some damage is unavoidable, but hundreds of millions of dollars per year could be saved through flood-prevention efforts.

About 60% of the flood damage shown in Table 18–3 occurs in upstream areas along small and moderate-sized streams where the drainage area above the site is less than 250,000 ac. One inference is that flood control in the upstream areas can do more good than large downstream structures. Upstream floods typically result from intense rainfall of limited duration. Terraces, contour tillage, and small upstream reservoirs can reduce the magnitude of upstream floods. Establishing protective vegetation and keeping buildings off floodplains are other means of reducing losses.

Downstream floods result from more widespread and longer-lasting storm systems than upstream floods. The downstream damage is more spectacular because it affects more large cities and is concentrated in occasional large events. There may be enough warning for sandbag levees to be erected and for people to be evacuated. Sometimes the damage is averted but other times the levees fail and major damage occurs. Levees provide some containment but they also cause the river to deposit sediment in its bed and thus raise its level to a more hazardous position. The lower reaches of rivers such as the Yellow River in China and the Mississippi River in the United States now flow some tens of feet above their floodplains for long distances, held only by their levees. Unfortunately, many people live on floodplains where a breach in a levee can permit the elevated river to inundate them. The surest way to reduce downstream flood disasters is to avoid building on floodplains.

Agriculture sustains more than half of the flood damage, as shown in Table 18–3. Most of the land where flood-control practices need to be applied is used for agricultural purposes. Flood control is therefore very significant to farmers and ranchers. Fortunately, soil and water conservation practices also help reduce flooding.

18–1.5 Environmental Benefits from Conservation

The soil conservation movement in the United States began in the 1930s out of concern for land that was suffering from erosion. Loss of land and reduced productivity were emphasized. Other environmental damage, including pollution caused by sediment, began to receive attention in the 1960s. Soil erosion damages property where the soil is deposited as well as where the soil loss occurred. The entire process is an environmental concern.

Several types of soil and water pollution can be reduced by soil conservation practices (see Chapters 16 and 17). Sediment and the chemicals often associated with it are prime examples. Soil conservation practices can keep most of the soil and chemicals in the fields where they are useful. Reduced runoff makes the water more useful. Landowners benefit from the productive value of the soil and water that are conserved. Downstream benefits occur when sedimentation, eutrophication, and chemical toxicity problems are reduced. Most of these benefits accrue to other landowners and to the general public, however, rather than to the person who built the terraces, changed tillage practices, or planted protective vegetation. Wildlife also benefit from vegetation planted to control erosion.

Dollar values can be attached to some, but not all, of the environmental benefits from conservation. For example, how much is it worth to be able to swim in clean water such as that shown in Figure 18–3? How much credit for reduced pollution should be given when a ton of composted sewage sludge is applied to a field instead of being allowed to enter a stream?

18–2 COSTS OF CONSERVATION PRACTICES

The costs of conservation practices are almost as hard to evaluate as the benefits. The prices of some practices are variable enough to make it difficult to obtain accurate averages. Furthermore, direct payments do not reflect the management input of the landowner and other persons who provide the required technology. A lot of experience and probably some research were needed to learn how to make the practice feasible.

Opportunity costs are a means of considering some of the less-obvious costs of conservation. For example, one must forgo the opportunity to grow a crop on the area where a grassed waterway is planted. The opportunity to make a profit is real, even though a gully might destroy it later.

Figure 18–3 Swimming in Lake Washington after new sewage-treatment plants were built and other pollution-control practices were established at a cost of $121,000,000. This beach is shown in its earlier, polluted condition in Figure 17–1. (Courtesy EPA-DOCUMERICA.)

The costs of conservation do not stop when a practice is installed. Opportunity costs remain like a ghost in the background. Someone will be tempted to plow up the waterways, fencerows, odd corners, and steep slopes, even if the erosion hazard is great. Educational programs are needed to teach new landowners and operators the lessons that their predecessors learned earlier. The cost of the educational programs is also chargeable to conservation.

Maintenance is an obvious continuing cost of conservation. Terrace channels and waterways must be cleaned and occasionally reshaped. Drainage systems must be kept open. Vegetation must be fertilized, mowed, reestablished, or otherwise tended.

18–2.1 Direct and Indirect Costs

Most conservation practices involve both out-of-pocket *direct costs* and less-tangible *indirect costs*. Building terraces involves large direct costs, usually with the expectation that more intensive use will increase profit and repay the costs. The alternative of less-intensive land use involves little or no direct costs but has indirect costs in the form of lower gross returns from the land.

Practices that involve less-intensive cropping tend to be used only where necessary because crop value is reduced year after year. Establishing permanent forage crops or other protective vegetation drastically reduces erosion but does not usually produce a high-value crop. Strip cropping and crop rotations have intermediate effects on both erosion and in-

TABLE 18–4 DIRECT COSTS OF SELECTED CONSERVATION PRACTICES

Practice	Cost (English units)	Cost (metric units)
Practices that produce a net short-term profit (direct costs compensated within a year):		
Basic good management:		
Adapted varieties and high-quality seed		
Proper timing of operations		
Optimum plant populations and spacings		
Fertilizing and liming as indicated by a soil test		
Conservation tillage:		
Contour tillage		
Reduced tillage		
No-tillage		
Practices with indirect costs but little or no direct costs:		
Land retirement		
Less-intensive cropping:		
Crop rotations		
Strip cropping		
Practices with large direct costs:		
Water management practices:		
Land smoothing	\$100—400/ac	\$250—1000/ha
Tile drainage	\$200—500/ac	\$500—1200/ha
Irrigation	\$500—1000/ac	\$1200—2500/ha
Land reclamation:		
Saline soils	\$50—200/ac	\$100—500/ha
Sodic soils	\$300—3000/ac	\$700—7000/ha
Sulfuric soils	\$300—3000/ac	\$700—7000/ha
Smoothing spoil heaps	\$300—600/ac	\$700—1500/ha
Establishing vegetation:		
Cover crops	\$50—150/ac	\$100—400/ha
Grassed waterways	\$100—300/ac	\$250—800/ha
Woodland or wildlife plantings	\$100—300/ac	\$250—800/ha
Terracing	\$1000—10,000/mi	\$600—6000/km
Fencing	\$1000—2000/mi	\$600—1200/km

come. None of these practices is very costly to install, but indirect costs sometimes prevent their use.

Large direct costs may not prevent a conservation practice from being used extensively if the costs can be spread over many years. Profit and convenience may justify the investment in a practice that is costly to install but inexpensive to maintain.

Typical direct costs of several conservation practices are shown in Table 18–4. The costs are presented as ranges to allow for variations in the work done and in local charges. Indirect costs are too variable to be generalized and can be estimated only when the circumstances of a specific situation are known.

The practices in the upper part of Table 18–4 do not show any direct costs because they either increase profits the year they are applied, or they involve little or no out-of-pocket expenses. They are legitimate conservation practices, however, because they reduce the amount of soil loss. Being profitable as well as soil conserving makes practices such as reduced tillage and proper fertilization especially easy to promote.

18–2.2 Conservation Practices as Investments

Long-lasting conservation practices are logically considered as investments. A drainage project, for example, is not expected to repay its total cost the first year. Sometimes the installation cost exceeds the gross returns from any one year's crop. It may take several years to repay the cost and make a net profit.

A conservation practice is a good investment if it produces a satisfactory rate of return over a long-enough lifetime. For example, an irrigation system with a lifetime of 20 years could be considered a good investment if the increased profits each year were as high as the annual payments on a 20-year mortgage with the same principal amount. A $20,000 system must increase the annual profit by $2037/yr to yield an 8% return on the investment and repay its costs in 20 years. Such investments may also be analyzed by comparing their *net present value* (NPV) to their costs (Note 18–2).

NOTE 18–2
CALCULATION OF NET PRESENT VALUE

A given amount of present benefits is generally preferred over an equal amount of future benefits. Interest rates can be used to determine either the future value of present benefits or the present value of future benefits. The general formula for calculating the present value as a fraction of the future value is

$$\frac{1}{(1 + x)^t}$$

where x is the interest rate in decimal form and t is the number of interest periods (usually years).

A practice with anticipated benefits worth $100 per year for a projected 10-year lifetime will serve as an example. An interest rate of 10% per year will be assumed.

Year (t)	$1/(1 + x)^t$	Anticipated benefits	Net present value of anticipated benefits
0	1.00	$ 100	$100
1	0.91	100	91
2	0.83	100	83
3	0.75	100	75
4	0.68	100	68
5	0.62	100	62
6	0.56	100	56
7	0.51	100	51
8	0.47	100	47
9	0.43	100	43
Totals		$1000	$676

The percentage of total returns counted as net present value (67.6% in the above example) is sensitive to both the interest rate and the time span involved. The above method can be used to calculate the following net present values as percentages of total returns:

	Interest rates (%)					
Time period (years)	4	6	8	10	12	15
5	92.6	89.3	86.2	83.4	80.7	77.1
10	84.4	78.0	72.5	67.6	63.3	57.7
20	70.7	60.8	53.0	46.8	41.8	36.0

Future costs can be discounted and summed in the same manner. Future costs and returns may be handled separately, or their differences may be calculated for each time period and discounted to obtain the net present value.

Most water management, land reclamation, and terracing projects are installed in anticipation of long-term profits. These practices are therefore good investments from economic considerations alone. Some other practices, such as plantings that provide good cover but little income, must be justified on the basis of nonmonetary values such as aesthetic considerations.

18–2.3 Costs of Soil Loss and Water Pollution Restraints

The first mandatory soil and water conservation law in the United States was passed in Iowa in 1971. This legislation authorized Iowa soil conservation districts to establish legal limits for soil-loss rates from each type of soil in the state. The U.S. Congress passed the Federal Water Pollution Control Act Amendments in 1972 with the objective of making all possible streams in the nation clean enough for swimming and fishing by 1983. The Food Security Act of 1985 added strong incentives for conserving soil and water. These laws and several other more recent laws place restraints on rates of soil loss. Landowners, taxpayers, and consumers of agricultural products will bear large direct and indirect costs in meeting legal standards that limit soil loss.

Heady and Nicol (1974) used a computer analysis to estimate the effects of a soil-loss limit of 5 tons/ac (11 mt/ha) on soil loss and crop prices in the year 2000. They concluded that if such a limit were enforced, the average annual soil loss from U.S. cropland would be reduced from 10 tons/ac (22 mt/ha) without the limit to 3 tons/ac (6 mt/ha) with the limit (Table 18–5). Crop prices under the 5 ton/ac limit in the year 2000 were estimated to be 7% higher for corn, 3% higher for hay, 4% higher for cattle, and 5% higher for hogs. Everyone's grocery bill would increase by about 4 or 5%.

The economic effects of soil-loss reductions depend greatly on how and where they are achieved. Taylor (1977) considered the effects that each of three approaches would have on farm income in the High Plains and in the Rolling Plains of western Texas. He concluded that terrace subsidies would increase farm income and reduce soil loss, but not enough to meet

TABLE 18–5 ESTIMATED AVERAGE ANNUAL SOIL LOSS IN REGIONS OF THE UNITED STATES IN THE YEAR 2000 (SOIL LOSS RESTRICTION TO BE MET BY PRACTICES SUCH AS CONTOURING, STRIP CROPPING, TERRACING, AND REDUCED TILLAGE, WITH CHANGES IN LAND USE WHERE REQUIRED)

	Estimated average annual soil loss			
			Where loss is limited to	
Region	Where loss is unrestricted		5 tons/ac	(11 mt/ha)
	tons/ac	mt/ha	tons/ac	mt/ha
National	9.9	22.2	2.8	6.3
North Atlantic	9.0	20.2	3.5	7.8
South Atlantic	21.5	48.2	3.3	7.4
North Central	9.2	20.6	2.8	6.3
South Central	15.1	33.8	3.6	8.1
Great Plains	3.2	7.2	1.5	3.4
Northwest	2.3	5.2	1.7	3.8
Southwest	3.3	7.4	2.5	5.8

Source: Calculated from Heady and Nicol, 1974.

tolerance limits. A tax on each ton of soil loss would reduce both net returns and soil loss. Establishing legal limits was the only approach that was projected to reduce soil loss below tolerance value in all places. Such limits would cause much land in the Rolling Plains to be idled. The acreage of cotton and wheat in that area would be reduced by 56% and net returns would be reduced by 63%. The projected effects in the High Plains were less severe—much grain sorghum would be replaced by wheat, and net returns would be reduced by 15%.

Soil-loss limitations that are applied only to certain areas or states place them at an economic disadvantage. Broader limits that restrict soil loss in the entire nation are likely to increase prices for the crops and may increase the net returns to the producers.

18–2.4 Costs of Erosion Control on Non-Agricultural Land

Very high erosion rates sometimes occur on constructions sites, mined areas, and other non-agricultural lands. Some soil-loss legislation is directed at erosion control for these areas. For example, the 1971 Iowa law mentioned in the previous section deals with the need for "soil and water conservation practices" on agricultural and horticultural lands and for "erosion-control practices" that apply to construction sites and commercial or industrial developments. Such legislation is likely to disregard profitability to the landowners or contractors involved because it is based on the public good. Perhaps a construction project should not be undertaken if it cannot support the costs of practices needed to control erosion and pollution. Similarly, mining is really not worthwhile if it cannot afford to contain and treat any acid runoff it generates. After all, if the persons involved do not handle the matter, the public will have to either suffer the environmental consequences or pay for repairing the damage. Furthermore, the persons involved often live in the vicinity and benefit from the control measures along with the rest of the public.

Erosion control practices are not always as burdensome as they may seem. Often there is some compensation in the form of associated values and public esteem. For example, Herzog et al. (2000) examined the economics of maintaining a vegetative cover on construction sites during a period of development. They surveyed a group of realtors and a group of potential homebuyers regarding the relative value of "green lots" with vegetative cover and "brown lots" without it. The realtors valued the green lots an average of $744 per lot higher than comparable brown lots, and the potential home buyers assigned them an average value $750 higher than the brown lots. The cost of seeding a lot was $100 to $300, so it was actually a profitable investment.

18–3 PAYING FOR SOIL AND WATER CONSERVATION

People are usually willing to pay for a conservation practice that produces short-term profits. Other practices that require long terms to repay their costs are adopted more cautiously, especially if large investments are required. The person paying for the practice must be assured of stable conditions for a long enough time to make the benefits worth more than the costs. An owner-operator is more likely to invest in such practices than is an operator with a short-term lease.

Sometimes an organization such as a drainage district or an irrigation district can be formed to handle worthwhile practices that are not profitable or practical for individuals. State or federal government may become involved through subsidies or other incentives. Careful planning may be required and detailed analyses made, but the practices may still be justified economically.

Aesthetic, legal, and moral considerations may motivate the use of some practices in spite of unfavorable economics. Some landowners may finance such practices, but group or public action is more common. A park, for example, might be supported by a community and made available to the public at little or no charge.

18–3.1 Payments by Landowners and Operators

Owners and operators pay most of the direct costs of practices that increase profits on their land. These include most crop management practices, such as the use of clean, viable seed of adapted crop varieties. Costs of conservation tillage and of fertilizing and liming also are usually paid by landowners and operators.

Investments in practices with large direct costs that require long terms to become profitable may be shared between landowners and public groups or government agencies. Operators with long-term leases might pay part of the costs, but those with short-term (one- to three-year) leases usually do not pay for long-term investments such as irrigation and drainage systems, land reclamation, and terracing, because these require several years to repay their costs.

Owners and operators divide costs in various ways according to lease arrangements. The costs of short-term practices are often shared in the same proportion as the profits from those practices. Owners pay most or all of the cost of long-term practices because these are investments that increase the value of their land.

18–3.2 Payments by Groups

Flood control, drainage projects, and irrigation developments are examples of practices that commonly require group action because they involve many people and properties. Typically, all of the people involved form a district or other legal entity that can act in their behalf. The district authority may include the right to design and apply needed practices, assess the costs to the individual members, obtain assistance from various agencies, borrow money to accomplish its purposes, and, in some cases, levy taxes. Counties or municipalities sometimes act in similar manners.

Groups are sometimes able to obtain assistance that would not be available to individuals. For example, an irrigation district needing to improve its water supply might obtain help from the Bureau of Reclamation. A flood-control district might obtain both technical assistance and cost-sharing from the small watershed program of the Natural Resources Conservation Service.

Group costs are much like those of individual landowners but on a larger scale. Usually, the costs must be recovered from the landowners by assessments, taxes, or selling of group services.

18–3.3 Government Participation in Soil Conservation

Districts, counties, and municipalities are all local units of government. State and federal governments represent broader areas and more people. Governmental units at any level may want to encourage conservation projects that benefit large numbers of people. The benefits from some projects are so diffuse that they probably would not be installed by individuals or small groups, but they may be feasible as government projects. Parks and preserves are examples.

Governments have several ways of encouraging the application of soil conservation practices. Many national governments provide education and technical assistance, and some allow tax credits or other benefits to those who apply the practices. Governments can provide low-interest loans or pay part or all of the costs themselves to finance the practices. Governments also have the option of bypassing economics by establishing legal requirements and enforcing them through their court systems.

The U.S. government pays part of the cost of selected conservation practices through the Environmental Quality Incentives Program (see Chapter 19). Payments under this program are authorized by the Farm Service Agency after the Natural Resources Conservation Service (NRCS) has certified that the practice was needed, practical, and properly installed. For the period from 1933 to 1988, the U.S. government spent about $15 billion on soil conservation practices (Colacicco et al., 1989).

The most ambitious U.S. soil conservation program to date was mandated by the Food Security Act of 1985. This act required that highly erodible lands be identified from soil maps and that farmers using such lands should have soil conservation plans for them by 1990. The act required these plans to be applied so they would reduce soil losses to tolerable levels by 1995. Models based on 1982 erosion rates indicated that without such action, the United States would lose the productive equivalent of 7.4 million acres (2.5 million ha) of cropland during the next 100 years (Putman et al., 1988).

The 1985 Food Security Act included provisions for a Conservation Reserve Program (CRP) designed to retire highly erodible land from cropping. Farmers could sign a 10-year contract with the government to plant perennial vegetation on such land in exchange for government payments (Osborn, 1993). By 1995, 36.4 million acres were enrolled in CRP, and the program was credited with having reduced soil erosion on this land by approximately 19 tons/ac-yr, or a total of nearly 700 million tons/yr. Additional benefits result from the gradual increase in organic matter in the soil under CRP. Gebhart et al. (1994) indicate that CRP land gains an average of 1.1 mt of carbon/ha-yr (about 1000 lb/ac-yr), thus sequestering about 45% of the 38 million tons of carbon that U.S. agriculture releases into the atmosphere each year. The 1996 farm bill provided for the continuation of the CRP.

The U.S. Natural Resources Conservation Service provides technical assistance without charge to farmers and others. This program is designed to conserve soil and water by helping people install conservation practices on their land. The technicians are paid by federal funds that are supplemented in some states by state and local funds.

Environmental concerns in the United States have resulted in the passing of several federal and state laws in recent years. For example, the 1972 Federal Water Pollution Control Act, as amended, includes provisions in Section 208 for reducing the sediment load in runoff water. Federal cost sharing is available through the Environmental Quality Incentives Program. Various states have developed supplemental programs to help keep pollutants out of streams and lakes.

Some state laws are more stringent than federal laws. The Iowa Soil Conservation Districts Law, as amended in 1971, provides means for declaring erosion a "nuisance" and for requiring landowners to limit soil loss to tolerable rates. It also requires that 75% cost sharing be available for mandated soil conservation practices and provides state funds to pay 25% of their cost in addition to the 50% available from the U.S. Environmental Quality Incentives Program.

18–3.4 Conservation Research and Education

Development and information dissemination must take place before conservation practices can be widely used. Research and education constitute important indirect costs of conservation. Much research is done through government agencies and universities with government support. Foundations, companies, and individuals also have research projects. In the United States, the Agricultural Research Service (ARS) and the Cooperative State Research, Education, and Extension Service (CSREES) of the USDA are the principal federal agencies involved, and the Hatch Act is one of the main federal sources of agricultural research funds in the universities. State governments and private companies also provide substantial funding for research. Many other nations are also involved through their own research, education, and extension agencies.

Educational activities dealing with soil conservation are part of the curricula of many schools at various grade levels. The educational effort is extended to farmers and others through an Extension Service. In the United States, the Extension Service is funded from federal, state, and sometimes county sources as a means of disseminating agricultural research findings from the land-grant universities. Educational work is also accomplished by the Natural Resources Conservation Service as a part of its technical assistance program and by such groups as the Soil and Water Conservation Society and the Soil Science Society of America. Various agribusiness companies also have programs that include conservation education. The largest expenditures for educational activities, however, are funded by the federal and state governments.

18–4 CONSERVATION INCENTIVES

People use conservation practices for various monetary, legal, convenience, aesthetic, moral, and other reasons (de Graaff, 2001). The relative importance of each factor varies from one person and situation to another.

The cost of a soil conservation practice is usually considered even when other factors influence the decision to install a practice. An economic analysis may calculate a *benefit/cost ratio* by dividing probable benefits by estimated costs. A benefit/cost ratio of 1.0 or larger may suffice to cause a practice to be installed. A ratio smaller than 1.0 is likely to end consideration of a project unless there are strong nonmonetary incentives in its favor. Fortunately, strong nonmonetary incentives often cause conservation practices to be installed in spite of benefit/cost ratios that are less than 1.0.

Legal requirements mean that a governmental body has decided that individuals must install a certain conservation practice. The legislators believe that important public benefits are involved—benefits that are more decisive for the public interest than for individual decision makers. The polluting effects of soil erosion, for example, are often ignored by the persons losing the soil. However, soil pollution has been the reason for the passage of several recent laws. The public benefits might be monetary savings such as not having to remove sediment and other pollutants from water, avoidance of public health problems, or aesthetic reasons such as the prevention of eutrophication. They might also be based on moral principles.

Convenience factors enter into many decisions. Wet spots, for example, cause inconvenience. The owner may drain a wet spot even if added returns are unlikely to repay the cost. Rock outcrops have been dynamited at great expense so the land could be tilled conveniently. Sometimes convenience is a secondary factor influencing the nature of the practice after the need has been established. For example, expensive parallel terraces might be installed for convenience even though cheaper, nonparallel terraces could hold the soil equally well.

Aesthetic appeal varies from person to person and may either help or hinder conservation. The feeling that crop residues left on the surface appear trashy is a negative effect. The desire for straight rows can make it hard for a person to accept contouring. However, many people see beauty in the patterns of contour strip cropping or in the uniform growth of crops on a field where soil has been reclaimed. Ponds and wildlife plantings are aesthetically appealing to many people and may be built or planted for that reason.

Moral reasons are another incentive for action. It is right to conserve soil rather than let it erode away. It is right to stop pollution even if it is expensive. It is right to provide some areas for wildlife even if crops could be grown there. Soil-stewardship programs emphasize morality by suggesting that soil should be conserved for the sake of future generations. The word *conservation* has a strong moral tone that helps to promote its cause.

SUMMARY

Soil and water conservation costs are usually immediate and obvious, whereas the benefits may be delayed, obscure, and dispersed. Government participation is justified to protect the productive potential of land and represent the interests of many people who are affected by the way soil is treated. Soil conservation emphases have shifted in recent years to include the effects of sedimentation and pollution as well as erosion.

Some conservation practices produce an immediate profit. Fertilization is so profitable that its conservation effects are often overlooked. Conservation tillage reduces energy-input costs. Water conservation increases yields on sloping land. Drainage of wet land and irrigation of dry land produce high returns on the investment. Effective pest control increases yields and reduces erosion.

Practices such as terracing, planting forage crops and trees, and land reclamation result in deferred profits. A conflict may develop between the need for short-term income and the need for long-term protection of the land. Environmental concerns also may conflict with practices that maximize income.

Eroded soil has a total plant nutrient value of about $5 per ton. Effective soil conservation may save $5 billion worth of plant nutrients each year in the United States and also reduce flood damage by hundreds of millions of dollars per year. Soil loss reduces yield potential by amounts ranging from 0 to 100% depending on the amount of soil lost, the nature of the soil, and its underlying material. Eroded spots in a field pose difficult management problems concerning future use of the field.

Direct costs represent only part of the total costs of conservation. Indirect costs such as research and education provide needed background. Often there are opportunity costs in the form of a profitable crop that could have been grown—reduced income from less-intensive cropping may be a stronger deterrent than large direct costs. Conservation practices are a good investment if they produce an adequate annual return over a period of years.

Legal soil-loss limits first became law in the United States in the 1970s. Such laws cause a competitive disadvantage if they apply only to certain areas. Applied to the entire nation, these laws tend to raise prices and increase the producers' income. Subsidies that pay much of the cost of installing conservation practices increase net farm income, whereas taxes on excessive soil loss would reduce it.

Owners and operators often share the costs of practices that produce short-term profits. Owners pay most or all of the costs when long-term profit is anticipated. Groups such as districts or other legal entities become involved when projects are too large for individuals. Governments can encourage the application of soil conservation practices by paying for them directly; by providing technical assistance, low-interest loans, tax credits, or other benefits; by taxing soil loss; or by passing conservancy laws. Governments can also help by providing good research and education programs.

Incentives for installing conservation practices include benefit/cost ratios larger than 1.0, legal requirements, convenience, aesthetic appeal, and moral considerations.

QUESTIONS

1. What are the costs of erosion to a nation?
2. Who suffers most from flood damage?
3. What are the effects of conservation tillage on income and expense for a particular crop?
4. How do direct and indirect costs influence the economic desirability of a conservation practice to a landowner?
5. Which conservation practices are likely to be evaluated on the basis of their suitability as investments?
6. What economic effects would mandatory soil conservation on farms have on a nonfarm family?

7. Should a building contractor be required to control erosion on a construction site? Explain.
8. What conservation costs should be paid by landowners? by operators? by groups? by the general public?

REFERENCES

ALLEN, R. R., B. A. STEWART, and P. W. UNGER, 1977. Conservation tillage and energy. *J. Soil Water Cons.* 32:84–87.

BATTE, M. T., 2000. Factors influencing the profitability of precision farming systems. *J. Soil Water Cons.* 55:12–18.

BORNSTEIN, J., and C. L. FIFE, 1973. Economic aspects of sloping land drainage. *J. Soil Water Cons.* 28:76–79.

BRANDLE, J. R., B. B. JOHNSON, and T. AKESON, 1992. Field windbreaks: Are they economical? *J. Prod. Agric.* 5:393–398.

BRUGGINK, D., 1996. No-till brings more time, more money. *No-Till Farmer,* Mid-April 1996, p. 4–5.

BULL, L., and C. SANDRETTO, 1996. *Crop Residue Management and Tillage System Trends.* USDA Economic Res. Serv. Stat. Bull. 930, 27 p.

COLACICCO, D., T. OSBORN, and K. ALT, 1989. Economic damage from soil erosion. *J. Soil Water Cons.* 44:35–39.

COUNTRYMAN, D. W., and J. C. MURROW, 2000. Economic analysis of contour tree buffer strips using present net value. *J. Soil Water Cons.* 55:152–160.

DE GRAAFF, J., 2001. The economic appraisal of soil and water conservation measures. In E. M. Bridges, I. D. Hannam, L. R. Oldeman, F. W. T. Penning de Vries, S. J. Scherr, and S. Sombatpanit (eds.), *Response to Land Degradation.* Science Pub., Inc., Enfield, NH, 510 p.

EL-SWAIFY, S. A., 1991. Effective resource conservation on hillslopes. In W. C. Moldenhauer, N. W. Hudson, T. C. Sheng, and S. W. Lee (eds.), *Development of Conservation Farming on Hillslopes.* Soil and Water Cons. Soc., Ankeny, IA, 332 p.

GEBHART, D. L., H. B. JOHNSON, H. S. MAYEUX, and H. W. POLLEY, 1994. The CRP increases soil organic carbon. *J. Soil Water Cons.* 49:488–492.

GUNATILAKE, H. M., and G. R. VIETH, 2000. Estimation of on-site cost of soil erosion: A comparison of replacement and productivity change methods. *J. Soil Water Cons.* 55:197–204.

HEADY, E. O., and K. J. NICOL, 1974. Models and projected results of soil loss restraints for environmental improvement through U.S. agriculture. *Agr. Environ.* 1:355–371.

HERZOG, M., J. HARBOR, K. MCCLINTOCK, J. LAW, and K. BENNETT, 2000. Are green lots worth more than brown lots? An economic incentive for erosion control on residential developments. *J. Soil Water Cons.* 55:43–49.

LARNEY, F. J., H. H. JANZEN, B. M. OLSON, and C. W. LINDWALL, 2000. Soil quality and productivity responses to simulated erosion and restorative amendments. *Can. J. Soil Sci.* 80:515–522.

LINGARD, J., 1994. The Economic Context of Soil Science. Ch. 18 in J. K. Syers and D. L. Rimmer (eds.), *Soil Science and Sustainable Land Management in the Tropics.* CAB International, Wallingford, U.K., 290 p.

NOWAK, P. J., 1988. The costs of excessive soil erosion. *J. Soil Water Cons.* 43:307–310.

OLSON, K. R., and E. NIZEYIMANA, 1988. Effects of soil erosion on corn yields of seven Illinois soils. *J. Prod. Agric.* 1:13–19.

OPIE, J., 1993. *Ogallala: Water for a Dry Land.* Univ. of Nebraska Press, Lincoln, NE, 412 p.

OSBORN, T., 1993. The conservation reserve program: Status, future and policy options. *J. Soil Water Cons.* 48:271–279.

PETERS, D. C., 1975. The value of soil insect control in Iowa corn, 1951–70. *J. Econ. Entomol.* 68:483–486.

PUTNAM, J., J. WILLIAMS, and D. SAWYER, 1988. Using the erosion-productivity impact calculator (EPIC) model to estimate the impact of soil erosion for the 1985 RCA appraisal. *J. Soil Water Cons.* 43:321–331.

RICHARDS, W., 1991. Restoring the land. *J. Soil Water Cons.* 46:409–410.

SANGHI, A. K., and R. KLEPPER, 1977. Economic impact of diminishing groundwater reserves on corn production under center-pivot irrigation. *J. Soil Water Cons.* 32:282–285.

SMART, J. R., and J. M. BRADFORD, 1999. Conservation tillage corn production for a semiarid, subtropical environment. *Agron. J.* 91:116–121.

SYLVESTER-BRADLEY, R., E. LORD, D. L. SPARKES, R. K. SCOTT, J. J. J. WILTSHIRE, and J. ORSON, 1999. An analysis of the potential of precision farming in northern Europe. *Soil Use and Management* 15:1–8.

TAYLOR, C. R., 1977. An analysis of some erosion control policies for the high and rolling plains of Texas. *J. Am. Soc. Farm Mgrs. and Rur. Appr. 41*(1):49–52.

TRIPLETT, G. B., JR., and D. M. VAN DOREN, JR., 1977. Agriculture without tillage. *Sci. Am.* 236:28–33.

URI, N. D., 2000. Agriculture and the environment—the problem of soil erosion. *J. Sust. Agr.* 16(4):71–94.

VAN DOREN, D. M., JR., G. B. TRIPLETT, JR., and J. E. HENRY, 1976. Influence of long term tillage, crop rotation, and soil type combinations on corn yield. *Soil Sci. Soc. Am. J.* 40:100–105.

WEERSINK, A., M. WALKER, C. SWANTON, and J. E. SHAW, 1992. Costs of conventional and conservation tillage systems. *J. Soil Water Cons.* 47:328–334.

WITTMUS, H., L. OLSON, and D. LANE, 1975. Energy requirements for conventional versus minimum tillage. *J. Soil Water Cons.* 30:72–75.

CHAPTER

19

SOIL AND WATER CONSERVATION AGENCIES IN THE UNITED STATES

Soil and water conservation depend on the combined efforts of many people and many organizations. These organizations may be as simple as a private agreement between landowners and operators in a particular locality, or as complex as a multi-nation treaty negotiated to protect important resources. Many of the private organizations are local, but some (such as the Ford Foundation and the Rockefeller Foundation) are large, endowed foundations that work on an international scale. Government agencies, likewise, represent various sizes of areas, including counties, states, and nations, and even continental or worldwide international organizations. The whole array is much too complex to describe in one chapter, or even in an entire book. The purpose of this chapter is to discuss the functions of the principal government agencies involved in soil and water conservation on a national scale in the United States.

In pioneer days, government had relatively little involvement in agriculture other than to register land ownership. Some people dislike some of the results, but things have changed and government now has a great deal of influence on the lives of everyone in the nation. The government agencies discussed in this chapter were established by laws that define their purposes and how they function. Some of the significant purposes recognized in these laws are:

1. To help farmers apply good farming methods, especially practices that conserve soil and water
2. To protect the environment and to control soil, water, and air pollution
3. To help farmers stay in business by stabilizing production and prices with government loans and insurance programs
4. To manage government-owned land

The laws provide specific means for accomplishing these purposes. Agencies such as the Agricultural Research Service conduct research to solve problems and discover what practices work best; colleges and the Extension Service function as educators to make this knowledge available to students and farmers; agencies such as the Natural Resources Conservation

Service provide technical assistance when the practices are installed on the land. The Farm Service Agency provides monetary services such as loans and cost-sharing and is involved in programs that restrict the production of certain crops. Pollution-control laws are administered by the Environmental Protection Agency. Taxation and financial penalties are also used sometimes to induce people to take certain actions when other incentives are inadequate.

Soil erosion officially became a matter of national concern in the United States in the 1930s. Four main factors were responsible for a widespread awareness of the erosion problem at that time: results from early field research, a worldwide depression, dust storms, and a dedicated leader named Hugh Hammond Bennett. The response of Congress and the people led to the development of a system of soil and water conservation that became a model for countries around the world.

19–1 EARLY WORK ON SOIL AND WATER CONSERVATION

The European settlers who established the colonies that became the United States brought European farming methods with them, but they soon found that the rainfall in the new land was more intense and erosive than that of Europe. The Native Americans had small patches of corn and other crops, but the new settlers soon began clearing more land so they could farm on a much larger scale. Erosion became a conspicuous problem in many fields, but the new nation centered its attention on other matters such as survival and expansion. Some individuals experimented with methods of conserving soil, but many had little concern if their land "wore out." They simply followed the frontier and began again on a new farm. This idea prevailed for more than a hundred years, until the frontier was gone and something had to change.

The first formal field research on soil and water conservation in the United States was established May 1, 1917, by M. F. Miller and F. L. Duley at the University of Missouri on the plots shown in Figure 19–1. The long record of data from these plots dating from 1917 to the present is valuable for understanding the erosion process and what can be done about it.

Figure 19–1 The first soil and water conservation field research plots in the United States were established by M. F. Miller and F. L. Duley on land that is now part of the campus of the University of Missouri at Columbia. The sign reads "Soil Erosion Experiment—Comparison of Different Crops and Methods of Tillage for Preventing Soil Washing. Begun May 1, 1917." The site is now a Registered National Historic Landmark. (Courtesy C. M. Woodruff, University of Missouri.)

Miller and Duley wanted to determine the reasons for declining soil productivity (Note 19–1). The first 14-year summary of the soil and water losses from the research plots is presented in Table 4–3. The table shows that the number of years required to erode an average 7-in. (18-cm) plow depth of soil varies from 24 years under fallow to 3043 years with continuous bluegrass. In 1941, 24 years after the erosion plots had been established, the surface of the continuous bluegrass plot was 8-in. (20-cm) higher than the one in continuous fallow.

NOTE 19–1
HISTORIC SOIL EROSION EXPERIMENT

These pioneering soil erosion research plots were designated by the National Park Service as a Registered National Historic Landmark in 1965. The inscription on the monument at the University of Missouri, Columbia, commemorating the first field research plots for studying soil and water conservation, reads as follows:

SITE OF FIRST PLOTS IN THE UNITED STATES FOR MEASURING RUNOFF AND EROSION AS INFLUENCED BY DIFFERENT CROPS

The research began in 1917. The first results were published in 1923. They provided the foundation for the soil conservation movement in the United States. The design of the experiment served as the prototype for future experiments by the USDA and land-grant universities throughout the United States and by other agencies abroad.

The investigations were initiated by M. F. Miller and F. L. Duley in their search for the causes of declining soil productivity. The original plots are now being used to investigate the renovation of eroded soil through the use of legumes, grass, and corn with treatments to supply the nutrients lost through erosion.

19–1.1 Buchanan Amendment

The U.S. Congress officially recognized soil erosion as a problem when it passed the Buchanan Amendment to the Agricultural Appropriation Bill for fiscal year 1930 (Public Law 70-769, February 16, 1929). The amendment provided $160,000 for establishing 10 soil-erosion experiment stations and 10 plant materials centers throughout the United States (Note 19–2).

19–1.2 Civilian Conservation Corps

The decade of the 1930s was a period of worldwide depression. Millions of people were unemployed and desperate for work. Seeking election as president of the United States, Franklin D. Roosevelt promised to do something about the job situation. He was inaugurated March 4, 1933, and instituted a large government spending program that same month. A law establishing the Emergency Conservation Work Agency was passed by Congress and signed by Roosevelt on March 31, 1933. Work for this and several other agencies was done

NOTE 19–2
EROSION EXPERIMENT STATIONS AND PLANT NURSERIES

The original 10 erosion experiment stations were located at

Guthrie, Oklahoma	Temple, Texas
Tyler, Texas	Hays, Kansas
Bethany, Missouri	Statesville, North Carolina
Pullman, Washington	Clarinda, Iowa
LaCrosse, Wisconsin	Zanesville, Ohio

The first 10 plant materials centers were located at

Mandan, North Dakota	Stillwater, Oklahoma
Cheyenne, Wyoming	Elsberry, Missouri
San Antonio, Texas	Shreveport, Louisiana
Ames, Iowa	Pullman, Washington
Belle Mina, Alabama	Safford, Arizona

The 10 plant materials centers have been increased to 23 centers located in 22 states.

by members of the Civilian Conservation Corps, better known as the CCC. (Author Roy Donahue was a forestry foreman in the CCC in 1933.)

About 3 million persons, mostly young men, were employed in government work programs from the time the Emergency Conservation Work Agency was established until the Civilian Conservation Corps was disbanded in 1942. They were stationed in camps throughout the nation and assigned to work for various government agencies. Many of them worked on soil and water conservation projects, such as planting trees and building dams.

19–1.3 Soil Erosion Service

The Soil Erosion Service was established as a temporary public works program in the U.S. Department of the Interior by a resolution of Congress adopted July 17, 1933 (Napier and Napier, 2000). Hugh Hammond Bennett was its director. Five million dollars were appropriated for soil-erosion control on both public and private lands. Contracts were signed with local governments and individual landowners that permitted persons employed by the Emergency Conservation Work Agency to plant trees, grasses, and legumes on eroding lands and to construct erosion-control dams in gullies and in other drainageways, as shown in Figure 19–2.

Forty soil-erosion-control projects were established throughout the United States as part of the growing conservation movement. Each project included an entire watershed for the purpose of making the public aware of soil erosion and how to control it. Landowners signed agreements with the Soil Erosion Service permitting the service to apply soil-erosion technology on their lands. All labor, machinery costs, materials costs, and technical assistance were supplied free to the landowners.

Figure 19–2 A masonry drop structure and tile outlet built by the Civilian Conservation Corps to prevent a gully from advancing up this Iowa waterway. (Courtesy USDA Natural Resources Conservation Service.)

The Soil Erosion Service had only a two-year life span before it was replaced by the Soil Conservation Service. Considering that the primary objective was to provide work for the unemployed, it had a tremendous impact on public awareness of the evils of erosion, as well as on erosion-control technology. Its work also led to the observation that farmers value conservation projects on their land more highly when they have contributed to their establishment. Consequently, more recent programs have been based on cost sharing between the government and the individual rather than the government paying the entire cost.

19–1.4 Soil Conservation Service (SCS)

Severe droughts from 1931 to 1938 coincided with the Great Depression and made the Great Plains of the United States a disaster area. Two giant dust storms made history on May 11, 1934, and March 6, 1935. These storms started in the powder-dry Great Plains and continued to the Atlantic seacoast and beyond. For several days, clouds of dust obscured the sun eastward for a distance of 2000 mi (3000 km). Total silt and clay as particulates in the air masses were estimated at 200 million tons (180 million mt). Dust from the Great Plains filtered into the offices and onto the desks of U.S. senators and representatives in Washington, D.C.

At this time, Hugh Bennett, Director of the Soil Erosion Service, was asking Congress to transfer the service from the Department of the Interior to the Department of Agriculture and to upgrade it from a temporary works program into a permanent soil conservation agency. The law establishing the Soil Conservation Service was passed without a dissenting vote on April 27, 1935, just 52 days after the start of the second disastrous dust storm. What senator or representative could vote against Public Law 74-46 while brushing the Great Plains dust from his desk? Bennett had won his political battle. He was named chief of the new agency, a position he held for the rest of his professional career, until his retirement in 1952.

The Soil Conservation Service developed into a large agency that serviced increasing numbers of soil conservation districts until they were established in nearly every county in the nation (see Section 19–3), and it became a pattern for similar work in other nations. Its early focus was on helping farmers plan and apply soil conservation practices on their land. With time, its responsibilities were expanded to include the national soil survey program, certification of soil conservation work that was to receive federal cost sharing through a sequence of agencies that is now represented by the Farm Service Agency (see Section 19–4), and several other specialized programs. Its emphasis also expanded from its initial objective of conserving soil to include other environmental concerns such as wildlife habitat and clean air and water (Johnson, 2000). The Soil Conservation Service performed these services for 60 years, until its name was changed to the Natural Resources Conservation Service.

19–2 NATURAL RESOURCES CONSERVATION SERVICE (NRCS)

The Natural Resources Conservation Service was established on June 3, 1995, by renaming the Soil Conservation Service. The new name was a recognition that this agency now deals with water conservation, mined resources, and pollution control along with soil conservation. Sentimental value attached to the old name was lost along with the familiar SCS initials, but the new name is a better fit for the multiple responsibilities that NRCS fulfills.

The National Cooperative Soil Survey program discussed in Chapter 7 is another major responsibility of the NRCS. Soil surveys are carried out under that program with the NRCS as the lead institution working in cooperation with personnel from the land-grant universities and several other federal and state agencies and institutions. The soil maps and descriptions serve as a base for several of the responsibilities of the NRCS that are discussed in the following sections. Additional information regarding the NRCS can be found on its Web site at http://www.nrcs.usda.gov.

19–2.1 Legal Mandates

The National Soil Conservation Act of 1935, Public Law 74-46 that established the Soil Conservation Service, together with subsequent amendments and laws, defines the soil conservation activities for the nation. Congress continues to budget much of the funding for NRCS soil and water conservation activities under this law.

The Natural Resources Conservation Service provides direct, formal national leadership and indirect, informal international leadership in soil and water conservation. Technical assistance concerning wise land use is given at no charge to individuals, groups, organizations, cities, towns, churches, schools, and county and state governments. The breadth of the assistance given can be judged by the kinds of specialists working for the NRCS. These include soil surveyors, soil scientists, soil conservationists, agronomists, foresters, plant materials specialists, land-use specialists, biologists, range-management

Figure 19–3 Minimum tillage makes erosion nearly zero in this Illinois field. Weeds were killed by a contact herbicide at the time the corn was seeded in the wheat stubble. (Courtesy USDA Natural Resources Conservation Service.)

specialists, wildlife specialists, geologists, economists, landscape architects, recreation specialists, ecologists, environmentalists, and the following types of engineers: agricultural, irrigation, hydraulic, civil, design, drainage, sanitary, and cartographic.

19–2.2 Assistance to Conservation Districts

Under Public Law 74-46 of 1935 and its amendments, the NRCS administers a broad program of assistance in soil and water conservation in cooperation with Conservation Districts (discussed in Section 19–3). Through district organizations, the NRCS provides these kinds of assistance at the request of farmers, ranchers, and other landowners or operators:

1. Mapping soils, publishing soil survey reports, and determining soil suitability guidelines for agriculture, housing, recreation, waste disposal, and road construction
2. Recommending the best management practices for soil erosion control (Figure 19–3)
3. Designing sediment interception systems
4. Designing water facilities, such as farm ponds

Figure 19–4 Terracing, contour strip cropping (eight rows of peanuts alternating with two rows of grain sorghum), winter cover crops, and subsoiling are used in combination to control erosion in this Texas field. (Courtesy USDA Natural Resources Conservation Service.)

5. Planning recreational facilities
6. Designing terrace, irrigation, and drainage systems
7. Developing cropping systems, such as that shown in Figure 19–4, to reduce erosion
8. Recommending pasture plantings
9. Developing range-management guidelines
10. Promoting wildlife conservation
11. Promoting woodland conservation
12. Supplying adapted plant materials for conservation plantings
13. Promoting surface-mine reclamation guidelines
14. Providing expertise on land-use planning

The NRCS now has about 12,000 career employees (Johnson, 2000) who provide on-the-land technical assistance for stabilizing the soil and water environment for the public good. It staffs approximately 2500 field offices located in almost every county throughout the nation (Natural Resources Conservation Service, 1996). Most of this service is provided through the conservation district offices. Specialists from the state and national offices are called upon when their expertise is needed.

19–2.3 Assistance to Individual Landowners and Operators

Individual landowners or operators obtain assistance from NRCS personnel by signing an agreement with the local conservation district (see Section 19–3.2). Land-use plans and designs for conservation practices recommended by the soil conservation technicians are based on soil maps interpreted according to the information in soil survey reports, as described in Chapter 7. Plans are made according to decisions of the cooperating landowner or operator in harmony with the capabilities and needs of the soils.

NRCS personnel make agronomic and engineering recommendations for each of the 13,000 soil series and map units recognized in the United States. Appropriate combinations of practices are recommended for the particular conditions existing at a specific site. The intent is to conserve soil and water and to protect the environment while using the land for the purposes desired by the owner or operator. The NRCS provides the technical assistance needed for planning, designing, and guiding the application of conservation practices.

Many types of assistance are available to conservation district cooperators without charge. Soil and crop specialists will develop cropping- and pasture-management systems to reduce erosion and sedimentation. Engineers design conservation structures including terraces, diversions, waterways, ponds, irrigation systems, drainage systems, and waste-disposal systems. Range conservationists assist ranchers with grazing management techniques for maximizing production with minimum erosion. Foresters recommend tree species and planting and harvesting techniques for woodlands and windbreaks. Plant materials specialists recommend special plant species for use in unusual sites such as acid mine spoils (Chapter 11). Many such plants are distributed each year from the 23 NRCS plant materials centers.

19–2.4 Surface-Mine Spoils Reclamation

Agronomists, soil scientists, plant materials specialists, and engineers work to help stabilize mine spoils. The help is available to mine operators, individuals, groups, or any governmental unit. Four phases of surface-mine reclamation are assisted by personnel of the NRCS:

1. Planning before mining, including the making of a soil survey
2. Applying conservation practices during mining
3. Applying soil and water conservation practices after mining to establish erosion-resisting perennial vegetation (Figure 19–5)
4. Assistance in reclaiming abandoned mine spoils

19–2.5 Watershed Surveys and Planning Program

Under Public Laws 78-534, passed in 1944, and 83-566, passed in 1954, the NRCS offers assistance to cities, towns, and rural groups organized to represent the inhabitants of a small watershed (not more than 250,000 ac or about 101,000 ha). The assistance may involve planning, cost estimates, and partial funding for flood reduction, erosion control, reducing siltation, and lowering maintenance costs for roads, bridges, and housing developments in an entire watershed area (Soil Conservation Service, 1992). This program recognizes that runoff and erosion problems are best treated on a watershed basis because they do not stop at individual property lines. The NRCS can provide both technical and financial assistance to groups it contracts with under this Small Watershed Program.

Figure 19–5 This area of coal mine spoils in Kentucky has been vegetated with shortleaf pine and wildlife food plants such as black cherry. (Courtesy USDA Natural Resources Conservation Service.)

The 1954 law (commonly called PL-566) also provided for flood prevention involving cooperation with federal, state, and local agencies to develop coordinated water resource programs through surveys and investigations of river basins. This function was merged with the small watershed program in 1996 under the title of Watershed Surveys and Planning Program. This program can deal with water reservoirs intended for multiple-use options such as a municipal water supply, fire protection, irrigation, and recreation.

There are more than 1600 watershed projects in operation in the United States, and the NRCS is currently providing assistance to more than 500 of them. The NRCS provides needed surveys and other natural resource information, assists the local group in planning, provides cost estimates and part of the funding (often half), and helps identify other possible sources of funding. NRCS technicians provide most of the agronomic, engineering, hydrologic, and other technical guidance needed for the project. The local group sponsors the project, seeks the help it needs from NRCS and other sources, makes the necessary plans, and works with landowners and contractors to implement the work.

19–2.6 Resource Conservation and Development (RC&D)

Most of the conservation districts organized according to Public Law 74-46 either coincide with a county or approximate a county in size. Public Law 87-703 of 1962 authorizes the NRCS to assist regional (multicounty) areas in aspects of physical resource conservation and development. The mandate includes conventional soil and water conservation activities, development of recreation facilities, fish and wildlife conservation, and reduction of air and water pollution. Assistance is also offered in land-use planning, preservation of scenic and historical sites, and industrial expansion. The goal is to improve the environment, economy, and living standards of people in designated areas. The program has completed more than 26,000 projects (U.S. Department of Agriculture, 1995). In 2002, it involved 315 RC&D areas averaging more than seven counties each, or 73% of the 3197 counties in the United States. Much of the work is accomplished by about 20,000 volunteers organized into RC&D councils and assisted by technicians from the NRCS and other agencies.

19–2.7 Environmental Quality Incentives Program (EQIP)

EQIP provides technical, financial, and educational assistance to resolve soil, water, and other resource problems on farms and ranches. It was formed in 1996 by combining several government programs that provided federal funds for conservation projects (see Section 19–4). The two largest predecessor programs were the Agricultural Conservation Program (ACP) and the Great Plains Conservation Program. EQIP includes a cost-sharing program that is jointly managed by the NRCS and the Farm Service Agency, much like the ACP was jointly managed by the Soil Conservation Service and the Agricultural Stabilization and Conservation Service. This cost-sharing function is discussed in Section 19–4.

As mandated by Public Law 84-1021 in 1956, the Soil Conservation Service administered the Great Plains Conservation Program. The Great Plains region is subject to severe climatic hazards such as periodic droughts and extremes in temperature. As a consequence, during periods of favorable weather, farmers and ranchers tend to cultivate marginal soils and to overstock ranges that, during dry years, are subject to devastating wind and water erosion.

Additional funds were set aside by Congress for exclusive use in the Great Plains to help intensify and accelerate the normal soil and water conservation activities of the Soil Conservation Service. Special emphasis was given to making a soils map of the entire Great Plains, converting marginal cropland to permanent pasture and range, and improving rangelands by encouraging more rational grazing management as shown in Figure 19–6. This program is now administered by the NRCS as a part of EQIP.

19–2.8 Land-Use Conversions and Reserve Programs

A large part of a nation's total soil loss comes from a relatively small part of its cropland. Some of this land could be protected with a system of terraces or some other set of structures and management practices that would permit it to continue to be cropped, but much land is simply being used more intensively than it should be. The most practical way to reduce erosion on such land and the associated sedimentation damage is to convert the land use to something less intensive, probably permanent grassland or woodland. The same may be true of some wetlands that are not subject to erosion but are subject to sedimentation, traffic problems, and/or salinization or are needed for pollution control and wildlife habitat (Farm Service Agency, 1997). Land-use conversions occur when farmers, ranchers, and other land managers are persuaded that such changes need to be made, either through their own observations or through the educational efforts of conservationists.

Figure 19–6 Controlled grazing is emphasized in the Great Plains Conservation Program as a part of EQIP. The tobosagrass in this Texas range shows the effects of severe overgrazing on the left, and proper grazing on the right. (Courtesy USDA Natural Resources Conservation Service.)

Based on a soil survey, NRCS soil scientists classify soils into Land-Use Capability classes identified by Roman numerals I through VIII. Class I land has the least hazard associated with use and Class VIII the greatest, as discussed in Chapter 7. This system helps to identify land that is suffering excessive erosion because of use beyond its long-term capability. For example, the land shown in Figure 19–7 was converted from cropland to rangeland because it was subject to severe wind erosion.

The 1985 Food Security Act made new provisions for land-use conversions under the Conservation Reserve Program (CRP). Under this program, farmers sign 10- to 15-year leases to set aside areas containing a large proportion of highly erodible land that has been used as cropland and to protect it with grass or trees in exchange for an annual rental payment from the government. CRP was popular enough to include 36.4 million acres by 1995. At that time, it was credited with having reduced total soil erosion for the nation by 22% and to have contributed significant water quality and wildlife habitat benefits (U.S. Department of Agriculture, 1995). The 1996 farm act therefore authorized the USDA to maintain 36.4 million acres of CRP land through the year 2002 (Osborn, 1996); the Farm Security and Rural Investment Act of 2002 continued it for another six years and allowed CRP and WRP (see next page) to expand to 39.2 million acres (National Association of Conservation Districts, 2002).

Land that is subject to excessive soil loss when used for intensive cropping is designated as *highly erodible land* (HEL) by the 1985 Food Security Act. Some of this land is now covered by CRP, but much of it can continue to be cropped if it is given adequate protection. The Food Security Act required that the SCS (now NRCS) assist farmers using such land to develop suitable conservation plans to protect it by 1990. Farmers had to have these plans in operation by 1995 in order to remain eligible for most government farm programs.

Figure 19–7 An example of land-use conversion. This former wheat field in Wyoming was subject to severe wind erosion until it was seeded to crested wheatgrass and converted to rangeland. (Courtesy USDA Natural Resources Conservation Service.)

These provisions, known as *conservation compliance,* also apply to any highly erodible land that is converted to cropland from other uses. The protective measures may be any combination of vegetative and mechanical practices that provide adequate erosion control as calculated by the soil-loss equations (Chapter 6). Erosion reductions from 16.2 tons/ac-yr down to 5.8 tons/ac-yr (a 64% reduction) were anticipated as an average result of these provisions (U.S. Department of Agriculture, 1995).

Another part of the 1985 Food Security Act designed to prevent wetlands from being converted to cropland became known as a "swampbuster." This provision caused a great deal of resentment by farmers and land developers who could no longer drain wetlands to convert them to cropland, housing, or other uses. The Food, Agriculture, Conservation, and Trade Act of 1990 therefore created a Wetland Reserve Program (WRP) as another way of protecting wetlands. WRP is authorized to contract for 975,000 acres of land with either permanent leases or long-term leases of 30 years or more (Lipton, 1996). This is a voluntary program with financial incentives for restoring wetlands and improving water quality and wildlife habitat. Swampbuster regulations still apply, but in a modified way that is intended to cause less interference with farming operations (Conrad, 2000).

The 1996 Federal Agriculture Improvement and Reform Act added a Conservation Reserve Enhancement Program (CREP) to the established CRP program. CREP is designed to coordinate federal and state efforts to reduce erosion, improve water quality, and enhance wildlife habitat in certain targeted areas. It requires measurable objectives chosen to fit local circumstances and annual monitoring to assure that the objectives are being achieved. Cooperating landowners and operators can receive incentive payments for installing designated conservation practices.

The 2002 farm bill extended or modified the CRP programs and extended the previously established Wildlife Habitat Incentive Program (WHIP) with cost-sharing incentives that encourage landowners to provide wildlife habitat areas on their land. It also established a new Conservation Security Program (CSP) whereby farmers can sign 5- to 10-year contracts agreeing to adopt and maintain specified conservation practices on cropland and a new Grassland Reserve Program (GRP) involving either 30-year or permanent agreements to restore grasslands that are to be grazed, hayed, or mowed only after the wildlife nesting season is over (National Association of Conservation Districts, 2002).

19–2.9 International Assistance

Several hundred foreign nationals come each year to the United States to study the organization and field operations of the NRCS (Soil Conservation Service, 1989). These educational experiences are usually financed by the U.S. Agency for International Development or by the Food and Agriculture Organization of the United Nations. Furthermore, many overseas governments request on-site services of experienced personnel from the NRCS. While studying the program of the NRCS, foreign nationals are invariably surprised to learn that much of its success is credited to the work of unpaid farm and ranch officials of the conservation districts and other volunteers.

Another popular form of international assistance is known as PL 480 or "Food for Peace" (Lipton, 1996). This program uses surplus U.S. production to provide needed wheat to other countries. Low interest loans, sometimes converted to gifts, are made to the other nation so it can buy wheat in the U.S. market. Repayment is in local currency that is then used to finance U.S. Agency for International Development programs. These programs often involve contracts with the NRCS as outlined in the preceding paragraph or with some of the land-grant colleges to provide technical agricultural assistance in developing nations.

19–3 CONSERVATION DISTRICTS

H. H. Bennett, the first chief of the Soil Conservation Service, has been called "the father of soil conservation" because of his influence in establishing both the Soil Conservation Service and the legislation that enabled the organization of soil conservation districts. Bennett was raised on a farm in the red Piedmont hills of Anson County, near Bollston, North Carolina. He witnessed serious sheet and gully erosion and the resultant widespread poverty and decline in crop yields. He became a professional soil scientist and for many years mapped soils in both southern and northern states. He realized that major input from farmers and ranchers was needed for a successful soil conservation program. As a consequence of his broad insight, he led the development of "A Standard State Soil Conservation Districts Law" to authorize the establishment of districts throughout the United States. For maximum acceptance among the states, Bennett persuaded President Roosevelt to send this model law to all state governors on February 27, 1937. The first Soil Conservation District actually organized was Brown Creek Soil Conservation District in North Carolina. It was approved on August 4, 1937, and included Bennett's birthplace.

By the end of 1937, 22 states had passed Soil Conservation District laws, and 10 years later all the states plus Puerto Rico and the Virgin Islands had enacted such legislation. Each state modified the model law to fit its land and the wishes of the people, but all of these laws included the principle that local citizens have the authority to establish policy and the mandate to accept responsibility for soil and water conservation in each district. During the 1960s, several states changed the names of their districts to "Soil and Water Conservation Districts" in recognition of the importance of water management in their programs. More recently, many of the names have been shortened to "________ Conservation District."

19–3.1 Present Scope

Conservation districts in all 50 states, Puerto Rico, and the Virgin Islands comprise 99% of all farms and ranches, 2.6 million cooperators, and 2183 million ac (883 million ha). Most conservation districts coincide with the boundaries of one of the nation's 3197 counties and territories. Each district has an elected governing board of three to five persons who serve without salary. In each state, a board or commission controls state appropriations and serves the conservation districts of the state in administrative, legal, and financial matters. Although the elected conservation district officials serve without pay, the states appropriate funds to be used for official travel, establishing an office, and employing a secretary.

The National Association of Conservation Districts (NACD) was organized in 1946; its permanent headquarters is in Washington, D.C. It holds annual meetings and publishes a bi-monthly newsletter called NACD News and Views. Its Web site is *http://www.nacdnet.org*.

19–3.2 Traditional Activities

Conservation districts are managed by private citizens elected by the residents of the district. Duties of the district supervisors (directors, board members, or commissioners) include the planning and directing of the soil and water conservation programs in the district. This involves the request for professional assistance from the USDA-NRCS, the USDA-Forest Service, the Land-Grant University Cooperative Extension Service, the Land-Grant Agricultural Experiment Station, the USDA-Bureau of Land Management, the Agricultural Research Service, and any other federal and state agencies that may be appropriate for the program of the particular district. Farmers and ranchers sign agreements with the local conservation district, and the district provides assistance through the personnel of the NRCS and other agencies with which it has cooperative agreements.

19–4 FARM SERVICE AGENCY (FSA)

The Farm Service Agency was formed in 1994 as a merger of the Agricultural Stabilization and Conservation Service (ASCS) and the Farmers Home Administration (FmHA). The changing emphasis of the ASCS is revealed by the names it has borne through the years. It began in the 1930s as the Agricultural Adjustment Agency (AAA) charged with helping agriculture recover from the depression, later it became the Production and Marketing Agency (PMA) to place more emphasis on farm markets, and next it was the Commodity Stabilization Service (CSS), indicating that recovery had progressed to where stability was desirable. It then became the Agricultural Stabilization and Conservation Service (ASCS) and was more closely allied with the Soil Conservation Service, and in 1994 it added the loan functions previously managed by the FmHA and received its new name—the Farm Service Agency. Information about this agency can be found on its Web site at *http://www.fsa.usda.gov*.

The Farm Service Agency finances part of the cost of designated soil and water conservation practices that are considered to be a joint responsibility of government and landowners. Cost sharing and loans are used to encourage cooperators to establish soil and water conservation practices on their land. The federal government often pays for about half of the cost of a conservation practice but may pay more or less, depending on the need to encourage the practice. This program provides a means for public funds to pay for public benefits resulting from soil and water conservation.

The Farm Service Agency, like its predecessors, is entirely a service and action agency. Rather than employing its own technical staff, it seeks and uses the advice of staff from other agencies in the U.S. Department of Agriculture. Personnel of the NRCS and the Forest Service are most often used to provide technical program guidance and on-farm assistance. Technical advice is also supplied by the State Agricultural Extension Service, the State Agricultural Experiment Station, and the State Forester. For example, the alliance with the NRCS (formerly SCS) involves joint administration of cost sharing through the Environmental Quality Incentives Program.

The Environmental Quality Incentives Program (EQIP) was formed in 1996 by consolidating the USDA cost-sharing programs previously known as the Agricultural Conservation Program (ACP), the Water Quality Incentive Projects (WQIP), the Colorado River Salinity Control Program (CRSCP), and the Great Plains Conservation Program (GPCP). Under this program, the NRCS supervises the work and verifies that it is done properly, and the Farm Service Agency pays the government's share of the cost through the Commodity Credit Corporation (CCC).

The original ACP program was established in 1936 to provide financial assistance to help farmers avoid bankruptcy. It, too, went through a series of name changes. It was known for some years as the Rural Environmental Assistance Program (REAP) and as the Rural Environmental Conservation Program (RECP). The emphasis has also shifted, and EQIP's primary objective now is to encourage landowners to protect natural resources from loss and pollution (Natural Resources Conservation Service and Iowa Soybean Promotion Board, 1996).

The Farm Service Agency also works with the Commodity Credit Corporation in administering the following types of agricultural loan programs (Farm Service Agency, 2000a):

- Farm ownership loans (to help rental farmers become farm owners)
- Farm operating loans (for established farmers)
- Emergency loans (when a county has been designated as a disaster area)
- Youth project loans (for rural or small town residents ages 10 to 20)
- Indian land acquisition loans (for tribes to purchase private lands inside their reservations)

The farm ownership loans, previously administered by the Farmers Home Administration, provide up to $200,000 of direct loan or $717,000 (adjusted annually for inflation) of guarantees for loans obtained from other sources to be used by persons who have at least three years of farming experience to become farm owners (Farm Service Agency, 2000a). Similarly, the farm operating loans provide these same amounts for established farmers when they are unable to obtain credit through commercial channels. These loans can be used for (and may require) the installation of needed soil and water conservation practices. Management advice and federal crop insurance are also available through the Farm Service Agency along with the loan programs.

One function of the farm operating loans is to serve as a price support system (Farm Service Agency, 2000b). A farmer can use a crop of wheat, corn, grain sorghum, barley, oats, rye, oilseeds, rice, tobacco, peanuts, milk, cotton, sugar, or honey as security for the loan. When the market price is high enough, the farmer can repay the loan and sell the crop at the higher price; if the market price stays low, the farmer can deliver the crop to the government and keep the loan money.

19–5 RESEARCH, EDUCATION, AND ECONOMICS

The U.S. Department of Agriculture has an Undersecretary of Agriculture for Research, Education, and Economics to supervise the activities of the Agricultural Research Service, the Cooperative State Research, Education, and Extension Service, the Economic Research Service, and the National Agricultural Statistics Service.

The Agricultural Research Service (ARS) is the federal agency responsible for solving problems and developing new knowledge relative to soil, water, and air resources, environmental pollution control, plant and animal productivity, agricultural commodity processing and marketing, and human nutrition. Much of the research work is done in collaboration with professors and graduate students at the land-grant universities throughout the United States.

The Cooperative State Research, Education, and Extension Service (CSREES, commonly known as the "Extension Service") makes agricultural information available to farmers, ranchers, and the general public through its many publications, meetings, and office consultations. Many people contact their local extension agent when they have a problem related to crop plants, animals, machinery, or farm buildings. State specialists are available for consultation on questions outside the expertise of the local agents. They are typically associated with the state land-grant university where they can be in contact with the latest research findings.

The Economic Research Service (ERS) and the National Agricultural Statistics Service (NASS) gather and publish a wide range of data relative to agricultural inputs, techniques, production, and marketing.

19–6 UNIVERSITIES AND COLLEGES

The U.S. Congress passed the first Morrill Act in 1862 to assist the states in establishing land-grant colleges to teach agriculture and applied science. Research at these and other institutions was assisted by the Hatch Act of 1887 and the Second Morrill Act of 1890. Statewide informal education from the colleges was authorized and assisted by the Smith-Lever Extension Act of 1914. These agricultural universities and colleges are all intimately involved in soil and water conservation. They conduct research, such as that shown in Figure 19–8, teach students about a variety of agricultural topics, and extend knowledge of soil and water conservation to land users in their respective states. Recently, some soil and water conservation research projects have been conducted at certain universities and colleges that are not part of the land-grant system.

Many students are taught principles of soil and water conservation in the agricultural universities before they are employed by such agencies as the USDA-NRCS. Research at the universities, combined with research from other sources, forms the basis for the soil and water management practices that are taught in classes and promoted throughout each state by the Extension Service and the NRCS. Furthermore, the National Cooperative Soil Survey is conducted in cooperation with the NRCS and the respective state agricultural university's agricultural experiment station. All 23 plant materials centers that are supervised by the NRCS are managed in close cooperation with the agricultural universities.

Many of the agricultural universities teach and conduct research in forestry. Much of this research is conducted in cooperation with the USDA-Forest Service and the state forestry agencies.

19–7 U.S. FOREST SERVICE AND STATE FORESTRY AGENCIES

The U.S. Forest Service is responsible for managing the 180 million ac (73 million ha) in the 150 national forests and the 3.7 million ac (1.5 million ha) of national grasslands. It also cooperates with agricultural universities and state forestry agencies in tree planting, fire protection, and other activities related to soil and water conservation. The Forest Service contact with private land management is usually through cooperative programs with the state forestry agencies and the conservation districts. The Farm Security and Rural Investment Act of 2002 established a new Forest Land Enhancement Program (FLEP) for the state forestry agencies to assist private landowners with various forest-related problems (National Association of Conservation Districts, 2002). Federal and state forest nurseries supply forest tree seedlings to plant on eroded and other lands.

The USDA-Forest Service is the principal agency in the United States conducting research on trees–soil–water relationships. This research applies to the 500 million ac (200 million ha) of commercial forests—more than 20% of the area of the United States.

Research relevant to this chapter includes that pertaining to the relationships among trees, water yield, and erosion. Some precipitation is intercepted by trees and evaporated into the atmosphere without reaching the soil. Another fraction infiltrates the soil, is absorbed by roots, and is transpired into the atmosphere through stomata of the leaves. A third segment becomes surface or subsurface runoff, and a final fraction moves downward to the water table.

The U.S. Forest Service has established a research project to find ways to increase runoff-water yield from a forested watershed. Treatments have included clearcutting the trees and establishing grass, as compared to cutting varying percentages of the trees. On some watersheds, an open stand of trees was found to yield more water than an all-grass watershed because the trees held additional snow. The open stand is also better than a clearcut watershed for erosion control.

Figure 19–8 Research conducted by Purdue University on the use of an aerated pond for cattle-waste disposal to minimize air and water pollution while using the waste for increasing pasture production: (A) aeration pond, (B) close-up of floating aerator in operation, (C) effluent being sprayed on pasture. (Courtesy A. C. Dale, Purdue University.)

Research on trees and watershed protection is continuing, especially that relating to establishing a reasonable balance between maximum water runoff for irrigation and soil-erosion losses. Another aspect of research, forest harvest methods and their relationship to erosion, is discussed in Chapter 12. Additional information about the Forest Service can be found on its Web site at *http://www.fs.fed.us*.

19–8 OTHER FEDERAL CONSERVATION AGENCIES

Almost all federal agencies have relevance to soil and water conservation; the agencies listed below are statutorily involved (National Wildlife Federation, 1989).

U.S. Army Corps of Engineers. The Corps manages navigable surface waters, wetlands, and recreation facilities adjacent to public reservoirs, builds dams for water conservation and power generation, conducts research on erosion and sedimentation, and dredges sediment-laden channels. Its Web site is *http://www.usace.army.mil*.

Bureau of Land Management, U.S. Department of the Interior. This agency administers 470 million ac (190 million ha) of U.S. public lands, mostly in the western states. This comprises about 60% of all public lands, and much of it is in small tracts. The principle of management is multiple-use, including livestock grazing, timber production, watershed protection, industrial mineral development, and outdoor recreation. It has a Web site at *http://www.blm.gov/nhp/index.htm*.

Bureau of Reclamation, U.S. Department of the Interior. The Bureau of Reclamation builds dams on federal lands in western states for power generation, irrigation, flood control, industrial development, fish and wildlife, and recreation, as shown in Figure 19–9. The Bureau of Reclamation Web site is at *http://www.usbr.gov/main/index.html*.

Bureau of Indian Affairs, U.S. Department of the Interior. The Bureau of Indian Affairs is responsible for soil and water conservation on all Native American lands. Personnel of this bureau advise the Native American land operators on useful and necessary conservation practices. When land is leased to nonreservation operators, the lease contains provisions for the conservation practices required of the lessees. The Bureau of Indian Affairs has a Web site at *http://www.ost.doi.gov*.

Environmental Protection Agency. This agency was established to reduce pollution of water, air, and soil by solid and liquid residues, toxic substances, pesticides, radiation, and noise. It is an independent agency functioning separately from major departments such as Agriculture and Interior. When eroded soil becomes a sediment pollutant of water or a particulate (dust) pollutant of air, its control is the official concern of this agency. The EPA administers the Water Pollution Control Act of 1972, Public Law 92-500, and amendments. It has a Web site at *http://www.epa.gov*.

Figure 19–9 This swimming facility in California was established and is managed by the U.S. Bureau of Reclamation. It is part of a multiple-use reservoir providing irrigation water and outdoor recreation. (Courtesy USDI Bureau of Reclamation.)

Tennessee Valley Authority. The Tennessee Valley Authority is an independent agency with headquarters in Knoxville, Tennessee. It was established in 1933 as an area-wide organization to develop the natural and human resources in the watershed of the Tennessee River, an area of 260 million ac (106 million ha) in seven states. It conducts research and demonstration projects to reduce erosion and sedimentation by applying all practical conservation measures. This includes a national research and demonstration laboratory for formulating new fertilizers and testing them in the Tennessee Valley and elsewhere, establishing model dams with hydroelectric generation, flood control techniques, and navigation development. Recreation has become one of its most popular uses of water and adjacent lands. There is a TVA Web site at *http://www.tva.gov*.

Food and Agriculture Organization. The agency with worldwide jurisdiction in soil and water conservation is the Food and Agriculture Organization of the United Nations, as explained in Chapter 20.

SUMMARY

In early 1929, the U.S. Congress responded to critical soil erosion problems by passing the Buchanan Amendment that established erosion experiment stations and plant nurseries. This was followed in 1933 by the establishment of the Soil Erosion Service, a public works program to provide employment during the Great Depression. Through the effective leadership of H. H. Bennett, assisted by two dust storm episodes, the permanent Soil Conservation Service was established in 1935. The successor of the Soil Conservation Service, known since June 3, 1995, as the Natural Resources Conservation Service, is the national and international leader in all phases of soil and water conservation.

Conservation Districts are now nationwide. They are legal subdivisions of state governments and are managed by a governing board elected by private citizens. They call for assistance from technical and other agencies, especially the Natural Resources Conservation Service.

The Farm Service Agency has many functions, one of which is to administer a cost-sharing program for soil and water conservation practices adopted by farmers and ranchers.

Research, education, and economic functions of the U.S. Department of Agriculture are directed by an undersecretary. Their assistance to states includes grants to the State Extension Services and to the State Agricultural Experiment Stations. Most of the research and extension work is done in collaboration with the land-grant universities. Some of the extension and research grants to the states are used to further soil and water conservation.

Teaching soil and water conservation in colleges is largely the province of the state agricultural universities. State funds are also used to support off-campus extension teaching and research on soil and water conservation.

The USDA-Forest Service practices and researches soil and water conservation on its national forests and national grasslands and offers management advice for private lands. It and the state forestry agencies promote conservation by supplying forest tree seedlings at concessional rates for planting.

Other federal agencies that practice conservation include the U.S. Army Corps of Engineers, Bureau of Land Management, Bureau of Reclamation, Bureau of Indian Affairs, Environmental Protection Agency, and the Tennessee Valley Authority.

QUESTIONS

1. Explain the factors that led to national awareness of soil erosion in the United States.
2. List the major mandates of the Natural Resources Conservation Service.
3. Enumerate the principal practices promoted by the Natural Resources Conservation Service.
4. Describe the origin and functions of Conservation Districts.
5. Explain cost sharing by the Farm Service Agency.
6. Agricultural universities have three main functions in soil and water conservation. Identify and explain their relevant work.
7. Name the other federal agencies that work in soil and water conservation.
8. What international agency has responsibility for soil and water conservation programs?

REFERENCES

Conrad, D., 2000. Implementation of conservation title provisions at the state level. Ch. 6, p. 63–75 in T. L. Napier, S. M. Napier, and J. Tvrdon (eds.), *Soil and Water Conservation Policies and Programs: Successes and Failures.* CRC Press, Boca Raton, FL, 640 p.

ECONOMIC RESEARCH SERVICE, 1996. *Provisions of the Federal Agriculture Improvement and Reform Act of 1996.* Agric. Info. Bull. 729, 147 p.

FARM SERVICE AGENCY, 1995. *Farm Service Agency Programs.* U.S. Government Printing Office, Washington, D.C., 8 p.

FARM SERVICE AGENCY, 1997. *The Conservation Reserve Program.* USDA PA-1603, May 1997, 40 p.

FARM SERVICE AGENCY, 2000a. *Producer's Guide to FSA Loan Programs.* USDA PA-1664, Revised July 2000, 20 p.

FARM SERVICE AGENCY, 2000b. *Serving America's Farmers and Ranchers.* USDA PA-1660, May 2000, 12 p.

JOHNSON, P. W., 2000. The role of the Natural Resources Conservation Service in the development and implementation of soil and water conservation policies in the United States. Ch. 4, p. 45–49 in T. L. Napier, S. M. Napier, and J. Tvrdon (eds.), *Soil and Water Conservation Policies and Programs: Successes and Failures.* CRC Press, Boca Raton, FL, 640 p.

LAL, R., (ed.), 1988. *Soil Erosion Research Methods.* Soil and Water Conservation Society, Ankeny, Iowa, 244 p.

LIPTON, K. L., 1996. Glossary of Terms. In *Provisions of the Federal Agriculture Improvement and Reform Act of 1996.* USDA Economic Research Service, Agric. Info. Bull. 729, 147 p.

NAPIER, T. L., and S. M. NAPIER, 2000. Soil and water conservation policy within the United States. Ch. 8, p. 83–94 in T. L. Napier, S. M. Napier, and J. Tvrdon (eds.), *Soil and Water Conservation Policies and Programs: Successes and Failures.* CRC Press, Boca Raton, FL, 640 p.

NATIONAL ASSOCIATION OF CONSERVATION DISTRICTS, 2002. The new farm bill—a giant step for conservation. In *NACD News & Views,* May/June 2002. 8 p.

NATIONAL RESEARCH COUNCIL, 1986. *Soil Conservation: Assessing the National Research Inventory.* Volume 1, 114 p.; Volume 2, 314 p.

NATIONAL WILDLIFE FEDERATION, 1989. *Conservation Directory,* 34th ed. Washington, D.C., 331 p.

NATURAL RESOURCES CONSERVATION SERVICE, 1996. *America's Private Land, A Geography of Hope.* USDA-NRCS Program Aid 1548, 80 p.

NATURAL RESOURCES CONSERVATION SERVICE and IOWA SOYBEAN PROMOTION BOARD, 1996. *The 1996 Farm Bill: Leading Conservation into the 21st Century.* 16 p.

OSBORN, C. T., 1996. Conservation. In *Provisions of the Federal Agriculture Improvement and Reform Act of 1996.* USDA Economic Research Service, Agric. Info. Bull. 729, 147 p.

SOIL CONSERVATION SERVICE, 1989. *Conservation Assistance Around the World.* SCS Program Aid 1431, 8 p.

SOIL CONSERVATION SERVICE, 1992. *Small Watershed Projects.* SCS Program Aid 1354, unpaged.

SOIL SURVEY STAFF, 1999. *Soil Taxonomy: A Basic System of Soil Classification for Making and Interpreting Soil Surveys,* 2nd ed. USDA Agric. Handbook 436, 869 p.

U.S. DEPARTMENT OF AGRICULTURE, 1995. *1995 Farm Bill: Guidance of the Administration.* Administration's Farm Bill Proposal, 94 p.

C H A P T E R

20

SOIL AND WATER CONSERVATION AROUND THE WORLD

The worldwide occurrence of soil erosion and the widespread deterioration of soil and water resources have prompted governments almost everywhere to respond in order to ensure the survival and well-being of their people. Most countries have made an effort, however uneven, to conserve soil and water (Figure 20–1). Many countries have developed a conservation agency very much like the Natural Resources Conservation Service in the United States within a national department or ministry of agriculture. Some good results have been obtained, but progress has been spotty and much remains to be done to reduce soil deterioration and water losses. This chapter presents examples of programs that have been developed in a number of countries.

20–1 WORLDWIDE NEEDS FOR SOIL AND WATER CONSERVATION

There are many soil and water conservation problems and solutions that do not stop at national boundaries. Conservation techniques developed in one country are often applicable in several other countries with similar climate and soils. Food, fiber, and wood products are shipped from one country to another in response to natural disasters, famines, and long-term needs and market conditions. Air and water pollution, acid rain, global warming, and the ozone hole are international concerns that have been important topics in a number of international conferences. Agreements have been made to reduce or eliminate the use of certain pesticides, chemicals such as chlorofluorocarbons (CFCs), and greenhouse gases. A single nation can do little to resolve these issues by itself, but a worldwide effort can accomplish much.

Global warming caused by greenhouse gases is a major environmental issue that has been the cause of much debate. Is it real or imagined? Is it natural or are humans causing it? Can we do anything about it? Does it matter, or will life go on anyway?

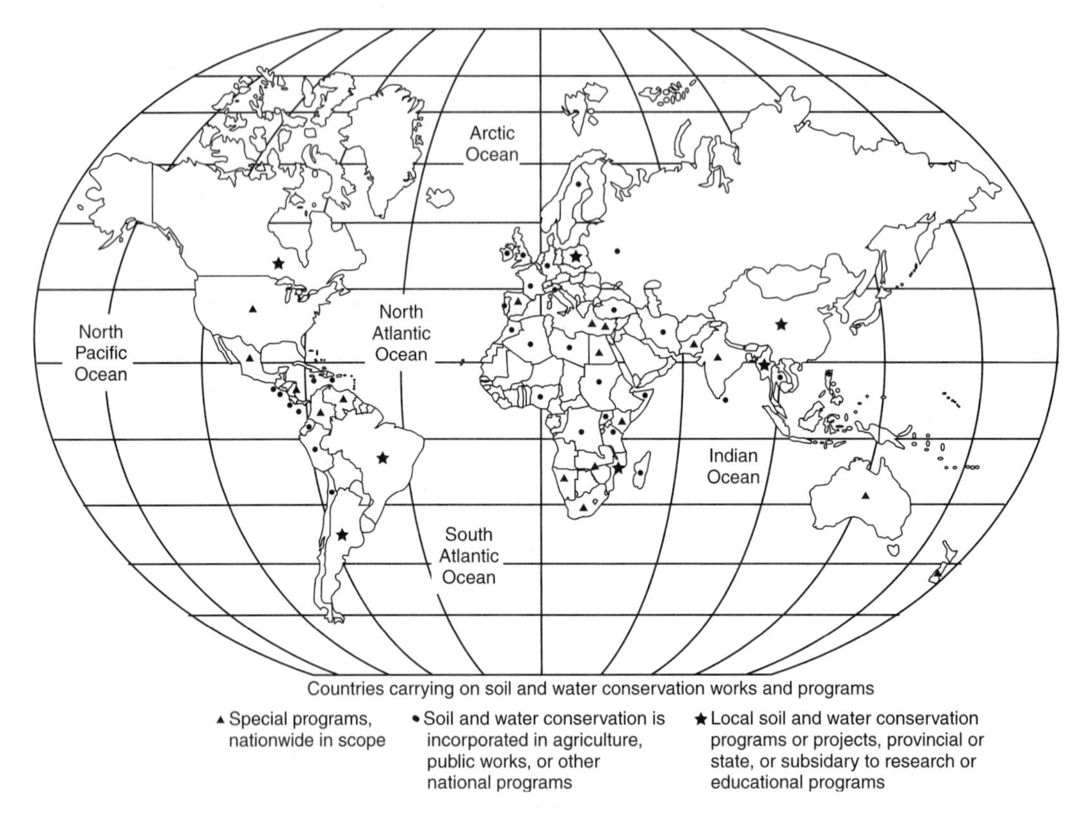

Figure 20–1 Almost every country has some form of soil and water conservation program. Many of them have been inspired and assisted by the work in the United States and by the Food and Agriculture Organization of the United Nations. (Courtesy USDA Natural Resources Conservation Service.)

Studies of sedimentary rocks deposited in past geologic eras show that the Earth's climate was variable long before there were enough humans to have influenced it. However, both theory and recent measurements indicate that current human activities are having a significant warming effect (Favis-Mortlock and Guerra, 2000). This could ultimately shift the temperature and precipitation bands around the Earth enough to relocate cropping zones, with some areas suffering reduced yields while others should show improvements. Another dramatic impact that is foreseeable is a rise in sea level. The world's glaciers have been receding rapidly in recent years and the water naturally flows to the oceans. Not only might productive delta areas be flooded within a few decades, but the inundation is likely to include many of the world's large cities that are built partly or entirely on lowlands near the present coastlines. The people of The Netherlands have shown that there are ways to battle the sea, but the cost is considerable and this would be on a much larger scale.

The decomposition of organic matter from the world's soils is a large source of the carbon dioxide that is a leading cause of global warming (Lal et al., 1995). The release of carbon dioxide as a result of clearing forests and cropland expansion to feed a growing population needs to be countered by practices that sequester carbon dioxide. Tree planting, restoring wetlands, and using less tillage (especially no-till) are helpful if they are applied on a grand scale. Reduced usage of fossil fuels would also help, but all of this may not be enough unless these techniques are accompanied by new practices that have yet to be developed.

20–2 TRANSFER OF CONSERVATION TECHNOLOGY

Some of the most difficult problems facing world development arise from geographic and cultural differences that impede technology transfer from one region to another. The majority of people in developed countries are willing, and many are eager, to help people in other countries with their food production problems, including soil and water conservation. However, it is very difficult to transfer ideas and techniques across cultural boundaries and into different physical and climatic environments (Hurni, 2000). Even the technicians of the USDA-Natural Resources Conservation Service and of the Food and Agriculture Organization of the United Nations have difficulty adapting effective techniques to other cultures, especially to problems as challenging as those in the tropics. The transfer of technology to the tropics and within the tropics will be emphasized here because many of the food shortages and the greatest unsolved problems of soil and water conservation are in developing countries with tropical and subtropical climates (Lal, 2000). Increased food production is essential to keep pace with their population growth. Soil degradation problems are common in developing nations where farmers try to get as much as possible from their land. For example, many farmers remove the entire plants (both crop plants and weeds, including the roots) from their fields to feed their livestock, leaving nothing to replenish the soil organic matter.

Some of the most serious mistakes in technology transfer have involved soil management. These mistakes have caused surface crusting, accelerated erosion, and a rapid decline in crop production. For example, large mechanization projects based on soil-management techniques of continuous cropping were introduced several years ago in Ghana and Tanzania. Rapid declines in crop yields in both countries resulted from irregular topography, small farms, fragile soils, plinthite (laterite) in the soil profiles, crusting of the soil surface, and serious wind and water erosion. The traditional shifting cultivation, often called "slash-and-burn," was a much more satisfactory cropping system than the substitutes that were instituted. However, traditional systems often break down under increasing population pressure. Shifting cultivation is inherently low intensity because a long fallow period must occur before the land can be cleared again.

Alternatives to shifting cultivation have become more feasible in recent times and will be discussed in Section 20–4. Starting in the 1960s, modern agricultural research centers have been established throughout the developing world. There are now 22 international agricultural research centers. The Consultative Group on International Agricultural Research supports 13 of them, and 9 are financed through other means. All of the centers were established primarily to increase food production, but indirectly they are concerned with the most crucial natural resources—soil and water. Much of the work is coordinated through the Food and Agriculture Organization of the United Nations.

20–3 FOOD AND AGRICULTURE ORGANIZATION

The Food and Agriculture Organization (FAO) was founded in 1945 and became an agency of the United Nations in 1946 with permanent headquarters in Rome, Italy. The FAO works worldwide but concentrates on helping developing countries. Hauck (1974) identifies the following specific projects of the Food and Agriculture Organization:

1. Sending consultants on soil and water conservation for specific work in response to government requests
2. Conducting seminars on a regional or country basis
3. Publishing bulletins on soil and water conservation
4. Conducting specific programs as requested in several countries

The FAO employs trained workers of various nationalities to work on agricultural problems around the world. They publish educational materials, conduct seminars, educate people at training centers, and work with local technicians and farmers to resolve a wide variety of problems. They do their own research work to supplement that available from other sources for various needs dealing with food production, erosion control, and reduction of pollution. They also have developed an FAO system of soil classification as a basis for working with world soil resources (Nachtergaele et al., 2000).

The FAO also gathers international information on food production, needs, and markets, including population data and trends (FAO, 1995). It works together with the World Health Organization (WHO) to address health and nutrition problems. The FAO responds to both long-term problems dealing with food production and nutrition and to disaster relief and other emergency assistance.

20–4 SHIFTING CULTIVATION AND CONSERVATION

About 250 million farmers in the humid tropics subsist by *shifting cultivation* ("slash-and-burn"). This kind of agriculture is common and sometimes dominant in parts of Asia, Africa, Central America, and South America. Close to 30% of the arable land in the world (about 400 million ha) is involved in shifting cultivation (Brady, 1996). The soils are generally permeable, are leached of many essential plant nutrients, and are commonly classified as Oxisols or Ultisols. The warm climate is favorable for the growth of many different crop plants. There are serious problems, however, including soil degradation, plant diseases, and the low natural soil fertility that is easily reduced to still lower levels under cropping conditions (Harwood, 1996).

Two contrasting kinds of shifting cultivation are used, depending on the soil and its vegetation. The most common type is practiced on humid-region, *forested,* low-fertility soils on steep slopes with high erodibility. The other type is used on humid-region, *grassed,* clay soils with high base saturation, gentle slopes, and moderate erosion hazard.

Figure 20–2 Shifting cultivation on forested soils requires that a dense stand of trees be cut down and burned. Bulldozers have been used experimentally, but they cause serious soil deterioration. Much of the work is done with an axe, as shown in this scene in the Ivory Coast. (Courtesy Roy L. Donahue.)

20–4.1 Forested Soils in the Tropics

Shifting cultivation on forested soils means partially clearing and burning a patch in a forest as shown in Figures 20–2 and 20–3, raising crops mostly by hoe-culture and to a lesser extent by animal-powered farming methods for a period of two to three years, then allowing the patch to revert to forest trees for 10 to 20 years while other tracts are cleared, burned, and farmed. After a decade or two of soil rejuvenation, the same patch is again cleared and cropped. The cropping period is usually extended for as long as the farmer considers crop yields satisfactory or until there are excessive infestations of weeds, insects, or diseases. This period seldom exceeds five years. Likewise, the period that forest trees are allowed to grow depends on how rapidly and completely trees occupy the cleared patches and how soon the farmer needs the area again for cropping. This system works well when the population pressure is low so the recovery period can be long, but it breaks down as population pressure increases and the forest period is shortened (Kleinman et al., 1995).

Trees are essential during the soil-rejuvenation stage to accelerate the weathering of soil minerals, to add organic matter, and to break the cycles of insects, diseases, and weeds.

Figure 20–3 A forest and grass fire in Ghana, western Africa. This is the second step in the shifting cultivation (slash-and-burn) system of farming. (Courtesy F. Botts for FAO.)

Soil productivity is renewed less rapidly where grasses dominate the vegetation during the rejuvenation cycle. This is especially true on soils with low-base status and plinthite. Trees maintain more uniform soil water and temperature conditions, which reduce crystallization of iron and aluminum into plinthite.

Plinthite forms in iron-rich soil layers that are saturated with water during part of the year. It becomes a rock-like and almost waterproof material called ironstone if it is subjected to excess drying, especially after it is exposed to repeated wetting and drying cycles and baking by the sun (Soil Survey Staff, 1999). The few inches up to three or four feet (from a few centimeters to a meter or more) of soil above the plinthite become saturated and are easily eroded during heavy rains.

Conditions conducive to soil erosion exist when a patch of steeply sloping land is cleared and burned for cropping. Most soils of the tropics are low in fertility, and their organic matter decomposes rapidly, especially when the land is cleared by burning and during the first year of cropping (García-Oliva et al., 1999). Precipitation is usually more intense than in temperate regions, and the resulting erosion is often much more rapid than it would be in temperate regions.

Many efforts have been made to find a viable alternative to shifting cultivation, but all failed until recently. An alternative that has met with some success is the use of fertilizers and pesticides combined with minimum tillage to leave organic residues on the surface of the soil. Such a system was studied in western Africa by the International Institute of

Tropical Agriculture at Ibadan, Nigeria (IITA, 1975). Western Africa is a humid, forested area and the dominant soils are high in plinthite and ironstone. Traditional shifting cultivation is the predominant cropping system. Research personnel at the institute hypothesized that modern scientific agriculture could replace shifting cultivation successfully. Field research the institute used to test this hypothesis included:

1. A surface mulch of crop residues and no-till cultivation to reduce extremes of soil moisture and soil temperature and to control soil erosion
2. Modern insecticides, fungicides, and herbicides to control pests
3. Chemical fertilizers and lime, based on a soil test, to increase soil fertility and crop productivity

Soybeans grown in this research lost only 1% of the precipitation received, and no soil was eroded. Plowed and cultivated soybeans caused a 15% loss of precipitation by runoff and suffered 17 tons/ac (38 mt/ha) of soil loss per year. Loss of water and soil from other crops was much higher.

One problem with this system is that fertilizers in many developing countries are commonly two to six times as expensive as they are in North America and Europe (Sanchez, 2002). Therefore, in many tropical countries, it makes more economic sense to rely on a *low-input* soil-management strategy that minimizes the use of expensive fertilizers and pesticides. This approach seeks to improve crop productivity by breeding or discovering plants that are better adapted to forested Oxisols and Ultisols. In low-input strategy, pasture grass and legume cultivars used in the forested tropics must be able to tolerate acidic soils high in aluminum saturation, to grow with low levels of available phosphorus, and to utilize low-cost rock phosphate. Progress is being made on all of these objectives (Table 20–1).

Even a low-input system requires some inputs to achieve good yields consistently. Nitrogen and phosphorus are usually deficient in these environments, and other nutrients may be needed as well. Sanchez (2002) proposes the use of alternative sources of these nutrients to limit costs in the subtropical parts of Africa that have wet and dry seasons. He recommends that leguminous trees be interseeded in corn and other crops to fix nitrogen in

TABLE 20–1 PROMISING GRASSES AND LEGUMES SELECTED/BRED FOR ACIDIC, INFERTILE, TROPICAL, FORESTED OXISOLS AND ULTISOLS

Grass cultivars	Legume cultivars
Signalgrass	Tick clover
(*Brachiaria decumbens*	(*Desmodium leonii*
Bluestem	Tick clover
(*Andropogon gayanus*	(*Desmodium ovalifolium*
Panicgrass	Zornia
(*Panicum maximum*)	(*Zornia latifolia*
Hyparrhenia	Stylosanthes
(*Hyparrhenia rufa*)	(*Stylosanthes capitata*
Crabgrass	Kudzu
(*Digitaria decumbens*)	(*Pueraria phaseoloides*

Source: Sanchez, 1986.

tropical fields. Growing trees such as *Sesbania, Crotalaria,* and other suitable species along with the crop and during an ensuing dry period can add 90 to 180 lb of N/ac (100 to 200 kg N/ha) when the leaves, pods, and small branches are hoed into the soil. He also recommends the use of indigenous rock phosphate to meet phosphorus needs. In his African work area, most of the soils are acidic enough to help dissolve this form of phosphorus.

The availability of all plant nutrients can be enhanced and the organic matter content of the soil increased by growing nutrient-accumulating shrubs such as *Tithonia diversifolia* along roadsides and as hedgerows and by adding their leaf biomass to the soil in the fields (Sanchez, 2002). This is a labor-intensive practice, so it is economically justifiable only on high-value crops. The *alley cropping* system described in Section 8–4 is a similar practice that uses lines of bushes grown on the contour to provide organic matter for the soil in the cropped alleys.

20–4.2 Grassland Soils in Humid Tropics and Subtropics

Vertisols are the most abundant grassland soils in humid tropical areas. The 5.78 million ac (2.34 million ha) that they occupy are mostly in tropical or subtropical areas with wet and dry seasons. They commonly occur where the annual rainfall is 24 to 40 in. (600 to 1000 mm). This is enough to support forest vegetation on other soils, but the Vertisols grow grasses. Two factors favor the grasses: the high base saturation and the high content of smectite clay. It takes at least 30% clay to a depth of 20 in. (50 cm) or more to give Vertisols their self-swallowing action. The shrink-swell potential of the smectite causes deep cracks to open in these soils during dry seasons, as shown in Figure 20–4. Granular surface soil falls into these cracks and causes the soil to shift and "churn" when it gets wet. The churning action distributes organic matter throughout the soil and produces a deep, dark-colored soil. It can also break deep roots of perennial plants and the foundations of buildings.

Traditional crop culture on Vertisols in Ethiopia includes cotton, grain sorghum, and sesame in a shifting cultivation that involves *burning the soil.* Only Vertisols are burned—never the adjoining red clay soils. The native tall-grass prairie is plowed with a village-made, tongue-type plow pulled by oxen (bullocks). The plowed rows are crooked, and the seedbed is always cloddy and full of large sod pieces with soil attached. The farmer gathers these sod pieces, puts them in piles, and adds cattle manure in the center of each pile, as shown in Figure 20–5. When the soil and sod pieces have dried, each pile is set on fire and may smolder for several days. Upon cooling, the residues from each burned pile are spread over the field. The soil has improved tilth after burning. After further plowings, seeds of cotton, grain sorghum, or sesame are sown. Field crops are then grown for a period of three to five years, after which the land is abandoned (fallowed). Weeds, annual grasses, and finally perennial native grasses invade and flourish during the 10- to 20-year fallow period; then the burning and cropping cycle is repeated.

During the cropping cycle, the soil becomes weedy, insect and disease populations build up, and soil tilth deteriorates. Soil fertility is decreased by crop removal, soil crust formation becomes more serious, and sheet and gully erosion are increased. By contrast, during the fallow cycle, the grass roots improve the physical condition of the soil, pests decrease, fertility increases by weathering of minerals, and erosion is controlled. The native grasses help to develop a fine crumb structure that makes a desirable physical, chemical,

Figure 20–4 A Vertisol in northwestern Ethiopia. This soil developed on high-lime materials, is dark colored, has a high percentage of smectite clay, and has wide, deep cracks for at least 90 consecutive days per year. (Courtesy Roy L. Donahue.)

and biological condition. Burning the black clay soil inside each pile of sod pieces destroys pests, eliminates cloddiness, and transforms the smectite clay into nonswelling particles that resemble brick dust. The surface soil that was a clay loam now acts like a loamy sand. The burning also increases soil pH, available phosphorus, and total carbonates but decreases organic carbon (Donahue, 1972).

Shifting cultivation on burned Vertisols supporting native grasses is an inefficient use of land, and it will not support rising populations with their increasing demand for land. Replacement of this traditional system with a more intensive system that will maintain production and control erosion awaits field research solutions. Perhaps a solution will be found that is comparable to that for forested areas, using minimum tillage, organic surface-residue management, pesticides, and chemical fertilizers (Sparovek and Schnug, 2001). In the meantime, Ethiopian farmers are using a system with no modern inputs that maintains yields and stabilizes the soil against excessive erosion.

20–5 SOIL AND WATER CONSERVATION IN SELECTED AREAS

Soil and water conservation are needed everywhere, and each area has its own unique needs, but only a few examples can be included here. Specific examples of soil and water conservation are cited here for arid northern Africa and the Middle East, Ghana and Liberia,

(A)

(B)

Figure 20–5 Traditional shifting cultivation on Vertisols in Ethiopia involves (A) plowing the grassland several times with a village-made plow and then (B) gathering sod pieces into piles. Cattle manure is added to the piles and they are allowed to dry; then they are burned and spread back over the field. (Courtesy Roy L. Donahue.)

South Africa, Greece, Russia and the former USSR, Iceland, India, the humid Amazon jungles of Peru, Brazil, Australia, and The Netherlands.

20–5.1 Arid Northern Africa and the Middle East

Soil and water conservation is especially difficult in arid regions. Aridity is common at latitudes between about 20 and 30° north and south of the equator. At these latitudes, global air circulation (Chapter 13) causes air masses to descend, become warmer, and increase their capacity to hold moisture. The result is low precipitation and little plant growth to protect the soil against erosion. Also, in areas receiving little precipitation, a single storm may bring all or most of the annual rainfall. The combination of torrential rain and scant vegetation results in severe water erosion, angular hills, and smooth alluvial lowlands. Wind erosion may be active during the long, dry periods.

Desert is commonly defined as an area that receives about 4 in. (100 mm) or less of annual precipitation—not enough to support protective vegetation. However, with an annual precipitation of 6 to 10 in. (150 to 250 mm), some plant species can be established to stabilize the soil and to supply some forage, firewood, and construction materials.

Special studies of techniques for successful establishment of productive and protective vegetation have been made in Tunisia, Morocco, and Algeria by the Food and Agriculture Organization of the United Nations (Bensalem, 1977). These results are considered to be equally applicable to many other countries with similar rainfall. Greater success is assured if the site selected for establishing vegetation is in a swale where additional rainwater collects, or if the soils are deep sands where plants root deeply and rainwater quickly infiltrates far enough to resist loss by evaporation. Deep-rooted, drought-resistant vegetation can be established in such areas.

About 120,000 ac (50,000 ha) of spineless cactus have been planted in Tunisia to stabilize the soil and to provide forage for livestock. Also in Tunisia, during a recent five-year period, spineless acacia were established on about 2500 ac/yr (1000 ha/yr). Windbreaks have been successfully established on fine-textured soils in Tunisia and Algeria by planting a mixture of native acacia species and eucalyptus trees from Australia. Two species of pines have proved satisfactory on deep sands: stone pine and cluster pine. New plantations must not be grazed by livestock for a period of two to five years. This restriction requires rigid enforcement by the local government. Trees and shrubs that have been successfully established in the arid and semiarid tropical and subtropical regions as windbreaks or woodlots are listed in Table 20–2.

TABLE 20–2 COMMON ENGLISH NAMES[a] OF TREE AND SHRUB SPECIES RECOMMENDED FOR PLANTING AS WOODLOTS, WINDBREAKS, OR SHELTERBELTS IN TROPICAL AND SUBTROPICAL REGIONS WITH ANNUAL PRECIPITATION OF 6 TO 10 IN. (150 TO 250 MM)[b]

Acacia	Mesquite tree
African locust	Mulga
Aleppo pine	Neem tree
Argan tree	Olive tree
Calligonum	Russian olive
Carob tree	Salsola shrub
Cluster pine	Shinus
Eucalyptus	Siris tree
Fourwing saltbush	Sissoo
Horsetail tree	Stone pine
Jerusalem thorn	Tamarix
Kassod tree	Tassili cypress

[a]Scientific names are given in Appendix B.

[b]All plants will grow in areas of the lower range of precipitation in swales and on sandy soils.

20–5.2 Ghana and Liberia

Ghana and Liberia are located in the humid part of western Africa about 5 to 10° north of the equator. The Soil Research Institute at Kwadaso-Kumasi in central Ghana has conducted outstanding research on soils, mechanization, erosion, and soil productivity. The institute made a soil survey of an area where attempts to use mechanized agriculture on soils with plinthite (laterite) had failed. The survey identified the problem soils, and it was recommended that all soils with plinthite be seeded to perennial pasture. Continuous mechanized cultivation was recommended for soils without plinthite. More detailed soil surveys made after this experience were used to make a map of Ghana that delineates areas suitable for the three kinds of cultivation: tractor-powered, ox-drawn, and hand-hoes, such as those shown in Figure 20–6.

Liberian agriculture consists mostly of shifting cultivation, with vegetables, upland rainfed rice, and cassava (tapioca) as the main crops. Many commercially successful rubber, cocoa, coffee, and oil palm plantations also exist. Erosion is serious on sloping croplands during extended periods of cropping, especially near large centers of population. Some soils apparently can be cropped without deterioration for five or more years, whereas others cannot support more than two or three crops without a serious decline in yields. To determine suitable soils for the most feasible land resettlement, the government of Liberia requested assistance in establishing a nationwide soil survey.

For three and a half years, the U.S. Agency for International Development financed the USDA-Soil Conservation Service to help establish a soil survey in Liberia. The survey is used to aid in scientific land-use decisions, including practices to maximize production and minimize erosion and sedimentation (Geiger, 1978).

20–5.3 Soil and Water Conservation Trials with Vetiver Grass

Mining spoils from the South African diamond mines were called *kimberlite* by Lewis in 1887. They consist of a mixture of minerals in a fine-grained ground mass of calcite ($CaCO_3$), olivine $(Mg,Fe)_2SiO_4$, and phlogopite $[K(Mg,Fe)_3AlSi_3O_{10}(OH,F)_2]$ (Grimshaw, 1996). Some kimberlite may be high in sodium (Na).

Piles of kimberlite mining spoils may be as high as 260 ft (79 m), cover areas as large as 500 ac (200 ha), and have slopes at the natural angle of repose of about 30°. These diamond mine dumps are in a semiarid climate that further increases the difficulty of stabilizing the spoils with vegetation (Bates and Jackson, 1980).

Vetiver grass was planted on the contour to stabilize diamond mine tailings in South Africa (Berry, 1996). The results are summarized as follows:

- Accelerated erosion and obvious sedimentation were arrested, even without slope contouring.
- Deep rooting of the vetiver grass was confirmed.
- The hostile environments included steep slopes, alkaline soils high in sodium, and surface soil temperature extremes from 32°F (0°C) to 122°F (50°C).
- As of 1996, a nursery was being established to propagate vetiver grass to plant on more diamond mine tailings.

(A)

(B)

Figure 20–6 Two kinds of tillage implements used in equatorial Africa. The village-made hoe, commonly used in subsistence agriculture, seldom disturbs the soil enough to cause serious soil erosion. The disk plow makes it possible to use tractor power in commercial farming and also increases the potential for soil erosion. (Courtesy Roy L. Donahue.)

The National Research Council (1993) lists 71 countries where vetiver grass has been used to control erosion and sedimentation for environmental improvement. This is a robust grass that grows as tall as 8 ft (2.4 m) in tropical regions of Africa and South America, yet it is hardy enough to withstand the mild frosts of central Europe and the southern United States. It also tolerates low soil fertility, acid or alkaline pH, and heavy metal concentrations that preclude growth of many other plant species. Its seeds are sterile, but it grows readily from cuttings.

20–5.4 Greece

Greece has a semitropical climate. Oranges are an important crop in the warmer parts of the country, and olives are grown over a more extensive area. Mean annual precipitation varies from 16 in. (400 mm) at Athens to 33 in. (850 mm) at Kalmia in the Peloponnesus. The dry summer season, ranging from about one month on the island of Crete to about four months at Athens, is a constraint on production of warm-season plants.

The soils of Greece are predominantly shallow over highly weathered limestone. Nearly all of the topography is hilly to mountainous. Shallow soils on steep slopes are conducive to erosion, even though the rains are usually gentle. Soil erosion is a serious handicap to the country because a deep soil is needed to hold sufficient available water for crop use during the dry summers, when temperatures are most favorable for plant growth.

According to a 20-year study by the Athens Soils Institute, serious erosion had occurred on 110,000 ac (45,000 ha) in the Peloponnesus. A study made by the Land and Water Reclamation Service in two states in northern Greece, Thrace and Macedonia, found the loss of agricultural production due to erosion amounted to $8 million per year.

Action programs to control erosion in Greece have included the use of terracing, contour farming, strip cropping, subsoiling, tree planting, grazing control, and gully reclamation. Terraces have been built to protect 22,000 ac (9000 ha) of land.

20–5.5 Russia and the Former Soviet Union

The former Union of Soviet Socialist Republics (Soviet Union) had 560 million ac (226 million ha) under cultivation. Its geographic location parallels that of Canada and the northern United States. The cropped area has a climate as variable as that from northern Alaska to southern California (Central Intelligence Agency, 1974). The Ukraine and nearby areas have warm, dry summers; Russian Siberia has a cold climate that limits crop production. About 60% of the Russian land area has permafrost (Gennadiyev, 2001).

Russia got an early start in the development of soil science in the 19th century through the work of Dokuchaev, Sibertsev, and others and has continued work in these areas since that time. They published a soil erosion map of the Soviet Union in 1968 and had a more detailed and updated version nearly completed in 1999 (Kastanov et al., 1999). These maps show that meltwater erosion is a major problem in the cold areas of northern Russia, rainfall erosion becomes more important in the middle latitudes, and wind erosion is dominant in the drier southern part. About two-thirds of the arable Russian farmland is subject to severe wind and water erosion. These natural processes have been accelerated in many areas by agricultural policy stressing production rather than conservation. Plowing of steep slopes, overgrazing, overcutting of trees, and other practices that remove vegetative cover have been particularly damaging in the forest-steppe and forest zones of European

Russia and in the Transcaucasus and central regions. Bare ground in the dry steppe and semidesert is frequently eroded by strong, dry winds.

Mountainous areas constitute another problem with their steep slopes and shallow soils (Il'ichev, 1999), and Russia has its share of mountains, including the Caucasus along its southern border and the Urals that divide Europe from Asia. Severe erosion results when humans harvest too many trees, graze too many cattle, and otherwise disrupt fragile mountain environments. The problem is even worse in mountainous areas that are more densely populated, such as those in Nepal, India, and Indonesia.

Thick loess deposits pose another problem. Loess is very susceptible to deep gullying, even on gentle slopes. Occasional downpours of summer rain produce vast networks of gullies and ravines, particularly in the deep loess deposits that occupy large areas of the Ukraine.

The severity of the erosion problem was acknowledged in a 1967 joint resolution of the USSR Communist Party and the Council of Ministers. The resolution called for increased use of erosion-control practices such as contour plowing, crop rotation, strip cropping, revegetation of steep slopes with grass, planting forest on uplands and in gullies, ravines, and along the shorelines of rivers and reservoirs, and building erosion-control and flood-control structures.

Windbreaks and shelterbelts are prominent among measures designed to combat both wind and water erosion. Concentrated on the steppe and forest-steppe of European Russia, these forest belts protect against wind erosion, increase the accumulation of snow, and check the erosive action of surface water. Trees and shrubs planted for windbreaks and shelterbelts include poplar, birch, black locust, Siberian elm, Siberian larch, Scotch pine, English oak, sea buckthorn, golden currant, Russian olive, caragana (Siberian peashrub), tartarian honeysuckle, tamarisk, and willow (Schroeder and Kort, 1989).

Contour plowing, crop rotations, fallowing, stubble-mulching, and terracing are used in many areas. Stubble-mulch tillage is a necessity for adequate erosion control in the New Lands area of western Siberia and northern Kazakhstan and in other moisture-deficient areas. Also called trashy fallow, this practice protects the soil from baking, contributes to lower soil temperatures in hot summer weather, decreases the depth of freezing in winter, impedes runoff and evaporation, and enables the soil to absorb more rainfall.

Various types of terraces are used in hilly terrain to reduce the slope length, especially in the mountains of Moldavia, the Caucasus, and Central Asia. Irrigation is widely practiced in the Ukraine and other nations in the southern part of the former Soviet Union. Water deficiencies and saline and sodic soils cause serious problems in these areas (Novikova, 1999).

Centralized planning in the time of the Soviet Union had the advantage of allowing the government to make large scale decisions regarding agricultural production, farming methods, and conservation practices. It also had the disadvantage of having decisions made by people who were thinking on the large scale and who were not on the site looking at specific problem areas. The result was that much soil degradation and erosion occurred (Gennadiyev, 2001). Land reforms in the 1990s allow individuals and local groups to own land and therefore place the decision making process under more local control. Conferences have been held and new technologies have been developed, but a lot of adjustment remains to be accomplished.

20–5.6 Iceland

Iceland is a large island (39,700 mi^2 or 102,800 km^2) located in the North Atlantic Ocean just south of the Arctic Circle. About 80% of the country is unpopulated and unsuited for agricultural use because it is covered with either glaciers or volcanic rock. The remainder has mostly poor soils and adverse climate, so it is used mostly for grazing livestock. Unfortunately, the land has been seriously overgrazed and so degraded that it is considered to be the nation's most serious environmental problem (Hannam, 2001). It was so serious that Iceland became the first nation in the world to establish a soil conservation agency in 1907. A land degradation survey made in the 1990s showed that 52% of the land was seriously degraded.

A more holistic ecological land management plan was adopted by the government of Iceland in 1996 with the objective of achieving a sustainable land management program by combining management of sheep grazing with soil conservation practices. The government is now offering incentives to land owners to drastically reduce grazing on degraded land (Hannam, 2001). Some of the damaged land is being converted to forest as a more appropriate alternative use.

The need for careful land management in Iceland is typical of other lands in cold climates. The soils in these marginal areas are fragile and have a very limited carrying capacity. Plants and animals show a remarkable ability to adapt to hostile environments, but their survival depends on an environment that is easily damaged. Once damaged, the environment is likely to require a long time to recover even if it is protected.

20–5.7 India

India has an extremely varied population of nearly a billion people with a mixture of racial and ethnic backgrounds, languages, and religions. Its topography ranges from coastal plains to rugged mountains, and the soils vary accordingly. About half of the soils in India are subject to deterioration by water and wind erosion, as shown in Figure 20–7. About 16.5% of the cropland area of India has extremely low productivity because of low precipitation, ironstone (hardened plinthite), riverine gullies, saline and sodic soils, sandy coastal land, and waterlogged soils (Das, 1977; National Commission on Agriculture, India, 1976) (Figures 20–8, 20–9, and 20–10).

Practices used to control erosion in India include contour cultivation, terracing, and establishing protective perennial vegetation. Most of the soil and water conservation work has been done to stabilize shifting sands or to reduce erosion and sedimentation in the watersheds above the reservoirs used to store water for hydroelectric-power generation and irrigation (Arakeri and Donahue, 1984).

20–5.8 Amazon Jungles of Peru

Research on intensive cropping of corn, peanuts, cowpeas, rice, soybeans, and cassava has been conducted at Yurimaguas, Peru, since 1969 (North Carolina State University, 1976). The soil is classified as a Typic Paleudult, a highly leached, acidic soil typical of the humid, lowland, forested tropics in the Amazon River Valley. Shifting cultivation is the dominant kind of traditional agriculture. The research treatments included applications of NPK fertilizers and lime, and mulching with kudzu. The results can be summarized in this way:

1. Mulching lowered soil temperatures, decreased bulk density and surface crust formation, controlled weeds, increased available soil moisture, and increased crop yields.

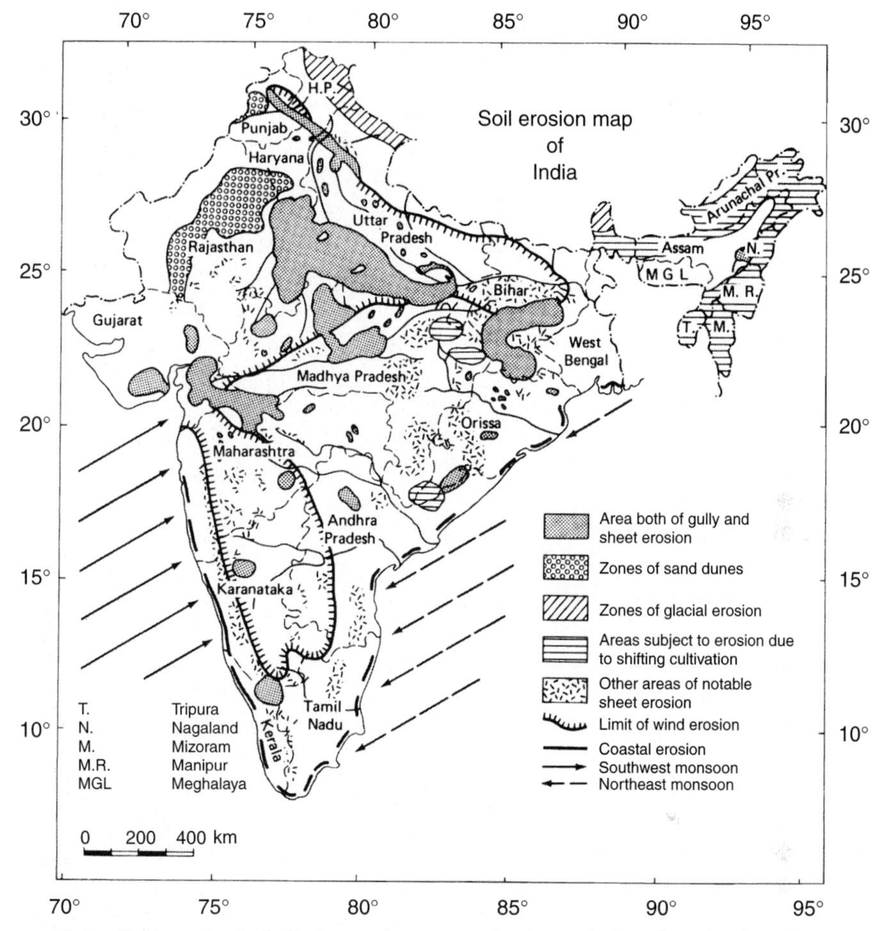

Figure 20–7 Soil erosion in India is a serious constraint for agricultural production. There are large areas of sheet, gully, and wind erosion plus areas of glacial erosion and erosion related to shifting cultivation. (*Source:* Food and Agriculture Organization of the United Nations, *FAO Soils Bull.* 33, D.C. Das, 1977.)

2. Fertilizing with 107 + 44 + 107 lb/ac (120 + 50 + 120 kg/ha) of N, P_2O_5, and K_2O, plus liming, resulted in yields equal to those of the mulched plots.
3. Leaving the grass and kudzu mulch on the soil surface resulted in higher yields of all crops than when the mulches were incorporated into the soil.

The researchers were not able to explain why the use of mulch alone on the soil surface gave crop-yield increases equal to the best chemical fertilizing and liming practices. Perhaps the chemical fertilizer and lime treatment added an excess of nutrients, or some of the nutrients may not have been available to the current crops. Another possibility is that the mulch may have supplied secondary and micronutrients and stimulated soil organisms to decompose soil minerals and bring slowly available plant nutrients into solution. Also,

Figure 20–8 This woody plant has held this 10-foot mound of soil in place in northern India while water erosion has devastated the surrounding soils. (Courtesy Roy Donahue.)

Figure 20–9 The U.S. Agency for International Development has helped India to establish soil conservation research and demonstration areas throughout the country. On the left is Roy Donahue and on the right is his Indian coworker in central India. (Note: The local script is Telugu.) (Courtesy USAID in India.)

the mulching would reduce the warmest soil temperatures, increase infiltration for cultivated crops, and be an effective soil and water conservation practice (Lal, 1974). Whatever the cause, surface mulching was very effective for increasing crop yields.

The temperate-region technique of clearing forested land with a bulldozer was tried and declared a failure in tropical Peru. The bulldozer scraped off too much of the most productive topsoil and compacted the remaining soil. The technique of clearing tropical jungle with hand tools and burning was far superior.

20–5.9 Brazil

Brazil is a large nation with a major area of tropical lowlands in the Amazon basin surrounded by highlands in the rest of the nation. A soil survey of 200 million ac (82 million ha) in the principal cropped areas in humid southern and western Brazil classified the soils as Oxisols, Ultisols, and Mollisols (Bloomfield, 1977). Annual precipitation in the region varies from about 48 to 60 in. (1200 to 1500 mm), with a dry season from April to August.

The survey also showed that 52% of the land in the 82-million ha (200-million ac) area had suffered moderate or severe erosion. The severity of past erosion indicated a need for an effective soil and water conservation program in Brazil. A proposal was made that soil and water conservation and soil survey be combined in an organization

Figure 20–10 Terracing technology reached India in the 1950s and was named "contourbund." The script to the right is in the state language, Telugu, of Andhra Pradesh (central India). (Courtesy Roy Donahue.)

similar to that of the USDA Soil Conservation Service (now the Natural Resources Conservation Service).

Erosion measurements indicated that a soybean field averaged about 9 tons/ac (20 mt/ha) of soil loss annually and water runoff loss was about 7% of the precipitation received. This compared with about 20 tons/ac (42 mt/ha) and 12% of the precipitation for castor beans, and 3 tons/ac (7 mt/ha) and 4% for a field planted to sweet potatoes. Fields in 10 other crops had soil and water losses between these extremes.

Erosion had been accelerated by the rapid increase in acreage of soybeans on slopes up to 12%. Soybean production in Brazil increased 385% from 3.2 to 15.5 million ac (1.3 to 6.3 million ha) during the period from 1970 to 1976, mostly on land formerly planted to coffee and pasture grasses.

Soybean acreage has continued to expand in Brazil, with much forest being cleared to make new cropland (Favis-Mortlock and Guerra, 2000). Large amounts of woody materials and other plant debris cleared from the area are burned, and the ashes are left to fertilize the soil. The organic-matter content of these tropical soils is low and becomes even lower with burning and cropping. Soil crusting is common on the exposed soils. Runoff and erosion are serious problems, especially under conventional tillage. No-till cropping increased rapidly in Brazil during the 1990s and reached about 6 million ha by 1996 (Bayer et al., 2001). The advantages of no-till for Brazil's tropical soils include not only erosion control but also a slower decomposition rate for soil organic matter.

20–5.10 Australia

Australia, a country about the size of the United States, lies 12 to 44° south of the equator. Whereas only about 10% of the United States is arid, two-thirds of Australia is arid, and there are no perennial streams. The arid lands receive less than 10 in. (250 mm) of rainfall a year, while coastal areas receive as much as 30 in. (750 mm). Temperatures vary from tropical in the north to semitropical in the south. Saline and sodic soils are common in the arid two-thirds of the country.

Wind erosion is a potentially serious problem in the arid rangeland of interior Australia. Control of wind erosion is accomplished by reducing the domestic livestock stocking rate. Fortunately the stocking rate can be controlled because the central government owns the rangeland and can dictate lease terms to private livestock owners.

Successful water-erosion control projects have included terracing, minimum tillage, and improved cropping systems.

One major soil degradation problem has been studied for which no solution has been found. It is called *dryland salinity.* Extensive areas of soils that were *never irrigated* are becoming salty. The principal theory is that salt is deposited in rainfall (Hallsworth, 1987). The problem is similar to that known as *saline seeps* in the Great Plains of the United States (see Section 14–3.7).

20–5.11 The Netherlands

Erosion is not a serious problem in much of The Netherlands, where half of the population lives below sea level, because these areas are nearly level and wind erosion is largely controlled by planting trees on the sandy soils. However, contamination of soil and groundwater in this setting could be catastrophic. It is very important to avoid excess applications of sewage sludge, animal manures, fertilizers, and pesticides. Understanding their proper use is crucial.

To minimize contamination, the groundwater is monitored in each soil-mapping unit by designating representative soil profiles and installing piezometers to measure depth to the water table. This research measures (1) the travel time for water to move from the soil surface to the mean water table, (2) soil cation-exchange capacity, and (3) the phosphate adsorption capacity of the soil.

The length of time it takes for water to reach the water table is a measure of the amount of soil depth that can adsorb pollutants. The cation-exchange capacity is a measure of the ability of the soil to adsorb fertilizer cations, heavy metals, and pesticide pollutants, and thus keep them out of the groundwater. The phosphate adsorption capacity of the soil determines how much phosphate is tied up and kept out of surface waters or groundwaters (Breeuwsma et al., 1986).

20–6 ONLY A SAMPLING

This is not the end of the story. Erosion and pollution problems exist in every country, and each country has its own approach to managing and conserving soil and water. This book has sampled only a few of the many examples that could have been cited. Many other problems and solutions could have served as well, but those chosen will convey something of the breadth and magnitude of the need for soil and water conservation not only in the United States but around the world.

Many things change with time, and the conservation story is no exception. People come and go, and they change their perspectives and ways of doing things. Even small changes in soil and crop management may have a great impact on the soil and water resources on which everyone depends. Governments and their programs are always limited by the Earth's finite land base. The human population, however, has never stopped increasing. The pressure on soil and water resources increases with the population, as their expectations increase for better lives and a cleaner, healthier environment.

SUMMARY

Soil and water conservation practices are essential around the world to retain the productive base of agriculture needed to support a rapidly increasing population with rising expectations. Most countries have established some kind of conservation agency, often based on the model and sometimes with the assistance of the USDA-Natural Resources Conservation Service. Progress to date, however, has been uneven, partly because the differences in culture, soil, and agriculture are so great that adaptation is necessary but usually has been inadequate. Massive mechanization schemes, as well as most temperate-region techniques of soil management, have failed in the tropics. One reason is that most soils in the tropics form surface crusts and become cloddy when cropped continuously, and crop yields soon decline below economic levels.

Coordinated worldwide response is needed to resolve worldwide problems such as global warming. Reforestation, restoring wetlands, no-till farming, and reduced use of fossil fuels are helping to reduce this problem, but more effort is needed. The Food and Agriculture Organization of the United Nations promotes international cooperation for such purposes.

Shifting cultivation is the traditional technique used in forested tropics to maintain yields and control the loss of soil and water. This consists of two to three years of cultivated crops followed by 10 to 20 years of wild trees and shrubs. Some Vertisols in the tropics are burned and cropped until yields decline, then are "rested in native grass" to rejuvenate them.

Terracing and ridging on the contour have failed in many tropical and subtropical regions because of physical deterioration of the soil. The use of annual cover crops has also failed because the additional tillage they require is conducive to erosion. Perennial crops in rotation with cultivated crops are satisfactory if the perennial crop is killed with an herbicide and the following crop is planted in the crop residue.

The most successful techniques of continuous cropping in the tropics have involved maintaining a mulch of plant residues on the soil surface at all times (often with no-till farming), controlling pests with appropriate chemicals, and supplying needed macro- and micronutrients along with enough lime to avoid excess acidity.

Arid regions, cold climates, mountainous areas, and lowlands each have unique conservation problems that affect the way the people live and how they manage their land. The conservation story continues with an endless variety of problems and solutions.

QUESTIONS

1. Describe three environmental concerns that require a worldwide response.
2. Why doesn't the technology of the U.S. Corn Belt work for raising corn in Nigeria?
3. Defend these statements: "Shifting cultivation is a successful technique of agricultural production." "Shifting cultivation is a failure and must be replaced by a more efficient system."
4. Explain how to stabilize the soil in the arid tropics against water and wind erosion and, at the same time, keep it productive for agriculture.
5. Tell how to control erosion in the humid, forested tropics while harvesting continuous agricultural crops.
6. What are the South African diamond mines doing to control erosion on kimberlite spoils?
7. What are the advantages of no-till cropping in a tropical setting?
8. How can the people of The Netherlands control soil salinity in soils that are below sea level?

REFERENCES

ARAKERI, H. R., AND R. L. DONAHUE, 1984. *Principles of Soil Conservation and Water Management.* Oxford & IBM Publishing Co., New Delhi, Bombay, and Calcutta, India, p. 186–207.

BATES, R. L., AND J. A. JACKSON (eds.), 1980. *Glossary of Geology,* 2nd ed. American Geological Institute, Falls Church, Virginia.

BAYER, C., L. MARTIN-NETO, J. MIELNICZUK, C. N. PILLON, and L. SANGOI, 2001. Changes in soil organic matter fractions under subtropical no-till cropping systems. *Soil Sci. Soc. Am. J.* 65:1473–1478.

BENSALEM, B., 1977. Examples of soil and water conservation practices in North African countries, Algeria, Morocco, and Tunisia. In *Soil Conservation and Management in Developing Countries.* Soils Bull. 33, FAO, Rome, p. 151–160.

BERRY, M. P. S., 1996. The use of vetiver grass in the revegetation of kimberlite mine spoils in respect to South African diamond mining. *Vetiver Newsletter* No. 15, April 1996. The Vetiver Network, 15 Wirt St. NW, Leesburg, VA, p. 25.

BLOOMFIELD, N. J., 1977. *An Evaluation of Soil Erosion in Southern Brazil, and a Proposal for an Integrated National Program of Soil Conservation and Soil Survey.* M. S. Thesis, University of Wisconsin, Madison, Wisconsin.

BRADY, N. C., 1996. Alternatives to slash-and-burn: A global perspective. *Agric. Ecosystems Environ.* 58:3–11.

BREEUWSMA, A., J. H. M. WOSTEN, J. J. VLEESHOUWER, A. M. VAN SLOBBE, and J. BOUMA, 1986. Derivation of land qualities to assess environmental problems from soil surveys. *Soil Sci. Soc. Am. J.* 50:186–190.

CENTRAL INTELLIGENCE AGENCY, 1974. *USSR Agriculture Atlas.* U.S. Government Printing Office, Washington, D.C.

DAS, D. C., 1977. Soil conservation practices and erosion control in India—a case study. In *Soil Conservation and Management in Developing Countries.* Soils Bull. 33, FAO, Rome, p. 11–50.

DONAHUE, R. L., 1972. *Ethiopia: Taxonomy, Cartography, and Ecology of Soils.* Monograph 1, African Studies Center and the Institute of International Agriculture, Michigan State University, East Lansing, MI, 44 p.

FAO, 1995. *Planning for Sustainable Use of Land Resources.* FAO Land and Water Bulletin 2, Food and Agriculture Organization of the United Nations, Rome, Italy, 60 p.

FAVIS-MORTLOCK, D. T., and A. J. T. GUERRA, 2000. The influence of global greenhouse-gas emissions on future rates of soil erosion: A case study from Brazil using WEPP-CO_2. Ch. 1, p. 3–31 in J. Schmidt (ed.), *Soil Erosion: Application of Physically Based Models*. Springer, Berlin, 318 p.

GARCÍA-OLIVA, F., R. L. SANFORD, JR., and E. KELLY, 1999. Effects of slash-and-burn management on soil aggregate organic C and N in a tropical deciduous forest. *Geoderma* 88:1–12.

GEIGER, L. C., 1978. Soil survey in Liberia. *Soil Cons. 43*(11):16–17.

GENNADIYEV, A., 2001. Conservation policy and socioeconomic development in Russia during the 20th century. Ch. 9, p. 414–421 in E. M. Bridges, I. D. Hannam, L. R. Oldeman, F. W. T. Penning de Vries, S. J. Scherr, and S. Sombatpanit, *Response to Land Degradation*. Science Pub., Inc., Enfield, NH, 510 p.

GRIMSHAW, R. G., 1996. *Vetiver Newsletter* No. 15, April 1996. The Vetiver Network, 15 Wirt St. NW, Leesburg, VA, p. 25.

GRIMSHAW, R. G., and L. HELFER (eds.), 1995. *Vetiver Grass for Soil and Water Conservation, Land Rehabilitation, and Embankment Stabilization.* A collection of papers and newsletters compiled by the Vetiver Network. The World Bank, Washington, D.C.

HALLSWORTH, E. G., 1987. Soil conservation down under. *J. Soil Water Cons.* 42:394–400.

HANNAM, I. D., 2001. A global view of the law and policy to manage land degradation. Ch 9, p. 385–394 in E. M. Bridges, I. D. Hannam, L. R. Oldeman, F. W. T. Penning de Vries, S. J. Scherr, and S. Sombatpanit, *Response to Land Degradation*. Science Pub., Inc., Enfield, NH, 510 p.

HARWOOD, R. R., 1996. Development pathways toward sustainable systems following slash-and-burn. *Agric. Ecosystems Environ.* 48:75–86.

HAUCK, F. W., 1974. Possibilities for assistance by FAO. In *Shifting Cultivation and Soil Conservation in Africa.* Soils Bull. 24. Swedish International Development Authority and FAO, p. 245–247.

HURNI, H., 2000. Soil conservation policies and sustainable land management: A global overview. Ch. 2, p. 19–30 in T. L. Napier, S. M. Napier, and J. Tvrdon (eds.), *Soil and Water Conservation Policies and Programs: Successes and Failures.* CRC Press, Boca Raton, FL, 640 p.

IITA, 1975. *Annual Report.* International Institute of Tropical Agriculture, Ibadan, Nigeria, 219 p.

IL'ICHEV, B. A., 1999. On the reproduction of loose mantles in mountain regions (by the example of northern Caucasus). *Eurasian Soil Sci.* 32:157–170. Translated from *Pochvovedenie* 1999, p. 182–194.

KASTANOV, A. N., L. L. SHISHOV, M. S. KUZNETSOV, and I. S. KOCHETOV, 1999. Problems of soil erosion and soil conservation in Russia. *Eurasian Soil Sci.* 32:83–90.

KLEINMAN, P. J. A., D. PIMENTEL, and R. B. BRYANT, 1995. The ecological sustainability of slash-and-burn agriculture. *Agric. Ecosystems Environ.* 52:235–249.

LAL, R., 1974. Soil erosion and shifting agriculture. In *Shifting Cultivation and Soil Conservation in Africa.* Soils Bull. 24, Swedish International Development Authority and FAO, Rome, p. 48–71.

LAL, R., 2000. Soil management in developing countries. *Soil Sci.* 165:57–72.

LAL, R., J. KIMBLE, E. LEVINE, and C. WHITMAN, 1995. World soils and greenhouse effect: An overview. Chapter 1, p. 1–7 in R. Lal, J. Kimble, E. Levine, and B. A. Stewart *Soils and Global Change*. Lewis Pub., Boca Raton, FL.

NACHTERGAELE, F. O., O. SPAARGAREN, J. A. DECKERS, and B. AHRENS, 2000. New developments in soil classification world reference base for soil resources. *Geoderma* 96:345–357.

NATIONAL COMMISSION ON AGRICULTURE, INDIA, 1976. *Report of the National Commission on Agriculture,* Part V. Resource Development, Government of India Press, New Delhi, p. 177–322.

NATIONAL RESEARCH COUNCIL, 1993. *Vetiver Grass: A Thin Green Line Against Erosion.* National Academy Press, Washington, D.C., p. 12.

NORTH CAROLINA STATE UNIVERSITY, 1976. *Tropical Soils Research Program, Annual Report for 1975.* Raleigh, NC, 312 p.

NOVIKOVA, A. F., 1999. Meliorative conditions and processes of soil degradation on irrigated lands of Russia. *Eurasian Soil Sci.* 32:558–569. Translated from *Pochvovedenie* 1999, p. 614–625.

SANCHEZ, P. A., 1986. A legume-based pasture production strategy for acid infertile soils of tropical America. In *Soil Erosion and Conservation in the Tropics.* Special Publ. 43, American Society of Agronomy and Soil Science Society of America, Madison, WI, p. 97–120.

SANCHEZ, P. A., 2002. Soil fertility and hunger in Africa. *Science.* 295:2019–2020.

SCHROEDER, W. R., and J. KORT, 1989. Shelterbelts in the Soviet Union. *J. Soil Water Cons.* 44:130–134.

SOIL SURVEY STAFF, 1999. *Soil Taxonomy: A Basic System of Soil Classification for Making and Interpreting Soil Surveys,* 2nd ed. USDA Agric. Handbook 436, 869 p.

SPAROVEK, G., and E. SCHNUG, 2001. Temporal erosion-induced soil degradation and yield loss. *Soil Sci. Soc. Am. J.* 65:1479–1486.

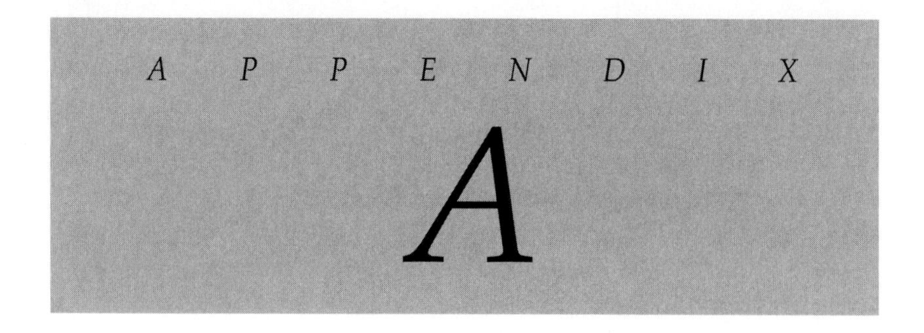

Conversion Factors

U.S. Public Law 94-168 in 1975 authorized the establishment of a board to plan for voluntary conversion to metric units by 1985. The first edition of this book, published in 1980, used mostly metric units in accord with the intent of the 1975 law. However, because only limited progress has been made on this conversion, the authors decided to use foot-pound-second units as the primary units for the subsequent editions. These are usually supplemented with approximate equivalents in the metric system. Metric units are retained as the primary system or even the only system for some uses where they are in common use in the United States. The intent is to make the book useful for readers both in the United States, where foot-pound-second units dominate, and in the rest of the world, where metric units dominate.

The metric equivalents indicated in the text in parentheses following foot-pound-second units (or occasionally the reverse set of equivalents) are usually only approximate. The degree of precision in matching these equivalents varies according to whether the measurements involved need to be expressed exactly or not. More exact equivalents are provided in this appendix along with many internal relationships for each of these systems of measurement. A few nonrelated units, such as atmospheres, also have been included for convenience. The units are grouped according to function.

Length, distance, depth, height (mile, yard, foot, inch, meter and derivatives)

1 mi = 5280 ft = 1609 m = 1.609 km

1 km = 1000 m = 3281 ft = 0.6214 mi

1 m = 100 cm = 1000 mm = 1.094 yd = 3.281 ft = 39.37 in.

1 yd = 36 in. = 3 ft = 0.9144 m

1 ft = 12 in. = 30.48 cm = 0.3048 m

1 in. = 2.54 cm = 25.4 mm

1 cm = 10 mm = 0.3937 in.

Area (length squared, acres, hectares)

1 mi^2 = 640 ac = 2.59 km^2 = 259 ha
1 km^2 = 100 ha = 0.3861 mi^2 = 247.1 ac
1 ha = 10,000 m^2 = 2.471 ac
1 ac = 43,560 ft^2 = 0.4047 ha
1 m^2 = 10.76 ft^2 = 10,000 cm^2

Volume (length cubed, gallons, quarts, liters)

1 ac-ft = 43,560 ft^3 = 12 ac-in. = 1233.5 m^3 = 12.335 ha-cm
1 ha-cm = 100 m^3 = 0.973 ac-in.
1 m^3 = 1000 l = 35.31 ft^3 = 264.2 gal (U.S.)
1 ft^3 = 7.48 gal (U.S.) = 28.32 l
1 gal (U.S.) = 4 qt (U.S.) = 3.785 l = 231 in.^{3}
1 l = 1000 ml = 1.057 qt (U.S.) = 61.02 in.^{3}
1 qt (U.S.) = 57.75 in.^{3} = 0.946 l

Weight (tons, pounds, metric tons, quintals, grams and derivatives)

1 mt = 1000 kg = 1.102 short tons = 0.984 long tons = 2204.6 lb
1 ton (short ton) = 2000 lb = 0.9072 mt = 907.2 kg
1 q = 100 kg = 220.46 lb
1 kg = 1000 g = 2.2046 lb
1 lb = 0.4536 kg = 453.6 g = 16 oz (avoirdupois)

Density (weight per unit volume)

1 g/cm^3 = 62.4 lb/ft^3 = 8.34 lb/gal

Rate, yield (weight per acre or hectare)

1 mt/ha = 10 q/ha = 0.446 short tons/ac = 892 lb/ac = 0.1 kg/m^2
1 ton/ac (short ton) = 2000 lb/ac = 2.2417 mt/ha = 2241.7 kg/ha
1 q/ha = 100 kg/ha = 1.49 bu/ac (soybeans or wheat at 60 lb/bu)
= 1.59 bu/ac (corn or sorghum at 56 lb/bu)
1 kg/ha = 0.892 lb/ac
1 lb/ac = 1.121 kg/ha
1 lb/ft^2 = 4.88 kg/m^2 = 0.488 g/cm^2

Pressure, tension (weight per unit area)

1 atm = 1.013 bars = 1.033 kg/cm^2 = 14.7 lb/in.^{2} = 10 megapascals
1 kg/cm^2 = 10 m of H_2O = 14.22 lb/in.^{2} = 0.968 atm = 9.68 megapascals

Energy, work (ft lb, kg m, joule, calorie)

1 j = 0.102 kg-m = 0.7377 ft lb = 0.239 cal = 10^7 ergs

Slope (horizontal distance:vertical distance or vertical distance/horizontal distance)

1:1 = 1 ft/1 ft = 100% = 45°
2:1 = 1 ft/2 ft = 50% = 26.6°
3:1 = 1 ft/3 ft = 33% = 18.4°
4:1 = 1 ft/4 ft = 25% = 14°
6:1 = 1 ft/6 ft = 17% = 9.5°
10:1 = 1 ft/10 ft = 10% = 5.7°

Velocity, permeability (distance per second, hour, or other unit of time)

1 mi/hr = 1.609 km/hr = 0.447 m/s = 1.4667 ft/s
1 km/hr = 0.6214 mi/hr = 0.27778 m/s = 0.9114 ft/s
1 m/s = 100 cm/s = 3.281 ft/s = 3.6 km/hr = 2.237 mi/hr
1 ft/s = 30.48 cm/s = 0.3048 m/s = 1.097 km/hr = 0.6818 mi/hr
1 m/hr = 1000 mm/hr = 100 cm/hr = 39.37 in./hr = 3.281 ft/hr

Flow rates (volume per second or hour)

1 m^3/s = 1000 l/s = 35.31 ft^3/s = 3600 m^3/hr
1 ft^3/s = 7.48 gal/s = 28.32 l/s = 101.9 m^3/hr
1 gal/min = 3.785 l/min = 227.1 l/hr = 0.2271 m^3/hr

Viscosity (poise)

1 p = 0.1 kg/m-sec

Temperature (degrees Fahrenheit, Celsius, Kelvin)

$°F = °C \times 1.8 + 32$
$°C = (°F - 32)/1.8 = °K - 273.15$

Chemical concentrations (normal, gram-equivalent, weight/volume of solution)

1 N = 1 g-eq/l = 1000 meq/l = 1 M/valence
1 g-eq = gram-weight of 1 mole/valence
1 mg/100 ml = 1000 μg/100 ml = 10 mg/l = 10 ppm (in water solution)
1 μg/100 ml = 10 μg/l = 10 ppb = 0.01 ppm (in water solution)
1 meq/100 g = 1 mg-eq/100 g of soil

Electrical conductivity (millimho/cm, siemen/cm)

1 mmho/cm = 0.001 mho/cm = 0.001 siemen/cm

Ratios (percent, parts per million)

1/1 = 100% = 1,000,000 ppm = 1,000,000,000 ppb
1% = 10,000 ppm

Scales (map distance:distance on ground)

1:15,640 = 4in.:1 mi
1:20,000 = 3.168 in.:1 mi = 5 cm:1 km
1:190,080 = 1 in.:3 mi
1:500,000 = 1 in.:7.9 mi = 1 cm:5 km
1:1,000,000 = 1 in.:15.8 mi = 1cm:10 km

APPENDIX

B

Common and Scientific Names of Plants Mentioned in the Text

The common English names of plants vary widely from one place to another. The following list is supplied to remove any ambiguity that would otherwise arise from the common names used in this book. The equivalent scientific names listed here have worldwide meaning.

Acacia—*Acacia farnesiana*
Alder:
  European black—*Alnus glutinosa*
  thinleaf—Alnus tenuifolia
Alfalfa—*Medicago sativa*
Alkali sacaton—*Sporobolus airoides*
Alkaligrass, nuttall—*Puccinellia airoides*
Apple—*Malus* species
Argan tree—*Argania spinosa*
Ash:
  green—*Fraxinus pennsylvanica*
  white—*Fraxinus americana*
Asparagus—*Asparagus officinalis*
Aspen, trembling—*Populus tremuloides*
Avocado—*Persea americana*

Bahiagrass—*Paspalum notatum*
Banana—*Musa* species
Barberry, creeping—*Berberis repens*
Barley—*Hordeum vulgare*

Bayberry—*Myrica cerifera*
Beachgrass:
  American—*Ammophila breviligulata*
  European—*Ammophyila arenaria*
Beachpea—*Lathyrus japonicus*
Bean:
  bush, climbing, wild—*Phaseolus* species
  faba or fava—*Vicia faba*
  field, garden—*Phaseolus vulgaris*
Bearberry—*Astostaphylos uva-ursi*
Beech, American—*Fagus grandifolia*
Beet:
  red, (garden)—*Beta vulgaris*
  sugar—*Beta vulgaris*
Bentgrass, creeping—*Agrostis stolonifera*
Bermudagrass:
  common or coastal—*Cynodon dactylon*
  African—*Cynodon transvaalensis*
Birch:
  river—*Betula nigra*
  white—*Betula alba*
  yellow—*Betula allegheniensis*
Bitterbrush:
  antelope—*Purshia tridentata*
  desert—*Purshia glandulosa*
Blackberry, common—*Rubus allegheniensis*
Bladdersenna—*Colutea arborescens*
Blueberry—*Vaccinium* species
Bluegrass:
  alpine—*Poa* species
  big—*Poa ampla*
  Kentucky—*Poa pratensis*
Bluestem:
  big—*Andropogon gerardii*
  broomsedge—*Andropogon virginicus*
  little—*Andropogon scoparius*
  sand—*Andropogon hallii*
  seacoast—*Andropogon littoralis*
  yellow—*Andropogon ischaemum*

Boxelder—*Acer negundo*
Broadbean—*Vicia faba*
Broccoli—*Brassica oleracea*
Bromegrass:
  California—*Bromus carinatus*
  mountain—*Bromus marginatus*
  smooth—*Bromus inermis*
Buckthorn—*Rhamnus* species
Buckwheat—*Fagopyrum esculentum*
Buffaloberry, russet—*Shepherdia argentea*
Buffalograss—*Buchloe dactyloides*
Bursage, white—*Franseria dumosa*

Cabbage—*Brassica oleracea*
Cactus—*Cactaea* species
Calligonum—*Calligonum arich*
Camellia—*Camellis japonica*
Canarygrass, reed—*Phalaris arundinacea*
Caper bush—*Capparis spinosa*
Caragana—*Caragana arborescens*
Carob tree—*Ceratonica siliqua*
Carrot—*Daucus carota*
Cashew—*Anacardium occidentale*
Cassava (tapioca)—*Manihot esculenta*
Cassia—*Cassia* species
Cattails—*Typha* species
Ceanothus:
  deerbrush—*Ceanothus interrimus*
  snowbrush—*Ceanothus velutinus*
  squawcarpet—*Ceanothus* species
  wedgeleaf—*Ceanothus cuneatus*
Cedar:
  red—*Juniperus virginiana*
  white—*Thuja occidentalis*
Cheatgrass—*Bromus tectorum*
Cherry:
  bessey—*Prunus besseyi*
  bitter—*Prunus emarginata*
  black—*Prunus serotina*

Nanking—*Prunus tomentosa*
wild red—*Prunus pennsylvanica*
Chickpea—*Cicer arietinum*
Chokecherry:
black—*Prunus virginiana melanocarpa*
common—*Prunus virginiana*
Cinquefoil—*Potentilla fruticosa*
Clover:
alsike—*Trifolium hybridum*
crimson—*Trifolium incarnatum*
red—*Trifolium pratense*
tick—*Desmodium leonii*
white (Dutch)—*Trifolium repens*
Cocoa—*Cacao nucifera*
Coffee:
arabica—*Coffea arabica*
robusta—*Coffea robusta*
Coffeetree, Kentucky—*Gymnocladus dioica*
Corn, field, pop, or sweet—*Zea mays*
Cotoneaster—*Cotoneaster apiculata*
Cotton, American, Egyptian—*Gossypium barbadense*
Cottonwood—*Populus deltoides*
Cowpea—*Vigna unguiculata*
Crabapple:
Siberian—*Malus baccata*
toringo—*Malus sieboldi*
Crabgrass—*Digitaria* species
Cranberry—*Vaccinium macrocarpon*
Creosotebush—*Larrea tridentata*
Crownvetch—*Coronilla varia*
Currant, golden—*Ribes* species
Cypress:
bald—Taxodium distichum
Tassili—*Cupressus dupreziana*

Dallisgrass—*Paspalum dilatatum*
Deerbrush—*Ceanothus interrimus*
Deertongue—*Panicum clandestinum*
Dogwood, redosier—*Cornus stolonifera*
Douglas-fir—*Pseudotsuga menziesii*

Elder, blueberry—*Sambucus cerulea*
Elm, Siberian—*Ulmus pumila*
Eriogonum, sulfur—*Eriogonum umbellatum*
Eucalyptus—*Eucalyptus* species

Fescue:
red and arctared—*Festuca rubra*
tall (reed)—*Festuca arundinacea*
Fir—*Abies* species
Flatpea—*Lathyrus sylvestris*
Flax—*Linum usitatissimum*
Forsythia—*Forsythia* species
Foxtail, creeping—*Alopecurus arundinaceus*

Grama:
black—*Bouteloua eripoda*
blue—*Bouteloua gracilis*
sideoats—*Bouteloua curtipendula*
Grape—*Vitus* species
Guar—*Cyamopsis tetragonolobus*

Hackberry—*Celtis occidentalis*
Hairgrass, Bering tufted—*Deschampsia caespitosa*
Hardinggrass—*Phalaris tuberosa*
Hawthorne, Arnold—*Crataegus* species
Hedgerose, Hanson—*Rosa* species
Hemlock:
eastern—*Tsuga canadensis*
ground—*Taxus canadensis*
western—*Tsuga heterophylla*
Hickory, shagbark—*Carya ovata*
Honeylocust, thornless—*Gledsia triacanthos*
Honeysuckle, tartarian—*Lonicera tartarica*
Horsetail tree (casuarina)—*Casuarina cunninghamia*
Hyparrhenia—*Hyparrhenia rufa*

Indiangrass—*Sorghastrum nutans, S. odorata*
Indian mustard—*Brassica juncea*
Indigo, false—*Amorpha fruticosa*
Indigobush—*Amorpha fruticosa*
Iris, wild—*Iris pseudacorus*

Jerusalem thorn—*Parkinsonia aculeata*
Johnsongrass—*Sorghum halepense*
Jujube—*Ziziphus jujuba*
Juniper:
  common—*Juniperus communis*
  creeping—*Juniperus horizontalis*
  one-seed—*Juniperus monosperma*
  Rocky Mountain—*Juniperus scopulorum*

Kassod tree—*Cassia siamea*
Kudzu—*Pueraria lobata*

Larch, European—*Larix decidua*
Leadplant—*Amorpha canescens*
Lentil—*Lens culinaris*
Lespedeza:
  bicolor—*Lespedeza bicolor*
  common (annual)—*Lespedeza striata*
  sericea (Chinese)—*Lespedeza cuneata*
  Thunberg—*Lespedeza thurnbergii*
Lettuce—*Lactuca* species
Lilac:
  common—*Syringa vulgaris*
  late—*Syringa villosa*
Linseed—*Linum usitatissimum*
Locoweed—*Astragalus* species and *Oxytropis* species
Locust:
  African—*Parkia clappertoniana*
  black—*Robinia pseudoacacia*
Lovegrass:
  Korean—*Eragrostis ferruginea*
  Lehmann—*Eragrostis lehmanniana*
  sand—*Eragrostis trichodes*
  weeping—*Eragrostis curvula*
Lupine, wild—*Lupinus perennis*

Manzanita, pinemat—*Arctostaphylos* species
Maple:
  amur—*Acer ginnala*

red—*Acer rubrum*
silver (white)—*Acer saccharinum*
sugar—*Acer saccharum*
Marramgrass:
Baltic—*Ammocalamagrostis baltica*
common—*Ammophila arenaria*
Medic, black—*Medicago lupulina*
Medusahead—*Taeniatherum asperum*
Melon—*Cucumis melo*
Mesquite—*Prosopis chilensis*
Milkvetch, cicer—*Astragalus cicer*
Millet, German—*Setaria italica*
Mulberry:
red—*Morus rubra*
white, silver—*Morus alba*
Mulga—*Acacia aneura*

Neem tree—*Azadirachta indica*

Oak:
bur—*Quercus macrocarpa*
English—*Quercus robur*
live—*Quercus virginia*
northern red—*Quercus rubra*
pin—*Quercus palustris*
water—*Quercus nigra*
Oat, common—*Avena sativa*
Oceanspray, bush—*Holodiscus discolor*
Olive:
autumn—*Elaegnus umbellata*
Russian—*Elaegnus angustifolia*
tree—*Olea europaea*
Onion—*Allium cepa*
Orange—*Citrus sinensis*
Orchardgrass—*Dactylis glomerata*
Osage-orange—*Maclura pomifera*

Palm:
date—*Phoenix dactylifera*
oil—*Elaeis guineensis*

Pangolagrass—*Digitaria decumbens*
Panicgrass, coastal—*Panicum amarulum*
Paspalum—*Paspalum* species
Pea:
  Austrian winter—*Lathyrus hirsutus*
  field, garden—*Pisum sativum*
Peanut—*Arachis hypogaea*
Pear, Oriental—*Pyrus pyrifolia*
Pearl millet—*Pennisetum americanum*
Peashrub, Siberian (caragana)—*Caragana arborescens*
Pea-tree, Siberian—*Caragana arborescens*
Pecan—*Carya illinoensis*
Penstemon—*Penstemon fruticosus*
Pepper, bell—*Capsicum annuum*
Pine:
  Aleppo—*Pinus halepensis*
  Austrian—*Pinus nigra*
  cluster—*Pinus pinaster*
  eastern white—*Pinus strobus*
  jack—*Pinus banksiana*
  loblolly—*Pinus taeda*
  lodgepole (shore)—*Pinus contorta latifolia*
  longleaf—*Pinus palustris*
  Monterey—*Pinus radiata*
  mugo (mugho)—*Pinus mugo*
  pitch—*Pinus rigida*
  ponderosa—*Pinus ponderosa*
  red—*Pinus resinosa*
  sand—*Pinus clausa*
  Scotch (Scots)—*Pinus sylvestris*
  shore—*Pinus contorta*
  shortleaf—*Pinus echinata*
  slash—*Pinus elliottii*
  stone—*Pinus monophylla, P. pinea*
  Virginia—*Pinus virginiana*
  western white—*Pinus monticola*
Plum:
  American—*Prunus americana*
  Assyrian—*Cordia myxa*
  beach—*Prunus maritima*

chickasaw—*Prunus angustifolia*
Poplar:
northwest, Norway (hybrid)—*Populus sargentii*
robusta (hybrid)—*Populus robusta*
white (silver)—*Populus alba*
yellow (tulip)—*Liriodendron tulipifera*
Potato:
Irish (white)—*Solanum tuberosum*
sweet—*Ipomoea batatas*

Rape—*Brassica napus*
Raspberry, red—*Rubus idaeus*
Redbud—*Cercis canadensis*
Redtop—*Agrostis alba*
Redwood—*Sequoia sempervirens*
Reed—*Phragmites* species
Rhodesgrass—*Chloris gayana*
Rice—*Oryza sativa*
Rosemary
bog—*Andromeda* species
common—*Rosmarinus officinalis*
Rose:
wild—*Rosa blanda*
woods—*Rosa woodsii ultramontana*
Rubber—*Hevea braziliensis*
Rush—*Juncus* species
Rye—*Secale* species
Ryegrass:
annual (Italian)—*Lolium multiflorum*
perennial—*Lolium perenne*

Safflower—*Carthamus tinctorius*
Sagebrush:
big—*Artemisia tridentata*
mountain—*Artemisia arbuscula*
Salsola shrub—*Salsola paletskiana, (S. richteri)*
Saltbush:
desert—*Atriplex polycarpa*
fourwing—*Atriplex canescens*

Saltgrass, seashore—*Distichlis spicata*
Sandcherry, western—*Prunus besseyi*
Sandgrass (purple)—*Triplasis purpurea*
Sassafras—*Sassafras albidum*
Scalebroom—*Lepidospartum squamtum*
Scotch broom—*Cytisus scoparius*
Sea-oatsgrass—*Uniola paniculata*
Sedges—*Carex* species
Serviceberry, Saskatoon—*Amelanchier alnifolia*
Sesame—*Sesamum indicum*
Sesbania—*Sesbania aculeata*
Shinus—*Shinus terebinthifolius, (S. molle)*
Signalgrass—*Brachiaria decumbens*
Silverberry—*Elaeagnus argentea, (E. commutata)*
Silvergrass, Chinese—*Miscanthus sinensis*
Siris—*Albizzia lebbek*
Sissoo—*Dalbergia sissoo*
Snowberry:
  common—*Symphoricarpos albus*
  creeping—*Gaultheria hispidula*
  mountain—*Symphoricarpos occidentalis*
Snowbrush—*Ceanothus velutinus*
Sorghum, grain—*Sorghum bicolor*
Soybean—*Glycine max*
Sphagnum (moss)—*Sphagnum* species
Spinach—*Spinacia oleracea*
Spirea, Douglas—*Spirea douglasii*
Spruce:
  Black Hills—*Picea glauca*
  Colorado—*Picea pungens*
  Norway—*Picea abies*
  red—*Picea rubens*
  Sitka—*Picea sitchensis*
  white—*Picea glauca*
Squawcarpet—*Ceanothus prostratus*
Strawberry—*Fragaria* species
Stylosanthes—*Stylozanthes capitata*
Sudangrass—*Sorghum sudanense*
Sugarcane—*Saccharum officinarum*
Sumac, Rocky mountain (skunkbush)—*Rhus glabra cismontana*

Sunflower—*Helianthus annuus*
Sunnhemp—*Crotalaria juncea*
Sweetclover:
white—*Melilotus alba*
yellow—*Melilotus officinalis*
Sweetgale—*Myrica gale*
Sweetgum—*Liquidambar styraciflua*
Switchgrass—*Panicum virgatum*
Sycamore (American)—*Platanus occidentalis*

Tamarix (tamarisk)—*Tamarix aphylla, (T. nilotica)*
Tapioca (cassava)—*Manihot esculenta*
Tea—*Camellia sinensis*
Thistle, Russian—*Salsola kali*
Timothy:
alpine—*Phleum alpinum*
common—*Phleum pratense*
Tobacco—*Nicotiana* species
Tobosagrass—*Hilaria mutica*
Tomato—*Lycopersicon esculentum*
Trefoil:
big—*Lotus uliginosus*
birdsfoot—*Lotus corniculatus*
Trisetum, spike—*Trisetum flavescens*
Tumbleweed—*Amaranthus albus*

Veldtgrass—*Ehrharta calycina*
Vetch:
hairy—*Vicia villosa*
reseeding—*Vicia* species
Vetiver grass—*Vetiveria* species
Virginia-creeper—*Parthenocissus quinquefolia*
Virginsbower, western—*Clematis virginiana*

Walnut:
black—*Juglans nigra*
English—*Juglans regia*
Water lily—*Nymphaea* species
Wax myrtle—*Myrica cerifera*
Wedgeleaf—*Ceanothus cuneatus*

Wheat, common—*Triticum aestivum*
Wheatgrass:
crested—*Agropyron desertorum*
intermediate—*Agropyron intermedium*
pubescent—*Agropyron trichophorum*
Siberian—*Agropyron sibiricum*
slender—*Agropyron trachycaulum*
tall—*Agropyron elongatum*
western—*Agropyron smithii*
Wildbean, trailing—*Strophostyles helvola*
Wildrye:
beardless—*Elymus triticoides*
Canada—*Elymus canadensis*
Russian—*Elymus junceus*
Volga—*Elymus giganteus*
Willow:
crack—*Salix fragilis*
desert—*Chilopsis linearis*
scouler—*Salix scouleriana*
white—*Salix alba*
Wintergreen—*Gaultheria procumbens*
Wormwood, Oldman—*Artemisia abrotanum*

Zornia—*Zornia latifolia*

REFERENCES

ANDERSON, K. L., and C. E. OWENSBY, 1969. *Common Names of a Selected List of Plants.* Tech. Bull. No. 117, Kansas State University, 62 p.

BAILEY, L. H., and E. Z. BAILEY, 1976. *Hortus Third: A Concise Dictionary of Plants Cultivated in the United States and Canada.* Macmillan, New York, 1290 p.

HALLS, L. K. (ed.), 1977. *Southern Fruit-producing Woody Plants Used by Wildlife.* General Tech. Report SO-16, USDA-Forest Service, 235 p.

HANSON, A. A., 1972. Grass Varieties in the United States. Agriculture Handbook No. 170, ARS, U.S. Dept. of Agriculture, 124 p.

SOIL CONSERVATION SERVICE, 1978. *Plant Performance on Surface Coal Mine Spoil in Eastern United States.* TP-155, U.S. Dept. of Agriculture, 76 p.

TERRELL, E. E., 1977. *A Checklist of Names for 3000 Vascular Plants of Economic Importance.* Agriculture Handbook No. 505. ARS, U.S. Dept. of Agriculture, 201 p.

Index